HANDBOOK OF RESEARCH ON SCIENCE LEARNING PROGRESSIONS

Gathering contributions from leading scholars around the world, this handbook offers a comprehensive resource on the most recent advances in research surrounding the theories, methodologies, and applications of science learning progressions.

Researchers and educators have used learning progressions to guide the design and alignment of curriculum, instruction, and assessment, and to help students learn scientific knowledge and practices in a coherent and connected way across multiple years. This handbook lays out the development and current state of research in this field across four sections: learning progression theories and methodologies; learning progressions to promote student learning; teachers' learning and use of learning progressions; and new technology in learning progression research.

Featuring internationally recognized experts in learning progression research as well as up-and-coming voices, the *Handbook of Research on Science Learning Progressions* offers a defining new resource for researchers, teachers and teacher educators, and curriculum and assessment developers in science education.

Hui Jin is an associate professor of science education at Georgia Southern University, USA.

Duanli Yan is the Director of Computational Research at ETS and an adjunct professor at Fordham University, USA.

Joseph Krajcik directs the CREATE for STEM Institute and is a University Distinguished Professor at Michigan State University, USA.

HANDBOOK OF RESEARCH ON SCIENCE LEARNING PROGRESSIONS

Edited by Hui Jin, Duanli Yan, and Joseph Krajcik

Routledge
Taylor & Francis Group
NEW YORK AND LONDON

Designed cover image: © Getty Images

First published 2025
by Routledge
605 Third Avenue, New York, NY 10158

and by Routledge
4 Park Square, Milton Park, Abingdon, Oxon, OX14 4RN

Routledge is an imprint of the Taylor & Francis Group, an informa business

ISBN: 978-0-367-77319-9 (hbk)
ISBN: 978-0-367-77318-2 (pbk)
ISBN: 978-1-003-17078-5 (ebk)

DOI: 10.4324/9781003170785

Typeset in Sabon
by MPS Limited, Dehradun

Access the Support Material: www.routledge.com/9780367773182

We dedicate this handbook to our children and grandchildren with the heartfelt hope that their education inspires and equips them to address the world's challenging problems.

Hui Jin, Duanli Yan, and Joseph Krajcik

CONTENTS

CONTRIBUTORS

Amy Adair is an NSF Graduate Research Fellow (DGE-1842213) and PhD candidate (Rutgers Graduate School of Education). Advised by Dr. Gobert, Amy works on AI for STEM learning and assessment. Prior to graduate school, she received a B.S. in mathematics and taught high school math and robotics.

Russell Almond is a statistician with extensive training and interest in computer science, particularly artificial intelligence. Together with R. Mislevy and L. Steinberg, he developed the early versions of evidence-centered assessment design (ECD) and wrote the book *Bayesian Networks in Educational Assessment* (published by Springer). His research interests include representations-building scoring systems for complex assessment systems (including simulations and games), knowledge representation for building educational assessments, applying machine learning techniques (particularly Bayesian networks) to educational measurement, and computational data analysis.

Alicia Alonzo is a Professor in the Department of Teacher Education at Michigan State University. Much of her research has focused on learning progressions as tools to support classroom assessment practices. She is the co-editor of *Learning Progressions in Science: Current Challenges and Future Directions* (Alonzo & Gotwals, 2012).

Tamer G. Amin is an Associate Professor in the Department of Education at the American University of Beirut. He earned his PhD in developmental psychology at Clark University. His research focuses on science concept learning, the acquisition of scientific expertise, and the implications of learning processes for science education. He is co-editor (with Olivia Levrini) of *Converging Perspectives on Conceptual Change* (Routledge, 2018).

Charles W. (Andy) Anderson is a former middle school science teacher and professor emeritus at Michigan State University. His research uses learning progressions for curriculum and assessment focused on environmental science literacy. He has served as president of NARST and co-editor of the Journal of Research in Science Teaching.

Angela Calabrese Barton, PhD, is a professor in Educational Studies at the University of Michigan. Her research focuses on equity and justice issues in STEM education in school/community settings. She takes an ecological and participatory approach to give witness to and learn with community partners in transforming teaching/learning towards socially just futures.

Kathryn M. Bateman is a Professional Learning and Research Manager with the Museum of Science, Boston. Her research applies post-structural theory to the intersection of teacher education and educational policies, employing asset-based approaches to science learning and transdisciplinary approaches to education, emphasizing geoscience learning K-16.

Julie A. Bianchini is a Professor of Science Education and Faculty Director of the CalTeach teacher recruitment and preparation initiative in the Department of Education at the University of California, Santa Barbara. Her research attends to equity and diversity in the teaching and learning of science at the secondary level.

Jason Y. Buell is a Postdoctoral Researcher at Northwestern University. A former middle and elementary school science teacher, Jason studies how to support teachers in learning to teach in increasingly equitable and sophisticated ways within complex systems. His current work is on using curriculum as a catalyst for teacher learning.

Todd Campbell is the Department Head of Curriculum and Instruction and a Professor of Science Education at the University of Connecticut. His research focuses on science teacher's professional learning and how science learners can be supported by model-based learning and justice-focused pedagogies in formal and informal STEM learning contexts.

Moraima Castro-Faix earned her PhD in Learning Sciences and MA in Education from Rutgers University and a BS in Genetics from the University of Puerto Rico. Her dissertation research focused on learning and teaching in genetics in K-12. She is also interested in supporting underrepresented students in STEM higher education.

Douglas H. Clements is a Distinguished University Professor and Kennedy Endowed Chair at the University of Denver, Colorado. Clements has published over 185 refereed research studies, 30 books, 107 chapters, and 300 additional works on the learning and teaching of early mathematics; research-based curricula; and taking interventions to scale. http://www.researchgate.net/profile/Douglas_Clements/

Beth A. Covitt is the Head of Science Education Research & Evaluation with the University of Montana's Broader Impacts Group. Her collaborative efforts in environmental and science education address the role education can play in preparing people to recognize the strengths and limitations of science for participation in discussions and decisions concerning socio-scientific issues.

Rachel Dickler, PhD (Learning Sciences, Rutgers University), obtained her PhD under Dr. Gobert's supervision. Rachel has an interdisciplinary research background spanning Artificial Intelligence in Education, Collaborative Learning Technologies, and Formal/Informal Science Education. Currently, Rachel is the Community Research and Evaluation Manager at the Denver Zoo, where she conducts social science research.

Jennifer H. Doherty is an Assistant Professor at Michigan State University in Lyman Briggs College and the Department of Physiology. She holds a PhD in Biology from the University of Pennsylvania. Her research focuses on how students develop principle-based reasoning in physiology, ecology, and evolution.

Ravit Golan Duncan, PhD, is a professor in Learning Sciences and Science Education at the Graduate School of Education at Rutgers University. Her research focuses on learning progressions in science, and designing instruction to promote student engagement with equitable and expansive scientific inquiry. She also prepares secondary life science teachers.

Richard A. Duschl is a TI Distinguished Professor, Lyle School of Engineering, Southern Methodist University; and Professor Emeritus, Penn State University. Member of the US National Academy of Education. Learning Sciences R&D includes (1) HPS in K-16 science education, (2) Design-Implementation of Argumentation/Assessment frameworks, and (3) DBIR/Improvement Science R&D in STEM formal/informal teaching and learning.

Kirsten D. Edwards is a doctoral candidate at Michigan State University and the Senior Director of K-8 Science at Achievement First Public Charter Schools. Her scholarship focuses on how secondary students engage in reading and writing as part of science learning and how curricular scaffolds can support that engagement.

David Fortus studies novel curricular approaches and their influence on adolescents' engagement with science, in and out of schools. He has been a high school physics teacher, the co-founder of a Waldorf high school, a physicist, and a project director in the aerospace industry. He holds the Chief Justice Bora Laskin Chair of Science Teaching at the Weizmann Institute of Science.

Allison Freed is an Assistant Professor of instructional technology and coordinator of the Instructional Technology (ITEC) graduate program at the University of Central Arkansas. Her teaching and research interests include internationalizing science teaching using citizen science, Collaborative Online International Learning (COIL), and action research.

Erin Marie Furtak is a Professor of STEM Education at the University of Colorado Boulder. A former high school science teacher, she studies how to support secondary science teachers to re-imagine formative assessments in their classrooms to better support student learning.

Janice Gobert, PhD (Cognitive Science, University of Toronto) is a Professor of Learning Sciences at Rutgers Graduate School of Education. Janice is also the CEO of Apprendis, which is productizing Inq-ITS and Inq-Blotter (inqits.com), which are patented, AI-based systems for the real-time assessment, scaffolding, and alerting on students' competencies at the practices of science.

Kristin L. Gunckel (she/they) is a Professor of Science Education at the University of Arizona. Kristin's research focuses on preparing teachers and students to understand and respond equitably to grand environmental challenges such as threats to freshwater and climate change.

Shannon Stark Guss is a Senior Research Associate at the Marsico Institute of Early Learning. Her research interests include the measurement of early learning outcomes, environments, and interactions. She is also interested in designing and evaluating school and home-based interventions that improve quality and outcomes for children from birth to age eight.

Christopher J. Harris is a Senior Director of Science and Engineering Education Research at WestEd. He designs and studies curricula and assessments that capitalize on innovative technologies and support teachers in creating engaging, interactive, and accessible learning environments. His efforts often involve collaborative work with districts and schools to inform research and practice.

Peng He is an Assistant Research Professor of Science Education at Michigan State University, CREATE For STEM Institute. His research interests include designing and testing learning environments with innovative technologies (e.g., artificial Intelligence), learning progression, classroom assessment, science measurement, and teacher knowledge and practices.

Kate Henson is a Research Scientist at the University of Colorado focusing on science education. She is co-director of the OpenSciEd High School Developers Consortium. Kate is a former high school science teacher and holds a BS and MS in Biology and a PhD in Curriculum and Instruction.

Yuxi Huang is a Ph.D. candidate in Science Education at the Mathematics, Science, and Social Studies Education Department, University of Georgia. Her primary research interest is teacher noticing. Yuxi has presented her work at the National Association of Research in Science Teaching (NARST) and the Association of Science Teacher Education (ASTE).

Hui Jin is an Associate Professor of science education at Georgia Southern University. Her research focuses on science learning progressions, science assessments, environmental literacy, mathematical thinking in science, and technology-enhanced data analysis in science.

Wendy R. Johnson taught high school science for 13 years and is currently a teacher support specialist with the National Center for Science Education. Her work focuses on engaging students' curiosity to drive science learning, improving understanding of evolution and climate change, and supporting teachers in implementing the NGSS.

Leonora Kaldaras is an Educational Researcher at Stanford University. She designs learning environments for developing complex cognitive skills such as knowledge application and blended Math-Science sensemaking across STEM disciplines. She leverages cutting-edge technologies such as artificial intelligence and interactive computer simulations to support learners from diverse backgrounds in learning science.

Michael T. Kane is the Messick Chair in validity at Educational Testing Service. Previously, he worked at the National Conference of Bar Examiners, the University of Wisconsin, and American College Testing. His research focuses on validity theory, generalizability theory, and standard setting. He has a Ph.D. in Education from Stanford University.

Joseph Krajcik directs the CREATE for STEM Institute and is a University Distinguished Professor at Michigan State University. Throughout his career, Joe has focused on working with colleagues and science teachers to design and test project-based learning environments to reform science teaching practices and to research student learning and engagement.

Okhee Lee is a Professor at the Steinhardt School of Culture, Education, and Human Development at New York University. She is committed to advancing research, policy, and practice that promote STEM and language learning with equity and justice for all students, particularly multilingual learners.

Cynthia Lima is an Assistant Professor at the University of Texas at San Antonio. Her research interests include science assessment for diverse populations and teacher preparation.

Xiufeng Liu is a Distinguished Professor at State University of New York at Buffalo. His research interest is the measurement and evaluation of STEM education. He has applied Rasch analysis to study students' long-term development of crosscutting concepts of matter and energy. His current research focuses on the measurement of STEM identity.

Yi-Meng Liu is a doctoral student at Beijing Normal University. Her research focus is on biology education.

Julie A. Luft is a Distinguished Research Professor and Athletic Association Professor of Mathematics and Science Education at the University of Georgia, Athens, Georgia. She has a long-standing research interest in newly hired science teachers. Over the years, she has published numerous books, chapters, and research articles on newly hired science teachers.

Seungho Maeng is an Associate Professor of the Department of Science Education at Seoul National University of Education, South Korea. His current research focuses on the epistemic process of Earth science learning with a specific interest in discursive practices. Learning progressions for geoscience and astronomy are also included in his research area.

Scott McDonald is a Professor of Science Education at The Pennsylvania State University and Director of the Krause Studios for Innovation. His research focuses on teacher professional vision, ambitious and equitable science teaching practices, and supporting student learning in Earth and Space Sciences using simulations and data visualizations.

Hannah K. Miller is an Associate Professor of education at Vermont State University. Her scholarship interrogates the process of social change in educational systems, with a focus on privilege and oppression. Her current CPAR project, TeachOut Vermont, examines how to build sustainable social networks with LQBTQ+ educators in Vermont.

An-Qi Mou is a teacher at Beijing No. 13 Middle School. Her research focus is on Geography Education.

Knut Neumann is the Director of Physics Education at the IPN–Leibniz Institute for Science and Mathematics Education. His research interests include how students develop competence in science, how to best support students in developing competence, and how to provide teachers with the professional competence to best support students.

Jeffrey Nordine is an associate professor of Science Education at the University of Iowa. He designs and investigates coherent approaches to science instruction and science teacher education. He has been a high school physics and mathematics teacher and chief scientist of a children's museum.

Jonathan Osborne is the Kamalachari Chair in Science Education, and Emeritus in the Graduate School of Education, Stanford University. His research focus is on the science curriculum and argumentation in science, and he chairs the expert group for the science assessments conducted by the OECD PISA Program in 2015 and 2025.

James W. Pellegrino is Professor Emeritus and Founding Co-director of UIC's Learning Sciences Research Institute. His research focuses on children's and adult's thinking and learning and the implications for assessment and instruction. He is a member of the National Academy of Education and the American Academy of Arts and Sciences.

Brian Riordan is the Principal AI Engineer at Outshift by Cisco. Prior to Cisco, he led natural language understanding for educational applications at ETS AI Labs. He has served as Principal Investigator on multiple NSF-funded projects and has authored more than 40 publications in natural language processing and machine learning.

Julie Sarama, Kennedy Endowed Chair and Distinguished University Professor, taught high school mathematics, computer science, middle school gifted mathematics, and early childhood mathematics. She has directed 30 projects (funded by NSF, IES, and others) and has authored over 80 refereed articles, 7 books, 60 chapters, and over 100 additional publications: https://www.researchgate.net/profile/Julie_Sarama.

Emily E. Scott earned her doctorate in Forest Ecology before shifting fields to investigate how students learn to reason about carbon-transforming processes and plant and animal physiology. She currently works for Washington State's Office of the Superintendent of Public Instruction as Assistant Director of Data and Implementation.

Jian-Xin Yao is an Associate Professor at the Department of Physics, Beijing Normal University. His research focus is on the science curriculum and assessment. He is a member of the expert group for national physics curriculum standards of China.

Shih-Ying Yao is a Lecturer at the Teaching and Learning Centre, Singapore University of Social Sciences. Her research primarily focuses on the development and validation of assessments and research instruments. In addition to educational measurement, she is also interested in the application of innovative educational technologies in higher education.

Namsoo Shin is an Associate Research Professor at Michigan State University and is interested in the research and development of individualized learning environments for diverse student needs. Her research is focused on evidence-based learning progressions and innovative learning technologies that enhance teaching and learning in everyday problem-solving skills.

Harleen Singh is an Assistant Professor of Education at the College of Education, Kinesiology, and Social Work at California State University, Stanislaus. Her research interest lies in teacher knowledge and teacher development. Specifically, her research centers on understanding the development of newly hired science teachers who are teaching out-of-field.

Carol L. Smith is a Professor Emerita in the Psychology Department at the University of Massachusetts/Boston. She earned her PhD in Personality and Development Studies at Harvard University. She is a cognitive developmental psychologist whose work on conceptual change in science education exemplifies part of the research basis for learning progressions.

Kenneth Steimel is an Associate AI Engineer at Educational Testing Service (ETS). He received his Master of Arts in Computational Linguistics from Indiana University and is currently a PhD candidate in Linguistics, with an emphasis on Computational Linguistics. His research interests lie at the intersection of Natural Language Processing and education.

Edna Tan, PhD, is Hooks Distinguished Professor of STEM education at the University of North Carolina at Greensboro. Her collaborative research is focused on long-termed, community-engaged partnerships with teachers and youth of color on the design and enactment of justice-oriented STEM teaching and learning across formal and informal settings.

Jian Wang is a Professor at the College of Biology, Beijing Normal University, P. R. China. His research focus is on Biology Education. He is a committee member of the National Association of Biology Education, Chinese Society of Education.

Lei Wang is a Professor at the College of Chemistry, Beijing Normal University. Her research focus is on Chemistry Education. She is the Chair of the Chemistry Education Committee of the Chinese Chemical Society.

Min Wang is a Professor at the Faculty of Geographical Science, Beijing Normal University. His research focus is on Geography Education. He is Chair of the Geography Olympic Competition Working Committee of the Geographical Society of China.

Quan Wang is a Doctoral Student at Beijing Normal University. His research focus is on Chemistry Education.

Mark Wilson is a Distinguished Professor at UC Berkeley. He teaches courses on measurement in the social sciences, and his research focuses on a new framework for measurement informed by the philosophy of measurement, including statistical models based on scientific models, and on developing instruments for new constructs.

Duanli Yan is the Director of Computational Research at ETS and adjunct professor at Fordham University, and has extensive experience in innovative psychometric research and development. She has published many books and received many awards including the 2016 AERA D Significant Contribution to Educational Measurement and Research Methodology Award, 2022 and 2023 NCME Bradley Hanson Award.

Xiaoming Zhai is an Associate Professor of Science Education and Artificial Intelligence and Director of the AI4STEM Education Center at the University of Georgia. He is interested in applying cutting-edge technology, such as AI, in assessment practices to facilitate science teaching and learning.

FOREWORD: PROGRESS IN THEORY, RESEARCH, AND APPLICATION OF LEARNING PROGRESSIONS IN SCIENCE EDUCATION

Much has changed over the past 20+ years regarding our expectations for student knowledge in the STEM disciplines. Nowhere is this more clearly articulated than in the frameworks and standards for STEM education. In science, the integration of scientific content and reasoning and the focus on depth rather than breadth of knowledge was most clearly articulated by the National Research Council's *Framework for K-12 Science Education* (NRC, 2012). The science *Framework* and the derivative *Next Generation Science Standards* (NGSS, 2013) include three dimensions that are conceptually distinct but are integrated into practice in the teaching, learning, assessment, and knowing of science and engineering. The *Framework* emphasizes that disciplinary knowledge and scientific practices are intertwined and must be coordinated in science and engineering education. By engaging in the practices of science and engineering, students gain new knowledge about the disciplinary core ideas and come to understand the nature of how scientific knowledge develops. The *Framework* and *NGSS* specify in detail what all students should know and be able to do in science by the end of high school and how they should be able to demonstrate their proficiency. The challenge with both the *Framework* and *NGSS* since their introduction during the last decade has been how to design curricular and instructional materials to support the acquisition of these important competencies and how to organize classroom instruction, including the design and use of assessment, to promote student attainment of the complex disciplinary objectives embodied by these contemporary STEM standards (see e.g., Pellegrino, et al., 2014).

Fortunately, the past 20+ years have included the publication of important syntheses regarding the general nature of learning and knowing, as well as the specifics of learning and knowing in the STEM disciplines (e.g., NRC, 2000; 2018), that have major implications for the design of curriculum, instruction, and assessment. Within that broader literature, it has been argued that sociocultural theory provides the most compelling general theory of knowing and learning because it accounts for how meanings, purpose, values, and motivation are jointly developed as part of learning and helps to understand the ecology of classroom instruction and learning (e.g., NRC, 2018; Shepard, Penuel, & Pellegrino, 2018). In this perspective, learning takes place as individuals participate in the practices of a community, using the tools, language, and other cultural artifacts of that community. Thus, learning is "situated" within, and emerges from, the practices in different settings and communities. In the case of school, a classroom constitutes such a community, and learning is situated in the practices of that community for the teaching, learning, and assessment of STEM disciplines like mathematics and science.

The sociocultural perspective has important implications for how academic disciplines such as science should be taught and learned in school. For example, the disciplines are distinct communities that engage in shared practices of ongoing knowledge creation, understanding, and revision. It is not surprising that a sociocultural perspective is reflected in the current *Framework* and *Standards* for primary and secondary science education. For example, the science *Framework* and *Standards* call for integrated development of science practices, crosscutting concepts, and core ideas of the discipline. They posit that proficiency involves much more than being able to recite core ideas; it requires the application of those ideas to explain phenomena in the world and solve problems using the kinds of science and engineering practices that disciplinary experts use.

While sociocultural theory establishes broad principles about the nature of knowing and learning in general, and classroom learning environments in particular, the specifics of curriculum, instruction, and assessment need to be supplemented by discipline-specific models of knowing and learning. In that respect, learning progressions have become the most prevalent example of the more detailed disciplinary learning models that Penuel and Shepard (2016a, 2016b) identified as socio-cognitive models of learning. Such models attend to the social nature of learning and to discipline-specific ways that core ideas and practices are developed over time with instruction.

As noted in the Introductory chapter, "the conceptualization of the LP approaches were initially presented in two reports by the National Research Council (NRC): *Systems for State Science Assessment* (NRC, 2006) and *Taking Science to School* (NRC], 2007). These reports explain how LPs can be used to guide the development and alignment of curriculum, instruction, and assessment systems that are developmentally, horizontally, and vertically coherent." Since that time a variety of definitions and instantiations of the learning progression (aka learning trajectory) construct have been presented in the literature, with substantial variability in focus and intent (see e.g., Confrey, 2008; Confrey et al., 2008, 2009; Corcoran, Mosher, & Rogat, 2009; Daro et al., 2011; Duncan & Rivet, 2018; Duncan & Hmelo-Silver, 2009).

Two of the most extensive early discussions of the learning progression/trajectory construct and its implications for curriculum, instruction, and assessment can be found in the Consortium for Policy Research in Education (CPRE) reports *Learning Progressions in Science* (Corcoran et al., 2009) and *Learning Trajectories in Mathematics* (Daro et al., 2011). As described therein, learning progressions (*trajectories*) are empirically grounded and testable hypotheses about how students' understanding of, and ability to use, core concepts and explanations and related disciplinary practices grow and become more sophisticated over time with appropriate instruction. These hypotheses describe the pathways students are likely to follow toward the mastery of core concepts and practices. They are based on research about how students' learning actually progresses. The hypothesized learning trajectories are tested empirically to ensure their construct validity – *does the hypothesized sequence describe a path most students actually experience given appropriate instruction?* – and ultimately to assess their consequential validity – *does instruction based on the learning progression produce better results for most students?* The reliance on empirical evidence differentiates learning progressions from traditional topical scope and sequence specifications typically based only on logical analysis of current disciplinary knowledge and on personal experiences in teaching.

There is little doubt that STEM education literature has embraced the concept of learning progressions – this volume's Introductory chapter shows relevant publication data related to science learning progressions spanning the period from 2009 to 2023. The available early literature on learning progressions was used in developing the *Framework for K-12 Science Education* and the *Next Generation Science Standards*. A considerable body of empirical research has emerged over the last dozen years on further developing and testing such progressions for various aspects of science. As such, the work varies considerably in scope, content, and specificity. Some progressions have been derived from cross-sectional data spanning multiple grade levels while others are based on longitudinal data from within a single grade level or instructional topic area. Consequently, progressions vary in their levels of detail and the specificity or "grain size" at which they describe elements along the progression with varying implications for instruction and assessment.

Where are we in terms of understanding the concept of a learning progression and providing evidence of its value for science education? What has been learned in the period spanning the last 15+ years since the idea was first introduced into the science education literature? What are the conceptual, empirical, and pragmatic issues associated with understanding, establishing, validating, and using the learning progression construct in science education? How does learning progression theory and research help us to establish learning environments that align with the challenging goals for science knowing and learning specified in the *Framework* and *NGSS*?

This *Handbook of Research on Science Learning Progressions* takes us a long way in answering these important questions. It is all about elucidating *Progress in Theory, Research and Applications of Learning Progressions in Science Education*. The four major Sections of this volume, and the 30 chapters contained therein, cover multiple aspects of the conceptual, empirical, and pragmatic space by synthesizing what we know, what we still need to know, why this is important to know, and what we can do with the current knowledge we have obtained. A lot has happened since the 2006/2007 introduction of the learning progression concept in science education. The work described in this volume gives us an appreciation of how important the concept can be for meeting the promise of advancing curriculum, instruction, and assessment in science education. The Co-editors of this volume deserve kudos for assembling such an authoritative and comprehensive set of contributions. The chapter Authors deserve equal kudos for presenting such thoughtful and important scholarly work. I have no doubt that the field of science education will benefit greatly from the comprehensive and collective work reported in this *Handbook*.

James W. Pellegrino
Founding Co-Director,
Learning Sciences Research Institute,
University of Illinois Chicago

1
AN INTRODUCTION TO SCIENCE LEARNING PROGRESSION RESEARCH

Hui Jin, Joseph Joe Krajcik, and Duanli Yan

The rapid advancement of technology is transforming every aspect of human life, reshaping not only what scientific knowledge students should learn but also how they learn it. Despite the progress made in the past decades, historically marginalized groups—Hispanic, Black, and American Indian or Alaska Native individuals—are still underrepresented in science and engineering degrees and occupations (National Center for Science and Engineering Statistics [NCSES], 2023). Data from large-scale assessments show persistent performance gaps for underrepresented students, which have been exacerbated by the COVID-19 pandemic (National Science Board [NSB] & National Science Foundation [NSF], 2021). All these underscore the pressing need to foster effective learning of useful scientific knowledge and practices for all students.

The Learning Progression (LP) research is powerful in tackling these challenges because it leverages advanced theories, ideas, and methodologies from multiple domains. First, Next Generation Science Standards (NGSS) (The NGSS Lead States, 2013) and their framework provide a vision of three-dimensional learning—"over multiple years of school, actively engage in scientific and engineering practices and apply crosscutting concepts to deepen their understanding of the core ideas" in science disciplines (National Research Council [NRC], 2012, pp. 8–9). When curriculum, instruction, and assessment are organized into a coherent system (National Research Council [NRC], 2006), they can significantly promote the effectiveness of three-dimensional learning. Second, scholars in educational measurement have proposed theories and methodologies that validate the interpretation and use of assessment results, ensuring classroom assessments offer valid and valuable information for instruction (e.g., Kane, 2013; Mislevy et al., 2003; Pellegrino, 2018; Wilson & Scalise, 2006). Third, researchers in cognition and learning sciences have uncovered the cognitive structure and mechanisms behind human thought (e.g., Kahneman, 2011; Pinker, 2007). Finally, the advancements in artificial intelligence, particularly in natural language processing, have made the automation of tasks such as scoring and providing feedback possible. The LP research, by utilizing all these resources, can effectively tackle the above challenges in the modern era.

Adopting a developmental perspective, LPs describe how students move from their existing understanding to more sophisticated and useful scientific understanding in meaningful ways. They guide the development and alignment of curriculum, instruction, and assessment to help students learn scientific knowledge and practices in coherent and connected ways over time so

DOI: 10.4324/9781003170785-1

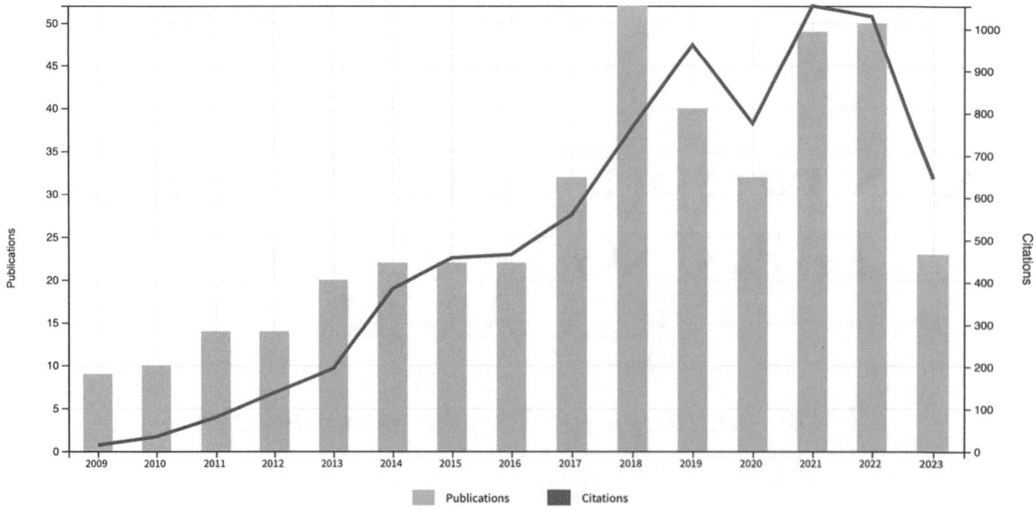

Figure 1.1 The number of LP articles published between 2009 and 2023 and their citations.

that knowledge is useable. During the past two decades, significant advances have been made in science LP research. Researchers have validated LP-based measurements (e.g., Jin et al., 2023; Kaldaras et al., 2021), developed LP-based instruction or the whole educational system (e.g., Anderson et al., 2018), studied teachers' use of LPs for classroom formative assessment (e.g., Alonzo, 2018), used automated scoring for LP-based assessments (Jescovitch et al., 2019), and so forth. A search of science LP research in the Web of Science resulted in 411 articles published between 2009 and 2023. These articles received a total of 7,557 citations. As shown in Figure 1.1, the interest in LP research grew exponentially from 2009 to 2021, with a sudden decline in 2020 likely attributed to the impact of the Pandemic.

In view of the substantial advancements in science LP research, it is important to review and analyze existing research to identify insights, challenges, and valuable lessons. Furthermore, LP research provides opportunities for exploring novel directions in curriculum, instruction, assessment, and teacher learning. To address this need, this handbook gathers contributions from leading LP scholars around the world; offers the latest developments in theories, methodologies, and applications of science LPs; and delves into future directions of LP research.

A Framework to Understand LP Research

In this introductory chapter, we present a framework that can be used to understand the development and the current state of LP research. As shown in Figure 1.2, the framework contains three components: the LP; the coherent curriculum, instruction, and assessment system; and teachers' learning and use of the LP and LP-based system.

LP: From Informal Reasoning to Disciplinary Logic

During the initiation of the LP research, science education researchers and developmental psychologists collaborated to develop a hypothetical LP for matter and its interactions (Smith et al., 2006). Unlike traditional approaches to matter, this LP addresses not only the

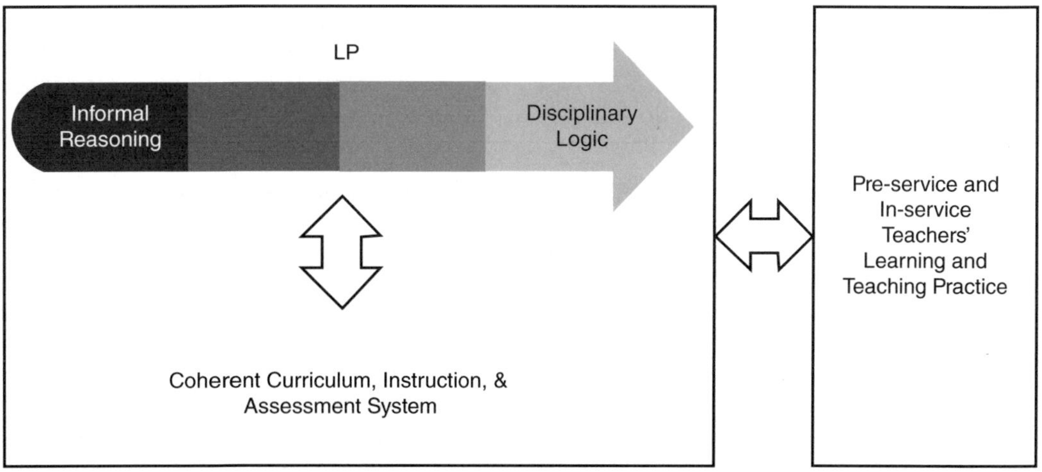

Figure 1.2 The framework for LP research.

disciplinary logic but also the informal reasoning. Here, disciplinary logic refers to the well-organized and integrated scientific knowledge and practices used in real-world situations, such as knowledge-in-use (Krajcik & Shin, 2023), three-dimensional learning (National Research Council [NRC], 2012), and styles of scientific reasoning (Kind & Osborne, 2017). It defines the upper anchor of the LP. Informal reasoning refers to the intuitive reasoning and nascent understanding that students hold at the starting point for a potential LP. Informal reasoning therefore defines the lower anchor of the LP. The levels that connect these two anchors depict how productive and effective learning can happen step by step.

In 2009, the Learning Progressions in Science (LeaPS) Conference was held at Iowa State University, Iowa City, IA. During that conference, researchers focused their discussions on four topics: defining LPs, developing LP-based assessments, psychometrics models used in LP research, and using LP for teaching and learning science. The product of the conference was an edited book, *Learning Progressions in Science: Current Challenges and Future Directions* (Alonzo & Gotwals, 2012). Also in 2009, a special issue of the *Journal of Research in Science Teaching* was devoted to science LPs. Duschl et al. (2011) reviewed this early work on science LPs and pointed out that LPs should not only address students' informal reasoning but also adopt a "work with it" view, as opposed to a "fix-it" view. From the "work with it" view, students' intuitive thinking and initial ideas should be used as resources for learning more advanced knowledge and progressing up the LP levels. Since then, researchers have conducted many empirical studies to develop LPs that guide students to move effectively from their initial ideas toward more sophisticated disciplinary knowledge.

An LP is a construct that depicts the tacit thinking and reasoning that students use to formulate responses. It is used to guide the development and use of coherent curriculum, instruction, and assessment systems. Therefore, both the LP and the use of the LP-based system should undergo validation to verify their suitability for scaling up and to ensure that teachers and students benefit from them. Given that LPs are tacit constructs and have intertwined relationships with curriculum, instruction, and assessments, design-based research methodology is often used to iteratively develop, revise, and validate LPs and LP-based systems. In this iterative process, the following questions should be considered: To what extent does the LP describe productive learning pathways that are justified with a theoretical rationale and supported by empirical data of student learning? Is the interpretation of LP-based assessment scores appropriate and to what extent does it offer meaningful insight into instruction? Is the use of the LP and LP-based assessment scores (e.g., to inform instruction) appropriate, and to what extent does it enhance student learning?

LP-based Curriculum, Instruction, and Assessment Systems

As noted by Lehrer and Schauble (Lehrer & Schauble, 2015), the conceptualization of the LP approaches was initially presented in two reports by the National Research Council (NRC): *Systems for State Science Assessment* (National Research Council [NRC], 2006) and *Taking Science to School* (National Research Council [NRC], 2007). These reports explain how LPs can be used to guide the development and alignment of curriculum, instruction, and assessment systems that are developmentally, horizontally, and vertically coherent. An LP-based system is developmentally coherent when it fosters the productive learning pathway described in the LP, begins with students' intuition, informal reasoning, or nascent understanding; and aims at the disciplinary logic. It is horizontally coherent when its curriculum, instruction, and assessment are designed to work in concert to support students in achieving the learning goals described in the LP (i.e., the upper anchor). It is vertically coherent when formative and summative assessments are employed at various levels, including classroom, school, school district, and national levels, to monitor students' progress toward shared learning goals (i.e., upper anchors).

A well-aligned LP-based system ensures students have coherent learning experiences, enabling them to learn disciplinary logic effectively and meaningfully. Beyond cohesion, the LP-based curriculum, instruction, and assessment systems should also address pressing issues in science education. For example, systems can be coherent but based on different instructional approaches. It is essential to determine which instructional approach is more effective and why. Furthermore, given the performance gaps and disparities, it is important to investigate questions such as: How can LP and LP-based systems be used in ways that promote social justice? How can they be designed in ways that provide effective support for underrepresented students who come from socially, culturally, and linguistically diverse backgrounds? As the chapters of this book show, while a few researchers examined different instructional approaches and proposed using LP with multiple learning pathways to fit the needs of diverse students, empirical investigations in these areas are insufficient. Furthermore, research utilizing LP approaches to promote social justice remains scarce.

Teachers' Learning and Teaching Practice

LPs and LP-based systems must be used by teachers with fidelity to achieve positive learning outcomes. Therefore, teachers' understanding and learning about the underlying theoretical ideas of LPs and LP-based systems, as well as their use of LPs and LP-based systems, are crucial aspects of LP research. In addition, the LP approaches can also be used to promote pre-service teachers' learning to teach. It is also important to recognize that the revision and enhancement of LPs and LP-based systems are also informed by teachers' use of them. To this end, the researcher-teacher partnership is critical. For example, to ensure the effectiveness and feasibility of the LP and LP-based systems, a team of researchers collaborated with teachers throughout the process of the development, revision, and use of LP and LP-based systems (e.g., Furtak & Heredia, 2014). Teachers play a central role in LP research because student progression relies heavily on teachers' effective use of LP approaches, LPs, and LP-based systems. However, compared with the research on the development of the LPs and LP-based systems, insufficient research attention has been given to teachers' learning and use of LPs.

Navigating the Handbook

In this handbook, Section I provides the theoretical and methodological foundation of LP research. Section II contains concrete examples of science LPs. Section III discusses the use of LPs to develop curriculum and instruction, as well as teachers' learning and use of LPs. The rapid advancement in technology is shaping the landscape of science education. Therefore, Section IV is devoted to the use of emerging technologies in LP research.

Section I discusses LP theories and methodologies. Chapters 2 and 3 provide theoretical foundations used to develop LPs. In Chapter 2, Smith and Amin discuss how theories from developmental psychology are utilized to develop the three key components of LP: lower anchor, upper anchor, and the processes that help students move from the lower anchor to achieve upper anchor. In Chapter 3, Jin and Krajcik respond to major critiques of the LP research: How can dynamic and complex science learning be presented as a sequence of a small number of LP levels? How can a single progression present the learning of many different students?

Chapters 4, 5, and 6 discuss validation in LP research. In Chapter 4, Kane discusses how an argument-based validation approach can be used to systematically validate interrelated claims about the theories, instructional design, and assessment used in LP research. In the subsequent chapters, two empirical studies are provided to illustrate the validation process for science LPs. In Chapter 5, Kaldaras and Krajcik discuss the validation of an LP for knowledge-in-use. Knowledge-in-use refers to the ability to apply organized and integrated scientific knowledge and practices to explain real-world phenomena and find solutions to real-life problems. Validating such a complex construct presents significant challenges. This chapter discusses the challenges and the strategies used to overcome those challenges. In Chapter 6, Wilson, Yao, and Osborne describe how the Berkeley Evaluation and Assessment Research (BEAR) Assessment System, which has been used in science LP research for a long time (Wilson, 2009), is used to develop and validate an LP for argumentation.

Chapters 7, 8, and 9 provide perspectives from international researchers and mathematics researchers. In Chapter 7, Neumann discusses a German approach, in which instruction systematically supports students in their progression toward building the envisioned competence. In Chapter 8, Wang and colleagues describe trends in the Chinese LP approach. They focus on the development of an integrated understanding of scientific knowledge and practices and the interaction between LPs and instructional practices. Mathematics education researchers have studied the learning progression/trajectory approaches much earlier than science education researchers (e.g., Simon, 1995). In Chapter 9, Clements, Guss, and Sarama, who are leading scholars in mathematics LPs, discuss how the various conceptions and methods used in mathematics LPs shed light on science LP research. In Chapter 10, the commentary author, Harris reviews the chapters in Section I and identifies challenges and opportunities for LP researchers and developers.

In Section II, scholars, who have been conducting longitudinal LP research for many years, provide insights on the effective utilization of LP to foster productive and meaningful learning in students. This section covers LPs in biology, chemistry, earth science, and environmental science.

Chapter 11 delves into the evolution of a genetics LP. In this chapter, Duncan and Castro-Faix detail how they and their colleagues employed various methods to overcome the challenges in the development and revision of the LP and associated assessment and instruction. An important contribution of their research is using an LP approach to compare the effectiveness of two instructional approaches, one used in molecular genetics and the other used in classical genetics. In Chapter 12, He, Shin, and Krajcik describe how they used design-based research methodology to develop three-dimensional LP for matter, interaction, and energy. The LP addresses a productive pathway toward using scientific knowledge to make sense of real-world phenomena and find solutions to real-world problems. In Chapter 13, Fortus and Nordine, who have been investigating energy in physics for years, propose an innovative LP-based instructional approach to energy. The traditional LP-based approach requires students to understand forms of energy, which often results in misconceptions and learning difficulties. Fortus and Nordine propose a system-transfer approach, which reflects a fundamental departure from this traditional approach. In Chapter 14, Duschl and colleagues point out the importance of LP research to reflect the most recent developments in geology and earth systems sciences. This involves understanding Earth as a dynamic adaptive system, acknowledging the role of new technology, and incorporating perspectives from trans-disciplinary geosciences research. Duschl and colleagues present two LPs to illustrate how LP research can address these recent trends. In Chapters 15 and 16, the researchers from the Carbon TIME project, including Scott, Miller, Anderson, et al., discuss how they develop LPs for matter in the context of environmental literacy. The LPs not only guide students' three-dimensional learning but also epistemic heuristics that prepare students for their future learning. In Chapter 17, Liu identifies key criteria for reviewing and comparing LPs based on existing literature. He applies these criteria to critically review the six LPs in this section, pinpointing areas requiring further research.

Section III features chapters that discuss how LPs and LP approaches are used in teaching, teacher education, and professional development. In Chapter 18, Gunckel, Covitt, and Bianchini discuss how LP-based curriculum materials can be designed with educative features to guide teachers in using LPs to bring coherence to curriculum, instruction, and assessments. In Chapter 19, Nordine and Lee identify the insufficient research base for LPs that explicitly use crosscutting concepts to guide curriculum, instruction, and assessment. Based on a review of relevant empirical research, they propose a path forward for conducting LP research that aligns with the special role that crosscutting concepts play in science learning.

Chapters 20, 21, and 22 focus on teachers' learning and use of LPs. In Chapter 20, Luft, Hang, and Singh focus on an important issue—using the LP approaches to promote teacher learning. Based on an examination of research on science teacher LPs, they identify three orientations in these LPs: LPs as resources, roadmaps, and representation. In Chapter 21, Alonzo tackles a critical issue in teachers' use of LP—Teachers frequently use LPs in ways that do not rely on LP levels. Alonzo further recommends future research and collaborative efforts with teachers to re-envision the use of LPs in enhancing student learning. In Chapter 22, Furtak, Henson, and Buell describe how an LP is used to support the teaching of energy across high school physics, chemistry, and biology classes. Specifically, the LP provides an orienting framework to guide teachers' professional learning. In Chapter 23, Barton and Tan tackle a critical issue that LP should address—social justice. They discuss how well the current LP research addresses youths' equitable presents and futures and offers fruitful directions for future LP research. In Chapter 24, Campbell reviews the chapters in this section. The review focuses on two critical questions about LP's use in teaching and teacher learning: What are the benefits and challenges of LPs supportive of science teacher learning and LP use? In what ways can science teachers learning with LPs, and their use, be (un)supportive of students' plural onto-epistemological reasoning?

Section IV features chapters on new technology used in LP research. In Chapter 25, Almond, Yan, and Jin provide an example of how dynamic Bayesian networks can be used to monitor student learning over time. In Chapter 26, Riordan and colleagues tackle a critical issue in automated scoring—How can scoring rubrics be used with Natural Language Processing Models to automatically score students' responses in LP-based assessment with validity? In Chapter 27, He and colleagues provide an example of integrating Artificial Intelligence (AI) technology into knowledge-in-use LPs. Specifically, AI technology is used for automatic scoring, personalized feedback, and instructional support in learning systems. In Chapter 28, Gobert and colleagues discuss how AI technology is used to develop a dashboard tool for teachers to assess and monitor students' progression along LPs. In Chapter 29, Zhai provides a critical evaluation of these chapters. He further points out that using technology in LP research is not merely a research trend; it suggests a shift in the research paradigm.

Conclusion

LPs, grounded in research and founded upon learning principles, present a promising avenue for affording every student equitable access to rigorous scientific concepts and practices. Although students, especially those originating from diverse backgrounds, exhibit varied learning modalities, LPs capture the predominant pathways that support students in learning. This facilitates understanding of challenging ideas and practices enabling learners to explain phenomena from their daily experiences, address challenges they face, and seek further knowledge when necessary.

Numerous chapters within this volume underscore that LPs are not panaceas, nor can they bring about change on their own. Instead, their potential can be fully harnessed only when complemented with student learning materials, pedagogical resources, classroom assessments, high-stakes evaluations, and continued professional development for educators. Furthermore, the emphasis of LPs on incremental knowledge accumulation necessitates policy endorsements. Many educational districts prescribe specific knowledge milestones; however, the heterogeneity in students' learning paces, with some necessitating extended exposures, renders it impractical to expect uniform content mastery timings. The pedagogical approach should eschew prematurely transitioning to new concepts, risking leaving some students behind. Such an approach perpetuates a detrimental cycle where certain students become disengaged from learning novel and challenging ideas and practices. The advocacy, therefore, is for policies and professional development that champion a longitudinal and development perspective on education.

Another pressing area for policy intervention remains elementary education. A significant portion of students, especially those in grades K-2, are scarcely exposed to learning science. This disparity results in inequitable scientific learning opportunities. Without foundational instruction in the early grades, students lag in grasping pivotal scientific ideas and practices. Grasping intricate ideas like adaptation, energy, and sustainability, as well as mastering practices like modeling and evidence-based explanations, develop as students encounter new experiences.

The chapters within this Handbook highlight the advancements in LP research that can positively impact student learning. Nonetheless, the onus remains on the professional community to amplify research efforts. It is our aspiration that readers find the chapters in this Handbook enlightening, spurring them to engage in research that ensures equitable learning opportunities for all children.

References

Alonzo, A. C. (2018). Exploring the learning progression–formative assessment hypothesis. *Applied Measurement in Education*, *31*(2), 101–103. https://doi.org/10.1080/08957347.2017.1408625

Alonzo, A. C., & Gotwals, A. W. (2012). *Learning Progressions in Science: Current Challenges and Future Directions*. Springer Science & Business Media.

Anderson, C. W., de los Santos, E. X., Bodbyl, S., Covitt, B. A., Edwards, K. D., Hancock, J. B., Lin, Q., Morrison Thomas, C., Penuel, W. R., & Welch, M. M. (2018). Designing educational systems to support enactment of the Next Generation Science Standards. *Journal of Research in Science Teaching*, *55*(7), 1026–1052. https://doi.org/10.1002/tea.21484

Duschl, R., Maeng, S., & Sezen, A. (2011). Learning progressions and teaching sequences: a review and analysis. *Studies in Science Education*, *47*(2), 123–182. https://doi.org/10.1080/03057267.2011.604476

Furtak, E. M., & Heredia, S. C. (2014). Exploring the influence of learning progressions in two teacher communities. *Journal of Research in Science Teaching*, *51*(8), 982–1020. https://doi.org/10.1002/tea.21156

Jescovitch, L., Doherty, J., Scott, E., Cerchiara, P., Jack, Wenderoth, M., Urban-Lurain, M., Merrill, J., & Haudek, K. (2019). Challenges in developing computerized scoring models for principle-based reasoning in a physiology context. In *annual meeting of the National Association for Research in Science Teaching*, Baltimore, MD.

Jin, H., Shin, H. J., & Cisterna, D. (2023). Systematic validation in science learning progression research. *International Journal of Science and Mathematics Education*. 10.1007/s10763-023-10359-w

Kahneman, D. (2011). *Thinking, fast and slow*. Macmillan.

Kaldaras, L., Akaeze, H., & Krajcik, J. (2021). Developing and validating Next Generation Science Standards-aligned learning progression to track three-dimensional learning of electrical interactions in high school physical science. *Journal of Research in Science Teaching*, *58*(4), 589–618. https://doi.org/10.1002/tea.21672

Kane, M. T. (2013). Validation as a pragmatic, scientific activity. *Journal of Educational Measurement*, *50*(1), 115–122. 10.1111/jedm.12007

Kind, P., & Osborne, J. (2017). Styles of scientific reasoning: A cultural rationale for science education? *Science Education*, *101*(1), 8–31. 10.1002/sce.21251

Krajcik, J., & Shin, N. (2023). Student concepts, conceptual change, and learning progressions. In N. G. Lederman, D. L. Zeidler, & J. S. Lederman (Eds.), *Handbook of Research on Science Education: Volume III*. Taylor & Francis.

Lehrer, R., & Schauble, L. (2015). Learning progressions: The whole world is NOT a stage. *Science Education*, *99*, 432–437. 10.1002/sce.21168

Mislevy, R. J., Almond, R. G., & Lukas, J. F. (2003). A brief introduction to evidence-centered design. *ETS Research Report Series*, *2003*(1), i–29. 10.1002/j.2333-8504.2003.tb01908.x

National Center for Science and Engineering Statistics [NCSES]. (2023). *Diversity and STEM: Women, minorities, and persons with disabilities 2023*. National Science Foundation; Special Report NSF 23-315. https://ncses.nsf.gov/pubs/nsf23315/

National Research Council [NRC]. (2006). *Systems for state science assessment*. National Academies Press.

National Research Council [NRC]. (2007). *Taking science to school: Learning and teaching science in grades K-8*. National Academies Press.

National Research Council [NRC]. (2012). *A framework for K-12 science education: Practices, crosscutting concepts, and core ideas*. National Academies Press.

National Science Board [NSB], & National Science Foundation [NSF]. (2021). *Science and engineering indicators 2022*. Elementary and Secondary STEM Education; NSB-2021-1. https://ncses.nsf.gov/pubs/nsb20211/

Pellegrino, J. W. (2018). Assessment of and for learning. In F. Fischer, C. E. Hmelo-Silver, S. R. Goldman, & P. Reimann (Eds.), *International Handbook of the Learning Sciences* (pp. 410–421). Taylor & Francis.

Pinker, S. (2007). *The stuff of thought: Language as a window into human nature*. Penguin.

Simon, M. A. (1995). Reconstructing mathematics pedagogy from a constructivist perspective. *Journal for Research in Mathematics Education*, *26*(2), 114–145. 10.5951/jresematheduc.26.2.0114

Smith, C. L., Wiser, M., Anderson, C. W., & Krajcik, J. (2006). Implications of research on children's learning for standards and assessment: A proposed learning progression for matter and the atomic-molecular theory. *Measurement: Interdisciplinary Research and Perspectives*, *4*(1–2), 1–98. 10.1080/15366367.2006.9678570

The NGSS Lead States. (2013). *Next Generation Science Standards: For States, by States*. National Academies Press. 10.17226/18290

Wilson, M. (2009). Measuring progressions: Assessment structures underlying a learning progression. *Journal of Research in Science Teaching*, *46*, 716–730. https://onlinelibrary.wiley.com/doi/abs/10.1002/tea.20318

Wilson, M., & Scalise, K. (2006). Assessment to improve learning in higher education: The BEAR assessment system. *Higher Education*, *52*(4), 635–663. 10.1007/s10734-004-7263-y

SECTION I

Learning Progression Theories and Methodologies

2

COGNITIVE FOUNDATIONS OF SCIENCE LEARNING PROGRESSIONS

Carol L. Smith and Tamer G. Amin

Introduction

A science learning progression (LP) is an *idealized model* of how students can develop understanding or skill in the use of disciplinary core ideas and science practices, given appropriate instruction, building on their initial conceptual and reasoning resources. Disciplinary core ideas typically involve a major theoretical framework in modern science such as atomic molecular theory or evolutionary theory: that is, unifying concepts, principles, and/or explanatory models. These are hard-won intellectual achievements involving counter-intuitive concepts. Science practices, such as modeling, designing and carrying out investigations, and argumentation, are what scientists do to construct and evaluate scientific knowledge. While they build on everyday sensemaking and reasoning processes, they constitute a historically recent and distinctive way of thinking that cannot be learned without instruction.

Initial interest in science LPs was driven by the dissatisfaction with the shallow science learning of students and the conviction that a deeper understanding of disciplinary core ideas and science practices could be developed *if* curricula and instruction supported their sustained development over a broad period drawing on learning processes needed to support meaningful learning and conceptual change (National Research Council (NRC), 2007). An extensive body of research showed that students shared many alternative ideas about science topics that they had developed from their everyday experiences that often remained unchanged after traditional instruction (Driver et al., 1994), leading to a wealth of research about how to redesign instruction to support conceptual change. It became clear that bridging the gap between student ideas and scientists' ideas involved multiple steps requiring coordinated changes in children's conceptual systems driven by instruction that emphasized modeling processes and drew on the knowledge and reasoning resources that children bring to school. The notion of an LP was proposed to show how this research could inform the rethinking of science standards and curriculum design. An LP typically includes a description of the targeted understanding of a disciplinary core idea or science practice ("the upper anchor"), the relevant precursor ideas and reasoning abilities available at the start of schooling ("the lower anchor"), and the productive intermediate levels of understanding and skill that can serve as targets of instruction between the lower and upper anchors, putting the next step in the LP within reach.

DOI: 10.4324/9781003170785-3

Constructing and evaluating LPs is challenging and requires interdisciplinary effort. Curriculum and instructional design specialists are needed to provide clear and practical characterizations of curricular sequences and instructional activities. Assessment experts are needed to help teachers design formative assessments to track students' progress in classroom settings and to build large-scale assessments. But these efforts must be grounded in a cognitive and developmental perspective—the focus of this chapter—that provides rich conceptual analyses of scientists' and students' conceptions, allowing us to identify early and intermediate forms of understanding that might be *causally connected* to building an understanding of the disciplinary core idea or practice, but are often overlooked. This perspective is also important for understanding the cognitive processes that instruction must engage to drive the desired changes.

Since the first LPs were proposed almost 20 years ago, many LPs for disciplinary core ideas and science practices have been suggested. This area of research is still new, so it is not surprising that approaches to their construction and evaluation have varied. Indeed, LP researchers have responded differently to various challenging questions: How conservative or transformative should we be when conceptualizing the upper anchor and intermediate levels of an LP (Lehrer & Schauble, 2015)? By what criteria does one judge an intermediate understanding as "more sophisticated" than another (Sikorski & Hammer, 2010)? How can we identify productive knowledge and reasoning resources that can serve as the lower anchor, when they may not look like those in the upper anchor or come from another domain (Metz, 2009)? How can one characterize a single or small number of idealized learning pathways when learners have diverse knowledge profiles and learning trajectories (Hammer & Sikorski, 2015)? What grain size is appropriate for describing levels within an LP (Lehrer & Schauble, 2015)? What is the relationship between developing an understanding of a disciplinary core idea and science epistemic practices (Smith & Wiser, 2015) and, more broadly, the relationship between an LP as a cognitive model of learning and instructional design (Duncan & Gotwals, 2015)?

How one conceptualizes the cognitive foundations of LPs has implications for how to address these challenges, so we conclude this chapter by discussing those implications. We clarify at the outset, though, that our approach is based on two assumptions (shared with NRC, 2007): that there is a complex interplay between maturation, experience, and instruction; and that cognitive development is not in any sense inevitable, but depends on learning experiences, and instruction can significantly alter developmental trajectories.

We organize this chapter around three questions: (a) What are widely shared student conceptions and reasoning abilities relevant to characterizing the lower anchor of a science LP at the start of schooling? (b) How do they compare with the target conceptions and reasoning abilities in the upper anchor? and (c) What are the processes that can be engaged through carefully designed instruction that allow students to build on, yet transform, conceptions in the lower anchor to achieve the target conceptions of the upper anchor? To answer these questions, we draw on research from cognitive development, the cognitive and learning sciences, science education research on conceptual change, and research on conceptual innovation in science.

We use the term "conception" here to refer to all kinds of knowledge resources (e.g., concepts, beliefs, models, image schemas). Indeed, we assume that when characterizing both intuitive and scientific conceptual systems, we need a systemic perspective, recognizing roles for multiple kinds of interacting knowledge elements that function at different levels of analysis. We use the term "concept" more narrowly to refer to a locally organized portion of a knowledge network, which draws on multiple resources and processes, to enable reference to, categorization of, and reasoning and communication about specific entities, relations, or processes (Amin, forthcoming; Amin et al., 2014; Wiser & Smith, 2016). Conceptual change is a protracted process that involves the addition and coordinated reorganization of different types of knowledge elements to enable change in both what the concept refers to and the reasoning it supports. Thus, in this chapter, we show how a systemic perspective is needed to characterize the lower anchor, the discontinuities and continuities between the lower and upper anchors, and the knowledge elaboration and bootstrapping processes that transform one to the other. We then consider the implications of this systemic perspective for the design of long-term LPs.

Characterizing Conceptual and Reasoning Resources for the Lower Anchor

We begin by surveying the diverse conceptual and reasoning resources available to young children of particular importance to their future learning of scientific concepts. We first consider the resources of *preverbal* infants that support their learning about everyday events; we then describe how language and other publicly shared symbol systems enrich and transform conceptual knowledge and reasoning in the preschool years. Understanding the range of resources, the contexts in which they develop, and the purposes they serve helps us understand the challenges, and potential, of repurposing those resources to support scientific understanding.

Conceptual and Reasoning Resources of Preverbal Infants

What resources are needed to get conceptual development started? Piaget (1952) argued that infant knowledge should be characterized exclusively in sensorimotor terms. In his account, *conceptual* representations—mental representations with content that goes beyond *perceptual* information such as color, size, shape, texture, and sound—developed in the preschool years. Generally, his view was that conceptual development proceeds from the concrete to the abstract: the capacity to represent and reason about entities that were not directly perceived was the result of a gradual process of forming and coordinating sensorimotor schemas. While developmental psychologists continue to accept that forming and coordinating sensorimotor schemas play an important role in the mental life of infants, and others, many no longer think these schemata are formed using only bottom-up learning mechanisms (such as association or abstracting statistical regularities from sensory input). Rather, it has been argued that conceptual development proceeds from the abstract to the concrete (Keil, 2007): infants are born with a range of domain-specific and domain-general learning devices, which already embody some abstract causal principles and meaningful *conceptual* content such as ideas about objects, agents, events, places, number, causes, and kinds (Carey, 2009). We highlight here some of the conceptual resources embodied in these devices and their importance for subsequent learning.

Researchers have hypothesized the presence at birth of core knowledge systems consisting of domain-specific core concepts and principles that enable infants to represent, reason, and learn about ecologically important entities in their environment. These knowledge systems are understood to be innate skeletal frameworks that are elaborated with learning, but never abandoned or revised. Evidence has been obtained with infants as young as two months (younger in some cases) using methods that rely on looking times, making fewer information processing demands than Piaget's action tasks. One paradigm takes advantage of infants' tendency to look longer at test events that violate rather than confirm their expectations, allowing researchers to infer the kinds of concepts and abstract principles that underlie these expectations. There is now extensive evidence for at least four different core knowledge systems (Carey, 2009; Kinzler & Spelke, 2007) allowing infants to represent and reason about physical objects, intentional agents, places, and approximate number. Research is also uncovering other possible core knowledge systems that underlie infants' understanding of non-solid substances (such as water and sand) (Hespos et al., 2016) and of animals as biological beings (Setoh et al., 2013).

An important part of this research is characterizing the types of perceptual cues exploited by innate *input analyzers,* the network of *concepts and principles* that are part of the skeletal framework for each domain (generated as output from these input analyzers), and the types of everyday *events* that the system can predict or explain. The core knowledge system dealing with physical objects is particularly relevant to our purposes. Infants not only possess the general concept of an *object*, but also distinguish *inert* and *self-propelled objects* (Lin et al., 2022). Based on spatiotemporal cues, infants segment the perceptual array into objects understood to be discrete, bounded, and cohesive entities that occupy space, move on continuous paths, and continue to exist when out of sight. In addition, core knowledge about inert objects includes understanding that unsupported objects fall and that objects remain at rest or once in motion don't change direction unless acted on by another object. In contrast, self-propelled objects are seen as containing an internal energy source, enabling them to start, stop, and change direction on their own. Crucially, the reasoning in this domain implicates the concept of permanent object persisting in time and space and causal notions of mechanical (contact) causality and internal energy which go beyond spatiotemporal information given in perception.

Another core knowledge system—one for predicting and explaining the behavior of agents—is centered on intentional rather than mechanical causality. This system (see review by Baillargeon et al., 2015) also includes a finer distinction between *human* and *non-human agents* (such as animals) and minimally includes the explanatory concepts of an agent's *goals* and *attentional focus*. Infants expect the actions of agents to be *rational*—i.e., consistent with their inferred goals— and to reflect relatively *efficient* paths to their goals. Significantly, the innate input analyzers in this case use patterns of an object's actions (e.g., contingent and reciprocal interactions with other objects; self-initiated, non-straight-line motion) to identify an agent and infer its goals, and cues from eye gaze to infer attentional focus.

Research has been uncovering how core knowledge systems support infant reasoning and learning about everyday events. As Lin et al. (2022) argue, core knowledge provides a basis for forming specific *event categories* in which objects causally interact in different ways such as occlusion, containment, support, and collision events. Infants use their (schematic) understanding of event categories to form specific event representations in working memory that encode features of the events they are observing that they deem causally relevant to making predictions and forming explanations. That is, they form an imagistic mental model of the event that embodies core principles to predict or imagine how it will unfold. There is strong evidence that infants as young as two months are forming and running such mental models. For example, infants are surprised (look longer) when a ball is rolled behind a screen and the screen is lifted to reveal the ball on the other side of an obstacle in its path, since the obstacle should have stopped it. Initially, infants include only basic categorical descriptors of the relations between objects in their event category knowledge. They then learn to include more relevant causal variables, treating them first qualitatively and then quantitatively. For example, in developing expectations about collision events, they initially expect an object will move if hit by another object, and only gradually refine their predictions to take account of how the relative sizes and weights of the colliding objects affect whether the object moves and how far.

This ability to simulate what should happen as events unfold allows infants to engage in a simple form of model-guided and explanation-based learning. According to Baillargeon and DeJong (2017), this learning process is triggered when infants encounter *unexpected contrastive outcomes*—situations that infants encode as the same but have different outcomes. This leads them to scrutinize the situations to identify a relevant variable that distinguishes them and that could explain the different outcomes in ways consistent with their existing knowledge. If they succeed and have a few more confirming examples, the newly discovered variable is incorporated into their event category. For example, infants who expect that objects should be stable when at least half their *bottom surface* is supported but then observe that an object sometimes falls in such situations could now pay attention to the *whole object* and note that when the object has an asymmetrical shape it is supported only when the bigger/heavier half is on the platform. Baillargeon and DeJong tested this analysis by exposing such infants to relevant contrastive outcomes and found they appropriately refined their expectations with exposure to just two or three contrastive experiences. Crucially, in another condition, infants who saw contrastive outcomes that would support a *reverse* expectation (e.g., The object is stable only when the smaller/lighter half is on the platform) and which cannot be plausibly explained given their existing knowledge, did not form that new expectation. So even at this very early point in conceptual development, generalizations infants form from experience are constrained by the cognitive representations and causal knowledge already in place.

In addition to domain-specific learning mechanisms, research suggests infants have a domain-general mechanism for forming causal representations that allow integrating causal information across domains (Carey, 2009) and an innately specified mechanism that provides a skeletal framework for representing any (lexicalized) kind concept (Prasada, 2016). The latter helps explain the ease with which infants form many basic-level kind concepts—such as kinds of objects (bottles, cups, trucks), animals (dogs, cats, birds) and substances (water, juice, milk)—by the end of their first year (Mandler, 2004); crucially, this mechanism enables them to represent the *principled* connection some properties have to being that kind (e.g., dogs have four-legs *because* they are dogs) and different instances as equivalent to other instances of that kind.

All this research clarifies the robust and subtle nature of the preverbal conceptual knowledge that develops over the first two years of life, in part, through a simple form of model and explanation-based reasoning. While detecting statistical regularities in perceptual and sensorimotor experiences contributes to the generalizations they form, the models they construct are highly constrained by their conceptual knowledge. Crucially, infants don't have to learn what an object, agent, goal, or cause is. Rather, because of their expanding experiences with contrasting cases as well as their increasing efficiency in encoding new information that allows them to attend to more without exceeding their processing limits, they gradually encode more detail about causal and spatial relations to support more sophisticated predictions about events. This knowledge continues to be a source of intuitive understanding for older children and adults.

Science education researchers have recognized the relevance of nonverbal, intuitive knowledge to science learning. Some focus on how this knowledge differs from canonical science knowledge and can interfere with learning (e.g., McCloskey, 1983). Others argue that this early developing knowledge can be repurposed and contribute productively to understanding scientific concepts (e.g., diSessa, 1993). We agree with *both* views and believe that research on infant knowledge helps us understand concept learning trajectories in two ways. First, recognizing that core conceptual representations are the output of innate input analyzers helps us anticipate the contexts in which they are likely to be activated and thus potentially interfere with learning scientific concepts. In these cases, instruction could help learners become metacognitively aware of this intuitive knowledge and develop their ability to inhibit the inferences automatically generated. Second, and crucially, this conceptual knowledge is the starting point of further conceptual development and the basis upon which scientific knowledge is ultimately built, so we need to understand how this knowledge contributes productively to subsequent learning.

Expanding the Range of Resources for Knowledge Building in the Preschool Years

Conceptual development in the preschool years builds on the resources of the preverbal infant but new resources appear. Characterizing what stays the same and what changes is actively debated. Here we present what is well-supported about the continuity in *content* of conceptual representations, and the ways this content becomes elaborated, including through dialog with others. We also discuss how acquiring language (and other publicly shared symbols) expands the available *formats* of conceptual representations, enabling fundamental changes in conceptual content and reasoning, and expanding the resources for knowledge building.

One area of continuity concerns much of the *content* of what preschoolers think and are now able to talk about. Many early words are learned through a process of "fast mapping" where a label is mapped directly onto conceptual representations already available to preverbal infants (Gelman & Brandone, 2010). There is also continuity in conceptual content at a more abstract level: just as core knowledge systems have been found to guide learning about specific entities in infancy, it is proposed that framework theories (consisting of interrelated concepts and causal principles that pick out ontologically important entities) guide learning in children beyond infancy. Wellman and Gelman (1992) describe three framework theories for which there is strong evidence in preschoolers: a naïve physics centered on the concept of *inanimate object* and an understanding of contact causality; a naïve psychology centered on a concept of an *intentional agent* with the understanding that mental states such as beliefs and desires guide behavior and that perceptual access or testimony from others affect beliefs (dubbed "a theory of mind"); and a naïve biology centered on the concept of *biological kinds* (initially limited to animals) with a rudimentary understanding that living things have internal and external parts that serve some function (e.g., movement, growth, health). For Wellman and Gelman, framework theories are more abstract with fewer worked-out details than specific theories but make possible and guide systematic inquiry and specific theory construction within a broad domain.

Exactly how infants' core knowledge systems relate to preschoolers' framework theories is an active area of research (Sodian, 2016). We find Carey's (2009) characterization of this relationship promising. She notes the considerable similarity in content between infants' core knowledge systems for objects and agents and preschooler's naïve physics and theory of mind, respectively. She explains this similarity by proposing that the conceptual *output* of innate input analyzers is likely the starting point for the construction of the first framework theories. Moreover, both support reference, inference, and learning. But she also lists several important differences. First, whereas core knowledge is represented nonverbally and most likely iconic in format, some aspects of framework theories are represented using publicly shared symbol systems such as language, which enable explicit propositional representations. In addition, the concepts and principles of core knowledge systems are innately specified and not revisable, whereas the linguistically encoded concepts and principles of framework theories are. These differences support differences in how these two kinds of concepts refer and in their inferential roles. Regarding reference, innate input analyzers determine how core knowledge systems pick out the members of distinct ontological kinds, whereas the mechanisms for sustaining reference of the concepts of framework theories are more complex: for example, once children map words to concepts, they can defer to others with greater expertise in establishing intended referents. Finally, hearing the ways words are interrelated in sentences opens up new inferential roles not innately specified and can lead to both enrichment and revisions in inferential networks.

Thus, learning language provides an important new resource for conceptual development allowing children to overcome multiple limitations of preverbal thought and ushering in uniquely human forms of reasoning, communicating, and learning. Let's now consider in more depth the multiple ways language both shapes and transforms children's concepts and reasoning. First, labels invite children to form conceptual categories more quickly or ones that they might not otherwise form simply based on firsthand observation (see review by Perszyk & Waxman, 2018). Labels also encourage children and adults to treat category members as if there is some underlying essence that explains their observable characteristics (Gelman, 2009). On Gelman's account, the essentialist assumption is already part of the representational repertoire of the preverbal child; the label simply triggers the essentialist interpretation of the category. More generally, Tomasello (1999) argues that when children learn a language, a rich perspectival system, they acquire the ability to name and construe situations flexibly in accord with the language of their community. For example, speakers can construe entities at different levels of generality ("animal," "dog," "poodle") and situations from different perspectives ("buy" versus "sell"); and they can construe events dynamically using verbs ("He *ran* home"), more statically using nouns ("He went for *a run*"), or with different force dynamic interpretations associated with different grammatical constructions ("He rolled the ball" versus "He kept the ball rolling").

The phenomenon of conceptual metaphor studied from an embodied cognition perspective by cognitive linguists not only provides another illustration of the perspectival nature of language, but also underscores the way more abstract concepts are systematically construed metaphorically in terms of image schemata emerging from sensorimotor experience (Lakoff & Johnson, 1980, 1999). Analysis of everyday language reveals widespread patterns where concepts from one (more abstract) ontological category are construed in terms of concepts from another (more concrete) one: Events Are Objects ("That was a really *big* party"); Time Is a Location ("We're *approaching* the holidays"); States Are Containers ("Are you *in* a bad mood?"), supporting the projection of inferences from image-schematic knowledge onto abstract concepts. Image schemas are understood as multimodal structured gestalts consisting of a small number of components. For example, the *container* image schema consists of an *exterior*, an *interior*, and a *boundary* (a schematization of recurrent experiences such as putting things in and taking them out of containers). As generalizations over sensorimotor experience, image schemas resemble event categories as discussed by Lin et al. (2022) and are likely constrained by innate core knowledge, explaining how they support inference and why the formation of some image schemas is more likely than others. Research has shown that the capacity for analogical reasoning has its roots in infants' abilities to detect relational similarity, and that even three-year-olds can engage in simple forms of analogical reasoning (Goswami, 2013). Language enhances this capacity leading to a powerful human form of relational reasoning, of which the phenomenon of conceptual metaphor is just one manifestation (Gentner, 2016).

These language-based construals can have both positive and negative consequences for future science concept learning. For example, a common label encourages children to generalize a new fact they have learned about a hidden property of one category member to other perceptually dissimilar members; it can also function as a placeholder to invite learning about a category, such as when one learns the name of a new type of animal. On the other hand, labels encourage children to treat category boundaries as fixed; this can impede the development of a population-based conception of species, for example, an idea central to understanding evolution by natural selection (Gelman, 2009). Similarly, the different "force dynamic" construals implicit in language, described extensively by Talmy (1988), can both guide and limit children's learning about motion. These construals involve simple categorical rather than quantitative distinctions and at most consider only two forces acting in the same or opposite directions: for example, *cause to happen* (e.g., cause to move), *enable/help* (e.g., reinforce motion with a secondary force in the same direction), and *prevent* (e.g., force opposing motion). Three- to six-year-olds were more successful in prediction tasks for situations that matched schematizations implicit in language than in more complex situations (George et al., 2019).

Our everyday metaphorical construals can also help or hinder learning scientific concepts. For example, Lacy et al. (2014) use the everyday metaphoric constructions "X *has* energy" and "Y *gives* Z energy" (construing energy as a possession and exchanged substance) as natural entry points for an LP on energy in the elementary years. Scientists also use metaphorical language, but their use is subtle and varies with context, presenting challenges to learners, especially when a metaphor misleads the learner by reinforcing a problematic ontological construal of a scientific concept (e.g., *heat as substance*; Brookes & Etkina, 2015).

Language has a still deeper influence by allowing the construction of entirely new sets of concepts via *Quinian bootstrapping* (Carey, 2009). In this process, language (or a public symbolic representation, more generally) encodes new symbols and their relations that are not yet understood and serves as a placeholder guiding the construction of new concepts. Carey's most developed example is the role of language in children constructing an understanding of natural number—a system that allows preschoolers to assign a precise numerical value to a set of any size. In this process, children first learn the count routine and the ordered set of words in the count list by rote, which serves as a placeholder for the conceptual structure to be built. They gradually fill in the meanings of the symbols in the structure: first mapping the word "one" to models of sets of one, then "two" to models of two items, and "three" to models of three items. They then make the analogical and inductive inference that *order* in the count list corresponds to the order of *numeric value* and the next word in the list is exactly *one more*. Thus, Quinian bootstrapping draws on a variety of reasoning and learning processes (e.g., associative learning, analogy, and induction) and involves integrating elements from previously distinct representational systems available to infants (e.g., representations of discrete objects, approximate number, sets, and linguistic representations).

Another example of conceptual bootstrapping in the preschool years involves the construction of the concept of a material kind as a separate explanatory level. Although even infants make an ontological distinction between solid objects and non-solid stuff, the innate input analyzers that trigger these categorizations have mutually exclusive referents. Wiser and Smith (2016) argue that language draws attention to the fact that solid objects can be construed as materials *as well as* objects by hearing the contrasting constructions for the same object (e.g., "This is a boat." vs. "This is wood."). They also hear constructions that relate both construals simultaneously (e.g., "This boat is made of wood.") and that invoke material-kind explanations (e.g., "The boat burned *because* it is *made* of wood."). In this way, language plays a role in bootstrapping a material kind construal distinct from a whole object construal and the emergence of a new framework theory for investigating and interrelating properties of objects and materials.

These two examples illustrate Quinian bootstrapping constrained by language acquisition in early childhood. Later, we offer a more detailed example of carefully designed instruction and discuss how it can account for profound conceptual changes that take place in science learning.

Language also enhances the capacity for logic and meta-representation (thinking about thinking) and the conscious metacognitive control and regulation of thinking. For example, language not only explicitly encodes propositions (in simple sentences) but connects them logically (with words like, "and", "or", "if/then" or "because"), marks their modal status (e.g., as *possible*, *impossible*, *likely*, or *necessary*), and expresses different epistemic attitudes towards them (e.g., *thinking*, *believing*, *knowing*, or *guessing*). During the preschool years, children acquire many words that express these subtle logical, modal, and epistemic concepts. For example, by four to five years of age, they can explicitly mark something as a possibility, consider multiple possibilities simultaneously, and distinguish certainty from uncertainty and partial from total ignorance (Leahy and Carey, 2020). Similarly, they can also explicitly consider and talk about their own and others' beliefs and recognize that people can have different beliefs based on what they have seen or heard from others (Sodian, 2016). Developing explicit metacognitive awareness about what they know enhances children's ability to monitor and control their learning in several ways. Preschool children not only ask many information-seeking questions, but also persist in asking questions if they judge responses to their questions not sufficiently informative, consistent with their monitoring whether their information-seeking goal has been achieved (Chouinard, 2007). With reliance on getting knowledge through testimony from others, preschoolers are also beginning to discriminate among more reliable, knowledgeable, and informative sources—an ability described as "epistemic vigilance" (F. Clement, 2016; Harris, 2012). Children's participation in conversation, such as dialogically rich classroom interactions, expands their ability to engage in metacognitive monitoring and improves self-regulatory learning in the early school years (Whitebread et al., 2014).

Preschool children also learn to use other public symbol systems (such as pictorial representations, maps, and 3D objects such as a globe) to represent and learn about the world and the ideas of their community. Indeed, by age 2½ to 3, children can form a dual representation of a picture or scale model as a thing in itself and as something that stands for something else (DeLoache, 2000). Once children treat pictures as representations of real-world entities, they can learn about these entities through picture books, combining both pictures and words.

Children can now draw on and integrate a wide range of resources into more complex knowledge structures. Consider children's understanding of the Earth during the early elementary years. Vosniadou and Brewer (1992) documented several models of the Earth that were constructed by children, many of which were non-canonical, but used coherently and consistently by a given child (e.g., a flattened spherical Earth, a fishbowl model with people living on a platform inside the bowl). They argued that children form these coherent models on the spot by integrating diverse resources: they assimilate the concept of the Earth to their ontological category of physical object which engages their commitment to the principle that unsupported things fall; they activate generalizations from everyday experience leading them to view the ground as flat and space as having absolute directions of up and down; they also receive input from others (verbally or in the form of 3-D models) that the Earth is a sphere. However, exposure to everyday and formal language can also interact with children's preverbal conceptual resources to disrupt the coherence of their developing concepts (see Masri & Amin, 2023 on the concept of force).

The expanding range of resources increases the importance of coordinating their activation and dealing with conflicting conceptions. This draws attention to the importance of children's developing metacognitive knowledge about their thinking and learning along with skill in monitoring and regulating these processes. But for any task—predicting, explaining, and communicating with others—the information processing demands can be considerable. Although we know that preschool and elementary school children's capacity to store and process information steadily increases, we also know that learning experiences and the internal and external representational resources that children are acquiring (especially language) are themselves sources of increased storage and processing efficiency. Thus, the developments in children's encoding efficiency and metacognitive knowledge and skill during this period along with their ability to use external representations and drawings to offload some storage demands, enable them, with instructional guidance, to reflect on their non-canonical concepts and models, compare them to the new ideas introduced to them, and consider evidence that could lead to model revision and greater coherence in their understanding (Hennessey, 2003).

In sum, conceptual development in the early years is the outcome of the interaction of multiple representational resources, some of which can be traced back to the preverbal infant (e.g., core knowledge, general representations of cause and kind, and event categories/image schemas) and some which appear in the preschool years (e.g., framework theories, linguistic, metacognitive, and epistemological resources). Compared to scientific understanding, the framework theories of early childhood lack mechanistic detail and are not put to rigorous empirical test. But by the start of elementary school, children have privileged starting points for developing LPs in at least five domains: naïve mechanics, some aspects of chemistry (e.g., initial ideas about materials and their properties), folk biology, folk psychology, and even earth systems and cosmology (NRC, 2007). Coming to understand scientific concepts will require significant alteration in the constraints and practices they use in model building as well as integration of resources from different domains (including mathematical ideas about number and space), but the knowledge structures and reasoning skills developed to this point are crucial resources for learning scientific concepts.

Comparing Conceptions of the Upper and Lower Anchors

By the end of 12th grade, students are expected to have developed some understanding of a broad range of contemporary scientific theories, as well as a more general understanding of science and its knowledge construction practices. These scientific theories differ in many ways from students' intuitive conceptions. At the same time, it is increasingly recognized that even scientific concepts are grounded in intuitive knowledge structures that are repurposed to support scientific understanding. Thus, any adequate account of how students learn scientific concepts must consider both discontinuities and continuities in the learning process.

Sources of Difference and Evidence for Discontinuities

Scientists, like children, are concerned with developing knowledge that supports explanation and problem solving. However, unlike children's skeletal framework theories, the theories of scientists are worked out, self-consciously held, formalized theories, with multiple levels of explanation and mechanistic detail. And while scientific theories explain everyday phenomena even young children observe, they also explain many phenomena that cannot be observed without special tools and instruments, which students will need to learn how to use. However, differences in detail, complexity, and phenomena considered do not imply fundamental discontinuity. Crucially, scientific theories are stated in terms of new lexicalized concepts (such as *mass, field, molecule, gene, tectonic plates*) that are entirely absent from young children's thinking as well as others (such as *force, weight, energy*) that share labels with radically different everyday concepts. Indeed, learning scientific concepts involves both changes in a concept's reference (what entities, properties, or relations in the world it picks out) and in inferential role (how it relates to other concepts to support inference).

Consider, for example, the contrasting concepts students and scientists use to reason about the motions of everyday (rigid) objects, both when thrown or in free fall. In their review, Nersessian and Resnick (1989) found that college students distinguished among *rest, motion, force, speed,* and *heaviness*, and that the following beliefs were widespread: there is a fundamental distinction between rest (a state) and motion (a process); motion (but not rest) requires explanation; motion is caused by a force (either imparted by an external agent or by some internal energy/power or property of the object); continuing motion is sustained by a stored force that can be used up as the motion continues and slows down (hence the amount of force is proportional to the object's speed); active and passive motions differ; downward motion is natural; and heavier objects fall faster than lighter ones. Many of the same beliefs are observed with elementary and secondary students (Driver et al.,1994). Indeed, many aspects of how students reason about force and motion can be traced to infants' physical reasoning which is reinforced by the force dynamic framework implicit in everyday language (discussed earlier).

In contrast, Newtonian mechanics rests on a very different network of concepts for describing and explaining the motion of objects—concepts that have been fully quantified and interrelated in an abstract model of point masses moving in an idealized Euclidean space (Nersessian, 1989). Within this model, different aspects of motion can be precisely quantified not only in terms of distance covered but also *average* and *instantaneous speed*, *velocity*, and *acceleration.* A variety of representations support describing motion in these terms including words, vector diagrams, numbers, position-time and velocity-time graphs, and equations. Given this more precise description, the motion of objects can be explained in terms of Newton's three laws. Within Newtonian mechanics, there are a variety of forces arising through direct interactions (e.g., normal, frictional, and applied forces) as well as through action-at-a-distance (e.g., gravitational force).

Thus, the concepts of students and scientists differ in numerous ways. First, they differ in what conceptual distinctions are made and hence what referents each concept picks out (e.g., scientists differentiate *average* and *instantaneous speed*, *velocity,* and *acceleration* whereas students primarily focus on whether an object is going *fast* or *slow*). Conversely, students differentiate concepts that scientists do not (e.g., treating *blocking* and *supporting* as distinct from *force*). Thus, scientists "see" forces that students overlook (e.g., frictional or normal forces) and students "see" forces that scientists do not (e.g., an *internal force* keeping an object in motion). Second, as new conceptual distinctions are made and others abandoned or reanalyzed, specific concepts take on new inferential roles. For example, force is related to *changes* in motion in the scientists' framework, but to *causing* motion in the students' framework. Further, students rely on simple linear causal patterns, where causes precede effects and come in physical contact with them. In contrast, causal interactions in Newtonian physics involve simultaneous, reciprocal (symmetric) interactions in a constraint-based system. (See Perkins & Grotzer, 2005, on how many causal patterns in modern science differ from those used by students.) Finally, there are differences in epistemological aims and ideals, as well as model-building skills. A central goal of Galileo and Newton was the construction of abstract entities defined in terms of mathematical relations, what Koyré called "the mathematization of nature" (Nersessian,1989, p. 174). Students, like Aristotle, start with more qualitative conceptual categories for describing motions and favor explanations that are closer to their perceptual experience. Therefore, students need to re-examine their epistemological assumptions about what makes a good model or explanation and develop more mathematical concepts and skills to use in model building.

Similar analyses of discontinuity between students' and scientists' conceptual systems have been provided for other major scientific theories. (See Wiser & Smith, 2013, 2016 for the atomic molecular theory; Evans, 2013, and Shtulman, 2006, for evolutionary theory).

In sum, the scale and depth of the differences between the conceptual systems of students and scientists means that Newton and students are not simply making *different claims* articulated over the *same concepts*. Rather, when students first encounter $f = ma$, they attempt to make sense of it using different concepts of *force*, *mass*, and *acceleration* than those assumed in the law. It is helpful to see these conceptual systems as "locally incommensurable," a phrase used by Kuhn (1982) to capture both the profound differences between some successive theories in the history of science (in explanatory structures, core concepts, and some methods) while accepting that there was enough in common (aspects of the phenomena of interest and some concepts) to allow meaningful comparison. Thus, Kuhn insisted that sometimes no direct "translation" between terms in one theory with those in another is possible. Instead, there could only be "interpretation," by reconstructing the world as seen by another, making use of what is common in the conceptual systems in question. It is important, therefore, to complement consideration of discontinuities in science concept learning with an examination of sources of continuity.

Sources of Similarity and Evidence for Continuities

Neglecting attention to continuity would be highly problematic in any constructivist account of learning as the changes that take place in the learner's conceptual system must draw on existing knowledge resources (Smith et al., 1993). Particularly relevant here is research on science concept learning that recognizes that knowledge structures emerging from sensori-motor experience continue to contribute to scientific understanding.

diSessa (1993) introduced the construct of a *phenomenological primitive* (or p-prim), a (nonverbal) structured schema that emerges from our experience interacting with the world. diSessa (1993, p. 126) describes the components of *Ohm's p-prim*, for example, as "an agent that is the locus of an impetus that acts against a resistance to produce some sort of result." He suggests this schema grounds intuitive inferences (e.g., The greater the resistance the more effort is needed to produce the same result) and identifies and describes dozens of other p-prims (e.g., *continuous force*, *force as mover*, *dynamic balance*) that constitute our intuitive "sense of mechanism." diSessa argues that developing an understanding of scientific concepts involves learning to activate many of these same p-prims in ways that enable us to formulate explanations and make predictions as a scientist would. Similarly, both Minstrell (1982) and Clement et al. (1989) have identified multiple "anchoring intuitions" readily activated by learners in specific contexts that could be used as starting points for learning sequences. Brown (1993) develops the idea further with his account of "core intuitions" which he describes as varying in their attributions of agency—e.g., an *initiating agent*, that has its own causal power; a *reactive agent*, which reacts with a degree of agency to an external agency acting on it; or an *instrumental responder*, which functions as a kind of medium to transmit agency. Core intuitions are integrated with other knowledge elements such as *implicit mental models*, conscious *verbal-symbolic knowledge*, and *conscious images* of specific situations, in the construction of more coherent and conscious *explanatory models*. Finally, Sherin (2001) suggests that physicists draw on intuitive conceptual resources to interpret physics equations. He introduced the construct "symbolic form" composed of a symbol template (e.g., □ - □; □ = □) and a conceptual schema, similar to a p-prim, (e.g., *opposing forces*, *balancing*), and argued that they help physicists meaningfully interpret the components of an equation and facilitate its application to the world.

Continuity in science concept learning can also be viewed from the perspective of embodied cognition positing that understanding abstract concepts, comprehending language, and reasoning all call for activating parts of our sensorimotor system (Gibbs, 2005). Applying this idea to science concept learning, Lindgren et al. (2016) designed mixed reality environments, in which learners *enact* the motion of physical objects (e.g., an asteroid orbiting a planet) with their bodies while interacting with a digital simulation of the physical system. They found that this supported learning about Newtonian force and motion better than "less embodied" experiences (e.g., using a regular desktop simulation). Analyzing the language of science also shows that many scientific concepts are grounded metaphorically in knowledge originating in sensorimotor experience (Beger & Smith, 2020). For example, energy forms and transformations are construed in terms of image schemas such as *containment* and *movement into/out of containers* (e.g., "The energy *went back and forth* between kinetic and potential energy"). From this perspective, learning science concepts is understood as (in part) appropriating the metaphorical mappings implicit in the language of science and learning how they support inference (Amin, 2020).

Although these researchers use different terminology (p-prims, core intuitions, conceptual schemata of symbolic forms, and image schemas), it is possible that they are all characterizing continuity in terms of the same kind of knowledge structure (a schematic knowledge structure emerging from patterns in sensorimotor experience) (Amin, 2021) and that many of these can be traced to the event categories developed during infancy.

Finally, there is also continuity in the reasoning *processes* that lead to conceptual restructuring, which we will take up in depth in the next section. The capacity for constructing analogical models to solve problems is a key feature of scientific creativity that draws on the basic human ability to construct and use mental models and engage in imagistic simulation (Nersessian, 2008). But much of the process of constructing scientific understanding goes beyond these basic human reasoning capabilities and the intuitive knowledge resources discussed above. We survey the processes that contribute to conceptual change next.

Characterizing the Learning Processes Underlying Conceptual Change

We turn now to two key challenges: (a) characterizing the learning mechanisms, and corresponding experiences, that transform learners' initial conceptual systems into those that support a greater understanding of disciplinary core ideas of science; and (b) characterizing how to approach the design of LPs in a way that is consistent with this account of learning processes. We first describe an "epistemic model" that integrates mechanisms of conceptual change that have been identified by historians and philosophers of science, developmental psychologists, and learning scientists; together these mechanisms enact *constraint-based modeling*. We then give a detailed analysis of an instructional sequence on force and motion that illustrates how constraint-based modeling leads to conceptual change. Finally, we argue that centering LPs around the construction of successively more powerful models is consistent with this epistemic model.

Constraint-based Modeling: An Epistemic Model of Conceptual Change

Earlier we noted that conceptual systems in the lower and upper anchors are often locally incommensurable, using Kuhn's term (1982). We also noted that Kuhn viewed conceptual change in science as a creative, *interpretive* process, not just simple translation. Clarifying what is meant by *interpretation* here is a useful entry point into the process of constraint-based modeling. Interpretation is an active, metacognitively guided process of meaning construction that involves satisfying multiple constraints. For example, when encountering a difficult poem, the meanings we initially attach to some words don't help us understand the text as a whole; we have to *construct* our understanding by using elements of the text as clues to (indeed, constraints on) what plausible interpretations might be; we also search beyond the meanings we originally attached to some words for alternative meanings and evaluate whether this produces a reading consistent with other parts of the text and more meaningful overall.

A scientist's construction of a new scientific idea or a learner coming to understand a science concept can both be seen as analogous to this process of text interpretation but with aspects specific to science. Crucially, the overarching goal of scientists (and learners) is to develop a deeper explanatory understanding of their *world*, not a text. But this is also a process of creative (re)interpretation of events with other observed phenomena and other knowledge of the world serving as constraints. According to some philosophers and historians of science, analogical models play a central role in scientific theories (Giere, 1988), and constructing a theory involves building and evaluating multiple models using model-based reasoning (Nersessian, 2008). A model is a representation at an intermediate level of abstraction between observed phenomena and theories formulated in linguistic or mathematical propositions, facilitating their application to the world. Scientists seek an idealized representation of an aspect of the natural world that would function to explain known observations and predict others. Coming up with the right idealized representation that supports inferencing in particular situations is a creative process—e.g., a scientist might recognize that what is known about the natural system under investigation resembles what is known in another domain; models and principles of that other domain can then serve as a starting point for model and theory construction in the new domain. Nersessian (2008) unpacks the details of this model building via analogical reasoning. She shows how the observed phenomena in a target domain, productive concepts already in place, and linguistic and mathematical propositions in an analog domain serve as constraints on a model-building process that often involves imagery, mental models, thought experiments, and limiting case analysis. The outcome of this process can be a profound restructuring of a conceptual system, with the concepts of the new explanatory model not clearly translatable to those in the prior explanatory models.

Developmental psychologist Susan Carey (2009) proposes a model of conceptual change in children (and learners more generally) that embraces the constraint-based modeling account just described. She highlights the process of Quinian bootstrapping in the construction of new concepts; this is a modeling process constrained by direct experience of the world, existing concepts, and, crucially, public symbolic representations, either linguistic or mathematical, introduced by others that may be poorly understood initially. For example, in the case of science learning in school, students might be told that "f=ma" or "All matter is made of a limited number of different kinds of atoms." These new, partially understood symbols and expressions serve as placeholders for the new network of concepts to be learned. The bootstrapping process involves filling in the meaning of these placeholders by modeling the phenomena in the domain using analogy, inductive and abductive inference, thought experiment, and limiting case analyses in ways that draw on and integrate existing conceptual resources (sometimes from another domain). Metacognition is needed to monitor comprehension, evaluate the consistency of models with the phenomena being considered, note conflicts among beliefs, and attempt to resolve them by bringing other knowledge to bear. Assembling and integrating multiple knowledge resources can strain information processing capacities, but, as Nersessian (2008) points out, much of this complex model-building process can be offloaded to external representations, reducing the demands on internal information processing capacities. Moreover, as we explain in the next sections, even profound conceptual changes can be decomposed into a series of more incremental steps and, in learning settings, can be scaffolded via instructional support.

Investigating both the creative reasoning of scientists and science learners, John Clement has unpacked the details of explanatory model construction. He has shown that when scientists are presented with challenging problems, they engage in cycles of model generation, evaluation, and modification (GEM cycles) (Clement, 2022). Revisions can involve the gradual elaboration and refinement of a model or more radical change, such as the abandonment of a model (and the generation of an entirely different one) or the development of a complementary model at a different level of analysis. Drawing on his work with scientists, Clement has created classroom environments that scaffold model construction, evaluation, and revision, with analogical reasoning that recruits physical intuitions in one domain central to the process of constructing explanatory models in another (Nunez-Oviedo & Clement, 2019).

Research by other learning scientists contributes to a more detailed account of how constraint-based modeling works. The knowledge-in-pieces perspective unpacks the subtleties of how p-prims are repurposed to construct scientific understanding (diSessa, 1993), guided by the constraints of observed phenomena (diSessa, 2014) and propositional representations, including language (Levrini & diSessa, 2008) and mathematics (Sherin, 2006). Differences in how individual learners activate intuitive knowledge elements in the context of conceptual change instruction have also been examined (Kapon & diSessa, 2012). Similarly, White (1993) has shown how computer-based microworlds can be designed to support learners in coordinating physical intuitions, analogical visual representations, and mathematical expressions to construct models to improve their understanding of Newtonian mechanics. Finally, drawing on the cognitive linguistic theory of conceptual metaphor, Amin (2009, 2018) has shown how conventional metaphorical expressions in the language of science can guide learners in making sense of abstract scientific concepts in terms of image schemas. He suggests that these implicit metaphors provide an initial layer of interpretation of otherwise poorly understood propositional representations functioning as symbolic placeholders in a Quinian bootstrapping process.

The processes we have just reviewed have parallels even in infant learning: an infant's construction of event representations constrained by core knowledge, event categories, and everyday observations is analogous to a scientist's construction of analogical models constrained by available concepts, image schemas/p-prims, and empirical phenomena. However, there are obviously important differences in the complexity of resources involved and the extent to which these resources can be repurposed to support the creation of novel conceptual frameworks. Significantly, scientists use propositional representations encoded in public symbols—linguistic and mathematical—to constrain scientific model-building and exploit the arbitrary nature of symbols to re-connect them to experience in novel ways. Further, scientists' understanding of mathematical concepts and the use of measurement tools, dramatically change the constraints on model building from those derived from everyday perceptual categories. Finally, scientists have greater metacognitive awareness of their knowledge, more formal logical reasoning abilities, and an epistemological understanding of the role of measurement tools and idealized models as sources of knowledge about the world. Developing these metacognitive and epistemological abilities in students supports their development of more coherent and robust knowledge structures, as well as their revision due to anomalies and inconsistencies detected through firsthand observation or thought experiments (Smith, 2018). Significantly, by first grade, children have developed enough conceptual, linguistic, mathematical, logical, and metacognitive resources to begin to engage in explicit modeling with instructional support (Hennessey, 2003).

An Example of Constraint-based Modeling: Developing Understanding of Newton's Laws

We present here Brown's (2017) modules for teaching pre-service elementary school teachers about forces and motion through guided firsthand inquiry, follow-up readings, and extensive class discussion because they nicely illustrate the application of the epistemic model we just outlined. The modules illustrate how learners are engaged in model-based reasoning constrained by observed phenomena and imagistic and propositional representations to achieve coordinated changes in concepts, principles, ontological and epistemological assumptions, and reference. Brown has not proposed an LP for force and motion, but his modules clarify the implications of a constraint-based modeling perspective for curricular sequencing.

The first module aims to restructure learners' motion concepts, a prerequisite to achieving a scientific understanding of the relationship between motion and force. Initial instruction builds on an understanding of *speed* as a relational concept (i.e., speed = distance/time)—itself an accomplishment of childhood and early math education that allows students to quantify average speed. The goal is to help them contrast *average* and *instantaneous speed* and develop a more precise and general understanding of *acceleration* differentiated from *speed*. Differentiating average and instantaneous speed is initially driven by contrasting quantities that they can readily think about: covering a certain distance in a given time (e.g., 40 miles in an hour) versus reading a speedometer that changes during a journey. These different representations of speed draw attention to the targeted conceptual distinction and place constraints on how many concepts are needed. Similarly, different combinations of constraints motivate greater precision and generalization of the concept of acceleration and its differentiation from "going very fast." The process of contrasting phenomena—such as hardly feeling motion in a plane moving at a high but steady speed, versus being pushed back in your seat in a car speeding up quickly—supports differentiating going very fast from acceleration and distinguishing uniform motion from acceleration. More precision in understanding acceleration is achieved by focusing on the speed of movement of the needle on a speedometer. Finally, learners are explicitly alerted to the difference between the everyday meaning of "acceleration" (as moving faster) and the more general use by scientists as any change in direction or speed (be it increase or decrease). Coalescing these in one concept is supported by the fact that changes in speed or direction all produce changes in bodily motion when riding in a vehicle, but uniform motion does not.

The goal of the next unit is to develop an intuitive understanding of the Newtonian concept of force as a symmetric interaction, as defined in Newton's third law. This departs from the typical introduction of Newton's laws in chronological order. Starting with the third law tackles at the outset an ontological obstacle: learners initially understand force either as a *property* of an object or as an *asymmetric relation* (an agent/object exerting a force on a passive recipient) rather than as a symmetric *interaction* between objects. By addressing this law first, students learn forces always come in equal and opposite pairs, which they can use to leverage subsequent learning.

Brown begins by introducing situations where the symmetric nature of forces is *intuitively compelling*: a hand pushing down or a book placed on a spring that pushes back. He applies the findings of classic research on "bridging analogies" where an anchoring intuition (the sense of force/agency associated with a compressed spring with a book on it) is projected to a situation where an upward force/agency is not intuitive (a book on a table) via consideration of various "bridging situations" (a book supported by a "springy" thin piece of wood) (Brown & Clement, 1989). The bridging situations, crucially, lend plausibility to the analogical reasoning the teacher is inviting the student to do. In the process, students learn to model interactions between objects—even static ones involving hard, rigid objects—using a microscopic explanatory model of particles undergoing springy interactions that incorporate an intuitive notion of force/agency.

Instruction then extends the springy interaction model to new situations, elaborating the model. For example, the model is extended to situations where the size of the forces can be directly compared and measured (e.g., two attached spring scales pulled in opposite directions providing identical readings); this increases the precision of the model and demonstrates that the forces are equal even where the objects are different in size. Then a challenging situation is addressed: two rigid objects of different masses that each behave very differently after colliding—e.g., a bowling ball colliding with a pin that it knocks down. Modeling this collision as a "springy" interaction allows learners to see that each object exerts an equal force on the other and to distinguish the size of the force from its effects. This sequenced extension of the "springy" interaction explanatory model gradually extends the range of situations the concept of force picks out (its reference) to include normal forces of rigid surfaces and less massive rigid objects often seen as passive *recipients* of force. Developing an epistemological understanding of the nature of explanatory models is important here as well: for example, understanding that a good model is one that can explain a pattern of results consistently and be fruitfully applied to new situations, rather than superficially resembling the situation in question.

It is in this context of making sense of this challenging situation that Newton's ideas are first explicitly presented. Brown begins by directly contrasting the everyday and Newtonian meaning of "force" noting that from the Newtonian perspective, you can't talk about the force of a single object; a force is always an *interaction* between two objects; he then presents the third law (expressed linguistically): "The force of A on B and B on A are always equal in size and in opposite directions." Introducing Newton's third law here is timely for two reasons. Prior experiences modeling different situations as springy symmetric interactions have prepared students to not only understand these statements, but also to find them plausible, and, expressing them linguistically helps students "hold on to" and apply the model in less intuitive situations.

With an understanding of force as a symmetric interaction and differentiated concepts of speed and acceleration, the next unit introduces Newton's second law, $f = ma$. This law not only relates force and acceleration, but also introduces a new conception of mass as inertial mass–something that resists change rather than as something that produces more change. (Students often initially think that more massive objects will accelerate more rather than less). Brown builds the law in stages, introducing it first as $a = f/m$, where the relationships among variables make intuitive sense using an *agent acting on an object to produce an outcome* schema. This schema is a p-prim and grounded in infants' core knowledge of objects; used to interpret the equation it grounds the understanding that as I increase the force on some object, it will accelerate more; if I exert the same force on an object that is more massive than another it will accelerate less. Demonstrating how spring scale readings increase when pulling a cart with more books further grounds understanding the mass as "sluggish," resisting the effects of a force. Brown then explicitly contrasts how the relationship between force and motion is understood in science (forces result in a *change* of motion–i.e., an acceleration) with everyday ideas (force is needed for motion) before extending the application of Newton's second law to less intuitive situations where the *recipient* of the force seems more like the *agent* (e.g., a fast-moving bowling ball colliding with a pin). Modeling the collision as a "springy" interaction allows students to see how the pin is also exerting a force on the ball in the opposite direction, causing it to slow down, reinforcing the conception of force as a symmetric interaction *between* objects that explains a *change* in motion.

Finally, the fourth unit aims to develop an understanding of the first law, the law of inertia (if there is no net force on an object its motion will not change); this is introduced as a special case—a logical implication—of the second law. Understanding the first law means giving up the naïve view that motion requires an explanation; instead, being at rest and being in uniform motion are equivalent. Although this is logically implied by the second law, logical implications that seem counterintuitive can be resisted, especially given, in the real world, things set in motion by a push slow down and eventually stop. Students need ways to understand why things sometimes stop to make the law of inertia empirically plausible.

The unit starts by inviting students to imagine what would happen to a puck set in motion on "perfectly slippery ice;" in this thought experiment, it makes sense for them to think that it would keep on moving. They are then asked to draw the forces acting on the puck when it is in motion on a frictionless surface and are led to see just two forces: the force of gravity downwards and the upward normal force exerted by the surface on the object. As both forces are equal, the net vertical force is zero. But this is identical to a stationary object at rest on some surface. So, students are explicitly told that this means that there is no distinction between rest and uniform motion in Newtonian physics in contrast to their everyday view that motion is different from rest and requires explanation. To strengthen the plausibility of this conclusion, the module includes the consideration of additional experiences about frames of reference (e.g., being in a stationary train seeing another move and thinking that you are moving).

Students then return to modeling what happens when the puck slides across a tile floor and why it slows down. This makes more sense if they understand the roughness of a surface as a source of force on the object (i.e., friction). So, students are invited to model the lower surface of the object and the surface along which it slides as jagged surfaces undergoing "springy" interactions (like two intermeshed hairbrushes). The smoother the surface, the more these interactions are reduced. They can then understand that an object sliding along a surface eventually stops *because* the force of friction results in acceleration (in this case a decrease in speed). But in the absence of any friction, the object would keep moving.

Throughout his instructional sequence, Brown encourages students to model new situations in ways that are *consistent* with various constraints— the phenomena themselves, their prior ideas, beliefs about other analogous situations, and verbal and mathematical formulations of some of Newton's laws—while offering plausible *explanations* of them. Further, as models are developed and elaborated to explain more phenomena, there is a progressive introduction and modification of constraints as more understanding of Newton's laws can be used to leverage new understandings, along with a greater understanding of the model-building process. Together, these small cycles of model building, elaboration, and revision result in an understanding of Newtonian concepts and laws grounded in a family of idealized models: elastic interaction of objects which grounds an interpretation of Newton's third law; an agent attempting to move, speed up or slow down an object despite its natural resistance which grounds interpretation of Newton's second law; and an object moving along a frictionless surface, with all elastic interactions removed, which grounds understanding of Newton's first law.

A similar approach to conceptual change instruction can be found in Smith's work on matter and density (Smith et al., 1997), White's (1993) on force and motion, Wiser's (Wiser & Amin, 2001) on heat and temperature, and Stewart et al.'s (2001) on genetics and evolution. All share the following features with Brown's approach: phenomena observed during inquiry activities create empirical constraints on models; visual representations (sometimes incorporated into computer-based simulations) or everyday situations that intuitively embody targeted scientific conceptual distinctions are introduced and guide mental model construction; multiple models are coordinated to make sense of phenomena; and there is explicit discussion of epistemic assumptions as well as the contrasting meanings of everyday and scientific terms.

This reveals one problem with traditional instruction where there is excessive focus on new terminology, tools, and formal procedures to solve problems without enough (or no) attention to the modeling processes that engage learners' sense-making resources. This also reveals a problem with instruction that emphasizes student-driven inquiry without sufficient guidance. Quinan bootstrapping emphasizes the importance of *changing constraints* in the course of model building *by* strategically selected phenomena, analogies, and symbolic placeholders.

Finally, we note that the question of whether to lead with everyday experiences, sense-making, and the building of models or with the introduction of placeholder structures cannot be answered in general. Learning the count words and procedure by rote plays an important role in driving conceptual change in the domain of number in early childhood. In contrast, in Brown's module, a substantial focus on making sense of everyday phenomena through analogical model building precedes the introduction of scientific terms and laws. However, both cases have in common complex interactions between experience, analogical sense-making, and symbol use, the essence of the epistemic model that accounts for conceptual change.

Implications for the Design of LPs

We turn now to the implications of this epistemic model for curriculum design, and by extension, for the design of science LPs. To do this, we emphasize two aspects of the account of concept learning just presented: (a) the systemic nature of conceptual knowledge, incorporating many different types of knowledge elements both iconic and propositional; and (b), the centrality of models in the representation of scientific knowledge and accounts of conceptual change.

Science curricula typically use a small number of declarative statements to articulate the grade band objectives for a disciplinary core idea. In our view, the characterization of knowledge states at a given point in an LP should capture the diverse kinds of knowledge elements (sensorimotor and imagistic as well as propositional, metacognitive, and epistemological) and forms of external inscription that need to be integrated, as well as how those knowledge states have changed from the previous level including changes in reference and the inferential roles of key concepts. This unpacking has a downside, however. A curricular document cannot be a treatise in cognitive science with detailed descriptions of heterogeneous knowledge elements and their transformations. Summary accounts are needed to serve as a practical guide for instruction and assessment. Given the centrality of models in the representation of scientific theories and modeling as a mechanism of knowledge change, we and others view models as prime candidates for efficiently characterizing successively more powerful knowledge states for a given disciplinary core idea (Rogat et al., 2011; Wiser & Smith, 2016).

Indeed, there are many reasons to organize LPs around the construction, evaluation, and revision of specific models. First, a particular model or (family of models) constitutes the core of a knowledge system, grounding understanding of concepts, linguistic and mathematical propositions, and epistemological assumptions. They are *integrative analogical representations* that capture important relations among concepts and are more readily applied to the world. They not only enable learners to run internal mental simulations of events but can also be coupled with external representations that support both making inferences and group discussion and critique. Second, models are intrinsically *simplifications* and *idealizations*; they are not true or false, but, rather, useful (or not) in a particular class of situations. Therefore, models are well suited to serve as pedagogical stepping-stones as well as support epistemic learning and discussion. Third, models can form a meaningful progression, both conceptually and epistemologically: they can be evaluated and revised on an empirical basis, thereby moving the knowledge network forward. Indeed, the drive to construct explanatory models is a general feature of human (even children's) reasoning that provides a powerful force for growth and change. Finally, modeling is an *integrative scientific practice* that involves constructing, evaluating, and revising representations that aid in solving specific problems. It constitutes a meaningful context for learning not only about specific conceptual content, but also related scientific practices, building on sense-making resources available to young children.

Wiser and Smith (2016) describe this approach as reducing the "incommensurability gap" between the upper and lower anchor, by identifying a series of conceptual stepping-stones, organized around models. They illustrate this approach within the context of an LP for matter. They suggest that curricula for young elementary school students could begin by drawing on their initial ideas about objects, materials, size, and weight that are initially organized around descriptions of the *perceptual* properties of large-scale objects and re-organize them into an initial *compositional model of matter*. In this model, any sample of a material can be imagined as being constituted of arbitrarily small pieces of that material; each piece has weight and size and the material's characteristic (macroscopic) properties, including its density. This model supports differentiating and inter-relating the quantities of weight, volume, and density *and* predicting and explaining diverse phenomena, such as why weight is extensive and doesn't change when an object is reshaped and why material identity is conserved when a solid is ground into a powder. It also highlights deeper similarities between solids and liquids as the model can be applied to samples of both. Further, it supports (and is supported by) learning about weight and volume measurement, both enriching students' available mathematical resources and leading to new epistemological insights about the importance of measurement and the limitations of unaided senses. Finally, although the model still rests on an experience-based understanding of matter as continuous, learners can now explain the properties of large-scale things in terms of underlying pieces that compose them. This model does have important limits, however: it can't explain many phenomena involving solids and liquids in which there can be changes in volume without changes in weight, what happens in phase changes, and most aspects of the behavior of gases.

These anomalous phenomena help motivate the construction of *a basic particle model*—a centerpiece of the next stepping-stone. In this model, matter is envisioned as being composed of discrete *preexisting* submicroscopic particles, with particle weight specific to a particular material; these particles are separated by empty space, are held together by bonds of varying strength that weaken with increasing temperature, and are in constant motion. Such a model can explain more phenomena (such as thermal expansion and phase change), supports reclassifying gases as matter, and distinguishes macroscopic and submicroscopic (particle) levels of description. The idea of particles that move on paths through space, maintain identity, and do not disappear builds on core ideas about objects; but, of course, these particles are not literally just little macroscopic objects. Particles are pre-existing pieces that have some, but not all, macroscopic properties of materials as well as new properties of their own. Understanding and accepting this model calls for developing a greater epistemological understanding that good models can involve conjectures about unseen events that provide compelling mechanistic explanations of observations. The next stepping-stone is now within reach: a simple version of *the atomic-molecular theory* in which atoms and molecules are distinguished ontologically and structurally, but atoms are spheres. This model begins to explain more phenomena, such as chemical reactions, and paves the way for more elaborated models of subatomic structure (Stevens et al., 2010).

Wiser and Smith (2016) discuss how the construction of each class of models takes extended time and how the models become elaborated as students extend the range of phenomena considered. They also describe the way each model builds on the conceptual and epistemological changes of the prior stepping-stone, while requiring additional conceptual and epistemological changes of its own. Other examples that involve identifying a progression of increasingly sophisticated models include White's (1993) work on force and motion and Plummer's (2012) work on models of celestial motion. Similarly, Lehrer and Schauble (2012) have proposed a sequence of four increasingly complex ways elementary students can model variation, change, and ecosystems to lay a foundation for evolutionary thinking.

Conclusion

In concluding, we return to the challenging questions in the LP literature that we raised in the introduction and discuss how our account of the cognitive foundations of science LPs suggests ways to begin to answer these questions.

We start with the questions of how to identify young learners' knowledge and reasoning resources that can serve as the productive starting point of an LP and what role they might continue to play in the upper anchor. Our review surveyed a broad literature that revealed many useful knowledge and reasoning resources: the concepts of core knowledge, event representations, abstractions from sensorimotor experience, framework theories, mental models, mathematically and linguistically formulated propositions, conceptual metaphors implicit in language, symbolic forms, analogical and imagistic reasoning, and the capacities for Quinian bootstrapping and metacognitive reflection. By unpacking the lower anchor and the upper anchor in terms of diverse knowledge elements and by characterizing the process of change in terms of constraint-based modeling, we can begin to see the resources that might help build the target structure, the changing roles they might play and the multiple ways to change the constraints considered in modeling.

With respect to the question of how conservative or transformative to be when conceptualizing an LP's upper anchor and the intermediate levels, our epistemic model implies our answer would fall along the transformative end of the continuum. First, by unpacking the targeted knowledge and intermediate states in the domain in terms of multiple kinds of coordinated knowledge elements, our epistemic model enriches our understanding of what it takes to understand a disciplinary core idea. Second, our epistemic model guides us in identifying some important non-canonical explanatory models as productive intermediate targets of instruction. Third, it makes the key (but often overlooked) scientific practices of modeling and model-based reasoning the centerpiece of the progression, avoiding teaching science as either a "rhetoric of conclusions" or as a generic process of decontextualized inquiry.

Our answer to another question–by what criteria should one judge an intermediate understanding as "more sophisticated" than another-falls directly out of this modeling approach: understanding at one level is more sophisticated than the one before in the range of phenomena that are explained by the model and how coherently interconnected the various knowledge elements are that make up the knowledge system, rather than by whether it exploits more pieces of canonical knowledge.

The design of LPs in terms of constructing increasingly more powerful explanatory models also directly addresses the questions of how to conceptualize the relationship between the development of understanding of a disciplinary core idea and the epistemic practices of science and between an LP as a cognitive model of learning and instructional design. As we have clarified, an important class of constraints on this modeling process are students' epistemic ideals, assumptions about reliable knowledge-generating processes, and skill at using a variety of knowledge abstraction techniques such as thought experiments and limiting case analyses. Thus, developing increasingly powerful models goes hand-in-hand with developing students' understanding of and skill in the epistemic practice of modeling. Further, since constraint-based modeling has been put forward as the learning mechanism that achieves the targeted conceptual changes, successful learning environments will be those that can enact the component processes of constraint-based modeling.

Hammer and Sikorski (2015) ask if it is wise to characterize a single or small number of idealized learning pathways for some core idea or practice when learners can have diverse knowledge profiles and learning trajectories. They also worry that LP-based instruction may (mistakenly) locate coherence in a curriculum rather than in the minds of students, undermining their sense-making agency (Sikorski & Hammer, 2017). These are valid concerns, but we would first like to point out that the main purpose of LPs is to provide a (testable) cognitive model of a series of knowledge states that incrementally bridge between the lower and upper anchors, not to describe individual learning pathways. An LP will be useful if it supports curricular redesign that allows a greater number of students to achieve the targeted understanding. Second, the research suggests there are many universal aspects to student starting points—commonalities that arise due to the existence of core knowledge, generalizations they form from similar everyday experiences, and the ways everyday language guides the construal of experience. These commonalities, in combination with what we know about learning mechanisms and the target conceptions to be achieved, place important constraints on any viable learning pathway. Finally, there are many excellent examples of curricula that combine a guided modeling approach while also promoting student agency in their sense-making and epistemological growth—see the work of White and Frederiksen (1998) and Sister Gertrude Hennessey (2003).

This leads to a final question—What grain size is appropriate for describing levels within an LP and how does this relate to the purposes an LP is expected to serve? Although "grain size" has often been thought of as the number of levels needed in an LP, our systemic account suggests a second relevant sense of "grain size": the degree of detail needed to characterize any given knowledge level. Here we suggest the level of detail needed depends on one's purpose: if the goal is designing instructional activities, attention to the specific knowledge elements (iconic and propositional, internal and external) one seeks to activate is important; but if the goal is to design curricular sequences and corresponding assessment, a coarser grain size (e.g., the explanatory model as a whole) might be sufficient. A theoretical framework is needed that allows moving easily between grain sizes given one's purposes. We suggest that our systemic account of concepts emphasizing models as a construct is just such a framework.

Acknowledgments

We thank David Brown, John Clement, Susan Carey, the anonymous reviewers, and editors for very helpful feedback on earlier drafts of this chapter.

References

Amin, T. (2009). Conceptual metaphor meets conceptual change. *Human Development*, *52*(3), 165–197. 10.1159/000213891

Amin, T. G. (2018). Representation, concepts, and concept learning. In T. G. Amin & O. Levrini (Eds.), *Converging perspectives on conceptual change: Mapping an emerging paradigm in the learning sciences* (pp.129–149). London: Routledge. 10.1063/9780735425477_011

Amin, T. G. (2020). Coordinating metaphor in science, learning and instruction: The case of energy. In A. Beger & T. H. Smith (Eds.), *How metaphors guide, teach and popularize science. Figurative thought and language series* (pp. 73–110). Amsterdam: John Benjamins. 10.1075/ftl.6.03ami

Amin, T. G. (2021). Understanding the role of image schemas in science concept learning: Can educational neuroscience help? In O. Levrini, G. Tasquier, T. G. Amin, L. Branchetti, & M. Levin (Eds.), *Engaging with contemporary challenges through science education research* (pp. 237–250). Cham, Switzerland: Springer. 10.1007/978-3-030-74490-8_19

Amin, T. G. (forthcoming). *Learning scientific concepts*. London: Routledge.

Amin, T., Smith, C., & Wiser, M. (2014). Student conceptions and conceptual change: Three overlapping phases of research. In N. Lederman & S. Abell (Eds.), *Handbook of research on science education, vol. II* (pp. 57–81). New York: Routledge. 10.4324/9780203097267.ch4

Baillargeon, R., & DeJong, G. F. (2017). Explanation-based learning in infancy. *Psychonomic Bulletin & Review*, *24*, 1511–1526. 10.3758/s13423-017-1334-4

Baillargeon, R., Scott, R. M., He, Z., Sloane, S., Setoh, P., Jin, K., Wu, D., & Bian, L. (2015). Psychological and sociomoral reasoning in infancy. In M. Mikulincer, & P. R. Shaver (Eds.), E. Borgida, & J. A. Bargh (Assoc. Eds.), *APA handbook of personality and social psychology: Vol. 1. Attitudes and social cognition* (pp. 79–150). Washington, DC: American Psychological Association. 10.1037/14341-003

Beger, A., & Smith, T. H. (2020). *How metaphors guide, teach and popularize science*. Amsterdam: John Benjamins. 10.1075/ftl.6

Brookes, D. T., & Etkina, E. (2015). The importance of language in students' reasoning about heat in thermodynamic processes. *International Journal of Science Education*, *37* (5-6), 759–779. 10.1080/09500693.2015.1025246

Brown, D. E. (1993). Refocusing core intuitions: A concretizing role for analogy in conceptual change. *Journal of Research in Science Teaching*, *30*(10), 1273–1290. 10.1002/tea.3660301009

Brown, D. E. (2017). *Forces and motion: Some ideas important for elementary teachers*. Unpublished manuscript. Department of Curriculum and Instruction, University of Illinois at Urbana-Champaign.

Brown, D. E., & Clement, J. (1989). Overcoming misconceptions via analogical reasoning: Abstract transfer versus explanatory model construction. *Instructional Science*, *18*(4), 237–261. 10.1007/bf00118013

Carey, S. (2009). *The origin of concepts*. Oxford, UK: Oxford University Press. 10.1093/acprof:oso/9780195367638.001.0001

Chouinard, M. (2007). Children's questions: A mechanism for cognitive development. *Monographs of the Society for Research in Child Development*, *72* (1), 1–112. 10.1111/j.1540-5834.2007.00416.x

Clement, F. (2016). Social cognition. In J. Greene, W. Sandoval, & I. Braten (Eds.), *Handbook of epistemic cognition* (pp. 86–99). New York: Routledge. 10.4324/9781315795225

Clement, J., Brown, D., & Zeitman, A. (1989). Not all preconceptions are misconceptions. *International Journal of Science Education*, *11*, 554–565. 10.1080/0950069890110507

Clement, J. (2022). Multiple levels of heuristic reasoning in scientific problem solving. *Frontiers in Psychology*, *13* (*Article 750713*), 1–26. 10.3389/fpsyg.2022.750713

DeLoache, J. S. (2000). Dual representation and young children's use of scale models. *Child Development*, *71*(2), 329–338. https://psycnet.apa.org/doi/10.1111/1467-8624.00148

diSessa, A. (1993). Towards an epistemology of physics. *Cognition and Instruction*, *10* (2/3), 105–224. 10.1080/07370008.1985.9649008

diSessa, A. (2014). The construction of causal schemes: Learning mechanisms at the knowledge level. *Cognitive Science*, *38*(5), 795–859. 10.1111/cogs.12131

Driver, R., Squires, A., Rushworth, P., & Wood-Robinson, V. (1994). *Making sense of secondary science: Research into children's ideas*. London: Routledge. 10.4324/9780203823583

Duncan, R., & Gotwals, A. (2015). A tale of two progressions: On the benefits of careful comparisons. *Science Education*, *99* (3), 410–416. 10.1002/sce.21167

Evans, M. (2013). Conceptual change and evolutionary biology. In S. Vosniadou (Ed.), *International handbook of research on conceptual change* (pp. 263–294). New York: Routledge. 10.4324/97802 03154472

Gelman, S. A. (2009). Essentialist reasoning about the biological world. In A. Berthoz & Y. Christen (Eds.), *Neurobiology of "Umwelt": How living beings perceive the world* (pp. 7–16). Springer. 10. 1007/978-3-540-85897-3_2

Gelman, S. A., & Brandone, A. (2010). Fast-mapping placeholders: Using words to talk about kinds. *Language Learning and Development*, *6*(3), 223–240 10.1080/15475441.2010.484413

Gentner, D. (2016). Language as cognitive tool kit: How language supports relational thought. *American Psychologist*, *71*(8), 650–657. 10.1037/amp0000082

George, N., Göksun, T., Hirsh-Pasek, K., & Golinkoff, M. (2019). Any way the wind blows: Children's inferences about force and motion events. *Journal of Experimental Child Psychology*, *177*, 119–131. 10.1016/j.jecp.2018.08.002

Gibbs, R. (2005). *Embodiment and cognitive science*. Cambridge, UK: Cambridge University Press. 10. 1017/CBO9780511805844

Giere, R. (1988). *Explaining science: A cognitive approach*. Chicago: University of Chicago Press. 10. 2307/2215515

Goswami U. (2013). The development of reasoning by analogy. In P. Barrouillet & C. Gauffroy (Eds.), *The development of thinking and reasoning* (pp. 49–70). London: Psychology Press. 10.4324/97802 03068748

Hammer, D., & Sikorski, T. (2015). Implications of complexity for research on learning progressions. *Science Education*, *99* (3), 424–431. 10.1002/sce.21165

Harris, P. L. (2012). *Trusting what you're told: How children learn from others*. Belknap Press of Harvard University Press. 10.4159/harvard.9780674065192

Hennessey, M. G. (2003). Metacognitive aspects of students' reflective discourse: Implications for intentional conceptual change teaching and learning. In G. M. Sinatra, & P. R. Pintrich (Eds.), *Intentional conceptual change* (pp. 103–132). Mahwah, NJ: Erlbaum. 10.4324/9781410606716-10

Hespos, S., Ferry, A., Anderson, E., Hollenbeck, E., & Rips, L. (2016). Five-month-old infants have general knowledge of how nonsolid substances behave and interact. *Psychological Science*, *27*(2), 244–256. https://www.jstor.org/stable/24763541

Kapon, S., & diSessa, A. A. (2012). Reasoning through instructional analogies. *Cognition and Instruction*, *30*(3), 261–310. 10.1080/07370008.2012.689385

Keil, F. (2007). Cognitive science and cognitive development. In W. Damon & R. Lerner (Eds.), *Handbook of child psychology, Vol. 2, Cognition, perception and language, 6th edition*. Hoboken, N.J.: Wiley. 10.1002/9780470147658.chpsy0214

Kinzler, K. D., & Spelke, E. S. (2007). Core systems in human cognition. *Progress in Brain Research*, *164*, 257–264. 10.1016/s0079-6123(07)64014-x

Kuhn, T. (1982). Commensurability, comparability, communicability. *PSA: Proceedings of the Biennial Meeting of the Philosophy of Science Association* (pp. 669–688). 10.1086/psaprocbienmeetp.1982. 2.192452

Lacy, S., Tobin, R., Wiser, M., & Crissman, S. (2014). Looking through the energy lens: A proposed learning progression for energy in grades 3-5. In R. F. Chen, R. F. A. Eisenkraft, A. F. Fortus, J. Krajcik, K. Neumann, J. Nordine, & A. Scheff (Eds.), *Teaching and learning of energy in K-12 education* (pp. 241–266). New York: Springer. 10.1007/978-3-319-05017-1_14

Lakoff, G., & Johnson, M. (1980). *Metaphors we live by*. Chicago: University of Chicago Press. 10.1017/ s0008413100023744

Lakoff, G., & Johnson, M. (1999). *Philosophy in the flesh: The embodied mind and the challenge to western thought.* New York: Basic Books. 10.1590/S0102-44502001000100008

Leahy, B., & Carey, S. (2020). The acquisition of modal concepts. *Trends in Cognitive Science*, 24 (1), 65–78. 10.1016/j.tics.2019.11.004

Lehrer, R., & Schauble, L. (2012). Seeding evolutionary thinking by modeling its foundations. *Science Education*, 701–724. 10.1002/sce.20475

Lehrer, R., & Schauble, L. (2015). Learning progressions: The whole world is NOT a stage. *Science Education*, *99* (3), 432–437. 10.1002/sce.21168

Levrini, O., & diSessa, A. (2008). How students learn from multiple contexts and definitions: Proper time as a coordination class. *Physical Review Special Topics Physics Education Research*, *4*, 1–18. 10.1103/physrevstper.4.010107

Lin, Y., Stavans, M., & Baillargeon, R. (2022). Infants' physical reasoning and the cognitive architecture that supports it. In O. Houdé & G. Borst (Eds.), *Cambridge handbook of cognitive development* (pp. 168–194). Cambridge, England: Cambridge University Press. 10.1017/9781108399838.012

Lindgren, R., Tscholl, M., Wang, S., & Johnson, E. (2016). Enhancing learning and engagement through embodied interaction within a mixed reality simulation. *Computers and Education*, *95*, 174–187. 10.1016/j.compedu.2016.01.001

Mandler, J. (2004). *The foundations of mind: Origins of conceptual thought.* Oxford, England: Oxford University Press. 10.1111/j.1467-7687.2004.00369.x

Masri, D. B., & Amin, T. G. (2023). The effect of language on the coherence of children's conceptions of force. *Journal of the Learning Sciences.* 10.1080/10508406.2023.2210138

McCloskey, M. (1983). Naïve theories of motion. In D. Gentner & A. L. Stevens (Eds.), *Mental models* (pp. 299–324). Hillsdale, N.J.: Erlbaum. 10.4324/9781315802725

Metz, K. (2009). Rethinking what is developmentally appropriate from a learning progression perspective. *Review of Science, Mathematics, and ICT Education*, *3*(1), 5–22. 10.26220/rev.119

Minstrell, J. (1982). Explaining the "at rest" condition. *The Physics Teacher*, *20*, 10–14. 10.1119/1.2340924

National Research Council (2007). *Taking science to school.* Washington, D.C.: The National Academies Press.

Nersessian, N. (1989). Conceptual change in science and science education. *Synthese*, *80*(1), 163–183. 10.1007/bf00869953

Nersessian, N. J., & Resnick, L. B. (1989). Comparing historical and intuitive explanations of motion: Does "naive physics" have a structure? *Proceedings of the Cognitive Science Society* (pp. 412–420). Hillsdale, NJ: Lawrence Erlbaum.

Nersessian, N. (2008). *Creating scientific concepts.* Cambridge, MA: MIT Press. 10.7551/mitpress/7967.001.0001

Nunez-Oviedo, M. C., & Clement J. J. (2019). Large scale scientific modeling practices that can organize science instruction at the unit and lesson levels. *Frontiers in Education*, *4*, 68. 10.3389/feduc.2019.00068

Perkins, D. N., & Grotzer, T. A. (2005). Dimensions of causal understanding: the role of complex causal models in students' understanding of science. *Studies in Science Education*, *41*(1), 117–165. 10.1080/03057260508560216

Perszyk, D. & Waxman, S. (2018). Linking language and cognition in infancy. *Annual Review of Psychology*, *69*, 231–250. 10.1146/annurev-psych-122216-011701

Piaget, J. (1952). *The origins of intelligence in children.* New York: W.W. Norton & Co. https://psycnet.apa.org/doi/10.1037/11494-000

Plummer, J. (2012). Challenges in defining and validating an astronomy learning progression. In A. Alonzo & A. Gotwals (Eds.), *Learning progressions in science* (pp. 77–100). Rotterdam: Sense Publishers. 10.1007/978-94-6091-824-7_5

Prasada, S. (2016). Mechanisms for thinking about kinds, instances of kinds, and kinds of kinds. In D. Barner & A. S. Baron (Eds.), *Core knowledge and conceptual change* (pp. 209–224). Oxford, England: Oxford University Press. 10.1093/acprof:oso/9780190467630.003.0012

Rogat, A., Anderson, C., Foster, J., Goldberg, F., Hicks, J., Kanter, D., Krajcik, J., Lehrer, R., Reiser, B., & Wiser, M. (2011). *Developing learning progressions in support of the new science standards: A RAPID Workshop Series.* CPRE. Teachers College, Columbia University. 10.12698/cpre.2011.lprapid

Setoh, P., Wu, D., Baillargeon, R., & Gelman, R. (2013). Young infants have biological expectations about animals. *Proceedings of the National Academy of Sciences, USA*, *110*, 15937–15942. 10.1073/pnas.1314075110

Sherin, B. (2001). How students understand physics equations. *Cognition and Instruction, 19* (4), 479–541. 10.1207/s1532690xci1904_3
Sherin, B. (2006). Common sense clarified: The role of intuition knowledge in physics problem solving. *Journal of Research in Science Teaching, 43*(6), 535–555. 10.1002/tea.20136
Sikorski, T., & Hammer, D. (2010). A critique of how learning progressions research conceptualizes sophistication and progress. *Proceedings of the 9th International Conference of the Learning Sciences, 1*, 1032–1039.
Sikorski, T., & Hammer, D. (2017). Looking for coherence in science curriculum. *Science Education, 101*, 929–943. 10.1002/sce.21299
Smith, C. L. (2018). Conceptualizing the interactions among epistemic thinking, metacognition, and content-specific conceptual change. In T. Amin & O. Levrini (Eds.), *Converging perspectives on conceptual change* (pp. 268–286). New York: Routledge. 10.4324/9781315467139-32
Smith, C. L., Maclin, D., Grosslight, L., & Davis, H. (1997). Teaching for understanding: A comparison of two approaches to teaching students about matter and density. *Cognition and Instruction, 15*(3), 317–393. 10.1207/s1532690xci1503_2
Smith, C. L., & Wiser, M. (2015). On the importance of epistemology-disciplinary core concept interactions in LPs. *Science Education, 99*(3), 417–423. 10.1002/sce.21166
Smith, J. P., diSessa, A. A., & Roschelle, J. (1993-1994). Misconceptions reconceived: A constructivist analysis of knowledge in transition. *Journal of the Learning Sciences, 3*(2), 115–163. 10.1207/s15327809jls0302_1
Sodian, B. (2016). Is false belief understanding continuous from infancy to preschool age? In D. Barner & A. S. Baron (Eds.), *Core knowledge and conceptual change* (pp. 301–320). Oxford, England: Oxford University Press. 10.1093/acprof:oso/9780190467630.003.0017
Stevens, S. Y., C. Delgado, & Krajcik, J. (2010). Developing a hypothetical multi-dimensional learning progression for the nature of matter. *Journal of Research in Science Teaching, 47*(6), 687–715. 10.1002/tea.20324
Stewart, J., Cartier, J., & Passmore, C. (2001). Developing understanding through model-based inquiry. In M. S. Donovan & J. D. Bransford (Eds.), *How students learn: Science in the classroom* (pp. 515–559). Washington, D.C.: The National Academies Press. 10.17226/11102
Shtulman, A. (2006). Qualitative differences between naïve and scientific theories of evolution. *Cognitive Psychology, 52*, 170–194. 10.1016/j.cogpsych.2005.10.001
Talmy, L. (1988). Force dynamics in language and cognition. *Cognitive Science, 12*, 49–100. 10.1207/s15516709cog1201_2
Tomasello, M. (1999). *The cultural origins of human cognition.* Cambridge, MA: Harvard University Press. 10.2307/j.ctvjsf4jc
Vosniadou, S., & Brewer, W. (1992). Mental models of the Earth: A study of conceptual change in childhood. *Cognitive Psychology, 24*, 535–585. 10.1016/0010-0285(92)90018-w
Wellman, H. M., & Gelman, S. A. (1992). Cognitive development: Foundational theories of core domains. *Annual Review of Psychology, 43*, 337–375. 10.1146/annurev.ps.43.020192.002005
White, B. (1993). Thinkertools: causal models, conceptual change, and science education. *Cognition and Instruction, 10*(1), 1–100. 10.1207/s1532690xci1001_1
White, B. Y., & Frederiksen, J. R. (1998). Inquiry, modeling, and metacognition: Making science accessible to all students. *Cognition and Instruction, 16*(1), 3–118. 10.1207/s1532690xci1601_2
Whitebread, D., Pino-Pasternak, D., & Coltman, P. (2014). Making learning visible: The role of language in the development of metacognition and self-regulation in young children. In S. Robson & S. F. Quinn (Eds.), *The Routledge international handbook of young children's thinking and understanding* (pp. 199–214). London: Routledge. 10.4324/9781315746043
Wiser, M., & Amin, T. (2001). "Is heat hot?" Inducing conceptual change by integrating everyday and scientific conceptions. *Learning and Instruction, 11*, 331–355. 10.1016/s0959-4752(00)00036-0
Wiser, M., & Smith, C. (2013) Learning and teaching about matter in the middle school years: How can the atomic-molecular theory be meaningfully introduced? In S. Vosniadou (Ed.), *International handbook of research on conceptual change* (2nd edition) (pp. 177–194). New York: Routledge. 10.4324/9780203154472-17
Wiser, M., & Smith, C. (2016). How is conceptual change possible? Insights from science education. In D. Barner & A. S. Baron (Eds.), *Core knowledge and conceptual change* (pp. 29–51). Oxford: Oxford University Press. 10.1093/acprof:oso/9780190467630.003.0003

3

ON THE CRITIQUES OF LEARNING PROGRESSION RESEARCH

Hui Jin and Joseph Krajcik

Introduction

Learning progressions (LPs) are "descriptions of successively more sophisticated ways of thinking about how learners develop key disciplinary ideas and practices within a grade level and across multiple grades" (Fortus & Krajcik, 2012, p. 784). Researchers made significant advances in science LP research during the past two decades. This body of literature is analyzed in two reviews, one covering earlier work (Duschl et al., 2011) and the other focusing on more recent research (Jin, Mikeska et al., 2019).

Several researchers raised critiques of the value of LPs (Shavelson & Kurpius, 2012; Sikorski & Hammer, 2010). In existing research, an LP typically contains one to several progress variables and a limited number of achievement levels per progress variable. According to critics, students' knowledge is fragmented, containing loosely connected pieces; the process of knowledge development is complex and dynamic; individual students may learn the same scientific idea and practice in very different ways. Then, how can science learning be presented as a sequence of a small number of achievement levels? How can a single progression present the learning of many different students? To validate an LP, researchers often use qualitative and quantitative methods to analyze students' responses; the analysis results show how well the achievement levels capture the patterns of students' responses and whether the achievement levels differentiate from each other in a progression (e.g., Jin, Delgado et al., 2019; Kaldaras et al., 2021). The critics point out that empirical validation, no matter how advanced, is problematic without a sound theoretical rationale. As Shavelson and Kurpius (2012) describe it, a danger exists in using empirical data to fit the researchers' arbitrary progressions, or in other words, their "procrustean beds" (Shavelson & Kurpius, 2012). This chapter provides the background and responses to these critiques.

Background of the Critiques of LP Research

The critiques of the LP research are rooted in a well-known debate between two perspectives on conceptual change: the knowledge-in-pieces perspective and the naïve-theory perspective. These two perspectives differ over (1) knowledge structure and (2) knowledge development.

DOI: 10.4324/9781003170785-4

A more detailed discussion of the naïve theory perspective as the cognitive foundation of LP research can be found in the chapter by Smith and Amin in this Volume.

Regarding knowledge structure, the debate focuses on the structure of students' knowledge and its coherence. From the knowledge-in-pieces perspective, students' knowledge can be represented as loosely connected fragments, known as "phenomenological primitives" or p-prim (diSessa, 2018). P-prims are "frequently fairly simple abstractions of familiar events, such as the fact that a pushed object moves parallel to the push" (diSessa, 2018, p. 39). diSessa (2002) identified many p-prims in physics. Some examples are:

- Ohm's p-prim: More effort begets more results, while more resistance begets less result.
- Force and mover: Things go in the direction you push them.
- Returning to equilibrium: Systems out of balance tend to return to equilibrium.

Because simple abstractions of familiar events result in students developing p-prims, generating new p-prims is effortless, and students hold many p-prims in their minds. When a student explains the phenomena about one topic, one or more p-prims in the student's mind become activated and used to form explanations. Which p-prims become activated is primarily determined by contextual characteristics. For example, students often believe that a spring exerts a force on the object on it, but a table does not. Since students' understanding is context-dependent and their minds contain many pieces of knowledge, some of which can be even in conflict, we cannot claim that students' knowledge is coherent. As will be elaborated later, researchers who adopt the knowledge-in-pieces perspective often identify and utilize certain knowledge pieces from students to assist them in constructing scientific understanding.

From the naïve-theory perspective, although students' knowledge is not as coherent as scientists' knowledge, it is theory-like and contains logically connected components (Gopnik & Wellman, 1994). Children can develop specific theories in naïve physics, naïve biology, and naïve psychology from a young age. Vosniadou's (2002) idea of "framework theory" gives an example of theory-like knowledge. According to Vosniadou, students tend to use a fundamental "framework theory" to construct *specific* theories, beliefs, and models about the behaviors of physical objects. This framework theory contains a set of ontological and epistemological presuppositions constraining physical objects' behaviors. Take the earth's shape as an example. A typical ontological presupposition is that gravity operates in an up-and-down fashion. A typical epistemological presupposition is that things are as they appear to be. Naïve students analyze their daily experiences based on these two presuppositions. They observe that objects always fall until they reach the ground or the lowest point. They also observe that the ground supports all objects (e.g., houses, cars, furniture, people, and plants). Based on their presuppositions and their life experiences, students make the following inferences: If the earth is a sphere, it can only support objects and people on top of it; things will fall when they move from the top to the side/bottom of the sphere; people live all around the earth, and they never fall; therefore, the Earth cannot be a sphere. It must be flat, like a disk or a rectangular board.

Similarly, students with naïve ideas often understand biological phenomena (e.g., living things growing and eating foods) in terms of vitalistic reasoning and teleological reasoning. According to vitalistic reasoning, living things gain life force from different sources and use it to sustain life (Inagaki & Hatano, 2004). According to teleological reasoning, the body organs of living things exist to sustain life (Carey, 1985). When using vitalistic reasoning and teleological reasoning, students see no need to reason about microscopic mechanisms of macroscopic changes, let alone reasoning about matter and energy.

Regarding knowledge development, the debate between these two perspectives focuses on how students transition from their intuitive thinking to scientific thinking. Knowledge-in-pieces researchers propose that students' knowledge can be used as productive resources (Hammer, 2000). They have developed instructional strategies that activate potentially productive p-prims in students' minds and guide students to build scientific concepts based on those p-prims. For example, students hold the intuitive idea that springs push things on top of them. Minstrell (1982) utilizes this knowledge piece as a resource for learning; he begins by guiding students in thinking about a table as an extremely stiff spring. From this analogy, he helps students gradually develop the scientific concept of normal force.

Naïve-theory researchers have identified several mechanisms of conceptual change, such as conceptual differentiation (Carey, 2011), belief revision, mental model transformation, and categorical shift (Chi, 2009). An example of conceptual differentiation is provided here: In science, density, an intensive quantity, does not depend on the amount of material present in an object; it is a property of materials. Mass, an extensive quantity, depends on the amount of material contained within a given object; it is a property of objects, not materials. However, central to students' knowledge structure about mass and density is the intuitive concept of "felt weight," which is "an unanalyzed perceptual property of entities that can be held" (p. 344, Smith, 2007). The concept of "felt weight" conflates the meaning of weight/mass and density. In other words, students mix characteristics of both weight and density; and they sometimes treat weight as an intensive property. For example, many students hold the idea that steel has more "stuff," while wood has less "stuff", so steel is heavier. For students to develop a scientific understanding, an instructional approach must be used to help students differentiate weight/mass and density. This process is called conceptual differentiation (Smith, 2007). It is important to note that fostering conceptual change in students is challenging. When learning science, students often modify their existing naïve theories to incorporate scientific ideas rather than completely restructuring their theories. Many "misconceptions" are synthetic conceptions or models (Vosniadou, 2002). For example, students' explanations of natural selection-related phenomena often blend features of both natural selection and need-based change (Furtak & Heredia, 2014).

Bootstrapping is an instructional approach that helps students make conceptual changes and restructure their naïve theories into scientific theories. In bootstrapping, researchers design activities in which students use their limited understanding as resources to generate in-depth understanding (Carey, 2011). For example, Smith (2007) designed a series of activities to help elementary students transition from the intuitive concept of felt weight to a differentiated understanding of mass, volume, and density. To help the students bootstrap from their perceptions of physical objects to mathematical quantification and measurement of objects, she began by engaging students in a thought experiment – imagining what happens as one divides an object into smaller and smaller pieces. This thought experiment helps students understand that certain pieces of matter may be too small to see and that what may seem light to us can be heavy to tiny creatures. This realization highlights the need for quantification and measurement, as relying solely on individual perceptions can result in varying judgments of an object's weight or mass. Introducing standardized measurements allows for a more objective understanding of weight and mass, free from the influence of individual perceptions. Next, she engaged the students in activities involving measuring an object's weight and volume. Through these activities, the students bootstrapped from mathematical quantification and measurement of objects to scientific understanding, in which they differentiate mass, volume, and density.

Table 3.1 presents a comparison between the knowledge-in-pieces and naïve-theory perspectives. The fundamental knowledge components of these two theories are p-prims

Table 3.1 Comparison of the knowledge-in-pieces perspective and the naïve-theory perspective

Perspectives	*Knowledge-in-pieces perspective*	*Naïve-theory perspective*
Knowledge Structure	*Components* • loosely connected knowledge pieces (i.e., p-prims) • many p-prims • a simple abstraction of familiar events *Using the knowledge to explain phenomena* • Depending on contextual characteristics, certain p-prims are activated and used to construct explanations. • Without instruction, the above process often generates inconsistent and incoherent explanations	*Components* • theory-like structure containing connected components • a small number of naïve theories • generalization of a wider variety of phenomena and experiences *Using the knowledge to explain phenomena* • Naïve theories are used to explain phenomena across contexts and topics. • The above process often generates relatively consistent and coherent explanations but does not match current scientific thinking.
Knowledge development	• P-prims constitute productive resources in students' minds. • Instruction is used to activate certain 'useful' p-prims and use them as the foundation for building scientific ideas.	• Typical mechanisms of conceptual changes (e.g., conceptual differentiation, synthetic model generation, belief revision, mental model transformation, and categorical shift) are identified. • Instruction helps students use their 'limited' understanding as a starting point for a specific conceptual change.

and naïve theories. While there are many p-prims, there are only a small number of naïve theories. While p-prims are products of simple abstraction of observations, naïve theories often reflect generalizations of a wider variety of phenomena and experiences. Due to these differences, these two perspectives provide different assumptions about how students use knowledge to explain phenomena. From the knowledge-in-pieces perspective, depending on specific context characteristics, some p-prims in students' minds are activated and used to explain the phenomenon. Without instruction or guidance, such a process often results in incoherent and inconsistent explanations. To foster effective learning, instruction can activate useful p-prims and guide students in building scientific ideas based on the activated p-prims. From the naïve-theory perspective, students' knowledge is theory-like, containing logically connected components. Students apply naïve theories across contexts and phenomena, producing relatively coherent and consistent explanations, but not necessarily scientific. Effective learning happens when students begin with their "limited" knowledge and experiences and go through the conceptual change mechanisms. Instruction is, therefore, designed to guide students to make the conceptual change. This comparison suggests a critical implication – although the naïve-theory researchers and the knowledge-in-pieces researchers disagree with each other on whether students' knowledge is relatively coherent, they use students' knowledge and experiences as resources for learning.

The above discussion of the two perspectives of student learning helps us understand the critiques of the LP approach. From the knowledge-in-pieces perspective, student knowledge is fragmented, containing many loosely connected pieces. When explaining phenomena, depending on contextual factors, certain knowledge pieces are activated to generate explanations. As such, it is inappropriate to represent student understanding of a scientific idea/practice in terms of a few LP levels. If this is true, the validation techniques used in LP research (e.g., interviews and IRT analysis of assessment scores) lack a theoretical basis. This is Critique 1, which is about the nature of student knowledge. The knowledge-in-pieces perspective asserts that learning is contextual and dynamic. Therefore, a single progression cannot accurately represent the learning of a diverse group of students. Due to variations in thinking, development, and learning styles, each student may have a unique learning trajectory. Thus, it is unlikely that one standardized progression can be applied to all students. This is Critique 2, which is about knowledge development, or in other words, the process of learning scientific knowledge.

Response to the Critiques

Researchers have used both naïve theories and knowledge-in-pieces to develop LPs. For example, when developmental psychologists and science educators initiated the LP approach, they collaborated on developing a hypothetical LP about matter (Smith et al., 2006). The development of that LP primarily draws upon ideas from naïve physics, such as material kind and conceptual differentiation. Science educators have used the matter LP as an exemplar, particularly during the earlier stages of science LP research. Researchers also used the knowledge-in-pieces perspective to develop LPs. For example, Stevens, Delgado, and Krajcik (2010) developed an LP for the nature of matter. The LP was produced based on diSessa's knowledge-in-pieces perspective, which posits that "students' knowledge structures may not be organized frameworks but pieces that are not put together in a systemic manner" (Stevens et al., 2010, p. 689). Building upon this perspective, the LP and its levels describe not only important matter concepts but also how the connections among the concepts build over time.

If substantial evidence suggests that students do not reason consistently across contexts or they do not use relatively coherent theories, LPs that use findings and ideas from naïve theories raise concerns. To address this concern, we first consider the empirical studies that examined whether students' knowledge is fragmented or theory-like. Next, we argue that science LPs are instruction-oriented; researchers should evaluate LPs for their effectiveness in promoting student learning.

Is Students' Knowledge Fragmented or Theory-like?

In a study on the meaning of forces, Ioannides and Vosniadou (2002) interviewed 105 Greek students from ages 4 through 15. They analyzed students' responses and identified seven separate meanings of force in students' responses. In the order of development, these force meanings are internal force, internal force affected by movement, internal and acquired force, acquired force, acquired force and force of push/pull, force of push/pull, and gravitational and other forces. They present a developmental trend from internal force to acquired force to gravitational force, with several synthetic force meanings in between. Ioannides and Vosniadou also found that 88.6% of the students consistently used one force meaning to construct explanations across the 27 interview tasks. Based on this finding, they further inferred a framework theory that students may use to generate the intuitive force meanings (i.e., internal force and acquired force). Apparently, if most students used a variety of force meanings at random during the interview, it would not be appropriate to infer the underlying "framework theory". This study provides empirical evidence supporting the naïve-theory perspective. However, in a quasi-replication of this study, diSessa and colleagues (diSessa et al., 2004) used the same interview tasks with 30 U.S. students. They found that only five students used internally consistent meanings of force, and the remaining 25 students used different force meanings to answer different interview questions. They argue that since the students did not use the force meanings consistently, it is inappropriate to infer a framework theory behind the intuitive force meanings.

Southerland and colleagues (Southerland et al., 2001) interviewed 96 students from 2nd, 5th, 8th, and 12th grades. In each interview, the student was asked to explain a series of graphics that depict biological events. The study generated two important findings. First, although the students came from different grade levels, the researchers identified the same reasoning categories from the students' responses. This result aligns with one idea from the naïve-theory perspective: Students use a small number of reasoning patterns to interpret domain phenomena. These reasoning patterns are non-scientific. Second, individual students often shift reasoning categories in explaining a single event or different events. This result aligns with one idea from the knowledge-in-piece perspective: When answering questions about similar phenomena, different knowledge pieces or ideas in students' minds are activated and used to construct explanations. Based on these results, Southerland and colleagues made the following inference about the nature of students' knowledge: If the assumption is that students use conceptual frameworks to understand the world, "the students have weak, poorly constructed frameworks for biological phenomena and these weak conceptions do not change substantially throughout years of schooling" (p. 343). If the assumption is that students hold loosely connected p-prims to explain phenomena, "the spontaneous nature of explanations based on p-prims would account for the tentative patterns, the shifting patterns, as well as the lack of development documented in students' explanations across grade levels" (p. 344). In sum, empirical studies do not provide enough evidence to confirm whether students' knowledge is fragmented or theory-like.

Nobelist Daniel Kahneman (2011) proposed a groundbreaking theory in his New York Times bestseller, *Thinking, Fast and Slow*. According to this theory, humans use two systems to process information and make decisions (pp. 20–21):

> System 1 operates automatically and quickly, with little or no effort and no sense of voluntary control. System 2 allocates attention to the effortful mental activities that demand it, including complex computations. The operations of System 2 are associated with the subjective experience of agency, choice, and concentration.

Examples of using System 1 to make decisions include detecting that one object is more distant than another, completing the phrase "bread and ...", finding the answer to 1 + 1, and driving a car on an empty road. Examples of using System 2 for decision-making include filling out a tax form, comparing two washing machines for overall value, and behaving appropriately in a social situation. System 1 and System 2 do not work in isolation. Instead, they interact frequently. Kahneman (2011) provides an example of how these two systems work together to answer the following question (p. 109):

> A bat and ball cost $1.10.
> The bat costs one dollar more than the ball.
> How much does the ball cost?

In his study, over 50% of Harvard, MIT, and Princeton students gave an intuitive incorrect answer – $0.10. Of course, most of these students have the capability of answering this simple problem. With a little math, they should be able to calculate that the correct answer is 5¢. But why did so many students provide an incorrect answer? Kahneman attributed the cause to the laziness of System 2. When answering the question, the students' System 1 quickly generates a plausible answer, $0.10. Overriding this answer requires the hard work of System 2. Students prone to place too much faith in their intuition and find cognitive effort at least mildly unpleasant often stick with the plausible answer generated by System 1. As humans, our System 2 tends to be lazy, and we tend to use as little cognitive effort as possible.

Kahneman's theory can explain why the empirical studies discussed above (diSessa et al., 2004; Ioannides & Vosniadou, 2002; Southerland et al., 2001) generated findings that support both the theory-like and the fragmented knowledge. In these studies, each participating student answered questions about one topic (e.g., earth's shape, biological events, etc.). Consider two *extreme* cases where two students, A and B, answer the questions in an interview. Student A uses System 1 to answer all questions. A's System 2 is extremely lazy during the interview; it proves all intuitive answers generated by System 1. When analyzing A's responses, the researchers are likely to find that A's responses are inconsistent; they may connect to each other but only loosely; there are also conflicts between some ideas. The researchers, therefore, infer that the students hold knowledge-in-pieces. In the other case, Student B is not tempted to accept any plausible answer that comes readily to mind. Instead, he uses System 2 to analyze each question deliberately, connect different questions, and identify and fill the gaps among the answers to different questions. In this case, when researchers analyze Student B's responses, the researchers are likely to find that B constructs relatively coherent explanations. The researchers infer that B demonstrates the use of theory-like ideas, although B's explanations don't necessarily match appropriate scientific ideas.

It is safe to assume that in the above-mentioned studies, most participating students would fall between these two extremes. The researchers examined students' responses to infer whether their knowledge was fragmented or coherent. However, the researchers have not considered a key component in the middle of this inferential chain. This component is the student's thought processes, i.e., how the student used System 1 and System 2 to answer the questions. A student who holds theory-like ideas may rely on System 1 to reason and does not use System 2 to deliberately analyze and evaluate the intuitive answers generated by System 1; as a result, the student will provide inconsistent responses in the interview. Therefore, the empirical studies on the coherence of students' knowledge do not provide enough information to determine whether the LPs using naïve theories have a sound theoretical rationale.

Linguistics research offers a potential solution to this issue. According to Pinker (2003), a vast body of evidence from linguistics research suggests that humans are born with limited innate mental modules and reasoning patterns. For example, human languages and children's speech patterns contain intricate grammar properties. Even in these different languages, errors are highly systematic. Additionally, instances exist where children who lived together without a pre-existing language developed their own language. Furthermore, force dynamics, a reasoning pattern about how entities interact concerning force, plays a structuring role in human languages. If students use specific reasoning patterns to learn and construct languages, they may utilize these patterns to analyze and explain natural phenomena in their everyday lives, which was confirmed in several empirical studies (Gunckel et al., 2012). In this sense, linguistics research provides indirect support for the naïve theory perspective and, therefore, for using the naïve-perspective to develop LPs.

In brief, a definitive conclusion regarding whether students hold pieces of knowledge or naïve theories cannot be drawn because the researchers do not know how students used System 1 or System 2 to construct their explanations in the interviews or the assessments. Thinking and reasoning are complex phenomena. It is likely that, in interviews and assessments, most students demonstrate consistent reasoning across some but not all contexts. They may reason consistently across contexts that share similar surface features but use different theories or ideas to explain contexts that have different appearances.

LP Researchers have employed qualitative and quantitative methods to validate the achievement levels in an LP. For instance, researchers frequently gathered think-aloud interview data to determine if students appropriately understood assessment items used to elicit their thinking (e.g., Gotwals & Songer, 2013). IRT analysis is often performed on assessment scores to generate Wright Maps, which reveal the extent to which the LP levels distinguish one level from another (e.g., Jin, Delgado et al., 2019; Kaldaras et al., 2020). Well-differentiated levels are empirical evidence for the LP levels. Researchers use validation methods to identify typical reasoning patterns of students. They are not meant to provide evidence that individual students use these reasoning patterns to consistently reason about phenomena across all or most contexts.

LPs Should Be Evaluated by Their Effectiveness in Promoting Student Learning

In an LP, the achievement levels are arranged sequentially to depict student development of scientific knowledge. According to Critique 2, presenting student learning as a progression ignores the dynamics of learning (Shavelson & Kurpius, 2012; Sikorski & Hammer, 2010). In response to this critique, we present two key points.

First, many existing studies developed LPs across grade levels. For these studies, the LPs were developed based on cross-sectional rather than longitudinal data. Researchers collect cross-sectional data from students at different grade levels at one time or several time points. Longitudinal data is collected from the same group of students as the students grow across grade levels. Researchers need to use longitudinal data to study individual student's trajectories as they move up levels within and across grades. Therefore, a simple response to Critique 2 is that an LP does not describe individual students' learning trajectories; rather, it outlines a typical pathway many students can follow. Furthermore, researchers have begun to explore multiple learning pathways. For example, Peng and colleagues, in this volume, present an LP for students developing an understanding of energy and the interaction of matter at the middle school level. What is unique about their work is the presentation of multiple pathways for students with different backgrounds to develop their understanding. Duncan and colleagues, in this volume, compared two other LPs and LP-based instructional materials for genetics.

Second, an LP is a model that describes how instructional environments can be designed to facilitate coherent experiences for students, enabling them to achieve a sophisticated understanding of scientific ideas or practices over time. In other words, LPs are not just models of learning but rather models of instruction-scaffolded learning. Such models use a developmental approach, requiring students to experience ideas and practices early in their education through planned, thoughtful instructional experiences. As students have further educational experiences, the knowledge and use of their knowledge become more sophisticated. Take, for example, learning Newton's laws of motion. When forces are balanced, there is no change in motion (although an object could be moving, its motion does not change, i.e., it is going at a constant speed). When forces are unbalanced, there is a change in the motion of an object (i.e., an object starts moving, slows down, speeds up, changes directions, or stops). If you see an object speeding up, a net force is acting on the object. Developing this level of scientific understanding is challenging for students because Newton's laws are counter-intuitive, and students receive no instructional experience related to them until middle grades.

Traditionally, students learn Newton's laws in a sequence: the first law, followed by the second, and then the third. Further, students typically study the first law until the middle grades. As a result, many students can memorize the scientific statements of Newton's laws, but they do not know how to apply them. Many students believe that to keep something moving at a constant speed, they need to apply force. This idea makes sense to students because it reflects their everyday experience: they must continually pedal their bikes to stay in constant motion. Students often fail to take into consideration other forces acting on the bike tires (e.g., friction exerted by the ground) because those forces are not obvious.

To help students develop a scientific understanding of Newton's laws and overcome their informal reasoning about forces and motion, it is essential to address three key points: 1) Forces are interactions between objects, not inherent properties of objects or 'impetus' stored within objects. 2) When we analyze the relationships between forces and motion, we must consider all individual forces exerted on the object, including those not readily apparent in everyday situations (e.g., friction). 3) Forces do not cause motion. Instead, forces are associated with changes in motion, and physicists have quantified the relationship between forces and motion as $F = ma$. Using an LP approach, instruction can utilize students' experiences and informal reasoning as resources to develop this level of scientific understanding. In early elementary grades, students are introduced to the idea of force as a push or a pull and have appropriate first-hand experiences. Later, they learn about forces that are not readily apparent such as friction and gravity. With the ability to identify various forces, students can effectively develop the idea of balanced and unbalanced forces. With this understanding, instruction can further help students understand that a balanced force means there will be no change in the motion of an object, and an unbalanced force means they will experience a change in motion. Then, in the middle grades, these ideas can be further developed, and in high school, a quantitative approach can occur. This is what is meant by a developmental approach. Students learn challenging ideas over an extended period through experiences appropriate for their age. Then, those ideas are built upon in subsequent units and grade levels to make their knowledge more sophisticated. This is why Lehrer and Schauble (Lehrer & Schauble, 2015) state that LPs can't be developed using traditional curriculum approaches because they don't take a developmental approach. LPs are not inevitable; they depend upon learners experiencing situations where they need to build and use appropriate ideas. Learning is a complicated process that is non-linear and has multiple pathways. LPs are developmentally appropriate, as they build over time based on learners' prior knowledge and experiences to build more sophisticated understanding.

Conclusion

Although the LP research has generated essential findings and products, it faces significant critiques. The critiques are rooted in the debate between the naïve-theory and the knowledge-in-pieces perspectives. While the naïve-theory perspective significantly influenced LP research, the critiques are based on the knowledge-in-pieces assumptions. Two major critiques are:

- Critique 1: Since students' knowledge contains many loosely connected pieces, it is inappropriate to represent student understanding of a scientific idea/practice at a few LP levels. If this is true, the validation techniques used in LP research (e.g., interviews and IRT analysis of assessment scores) lack a theoretical basis.
- Critique 2: Since learning is dynamic and context-dependent, one progression cannot represent many students' learning.

In response to Critique 1, we investigated whether empirical studies support the assumption that students' knowledge is fragmented rather than theory-like, as proposed by the knowledge-in-pieces perspective. We found that these studies have presented evidence that supports both viewpoints. Kahneman's (2011) dual-system theory offers a framework to comprehend these results. When generating a response, System 1 provides rapid and instinctive answers, whereas System 2 makes efforts to conduct logical analysis and evaluate the response generated by System 1. In these studies, researchers inferred the cohesion of students' ideas from their responses. What is absent in their inferential chain is how System 1 and System 2 collaboratively generate responses. Without considering this critical component, researchers cannot make a claim about the cohesion or fragmentation of students' ideas. Moreover, linguistic research provides ample evidence that humans tend to use similar ways of thinking to construct language (Pinker, 2003), indirectly suggesting that it is feasible to capture the knowledge of many students by employing only a few reasoning patterns (i.e., LP levels). Given this theoretical rationale, it is appropriate to verify whether the levels of achievement in students' responses "exist" by utilizing techniques such as think-around interviews and IRT analysis.

In response to Critique 2, we clarified that most LP were developed based on cross-sectional rather than longitudinal data. Therefore, they do not describe individual students' learning trajectories. We further emphasize that LPs are models that describe instruction-facilitated learning. These models use a developmental approach in which students learn challenging ideas over an extended period of time. This is achieved by purposefully designing instruction and learning environments, which provide experiences appropriate for students with a given background and developmental age and leverage students' existing ideas to build more sophisticated ideas. Moreover, these models can have multiple learning pathways to address the needs of diverse students.

References

Carey, S. (2011). The origin of concepts: A précis. *The Behavioral and Brain Sciences*, *34*(3), 113.

Carey, S. (1985). *Conceptual change in childhood.* MIT Press.

Chi, M. T. H. (2009). Three types of conceptual change: Belief revision, mental model transformation, and categorical shift. In *International handbook of research on conceptual change* (pp. 89–110). Routledge.

diSessa, A. A. (2002). Why "conceptual ecology" is a good idea. In *Reconsidering conceptual change: Issues in theory and practice* (pp. 28–60). Springer.

diSessa, A. A. (2018). A friendly introduction to "knowledge in pieces": Modeling types of knowledge and their roles in learning In G. Kaiser, H. Forgasz, M. Graven, A. Kuzniak, E. Simmt, & B. X. Cham (Eds.) *Invited lectures from the 13th International Congress on Mathematical Education (ICME-13)* (pp. 65–84). Cham, CH: Springer Open. 10.1007/978-3-319-72170-5.

diSessa, A. A., Gillespie, N. M., & Esterly, J. B. (2004). Coherence versus fragmentation in the development of the concept of force. *Cognitive Science*, *28*(6), 843–900.

Duschl, R., Maeng, S., & Sezen, A. (2011). Learning progressions and teaching sequences: a review and analysis. *Studies in Science Education*, *47*(2), 123–182. https://doi.org/10.1080/03057267.2011.604476

Fortus, D., & Krajcik, J. (2012). Curriculum Coherence and Learning Progressions. In B. J. Fraser, K. Tobin, & C. J. McRobbie (Eds.), *Second International Handbook of Science Education* (pp. 783–798). Springer: Netherlands. https://doi.org/10.1007/978-1-4020-9041-7_52

Furtak, E. M., Morrison, D., & Kroog, H. (2014). Investigating the link between learning progressions and classroom assessment. *Science Education*, *98*(4), 640–673. 10.1002/sce.21122

Gopnik, A., & Wellman, H. M. (1994). The theory theory. *Mapping the mind: Domain specificity in cognition and culture*, 257.

Gotwals, A. W., & Songer, N. B. (2013). Validity Evidence for Learning Progression-Based Assessment Items That Fuse Core Disciplinary Ideas and Science Practices. *Journal of Research in Science Teaching*, *50*(5), 597–626. https://doi.org/10.1002/tea.21083

Gunckel, K. L., Covitt, B. A., Salinas, I., & Anderson, C. W. (2012). A learning progression for water in socio-ecological systems. *Journal of Research in Science Teaching*, *49*(7), 843–868. 10.1002/tea.21024

Hammer, D. (2000). Student resources for learning introductory physics. *American Journal of Physics*, *68*(S1), S52–S59.

Inagaki, K., & Hatano, G. (2004). Vitalistic causality in young children's naive biology. *Trends in cognitive sciences*, *8*(8), 356–362.

Ioannides, C., & Vosniadou, S. (2002). The changing meanings of force. *Cognitive Science Quarterly*, *2*(1), 5–62.

Jin, H. Delgado, C. Bauer, M. I. Wylie, E. C. Cisterna, D. & Llort, K. F. (2019). A hypothetical learning progression for quantifying phenomena in science. *Science & Education*, *28*, 1181–1208.

Jin, H. Mikeska, J. N., Hokayem, H. & Mavronikolas, E. (2019). Toward coherence in curriculum, instruction, and assessment: A review of learning progression literature. *Science Education*, *103*(5), 1206–1234.

Kahneman, D. (2011). *Thinking, fast and slow*. Macmillan, London.

Kaldaras, L., Akaeze, H., & Krajcik, J. (2021). Developing and validating next generation science standards-aligned learning progression to track three-dimensional learning of electrical interactions in high school physical science. *Journal of Research in Science Teaching*, *58*(4), 589–618. https://doi.org/10.1002/tea.21672

Lehrer, R., & Schauble, L. (2015). Learning progressions: The whole world is NOT a stage. *Science Education*, *99*(3), 432–437.

Minstrell, J. (1982). Explaining the "at rest" condition of an object. *The Physics Teacher*, *20*(1), 10–14. https://doi.org/10.1119/1.2340924

Pinker, S. (2003). Language as an adaptation to the cognitive niche. *Studies in the Evolution of Language*, *3*, 16–37.

Shavelson, R. J., & Kurpius, A. (2012). Reflections on learning progressions. In *Learning progressions in science* (pp. 13–26). Brill Sense.

Sikorski, T.-R., & Hammer, D. (2010). A critique of how learning progressions research conceptualizes sophistication and progress In Gomez, K., Lyons, L., & Radinsky, J. (Eds.), *Learning in the Disciplines: Proceedings of the 9th International Conference of the Learning Sciences (ICLS 2010) - Volume 1, Full Papers* (pp. 1032–1039). Chicago IL: International Society of the Learning Sciences.

Smith, C. L., Wiser, M., Anderson, C. W., & Krajcik, J. (2006). Implications of research on children's learning for standards and assessment: A proposed learning progression for matter and the atomic-molecular theory. *Measurement: Interdisciplinary Research & Perspective*, *4*(1-2), 1–98.

Smith, C. L. (2007). Bootstrapping processes in developing students' commonsense matter theories: Using analogical mappings, thought experiments, and learning to measure to promote conceptual restructuring. *Cognition and Instruction*, *25*(4), 337–398.

Southerland, S. A., Abrams, E., Cummins, C. L., & Anzelmo, J. (2001). Understanding students' explanations of biological phenomena: Conceptual frameworks or P-Prims? *Science Education*, *85*, 328–348. 10.1002/sce.1013

Stevens, S. Y., Delgado, C., & Krajcik, J. S. (2010). Developing a hypothetical multi-dimensional learning progression for the nature of matter. *Journal of Research in Science Teaching*, *47*(6), 687–715.

Vosniadou, S. (2002). On the nature of naive physics. In *Reconsidering conceptual change: Issues in theory and practice* (pp. 61–76). Springer.

4

VALIDATION OF LEARNING PROGRESSIONS

Michael T. Kane

Chapter 4: Validation of Learning Progressions

Learning progressions (LPs) have several appealing characteristics (National Research Council, 2001, 2005, 2007). First, they emphasize the mastery of big ideas and broadly applicable foundational theories and practices, rather than recall of descriptive information and mastery of algorithmic skills, and in doing so, they define achievement in terms of what students can do. In particular, they define achievement and progress in terms of qualitatively different achievement levels, representing increasing levels of sophistication in some topic or domain. Second, they focus on instruction, learning, and progress in the domain. Third, the definitions of the levels and the instructional methodology are based on research on how students achieve increasing sophistication in a domain. So, they can be well-grounded and can provide specific, helpful guidance to teachers. This chapter focuses on the validation of LPs in science education, but most of the discussions are generally applicable to LPs.

The validation of LPs presents an unusually complex set of challenges because work on LPs is relatively new, with most of the literature involving reports on efforts to develop effective LPs, general discussions of the potential benefits of LPs and how they might be developed, and a few papers on methodological issues. Most of the learning progressions in science that now exist focus on specific topics at specific grade levels, and the different LPs include a variety of topics and a variety of grade levels. That is, there is not a single well-defined target for validation. In addition, some proposed LPs involve ambitious, long-term curricula development efforts spanning multiple grades and topics, rather than a single topic and a single grade level.

In addition, the standards to be relied on in validating LPs are not so clear. In one sense, a learning progression can be viewed as a local theory of learning specific to some domain and grade level, and as such it would be evaluated against the standards for validating a scientific theory. It can also be viewed as an instructional program to be evaluated in terms of its effectiveness in promoting student progress. Finally, it can be viewed as an assessment program yielding results to be interpreted and used in decision-making and evaluated in terms of professional standards for assessments. It may seem like a heavy burden to require LPs to meet three separate sets of requirements for validation, but it is not too heavy, because the three sets of requirements are elaborations of one basic requirement: that public claims be supported by adequate evidence.

DOI: 10.4324/9781003170785-5

Validation

There are several distinct but related ways to think about validation for learning progressions. First, we can think about validating the LP as a scientific theory. After a theory has been proposed as a basis for describing and explaining some phenomena, it needs to be "validated" before it is accepted by the scientific community. Although there are plenty of disputes about how exactly to do this (Kuhn, 1970; Lakatos, 1970; Popper, 1962; Toulmin, 1953), theory validation generally involves at least three major steps. Step 1, the theory is spelled out in some detail and shown to be internally consistent and reasonable as a theory of the phenomenon in question, and at least some of the terms in the theory are connected to aspects of the observable phenomena (Suppe, 1977). Step 2, the theory is used to make predictions about the phenomena of interest, and these claims are empirically checked against observations of the phenomena. If the observations are generally consistent with the predictions, confidence in the theory increases, and if observations are not consistent with the predictions, confidence in the theory decreases. Step 3, if the theory is partially confirmed as a model of the phenomena, but not completely confirmed, it is likely to be revised in some way. As the theory develops, it can generate new predictions which can be evaluated against new data. To the extent that a theory continues to account for the data, as it is evaluated against empirical checks, it tends to gain greater acceptability (Lakatos, 1970). Note that the empirical validation of theories can also provide a basis for validating the indicators used to connect the theory to the measurable properties of the phenomena (Cronbach and Meehl, 1955).

Second, we can think about validating the effectiveness of the LP as an educational program. Educational programs are designed to provide instructional support to teachers and students that will promote learning, and the program is validated mainly by showing that it achieves its stated goals without causing any serious negative consequences. Lehrer and Schauble (2015) have emphasized this usage of the term, validation, in suggesting that LPs be evaluated in terms of their effectiveness as tools:

> An LP, then, is not an object for validation in the traditional sense of a test or even in the broader sense as a theory, but as a tool validated by the pragmatics of long-term viability and utility for education. (p.437)

A model specifies how the program operates and achieves its intended effects. Some parts of this model may be formally described in terms of mathematical models, but most of it tends to be in narrative form. According to Cronbach (1982), these evaluative interpretations are "held in place by chains (or cobwebs) of reasoning" (p.76) and the interpreter "assembles information and assumptions to make a persuasive case" (p.77). To evaluate the effectiveness of the program is to evaluate these claims, and this is done by evaluating how well the program promotes learning and by evaluating the various assumptions, inferences, and claims in the narrative. It is especially important to evaluate challenges to the assumptions, inferences, and claims in the narrative. If the program fails to achieve its goal (promoting student learning) or has negative side effects on individual students or the school community, or any of the underlying assumptions fail, the narrative as a whole loses credibility, and confidence in the instructional program decreases (Cronbach, 1982).

Third, we can think about validating the LP as a framework for assessing student progress in a domain. Within educational measurement, validation can be thought of as an evaluation of the coherence and completeness of an interpretation/use argument (IUA) specifying the proposed interpretation and use of assessment results and the plausibility of the inferences and assumptions supporting these claims (Kane, 2006, 2013). The IUA can be thought of as a local theory specifying the interpretation and uses of the assessment results, and therefore, validation of the assessments can be thought of as an example of construct validation based on this local theory (Cronbach and Meehl, 1955).

If the assessments do not work as intended within the context of the LP, this failure can cast doubt on the assessments, on the local theory associated with the LP, or on the implementation of the instructional program. That is, the consequences associated with the implementation of the LP play a major role in our confidence in the program and all of its components (Cronbach, 1971, Kane, 2006, Messick, 1989).

Note that although each of these kinds of validation is a distinct activity with its own particular methodologies, standards for evidence, and literature, they have a basic structural similarity. They all involve putting forward an integrated set of claims or hypotheses about some domain of interest (scientific phenomena, score interpretations and uses, an instructional program) and a subsequent conceptual and empirical evaluation of the claims being made. In the case of theory validation, the basic claim is that the theory adequately explains the phenomena. In the case of assessments, the interpretation and use of assessment scores constitute the claims to be evaluated. And in the case of program evaluation, the claims to be evaluated involve the intended and unintended outcomes of the program and its efficacy in promoting desirable outcomes. This general validation strategy of explicitly stating a coherent set of claims and then evaluating these claims and their underlying assumptions critically and empirically is widely applicable. As Cronbach (1980) paraphrased Popper (1962),

> The job of validation is not to support an interpretation, but to find out what might be wrong with it. A proposition deserves some degree of trust only when it has survived serious attempts to falsify it" (Cronbach, 1980, p. 103).

Argument-based validations require that the inferences and assumptions inherent in the claims being made be specified in enough detail that they can be challenged.

In addition to their structural similarities, the three kinds of validations are intertwined for LPs. As noted earlier, LPs are essentially local theories of instruction and learning. The assessments depend on the assumptions built into the LP in a way that provides a working model of Cronbach and Meehl's (1955) conception of construct validity, in that the validation of the assessments depends on the validation of the local theory. Finally, the LP's instructional strategies are based on the local theory and the assessment results. As a result, the expectations for the success of the instructional program tend to be supported by the theory and the assessments, and at the same time, the success of the instructional program in supporting student progress tends to support the validity of both the local theory and the assessments.

Formative and Summative Assessment

The *Standards for Educational and Psychological Testing* (AERA, APA. NCME, 2014) provides professional guidelines for the validation of standardized tests, but these standards are not fully applicable to classroom assessments. As the *Standards* put it:

> Rigorous application of the *Standards* ... to instructor-made tests that are used to evaluate student performance in education and training is generally not possible. (AERA, et al., 2014, p.2)

The technical prescriptions in the *Standards* are not designed for classroom assessments and are not intended to apply to most assessments embedded in classroom instruction, but many of the basic issues raised in the *Standards* would be applicable, in general terms, to any educational assessment (Ferrara, Maxey-Moore, and Brookhart, 2020).

A useful distinction can be drawn between *formative assessment* and *summative assessment*, both of which are concisely defined in the glossary of the *Standards*:

> formative assessment: An assessment process used by teachers and students during instruction that provides feedback to adjust ongoing teaching and learning with the goal of improving students' achievement of intended instructional outcomes. (AERA et al., 2014, p. 219)

and

> summative assessment: The assessment of a test taker's knowledge and skills typically carried out at the completion of a program of learning, such as the end of an instructional unit. (AERA et al., 2014, p. 224)

Note that this distinction was originally proposed in the context of program evaluation as formative evaluation and summative evaluation (Scriven, 1967).

The distinction between formative and summative assessment is one of purpose and use, but it has implications for the structure of an assessment. Summative assessments are often used to make high-stakes decisions with long-term consequences that are difficult to change, and therefore the generalizability (over tasks, contexts, raters, and occasions) of the results and the accuracy of the decisions are major concerns (Brennan, 2001). They tend to summarize the test taker's level of achievement in a domain and pay little attention to the steps leading to that level of achievement. Their design and development tend to rely on input from content experts on the knowledge, skills, and competencies deemed central to mastery of the domain. Summative assessment can benefit from a detailed analysis of how students learn, but they generally do not rely on this kind of theory for their design.

Assessments used for formative purposes are typically intended to support short-term instructional decisions by particular teachers in particular contexts, therefore, the generalizability of the results is not a major concern, and the consequences of an inaccurate decision can be significant but tend to be short-term (Kane and Wools, 2020). Formative assessment is used to support instruction by providing effective guidance on the next steps (Bennett, 2011; Shepard, Penuel, and Davidson, 2017). To do this, the teacher needs to have a clear understanding of what the complete, intended performance would look like, and the assessment evidence can indicate the student's current understanding of the specific topics and skills being taught. Using the information about gaps between current and intended performance, the teacher can decide how to help the student improve their performance, and therefore:

> Formative assessment should be based on cognitive theories about how people learn particular subject matter to ensure that instruction centers on what is most important for the next stage of learning, given a learner's current state of understanding. (Pellegrino, 2016, p.71)

Given their role in guiding instruction, the assessments included in LPs would generally be considered instances of formative, classroom assessment, and therefore not generally be subject to many of the requirements in the *Standards*. However, if the same tasks were incorporated in a standardized test used to make high-stakes decisions, they would be subject to the *Standards*.

Learning Progressions

To consider how to validate learning progressions, we must first come to grips with what they are, and this is a bit complicated, because the term, "learning progressions", is applied to a multitude of possible instructional/assessment programs, all of which share some basic characteristics, but otherwise are quite varied. The central characteristic is the specification of levels of achievement; learning progressions are commonly thought of as being defined in terms of a sequence of qualitatively distinctive levels representing "successively more sophisticated ways of thinking about a topic" (NRC, 2007, p.214), but the notion has evolved.

Corcoran, Mosher and Rogat (2009) edited a report of a panel of experts in science education and cognitive scientists commissioned by the Center on Continuous Instructional Improvement to review the state of the then-current work on LPs. After reviewing the available work on LPs, the panel decided on the following definition:

> Learning progressions are hypothesized descriptions of the successively more sophisticated ways student thinking about an important domain of knowledge or practice develops as children learn about and investigate that domain over an appropriate span of time. (Corcoran, et al., 2009, p. 37)

Note that this definition associates LPs with descriptions of how student thinking about a domain develops over time. It assumes the existence of a domain of knowledge or practice and focuses on how children can achieve successively more sophisticated ways of thinking about the domain, but it does not focus on assessment. That is, the Corcoran et al. (2009) definition suggests that an LP is a particular kind of domain-specific theory of learning.

There is some ambiguity in talking about the development of student thinking "over an appropriate span of time" (Corcoran et al., 2009, p. 37). For some LPs the span is a unit of instruction lasting days or a few weeks (Gotwals and Songer, 2013), and for others (Briggs and Furtak, 2020; Mohan, Chen, and Anderson, 2009), the program can last years. The unit-level LPs can rely on formative methodology, but the longer-term LPs will involve both formative and summative methods in most cases (Alonzo, 2018). The differences in grain size are very large across this continuum, and these differences imply different approaches to validation. In this chapter, I focus on the validation of unit-level LPs, but I will also mention how the issues would evolve as the length and complexity of LPs increase.

Corcoran, et al. (2009, p. 37–38) went on to specify some structural, and operational requirements for LPs, and I will adopt their definitions of the core elements in LPs and their terminology. Corcoran, et al. (2009) suggest that LPs must include the following elements:

1. Target performances or learning goals, which are the endpoints of a learning progression and are defined by societal expectations, analysis of the discipline, and/or requirements for entry into the next level of education;
2. Progress variables, which are the dimensions of understanding, application, and practice that are being developed and tracked over time. These may be core concepts in the discipline or practices central to scientific work;

3 Levels of achievement that are intermediate steps in the developmental pathway(s) traced by a learning progression. These levels may reflect levels of integration or common stages that characterize the development of student thinking. There may be intermediate steps that are non-canonical but are stepping stones to canonical ideas;
4 Learning performances, which are the kinds of tasks students at a particular level of achievement would be capable of performing. They provide specifications for the development of assessments by which students would demonstrate their knowledge and understanding; and,
5 Assessments, which are the specific measures used to track student development along the hypothesized progression. Learning progressions include an approach to assessment, as assessments are integral to their development, validation, and use. (pp. 37–38)

The first four elements define the basic structure of LPs. We have *target performances* that specify the goals of the LP, and *progress variables* that specify the dimensions along which progress is to be recorded. In addition, we have intermediate *achievement levels*, which are defined in terms of particular levels of the *progress variables* and learning performances on tasks that are used to define the achievement levels. Finally, the last element assumes that we have assessments used to track student progress on the progress variables across the achievement levels. Note, however, that the definitions of LPs in the literature vary, and in some cases, the assessments are not considered part of the LP.

LPs can be thought of as domain-specific theories of learning, but when thought of as theories, they have some unusual characteristics that are worthy of note. First, they are explicitly designed to achieve certain goals, rather than to describe and explain phenomena, and in this vein, they are more characteristic of engineering or technological theories than of basic scientific theories. Second, they prescribe a specific structure for the theory, involving target goals and intermediate achievement levels defined in terms of progress variables. Third, they include assessments defined in terms of the progress variables that "are integral to their development, validation, and use".

Corcoran, et al. (2009) suggest that the panel believed that LPs have several additional characteristics, which I will paraphrase rather than quote. They suggest that, like other scientific theories, LPs are based on research and are empirically testable. The LPs are to be based on foundational ideas and practices. Like all theories, LPs are expected to be conceptually coherent, and the achievement levels are to describe successively more sophisticated levels of student thinking as they make progress.

However, they suggest that the order in which the levels are achieved is not necessarily fixed. Curricula and instructional contexts may need to be consistent with the assumptions built into the LP for it to provide accurate descriptions of student progress, and even within a particular instructional program, students may follow different paths as they progress. That is, unlike Piaget's stages, the achievement levels are not generally fixed in their order, and the ordering of the levels in an LP will generally depend on the curriculum and teaching practices as well as the target domain. Some LPs may be robust in the sense that they tend to hold up well across educational contexts that vary in their curriculum contexts and approaches to instruction, but many are specific to their associated instructional programs.

LPs can be viewed as "local theories" of learning (National Research Council, 2001; Shepard, Penuel, and Davidson, 2017; Shepard, Penuel, and Pellegrino, 2018). These "theories" do not make strong claims about the generalizability of the achievement levels and certainly not the universality of the achievement levels, or the paths followed by students. Rather, the LP claims that most students will follow the prescribed path in a particular instructional context, and it suggests instructional strategies that will promote student progress from one achievement level to another. To the extent that such local theories are successful in these efforts, they can be useful in teaching.

The general model for LPs outlined above tends to work best as a local theory for LPs that are based on a single topic and that describe a unit of instruction on that topic (Covitt, Gunckel, Syswerda, and Caplan, 2018; Alonzo and Steedle, 2009). For brevity, I will refer to this model as the Corcoran-Mosher-Rogat (CMR) model (Corcoran, et al., 2009). The label "learning progression" has also been applied to longer chunks of the science curriculum (e.g., Briggs and Furtak, 2020), and although the general principles that have inspired the growing interest in LPs in science are applicable to these more ambitious efforts to varying degrees, the more detailed assumptions built into the CMR model tend to unravel as the LP gets more extensive and complex. In the discussion of validation presented below, I start with an analysis of the main claims included in the CMR model, and then examine how increasing length and complexity in the proposed instructional program can introduce complexity of various kinds.

Validation of Claims for Learning Progressions

The central idea across the many uses of the term, "validation" is that the claims being made need to be supported by appropriate evidence. Given the structure of the CMR model for LPs, the main claims that need to be examined can be divided into four clusters: the general claims inherent in the LP as a local, domain-specific theory of learning, those associated with the assessments as appropriate measures of the performance variables, those associated with achievement levels as plausible stepping stones leading to the target performances, and those associated with the effectiveness of the instructional strategies associated with the LP. These four sets of claims build on each other and are therefore not independent of each other, but for exposition, it will be useful to consider them separately. Doing so will allow for a more detailed analysis of each set of claims and will help to highlight the relationships among the different claims.

The development of the learning progression is necessarily an iterative process (Breslyn, McGinnis, McDonald, and Hestness, 2016; Jin, van Rijn, Moore, Bauer, Pressler, and Yestness, 2019; Shepard, 2018). It begins with a general local theory about how the LP is to work, including the target behavior for the learning progression, the achievement levels, the progress variables and their assessments, and instructional strategies based on each student's estimated achievement level. In the first iteration, a rough draft of the learning progression and the local theory are developed, along with preliminary versions of the progress variables and their assessments, the achievement levels, and instructional strategies.

The learning progression can then be evaluated empirically by seeing whether the assumptions built into the learning progression hold, at least approximately, for the student performances that are observed, and as the bottom line, whether the LP-based instruction has a positive impact on learning. The first iteration of this evaluation process, involving initial specifications of the progress variables, assessments, achievement levels, and learning strategies, is not likely to be fully satisfactory. Some of the assessments may not work as intended, perhaps because their progress variables are not clearly defined, some of the instructional strategies may not work very well, etc. To the extent that a particular component of the LP is not working well, it can be revised. In its early stages, the iterative evaluation of the LP should be formative (Scriven, 1967) in the sense that its main goal is to improve and then fine-tune the LP.

However, later in the process, the evaluation needs to adopt a more critical stance and think about what might go wrong in implementing the LP. Identifying potential problems of implementation can be helpful in designing an LP so that it avoids or mitigates these problems. Second, this critical stance can identify the most questionable assumptions in the theory and can be used to identify empirical challenges that can be used in a summative evaluation (Scriven, 1967) of the LP. In the very earliest stages of developing the LP, it would be appropriate to be tolerant about working assumptions, but serious work on validating the local theory and its components should begin fairly early in the process.

Initial Validation of the Learning Progression as a Local Theory

As noted earlier, an LP can be interpreted as a local domain-specific theory of learning consisting of a sequence of achievement levels defined by performance variables, by a set of assessments for measuring the performance variables, and by instructional guidelines provided by the learning progression. It purports to describe student learning in terms of their progress through successively more sophisticated levels of achievement in some part of a discipline. In addition to being domain-specific, the LP is also likely to be context-specific in that it assumes that the student progress occurs in the context of a particular approach to instruction, or at least, a kind of approach to instruction. The theory is a local theory, specific to a topic and to an instructional approach, and it does not purport to be a general theory of learning.

As a first step in validating the local theory for an LP, its structure would be spelled out in some detail and its conceptual coherence and plausibility would be evaluated, given previous research and the nature of the target performances to be achieved. This may seem obvious, but a thorough, critical review of existing research can provide a substantial base of validity evidence for the LP. The target performances should be significant enough to provide the endpoint of the LP, and the achievement levels should be well defined, represent increasing levels of sophistication, and provide plausible stepping stones leading to the target performances. The progress variables should be well enough defined to provide a dependable basis for assigning each student to a particular achievement level in the LP, and the instructional strategies based on these assignments should be supported by the research base. The literature on LPs suggests that this kind of preliminary work is being conscientiously done, with attention being given to current findings in cognitive science, more general work on teaching and learning, and prior experience with LPs.

A preliminary definition of the local theory requires that we develop preliminary, working definitions of its components, beginning with a clear conception of the target performances that constitute the goals of the learning progression. We would also need definitions of the progress variables that are to be used to indicate progress toward the target performances, along with specifications for the assessments that are to be used to determine each student's level of achievement on each of the progress variables. We would need to hypothesize achievement levels that are thought of as intermediate steps on the way to the target performances, defined in terms of values of the progress variables, and we would need to develop instructional strategies for helping students to get from one achievement level to another. Finally, we would need to have a preliminary sense of the most likely paths through the achievement levels to the target performances given any starting point in the learning progression.

The preliminary work involved in developing the initial version of the learning progression can provide a base of support for the validity of the learning progression as a local theory by establishing its overall plausibility, given the results of research on general learning theories, more specific research on learning particular topics, and the properties of the assessments that are being used by the learning progression. That is, this preliminary work can assemble various kinds of evidence for the plausibility of the assumptions built into the learning progression.

The development of this initial version of the learning progression, based on prior research and experience, does not offer many opportunities for LP to be seriously challenged empirically, because as it is being developed, the components of the LP are also being developed, and little or no data are available. As work on defining the specific components of the LP (i.e., the progress variables and their assessments, the achievement levels, and the instructional strategies) gets started, it will be possible to check whether the empirical properties of these components are consistent with expectations and to adjust, if necessary. Based on observations of how well this preliminary version of the LP works, we can revise the local theory or specific components in the LP.

As the LP gets developed and the components become more clearly defined, it should become possible to generate serious empirical challenges to the LP's claims. For example, to empirically check the assumption inherent in LPs, that the achievement levels are generally attained in the specified order, it would be necessary to determine when students have attained the different achievement levels, and this requires that we have confidence in the definitions of the progress variables, the assessments, the achievement levels, and the instructional strategies being implemented. This is a common problem in theory development and validation in the sciences and can be addressed by "bootstrapping". We begin by hypothesizing an LP and its components and then examine how well these elements seem to work to support the desired outcome and achievement of the target performances, then, we iterate to improve the LP and its components.

Claims about Progress Variables and Their Assessments

Starting with the preliminary description of the learning progression, we can begin to investigate how well the LP seems to work in more detail. A basic assumption in LPs is that students' levels of performance on the progress variables improves over time while they are in the instructional context associated with the LP, because they represent progress on the cognitive competencies that the learning progression is intended to promote. If the assessments involve the kinds of performance specified for the progress variables and are not overly influenced by construct-irrelevant variance, there is not much of a leap from the assessment results to the progress variables.

The direct assessment of our most important educational goals seems to require that complex performances be evaluated (Jin and Anderson, 2012). For example, if we want to assess students' abilities to use core scientific concepts and practices, it seems essential that we give them the opportunity to do so. If we wish to evaluate students' abilities to design experiments and draw conclusions, the most direct approach to assessment would be to observe their performances when they are asked to do so. Inferences from observed performances on a sample of tasks to conclusions about the ability to perform this kind of task are intrinsically plausible. The directness of the interpretation makes many competing interpretations (e.g., in terms of method effects) implausible (Kane, Crooks, and Cohen, 1999).

Think-aloud protocols and follow-up interviews can provide effective checks on this assumption by illuminating the thinking processes employed by students while performing the assessment tasks (Jin et al., 2019; Gotwals and Songer, 2013). If these processes correspond to the kinds of processes being developed by the LP, that will tend to support the learning progression as a whole and also the validity of the assessment results as measures of the target performances for the LP. On the other hand, if the thought processes reflected in the assessments do not fully reflect the cognitive processes in the target processes, the assessments would probably need to be revised, or the definitions of the progress variables would need to be reconsidered. For example, based on interviews following an assessment, Gotwals and Songer (2013) found that the assessment process being used was not sensitive to the full range of student achievement, because "some students may understand how to support their claims through evidence and reasoning, but not make it explicit" (p. 620).

Most of the problems unearthed by think-aloud protocols and follow-up interviews would be considered sources of systematic errors (i.e., errors that tend to recur over instances of the assessment), as in the case reported by Gotwals and Songer (2013), but the think-aloud protocols and especially the interviews would also provide indications of the frequency of random errors (e.g., guesses and slips of attention that are not likely to recur over instances of the assessment). In psychometric theory, random errors are evaluated in terms of reliability or generalizability (Brennan, 2001), but issues of generalizability tend to play a smaller role in formative assessment than in summative assessments because the claims based on the assessment results are local and limited (Kane and Wools, 2020).

The empirical data generated by the assessments will also provide evidence for or against assumptions built into the LP. Bennett (2015) summarized the main psychometric analyses that might be applied to LP assessments:

> items meant to measure lower levels should be easier than items mapped to higher levels, and individual examinees should generally get items below their level correct and items above it incorrect. Furthermore, that assessment should place individuals into levels consistently; parallel forms of the test taken at the same point in time should produce the same placement. Finally, placements suggested by the test should generally agree with independent indicators of those assignments (e.g., the judgments of expert teachers). (p. 381)

van Rijn, Graf, and Deane (2014) provide examples of these kinds of empirical analyses.

In performance testing, it is usually necessary to have a rater (or raters) evaluate the student's performance against a rubric of some kind, and multiple ratings of the same performance are typically not in perfect agreement, resulting in less-than-perfect inter-rater reliability (AERA et al., 2014). In using assessments in classrooms, the teacher is usually the rater, the person who interprets the rating, and the person who makes decisions about what to do next. Within this kind of closed loop, the issues associated with generalizing the ratings across raters are much less severe than they can be in rating performances within large-scale, summative assessments. Inter-rater reliability tends to become more serious as the performances and their rubrics become more complicated, but it is generally not a very serious concern as long as the ratings are "trustworthy" in that they are accurate enough to support the decisions based on the assessment results (Alonzo, 2020).

To the extent that the progress variables and their assessments are defined in terms of performance domains that include multiple kinds of performances, rather than a specific kind of performance, as is generally the case for assessments that focus on performance across some extended domain, generalizations over tasks, occasions, and contexts tend to be concerns. In formative assessment, inferences tend to be based on observations of the kind of performance of interest, and decisions about the next steps are made based on these observed performances. The tasks used to monitor student progress in many LPs are closely tied to the content and format of the instructional program and would not generally fit well in different instructional programs (Shepard, 2018).

In summative assessments, the inferences to be made concern the expected level of performance over a larger domain of tasks. If the domain is large, these summative assessments will necessarily involve sampling from the domain, and therefore questions arise about the dependability of the results over samples from the domain. The generalizability of the results can be evaluated using a G coefficient or reliability coefficient (Brennan, 2001). In addition, as the LPs get larger (e.g., covering multiple topics over a school year or years, and multiple teachers), the need for sampling and evaluating the generalizability of the assessment results will also tend to increase.

Note that the analyses discussed in this section have focused on direct evaluations of the assessments as measures of the progress variables. They provide preliminary evidence for the psychometric validity of the assessments. Additional (and perhaps stronger) evidence for psychometric validity can be developed by examining how well the assessments support the LP as a whole. To the extent that the assessments of the progress variables lead to sensible assignments to achievement levels, effective instruction, and a parsimonious model for progress through the achievement levels, confidence in their psychometric validity will be more strongly confirmed (Cronbach and Meehl, 1955).

The validation of the claims about student progress based on the assessments highlights the importance of theory-based interpretations and uses. Validation of educational measures typically requires investigations of the accuracy and dependability of proposed interpretations of the results on the assessments (qualitative or quantitative), and investigations of the consequences of using these results in certain ways (Cronbach, 1971; Messick, 1989; Kane, 2006).

Claims about the Achievement Levels

For LPs that can be represented as a CMR model, the achievement levels can be defined in terms of the specific tasks that are the focus of the LP, and more specifically in terms of how well or how completely students at each level can perform. Students at more advanced achievement levels can perform the assessment tasks without scaffolding, and students at lower levels can't perform the tasks (or can't perform them as well or as completely, or without certain scaffolding). Initially, the achievement levels are defined in terms of the initial progress variables, or dimensions of understanding and practice leading to the target performance, and students' achievement levels are identified in terms of results on the initial versions of the assessments.

If the achievement levels make sense in general terms as intermediate stages on the way to the target performances, the main issue in evaluating the achievement levels is the question of how well the progress variables define them. If the achievement levels are defined by a single progress variable, which is closely aligned with the definitions of the achievement levels, then most students can be unambiguously assigned to achievement levels based on the assessment of that progress variable. For example, if the difference between adjacent achievement levels is specified in terms of a common misconception that occurs in students at the lower level but does not occur at the next higher level, or in terms of the scaffolding required for effective performance at one level but not at the next higher level, and the progress variable and assessment are based on these differences, there may be little uncertainty in the assignment of students to these achievement levels (Kane et al., 1999).

Note, however, that the structure of LPs invites a kind of bias associated with expectations. If a student has been performing at a certain level, and a current performance reflects a lower level, the teacher/assessor might tend to provide the students with additional support to see whether they can perform at the expected level. On the other hand, if a student is performing at a much higher level on the assessment than they have been performing, that might reflect a dramatic improvement or an error in the current assessment (or the earlier assessments). The expectation that students will proceed in an orderly way through the achievement levels provides a check on the achievement levels and the assessments.

A second issue to be considered in evaluating the achievement levels involves their utility for instruction. That is, does the definition of the achievement level clearly indicate the kinds of instructional strategies that will help students get to the next higher achievement level, or more generally, to a more advanced achievement level? For example, if an achievement level is specified in terms of a misconception that is characteristic of the current level but does not occur at the next higher level, instructional strategies that tend to clarify the misconception (e.g., by suggesting a counterexample that contradicts the misconception) would be appropriate. This assumption can be evaluated empirically by examining whether the instructional strategies associated with the achievement levels are effective in promoting progress to higher achievement levels.

For relatively simple LPs, involving a few achievement levels defined in terms of one or two progress variables, the local theory defining the LP may be highly plausible a priori. However, if the achievement levels are defined in terms of more than one progress variable, the assignments to specific achievement levels are likely to be ambiguous, perhaps very ambiguous, if the progress variables do not agree with each other. As the saying goes, a person with one watch knows what time it is, but a person with two watches is never quite sure. The different progress variables are not likely to be in perfect agreement, but they do not need to be entirely consistent to work well in defining the achievement level. Nevertheless, any disagreements among the progress variables (especially those that tend to affect achievement-level assignments) add uncertainty to the system.

If there is only one progress variable used to define the achievement levels in the LP, the problems associated with evaluating the assessment of this progress variable are relatively simple. If more than one progress variable is used to define the achievement levels within the learning progression, this problem becomes a bit more difficult (van Rijn, Graf, and Deane, 2014). In addition to verifying that the assessments accurately reflect levels of performance on the progress variables, and the fact that students make regular progress on these variables during instruction, we would need to evaluate how well the progress variables hang together in defining the achievement levels.

Claims about Likely Paths through the Achievement Levels

A basic assumption in CMR LPs is that, for most students, the target performances or learning goals are achieved via a hypothesized pathway (or via one of a small number of likely pathways) through the achievement levels, which serve as intermediate steps in getting to the target performances. This assumption is central to the general conception of an LP as a "progression" from some entry point in the LP to the target performances.

Once the LP is well-developed, with well-defined progress variables, assessments, achievement levels, and instructional strategies, this claim can be checked empirically by examining how students perform on the associated assessments. If most students make adequate progress, and do so along the prescribed path, the claim can be accepted. If some students are running into snags because of defects in some component in the LP (e.g., unclear instructions in an assessment, inadequate instruction), the LP can be revised, and the new LP can be evaluated with a new sample of students and teachers. That is, the process is iterative and formative (Scriven, 1967). If substantial numbers of students are deviating from the intended paths, due to differences in learning styles or backgrounds, it might be reasonable to explicitly allow for multiple paths.

For LPs that are relatively limited in scope, covering say a single unit within a course, the CMR model is likely to hold and the claim that there is only one preferred path, or a small number of likely paths, through the achievement levels would be likely to be supported for several reasons. The achievement levels are designed to represent a more-or-less natural sequence of stages leading to the target performance, with the learning opportunities provided to the students at each stage designed to facilitate the achievement of the next stage, or achievement level. For example, if the current achievement level is defined in part by certain limitations in the student's understanding or certain misconceptions that a student seems to have, the instructional strategies for that achievement level would focus on clarifying the misunderstandings or less-than-complete understanding, thus leading to an achievement level that does not include that misconception in its definition. That is, as learning progressions are structured, the instruction that occurs for students at any particular achievement level is explicitly aimed at getting them to the next level along the hypothesized path, and therefore, the instruction, if effective, is likely to limit the tendency of students to follow other paths.

Also, for LPs with a limited scope and a limited number of achievement levels, there would be a relatively small number of possible paths through the achievement levels, and an even smaller number of plausible paths through the achievement levels. Some paths would be very unlikely. For example, if a student were assigned to a particular achievement level characterized by a complete understanding of some aspect of the topic under consideration, and a new assessment were to indicate that the student should be assigned to a lower achievement level, characterized by a very limited understanding of that topic, we would probably not simply assign the student to the lower achievement level; rather, we would question the new assessment results or the earlier results. If this kind of glitch occurred frequently, we would tend to question the accuracy of the assessments (Cronbach and Meehl, 1955).

For longer, more complex LPs, there may be a large number of achievement levels and a large number of possible paths, and a substantial subset of these paths may be fairly plausible. As the number of plausible paths increases, the question of whether most students' progress tends to follow a single path becomes more salient, because the number of opportunities for getting off the path would increase substantially. Of course, the teachers implementing the LP can deal with some variability in the paths taken, but if we try to apply the CMR model to curricula spanning a full semester or more, the number of possible achievement levels and paths will be quite large.

Many of the possible paths will not be all that likely to occur in practice, but in a diverse population of students, some variations in the paths taken by students are almost inevitable. Indeed, as Lehrer and Schauble (2015) pointed out:

> Children's differing histories make it likely that the development of some individuals will deviate for even a carefully constructed and well tested model – indeed, any child is highly unlikely to conform to the model in all respects. (p. 436)

As the LPs get more ambitious, covering more content and longer periods of instruction, the definitions of the achievement levels are likely to become more complex, involving assessments of several (rather than one or two) progress variables, and as a result, fuzzier, and deviations from the intended path will become more common.

In addition, as the length and complexity of an LP increases, some practical issues will complicate its implementation further. In the United States, the students in a classroom can change drastically over the course of a year, as students enter and leave the school for various reasons. Students coming from a different school with different curricula and different sequencing of topics, may not fit easily into the LP.

For any LP, but especially for longer, more comprehensive instructional programs, assuming that the number of paths to be followed by students should be small may be counter-productive, even if this goal can be achieved. As indicated above, this goal will probably be unrealistic for more comprehensive programs, and adopting this goal may interfere with instructional effectiveness for some students. Penuel and Shepard (2016) suggest that LPs may not allow for equitable access to instruction if they do not accommodate differences in socio-cultural and instructional backgrounds by allowing for flexibility in paths through the LP. Some latitude in how students make progress is not a problem if they are making substantial progress.

Claims about Instructional Effectiveness

The evaluation of the LP as a local theory will generally provide substantial evidence for the validity of the local theory's claims about the efficacy of its instructional practices, based on prior empirical research and theory indicating that these practices are likely to be effective, but this claim needs to be checked empirically. As the iterations proceed and the local theory, the progress variables, assessments, achievement levels, and the instruction program are refined and improved, the iterative validation process will generate additional evidence supporting the claims being made by the LP (or the initial version of the LP will be revised or abandoned). Given that LPs are designed to help students achieve sophisticated ways of thinking about some topic or domain, the instructional effectiveness of the LP, as implemented, is particularly important (Pellegrino, diBello, and Goldman, 2016).

To be effective in improving instruction and learning, an LP would need to provide discipline-specific or topic-specific guidance on how "partial understandings might mature into more sophisticated conceptions" (Shepard, 2018, p.170), and on how misconceptions might be effectively challenged. Within a unit-sized LP with a single path, the target performances, a student's current achievement level, and the next achievement level to be achieved by the student are all known:

> Learning progressions benefit student learning by illuminating student thinking and by providing insights about instructional next steps. (Shepard, 2018, p.166)

In addition to providing a basis for assigning a student to an achievement level, the assessments should illuminate some detailed patterns of student thinking, and the local theory should suggest instructional strategies. It could, for example, provide suggestions on the feedback to provide to the student as well as materials and challenges that might help the student to progress to the next level. That is, the assessment results should point to explicit next steps in instruction and learning:

> It does little good to tell fourth grader that his essay is a "2" or that he needs three more items correct to reach mastery on a numeracy test. Feedback is more likely to be

beneficial when it helps students see *how* to improve, and this invariably calls for information that is qualitative rather than quantitative. [italics in original] (Shepard, 2018, p.169)

The preferred outcome of a formative assessment is not generally a score, but rather, suggestions for what might be tried next, together with the opportunity to apply the feedback.

The instructional strategies incorporated in an LP need to be evaluated at two levels. A fundamental concern in validating an LP is whether it achieves the goal of helping students achieve the target performances. The evidence supporting this summative claim can be based on assessments of various kinds administered at the end of the unit of instruction. In addition to the assessments within the LP that can be used for this purpose, it would generally be good to get a second opinion based on externally developed assessments of the same kinds of performances (Furtak, Heredia, and Circi, 2018). Given the emphasis on developing proficiency in science practices as well as content in many science LPs, it could also be illuminating to look for transfer across LPs in the development of these science practices.

Given the iterative process that is likely to be necessary for most LPs, a more detailed analysis of performance at successive achievement levels along the most likely paths through the LP would also be desirable. That is, data can be collected on the effectiveness of the instructional strategies employed with the students at each achievement level to evaluate their effectiveness, and as part of the iterative process, strategies that are found to be relatively ineffective can be dropped or revised.

These achievement level-specific analyses can also provide additional evaluations of the definitions of the achievement levels and of the assessments used to assign students to achievement levels. For example, if the strategies are found to be more effective with some students (or groups of students), and other strategies are more effective with other students (or groups), this added complexity could be reflected in the LP, and the achievement levels or the assessments might need to be reconsidered.

After enough iterations, these local evaluations may indicate that the more specific strategies are working well and can provide additional support for the overall effectiveness of the LP. And working backwards, summative assessments of the effectiveness of the LP can provide indications of specific weaknesses in the LP and suggestions for further improvement. For example, Furtak, Heredia, and Circi (2018) examined item-level performance in a pre-post design and found that there was variability in the observed improvement across item sets, indicating that some target performances were not adequately achieved, and thereby, they identified goals for improvement of the LPs.

As noted earlier, there is necessarily, a substantial difference in grain size between a CMR model and a larger-scaled instructional program or a summative assessment program. In the context of an LP covering a unit of instruction, it could be possible to suggest specific instructional strategies to help students at a particular achievement level to achieve the next higher level. On a larger scale, the achievement levels will necessarily be described in more general terms and the instructional strategies that are suggested will also be more general.

With a single path through an LP with a small number of achievement levels, the teacher knows what to do next. If the student is at the fourth achievement level, AL4, in the path, the teacher needs to help the student get to the fifth achievement level, AL5. This may not be easy to achieve for the student or the teacher, but the path is clear. If AL4 is on several plausible paths, the next step is less clear; the teacher could try to help the student to reach AL5, but if the student has been following a different plausible path on which AL5 is generally achieved before AL4 a different strategy would be called for. At some point, as the length and complexity of LPs increase, and the number of achievement levels and paths increase, the structure envisioned by Corcoran et al. (2009) will become unwieldy, and more complex frameworks will be needed.

Long-term LPs

Many of the components in LPs could be retained and expanded in more broadly defined and long-term instructional programs. The emphasis on core concepts in a domain and on the use of local theories of learning to guide instruction and assessment could be retained, along with an emphasis on the intersection of core ideas, scientific practices, and crosscutting concepts (NRC, 2014). However, the conception of an LP with a modest number of achievement levels and one or a few paths followed by most students, highly dependable assignment of students to achievement levels, and unambiguous instructional guidance for the students at each achievement level will fail at some point as the length and complexity of the instructional program increases. That is, the CMR model (Corcoran et al., 2009) will have to be abandoned at some point, at least in its strong form. For example, Briggs and Furtak (2020) describe a program on modeling energy flows for grades 9–12, covering topics in physics, chemistry, and biology. The structure of the proposed LP is specified in terms of 11 performance expectations, which are necessarily described in general terms and associated with a more detailed curriculum. Briggs and Furtak emphasize the role of formative and summative assessments in their discussion of the energy-flow LP, but a full description of this LP could not be fit neatly in a simple structure with one or a few paths describing student progress through the curriculum. For such long-range instructional programs, LPs are likely to function less as prescriptions and more as guidelines or metaphors for planning the program.

This is not a problem as long as the differences between small to moderate-sized LPs emphasizing the systematic use of formative assessment to guide progress toward "successively more sophisticated ways of thinking about a topic" (NRC, 2007, p.214) and instructional programs covering a substantial number of related topics are recognized, but the methodology that would be most appropriate for validating programs at different points on this continuum would be considerably different. This chapter has focused on the validation of LPs that can be viewed as exemplars of the CMR model. For larger-scaled programs, it will probably be necessary to include summative evaluations as well as assessments that can be used for formative purposes, and to give more attention to the methodology of program evaluation.

Work has begun on various approaches to extending LP models to larger-scale programs and summative assessments (Duncan, Rogat, and Yarden, 2009; Mohan, Chen, and Anderson, 2009). Briggs and Furtak (2020) proposed a direct extension of the basic LP model, with some loosening of assumptions. Alonzo and Ke (2016) suggested embedding more ambitious assessment tasks in classroom instruction as embedded assessments. Wilson (2009) and Black, Wilson, and Shih-Ying (2011) proposed embedding LPs, construed as construct maps, in larger structures within curricula.

Conclusions

Although there have been a number of serious efforts to develop and validate "learning progressions", this concept is relatively new and is still only loosely defined. I assumed the kind of structure described by Corcoran et al. (2009) as a working framework throughout this chapter, because it seems clear and well-defined and because it arose from a consensus effort.

In any case, I suggest that the assumptions built into an LP, however it is defined, be specified clearly and completely and that they be evaluated critically.

There are two basic sources of ambiguity in validating LPs. The first is inherent in our uses of the word "validate" in multiple senses across multiple contexts. In this chapter, I have adopted a broad definition of validation, one that encompasses the evaluation of the local theory for the LP, the appropriateness of its assessments and achievement levels, and the effectiveness of its learning strategies. In its early stages, validation of an LP is iterative and formative, with an emphasis on improving how well the LP works; in its latter stages, the focus shifts to a more summative evaluation of how well the LP works. Central to this broad conception of validation is a critical stance, which focuses on identifying and challenging the most questionable assumptions in the LP and its components. If the assumptions inherent in the LP can stand up to serious criticism, it deserves our trust.

The second source of ambiguity results from the broad usage of the expression "learning progression", which has been applied to a wide range of programs, ranging from a specific unit of instruction lasting perhaps a week or two to a multiyear sequence of courses. In validating any educational program, it is necessary to evaluate the claims made by the program, and since these claims will vary from program to program, the validation model needs to be flexible. This chapter has proposed an iterative, argument-based approach to validation for LPs that reflects a version of the CMR model:

> Learning progressions in science are empirically grounded and testable hypotheses about how students' understanding of, and ability to use, core scientific concepts and explanations and related scientific practices grow and become more sophisticated over time, with appropriate instruction (NRC, 2007). These hypotheses describe the pathways students are likely to follow to the mastery of core concepts. They are based on research about how students' learning actually progresses - as opposed to selecting sequences of topics and learning experiences based only on logical analysis of current disciplinary knowledge and on personal experiences in teaching. These hypotheses are then tested empirically to ensure their construct validity (Does the hypothesized sequence describe a path most students actually experience given appropriate instruction?) and ultimately to assess their consequential validity (Does instruction based on the learning progression produce better results for most students?). (Corcoran et al., 2009, p.15)

LPs that satisfy this model will generally be specific to the learning of a particular topic, and they will be moderate in duration (lasting weeks). For longer programs of instruction (lasting semesters), effective validation will need to address some of the issues discussed in this chapter, but the CMR model will probably not work well, and validation will generally require a more complex model. That is, I am not optimistic that the basic assumptions in the CMR model can be scaled up, without being seriously diluted. Longer programs may take inspiration from the CMR model, but these more ambitious efforts will need to be specified at a much larger grain size than most current LPs and probably need to involve summative evaluations as well as formative evaluations. However, even if one accepts this conclusion, LPs focused on a unit of instruction may prove very powerful in improving current instruction and learning and in promoting progress in learning theory, and these would not be small achievements.

The interpretive claims based on LPs could initially be investigated under the heading of theory validation, because the claims inherent in the LP are based on the local theory. The progress variables and assessments, the achievement levels, and instructional strategies incorporated in the LP can be developed and refined in an iterative process. If an implementation of a particular version of the LP is found to not work, one or more of these components can be revised until they work well together. As Shepard put it:

> Learning progressions have a certain amount of validity "built in" not just because of their initial conceptual theorizing, but because they are developed recursively with explicit testing of proposed orderings and effectiveness of specific instructional supports. (p.172)

This kind of iterative process can lead to a high level of horizontal coherence between instruction and assessment and can therefore lead to assessments that have a more positive impact on instruction and learning than most educational tests have managed to achieve (Bennett, 2015; Penuel and Shepard, 2016; Shepard, 2016, 2018), and it could make the promise inherent in formative assessment more of a reality. The bottom line for the validation of LPs is the evaluation of the extent to which they promote effective teaching and learning.

References

Alonzo, A. (2020). Defining trustworthiness for teachers' multiple uses of classroom assessment results. In S. Brookhart and J. McMillan (Eds.), *Classroom assessment and educational measurement* (pp. 120–145). The NCME Applications of Educational Measurement and Assessment Book Series. Routledge.

Alonzo, A. C. (2018). An argument for formative assessment with science learning progressions. *Applied Measurement in Education*, *31*(2), 104–112.

Alonzo, A. C., & Ke, L. (2016). Taking stock: Existing Resources for Assessing a New Vision of Science Learning. *Measurement: Interdisciplinary Research and Perspectives*, *14*(4), 119–152.

Alonzo, A. C., & Steedle, J. T. (2009). Developing and assessing a force and motion learning progression. *Science Education*, *93*, 389–421. doi:10.1002/sce.v93:3

American Educational Research Association, American Psychological Association, & National Council on Measurement in Education (2014). *Standards for educational and psychological testing*. American Psychological Association.

Bennett, R. E. (2015). The changing nature of educational assessment. *Review of Research in Education*, *39*, 370–407.

Bennett, R. E. (2011). Formative assessment: A critical review. *Assessment in Education: Principles, Policy, and Practices*, *18*(1), 5–25.

Black, P., Wilson, M., & Shih-Ying, Y. (2011). Road maps for learning: A guide to the navigation of learning progressions. *Measurement: Interdisciplinary Research and Perspectives*, *9*, 71–123.

Brennan, R. (2001). *Generalizability theory*. Springer-Verlag.

Breslyn, W., McGinnis, J., McDonald, R., & Hestness, E. (2016). Developing a Learning Progression for Sea Level Rise, a Major Impact of Climate Change. *Journal of Research in Science Teaching*, *53*, 1471–1499.

Briggs, D., & Furtak, E. (2020). Learning progressions and embedded assessment. In S. Brookhart and J. McMillan (eds.), *Classroom assessment and Educational Measurement* (pp. 147–169). The NCME Applications of Educational Measurement and Assessment Book Series, Routledge.

Corcoran, T., Mosher, F. A., & Rogat, A. (2009). *Learning progressions in science: An evidence-based approach to reform (CPRE Research Report #RR-63)*. Consortium for Policy Research in Education.

Covitt, B., Gunckel, K. L., Syswerda, S., & Caplan, B. (2018). Teachers' use of learning progression-based formative assessment in water instruction. *Applied Measurement in Education*, *31*(2), 128–142.

Cronbach, L. J. (1982). *Designing evaluations of educational and social programs*. Jossey-Bass.

Cronbach, L. (1980). Validity on parole: How can we go straight? *New Directions for Testing and Measurement: Measuring Achievement Over a Decade*, *5*, 99–108.

Cronbach, L. J. (1971). Test validation. In R. L. Thorndike (Ed.), *Educational measurement* (2nd ed., pp. 443–507). American Council on Education.

Cronbach, L. J., & Meehl, P. E. (1955). Construct validity in psychological tests. *Psychological Bulletin*, *52*, 281–302.

Duncan, R. G., Rogat, A. D., & Yarden, A. (2009). A learning progression for deepening students' understandings of modern genetics across the 5th–10th grades. *Journal of Research in Science Teaching*, *46*, 655–674. doi:10.1002/tea.v46:6

Ferrara, S., Maxey-Moore, K., & Brookhart, S. (2020). Guidance in the standards for classroom assessment. In S. Brookhart and J. McMillan (eds.), *Classroom assessment and educational measurement* (pp. 97–119). Routledge.

Furtak, E., Heredia, S., & Circi, R. (2018). Exploring alignment among learning progressions, teacher-designed formative assessment tasks, and student growth: Results of a four-year study. *Applied Measurement in Education*, *31*(2), 143–156.

Gotwals, A. W., & Songer, N. B. (2013). Validity evidence for learning progression-based assessment items that fuse core disciplinary ideas and science practices. *Journal of Research in Science Teaching*, *50*, 597–626.

Jin, H., van Rijn, P., Moore,J., Bauer, M., Pressler, Y., & Yestness, N. (2019). A validation framework for science learning progression research. *International Journal of Science Education*, *41*(1), 1–23.

Jin, H., & Anderson, C. (2012). Development of assessments for a learning progression on carbon cycling in socio-ecological systems. In A. Alonzo and A. W. Gotwals (Eds.), *Learning progressions in science: Current challenges and future directions* (pp. 151–182). Sense Publishers.

Kane, M. (2006). Validation. In R. Brennan (Ed.), *Educational Measurement* (4th ed., pp. 17–64), American Council on Education; Praeger.

Kane, M. (2013). Validating the interpretations and uses of test scores. *Journal of Educational Measurement*, *50*(1), 1–73.

Kane, M. T., Crooks T. J., & Cohen, A. S. (1999). Validating measures of performance. *Educational Measurement: Issues and Practice*, *18*(2), 5–17.

Kane, M., & Wools, S. (2020). Perspectives on the Validity of Classroom Assessments. In S. Brookhart and J. McMillan (eds.), *Classroom assessment and educational measurement* (pp. 11–26). Routledge.

Kuhn, T. S. (1970). *The structure of scientific revolutions* (2nd ed.). University of Chicago Press.

Lakatos, I. (1970). Falsification and the methodology of scientific research programs. In I. Lakatos and A. Musgrave (Eds.), *Criticism and the growth of knowledge* (pp. 91–196). Cambridge University Press.

Lehrer, R., & Schauble, L. (2015). Learning progressions: The whole world is NOT a stage. *Science Education*, *99*, 432–437.

Messick, S. (1989). Validity. In R. L. Linn (Ed.), *Educational Measurement* (3rd ed., pp. 13–103). American Council on Education. Macmillan.

Mohan, L., Chen, J., & Anderson, C. W. (2009). Developing a multi-year learning progression for carbon cycling in socio-ecological systems. *Journal of Research in Science Teaching*, *46*, 675–698. doi:10.1002/tea.v46:6

National Research Council. (2001). *Knowing what students know: The science and design of educational assessment*. Committee on the Foundations of Assessment. J. Pellegrino, N. Chudowski, and R. Glaser, (Eds.). The National Academies Press.

National Research Council (2005). *Systems for state science assessments. Committee on Test Design for K-12 Science Achievement*. M. R. Wilson, and M. W. Bertenthal (Eds.), The National Academies Press.

National Research Council (2007). *Taking science to school: Learning and teaching science in grades K-8. Committee on Science Learning, Kindergarten through eighth grade*. R. A. Duschl, H. A. Schweingruber, and A. W. Shouse (Eds.). The National Academies Press.

Pellegrino, J. W., DiBello, L. V., & Goldman, S. R. (2016). A framework for conceptualizing and evaluating the validity of instructionally relevant assessments. *Educational Psychologist*, *51*(1), 59–81.

Penuel, W. R., & Shepard L. A. (2016). Assessment and teaching. In D. H. Duschl & C. A. Bell (Eds.), *Handbook of research on teaching*. American Educational Research Association.

Popper, K. R. (1962). *Conjecture and refutation: The growth of scientific knowledge*. Basic Books.

Scriven, M. (1967). The methodology of evaluation. In R. W. Tyler, R. M. Gagne, & M. Scriven (Eds.), *Perspectives on curriculum evaluation* (Vo1. 1, pp. 39–83). Rand McNally.

Shepard, L. A. (2016). Evaluating test validity: Reprise and progress. *Assessment in Education: Principles, Policy & Practice*, *23*, 268–280.
Shepard, L. A. (2018). Learning progressions as tools for assessment and learning. *Applied Measurement in Education*, *31*(2), 165–174.
Shepard, L. A., Penuel, W. R., & Davidson, K. L. (2017). Design principles for new systems of assessment. *Phi Delta Kappan*, *98*(6), 47–52.
Shepard, L. A., Penuel, W. R., & Pellegrino, J. W. (2018). Using learning and motivation theories to coherently link formative assessment, grading practices, and large-scale assessment. *Educational Measurement: Issues and Practice*, *37*(1), 21–34.
Suppe, F. (1977). *The structure of scientific theories* (2nd ed). University of Illinois Press.
Toulmin, S. (1953). *The philosophy of science: an introduction*. Harper and Row.
van Rijn, P., Graf, A., & Deane, P. (2014). Empirical recovery of argumentation learning progressions in scenario-based assessments of English language arts. *Psicologia Educativa*, *20*, 109–115.
Wilson, M. (2009). Measuring progressions: Assessment structures underlying a learning progression. *Journal of Research in Science Teaching*, *46*, 716–730.

5

DEVELOPMENT AND VALIDATION OF KNOWLEDGE-IN-USE LEARNING PROGRESSIONS

Leonora Kaldaras and Joseph Krajcik

Development and Validation of Knowledge-In-Use Learning Progressions

Introduction

Learners demonstrate deep science understanding by applying scientific knowledge to explain phenomena and solve real-life problems (Duschl, Schweingruber, Shouse, 2007). Supporting all learners in developing deep science understanding has been studied for many years and remains a challenging and complex task yet to be accomplished in science classes. Factors contributing to the development of deep science understanding have been identified, and they stem from years of research on the developmental nature of student understanding (Smith, Wiser, Anderson, & Krajcik, 2006) and situated learning (Duschl et al., 2007), both of which represent important elements of constructivist learning theories.

The developmental nature of student understanding in the context of science education has been discussed in detail in several NRC reports, including *The Framework for K-12 Science Education* (NRC, 2012), *Taking Science to School* (Duschl et al., 2007), and *How People Learn* (NRC, 2000). The factors influencing a developmental approach include learning as a complex phenomenon that requires supporting learners over a broad period of time, aligning curriculum, instruction, and assessment to ensure coherence of the learning process, and providing appropriate scaffolding to support students in progressing to higher levels of understanding (Duschl et al., 2007; NRC, 2000; Smith et al., 2006). Research shows that cognitive development does not occur in specific age-defined stages, and it is not uniform across individuals (Duschl et al., 2007). Rather, developmental appropriateness is not a simple function of grade or age but is largely dependent on prior opportunities to learn (Duschl et al., 2007). Grounded in prior research on student learning and the logic of the discipline among other factors, learning progressions outline increasingly sophisticated ways of thinking about a construct (Duschl et al., 2007). In accordance with the developmental approach, LPs do not assume that all students progress in a single linear, defined set of levels (Shepard, 2018). Rather, LPs aim to describe the most frequent orderings of ways in which students can think about a topic grounded in empirical research and the logic of the discipline among others (Shepard, 2018).

Learning progressions embody a developmental vision of student understanding by allowing accurate diagnosis of the current level of understanding of individual learners and by providing

 DOI: 10.4324/9781003170785-6

a guide to educators on how to support learners in transitioning to higher levels. Therefore, learning progressions shift the goal of the learning process from focusing on what students should be learning and doing based on their age or some other external measure of cognitive development to focus on creating learning opportunities for students based on their individual level of understanding, which is contingent upon their prior learning opportunities. By enabling this shift, LPs represent a powerful tool for helping to adjust instruction to the needs of individual learners by guiding the development of meaningful learning opportunities for students at different levels of understanding.

Situated learning assumes that students learn best when they need to apply science content to relevant situations (Duschl et al., 2007). The situated view on learning considers not only what is in someone's head but also recognizes the importance of the relationship between the individual and the learning environment, emphasizing that learning is always situated in the context of the activity that an individual is engaged in, including social, cultural, and physical contexts (Gee, 2008). Therefore, effective learning environments should engage students in activities that support using content ideas to make sense of meaningful situations. That way, students will develop a deeper understanding of the content. Learning progressions embody this situated learning vision by describing increasing levels of sophistication in terms of learning performances; that is, LPs focus on outlining what students should be able to do with the content under study at various levels of sophistication (Duncan & Hmelo-Silver, 2009). Therefore, LPs aim to shift the goal of the learning process from focusing on content memorization to focusing on supporting students in developing proficiencies related to applying the content in various contexts. It is important to point out that effective situated learning can only occur when individuals can perceive the provided learning opportunities as feasible (Duschl et al., 2007). In other words, it is important to situate learning in the types of activities that can allow for meaningful engagement to students at all levels of cognitive development and from diverse academic and cultural backgrounds (Duschl et al., 2007). Learning Progressions could potentially guide the development of such learning opportunities assuming the various ways of thinking and knowing were considered at the stage of LP development and validation. To summarize, learning progressions (LPs) in science embody the aspects of both the developmental nature and situated cognition theory and represent a promising tool for aligning curriculum (learning materials), instruction (teaching strategies), and assessment development.

In general, LPs describe how students progress toward a more sophisticated understanding of a big idea of science over time (Duschl et al., 2007, Smith et al., 2006). Therefore, in theory, LP represents a "roadmap" of how students can potentially move toward a more sophisticated understanding of science over a broad, defined period (Alonzo & Gotwals, 2012; Duschl et al., 2007; Smith et al., 2006). The current most concrete definition of an LP stems from the domain of assessment. It focuses on building assessment systems combining validated LPs for relevant science constructs and associated assessment items that probe various levels of understanding to track student progress (Brown & Wilson, 2011). According to this comprehensive definition, LPs should combine four structural and theoretical considerations (Duncan & Hmelo-Silver, 2009). First, LPs should incorporate elements of both content and practice. They should reflect complex science constructs beyond describing factual knowledge to outline increasingly sophisticated ways of applying the understanding by demonstrating the ability to integrate relevant content and practices consistent with situated learning theories. Second, LPs are bound by an upper anchor that describes what students are expected to be able to do at the end of the progression and a lower anchor that describes students' prior knowledge and skills, which is consistent with the developmental nature of student understanding. The upper and lower anchors are determined by research-based standards. Third, intermediate levels are expressed as learning performances defining what students know and can do at different levels of sophistication, consistent with the developmental approach. Learning performance expresses what students should know and be able to do. Fourth, LPs are mediated by curriculum, instruction, and formative assessment that provide developmentally appropriate context and scaffolding, consistent with the developmental approach and situated cognition theories.

Importance of Learning Progressions in Education

An assessment system consists of three elements that include a model of cognition, tasks or situations that allow for observing student performance, and an interpretation method that would allow drawing conclusions regarding student performance based on the observations (National Research Council [NRC], 2001). Cognition models (such as LPs) describe how students represent knowledge and develop competence in a subject domain (NRC, 2001). Using cognition models as an organizing tool for aligning curriculum, instruction, and assessment can significantly benefit all three (NRC, 2001). In the context of curriculum development, grounding the curriculum in relevant cognition theories allows for developing sequences to help students learn scientific ideas in a more meaningful and efficient manner. For instruction, the model of cognition helps interpret the assessment results more meaningfully and guides teachers in supporting students in attaining higher proficiency levels (Brown & Wilson, 2011; Kaldaras, Akaeze, & Krajcik, 2021a). Finally, for assessment, cognition models tell us where students are with respect to understanding specific big ideas and what support they need to move up the LP levels. Additionally, cognition models ensure that what is being measured on the assessment goes beyond memorization-based strategies to include complex constructs that focus on measuring knowledge application, which is indicative of deep science understanding (Brown, Wilson, 2011; Kaldaras, Akaeze, & Krajcik, 2021b).

A model of cognition was historically missing from traditional assessment systems (Brown & Wilson, 2011). Standard test theory, which has historically been the pillar of traditional assessment systems, allows obtaining information regarding student proficiency in a domain but generally does not allow defining proficiency or how it develops (Mislevy, 1996). This lack of information regarding proficiency and its development severely limits the validity of test scores (Mislevy, 1996). Information about proficiency is an important part of construct validity, which is a concept introduced as part of the current test validity theory and reflects a unifying concept that includes various types of validity (Jin, Shin & Cisterna, 2023). LPs in turn outline what proficiency looks like in a domain and how it develops across time, therefore, allowing for empirical testing and valid interpretation of assessment results, contributing to the validity of the assessment system. A valid assessment system, in turn, can inform revision and alignment between curriculum, instruction, and assessment and better support students in developing deep science understanding. Therefore, obtaining validity evidence for a specific LP helps revise and align curriculum, instruction, and assessment to better support students in developing a deeper understanding of science (Neumann, Viering, Boone, & Fischer, 2013). However, these benefits of validated LPs will only be relevant in practice if students have opportunities to learn the big ideas and scientific practices in the same way as specified by the LP.

It is important to point out that LPs are not developmentally inevitable and depend on curriculum, instruction, and student prior knowledge, among other factors (Smith et al., 2006). In other words, even though LPs are often represented as linear learning pathways, no single path can successfully account for all possible individual learning trajectories. In practice, there might be several viable paths to describe how students can develop toward a higher-level understanding of a given big idea. Nevertheless, the development and validation of specific LPs is still a valuable and practical endeavor because validated LPs represent cognition models reflecting the most frequent orderings in which understanding of science constructs develops in practice and the associated frequent conceptual challenges to students' experience (Shepard, 2018). Examples of such common orderings related to big ideas in science include understanding that similar charges repel, and opposite charges attract prior to discussing Coulomb's law (Kaldaras, Akaeze, & Krajcik, 2021a), and perceiving that matter is not continuous before discussing the atomic nature of matter (Smith et al., 2006). Examples of frequent conceptual challenges related to big ideas in science include the idea that applied force is proportional to an object's acceleration rather than speed (Alonzo & Steedle, 2009) or that the causal driving factor behind chemical bond formation is a balance of attractive and repulsive forces and energy minimization rather than achieving octet by an atom (Cooper et al., 2012). In the context of complex science constructs, LPs combine elements of big ideas and associated science practices. Therefore, from the perspective of the developmental approach, LPs that incorporate disciplinary knowledge and scientific practices should describe scientific practices that weave together with the common orderings of core ideas (and challenges) such that they form the levels of progression. Supporting students in progressing up the levels of such an LP would therefore involve engaging them in practices integrated with the common orderings of big ideas in the increasingly sophisticated fashion described by the LP. Validated LPs for complex constructs, therefore, should take into consideration how relevant science practices develop in the context of specific big ideas. Few such integrated LPs exist, and there is a growing need to conduct research focused on the validation of LPs for complex constructs focused on describing and supporting knowledge application ability in practice (Kaldaras, Akaeze & Krajcik, 2021a; Harris et al., 2019).

Learning Progressions in Science Education Literature

So far, the LPs reported in the literature can be characterized largely based on their focus: some focus on describing proficiency in the development of explanations of relevant disciplinary ideas, others only on practices, and some combine elements of content and practice. Additionally, some LPs are purely hypothetical, and others have varying degrees of validity evidence supporting the LPs resulting from one or more validation approaches discussed in the next section. We will further discuss each type of LP in more detail and provide examples.

Dimensions of Learning Progressions

Learning Progressions for the Development of Disciplinary Ideas

Most LPs developed early during the timeline of LP research focused on defining proficiency in building explanations of relevant disciplinary ideas. These LPs were often grounded in the expert-like vision of a given discipline, built following the evaluation of relevant educational standards or textbooks and the review of existing curriculum and research on students' understanding of a given science domain. Examples of such LPs include the atomic molecular nature of matter LP (Smith et al., 2006), genetics (Duncan & Yarden, 2009), and celestial motion (Plummer & Krajcik, 2010) among others. Such LPs often were purely hypothetical, suggested possible, sometimes novel learning pathways, and outlined potential student difficulties in developing higher levels of understanding following these pathways. For example, Duncan and Yarden (2009) focused on challenges associated with supporting students in building molecular-level mechanistic explanations of phenomena involving genetics ideas. Plummer and Krajcik (2010) identify important ideas students need to understand to develop a deep knowledge of celestial motion. These ideas include understanding how our perspective changes as the Earth rotates on its tilted axis while slowly orbiting the Sun and understanding how the apparent motion and appearance of the Moon changes as it slowly orbits the Earth once a month (Plummer & Krajcik, 2010). Some of the initially proposed hypothetical LPs were later validated to a certain degree by analyzing student response data from LP-aligned assessments. These include further development and validation of a construct map for changing seasons (Plummer & Maynard, 2014) and validation of the LP for genetics (Castro-Faix, Duncan, & Choi, 2021), while others remained purely hypothetical (Smith et al., 2006). As the LP research field developed, more empirically validated LPs emerged.

The degree to which these LPs were grounded in empirical evidence varies. For example, some LP validation studies presented initial validity evidence for the proposed LP by analyzing student response data from a large number of students using an LP-aligned assessment instrument, such as an energy LP (Neumann, Viering, Boone, & Fischer, 2013) and then refined the LP by revising the levels and administering the revised instrument to a new sample of students (Yao, Guo, & Neumann, 2017). Similarly, an LP for force and motion offers validity evidence based on cognitive interview analysis with selected students and quantitative item score analysis of a larger student sample (Alonzo & Steedle, 2009). Another extensively validated LP includes an associated assessment instrument designed to score student responses automatically, instructional materials, and a professional development (PD) package is an LP for carbon transforming processes based on Carbon Time curriculum materials (Anderson et al., 2018). This educational package resulted from years of research on LP validation through iterative cycles of LP and curriculum revisions for disciplinary ideas related to carbon-transforming processes (Gunckel, Covitt, Salinas, Anderson, 2012) and energy (Jin & Anderson, 2012) in the context of the carbon cycle.

Learning Progressions for Scientific Practices and Student Reasoning

Another important category includes LPs for constructs other than disciplinary ideas, including scientific practices and various types of student reasoning. Examples of LPs for scientific practices include scientific argumentation (Osborne, Henderson, MacPherson, Szu, Wild & Yao, 2016), scientific modeling (Schwarz, Reiser, Davis, Kenyon, Achér, Fortus, Shwartz, Hug & Krajcik, 2009), and scientific explanation (Yao & Guo, 2018). Examples of LPs for general scientific reasoning constructs include statistical reasoning (Lehrer, Kim, Ayers & Wilson, 2014) and blended Math-Science sensemaking (Kaldaras & Wieman, 2023a,b).

Learning Progressions that Integrate Core Ideas with Scientific Practices

Finally, another important category relates to more sophisticated LPs that describe the integration of relevant disciplinary ideas and practices. These LPs are grounded in situated cognition theories stating that deep science learning can only be achieved if students engage in authentic activities that require them to apply their understanding of relevant disciplinary ideas (Smith et al., 2006). From this perspective, engaging students in authentic scientific practices such as developing and using models or constructing scientific explanations when learning relevant disciplinary ideas results in developing a deep understanding that students can apply to explain various phenomena and solve problems. Examples of such LPs include developing scientific explanations for biodiversity (Gotwals & Songer, 2013) and developing models and scientific explanations for the interaction of matter (Kaldaras, Akaeze, & Krajcik, 2021a and b).

Validation Approaches for Learning Progressions

Hypothetical learning progressions require validation. The overarching approach to LP validation is grounded in the vision outlined in the Standards of Psychological and Educational Testing (Eignor, 2013; Messick, 1980), which refers to validity as appropriateness, meaningfulness, and usefulness of specific inferences made from the test scores. In the context of LP validation, test scores refer to student performance on an assessment designed to probe various LP levels. The standards further state that test validation occurs through a process of accumulating evidence to support the inferences. The accumulated validity evidence reflects the degree to which the LP-aligned assessment items elicit the ideas at each level of the LP and the degree to which student responses on the items fall into the hypothetical suggested LP. Accumulating this evidence is an iterative process. At each stage, valuable evidence is collected to inform the refinement of the LP levels, revision of the associated assessment instrument, and inform curriculum development and instruction.

Multiple approaches to validating LPs in practice can be used depending on how far along a given research study is in the validation cycle. In general, the validity studies can be pursued in three broad directions: 1) developing purely hypothetical LPs grounded in a review of relevant literature with no further validation conducted (no assessment instrument or associated curriculum); 2) using existing hypothetical LP and assessment instrument probing the LP levels to investigate how well existing curriculum supports student learning along the progression (includes associated assessment instrument but doesn't have LP-aligned curriculum designed); 3) using previously developed hypothetical LP with LP-aligned assessment and curriculum to investigate how well the curriculum supports student learning along the LP levels.

The starting point (direction #1 above) is a hypothetical LP developed based on existing research and content analysis of a given domain (Smith et al., 2006; Duncan, Rogat, & Yarden, 2009). Purely hypothetical LPs usually do not have an assessment instrument associated with probing different levels of the LPs. To validate these types of LPs, researchers need to develop associated curriculum, assessment, and professional learning (PL) programs that will provide validity evidence for the LPs.

Another approach (direction #2 above) represents cross-sectional studies that describe the progression of student understanding of ideas in a domain across multiple grades (Herrmann-Abell & DeBoer, 2018). In this case, assessment instruments are designed to probe various levels of the LP and can investigate how the existing curriculum impacts student learning. This approach requires some alignment between the assessment instrument probing the LP levels and the existing curriculum. For further validation studies, researchers need to design instructional interventions that align with these LPs to provide further evidence for the validity of the proposed LP and the associated assessment instrument.

Finally, the most advanced type of LP validation study (direction #3 above) represents the validation of an LP in the context of a carefully designed curriculum aligned to the LP levels across multiple grades (Kaldaras, Akaeze, & Krajcik, 2021a; Kaldaras, Akaeze & Krajcik,

2023). This approach provides validity evidence for the LP in the context of a carefully designed LP-aligned curriculum accompanied by the relevant professional learning programs to ensure fidelity of implementation of the curriculum, which in turn will facilitate student progression along the LP levels.

It is important to point out that it is not necessary to validate LPs in the context of a specific curriculum to produce an extensively validated LP. In fact, depending on the intended use, LPs validated in the context of a specific curriculum might not be the most useful. For example, if the intended use of the LP is for standards development and monitoring assessments, developers need information about broader developmental trends rather than curriculum-specific results. In this case, an LP validation study involving multiple contexts and curricula is needed. Therefore, validation is always context-dependent. A measure validated in one context may not be valid in another context. Therefore, researchers should determine what theoretical rationales and empirical evidence should be collected to validate the LP and its use in science education.

However, regardless of the intended use of the LP, students need to have opportunities to learn the content outlined in the LP, and teachers need sufficient support in helping students transition to higher LP levels. In this regard, LPs such as those described by Songer and colleagues (2009) and Kaldaras and colleagues (2021a) represent examples of more advanced stages of LP validation.

Validation of Three-Dimensional Learning Progressions

Multi-Dimensional Learning: an example of three-dimensional learning

According to *the Framework for K-12 Science Education* (NRC, 2012), deep science understanding is reflected in the ability to apply relevant disciplinary ideas to explain phenomena and solve real-life problems (NRC, 2012). This knowledge application ability is also called knowledge-in-use (Harris et al., 2019). Supporting the development of knowledge-in-use requires time and appropriate scaffolding reflected in timely and useful instructional support provided to learners (NRC, 2012). To better support the development of knowledge-in-use, *the Framework* outlines two important ideas: 1) it defines the three dimensions of scientific knowledge that developers need to integrate into the curriculum, instruction, and assessment to help students develop deep science understanding reflected in knowledge application ability; and 2) it promotes the need to organize the learning process along validated LPs that combine the three dimensions of scientific knowledge. The three dimensions include disciplinary core ideas (DCIs), scientific and engineering practices (SEPs), and crosscutting concepts (CCCs). DCIs represent big ideas in each discipline (including Physical Science, Life Science, and Earth Science) that hold significant explanatory power and can be used to explain a wide range of phenomena (NRC, 2012). Examples of DCIs include "Matter and Energy", "Forces and Interactions", "Earth's Systems", "Heredity, Biological Evolution". SEPs refer to the scientific practices used by scientists to generate and revise scientific knowledge in real-life settings. Examples of SEPs include "Developing and Using Models", "Analyzing and Interpreting Data", and "Planning and Carrying out Investigations". CCCs are lenses one can use to explain phenomena. Examples of CCCs include "Patterns", "Cause and Effect", "Structure, and Function".

Science educators refer to the ability to integrate the three dimensions of science (DCIs, SEPs, and CCCs) when making sense of phenomena and solving real-life problems as three-dimensional (3D) learning (NRC, 2012). Demonstrating the ability to integrate the three dimensions when making sense of phenomena indicates knowledge application ability and is, therefore, an indicator of deep science understanding (NRC, 2014). Both *the Framework* and Next Generation Science Standards (NGSS, NGSS Lead States, 2013) outline large-grain, grade band-specific hypothetical LPs for each of the three dimensions separately. However, currently, there is very little empirical evidence that demonstrates the feasibility of developing NGSS-aligned LPs that integrate the three dimensions of scientific knowledge (DCIs, SEPs, CCCs). Examples of 3D LPs validated using qualitative analysis of student response patterns include patterns of evolution (Wyner & Doherty, 2017) and solar system (Plummer et al., 2020). Examples of 3D LPs validated using both qualitative and quantitative analysis of student response patterns include a 3D LP for electrical interactions described by Kaldaras, Akaeze, and Krajcik (2021a). This study also demonstrates the feasibility of developing NGSS-aligned LP that can place individual students on an LP level with a high degree of accuracy, suggesting immediate applicability for pedagogical use.

However, empirical research on the development and validation of 3D LPs remains minimal, and, we believe, should be placed at the center of research initiatives that promote the successful implementation of NGSS. Our reason for this claim includes that LPs represent cognition models that are at the heart of the new vision for science education put forth by both NGSS and *the Framework*. The development of empirically validated, NGSS-aligned LPs that describe changes in student thinking at both larger and smaller grain sizes is a necessary initial step before any high-quality NGSS-aligned curriculum and assessment materials can be developed. In this chapter, we discuss the challenges of validating NGSS-aligned LPs measuring knowledge-in-use and provide examples of the current work in this area of research. For a broader discussion of LP validation, see Chapter 6 by Dr. Michael Kane in this book.

Validation of Three-Dimensional Learning LPs

According to the Standards of Educational and Psychological Testing (American Educational Research Association [AERA, 2018]), validation begins with an explicit statement of the proposed interpretation of test scores along with a rationale for the relevance of interpretation to the proposed use. 3D LPs describing knowledge-in-use are built on a set of theoretical assumptions that include a) the assumption that student proficiency is reflected in their ability to integrate the three dimensions of NGSS (DCI, CCC, SEP) when explaining phenomena, and b) the assumption that this ability develops in a manner described by the levels of the 3D LP. The goal of the validation process, therefore, is to evaluate to what degree student response data support these theoretical assumptions. If sufficient evidence is generated for each of these assumptions, this would suggest that the 3D LP and the associated assessment instrument can evaluate student proficiency in a given construct with respect to LP levels. That is, the researchers and teachers can use scores on assessments aligned to the 3D LP to draw inferences regarding student proficiency with respect to the 3D LP levels. In this context, a 3D LP includes a description of LP levels reflecting increasingly sophisticated ways of integrating the three dimensions to make sense of phenomena for a specific construct of interest, associated assessments probing the levels, and rubrics that explicitly relate student responses on the assessments to the levels of 3D LP. To achieve the abovementioned goals of the validation process, the assessments that align to the 3D LP require the following properties: a) measure student ability to integrate the three dimensions of science for a given context; b) measure students ability at varying levels of sophistication, and c) student response data should support these levels. If enough evidence exists to conclude that the scores on 3D

LP-aligned assessments have these characteristics, then we can conclude that the 3D LP exhibits the validity properties for each of the theoretical assumptions. Note that these assumptions relate to the overarching goal of using the scores on 3D LP-aligned assessments to characterize student proficiency in integrating the three dimensions of NGSS with respect to the levels outlined in the LP. In this context, it is not of primary concern whether the assessments are used for summative or formative purposes, since in both cases the central goal is accurate LP-level placement of students. Accurate and valid LP level placement is central to achieving the goals of both types of assessments, irrespective of whether the subsequent use of these results would adjust instruction (in the case of formative) or provide some final evaluation measure of student progress (in the case of summative assessment). We further discuss the types of validity evidence that researchers can use for evaluating the above-mentioned assumptions.

Response-process-based validity (American Educational Research Association [AERA, 2018]) provides evidence about the cognitive processes that the test takers engage in when answering the test question (AERA, 2018). Since 3D LPs represent complex models that describe specific cognitive processes related to student ability to integrate relevant DCIs, SEPs, and CCCs at increasing levels of sophistication, evaluating response-process-based validity is essential for assessing the viability of 3D LPs to reflect student 3D understanding of relevant cognitive constructs. Response-process-based validity provides evidence concerning the fit between the theoretical construct and the types of responses provided by learners on tests designed to measure that construct (AERA, 2018). In the context of complex 3D constructs that combine elements of the three dimensions of science (DCIs, SEPs, CCCs) response-process-based validity provides evidence about the fit between the cognitive model describing how proficiency in a given 3D construct develops-the LP, and the types of responses students demonstrate on a test designed to probe the LP levels (Kaldaras, et al., 2021a; Kaldaras, Akaeze, Krajcik, 2023). Therefore, in the context of 3D LP validation, response-process-based validity provides evidence on whether students engage in the cognitive processes of integration of the three dimensions of NGSS described at the corresponding LP levels. Importantly, this type of validity evidence goes beyond simply evaluating whether the test taker understood the question. Rather, evaluating this type of validity evidence should focus on evaluating whether sufficient evidence exists to conclude that student responses to the assessments reflect the types of cognitive processes that relate to the construct of interest, which the theoretical 3D LP describes. As pointed out in the Standards for Educational and Psychological Testing (AERA, 2018, p.16), evaluating response-process-based validity would involve evaluating the actual path students took to answering the problem and the types of cognitive processes students engaged in, not whether the student answered the problem correctly. Since 3D LPs reflect specific path students can follow in developing higher proficiency in a given construct and specific cognitive processes students should engage in (integration of DCIs, SEPs, and CCCs), evaluating response process-based validity builds a compelling validity argument for supporting the assumption that student proficiency reflects their ability to integrate the three dimensions (assumption "a" above).

In prior studies, response process-based validity of LP-aligned assessments for NGSS has been evaluated qualitatively using cognitive think-aloud interviews or qualitative analysis of student responses to items aligned to 3D LPs (Kaldaras et al., 2021a; Kaldaras, Akaeze & Krajcik, 2023; Wyner & Doherty, 2017). For this validation approach, the qualitative analysis of student response patterns should focus on investigating whether the 3D LP-aligned items elicit the integration of the three dimensions of NGSS in a manner described by the 3D LP. If sufficient evidence exists to conclude that this integration happens and, in a manner, described by the 3D LP, researchers can conclude that the 3D LP and the associated assessment instrument exhibit response-process-based validity.

While this type of validation study provides essential evidence towards the validity of the proposed 3D LP with respect to the assumption "a" stated above, this evidence is not enough to use the LP and the associated assessment instrument to accurately place students on LP levels placement. Response-process-based validity provides only preliminary evidence on whether the cognitive model reflects a progression. For example, if we see a clear pattern in the student interviews demonstrating understanding consistent with all LP levels up until a certain highest assigned level, this suggests that the order of the levels outlined in the LP is possible (Kaldaras & Wieman, 2023a). To further confirm this assumption, LP validation requires quantitative studies, which will allow to further confirm or disprove this assumption with a larger number of responses and yielding standardized item parameters that allow for more accurate quantitative evaluation.

Quantitative studies of LP validation need to use LPs and the associated assessment instruments as standardized, quantitative measures for diagnosing student proficiency. Quantitative LP validation studies involve using one of the latent variable modeling (LVM) methods, such as item response theory (IRT), to develop a standardized scale for the LP-aligned assessment instrument. These LVM approaches help to establish a quantitative link between LP-aligned item properties, student responses to the items, and the latent trait being measured. The latent trait in this context refers to student understanding of a given topic at different levels of LP. IRT assumes that an unobservable continuum exists along the latent trait and LP-aligned assessment items.

The primary purpose of IRT analysis is to analyze the developmental nature of the learning progression through a visual interpretive map called the Wright Map. This analysis will provide evidence for the validity of internal latent structure reflected in the developmental levels described for a given LP (Bernbaum Schoenfeld, Wilson, Champney & Zahner, 2011). Internal latent structure in this context refers to the order of the levels in which proficiency for a given construct develops as described by the LP. If IRT analysis demonstrates that evidence exists to support this order, we can conclude that the LP exhibits internal latent structure-based validity. Figure 5.1 shows a sample Wright Map.

The example shown in Figure 5.1 demonstrates an assessment instrument comprised of 8 items, each probing both levels 1 and 2 of the 3D LP Electrical Interactions. In this example, student responses are scored directly into the corresponding LP level according to proficiency described for that level (Kaldaras et al., 2021a). These responses serve as a basis for IRT model parameter estimation. The Wright Map shows a visual representation of estimated student proficiency on the left and estimated difficulty for each item in Logits (an ability scale) corresponding to the ability level. The interpretation of the Wright Map is meaningful psychometrically because each of the difficulties on the Wright Map relates to the levels of the LP (Bernbaum et al., 2011). Based on the design of the hypothetical 3D LP, we expect that the levels of the construct will occur in a certain order (Kaldaras et al., 2021a). As the map shows, difficulty parameters corresponding to level 1 are well separated from those corresponding to level 2 of the 3D LP. Additionally, level 1 difficulty parameters are consistently lower on the ability scale (logits) compared to level 2 difficulty parameters, suggesting that students at lower ability levels are less likely to demonstrate the types of responses consistent with level 2 of the LPs. Therefore, the estimated order for each of these levels in the Wright Map suggests an alignment between the theoretical expectations in the learning progression and the empirical results in the Wright Map. Thus, the theoretical framework as a 3D developmental LP exhibits internal latent structure validity (Bernbaum et al., 2011; Kaldaras et al., 2021a,b).

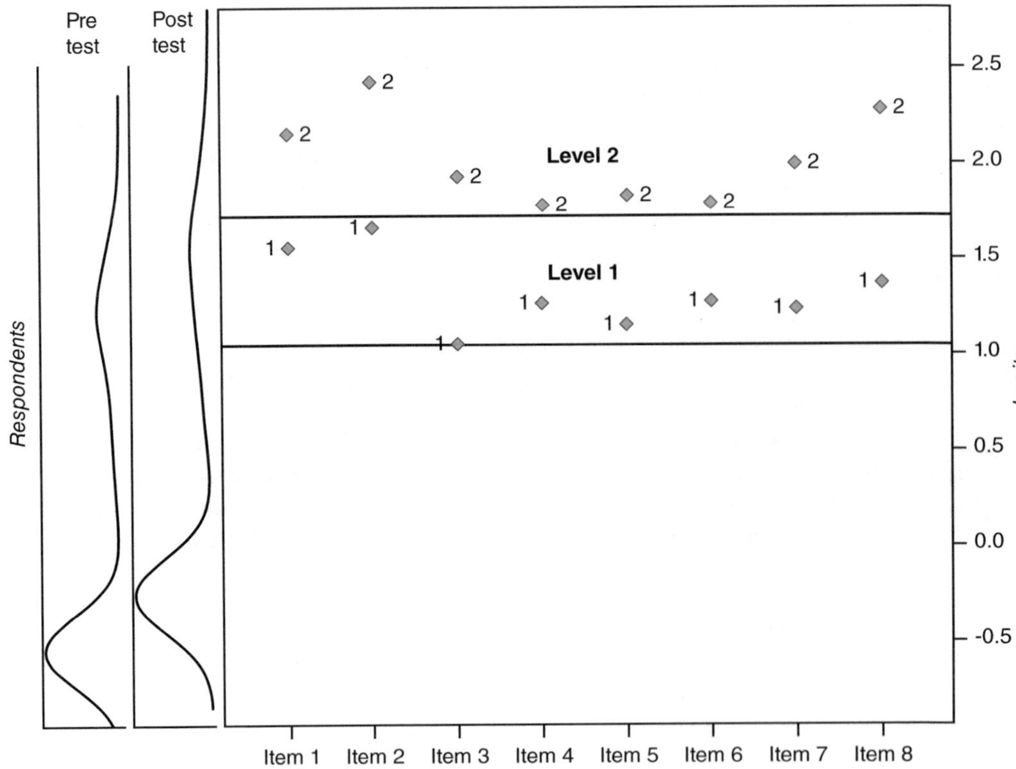

Figure 5.1 Wright map for 3D learning progression for electrical interactions (adopted from Kaldaras, Akaeze, Krajcik, 2021a).

Establishing internal latent structure-based validity focusing specifically on the order of the LP levels is an essential piece of validity evidence in the context of LP validation because it allows researchers and teachers to put trust into the developmental levels described by the LP. Since the main feature of the LP as a cognitive model describes developmental levels that students can follow in attaining higher proficiency in a given construct, obtaining validity evidence for the order of these developmental levels is critical for establishing the validity of the LP. Establishing the validity of the order of the LP levels also allows the use of the LP in the accurate diagnosis of group and individual student proficiency with respect to the LP levels. For example, Figure 5.1 shows that some learners progressed from low level 1 to high level 1 and some transitioned to level 2 between pre and post-test. A more detailed description of student learning gains can be found in Kaldaras and colleagues (2021a). Similarly, some LPs might allow for an accurate LP-level placement of individual students. A modified Wright Map shown in Figure 5.2 from the same study by Kaldaras and colleagues, (2021a) shows how the same Wright Map shown in Figure 5.1 designed to establish the order of the LP levels can be reorganized in the order of increasing ability to demonstrate estimated proficiencies of individual students. Depending on the accuracy of the proficiency estimates, individual student-level placements can be determined with varying degrees of accuracy (Kaldaras et al., 2021a). It is important to point out that ability estimates for each student shown in Figure 5.2 are based on the student's overall response pattern on all eight assessment items designed to probe the LP levels. A single item will not

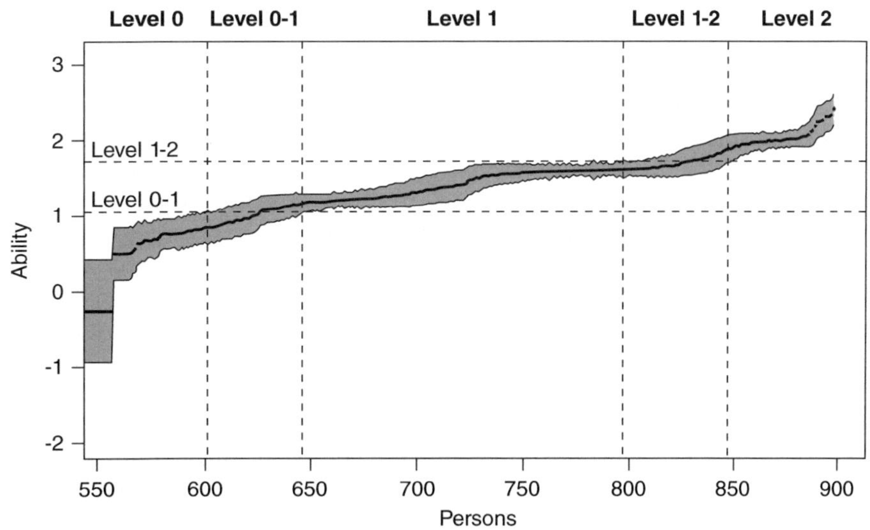

Figure 5.2 Modified Wright map showing student proficiency estimates and standard error bands from lowest to highest (adopted from Kaldaras et al., 2021a).

provide sufficient information for conducting IRT analysis and accurate estimation of student LP placement.

Therefore, IRT analysis of student responses helps obtain the critical piece of validity evidence to establish the LP and the associated assessment instrument as a standardized scale that can accurately place individual students on an LP level. This type of validity evidence is critical to LP validation because it allows researchers to go beyond establishing qualitative accounts of student learning to quantifying the degree of student progress with a high degree of accuracy. The more accurately researchers can diagnose student proficiency on a given construct and measure student progress, the better researchers, developers, and teachers can organize further instruction to support students in developing higher proficiency in a topic.

In the context of NGSS-aligned 3D LPs, very limited research has been conducted on demonstrating the feasibility of developing 3D LPs validated using both qualitative and quantitative Item Response Theory (IRT) based approaches. For example, the study by Kaldaras and colleagues (2021a) introduces an NGSS-aligned 3D LP for electrical interactions validated using both qualitative analysis of student interviews and IRT modeling to show that the 3D LP and the assessment instrument can place individual students on a level with a high degree of accuracy (Kaldaras et al., 2021a). Further, a similar study was conducted by Kaldaras, Akaeze & Krajcik (2023) to develop an NGSS-aligned 3D LP for chemical bonding. Both studies were conducted in the context of the previously designed NGSS-aligned curriculum for 9th-grade Physical Science. These studies represent valuable examples of the methodological and conceptual approaches to NGSS-aligned LP validation of finer-grained LPs capable of providing student-specific level placement. However, additional studies in different contexts and with a larger number of items are needed to develop both finer and larger grain-sized NGSS-aligned LPs that could directly impact educational settings. We will further discuss challenges associated with latent structure validation for NGSS-aligned 3D assessments and why this step is important for validating 3D LPs.

Challenges of Validating 3D Learning Progressions

One of the central issues for validation studies of 3D LPs is studying how the three dimensions of NGSS (DCIs, SEPs, CCCs) manifest in the context of the current measurement approaches used to validate the LP levels quantitatively (Kaldaras et al., 2021b). In the context of 3D LP validation, dimensionality represents a different aspect of latent structure validity (Kaldaras et al., 2021b). Two primary reasons exist for studying the dimensionality of 3D constructs: a) dimensionality is a prerequisite for choosing appropriate psychometric models to quantitatively evaluate the item and student performance to obtain information about student growth (Kaldaras et al., 2021b); b) understanding how the three dimensions of NGSS manifest in practice will help improve assessment and instructional practices in order to better support students in engaging in 3D learning as will be further discussed below.

The theoretical premise of *the Framework* and NGSS is that the three dimensions of science should not be separated but rather integrated together in curriculum, instruction, and assessment to ensure that students engage in authentic 3D learning as defined by *the Framework* (NRC, 2012). This premise, grounded in situated learning, states that students cannot learn the content separately from the context (Duschl et al., 2007), and the developmental approach stating that deep understanding takes time and appropriate scaffolding to develop (Smith et al., 2006). In the context of educational measurement, integration of the three dimensions discussed in *the Framework* suggests that these three dimensions are components of one complex construct consisting of relevant DCIs, SEPs, and CCCs. The LPs describing such complex constructs should outline an increasingly sophisticated ability to integrate the relevant DCIs, SEPs, and CCCs for a specific complex construct. Assessment capable of probing the levels of such 3D LP should aim to elicit relevant DCIs, SEPs, and CCCs in an integrated fashion at different levels of sophistication consistent with the 3D LP to which it aligns. A principled approach to NGSS-aligned assessment development that ensures stepwise, NGSS-driven assessment development focused on evaluating knowledge-in-use has been described by Harris, Krajcik, and Pellegrino (2019), and successfully employed by several studies focused on NGSS assessment development (Kaldaras et al., 2021a,b). This approach builds on argument-based assessment development methods such as evidence-centered design (Mislevy, Haertel, 2006) to design items for measuring knowledge-in-use.

Historically, item response theory (IRT) models, and Rasch models, in particular, have been used to derive standardized item parameters for LP-aligned assessments and track individual student progress along the LP levels (Wilson, 2004). These traditionally used IRT approaches assume that the underlying construct measured on a given assessment is unidimensional. Therefore, only student understanding of the construct measured on the test affects the observed item response patterns (Wilson, 2004). This assumption seems logical since most LPs were designed to measure only one construct or two closely related ones (such as a disciplinary idea and a practice, as in Songer et al., 2009). However, with multiple dimensions of science understanding (DCIs, SEPs, and CCCs) defined by the *Framework* and NGSS, the assumption of unidimensionality needs to be empirically checked. It is possible that in the context of NGSS, that assumption is potentially violated (Kaldaras et al., 2021b). In cases where the unidimensionality assumption is violated, using traditional unidimensional IRT models for LP validation will yield inaccurate parameters and compromise the validity of the LP. In these situations, researchers cannot draw accurate conclusions about student learning with respect to LP levels (Kaldaras et al., 2021b).

Further, it is also important to evaluate the dimensionality of assessments used for classroom assessment purposes because this information can help educators better understand how to support students in learning to integrate the three dimensions of NGSS. For example, if sufficient evidence exists to conclude that 3D tasks better describe unidimensional models, that would suggest that a 3D construct is a complex construct that combines the three dimensions of scientific knowledge instead of three separate constructs. It would therefore imply that the 3D construct is a complex conceptual dimension that manifests as a single psychometric dimension. This would be consistent with the situated cognition premise of the *Framework for K-12 Science Education* and *NGSS* that states that learning content (DCIs) is inseparable from engaging in SEPs and CCCs. This has an important instructional implication because it suggests that students cannot gain a deep understanding of the content (DCIs) without a context that allows for the simultaneous use of CCCs and SEPs. For instance, if a student can construct a model in Physical Science, it would not mean that the student can construct a model in the context of the Biology discipline since the content in the two cases would be different. As a result, for designing classroom assessments, this finding suggests that content should not be evaluated separately from the context and that the three dimensions should be integrated into classroom instruction and assessment (Kaldaras et al., 2021b). In contrast, if sufficient evidence exists to conclude that 3D tasks are better described by multidimensional models, further studies need to determine which dimensions manifest as separate constructs. This would provide important insights into the potential transferability of skills across contexts. For example, if researchers find that a SEP manifests as a separate latent construct, it would suggest that students can transfer their modeling skills across disciplines, which would in turn have implications for organizing instruction and designing classroom assessments.

At present, not enough empirical evidence exists to assume that complex NGSS constructs combining elements of DCIs, SEPs, and CCCs manifest as a single latent construct. Very little research has been conducted on evaluating whether unidimensional or multidimensional models better describe complex NGSS constructs. For example, Kaldaras and colleagues (2021b) present an extensive internal latent structure validation study for two NGSS-aligned 3D assessment instruments that supports the theoretical premise put forth by the *Framework* and demonstrates that the three dimensions (DCIs, SEPs, CCCs) manifest as a single latent construct. This study was conducted in the context of a previously designed NGSS-aligned curriculum where students had opportunities to engage in the DCIs, SEPs, and CCCs measured on both assessment instruments. Therefore, the study provides a good example of latent trait analysis in the context of a well-aligned, tightly integrated system consisting of curriculum, instruction, and assessment components. However, it is unclear whether similar conclusions regarding the latent dimensionality of 3D constructs would hold in the instructional context that is not as tightly aligned to the assessment as in the proposed study. Specifically, if students do not have opportunities to engage with DCIs, SEPs, and CCCs measured on the test, the dimensionality of the instrument might be different. Therefore, more research needs to be conducted on how different degrees of engagement with the relevant DCIs, SEPs, and CCCs affect the latent dimensionality of the complex constructs described by the NGSS LPs. Similar research studies should also be conducted with multiple DCIs, SEPs, and CCCs across different disciplinary contexts.

Further, another important reason why the unidimensionality of the resulting 3D construct might not be supported in practice relates to educational settings where students do not have opportunities to engage with the same types of SEPs and CCCs in the context of DCIs described by the 3D LP. For example, suppose a validated 3D LP described increasing levels of understanding related to constructing scientific explanations of electrical interactions. In that case, the question becomes, can this LP be used to track students' ability to develop and use scientific models to explain phenomena involving electrical interactions? To our knowledge, no research on these types of differences regarding various combinations of SEPs and CCCs related to the same DCIs described by the 3D LP has been conducted. The different sets of SEPs and CCCs could contribute differently to the resulting latent dimensionality, possibly resulting in a multidimensional construct. Alternatively, the different dimensions might result in entirely different ability levels and potentially the need for the development and validation of a different (but related) 3D LP. More research is needed on how different combinations of DCIs and SEPs affect the resulting latent dimensionality of complex 3D constructs focused on specific DCIs. In the context of LP validation, this also bears more thought regarding how to define LPs with respect to the dimensions of NGSS. Specifically, would LPs that focus on the same set of DCIs, but a different set of SEPs and CCCs, show different results? If yes, is this difference significant, and in what way? What does that mean for the latent variable modeling and the definition of proficiency on the complex construct described by those related but different LPs? These are important questions to pursue for future research studies in the field of NGSS-aligned LP validation. This research would provide evidence for using unidimensional or multidimensional LVM approaches for modeling NGSS constructs.

Another study comparing unidimensional and multidimensional IRT model fit for NGSS-aligned items showed that although a 3D IRT model provided a slightly better overall model fit than the 1D IRT model, the item-level fit of the 3D model was poor (Chen, Lee, Nichols, Schneider, 2021). This suggests that in the case of complex NGSS constructs, it is essential to check both the overall model fit and the model assumptions, including person and item fit and possibly local independence. For example, as shown by Chen et al. (2021), some items might not exhibit good model fit even when the overall model fit is acceptable, suggesting a possible need for revision of those items. Suppose poorly fitting items are also aligned to a relevant 3D LP. In that case, this could indicate potential conceptual problems with the LP and the item itself, all of which should serve as red flags for researchers and assessment developers. Similarly, the local independence assumption is critical to check to ensure that only one underlying construct contributes to the student response pattern on the assessment. This is an easy step that could be conducted in the context of IRT modeling and checking the overall model fit, item, and person fit. Finally, yet another explanation could be that methodologically multidimensional IRT is in fact a better choice for modeling 3D assessments, and poor item fit depends on the ability level of the respondents: multidimensional for lower-level responses and unidimensional for higher-level responses. This is because as learners progress towards higher levels of the 3D LP, they learn to better integrate the three dimensions in their responses. All these possible explanations require further research on the topic.

Conclusion

This chapter presents a brief review of key directions focused on developing and validating LPs that describe knowledge-in-use. LPs are an essential component of the new vision of science education because they represent cognitive models grounded in evidence that can drive the development of curriculum, assessment, instruction, and professional learning aligned to current learning theories reflected in the *Framework* and NGSS. It is, therefore, crucial to develop scientifically sound solutions for validating LPs focused on describing knowledge-in-use (such as NGSS-aligned LPs). The new vision of science education is driven by the need to help students develop deep science understanding reflected in the ability to apply knowledge to explain the world. This deep science understanding is reflected in the ability to integrate the three dimensions of scientific knowledge when explaining the world (the NGSS Lead States, 2013). According to the *Framework*, the three dimensions can't be separated and should be integrated into the curriculum, instruction, and assessment (National Research Council, 2012). The LP approach is also the cornerstone of the new vision for science education (National Research Council, 2012). This approach leads to the need to develop three-dimensional (3D) LPs for the complex constructs combining relevant DCIs, SEPs, and CCCs. There is currently minimal research on building and validating 3D LPs. Several future research avenues have been discussed, including designing psychometrically sound approaches to study how the three dimensions of scientific knowledges manifest in practice in different assessment and instructional scenarios. Further, the need to address the issues related to latent dimensionality during 3D LP validation has been discussed. Finally, the need to develop finer and larger grain-sized 3D-aligned LPs has also been emphasized. Importantly, using validated LPs to track student progress and support student learning in classrooms will only be feasible if students have opportunities to engage with the content in the way envisioned by *the Framework* and NGSS. It is therefore important to support teachers through professional development efforts and design instructional and curricular supports to ensure that students have opportunities to learn the content and develop the ability to integrate the three dimensions of scientific knowledge described by *the Framework* and NGSS.

References

Alonzo, A. C., & Gotwals, A. W. (Eds.). (2012). *Learning progressions in science: Current challenges and future directions*. New York: Springer Science & Business Media.

Alonzo, A. C., & Steedle, J. T. (2009). Developing and assessing a force and motion learning progression. *Science Education*, *93*(3), 389–421.

American Educational Research Association. (2018). *Standards for educational and psychological testing*. American Educational Research Association.

Anderson, C. W., de los Santos, E. X., Bodbyl, S., Covitt, B. A., Edwards, K. D., Hancock, J. B., ... & Welch, M. M. (2018). Designing educational systems to support enactment of the Next Generation Science Standards. *Journal of Research in Science Teaching*, *55*(7), 1026–1052.

Bernbaum Wilmot, D., Schoenfeld, A., Wilson, M., Champney, D., & Zahner, W. (2011). Validating a learning progression in mathematical functions for college readiness. *Mathematical Thinking and Learning*, *13*(4), 259–291.

Brown, N. J., & Wilson, M. (2011). A model of cognition: The missing cornerstone of assessment. *Educational Psychology Review*, *23*(2), 221.

Castro-Faix, M., Duncan, R. G., & Choi, J. (2021). Data-driven refinements of a genetics learning progression. *Journal of Research in Science Teaching*, *58*(1), 3–39.

Chen, J., Lee, J., Nichols, P., & Schneider, M. C. (2021). Psychometric models for a new state science assessment aligned to the next generation science standards. In *Quantitative Psychology* (pp. 407–416). Cham: Springer.

Cooper, M. M., Underwood, S. M., Hilley, C. Z., & Klymkowsky, M. W. (2012). Development and assessment of a molecular structure and properties learning progression. *Journal of Chemical Education*, *89*(11), 1351–1357.

Duncan, R. G., & Hmelo-Silver, C. E. (2009). Learning progressions: Aligning curriculum, instruction, and assessment. *Journal of Research in Science Teaching*, *46*(6), 606–609.

Duncan, R. G., Rogat, A. D., & Yarden, A. (2009). A learning progression for deepening students' understandings of modern genetics across the 5th–10th grades. *Journal of Research in Science Teaching*, *46*(6), 655–674.

Duschl, R. A., Schweingruber, H. A., & Shouse, A. (Eds.). (2007). *Taking science to school: Learning and teaching science in grades K-8*. Washington, DC: National Academy Press.

Eignor, D. R. (2013). The standards for educational and psychological testing.

Gee, J. P. (2008). A sociocultural perspective on opportunity to learn, *Assessment, Equity, and Opportunity to Learn*, 76–108. Chicago.

Gotwals, A. W., & Songer, N. B. (2013). Validity evidence for learning progression-based assessment items that fuse core disciplinary ideas and science practices. *Journal of Research in Science Teaching*, *50*(5), 597–626.

Gunckel, K. L., Covitt, B. A., Salinas, I., & Anderson, C. W. (2012). A learning progression for water in socio-ecological systems. *Journal of Research in Science Teaching*, *49*(7), 843–868.

Harris, C. J., Krajcik, J. S., Pellegrino, J. W., & DeBarger, A. H. (2019). Designing knowledge-in-use assessments to promote deeper learning. *Educational measurement: issues and practice*, *38*(2), 53–67.

Herrmann-Abell, C. F., & DeBoer, G. E. (2018). Investigating a learning progression for energy ideas from upper elementary through high school. *Journal of Research in Science Teaching*, *55*(1), 68–93.

Jin, H., & Anderson, C. W. (2012). Developing assessments for a learning progression on carbon-transforming processes in socio-ecological systems. In *Learning progressions in science* (pp. 149–181). Brill Sense.

Jin, H., Shin, H. J., & Cisterna, D. (2023). Systematic validation in science learning progression research. *International Journal of Science and Mathematics Education*, 1–21.

Kaldaras, L., & Wieman, C. (2023a) Cognitive framework for blended mathematical sensemaking in science. *International Journal of STEM Education*, *10*(1), 1–25.

Kaldaras, L., & Wieman, C. (2023b). Instructional model for teaching blended math-science sensemaking in undergraduate science, technology, engineering, and math courses using computer simulations. *Physical Review Physics Education Research*, *19*(2), 020136.

Kaldaras, L., Akaeze, H. O., & Krajcik, J. (2023). Developing and validating an next generation science standards-aligned construct map for chemical bonding from the energy and force perspective. *Journal of Research in Science Teaching*.

Kaldaras, L., Akaeze, H., & Krajcik, J. (2021a). Developing and validating next generation science standards-aligned learning progression to track three-dimensional learning of electrical interactions in high school physical science. *Journal of Research in Science Teaching*, *58*(4), 589–618.

Kaldaras, L., Akaeze, H., & Krajcik, J. (2021b). A methodology for determining and validating latent factor dimensionality of complex multi-factor science constructs measuring knowledge-in-use. *Educational Assessment*, *26*(4), 241–263.

Lehrer, R., Kim, M. J., Ayers, E., & Wilson, M. (2014). Toward establishing a learning progression to support the development of statistical reasoning. *Learning over time: Learning trajectories in mathematics education*, 31–60.

Messick, S. (1980). Test validity and the ethics of assessment. *American Psychologist*, *35*(11), 1012.

Mislevy, R. J. (1996). Test theory reconceived. *Journal of Educational Measurement*, *33*, 379–416.

Mislevy, R. J., & Haertel, G. D. (2006). Implications of evidence-centered design for educational testing. *Educational Measurement: Issues and Practice*, *25*(4), 6–20.

National Research Council. (2000). *How people learn: Brain, mind, experience, and school: Expanded edition* (Vol. 1). National Academies Press.

National Research Council. (2001). *Knowing what students know: The science and design of educational assessment*. National Academies Press.

National Research Council. (2012). *A framework for K-12 science education: Practices, crosscutting concepts, and core ideas*. Washington, DC: National Academies Press.

Neumann, K., Viering, T., Boone, W. J., & Fischer, H. E. (2013). Towards a learning progression of energy. *Journal of research in science teaching*, *50*(2), 162–188.

NGSS Lead States. (2013). *Next Generation Science Standards: For States, By States*. Washington, DC: The National Academies Press.

Osborne, J. F., Henderson, J. B., MacPherson, A., Szu, E., Wild, A., & Yao, S. Y. (2016). The development and validation of a learning progression for argumentation in science. *Journal of Research in Science Teaching*, *53*(6), 821–846.

Plummer, J. D., & Krajcik, J. (2010). Building a learning progression for celestial motion: Elementary levels from an earth-based perspective. *Journal of research in science teaching*, *47*(7), 768–787.

Plummer, J. D., & Maynard, L. (2014). Building a learning progression for celestial motion: An exploration of students' reasoning about the seasons. *Journal of Research in Science Teaching*, *51*(7), 902–929.

Plummer, J. D., Palma, C., Rubin, K., Flarend, A., Ong, Y. S., Ghent, C., ... & Furman, T. (2020). Evaluating a learning progression for the solar system: Progress along gravity and dynamical properties dimensions. *Science Education*, *104*(3), 530–554.

Schwarz, C. V., Reiser, B. J., Davis, E. A., Kenyon, L., Achér, A., Fortus, D., Shwartz, Y., Hug, B. & Krajcik, J. (2009). Developing a learning progression for scientific modeling: Making scientific modeling accessible and meaningful for learners. *Journal of Research in Science Teaching*, *46*(6), 632–654.

Shepard, L. A. (2018). Learning progressions as tools for assessment and learning. *Applied Measurement in Education*, *31*(2), 165–174.

Smith, C. L., Wiser, M., Anderson, C. W., & Krajcik, J. (2006). FOCUS ARTICLE: Implications of research on children's learning for standards and assessment: A proposed learning progression for matter and the atomic-molecular theory. *Measurement: Interdisciplinary Research & Perspective*, *4*(1–2), 1–98.

Songer, N. B., Kelcey, B., & Gotwals, A. W. (2009). How and when does complex reasoning occur? Empirically driven development of a learning progression focused on complex reasoning about biodiversity. *Journal of Research in Science Teaching*, *46*, 610–631.

Wilson, M. (2004). *Constructing measures: An item response modeling approach*. Routledge.

Wyner, Y., & Doherty, J. H. (2017). Developing a learning progression for three-dimensional learning of the patterns of evolution. *Science Education*, *101*(5), 787–817.

Yao, J. X., & Guo, Y. Y. (2018). Validity evidence for a learning progression of scientific explanation. *Journal of Research in Science Teaching*, *55*(2), 299–317.

Yao, J. X., Guo, Y. Y., & Neumann, K. (2017). Refining a learning progression of energy. *International Journal of Science Education*, *39*(17), 2361–2381.

6

COORDINATING ASSESSMENTS WITH A LEARNING PROGRESSION

Mark Wilson, Shih-Ying Yao, and Jonathan Osborne

Coordinating Assessments with a Learning Progression

The concept of a learning progression has been implicated in curriculum and instructional ideas that have existed in the educational instruction and assessment literatures over many decades.[1] One early definition is as follows:

> Learning progressions are descriptions of the successively more sophisticated ways of thinking about an important domain of knowledge and practice that can follow one another as children learn about and investigate a topic over a broad span of time. They are crucially dependent on instructional practices if they are to occur. (CCII, 2009)

The focus of a learning progression is on the development of the learner's understanding: thus, curriculum activities must fit with the structure of the learning progression—and the topics in a textbook should broadly follow the map identified by the learning progression, as should the curriculum standards. The learning progression must document the critical developmental points in a student's growth that are key to designing instruction to bring about change. The developmental points can be seen as defining a "roadmap" of important forms of student thinking (Black et al., 2011). In the long run, as for any other model, a learning progression must be judged by its utility. Its descriptions are ideally situated at a "meso-level" without being either so broad as to provide very little guidance for instruction and assessment (i.e., the "macro-level" that would be suitable for state and international testing, etc.), or so very fine-grained as to make it difficult for a teacher to use them in planning for a classroom setting (i.e., "micro-level") (Wilson, 2020).

A learning progression is a conceptual device intended to help address a foundational problem in education that is difficult to grasp due to its complexity: that is the interaction of disciplinary knowledge, student learning, the accompanying appropriate instruction, and assessment to achieve an educational aim. That is, a learning progression is a contribution to the (complex) answer to the four-part question of how we are to establish the *necessary coherence* among: the goals of the education expressed as substantive disciplinary content, whether it is a part of language arts, mathematics, or science (as is the focus here); the principal steps in the development of a more sophisticated understanding of a given concept

 DOI: 10.4324/9781003170785-7

(i.e., signposts along the prime roads to understanding); the focus of instruction and ordering to align with the learning progression; and the foci for the assessment of the progress of students understanding as they learn (i.e., where they are on the "roadmap," and where they are going). In addition, in any educational approach based on a learning progression, it is crucial to provide professional development for teachers on how to use the learning progression, and this should be based on the same quadripartite view—teachers must "know the discipline," "know how students learn," "be able to design and carry out instruction to assist that learning,"[2] and "know how to assess the success of their students."

In this chapter, we will focus on the fourth aspect of the use of learning progressions, the assessment aspect. To this end, we will not only explore this role for assessment, but also attempt to support our basic proposition. This is that: *Good measurement can provide important evidence in support of a learning progression, and the outcomes can be helpful for teachers, but it requires a systematic and rigorous effort by assessment developers to create materials that can help teachers in the classroom.* We will make this argument, in part, by using an example of an assessment of a specific topic, scientific argumentation. As we work through the example below, we will follow these steps:

i create a progression as a sequence (or multiple sequences) of waypoints of student understanding (i.e., qualitatively distinct ways of understanding, seen as leading to a particular goal);
ii develop ways to assess these waypoints in each construct in this learning progression;
iii develop reports (and associated professional development activities) to help teachers use the information from the assessments for instructional design and within the activities in a classroom; and
iv study whether the empirical information from students taking the assessments is consistent with the hypothesized progression.

One of the challenges of this process is that to do (i) well, one must already have succeeded (at least in part) in (ii) through (iii)—that is, this is a process that is not linear, as it might appear from the labeling above. Rather, (i) through (iv) is an iterative process, and frequently it involves other non-linearities where, for example, steps are skipped, and some may be fed back or re-visited.

For the next three sections of this chapter we will illustrate, using the example of scientific argumentation, how assessments based on a learning progression can help teachers (a) to understand the waypoints of the learning progression, (b) to make design decisions, and (c) to make practical every-day decisions about learning.

First, we lay out foundational research and ideas for understanding students' growth in scientific argumentation. Following that, we describe (a) the development of an assessment that attempts to measure students' argumentation competency in the context of middle school science, followed by (b) an account of one critical aspect of the validation study of this assessment. The chapter concludes with a discussion of the challenges that are involved in developing assessments that align with a learning progression.

An Exemplary Context: Scientific Argumentation

In the past two decades, there has been an increasing emphasis on incorporating scientific argumentation into the practice of science education (e.g., Driver et al., 2000; Duschl & Osborne, 2002). Extensive research suggests that engaging in the process of argumentation supports students' understanding of scientific content knowledge (e.g., Mercer et al., 2004; Schwarz et al., 2000; Venville & Dawson, 2010; von Aufschnaiter et al., 2008; Zohar & Nemet, 2002). The need for engaging students in argumentation is addressed explicitly in the national science standards in the US. Specifically, the *Framework for K–12 Science Education* (National Research Council, 2012), along with the Next Generation Science Standards (NGSS Lead States, 2013), specifies "engaging in argument from evidence" as one of the eight essential science practices in grades K-12.

Following the logic of Henderson et al. (2014), we consider argumentation as a competency demanding both

i knowledge, i.e., content knowledge, which is required to adjudicate multiple pieces of evidence, and
ii practice, i.e., epistemic knowledge to construct and critique various forms of reasoning.

Thus, with an increasing focus on incorporating argumentation into the teaching and learning of science, an emerging challenge is how to assess students' competency with scientific argumentation. For this work, we take advantage of previous research on how to assess scientific argumentation to provide the waypoints of the learning progression (Henderson, et al., 2014; Osborne et al., 2016). Without valid and reliable assessments, science educators will not be able to readily assess students' argumentation competency.

Toulmin's Model of Argument

Osborne and colleagues' initial hypothesis about how scientific argumentation is learned (Henderson et al., 2014; Osborne et al., 2016) is based in part on Toulmin's (1958) model of practical argument. In contrast to the logic of more traditional deductive arguments that start from first principles to logically deduce or prove a claim, practical arguments instead begin with claims and seek justification and evidence to support those claims. This justificatory view of argumentation is more in line with how argument is used in actual scientific practice – scientists begin with a claim (i.e., a hypothesis) and test its veracity against empirical data.

The Toulmin (1958) model begins with a claim as a "conclusion whose merits we are seeking to establish" (p. 90). It follows that in seeking to establish the merit of a claim, one turns to evidence that may lend support. In turn, the relation between the evidence and the claim is provided by a warrant that forms the substance of the justification for the claim. The Toulmin conceptions of claims, warrants, and evidence play a key role in the development of the Osborne et al. construct, just as it has formed the basis of many schemas used in research for analyzing student discourse (Cavagnetto, et al., 2010; Jimenez-Aleixandre et al., 2000; Osborne et al., 2004; Zohar & Nemet, 2002).

The Role of Critique

In addition to Toulmin's (1958) focus on the justificatory role of argument, the process of critique is also incorporated into the Osborne et al. (2016) construct map for scientific argumentation. As Ford (2008) has argued, the goal of scientific reasoning is knowledge construction, achieved by a dialectic consisting of both construction and critique.

More specifically, argument is used to justify the validity of explanatory hypotheses, experimental designs, and interpretations of a given data set. Critique is thereby essential to identifying flaws in such arguments. Without critique, adjudication of competing arguments is not possible.

The ability to engage in critique, however, requires a somewhat different competency than construction. For instance, rather than constructing a claim, critique requires the ability to identify what is the claim in any given argument, what elements constitute the data that is used to support the argument, and what kind of reason is used to relate the data to the claim. Such competency is reliant on at least a tacit meta-knowledge of the features of an argument and the ability to distinguish its component elements. Taken a step further, critique requires the ability to construct a rebuttal that would explain why the reasoning in a given argument is flawed.

Commonly, this may require the cognitive performance of comparing and contrasting the relative merits of two arguments or constructing an argument for why some evidence has higher epistemic validity than other evidence.

Accounting for Cognitive Load

Taken as a whole, therefore, a competency with scientific argumentation demands a complex orchestration of construction and critique of claims, warrants, and evidence in situations that require scientific knowledge to resolve. Unfortunately, opportunities to engage in argumentation rarely occur in traditional science classrooms in K-12 education (Banilower et al., 2013; Newton et al., 1999; Weiss et al., 2003). Thus, the multitude of information that must be processed to construct and/or critique a scientific argument, argumentation is both a novel and demanding task for many students that can – depending on the complexity of the argument – make substantial cognitive demands.

Given that students have little experience with scientific argumentation, any construct map for scientific argumentation should also account for limitations in human working (i.e., short-term) memory. This is because working on novel tasks places a high cognitive demand on human working memory. These limitations in working memory are perhaps most famously documented by Miller (1956) who described how, across multiple cognitive experiments, the number of entities that could be simultaneously held in human working memory is seven +/–2.

These findings have served as a foundation upon which cognitive load theory (Chandler & Sweller, 1991; Paas et al., 2003; Sweller, 1994) has been built.

Pollock et al. (2002) summarize two basic sources of cognitive load. The first source – intrinsic cognitive load – "is determined by the extent to which various elements interact ... An element is the information that can be processed by a particular learner as a single unit of working memory" (p. 62). In contrast, extraneous cognitive load "is generated by the manner in which information is presented to learners and is under the control of instructional designers" (p. 62).

All of these elements derived from the literature on the competency of scientific argumentation were used to develop the first of the BAS building blocks—the construct map of the entity to be assessed, as described in the next section.

Constructing the Assessment of Scientific Argumentation

The work reported here was part of a larger research project designed to examine how students' understanding of the structure of matter and competency in argumentation in the scientific context develop. The development of this scientific argumentation assessment is based on a construct modeling approach (Brown & Wilson, 2011; Wilson, 2004, 2005, 2009, 2012, 2023), known as the BEAR Assessment System (BAS), which consists of four building blocks: construct

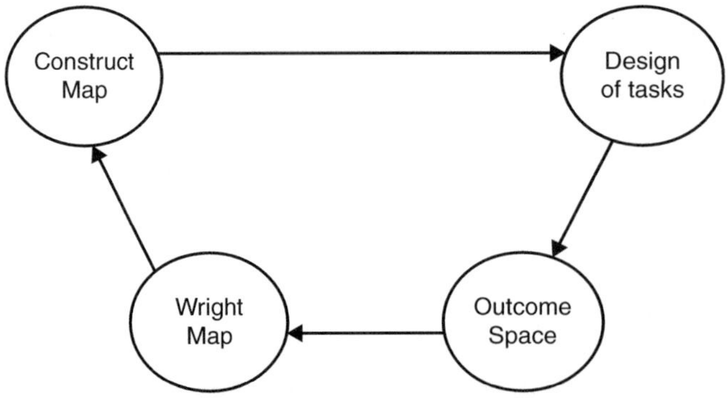

Figure 6.1 A representation of the BEAR assessment system (BAS).

map, item design, outcome space, and measurement models. The BAS is a general approach and can be applied to both achievement test development (K-12 and higher education) as well as to attitude scales (Wilson, 2005, 2023). How each of these four building blocks contributes to the development of the assessment will be explained in more detail below.

This section describes each of the four BAS building blocks (see Figure 6.1) in the development of the scientific argumentation assessment. The first building block describes the theory underpinning a proposed learning progression for scientific argumentation upon which our assessment is based. The BAS proposes a simple form for the learning progression, one that can be made up of a set of *construct maps*. A construct map consists of a series of increasingly sophisticated ways of understanding a certain idea or process, with a qualitative definition of each successive node in that succession—these nodes are called waypoints (Wilson, in press). The simplest form is where there is just a single ordered set of waypoints, but more complex patterns may be a better representation of some hypothesized structures (see Wilson (2009) for multiple examples of these more complicated structures). The second and third building blocks illustrate the process of designing assessment items and scoring rubrics focused on probing the various waypoints of the hypothesized construct map. The fourth building block describes the measurement model of choice.

Building Block One: A Scientific Argumentation Construct Map

As noted above, Osborne et al. (2016) use cognitive load theory as a lens through which the competency of scientific argumentation can be operationalized as a single construct. More specifically, given that Toulmin (1958) resolves argumentation into fundamental components such as claim, warrant, and evidence, the process of constructing or critiquing arguments can be viewed as an orchestration of various combinations of these elements of an argument. It follows that for tasks that increase the number of elements that must be processed, the intrinsic cognitive load on the working memory increases, thereby making it more difficult to demonstrate argumentative competency.

Hence, conceptualizing the elements of argument as a source of intrinsic cognitive load guided Osborne et al. (2016) as they hypothesized certain levels of sophistication of argumentation to be more difficult than others in their proposed learning progression for scientific argumentation. Consequently, Osborne et al.'s (2016) hypothetical learning progression for scientific argumentation (see Figure 6.2,[3]) consists of three broad sets of waypoints of argumentation differentiated by intrinsic cognitive load (and an initial waypoint

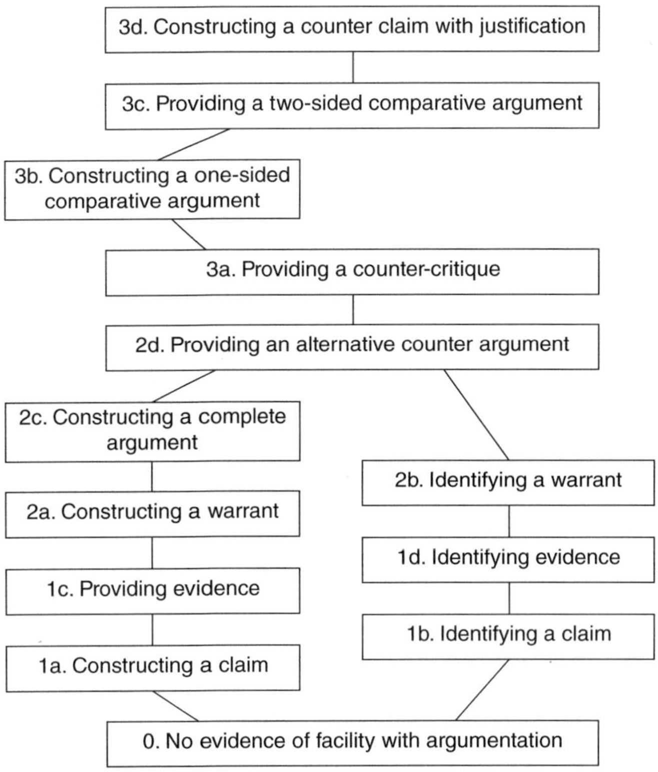

Figure 6.2 The hypothesized construct map for argumentation.

with no evidence of argumentation ability), where each higher waypoint is seen as requiring more connections to be made between claims and pieces of evidence.

The lowest waypoint of the construct map (at the bottom of Figure 6.2) is the waypoint where no part of an argument is identifiable in the student's response, and this is labeled "0".

The next set of waypoints are prefixed with the number "1"—this denotes that assessment items probing these stages do not ask for explicit connections between claim and evidence to be made or recognized. These connections – warrants under the Toulmin model – are not specifically asked for. Rather the question is whether the student can identify the individual elements of the argument and demonstrate competency with the identification/critique of a claim and/or evidence without making a logical connection between them. Thus, the 1 prefix for these waypoints indicates that the response is unitary and hence does not show any degrees of coordination. Construction of an explicit claim (Waypoint 1a) is hypothesized to be the most basic demonstration of argumentation competency, as doing so does not technically require any additional knowledge of the features of an argument. Indeed, it is possible for a student to advance a claim without relating it to either evidence or a warrant that might offer justification. Waypoint 1b is the matching ability to recognize a claim.

Items that do indeed require explication of warrants mark the transition from the set of Waypoints 1 to the set of Waypoints 2 in the construct map. These intermediate waypoints are prefixed with the number "2" to denote that they do involve coordination between two elements of an argument – i.e., students need to make at least one explicit logical connection between claim and evidence with a warrant. Thus, Waypoint 2a builds on the set of Waypoint 1 of the construct map, as it requires understanding not only what constitutes a claim or a piece of evidence, but also how to construct or critique a relationship between claim and evidence.

Assessments probing Waypoint 2a explicitly ask for such a warrant to be constructed. Specifically, at Waypoint 2a, students can construct an explicit warrant that connects a claim to evidence. Alternatively, Waypoint 2b requires the ability to identify the warrant that another person provided. Waypoints 2c and 2d are concerned with a student's competency in constructing a complete argument or providing an alternative counterargument to rebut a different claim.

The most advanced waypoints of the construct map are prefixed with the number "3" to denote coordination between more than two parts of an argument. At these waypoints, the student not only has to construct their own argument but also identify the elements in another argument and compare and contrast the two. Waypoint 3 items require students to explicate two or more warrants. In particular, assessments probing Waypoint 3a require students to critique another person's argument (e.g., explain why an argument is flawed). Waypoints 3b and 3c require students to evaluate the merits of two competing arguments. At Waypoint 3b, students are able to justify the value of one argument and yet do not provide a warrant to show why the other argument is less (or more) superior. Alternatively, at Waypoint 3c, students are not only able to evaluate two competing arguments but also able to explicitly argue why one argument is considered stronger than the other. Lastly, Waypoint 3d requires the ability to construct a counterclaim. This is the most proficient waypoint since it requires all of the capabilities. Specifically, it requires the individual to construct an alternative explanatory hypothesis, compare it with the existing argument, and evaluate why it is superior. The latter requires a reason for why it offers greater explanatory coherence and refutes alternative theories or arguments. Although the descriptions of the successive waypoints in the paragraphs above have not made explicit references to the relative cognitive loads involved at each waypoint, a clear succession of increasing pieces of the Toulmin argument framework is evident.

Building Block Two: The Items Design

To investigate the validity of the hypothesized construct map described in Building Block One, multiple paper and pencil assessment items were designed to elicit student responses commensurate with the various waypoints of our construct map. Each argumentation assessment in this study began by establishing a scientific scenario relevant to how matter is structured.

Students were then presented with a set of items related to the specific scenario. Within each scenario, multiple items were designed to span a range of argumentation skills included in the hypothesized construct map (Figure 6.2). Argumentation items sharing a common scenario will be referred to as item bundles, following Rosenbaum's (1988) use of the term to denote a subset of items that share a common stimulus. Figure 6.3 presents an example of an argumentation item bundle called "Sugar in the Water." The choice of specific domain knowledge that is involved in the argumentation items is, of course, a crucial decision. The choice in the examples illustrated here flowed from a decision to investigate the Structure of Matter. See Black et al. (2011)—for a discussion of the Structure of Matter learning progression that the items are based upon.

Note that the design of these items was informed by peer review from both educational researchers and teachers. Additionally, cognitive think-aloud interviews (Wilson 2005, 2023) with a group of grade eight students in a local district provided a gauge on whether each item elicited the kinds of cognitive response processes that were intended when the items were originally written. Utilizing an iterative process of item refinement when items did not perform as expected, the resulting modifications were again subjected to additional peer

Item A2. What Happens to Sugar in the Water?

Three students are discussing what happens to sugar grains when the sugar is added to a cup of hot water.

Paul says: *The sugar breaks up and disappears for good.*

Mary says: *The sugar breaks up but is still there and has not changed.*

Laura says: *The sugar has changed to make a new substance.*

They have several pieces of evidence for their argument.

1. The water tastes sweet.
2. After stirring the sugar can no longer be seen.
3. They leave the cup for several days. All the liquid evaporates leaving a sticky, solid substance at the bottom of the cup. This tastes sweet but is one large lump rather than separate grains.
4. They weigh the cup of water and the sugar separately. The weight of the cup of water and the sugar added together is the same as the weight of the cup of water with the sugar added.
5. They have been told that you cannot destroy or make new matter.

5a. Which evidence do you think best supports Paul's argument?
(a) Number ____
(b) Explain why you think this evidence supports Paul's argument.

5b. Which evidence challenges Paul's argument?
(a) Number____
(b) Explain why you think this evidence challenges Paul's argument.

5c. Which piece of evidence best supports Mary's argument?
(a) Number ____
(b) Explain why you think this evidence supports Mary's argument.

5d. Laura claims that the evidence available is not enough to decide whether she or Mary is right. Do you agree? Why?

Figure 6.3 The "Sugar in the Water" argumentation item.

review and think-aloud interviews. It is important to note that the quality of the items is also very important not only in promoting teacher usage of the assessments and the student outcomes from those assessments, both in terms of teachers understanding how the items reflect the waypoints of the construct map, and how the findings from assessments using those waypoints can be useful to them, but also in promoting teacher acceptance of the assessment.

Building Block Three: The Outcome Space

Given that the preponderance of argumentation items required open-ended written responses, it was important to develop scoring rubrics that facilitated reliable human scoring of written excerpts. More specifically, the design of the scoring rubrics was initially based on mock student responses that were hypothetically anticipated for each item. Researchers then used these preliminary rubrics to assign scores on pilot samples of actual student responses for each item.

Working in groups of no less than three researchers, each researcher independently scored the same sample of pilot tests. For consistency, researchers focused on scoring only one specific part of one specific item at a time. Each researcher then shared their scores on a specific part of a specific item with the rest of the group. If researchers did not reach a unanimous scoring decision for each pilot test, a discussion would ensue, and the scoring rubrics were modified based on this peer review. This moderation process was repeated until the revised scoring rubrics indeed yielded unanimous scores for each pilot test. Along the way, empirical samples of student responses were added to the rubrics to complement the hypothetical mock responses. The materials developed to help the researchers make these judgments are retained as training materials for others who will use these items, as well as the arrangements for the training procedures—these are all considered a part of the outcome space.

In later iterations of this developmental work (Morell et al., 2020), these recorded open-ended responses were mined to use as sources of options for multiple-choice and technology-enhanced items (TEIs). This results, eventually, in a two-part item bank—an open-ended portion, which can be used by teachers as rich sources of stimuli for formative assessment conversations, and a selected response portion which can be used for teachers to get quick feedback from their students. The remainder of this account focuses on the former portion, the open-ended items.

As described previously, each argumentation item is designed to assess one of the three main waypoints (i.e., Waypoints 1, 2, or 3) in the hypothesized scientific argumentation construct map. Note that partial credit was given to responses where reasoning was largely, yet not entirely correct. In other words, responses with full credit are regarded as fully demonstrating the argumentation construct waypoint that the item was designed to assess.

Alternatively, responses receiving partial credit are considered as demonstrating the construct waypoint in only a partial and incomplete fashion. This logic applies to the scoring of all items.

As an example, Table 6.1 presents the final scoring guide of item 5b in the "Sugar in the Water" item bundle (see Figure 6.3). This item asks students to identify a piece of evidence that challenges the claim made by a hypothetical student, Paul, in the prompt, and provides a justification for how the selected piece of evidence challenges Paul's claim. As described previously, each argumentation item is designed to assess one of the three main waypoints, i.e., Waypoint 1, 2, or 3, in the hypothesized scientific argumentation learning progression. This particular item was developed to assess Waypoint 2 in the hypothesized argumentation learning progression, which is to identify evidence that contradicts a given claim and build an alternative argument about why that claim may be flawed. Note that partial credit was given to responses, where the reasoning is largely yet not entirely correct. Responses with full credit, i.e., code 2 in the current example, are regarded as fully demonstrating the proficiency waypoint an item is designed to assess, i.e., Waypoint 2 in this example. Alternatively, responses with partial credit, i.e., code 1 in the current example, are considered as demonstrating the proficiency waypoint partially. This logic applies to the scoring of all items.

Building Block Four: The Measurement Model

The fourth building block is represented by measurement models, which allow one to mediate the information from a sample of students back to the original intent in terms of the estimated parameters of the model (both the student and the item parameters). Through a measurement model, scored student responses were calibrated and related to the cognitive model (Wilson 2005, 2023). There have been many measurement models proposed and used in the field of

Table 6.1 The scoring guide of the item 5b in the "sugar in the water" item bundle

Code	*Descriptions*	*Mock student responses*	*Empirical student responses*
2	Correct answer with appropriate reasoning. *A correct piece of counter-evidence is identified and reasoning is provided as to how it counters Paul's claim.*	"Number [anything but 2] challenges Paul's argument, as he claims the sugar disappears for good, but in fact [any piece of evidence besides suggests that something remains]."	[PICKED 5] "Paul said the sugar was gone for good but due to the law of conservation, the sugar is still there just that it has dissolved." (Note that this response tied this evidence back to Paul's claim)
1	Reasoning is largely correct, i.e., correct answer with inadequate reasoning or good reasoning that is used to justify an incorrect answer *A correct piece of counter-evidence is identified, but reasoning is not provided as to how it counters Paul's claim.*	"Number [anything but 2] challenges Paul's argument, as he claims the sugar disappears for good, but in fact [any reasoning that does not justify why the selected evidence does not support Paul]."	[PICKED 5] "Because you cannot destroy or make new matter." (Note that this response did not tie this back to Paul's claim) [PICKED 1] "Followed by what Paul said, the sugar breaks up and mixed with the water."
	Alternatively, *A correct piece of counter-evidence is not identified, but at least reasoning is provided as to how the student thinks it counters Paul's claim.*	*Alternatively,* "Number [2] challenges Paul's argument, as he claims the sugar disappears for good, but in fact [any reasoning that tries to justify why the selected evidence does not support Paul]."	[PICKED 4] "Because number 4 said that the sugar 0
0	Answer fails to address the question. *The student not only fails to identify a piece of evidence that challenges Paul's argument, but no justification is provided for why in fact they believe the piece of they chose counters Paul.*	"Number [2] challenges Paul's argument, as he claims the sugar disappears for good, but in fact [any reasoning that does not justify why the selected evidence does not support Paul]."	[PICKED 2] "Because Paul says the sugar breaks apart and disappears, not stirring the water." [PICKED 2] "Tells more details, and makes his statement sort of clear."
B	No Response		

educational and psychological testing. In describing the philosophy of the construct modeling approach (i.e., the BAS approach), Wilson (2005, 2023) focused on the special role the Rasch models (Rasch, 1960) play in understanding a construct of interest. In this study, both exploratory factor analysis and Rasch partial credit model were applied to the investigation of the function of the assessment of interest, which are illustrated in detail in the subsequent section.

Among the various aspects of validity evidence, this study focuses on the evidentiary strand of internal structure validity for the assessment of scientific argumentation– the focus of the study. According to the *Standards for Educational and Psychological Testing*, investigation of the empirical manifestations of the internal structure of a test can provide empirical evidence about "the degree to which the relationships among test items and test components conform to the construct on which the proposed test score interpretations are based" (American Educational Research Association, American Psychological Association, & National Council for Measurement in Education, 2014). The assessment investigated here was designed to measure a unidimensional construct of argumentation. Whether empirical evidence supports this assumption of unidimensionality is, therefore, the first question of interest.

In addition, as each item was designed to target a specific waypoint on the hypothesized construct map of argumentation, the extent to which the empirical pattern of item parameters supports this item design feature constitutes a second focus of interest.

Methods

Data

Twenty items were developed to assess students' argumentation competency, each specifically designed with one waypoint on the argumentation construct map in mind; the items were embedded in four scenarios contextualized in the domain-specific area of Structure of Matter for grade eight. Based on our expectations of the approximate ability distribution of the students, 11 items were designed to measure Waypoint 1, seven items were designed to measure Waypoint 2, and two items were designed to measure Waypoint 3 of the hypothesized argumentation construct map. The 20 argumentation items were administered in paper and pencil format. The students were given what was considered ample time to respond to the 20 items— most were finished well before the available time, though a few were still working at the end.

The data used in this study were collected from 347 grade eight students in a local school district in the US. Non-responses were coded as inaccurate responses. Information about five students was incomplete. The remaining sample consists of 188 girls (54.97%) and 154 boys (45.03%).

According to the results of the 2011 California Standards Test in Science (California Department of Education, 2013), 195 of these students (57.02%) were classified into the "Advanced" category, 61 students (17.84%) were in the "Proficient" category, 57 students (16.67%) were in the "Basic" category, 20 students (5.85%) were in the "Below Basic" category, and nine students (2.63%) were in the "Far Below Basic" category.[4]

Data Analysis

To assist the investigation into the dimensionality of the assessment, exploratory factor analysis (EFA) was performed. In a study of dimensionality assessment of polytomous items,

Timmerman and Lorenzo-Seva (2011) suggested that the parallel analysis with minimum rank factor analysis (PA-MRFA) performed consistently well at identifying the number of major factors when minor factors were present in the simulated data. In the current study, the PA- MRFA procedure described by Timmerman and Lorenzo-Seva (2011) was performed with the software FACTOR (Lorenzo-Seva & Ferrando, 2006). For a detailed account of the PA-MRFA procedure implemented, please refer to Yao (2013). To evaluate the number of factors, the scree plot (Cattell, 1966) was also examined.

The Rasch partial credit model (Masters, 1982) was used as a confirmatory model to evaluate the measurement properties of the instrument as a whole and of each individual item. Under the Rasch partial credit model, the probability of student responses is formulated as follows:

$$P_{pix}(\theta) = \frac{\exp\left[\sum_{j=0}^{x}(\theta_p - \delta_i - \tau_{ij})\right]}{\sum_{r=0}^{m_i}\left[\exp\sum_{j=0}^{r}(\theta_p - \delta_i - \tau_{ij})\right]},\ x = 0, \ldots, m_i, \quad \text{where} \ \sum_{j=0}^{0}(\theta_p - \delta_i - \tau_{ij}) \equiv 0,$$

where θ_p stands for the latent ability of student p. δ_i represents the overall difficulty of item i, and τ_{ij} refers to the deviation step parameter associated with category j of item i.

The partial credit model was estimated with the software, *ConQuest* (Adams et al., 2020), using the marginal maximum likelihood estimator (Gauss-Hermite Quadrature with 15 nodes), where the ability parameter θ_p is assumed to be normally distributed as $N(\mu_\theta, \sigma_\theta^2)$. Item parameters, δ_i and τ_{ij}, are assumed to be fixed parameters. For model identification, the mean of the ability distribution is fixed as 0 and $\sum_j \tau_{ij} = 0$ for each item i. In the Conquest outputs used below, the reported Wright maps utilize a calculated value called a "Thurstonian threshold" (Adams et al., 2020). For the cases shown, this value represents the location at which the probability of observing the highest category (in terms of the construct map waypoints shown in Figure 6.2) is 0.50.

Results

Unidimensionality–Results from the Exploratory Factor Analysis

The EFA results first provide statistical evidence for whether the studied argumentation items measured one salient latent construct as intended. Figure 6.4 is the scree plot of the

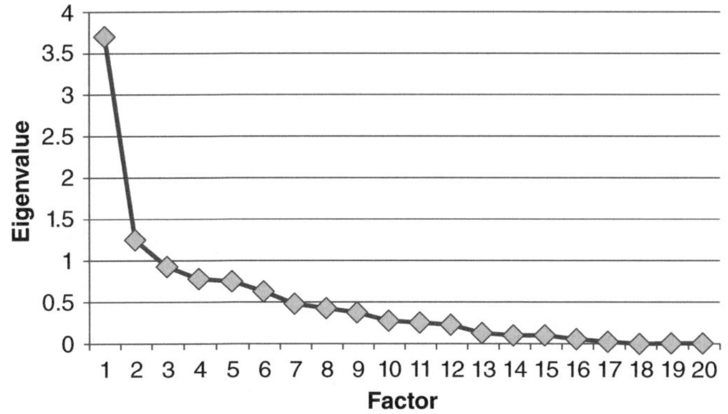

Figure 6.4 Scree plot for the EFA analysis.

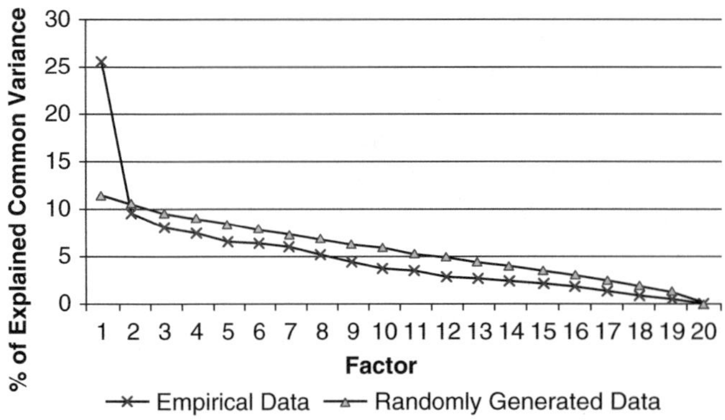

Figure 6.5 Results of the parallel analysis.

eigenvalues of the reduced correlation matrix (Timmerman & Lorenzo-Seva, 2011). The plot shows that a sharp bend occurs around the second eigenvalue. Although the first two factors both have eigenvalues larger than one, the eigenvalue of the first factor is about three times as large as the eigenvalue of the second factor and the eigenvalue of the second factor is only slightly larger than one. This evidence points to the presence of one major factor.

The presence of one salient factor is further observed in the result of parallel analysis of the data (Timmerman & Lorenzo-Seva, 2011). The description of this analysis and corresponding terminologies is beyond the scope of this chapter. Please refer to Timmerman and Lorenzo-Seva (2011) for details. As shown in Figure 6.5, only Factor 1 has the proportion of explained common variance from the empirical data larger than that from the randomly generated data. This result clearly points to a one-factor solution.

To evaluate the fit of the one-factor model, the value of the root mean square of residuals was also checked. The values of the root mean square residuals should be below 0.08, with lower values indicating better model fit (Hu & Bentler, 1999). The one-factor model has the value of the root mean square of residuals equal to 0.08, which suggests a reasonable fit. For estimates of factor loadings, please refer to Appendix A.

Results from the Rasch Analysis

The person separation reliability[5] was estimated as 0.81, and the expected a posteriori reliability was estimated as 0.82. This evidence suggests a reasonable reliability result for the test.

Item Fit Statistics

The weighted mean square statistic was used to evaluate the fit between the partial credit model and data on each item. The weighted mean square statistic has an expected value of one. Items with values of the weighted mean square statistic larger than one are those with observed variance greater than the expected and contribute less toward the overall estimation of the latent variables. Items that have a weighted mean square statistic outside the range of 0.75 and 1.33 and have the absolute value of the weighted t statistic larger than 1.96 should be considered problematic (Wilson, 2005; p.129). All of the investigated argumentation items have weighted mean square statistics within the range of 0.75 and 1.33, which provides support for the use of the partial credit model in data calibration. This result should not be surprising, as the 20 items used in this study had already been calibrated once before in an earlier round of data used for item selection and improvement. and found to be reasonably well-fitting in that earlier analysis.[6]

Item discrimination

Students who demonstrate a higher proficiency waypoint on the latent construct should also tend to score higher on each item than their counterparts—this is evidence for the discrimination of the item. To check this consistency between each single item and the whole instrument, Wilson (2005) recommended examining the mean ability of students within each score category for a given item. Specifically, the mean ability of students within each score category is expected to *increase* as the score increases. All of the studied argumentation items met this expectation, except for item A2_5d. Figure 6.3 presents the prompt for this item and Table 6.2 presents the information about each score category of this item. As shown in the Table, the average ability estimate increases as the score increases for item A2_5d, except for the category of score 2. For this item, categories of score 1 and score 2 are responses that earned partial credits. According to the scoring guide for this item, responses as described in the category of score 2 (i.e., recognize uncertainty but DO NOT provide reasoning) got a higher overall estimate than responses as described in the category of score 1 (i.e., provide reasoning but DO NOT recognize uncertainty), because recognizing uncertainty in evidence is considered to be more sophisticated than providing reasoning to back a claim on the hypothetical argumentation progress map. The empirical evidence, however, shows that the average ability of the students who scored 2 (n = 80) on this item is lower than the average ability of the students who scored 1 (n = 82). The reason why item A2_5d did not show as displaying a large misfit statistic is probably because there are only 80 students in the misbehaving response category (i.e., score 2). Nevertheless, the disordering in the average

Table 6.2 Information about each score category of Item A2_5d

Score	*Descriptions*	*Mean ability (logit)*	*Number of students*
3	Recognize uncertainty AND provide reasoning. *The student recognizes that the evidence is not conclusive and makes a case for why this uncertainty does not allow them to side with either Laura or Mary*	0.38	28
2 .	Recognize uncertainty but DO NOT provide reasoning. *The student recognizes that the evidence is not conclusive but does not make a case for why this uncertainty does not allow them to side with either Laura or Mary.*	0.05	80
1	Provide reasoning but DO NOT recognize uncertainty. *Here students, failing to recognize the uncertainty in the evidence, conclusively side with either Laura or Mary, but at least they provide a plausible reason why they chose to side with Laura or Mary.*	0.30	82
0	DO NOT provide reasoning/ DO NOT recognize uncertainty. *The student not only fails to see the uncertainty in the evidence, but they are unable to correctly reason for the claim (i.e., side with either Laura or Mary) in which they feel the evidence is indeed sufficient.*	–0.27	157

ability estimates of the students across response categories is indeed manifested in the weighted mean square statistic. Specifically, the weighted mean square statistic of item A2_5d is 1.21, which is close to the upper boundary of the acceptable range (i.e., 1.33). When the value of the weighted mean square statistic is larger than one, it suggests that the observed residual variance is greater than expected. This is the only argumentation item that assessed students' ability to recognize uncertainty in evidence. More items of this kind are needed in the future to investigate whether recognizing uncertainty is indeed more sophisticated than providing reasoning to back a claim as hypothesized.

The Wright Map

A Wright map displays the student ability distribution and item locations on the same logit scale. When we produced the full set of item locations for the Wright map for these items, including those related to the partially correct responses (e.g., partially correct at Waypoint 2), the pattern of results was not clear—the conclusion from this analysis was that these partially-correct categorizations were not as consistent across items as we would desire them to be. Hence, it was decided to focus attention on only the "full credit" categorizations. With this focus, we found that the pattern of results was much clearer.

Thus Figure 6.6 presents the Wright map based on the *full-credit response category* for each item. The first column on the left of Figure 6.6 shows values of the logit scale. The second column on the left represents the distribution of the estimated student ability from the most able at the top to the least able at the bottom of the map, with a mean equal to 0 (by convention) and variance equal to 0.39. Each "x" in this column represents approximately 2.1 students and the location of the "x"s (with respect to the logit scale) indicates the waypoint of their ability estimate. The third to the fifth columns present the item Thurstonian thresholds (Wilson 2005, 2023), which are arranged in columns 3, 4, and 5 depending on the construct waypoint an item step was designed to measure. The jth Thurstonian threshold represents the point at which the probability of scoring below j is equal to the probability of scoring j and above. If the ability estimate of a student and the full-credit Thurstone threshold of an item are located at the same position on the map, the probability of this student obtaining full credit on the item is 0.5.

Checking the Internal Structure Validity Evidence

Each item is hypothesized to measure one specific construct waypoint. Items measuring a lower construct waypoint are expected to have lower difficulty waypoints, compared to items measuring a higher construct waypoint. Ideally, item thresholds tapping the same construct waypoint will be located at similar positions on the Wright map. When the locations of the thresholds for a certain construct waypoint fall close together, we can specify this by a partitioning of the scale on the Wright map; this is called "banding." Once a set of "bands" representative of construct waypoints are drawn, the empirical ordering of these "bands" can then be compared to the theoretical ordering of the waypoints as hypothesized in the construct map. Examining Figure 6.6, we can readily see that the locations of the thresholds for a given construct waypoint do not fall closely together on the Wright map, and thus a set of "bands" representative of construct waypoints cannot readily be drawn in Figure 6.6.

Given that the empirical pattern of the locations of the item thresholds on the Wright map does not initially support the expected ordinal structure of the construct waypoints, one

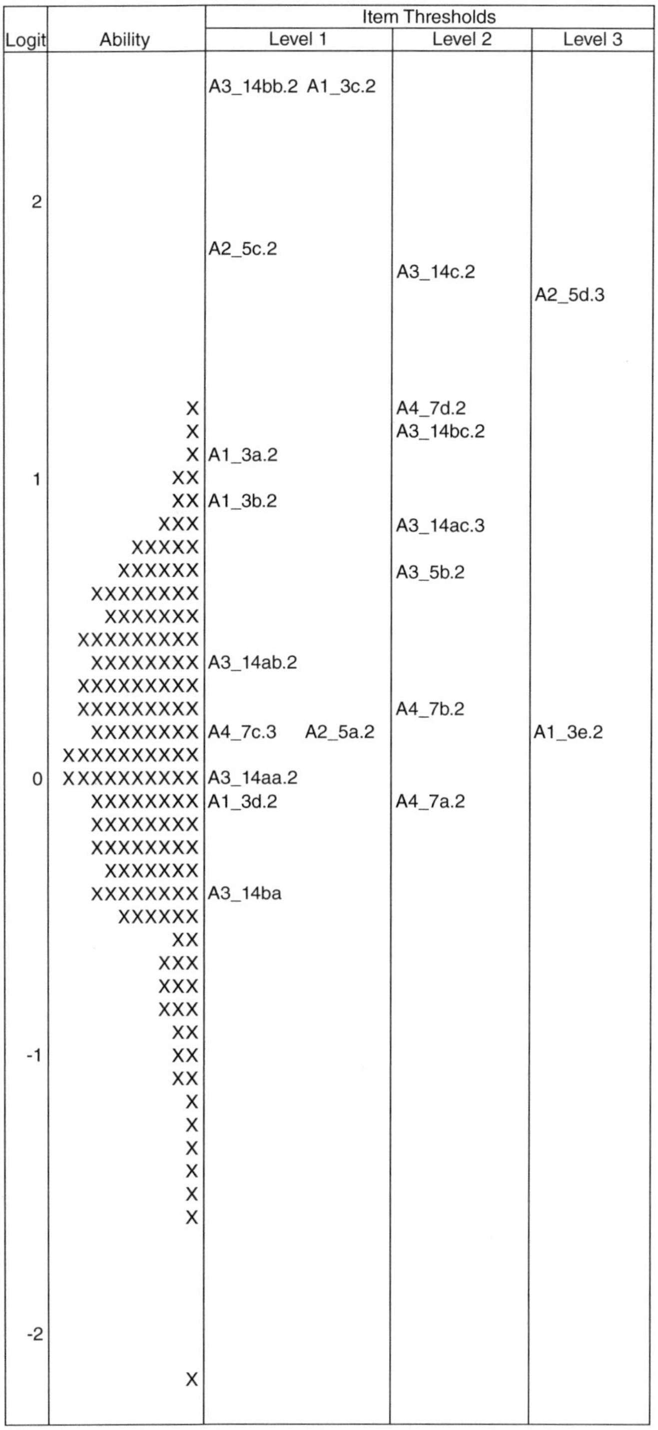

Figure 6.6 Wright map with item thresholds of the full-credit response categories.

plausible step is to identify items with threshold locations that do not conform to the expectation and investigate possible causes for their unexpected functions.

One reason for this lack of clear structure may be that some items did not perform as well as others on measuring the common latent argumentation competency, and thereby brought "noise" into the Wright map in Figure 6.6. Five of the items that do not conform to the banding are among those that have small factor loadings (please see Appendix A): items A1_3b, A3_14c, A3_14bb, A4_7a, and A4_7b. To focus on items that more clearly are centered on the argumentation construct, it was decided to drop these five items from the remaining analysis.

Figure 6.7 presents the banding of the Wright map based on the smaller set of items, where these five items with small factor loadings are excluded. Although not perfect, an ordinal structure among locations of item thresholds associated with different construct waypoints more readily emerges in Figure 6.7.

Waypoint-1 Items

First, all Waypoint-1 item thresholds fall in the range between –0.47 logit and 0.41 logit on the Wright map, except for the thresholds of the following three items: items A1_3c, A1_3a, and A2_5c. Getting full credit on these three items appears to be much more difficult than getting full credit on the other Waypoint-1 items. An examination of item A1_3c (see Figure 6.8) reveals a potential influence of language on the difficulty waypoint of the item. Bundle A1 establishes a scenario with three paragraphs: the first paragraph presents the issue of California being short of fresh water, and each of the remaining two paragraphs presents a method of desalination to obtain fresh water from the oceans. Hypothetically, items A1_3c and A1_3d (see Figure 6.8) are expected to have similar difficulty waypoints, since they ask students to identify an argument against each of the two desalination methods. However, item A1_3c appears to be much more difficult than item A1_3d in the empirical data. One possible reason is the ambiguity of the language. Item A1_3d explicitly specifies the desalination method students should identify an argument against, which helps students locate the critical information in the prompt. In contrast, item A1_3c does not clearly specify what students should look for.

An examination of item A1_3a provides another implication for the assessment of scientific reasoning. This item asks students to identify the main claim that is being made in the scenario of Bundle A1. According to the scoring guide, students received full credit only if their responses referred to both (a) the water shortage and (b) how the water shortage is to be addressed. 55% of the students mentioned only one of the two pieces in their responses and, thus, were awarded partial credit. These pieces of evidence suggest that identifying parts of a claim may be easy as hypothesized yet it is difficult to identify *all* pieces in each claim. This finding calls for reflection on two general assessment issues: (a) what waypoint of sophistication of student responses should be sufficient to be considered as demonstrating performance at a particular construct map waypoint, and (b) how item design can be improved to prompt students to demonstrate their understanding and proficiency waypoint to the fullest.

Osborne and colleagues (2016) have provided a discussion about why item A2_5c was surprisingly difficult. As shown in Figure 6.3, item A2_5c asks students to identify a piece of evidence that best supports the claim the hypothetical student Mary made and to provide an appropriate explanation. Item A2_5a is identical to item A2_5c, except that item A2_5a is about identifying evidence for the claim the other student Paul made. By design, these two items were both regarded as requiring Waypoint-1 competency, as they only ask for the identification of evidence supporting a given claim. However, the empirical evidence suggests that it is much more difficult to get full credit on item A2_5c than on item A2_5a. An examination of the

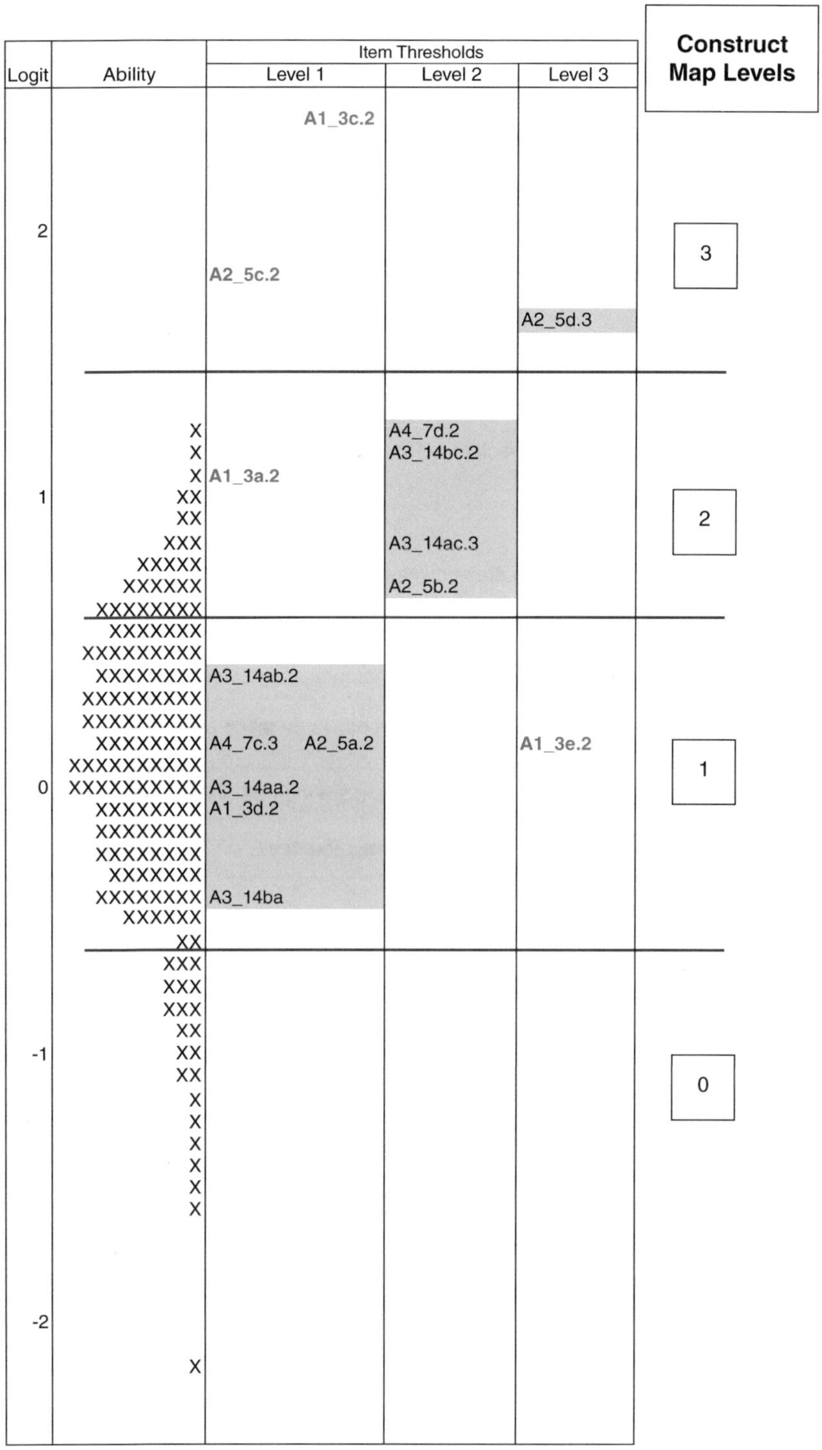

Figure 6.7 Edited version of the Wright map with item thresholds of the full-credit response categories.

Item A1. Desalinating Water
California is increasingly short of fresh water. Lack of fresh water limits the amount of food that can be grown. Clearly, there is a need to better manage this increasingly valuable resource. 97% of the world's water is in the oceans. The trouble is that it is salty and cannot be drunk. The process of removing salt from water is called *desalination.* One method of desalination is to use energy to heat seawater so that it evaporates leaving the unwanted salt behind. The water vapor is then passed over a cold object where it turns back to pure water. A great deal of energy is lost during this process. Another method involves pushing the seawater through a sheet with tiny holes in it called a membrane. Only the water particles pass through leaving all the salt on the other side. This uses half the energy, but the membranes get clogged and have to be cleaned regularly. All the sludge produced then has to be taken away.
3c. What is the main argument against the standard method of desalinating water? 3d. What is the main argument against using membranes?

Figure 6.8 Items A1_3c and A1_3d.

question shows that only one of the five pieces of evidence given in the prompt could support Paul's claim, while four pieces of evidence could support Mary's claim. Osborne and colleagues (2016, p.11) argued that "in the case of Paul, the question merely demanded an assessment of which single element was the most significant piece [of] evidence for his claim. However, in the case of Mary [,] the question demanded the assessment of the relative significance of multiple pieces of evidence in relation to her claim." As described previously, the competency for scientific argumentation is conceptualized as a complex orchestration of various elements of argument (e.g., claim, evidence, etc.) in the current construct map. Although item A2_5c and item A2_5a were initially designed to assess the same type of coordination (i.e., identifying evidence supporting a claim), item A2_5c imposes a higher waypoint of the intrinsic cognitive load as the pieces of information students need to process is larger.

Waypoint-2 Items

As to Waypoint-2 item thresholds, the four Waypoint-2 item thresholds cluster between the range of 0.70 logit and 1.30 logit and are located higher on the Wright map than most of the Waypoint-1 item thresholds. This pattern matches with the expectation that item thresholds targeting the same construct waypoint have similar difficulty waypoints and item thresholds of a higher construct waypoint should be located higher on the Wright map than those of a lower construct waypoint. Development of more Waypoint-2 items will be needed in the future to obtain more evidence about this pattern.

Waypoint-3 Items

For Waypoint 3, item A1_3e is surprisingly easy. Also embedded in Bundle A1, item A1_3e asks students to decide which desalination method is the best and explain their choice. It was hypothesized that this item assessed a student's ability to evaluate and compare competing arguments. However, according to the scoring guide, students get full credit if they successfully give a claim (i.e., choose one of the two desalination methods) and a reason provided in the prompt. Cognitively, succeeding on this item thus only requires the ability to identify a claim and a relevant piece of supporting evidence. We would therefore re-classify item A1_3e as Waypoint 1. This item is a good example of how the way an item is written may affect the cognitive demand it requires. Also, more Waypoint-3 items will be needed for a further examination of the "banding" and establishing the typical difficulty locations of the Waypoint-3 items.

Discussion of Results

The overall picture of alignment, following the investigation of the non-aligned items above, as shown in Figure 6.7 (on the far right-hand side) allows one to see a potential banding for the scientific argumentation construct. The banding was based on an application of the "construct mapping" standard-setting technique. (Draney & Wilson, 2011; Wilson & Draney, 2002). This is shown by the horizontal lines through the Wright map separating the scale into four segments, each labeled on the right-hand side by the relevant Waypoint of the construct map (Figure 6.2).

Bearing in mind that the location of a student at the same point as an item means that a student tends to succeed about 50% of the time, these bands can now be interpreted as indicating the regions of the scale where a student is learning at a certain Waypoint of the construct map. Thus, for example, a student at, say, 0.0 logits, is currently learning about claims and data (i.e., Waypoint 1 on the construct map), while a student at, say, 1.0 logits, is learning about arguments and rebuttals (i.e., Waypoint 2). In contrast, students at, say, –1.0 logits, are not yet at the point where they can consistently make a claim. And, if any student were located at 2.0 logits, then they would be working on articulating different arguments (but no students were consistently doing that!). This banding shows the way that the foundational effort to create the construct map waypoints can then be wound around the BAS to allow one to interpret the quantitative estimates on the logit scale.

In summary, we chose to use this example of how to calibrate a learning progression not because the results are clean and perfect, but as discussed above, because the results show the somewhat consistent, and somewhat inconsistent patterns that we typically find as we start to explore applications of this new paradigm. What one would need to do next is to *iterate* the procedures carried out above, with new items (as well as the old ones), and seek to confirm that the patterns of results can be replicated. One important part of this strategy is to examine the items that did not conform to the overall pattern (such as the "non-aligned" items mentioned above, as well as the five items that did not load well on the construct) and seek to understand what it was about the items that brought this about.

Reporting Results to Teachers

It is crucial for the success of the project that the assessment results be reported to teachers in ways that allow them to make efficient use of the information. The work to develop these is currently underway and has been partially reported in Morell et al., 2019; Morell et al., 2021; Wilson et al., 2019, and is currently under submission to appropriate journals.

Examples of the reports we have developed are shown and discussed in this subsection.

First, Figure 6.9 shows a display of results from a (fictitious) set of students—this is called the "Group Report." The waypoints of the construct are shown advancing from left to right across the page, from Waypoint 0 to Waypoint 3. Students are shown down the page, ordered from the lowest performers at the bottom to the highest at the top. Each student has a separate row, and his/her location is indicated by the black dot. Stretching to the left and right of the student are 67% confidence intervals for that student, expressed as the thin black bars. This report can give teachers several useful pieces of information, that is:

a the range of the students, in this case from Waypoint 0 ("ARG 0" in Figure 6.9) to Waypoint 3 ("ARG 3)";
b a quick indication of the relatively more frequent Waypoints in the class (in this case Waypoints 1 and 2);
c identification of the students at the top (in this case 3 students), and at the bottom (in this case 2 students);
d some indications of uncertainty about specific student locations (for example, note that the two students at the bottom are relatively wide, probably associated with some missing data).

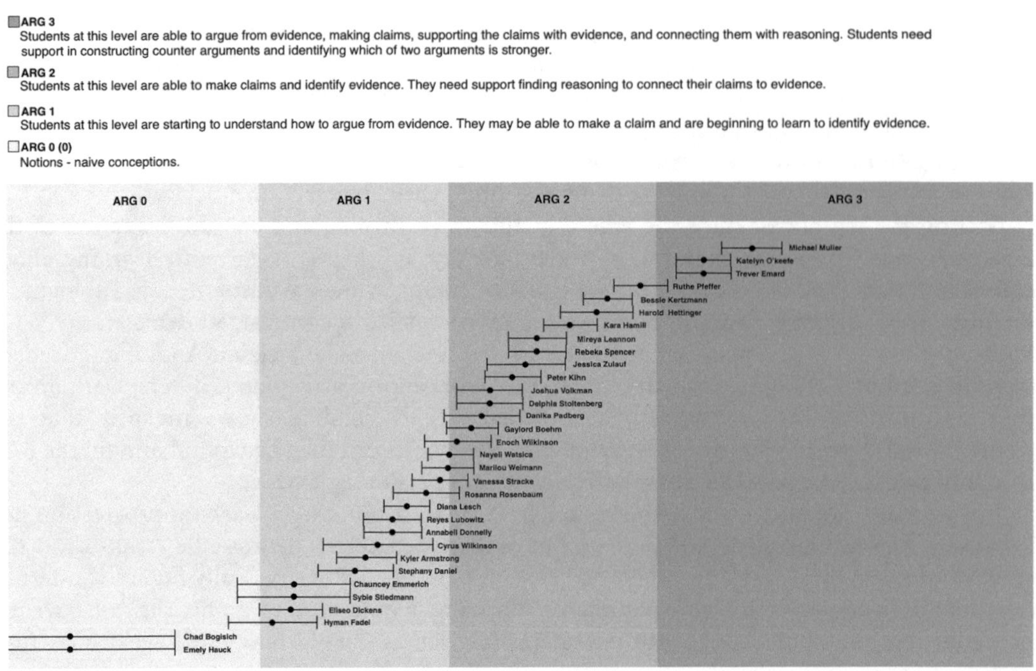

Figure 6.9 Graphic display of the locations of a class of students on the argumentation learning progression.

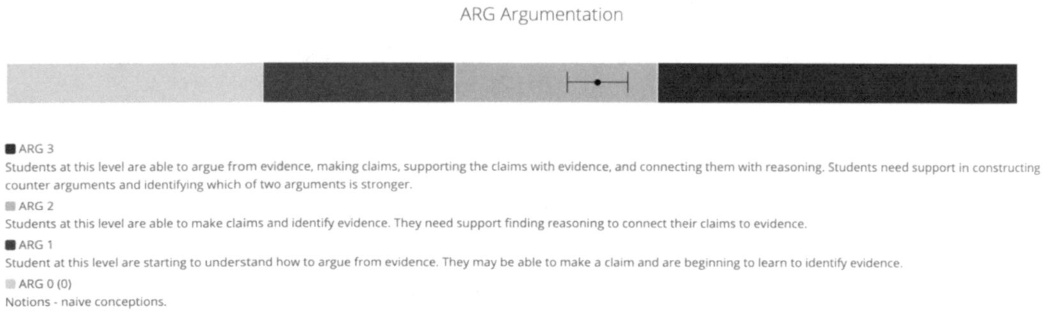

Figure 6.10 Graphic display of an individual student's location on the argumentation learning progression.

In our discussions with teachers, they have indicated that the Group Report (as shown in Figure 6.9) would be most useful in grouping students for subsequent instruction (Morell et al., 2021).

The information in this graph is also augmented by tables, which show all the above information in explicit detail.

Teachers can also see, for example, a report on an individual student, as shown in Figure 6.10—this is called the "Individual Report." Here the layout and graphical features are the same as for Figure 6.9, but the focus is on just one student. In this case, the student is performing solidly in Waypoint 2, with sufficient information that the student is beyond Waypoint 1 and not yet at Waypoint 3 (as indicated by the observation that the error bars do not overlap with those regions of the graph). Thus, a teacher observing this performance could interpret that the student was making reasonable progress in constructing an argument. Further information is also available in the "Scores Report," which shows each student's responses on each item that was attempted.

Conclusion

With the formal inclusion of argumentation as an essential practice in science classrooms in national standards (e.g., NGSS Lead States, 2013), how to evaluate a student's scientific argumentation proficiency becomes a necessity. And, without valid argumentation assessments, science educators will not be able to diagnose their students' learning needs readily nor provide appropriate instructional interventions. Consequently, argumentation may not be adequately incorporated into science classrooms as desired. This study presented the development of an assessment for a learning progression based on scientific argumentation for middle school students and illustrated the procedure of assessment validation. Specifically, (i) we used the scholarship of Osborne et al. (2016) as a basis for a progression as a sequence of waypoints of student sophistication in argumentation; (ii) we applied the BEAR Assessment System (BAS) to develop ways to assess the waypoints in this progression; (iii) we developed reports intended to help teachers use the information from the assessments for instructional design and also within the activities in a classroom; and (iv) we used both exploratory factor analysis and Rasch modeling approach to study whether the empirical information from students taking the assessments is consistent with the hypothesized progression. The process of developing assessments that align to a learning progression shaped as a set of construct maps, has been described in several other contexts. Recent examples illustrating this can be found in Dray et al. (2019), Junpeng et al. (2018), Metz et al. (2019), Morell et al. (2017), Siddiq et al. (2017), and Wilson (2018).

Findings of the Internal Structure Validity Evidence

In the final sections, we discuss the findings and challenges of the current research.

Comparing the empirical pattern of the item parameter estimates to the theoretical structure of the progression waypoints, the empirical pattern of item thresholds on the original Wright map did not support the expected structure of the learning progression waypoints. As Wilson (2005) has discussed, negative evidence for internal structure validity could originate from at least three sources: (1) the theory of the construct (i.e., learning progression in the current case) which was inaccurate in some way, (2) the items did not work as intended, or (3) the outcome space was not developed appropriately. In this study, we traced the sources of the problems through the examination of the result of exploratory factor analysis and through the content analysis of the items. Items that did not conform to the expected structure were found to either measure the argumentation construct poorly (i.e., have small factor loadings in the exploratory factor analysis) or measure nuisance variables in addition to the argumentation construct of interest.

Except for these erratic results in the initial development of items, the remaining items in this pilot study were found to have locations of thresholds that conformed to the expected pattern: (a) item thresholds of a lower progression waypoint have lower difficulty locations than item thresholds of a higher progression waypoint, and (b) item thresholds of the same progression waypoint located at similar positions on the Wright map. This we see as supporting the Osborne et al. learning progression.

Challenges of Developing Learning-Progression-Based Assessments

Challenge 1: Designing Items Capable of Eliciting Student Performance Indicative of Progression Waypoints

To accomplish the goal of diagnosing where a student stands in the path of a learning progression, it is fundamental to develop a set of items that are capable of eliciting student performance indicative of progression waypoints on the hypothetical learning progression. In this study, we attempted to assess a student's scientific argumentation competency with items in the paper and pencil format. We found that it was most challenging to develop items that could elicit student performance indicative of high-level competency. For example, item "A1_3e" was originally expected to assess a student's ability to evaluate and compare competing arguments, which was regarded as the Waypoint-3 performance. However, it was found that succeeding on this item requires only the ability to identify a claim and a relevant piece of supporting evidence, which was essentially only Waypoint-1 performance. That is, this item was not capable of differentiating students at Waypoint 3 from students at lower locations as expected and appeared as a surprisingly easy item in the analysis output. This example illustrates how the way an item is written may affect the cognitive demand it requires and shows the importance of evaluating students' cognitive or test-taking strategies in the process of responding to a specific item. Thus, although it is not too hard to write very demanding open-form items, it can be difficult to write items that assess the specific higher-order waypoints of a competency.

Challenge 2: Linking Student Performance on an item to a Specific Progression Waypoint

Wilson (2005) has suggested that "Although qualitative levels [on a learning progression] are definable, we assume that the respondents can be at any point in between—-that is, the underlying construct is continuous." (pp.6–7). Based on the hypothesis that a learning progression is essentially a latent continuum, variation in student performance *within* a learning progression waypoint should be expected. Following this idea, in this study student responses demonstrating partial understanding of a Waypoint, which could not be mapped to a specific lower progression Waypoint, were regarded as incomplete or imperfect performances at that Waypoint, and hence given an intermediate score. However, the Wright map results from this tactic were not conclusive. Instead, it was found that only the results from the full-credit scores gave a clear pattern of results. Our conclusion from that evidence is that we should not code out partial success at a Waypoint but stick to full credit in the development of future items.

Challenge 3: Handling Unexpected Variables that Affect Student Performance on an Item

As discussed previously, an assessment is never strictly unidimensional. While some variables, such as the use of language in item design, may be regarded as "nuisance" variables, other variables may systematically influence student success on items. For example, as discussed in detail above (in the section *Checking the internal structure validity evidence*), items "A2_5a" and "A2_5c" were both initially expected to assess Waypoint-1 competency. However, they differed significantly in terms of cognitive demand due to the specifics of the item design beyond the amount of detail provided in the description of Waypoint 1 of the Argumentation Construct Map. In this case, the item design issue had to do with the nature of the options in a multiple-choice item—most likely this would never rise to the importance of being a specific part of a Waypoint descriptor but is nevertheless important in understanding the relationship between the item design and its difficulty.

A Broader Perspective

Our argument has been that a rigorous and systematic effort is needed to investigate the alignment of a learning progression with empirical data from an assessment. The question remains, however, about whether the alignment illustrated in Figure 6.7 constitutes useful evidence for the learning progression hypothesized in Figure 6.2. Specifically, what we have found is that over half of the items in this pilot study have locations of thresholds that conform to the expected pattern: (a) item thresholds of a lower progression waypoint have lower difficulty locations than item thresholds of a higher progression waypoint, and (b) item thresholds of the same progression waypoint locate at similar positions on the Wright map. In fact, we see this finding as a critically important finding, pointing towards a confirmation of a learning progression in the area of scientific argumentation. This finding needs the usual scientific support in the shape of further data from new samples (which we are in the process of gathering; Morell et al., 2020; Wilson et al., 2019). We expect that there will be adjustments to the progression as we add new data and studies, but we are also expecting that the major outline of the learning progression will prove to be resistant to changes in assessment formats, such as differences between selected response and constructed response items (Wilson, Morell, Osborne, Dozier & Suksiri, 2019). As we move forward with this research, we will also engage directly with the teacher's use of the construct map and the assessment results (Morell et al., 2021), which will add to our understanding of the implications of the development of this learning progression and its value for improving the quality of information that we can provide about student learning and progression to teachers. Looking beyond its application in science education, it would be interesting to try the learning progression out in other educational areas: in particular, argumentation can be important in mathematics education, but it is not clear that the same structure will be generated by the sorts of argumentation tasks that mathematics teachers ask of their students. As Hattie has argued, high-quality feedback is one of the most effective tools for improving teaching and learning (Hattie and Timperley, 2007; Hattie, 2008). We see our work as a contribution to this goal.

Notes

1 It is close to the concept of a learning *trajectory* as is often used in mathematics education (Clements & Sarama, 2009).
2 Although the specifics of that instruction are usually not contained in the learning progression.
3 Note that the numbering of the waypoints in Figure 2 has been changed from that given in Osborne et al. (2016) to make the scoring clearer—the Waypoints "0a" etc. have been re-labeled "1a" etc., and similarly for Waypoints 1a through 1d and 2a through 2d.
4 This data was used as part of an earlier analysis that focused on the relationship between the argumentation dimension and the content dimension (Yao et al., 2015).
5 The person separation reliability is the equivalent of Cronbach's alpha, following the translation of the raw scores (used in Cronbach's alpha) into estimates on the logit scale, based on the Rasch model calibration (Masters, 1982).
6 The results of this earlier data collection and analysis were formative of the instrument, and hence were not published at that time.

References

Adams, R. J., Wu, M. L., Cloney, D., & Wilson, M. R. (2020). *ACER ConQuest: Generalised Item Response Modelling Software, Version 5 [Computer software]*. Australian Council for Educational Research.

American Educational Research Association, American Psychological Association, & National Council for Measurement in Education. (2014). *Standards for educational and psychological testing*. American Educational Research Association.

Banilower, E. R., Smith, P. S., Weiss, I. R., Malzahn, K. M., Campbell, K. M., & Weis, A. M. (2013). *Report of the 2012 National Survey of Science and Mathematics Education*. Horizon Research.

Black, P., Wilson, M., & Yao, S.-Y. (2011). Road maps for learning: A guide to the navigation of learning progressions. *Measurement: Interdisciplinary Research and Perspectives*, 9(2-3), 71–123.

Brown, N. J. S., & Wilson, M. (2011). A model of cognition: The missing cornerstone of assessment. *Educational Psychology Review*, *23*(2), 221–234.

California Department of Education. (2013). STAR CST blueprints. Retrieved from https://www.cde.ca.gov/re/pr/star13rfl.asp

Cattell, R. B. (1966). The scree test for the number of factors. *Multivariate Behavioral Research*, *1*(2), 245–276.

Cavagnetto, A., Hand, B. M., & Norton-Meier, L. (2010). The nature of elementary student science discourse in the context of the science writing heuristic approach. *International Journal of Science Education*, *32*, 427–449.

Center for Continuous Instructional Improvement (CCII). (2009). *Report of the CCII Panel on Learning Progressions in Science*. CPRE Research Report, Columbia University.

Chandler, P., & Sweller, J. (1991). Cognitive load theory and the format of instruction. *Cognition and Instruction*, *8*(4), 293–332.

Clements, D. H., & Sarama, J. (2009). *Learning and Teaching Early Math: The Learning Trajectories Approach*. Routledge. doi: 10.4324/9780203883389

Draney, K., & Wilson, M. (2011). Understanding Rasch measurement: Selecting cut scores with a composite of item types: The Construct Mapping procedure. *Journal of Applied Measurement*, *12*(3), 298–309.

Dray, A. J., Brown, N. J. S., Diakow, R., Lee, Y., & Wilson, M. (2019). A construct modeling approach to the assessment of reading comprehension for adolescent readers. *Reading Psychology*, *40*(2), 191–241. DOI: 10.1080/02702711.2019.1614125

Driver, R., Newton, P., & Osborne, J. (2000). Establishing the norms of scientific argumentation in classrooms. *Science Education, 84*, 287–312.

Duschl, R. A., & Osborne, J. (2002). Supporting and promoting argumentation discourse in science education. *Studies in Science Education, 38*(1), 39–72.

Ford, M. (2008). Disciplinary authority and accountability in scientific practice and learning. *Science Education, 92*(3), 404–423. doi:10.1002/sce.20263

Hattie, J. (2008). *Visible Learning: A Synthesis of over 800 Meta-Analyses Relating to Achievement*. Routledge.

Hattie, John, & Timperley, H. (2007). The power of feedback. *Review of Educational Research, 77*(1), 81–112. 10.3102/003465430298487. http://rer.sagepub.com/cgi/content/abstract/77/1/81.

Henderson, J.B., Osborne, J., MacPherson, A., & Szu, E. (2014). A new learning progression for student argumentation in scientific contexts. In C. P. Constantinou, N. Papadouris & A. Hadjigeorgiou (Eds.), *E-Book Proceedings of the ESERA 2013 Conference: Science Education Research for Evidence-based Teaching and Coherence in Learning: Part 7* (co-ed. M. Evagorou & K. Iordanou) (pp. 26–42). European Science Education Research Association.

Hu, L., & Bentler, P. M. (1999). Cutoff criteria for fit indexes in covariance structure analysis: Conventional criteria versus new alternatives. *Structural Equation Modeling: A Multidisciplinary Journal, 6*(1), 1–55.

Jimenez-Aleixandre, M. P., Rodriguez, A. B., & Duschl, R. A. (2000). "Doing the lesson" or "doing science": Argument in high school genetics. *Science Education, 84*, 757–792.

Junpeng, P., Inprasitha, M., & Wilson, M. (2018). Modeling of the open-ended items for assessing multiple proficiencies in mathematical problem solving. *The Turkish Online Journal of Educational Technology*, Special Issue for INTE-ITICAM-IDEC 2018 (Volume 2), 142–149.

Lorenzo-Seva, U., & Ferrando, P. J. (2006). FACTOR: A computer program to fit the exploratory factor analysis model. *Behavior Research Methods, 38*(1), 88–91.

Masters, G. N. (1982). A Rasch model for partial credit scoring. *Psychometrika, 47*(2), 149–174.

Mercer, N., Dawes, L., Wegerif, R., & Sams, C. (2004). Reasoning as a scientist: Ways of helping children to use language to learn science. *British Educational Research Journal, 30*(3), 359–377.

Metz, K. E., Cardace, A., Berson, E., Ly, U., Wong, N., Sisk-Hilton, S., Metz, S. E., & Wilson, M. (2019). Primary grade children's capacity to understand microevolution: The power of leveraging their fruitful intuitions and engagement in scientific practices. *Journal of the Learning Sciences, 28*(4-5), 556–615. DOI: 10.1080/10508406.2019.1667806

Miller, G. A. (1956). The magical number seven, plus or minus two: Some limits on our capacity for processing information. *Psychological Review, 63*(2), 81–97. 10.1037/h0043158

Morell, L., Collier, T., Black, P., & Wilson, M. (2017). A Construct-Modeling approach to develop a learning progression of how students understand the structure of matter. *Journal of Research in Science Teaching, 54*(8), 1024–1048. doi:10.1002/tea.21397.

Morell, L., Suksiri, W., Dozier, S., Osborne, J., & Wilson, M. (2019, January). *Addressing the NGSS practice of Arguing from Evidence using forced-choice item formats: Challenges and successes*. Paper presented at the IES Annual Principal Investigators Meeting, Washington, DC.

Morell, L. Dozier, S., Suksiri, W., Osborne, J., & Wilson, M. (2021, Feb). *Gaining insight into teachers' interpretations of computer-based feedback for the purpose of valid inferences*. Paper presented at the IOMW 2020 Virtual Conference.

Morell, L., Suksiri, W., Dozier, S., Osborne, J., & Wilson, M. (2020, September*). An exploration of selected-response items compared to constructed-response item types in science education*. Presented at the National Council on Measurement in Education (NCME), online Annual Meeting.

National Research Council. (2012). *A framework for K-12 science education: Practices, crosscutting concepts, and core ideas*. The National Academies Press.

Newton, P., Driver, R., & Osborne, J. F. (1999). The place of argumentation in the pedagogy of school science. *International Journal of Science Education, 21*(5), 553–576.

NGSS Lead States. (2013). *Next generation science standards: For states, by states*. The National Academies Press.

Osborne, J., Erduran, S., & Simon, S. (2004). Enhancing the quality of argumentation in school science. *Journal of Research in Science Teaching, 41*, 994–1020.

Osborne, J. F., Henderson, J. B., MacPherson, A., Szu, E., Wild, A., & Yao, S. Y. (2016). The development and validation of a learning progression for argumentation in science. *Journal of Research in Science Teaching, 53*(6), 821–846. 10.1002/tea.21316

Paas, F., Renkl, A., & Sweller, J. (2003). Cognitive load theory and instructional design: Recent developments. *Educational Psychologist, 38*(1), 1–4.

Pollock, E., Chandler, P., & Sweller, J. (2002). Assimilating complex information. *Learning and Instruction*, *12*(1), 61–86.

Rasch, G. (1960). *Probabilistic models for some intelligence and attainment tests*. Danmarks Paedagogiske Institut.

Rosenbaum, P. R. (1988). Item bundles. *Psychometrika*, *53*(3), 349–359.

Schwarz, B. B., Neuman, Y., & Biezuner, S. (2000). Two wrongs may make a right ... If they argue together! *Cognition and Instruction*, *18*(4), 461–494.

Siddiq, F., Gochyyev, P., & Wilson, M. (2017). Learning in digital networks—ICT literacy: A novel assessment of students' 21st century skills. *Computers & Education*, *109*, 11–37.

Sweller, J. (1994). Cognitive load theory, learning difficulty, and instructional design. *Learning and Instruction*, *4*, 295–312. 10.1016/0959-4752(94)90003-5.

Timmerman, M. E., & Lorenzo-Seva, U. (2011). Dimensionality assessment of ordered polytomous items with parallel analysis. *Psychological Methods*, *16*(2), 209–220.

Toulmin, S. (1958). *The uses of argument*. Cambridge University Press. Retrieved from http://www.dlpdfs.com/pdf03/uses%20of%20argument.pdf

Venville, G. J., & Dawson, V. M. (2010). The impact of a classroom intervention on grade 10 students' argumentation skills, informal reasoning, and conceptual understanding of science. *Journal of Research in Science Teaching*, *47*(8), 952–977.

Von Aufschnaiter, C., Erduran, S., Osborne, J., & Simon, S. (2008). Arguing to learn and learning to argue: Case studies of how students' argumentation relates to their scientific knowledge. *Journal of Research in Science Teaching*, *45*, 101–131.

Weiss, I.R., Pasley, J. D, Smith, P. S., Banilower, E. R., & Heck, D. J. (2003). *A Study of K–12 Mathematics and Science Education in the United States*. Horizon Research.

Wilson, M. (2004). A perspective on current trends in assessment and accountability: Degrees of coherence. In, M. Wilson, (Ed.). *Towards coherence between classroom assessment and accountability. 103rd Yearbook of the National Society for the Study of Education, Part II* (pp. 272–283). University of Chicago Press.

Wilson, M. (2005). *Constructing measures: An item response modeling approach*. Lawrence Erlbaum Associates, Inc.

Wilson, M. (2009). Measuring progressions: Assessment structures underlying a learning progression. *Journal of Research in Science Teaching*, *46*(6), 716–730.

Wilson, M. (2012). Responding to a challenge that learning progressions pose to measurement practice: Hypothesized links between dimensions of the outcome progression. In A. C. Alonzo & A. W. Gotwals (Eds.), *Learning progressions in science: Current challenges and future directions* (pp. 317–343). Sense Publishers.

Wilson, M. (2018). Making measurement important for education: The crucial role of classroom assessment. *Educational Measurement: Issues and Practice*, *37*(1), 5–20.

Wilson, M. (2020, April). *Promises and perils for classroom assessments in the digital era*. E. F. Lindquist Award presentation, AERA Annual Meeting San Francisco, CA http://tinyurl.com/wnxmsxl (Conference Canceled).

Wilson, M. (2023). *Constructing measures: An item response modeling approach*, Second Edition. New York: Routledge.

Wilson, M., & Draney, K. (2002). A technique for setting standards and maintaining them over time. In S. Nishisato, Y. Baba, H. Bozdogan, & K. Kanefugi (Eds.), *Measurement and multivariate analysis* (Proceedings of the International Conference on Measurement and Multivariate Analysis, Banff, Canada, May 12-14, 2000), pp 325–332. Springer-Verlag.

Wilson, M., Morell, L., Osborne, J., Dozier, S., & Suksiri, W. (2019, April). *Assessing higher order reasoning using technology-enhanced selected response item types in the context of science*. Paper presented at the 2019 Annual Meeting of the National Council on Measurement in Education in Toronto, Ontario, Canada.

Yao, S.Y. (2013). Investigating the validity of a scientific argumentation assessment using psychometric methods. Doctoral dissertation, UC Berkeley.

Yao, S-Y., Wilson, M., Henderson, J. B., & Osborne, J. (2015). Investigating the function of content and argumentation items in a science test: A multidimensional approach. *Journal of Applied Measurement*, *16*(2), 171–192.

Zohar, A., & Nemet, F. (2002). Fostering students' knowledge and argumentation skills through dilemmas in human genetics. *Journal of Research in Science Teaching*, *39*(1), 35–62.

Appendix A: Pattern of the Factor Loadings

Table A1 below presents the pattern of factor loadings for the 20 argumentation items. As a common criterion, 0.3 is used to determine the strength of the relationship between an item and a factor. To make the factor loading pattern more readable and interpretable, factor loadings with absolute values lower than 0.3 are reported in gray scale. As shown in the Table, all of the argumentation items load positively on the factor as desired. Seven out of the 20 argumentation items have loadings smaller than 0.3.

Table A1 Factor loadings obtained from the one-factor model

Variable	*Factor loading*
A1_3a	0.23
A1_3b	0.28
A1_3c	**0.39**
A1_3d	**0.53**
A1_3e	**0.42**
A2_5a	**0.59**
A2_5b	**0.50**
A2_5c	0.29
A2_5d	0.22
A3_14aa	**0.64**
A3_14ab	**0.49**
A3_14ac	**0.54**
A3_14ba	**0.41**
A3_14bb	0.13
A3_14bc	**0.54**
A3_14c	0.28
A4_7a	**0.30**
A4_7b	0.28
A4_7c	**0.52**
A4_7d	**0.54**

7

LEARNING PROGRESSION APPROACHES USED IN GERMANY

Knut Neumann

As large-scale assessments repeatedly showed that German students' achievement was not competitive with achievement internationally, Germany initiated a major transformation of its education system: the introduction of National Education Standards (Neumann et al., 2010). Historically, the German education system had rested on a description of the contents to be taught, teacher education focused on content knowledge and a particular freedom of teachers in designing their instruction. The National Education Standards introduced mandatory educational goals that could be assessed through standardized tests and that teachers, schools, and administration could be held accountable against (Klieme et al., 2003).

Standards were to be developed for each subject (e.g., biology, chemistry, and physics) and educational level: primary level (grades 1–4), lower secondary level (grades 5–10), and upper secondary level (11–12). For each subject and educational level, the standards were to articulate the competencies that students are expected to have developed by the end of the respective level; each of these competencies reflects the integration of the different knowledge, skills, and abilities required to identify and solve problems typical of the respective domain (Weinert, 2001). For example: "Students describe phenomena and link them back to known principles in physics" (Sekretariat der Ständigen Konferenz der Kultusminister der Länder in der Bundesrepublik Deutschland [KMK], 2005c, p. 11). Since the competencies reflect complex patterns of students' resources (i.e., the integration of a range of different knowledge, skills, and abilities), their assessment required further delineation (Mislevy et al., 2002).

To delineate competencies into different students' resources, researchers proposed so-called competency models. A competency model identifies the knowledge, skills, and abilities that constitute a competency or area of competency in a particular domain (e.g., competency in the planning, performance, and analysis of investigations, Neumann et al., 2019). After an extended period of research on the empirical validation of these models, researchers eventually turned to the development of competencies. Following the same iterative procedure, models of competency were refined into models explicitly accounting for how students are expected to progress in developing the competencies in question. These models of competency development were then validated in large-scale cross-sectional studies with students from different grades and grade bands (e.g., Neumann et al., 2013). The models were supposed to help break down the standards to performances expected from students at each

 DOI: 10.4324/9781003170785-8

grade to guide the design of more coherent instruction; that is, instruction that systematically supports students in their progression towards building the envisioned competence.

The present chapter will continue with a brief outline of the German education system and the state of science education in Germany to provide the background for discussing the German approach to learning progressions; that is, models of competence development. Drawing on different examples of scholarly work on competency development models in science, the chapter will highlight the characteristic features of the German approach. The chapter will conclude by discussing the contribution this work can make to the international debate about learning progressions: namely, careful delineation of how students develop competence in a domain into respective models; validation of these models in large-scale studies involving students from different grades and grade bands; and the use of specifically designed assessments and elaborated statistical procedures in the validation process.

Science Education in Germany

The German Education System has traditionally focused on content. Even at the dawn of the 21st century, the goals of science education were essentially long lists of content students were expected to know. Teacher education was based on a strong education in the discipline. Even today, the majority of the credits in teacher education programs are to be earned in the disciplines (e.g., 120 credits in physics vs. 30 credits in physics education in the teacher education program at the University of Kiel) (Neumann et al., 2017). Correspondingly, video studies revealed a teacher-centered, direct teaching approach which puts introducing and working with new content at the forefront (Seidel et al., 2007).

History of the German Education System

The strongly academic perspective on teaching and teacher education is deeply rooted in the origins of the German education system. As compulsory schooling was introduced in Germany, in parallel to the *Volksschule* (people's school) a second school type emerged, the Gymnasium. The Volksschule provided the average citizen with development in the most basic skills: reading, writing, and some basic mathematics. The Gymnasium, on the contrary, was dedicated to preparing the future academic elite (for details see Neumann et al., 2010; Neumann et al., 2017). Both types of schools still exist today. The Volksschule, now often referred to as *Gemeinschaftsschule* (comprehensive school), is a track open to all students, and the Gymnasium is the track for students apt to more academic learning.

Teachers for the Volksschule were originally recruited from people with expertise in reading, writing, or basic mathematics and educated in a master-apprenticeship approach. Gymnasium teachers were typically recruited from university faculty trained in a respective discipline, who were often already working as private teachers of that discipline. Over time, teacher education developed into a two-phase system effectively combining both approaches. That is, in the first stage (future) teachers are going through a five-year university program with a strong disciplinary focus. As part of the program, they take classes in the discipline (e.g., physics), education in the discipline (e.g., physics education), and pedagogy. In many cases, teacher students are taking the same classes as students of the discipline (e.g., students majoring in physics). In the second stage, after completing the university program, (future) teachers go through a two-year in-service training program. In this program, they train in the planning, implementation, and reflection of instruction. During the program, they are mentored by an experienced teacher in a master-apprentice style relationship.

The intensive education in the discipline and solid practical training of teachers came along with strong confidence in their professional competence and respective freedom for teachers in the planning of, implementing of, and reflecting on their instruction. That is, there were no standards or other specifications of what was expected from students, but a *Lehrplan* defining what aspects of a subject had to be covered. Each subject and level of education (i.e., primary, lower secondary, and upper secondary level) had its own Lehrplan. For Science, there was one Lehrplan at the primary level. However, since science is taught as biology, chemistry, and physics at the secondary level, there was one Lehrplan each for lower and upper secondary levels. Each Lehrplan typically included a general pre-amble about the role of science education (e.g., preparing students to understand the world of science), instructional principles (e.g., active learning), and the actual content to be covered (e.g., knowledge of Ohm's Law) at the respective level often broken down by grades or pairs of grades (e.g., grades 5/6, 7/8, and 9/10 for the lower secondary level).

Goals of Science Education in Germany

In Germany, the goals of science education were historically defined by those teaching it. The Lehrplan was written by selected teachers and teachers were largely free to decide what specifically students were to learn about the contents specified in the Lehrplan. As in the early days of science education, teachers were university scholars, and what was taught and how it was taught reflected to a great extent the organization of the respective disciplines at the university level. School physics was, for example, organized into areas like mechanics, electricity, optics, or thermodynamics. The content covered was often simply reflecting the content at the university level – reduced in terms of scope and complexity. In fact, the *Elementarisierung* (elementarization) of disciplinary content (i.e., how to reduce disciplinary content in scope and complexity appropriate for a particular grade) was a standard procedure for teachers to learn during their professional training as a preparation for planning their own instruction.

When German students repeatedly showed mediocre science performance in international large-scale assessments (Beaton et al., 1996; Organisation for Economic Co-operation and Development [OECD], 2001), the strong focus on content was highlighted as one reason. Students were considered to not move beyond the knowledge of isolated facts because instruction did not provide enough opportunities to engage in the practices of science and learn about the nature of science (Prenzel et al., 2001). Another reason discussed in the literature was a lack of coherent instruction (Baumert et al., 1997). The findings resulted in an intense discussion about the goals of science education and their implementation in instruction. This discussion eventually led to a fundamental reform of (science) education: the introduction of National Education Standards for science (as well as other) subjects. For the end of primary, lower secondary, and upper secondary levels, the Standards were supposed to define common goals of education across states. Subsequently, the states were to develop so-called Kerncurricula (core curricula) that would break down the goals for each level into goals for individual grades or pairs of grades within that level (Klieme et al., 2003).

The vision of students developing *Naturwissenschaftliche Grundbildung* served as a conceptual foundation for the Standards. This vision derives from Bybee's (1997) idea of scientific literacy, which has also served as the foundation for the PISA Framework (Organisation for Economic Co-operation and Development [OECD], 2006). Naturwissenschaftliche Grundbildung refers to the capacity to use scientific knowledge to make sense of phenomena and solve problems, to understand and be able to engage in the practice of scientific inquiry, to communicate scientific results, and to understand and assess science-and technology-related issues and decisions. Such Naturwissenschaftliche Grundbildung is supposed to help students act as informed citizens and prepare them for occupational training (e.g., KMK, 2005c). As in PISA, the Standards operationalize the vision of Naturwissenschaftliche Grundbildung into selected competencies that students are expected to develop as a function of science education. Subsequent benchmarking was supposed to provide guidance about the level of proficiency that students should be expected to reach in the respective competencies.

Standards for Science Education

The first set of Standards for Science Education was published in 2005 for the lower secondary level (KMK, 2005a, 2005b, 2005c). Although the Standards for Biology, Chemistry, and Physics were published separately, they all build on the same vision and follow the same structure. A preamble outlines the contribution of the respective subject to students' *Naturwissenschaftliche Grundbildung*. Then, four areas in which students are expected to develop competency by the end of the lower secondary level are described: *content knowledge* (knowledge of phenomena, concepts, principles, facts, and laws and the ability to use this knowledge), *scientific inquiry* (using models as well as experimental and other methods of investigation), *communication* (communicating information in accordance with the content and audience), and *evaluation* (recognizing and evaluating relevant situations in different contexts based on scientific facts and principles). Finally, for each area of competency, between four and ten standards are listed to characterize the competency the average student is expected to hold in the respective area.

Table 7.1 shows two examples of standards for each area of competency. Note that, regarding content knowledge, students are not just expected to hold a broad range of knowledge, but are expected to develop a profound, well-structured knowledge organized around basic (i.e., core) concepts. According to Liu (2009), organizing standards around core concepts means "that the same ... concept may be learned again and again at different grades with increasing complexity and thus increasingly higher expectations of proficiency" (p. 8). In the context of the Standards, the basic concepts serve two functions. First, basic concepts are supposed to help teachers move away from teaching a broad range of content, instead focusing on the most important concepts of the domain. Second, by organizing science education around basic concepts, students are supposed to be supported in the development

Table 7.1 Example competencies from the German science education standards for Physics

Area of competency	*Standards* *Students ...*
Content knowledge	• ... possess a profound, well-structured knowledge based on the basic concepts. • ... use analogies to solve tasks and problems.
Scientific inquiry	• ... formulate hypotheses for ordinary examples. • ... plan, perform, and analyze ordinary experiments.
Communication	• ... research a topic using different sources. • ... present results of their work group related to a target problem.
Evaluation	• ... compare and judge alternative technical solutions considering physical, economic, social, and ecological aspects. • ... name implications of physics-related findings with respect to historical and social contexts.

of knowledge that goes beyond the knowledge of isolated facts. For Biology the basic concepts are *system*, *structure and function*, and *evolution;* for Chemistry, *matter-particle relations*, *structure-property relations*, *chemical reactions*, and *energy*; for Physics, *matter*, *interaction*, *system*, and *energy*.

Other than the Next Generation Science Standards (NGSS), the National Education Standards for Biology, Chemistry, and Physics do not derive from an underlying framework. Instead for each area of competency, the competency an average student is expected to develop is characterized in broad strokes by a set of somewhat general standards. Hence, to delineate the competency students are expected to develop in the different areas, researchers have begun to develop models that (like the Framework of K-12 Science Education) identify the knowledge, skills, and abilities that constitute the respective competency. These models are commonly referred to as models of competency or competency models. The process is referred to as competency modeling.

Models of Competency in the Sciences

In the aftermath of the publication of the Standards, a variety of competence models has been developed in biology (e.g., Mayer et al., 2008), chemistry (Bernholt & Parchmann, 2011), and physics (Kauertz & Fischer, 2006; Schecker & Parchmann, 2006). These models organize the knowledge, skills, and abilities that constitute a competency into areas and levels of competency. Areas of competency are distinct aspects of a competency such as the planning, performance, and analysis of experiments. Levels of competency describe different levels of student proficiency with respect to the areas of competency. As such, competency models are the missing link between the abstract standards and benchmarks of the standards. A competency model makes it possible to determine the extent to which a student has mastered the competency in question. Models of competency therefore help in the difficult task of assessing standards.

The development of a competency model often involved extensive efforts to empirically validate it. This work was strongly influenced by scholarly work on modeling scientific literacy in the scope of TIMSS and PISA, in particular, work focused on theoretically justifying and empirically validating proficiency levels with respect to scientific literacy (Klieme et al., 2000). Validation of a model included developing items targeting each combination of dimensions, grouping the resulting (large number of) items into different booklets, administering them to a larger number of students, and using psychometric methods to link student responses back to the model (for an overview, see, Leutner et al., 2017).

Several researchers were successful in validating a model for the competency area of content knowledge. More specifically, these researchers were able to provide evidence that competency levels can be described based on the complexity of the knowledge that needs to be applied in a given situation (Bernholt & Parchmann, 2011; Kauertz & Fischer, 2006). Kauertz and Fischer (2006), for example, provide evidence that a system of six hierarchical levels of complexity from factual knowledge to conceptual understanding successfully describes students' increasing competency in using their content knowledge. As this model was the first in which the theoretically hypothesized levels could be empirically validated, it served as a point of departure into the development of a competency model used to benchmark and subsequently adjust the German National Education Standards for science subjects in the lower secondary level (i.e., grades 5 to 10).

Interestingly, most competency models, albeit formulating hierarchical levels of competence, do not take a developmental perspective (Leutner, 2017). That is, the postulated levels describe increased competency, but do not necessarily reflect students' progression in developing such competency. Hence after an initial phase of research for the development and validation of competency, researchers turned to how students develop the competencies in question. To do so, researchers either revised existing models to incorporate a developmental perspective and validated it, or they investigated how students' competency (structure) changes; both to obtain models of competency development that can guide the process of breaking down the standards into goals for individual grades.

Learning Progression Approaches in Germany

Models of competency development represent the German approach to learning progressions. Since models of competency development were largely based on previously developed models of competency, many features of the development process carried over as well. This, first and foremost, includes the focus on either concept or practice-related competencies and the delineation of these competencies into complex, multi-dimensional models. Next, the German Standards specify only the level of competency expected from students by the end of an educational level and provide no guidance on how to support students in developing the competencies. Hence, in the delineation and validation of competency development models, a long-term perspective was taken focusing on students' progression across grades and grade bands. At last, in the validation process, a particular emphasis was put on the model-based creation of assessment items and the use of tailored psychometric models to link the empirical data obtained through these assessments back to the model, and hence to align standards and assessment. These features will be discussed in greater detail in the following sections.

Modeling Competency Development

A first model of competency development (or: learning progression) was proposed by Neumann et al. (2013) for the concept of energy. The model extended the competency model used for benchmarking the Standards (Walpuski et al., 2008). This model used for benchmarking the Standards describes different levels of students' competence regarding their ability to use their knowledge about the four basic concepts of physics (energy, matter, interactions, and systems). Drawing on theories of knowledge organization (Bernholt & Parchmann, 2011; Clark & Linn, 2013; Kauertz & Fischer, 2006), the model identifies five hierarchical levels of increasingly complex knowledge: facts, multiple facts, relationships, connected relationships, and conceptual understanding. However, students do not progress in their understanding of energy by learning all the facts at the primary level, establishing connections at the lower secondary level, and reaching conceptual understanding only at the end of the upper secondary level. Instead, conceptual change theory suggests that students can develop conceptual understanding very early (which may, however, not always correspond with the scientific understanding, Posner et al., 1982). Hence, Neumann et al. (2013) extended the model originally developed for benchmarking the standards to obtain a model more accurately describing students' progression in developing competency in using their knowledge about energy to explain phenomena or solve problems spanning the lower (and potentially upper secondary) level.

Integrating theories of knowledge organization with findings from research on students' (conceptual) understanding of energy (for an overview see Driver et al., 1994; Liu & McKeough, 2005), Neumann et al. (2013) identified two dimensions in which students would make progress in developing competency in using their knowledge about energy: aspects of energy and complexity of knowledge about these aspects. The authors hypothesized that this progression would be marked by students incorporating new knowledge about energy into their conceptual ecology: starting from everyday understandings, students would acquire new knowledge about how energy manifests itself in different forms, effectively moving towards an understanding of energy forms. In the same way, by integrating new knowledge about further aspects of energy (i.e., energy transformation, dissipation, and conservation) into their knowledge base, students are expected to continue developing an understanding of how these forms are transformed into one another; how during each transformation some energy is converted into thermal energy, which is dissipated even though overall the total energy of an (isolated) system is conserved (see also the chapter by Fortus & Nordine, this volume).

Many other researchers took a similar approach to modeling the development of competency. Combining a review of the literature on students' conceptual understanding of matter with theories of knowledge organization, Hadenfeldt et al. (2016), for example, proposed a model of competence development for the matter concept spanning grades 5 to 12. The model differentiates between four aspects of matter – structure and composition, physical properties and changes, chemical properties and change, chemical reaction, and conservation – and describes five levels describing the complexity of the knowledge about each aspect from everyday knowledge to knowledge about the relation of the submicroscopic structure of matter with its observable, macroscopic properties. Based on an in-depth domain analysis, Emden et al. (2018) proposed a more detailed version of this model for grades 7 to 9 in the form of a strand map that identifies two progress variables – Structure of Matter and Chemical Reactions – and three interdependent stages of conceptual development for each progress-variable – everyday understanding, Daltonian model, to the refined electron shell model of Bohr. For evolution, Tibell and Harms (2017) proposed a two-dimensional model that links students' progression in understanding the concept of evolution to so-called threshold concepts. Tibell and Harms (2017) argue that understanding evolution through natural selection requires mastery of so-called threshold concepts; that is, understanding the different threshold concepts (i.e., randomness, probability, spatial scale, and temporal scale) is a pre-requisite for building an understanding of key concepts about evolution (i.e., selection, inheritance, and variation).

Others have taken existing models of competency as a starting point to establish a learning progression (or model of competence development). One such example is the work of Upmeier zu Belzen and Krüger (2010) who have proposed a model of students' competence in developing and using models in science (more specifically: biology). The model distinguishes five sub-competencies related to different aspects of models or modeling: properties of models, alternative models, using models, testing models, and revising models. For each sub-competency, three different levels of proficiency are described. For example, for the use of models these are: using a model to i) describe something, ii) explain connections or correlation between variables, and iii) predict new insights based on connections or correlations between variables (Patzke et al., 2015). Eggert and Bögeholz (2006), in a similar approach, described a model of students' competency in (socio-)scientific reasoning and decision-making. The model differentiates four sub-competencies: i) generating and reflecting factual information; ii) evaluating, deciding, and reflecting; iii) knowing and understanding values and norms; and iv) knowing and understanding sustainable development. For each sub-competency, the authors describe a series of proficiency levels. For 'evaluating [information], deciding and reflecting' the authors, for example, propose a set of four proficiency levels, with the lowest level characterized by a focus on select criteria and intuitive decision-making strategies. The highest level refers to students being able to consider all relevant criteria, apply compensatory decision-making strategies, reflect normative aspects of the decision, and recognize the limitations of decision-making strategies.

In summary, the German approach to learning progressions is characterized by a focus on the delineation of complex, multidimensional models; so-called models of competence development. These models are often based on models of competency that were used as such or were extended by additional dimensions, incorporating findings from previous research on students' progression in developing the competencies in question. Separate models have been developed for core (i.e., basic) concepts and practices. Interestingly, models focusing on core concepts are often extended versions of a model describing increased competency by the availability of more complex knowledge, the extension typically building on findings from conceptual change research. Scholarship focusing on core practices often builds on models of competency without further refinement, likely because there has been little research on students' progression in developing competency in the practices across grades or grade bands.

Development Across Grades and Grade Bands

A second feature of the German approach to learning progression is that it emphasizes the long-term perspective. Neumann et al. (2013), for example, initially proposed their learning progression for the lower secondary level, validating it with students from grades 5 to 10. In later work, the authors extended the scope of application to the upper secondary level (Opitz et al., 2019). The work by Hadenfeldt et al. (2016) also spans grades 5 to 13; that is lower and upper secondary level. Emden et al.'s work (2018) is focused on grades 6 to 8, but since it is situated within the larger learning progression by Hadenfeldt et al. (2016), effectively differentiating the lower levels, it also implements a longer-term perspective. Interestingly, the long-term perspective in most cases has implications for the kind of work carried out on the validation of the proposed learning progressions. A substantial amount of work has been carried out to validate the model of competency (development) in the practice of modeling proposed by Upmeier zu Belzen and Krüger (2010) in a cross-sectional approach including students from grades 7 to 10 (e.g., Grünkorn et al., 2014; Krell, 2013; Terzer, 2013). Eggert and Bögeholz (2010) also report findings from two studies including students from grades 6 to 12 with respect to their model of competency in (socio-)scientific reasoning and decision-making. Notably, all of this validation work took a large-scale cross-sectional approach, involving different groups of students for each grade or combination of grades (e.g., grades 7/8) covered by the learning progression. This is likely due to the fact, that running a longitudinal study across the grades covered (i.e., 5–8 years) presents a particular effort and comes with substantial methodological challenges (e.g., panel mortality).

Building on the results from the large-scale cross-sectional validation studies, later work has, however, adopted a longitudinal approach. Weßnigk et al. (2017), for example, report on a longitudinal study based on the model proposed by Neumann et al. (2013), tracking students from grade 6 to grade 9. Another example includes a two-cohort longitudinal study reported by Patzke et al. (2015). Based on the model of competency development on the modeling practice proposed by Upmeier zu Belzen and Krüger (2010), Patzke et al. (2015) investigated students' progression in modeling competence with assessments at the end of each grade from grade 7 to 9 (first cohort) and 8 to 10 (second cohort). A few studies are particularly noteworthy in this context. One such study is by Zabel and Gropengiesser (2011), who attempt to develop a learning progression "from scratch" by tracking students' learning over the course of a teaching unit on Darwin's theory of natural selection. From writing assignments before and after the five-week (10 class period) unit, the authors derive a so-called conceptual landscape similar to strand maps used for example by Emden et al. (2018). This conceptual landscape mapped students' explanatory patterns against the background of key concepts of Darwin's theory of natural selection and identified pathways indicating changes in students' explanatory patterns as a result of the instructional unit. A second study of note, run by Bernholt et al. (2020), investigates students' progress in developing an understanding of the core concepts of energy, matter, and chemical reactions. Bernholt et al. (2020), building on the prior work by Neumann et al. (2013; see also Opitz et al., 2017, 2019), Hadenfeldt et al. (2016) and own work on chemical reactions, identify one progress variable for each concept. In a large-scale longitudinal study covering grades 9 to 12, the authors examine students' progression with respect to each progress variable and potential interrelations between them as a function of the existing curriculum in one German state (i.e., Schleswig-Holstein).

In summary, the German approach to learning progressions is characterized by a long-term perspective spanning multiple grades and even grade bands. One reason for this was the fact that the Standards were formulated for the end of the primary, lower, and upper secondary levels. Hence, to guide the development of so-called core curricula which would break down the standards to individual grades or combinations of grades, models of competency development incorporated a long-term perspective. The development of these models followed a standard scheme: the point of departure were models of competency. These models were either developed into models of competency development or used as such. Based on the models, the competency of students in different grades was assessed in large-scale cross-sectional studies. In some cases, the findings were further substantiated by longitudinal studies, sometimes employing a multi-cohort design to reduce effort. Interestingly, the development and validation of competency development models were mostly based on the existing curriculum. In very few studies, students' competency development was investigated based on newly designed curricula (e.g., Scheuch et al., 2019).

Alignment of Standards and Assessments

The third notable feature of the German approach to learning progressions is an emphasis on the alignment of standards and assessment over the alignment of standards and curriculum or curriculum and assessment. One important part of the process of developing and validating a model of competency development was the operationalization of the model into tasks. Each cell (i.e., each combination of the different dimensions of the often-multi-dimensional models) was carefully described. Since the competencies were often delineated into fine-grained dimensions, the cells represented a very specific aspect of the competency in question and allowed for a straightforward operationalization into tasks. Neumann et al. (2013), for example, worked on a two-dimensional description of four aspects of energy (forms, transformation, dissipation, and conservation) and five levels of knowledge about each of them (one fact, multiple facts, one relation, multiple relations, overarching concept). Following a strict procedure, the authors first developed items assessing a given aspect at the level of the overarching concept, and then developed further items by reducing the initial item complexity step by step. In many cases, the items utilized were multiple-choice. Hadenfeldt et al. (2016) used a specific sub-type of items, so called ordered multiple-choice items. Instead of one correct, and several incurred response options, these items include response options representing different levels of proficiency (for details, see Briggs et al., 2006). Some authors utilized open-ended items (e.g., Grünkorn et al., 2014) or combinations of multiple-choice and open-ended items (e.g., Patzke et al., 2015). A method used more scarcely was interviews. Scheuch et al. (2019), for example, used interviews to assess students' learning as a function of instruction designed based on a learning progression for evolution. More often, however, interviews were used for validation purposes, for example, by Hadenfeldt et al. (2014), who used think-aloud interviews to validate students' responses to the ordered multiple-choice items utilized.

Since commonly 3 to 5 items were developed for each cell and models included 20 or more cells, the resulting set of items became too large for a single student to answer. Hence researchers employed a so-called matrix design. In a matrix design, items are spread across multiple booklets following a specific procedure (Frey et al., 2009). To match the data from students who received different booklets, procedures based on Item Response Theory (IRT) were used. These procedures allow for locating item difficulty and student ability on the same (latent) continuum. This, in turn, allows for interpreting students' performance in relation to the continuum of tasks. Assuming the tasks can be arranged such that they reflect the development of competence, conclusions can be drawn about where students are in their progression in developing competence in a domain (e.g., understanding of a core concept). One of the most commonly used IRT models is the Rasch model. Neumann et al. (2013) used the Rasch model, for example, to analyze how the performance of N = 1856 students from grades 6, 8 and 10 maps onto their hypothesized learning progression. Hadenfeldt et al. (2016) used a multi-dimensional extension of the Rasch model to examine the extent to which the identified aspects of the matter concept mark dimensions of a higher construct (i.e., the understanding of matter).

Later work tended to use even more complex psychometric models to better account for the complex assumptions about students' progression in developing an understanding of a particular concept or competence in using a particular practice. Neumann et al. (2013), for example, hypothesized that students will learn about energy in four stages: forms, transformation, degradation, and conservation; in each stage developing an increasingly complex knowledge from facts to conceptual understanding. The authors initially allowed for some overlap between stages, but acknowledged in later work that students can learn about each of the four aspects (e.g., forms or transformation) in parallel. That is, students can learn about energy conservation early on, but for a true understanding of conservation need to integrate knowledge across all four aspects. To test this complex assumption, Neumann and Nagy (2013) used a multi-group two-parameter IRT model (in the threshold parametrization, Muthén, 1979). In contrast to the Rasch model, this model includes, in addition to a person ability parameter and an item difficulty, an item discrimination parameter indicating the extent to which the item measures the underlying trait. The multi-group version allows for parameters to vary across groups. Neumann and Nagy (2013) demonstrated that while in grade 6 mainly items measuring knowledge of forms exhibit high discrimination, with increasing grade discrimination for items measuring knowledge of transformation, degradation and conservation increases as well, suggesting the mechanism underlying students' progression is indeed a knowledge integration. As a last example, Bernholt et al. (2020) in their analysis of the interaction of students' progression in developing competence about the three major concepts of chemistry used cross-lagged panel analysis. Cross-lagged panel analysis is a method based on structural equation models to analyze repeated measurement of variables and how the measurement of one concept (e.g., energy) at one point in time (e.g., grade 9) affects that of another concept (e.g., matter) at the subsequent point in time (e.g., grade 10). The authors found out that students' progression in mastering the matter concept determines their progression in mastering chemical reactions.

In summary, the development and validation of models of competency development have followed a rigorous process of operationalizing the model through tasks, administering the resulting pool of tasks to students from different grades, and analyzing the resulting data against the model; effectively implementing the assessment triangle (Pellegrino et al., 2001). In doing so, researchers tended to use increasingly sophisticated methods to model the theoretical assumptions about students' progression in developing competency more accurately. Hence, the German approach to learning progressions was very much driven by the idea of testing the progression hypothesized (based on the existing curriculum) against data collected from students (who learned according to the existing curriculum); with the aim to identify opportunities to improve the existing curriculum. However, this last step has rarely been taken so far.

Summary and Conclusions

The German approach to learning progressions—that is, modeling the development of competency in a domain—comes in a variety of slightly different approaches. There is work that focuses on core concepts (e.g., Neumann et al., 2013) and work that focuses on core scientific practices (e.g., Krell, 2013). Some of this work investigates students' progression in developing competence (e.g., Zabel & Gropengiesser, 2011) to build a model of competency development; other work aims to validate a hypothetical model derived from theory (e.g., Hadenfeldt et al., 2016). Some of this work even builds explicitly on the actual definition of learning progressions as expressed for example in Duschl et al. (2007): delineating an idealized, hypothetical progression of students' mastery of a domain across grades or even grade bands, as a means to align standards, curriculum, and assessment. Despite the differences between the various works, there are also commonalities shared by most of the work.

First, almost all approaches to develop learning progressions in Germany build off a model of competency. The National Education Standards formulated competencies students were expected to develop but provided little guidance for the assessment and development of these competencies. Hence, competency modeling (i.e., the development and validation of competency models) became a growing area of research (for an overview see Leutner, 2017). Competency models were supposed to organize the knowledge, skills, and abilities required to be considered competent in a domain into areas of competency and levels of competency. These multi-dimensional models were supposed to serve as a foundation for assessing the Standards and as a guidance for designing instruction aligned with the Standards. However, the models often did not include a developmental perspective. Hence, researchers turned to developing (and validating) models of competency development. In some cases, this work built on existing models and simply investigated students' competency across grades (e.g., Patzke et al., 2015). In the majority of cases, however, researchers proposed a refined model of competence development (i.e., a model that, based on a model of competence, describes how students would progress in developing that competence; e.g., Neumann et al., 2013; Weber et al., 2016). Either way, this work resulted in often complex, multi-dimensional models.

Second, most of the work on modeling competency development took a multi-year perspective spanning multiple grades or even grade bands. The Standards provided little guidance regarding how students were expected to progress in developing the envisioned competency across grades, not to mention how instruction could support this progression. Models of competency development were supposed to provide exactly this guidance; for example, to break down the Standards that were formulated for the end of primary, lower secondary, and upper secondary levels to individual grades. Due to the effort and cost coming with multi-year longitudinal studies, much of this work drew on so-called quasi-longitudinal studies; that is, cross-sectional studies including students from different grades. Due to the strong focus on validation, these studies were often large-scale in nature, involving hundreds of students from each grade; and while in some cases, the results were later corroborated or extended in longitudinal studies, very little work actually made it to the point where the curriculum was revised and students' progression was re-investigated to see if the revised curriculum would lead to a better competency development as promised (see, for example, Neumann et al., 2013).

Third, since the work on competency development models followed the same iterative procedure of development and validation of competency models, researchers put great emphasis on the operationalization of the models into items. The resulting extensive item pools were grouped into multiple booklets, administered to students, and analyzed using sophisticated statistical procedures. This work examined students' progression in developing competence as a function of the current instruction to validate the model but at the same time identify issues that specific groups of students (i.e., low-achieving students) may exhibit in their progression as a basis for optimizing (current) instruction. While there is work that investigated students' competence development as a function of an intervention; this work was often not aimed at describing or understanding *how* students progress in developing increasing competence but more generally focused on *whether* and to what extent students' progress as a function of an intervention (e.g., Gresch et al., 2013). That is, while this line of work investigated students' competence development, it was not so much interested in modeling the actual progression, but to confirm there is progression.

One conclusion from this review of the German approach to learning progressions, therefore, is that there is little evidence that the competency development models have indeed guided the development of (better) curriculum or instruction. The international approach to learning progression, in contrast, has focused much more on using learning progression as a basis for developing curriculum and instruction. This is one area in which the German approach differs from the international approach. Another area of complementarity is the alignment between standards, curriculum, and assessment. Internationally, the focus has been on aligning standards, curriculum, and assessment – with the purpose of supporting students in reaching educational goals. In Germany, a particular emphasis has been placed on aligning standards and assessment, with the purpose of understanding how much the current curricula support students' progression in meeting the educational goals (or not). A last difference is the use of more sophisticated (as in more fitting to the theoretical assumptions) methodological approaches to analyzing the data collected using these instruments. These methods should not only provide deeper insights into how the inner structure of students' competence changes as they progress (e.g., Neumann & Nagy, 2013) but also allow a link of students' learning in a day-to-day curriculum with their overarching progression across grades or grade bands.

One thing to keep in mind when comparing the German approach to learning progressions to the international approach is that it did not take place in an isolated world. Some researchers quickly realized parallels between the scholarship on modeling competence development and learning progression research, for example, the goal of proposing and validating a model of students' progression in developing competence in a domain. Based on the tradition of developing and validating models of competence, the German approach largely took the work to the large scale, aiming to obtain evidence that students indeed progress as hypothesized based on theoretical considerations (i.e., a combination of curriculum analysis and analysis of prior research on students' learning about the domain).

In doing so, the German approach has yielded findings that complement the findings that have been made internationally about student competence. As a result, it may be fruitful to follow an approach recently taken by the field of research on teacher professional knowledge: synthesizing existing work into a consensus model and continually revising it as new evidence emerges (Berry et al., 2015; Borowski et al., 2019).

References

Baumert, J., Lehmann, R., Lehrke, M., Schmitz, B., Clausen, M., Hosenfeld, I., & Köller, O. (1997). *TIMSS. Mathematisch-naturwissenschaftlicher Unterricht im internationalen Vergleich. Deskriptive Befunde*. Leske+Budrich.

Beaton, A. E., Martin, M. O., Mullis, I. V. S., Gonzalez, E. J., Smith, T. A., & Kelly, D. L. (1996). *Science Achievement in the Middle School Years: IEA's third International Mathematics and Science Study (TIMSS)*. Center for the Study of Testing, Evaluation, and Educational Policy, Boston College.

Bernholt, S., & Parchmann, I. (2011). Assessing the complexity of students' knowledge in chemistry. *Chemistry Education Research and Practice*, *12*(2), 167.

Bernholt, S., Höft, L., & Parchmann, I. (2020). Die Entwicklung fachlicher Basiskonzepte im Chemieunterricht—Findet ein kumulativer Aufbau im Kompetenzbereich Fachwissen statt? *Unterrichtswissenschaft*, *48*, 35–48. 10.1007/s42010-019-00065-4

Berry, A., Friedrichsen, P. J., & Loughran, J. (Eds.). (2015). *Re-examining pedagogical content knowledge in science education*. Routledge.

Borowski, A., Cooper, R., & Hume, A. (Eds.). (2019). *Repositioning Pedagogical Content Knowledge in Teachers' Knowledge for Teaching Science* (1st ed. 2019). Springer Singapore: Imprint: Springer. 10.1007/978-981-13-5898-2

Briggs, D. C., Alonzo, A. C., Schwab, C., & Wilson, M. (2006). Diagnostic assessment with ordered multiple-choice items. *Educational Assessment*, *11*(1), 33–63.

Bybee, R. W. (1997). Toward an understanding of scientific literacy. In W. Gräber & C. Bolte (Eds.), *Scientific literacy. An international symposium* (Vol. 154). IPN.

Clark, D. B., & Linn, M. C. (2013). The knowledge integration perspective: Connections across research and education. *International Handbook of Research on Conceptual Change*, 520–538.

Driver, R., Leach, J., Scott, P., & Wood-Robinson, C. (1994). Young people's understanding of science concepts: Implications of cross-age studies for curriculum planning. *Studies in Science Education*, *24*, 75–100.

Duschl, R. A., Schweingruber, H. A., & Shouse, A. W. (2007). *Taking science to school: Learning and teaching science in grades K-8*. National Academies Press.

Eggert, S., & Bögeholz, S. (2006). Göttinger Modell der Bewertungskompetenz—Teilkompotenz„Bewertung, Entscheiden und Reflektieren" für Gestaltungsaufgaben Nachhaltiger Entwicklung. *Zeitschrift Für Didaktik Der Naturwissenschaften*, *12*, 177–198.

Eggert, S., & Bögeholz, S. (2010). Students' use of decision-making strategies with regard to socioscientific issues: An application of the Rasch partial credit model. *Science Education*, *94*(2), 230–258.

Emden, M., Weber, K., & Sumfleth, E. (2018). Evaluating a learning progression on 'Transformation of Matter' on the lower secondary level. *Chemistry Education Research and Practice*, *19*(4), 1096–1116. 10.1039/C8RP00137E

Frey, A., Hartig, J., & Rupp, A. A. (2009). An NCME Instructional Module on Booklet Designs in Large-Scale Assessments of Student Achievement: Theory and practice. *Educational Measurement: Issues and Practice*, *28*(3), 39–53.

Gresch, H., Hasselhorn, M., & Bögeholz, S. (2013). Training in decision-making strategies: An approach to enhance students' competence to deal with socio-scientific issues. *International Journal of Science Education*, *35*(15), 2587–2607. 10.1080/09500693.2011.617789

Grünkorn, J., zu Belzen, A. U., & Krüger, D. (2014). Assessing students' understandings of biological models and their use in science to evaluate a theoretical framework. *International Journal of Science Education*, *36*(10), 1651–1684. 10.1080/09500693.2013.873155

Hadenfeldt, J. C., Liu, X., & Neumann, K. (2014). Framing students' progression in understanding matter: A review of previous research. *Studies in Science Education*, *50*(2), 181–208. 10.1080/03057267.2014.945829

Hadenfeldt, J. C., Neumann, K., Bernholt, S., Liu, X., & Parchmann, I. (2016). Students' progression in understanding the matter concept. *Journal of Research in Science Teaching*, *53*(5), 683–708. 10.1002/tea.21312

Kauertz, A., & Fischer, H. E. (2006). Assessing students' level of knowledge and analysing the reasons for learning difficulties in physics by Rasch analysis. In X. Liu & W. J. Boone (Eds.), *Applications of Rasch Measurement in Science Education* (pp. 212–246). JAM Press.

Klieme, E., Avenarius, H., Blum, W., Döbrich, P., Gruber, H., Prenzel, M., Reiss, K., Riquarts, K., Rost, J., Tenorth, H.-E., & Vollmer, H. J. (2003). *Zur Entwicklung nationaler Bildungsstandards—Eine Expertise*. Bundesministerium für Bildung und Foschung (BMBF).

Klieme, E., Baumert, J., Köller, O., & Bos, W. (2000). Mathematische und naturwissenschaftliche Grundbildung: Konzeptuelle Grundlagen und die Erfassung und Skalierung von Kompetenzen. In J. Baumert, W. Bos, & R. Lehmann (Eds.), *TIMSS/III* (Vol. 1, pp. 85–133). Leske + Budrich.

Krell, M. (2013). *Wie Schülerinnen und Schüler biologische Modelle verstehen: Erfassung und Beschreibung des Modellverstehens von Schülerinnen und Schülern der Sekundarstufe I*. Logos-Verlag.

Leutner, D. (2017). *Competence assessment in education: Research, models and instruments*. Springer Berlin Heidelberg.

Leutner, D., Grünkorn, J., Fleischer, J., & Klieme, E. (2017). *Competence assessment in education: Research, models and instruments*. Springer Berlin Heidelberg.

Liu, X. (2009). *Linking competence to opportunities to learn: Models of competence and data mining*. Springer.

Liu, X., & McKeough, A. (2005). Developmental growth in students' concept of energy: Analysis of selected items from the TIMSS database. *Journal of Research in Science Teaching*, *42*(5), 493–517.

Mayer, J., Grube, C., & Möller, A. (2008). Kompetenzmodell naturwissenschaftlicher Erkenntnisgewinnung (im Druck). In U. Harms & A. Sandmann (Eds.), *Ausbildung und Professionalisierung von Lehrkräften*.

Mislevy, R. J., Steinberg, L. S., & Almond, R. G. (2002). Design and analysis in task-based language assessment. *Language Testing*, *19*(4), 477–496. 10.1191/0265532202lt241oa

Muthen, B. (1979). A structural probit model with latent variables. *Journal of the American Statistical Association*, *74*, 807. 10.2307/2286403.

Neumann, K., Fischer, H. E., & Kauertz, A. (2010). From PISA to educational standards: The impact of large-scale assessments on science education in Germany. *International Journal of Science and Mathematics Education*, *8*(3), 545–563.

Neumann, K., Härtig, H., Harms, U., & Parchmann, I. (2017). Science teacher preparation in Germany. In J. Pedersen, T. Isozaki, & T. Hirano (Eds.), *Model science teacher preparation programs* (pp. 29–52). Information Age Publishing.

Neumann, K., & Nagy, G. (2013). *Students' progression in understanding energy*. Paper presented at the Annual Conference of the National Association for Research in Science Teaching (NARST), Rio Grande, Puerto Rico.

Neumann, K., Schecker, H., & Theyßen, H. (2019). Assessing complex patterns of student resources and behavior in the large scale. *The ANNALS of the American Academy of Political and Social Science*, *683*(1), 233–249. 10.1177/0002716219844963

Neumann, K., Viering, T., Boone, W. J., & Fischer, H. E. (2013). Towards a learning progression of energy. *Journal of Research in Science Teaching*, *50*(2), 162–188. 10.1002/tea.21061

Opitz, S., Neumann, K., Bernholt, S., & Harms, U. (2017). How do students understand energy in Biology, Chemistry, and Physics? Development and validation of an assessment instrument. *EURASIA Journal of Mathematics Science and Technology Education*, *13*(7), 3019–3042. 10.12973/eurasia.2017.00703a

Opitz, S., Neumann, K., Bernholt, S., & Harms, U. (2019). Students' energy understanding across Biology, Chemistry, and Physics contexts. *Research in Science Education*, *49*, 521–541. 10.1007/s11165-017-9632-4

Organisation for Economic Co-operation and Development [OECD]. (2001). *Knowledge and Skills for Life*. Organisation for Economic Co-operation and Development. http://public.eblib.com/choice/publicfullrecord.aspx?p=533314

Organisation for Economic Co-operation and Development [OECD]. (2006). *Assessing Scientific, Reading and Mathematical Literacy: A Framework for PISA 2006* (1st ed.). OECD Publishing.

Patzke, C., Krüger, D., & Umpeier zu Belzen, A. (2015). Entwicklung von Modellkompetenz im Längsschnitt. In M. Hammann, J. Mayer, & N. Wellnitz (Eds.), *Lehr-und Lernforschung in der Biologiedidaktik. Band* 6 (pp. 43–58). Studien Verlag GmbH.

Pellegrino, J. W., Chudowsky, N., & Glaser, R. (2001). *Knowing what students know: The science and design of educational assessment*. National Academy Press.
Posner, G. J., Strike, K. A., Hewson, P. W., & Gertzog, W. A. (1982). Accommodation of a scientific conception: Toward a theory of conceptual change. *Science Education*, *66*(2), 211–227.
Prenzel, M., Rost, J., Senkbeil, M., Häußler, P., & Klopp, A. (2001). Naturwissenschaftliche Grundbildung: Testkonzeption und Ergebnisse. In J. Deutsches PISA-Konsortium, E. K. Baumert, M. Neubrand, M. Prenzel, U. Schiefele, W. Schneider, P. Stanat, & K.-J. Tillmann (Eds.), *PISA 2000* (pp. 191–248). Leske+Budrich.
Schecker, H., & Parchmann, I. (2006). Modellierung naturwissenschaftlicher Kompetenz. *Zeitschrift Für Didaktik Der Naturwissenschaften*, *12*, 45–66.
Scheuch, M., Scheibstock, J., Amon, H., Bauer, H., Harms, U., & Reiss, M. (2019). Teaching Evolution Along a Learning Progression. In *Evolution Education Re-considered* (pp. 81–99). Springer.
Seidel, T., Prenzel, M., Rimmele, R., Herweg, C., Kobarg, M., & Schwindt, K. (2007). Science teaching an learning in German physics classrooms. In M. Prenzel (Ed.), *Studies in the educational quality of schools. The final report on the DFG Priority Programme* (pp. 79–99). Waxmann.
Sekretariat der Ständigen Konferenz der Kultusminister der Länder in der Bundesrepublik Deutschland [KMK]. (2005a). *Bildungsstandards im Fach Biologie für den Mittleren Schulabschluss. Beschluss vom 16.12.2004*. Luchterhand.
Sekretariat der Ständigen Konferenz der Kultusminister der Länder in der Bundesrepublik Deutschland [KMK]. (2005b). *Bildungsstandards im Fach Chemie für den Mittleren Schulabschluss. Beschluss vom 16.12.2004*. Luchterhand.
Sekretariat der Ständigen Konferenz der Kultusminister der Länder in der Bundesrepublik Deutschland [KMK]. (2005c). *Bildungsstandards im Fach Physik für den Mittleren Schulabschluss. Beschluss vom 16.12.2004*. Luchterhand.
Terzer, E. (2013). *Modellkompetenz im Kontext Biologieunterricht*.
Tibell, Lena A. & Harms, Ute. (2017). Biological Principles and Threshold Concepts for Understanding atural Selection. *Science & Education*, *26*, 953–973. 10.1007/s11191-017-9935-x
Upmeier zu Belzen, A. zu, & Krüger, D. (2010). Modellkompetenz im Biologieunterricht. *Zeitschrift Für Didaktik Der Naturwissenschaften*, *16*(1), 41–57.
Walpuski, M., Kampa, N., Kauertz, A., & Wellnitz, N. (2008). Evaluation der Bildungsstandards in den *Naturwissenschaften. Der Mathematische Und Naturwissenschaftliche Unterricht*, *61*(6), 323–326.
Weber, K., Emden, M., & Sumfleth, E. (2016). Development of a Learning Progression on Chemical Reactions. In J. Dolin & P. M. Kind (Eds.), *Electronic Proceedings of the ESERA 2015 Conference. Science education research: Engaging learners for a sustainable future, Part Evaluation and assessment of student learning and development* (pp. 1589–1597).
Weinert, F. E. (2001). Concept of competence: A conceptual clarification. In D. S. Rychen & L. H. Salganik (Hrsg.), *Defining and selecting key competencies* (S. 45–65). Hogrefe.
Weßnigk, S., Neumann, K., Viering, T., Hadinek, D., & Fischer, H. E. (2017). The development of students' physics competence in middle school. In *Competence Assessment in Education* (pp. 247–262). Springer.
Zabel, J., & Gropengiesser, H. (2011). Learning progress in evolution theory: Climbing a ladder or roaming a landscape? *Journal of Biological Education*, *45*(3), 143–149. 10.1080/00219266.2011.586714

8

LEARNING PROGRESSION APPROACHES USED IN CHINA

Lei Wang, Jian-Xin Yao, Quan Wang, Yi-Meng Liu, An-Qi Mou, Jian Wang, and Min Wang

Introduction

At the beginning of the 21st century, the essential characteristics and key abilities required for students' lifelong development were sorted out and condensed, with an integrated perspective, into the core competencies system in education (C. D. Lin, 2016). To achieve the development of students' core competencies, many countries have developed relevant curriculum standards that emphasize the basis of learning progressions (LPs), such as national science curriculum standards (Ministry of Education, P. R. China [MOE], 2022) and *The Next Generation Science Standards* (NGSS Lead States, 2013).

Learning progressions are blueprints describing students' domain-specific cognitive development of core ideas and key practices, some of which provide targeted solutions for instruction improvement (National Research Council [NRC], 2007; Jin et al., 2019; Yao & Y. Y. Guo, 2014, 2018a). LPs have gained widespread attention from educational researchers and practitioners, resulting in the conduction of many empirical and theoretical studies about LPs. The development of domain-specific cognitive development of core ideas and abilities is assessed and described by the LP studies. LP studies also provide a cognitive basis for the systematization of science curriculum planning and assessment design. The studies also provide a research foundation for the implementation of the 21st-century science education idea of "integrated development" (e.g., Harlen, 2010; Y. Y. Guo et al., 2013). Chinese scholars have performed LP studies for more than 10 years. Apart from some unique features, Chinese LP studies are comparable to international research on LPs. In this chapter, we review the theoretical foundation, progress variables, research paradigms, representative results, and applications of LP research.

Theoretical Foundation

In 2004, the idea of "learning progression" was formally introduced in science education. Although learning progression is a newly introduced term in science education, the idea has long been rooted in education, with Piaget's genetic epistemology and Vygotsky's zone of proximal development (ZPD) revealing that a systematic study of student development can bring critical changes in teaching and learning. At the end of the 20th century, scholars began to reflect on the

DOI: 10.4324/9781003170785-9

Alternative Conceptions Movement. One of the important points of reflection was that many studies focused on students' conceptual states at a certain stage and failed to include a developmental variable to systematically portray students' conceptual development. In response to this reflection, some scholars began to analyze students' "conceptual trajectory" and "conceptual development." For example, when proposing conceptual trajectories, Driver et al. (1994) pointed out that building a bridge between children's science and formal science involved some "intermediate understanding" or "intermediate concepts." Therefore, the progression process would follow a conceptual trajectory that depicted the key steps in the development of knowledge in a specific domain. The guidelines for LPs have included a starting point of the progression (children's science), a goal of the progression (formal science), and the intermediate nodes (intermediate understanding/concepts) ever since.

The framework proposed by NRC (2007) provided an early common foundation for LP research. However, to date, researchers have presented many different representations and interpretations of LP definitions and connotations and applied different research philosophies and research methods (Duschl et al., 2011; Yao & Y. Y. Guo, 2014). Initially, defined LPs were descriptions of typical paths for the development of students' understanding of core ideas, which can be drawn from curriculum standards and research on students' conceptions. Soon, assessment-based approaches became the major focus of LP studies. There are many representative LP studies from this time. Some studies focus on disciplinary core ideas and crosscutting concepts. Other studies focus on scientific practices or scientific thinking.

In recent years, studies on LP have addressed two challenges: first, the response of LPs developed using the large-scale assessment approach to the basic principle of "teaching-assisted development" in science education; second, how to develop integrated LPs that meet the multidimensional objectives of science education, such as integrating LPs of core ideas and key practices (e.g., Lehrer & Schauble, 2015; Y. Y. Guo & Yao, 2016). Despite these challenges, the positive impact of evidence-based LPs in curriculum and assessment is undeniable. The ongoing discussion on the definition and connotation of LPs reflects the different positions of different schools of thought in science education and contributes to deepening the understanding of the nature of LP. Moreover, facing these challenges, LP research has demonstrated its vitality and developmental potential, with continual self-reflection and transformation to allow paradigm evolution (Duschl et al., 2011; Jin, Mikeska, et al., 2019; Yao & Y. Y. Guo, 2018a).

Progress Variables of LPs

In essence, LPs are a series of models of students' cognitive development. These models need to be tested by representative and stable observational evidence. Before accumulating data through tests, interviews, teaching experiments, etc., a well-designed LP hypothesis plays an overarching role. The LP hypothesis is generally composed of a progression starting point (lower anchor), a progression goal (upper anchor), and intermediate levels, which are linked by one or multiple progress variables. The progress variable helps LPs describe what precisely students learned. The progress variable establishes the theoretical framework for the LP hypothesis, and its selection and design are the core work of proposing an LP hypothesis (Jin, Mikeska, et al., 2019; Yao & Y. Y. Guo, 2018a).

Using the ontology of scientific knowledge as a progress variable was dominant in the early stages of LP studies on disciplinary core ideas. In these studies, the LP levels were delineated based on the analysis of the structure of scientific knowledge, curriculum materials, and previous conceptual learning research. Later studies used theories of cognitive psychology and

Table 8.1 Developmental hierarchical model of conceptual understanding of science (Y. Y. Guo & Yao, 2016)

Development levels	*Level description*
Integration	Students are able to unify their understanding of a scientific idea (e.g., the idea of matter, the idea of energy, etc.) from core concepts and construct connections between scientific ideas and crosscutting concepts (e.g., systems, scales, etc.).
System	Students are able to reconcile the self-varying and covarying relationships of the variables in a multifactor structure at the system level.
Relation	Students are able to construct relationships between abstract terms and several observable concrete features of things.
Mapping	Students are able to construct mapping relationships between concrete features of things and abstract terms.
Experience	Students have everyday experiences and fragmented facts that are not yet interrelated.

learning sciences as progress variables. Taking the LPs in physics as an example, for the development of LPs of disciplinary core ideas, a hierarchical model for the development of conceptual understanding in science (Table 8.1) was proposed based on cognitive theories such as hierarchical complexity and knowledge integration. Using the levels in this model as progress variables, the performance expectations of core concepts such as energy, mechanical motion and force, electricity, and magnetism were described at various levels (Y. Y. Guo & Yao, 2016).

For research on LPs in chemistry, one of the most frequently used progress variables for Chinese scholars is abstracted from the Subject Competency Framework (SCF, Figure 8.1, Lei Wang et al., 2022). In SCF, the subject core knowledge is a fundamental but insufficient component for subject competency. Subject core knowledge can help students develop subject competency only when it is transformed into cognitive perspectives (e.g., analyze the substance from elemental and particle perspectives, explore the substance property from category and valence perspective) and cognitive mode (Lei Wang, 2016; Lei Wang et al., 2022). As a result, cognitive mode has become a frequently used progress variable (Lei Wang, 2016). Scholars have used cognitive mode or its subordinate elements as the progress variables to delineate levels in their LPs. For example, in W. Y. Wang's LP study on the theme of the periodic law of elements, she took the research objects, cognitive perspectives, and

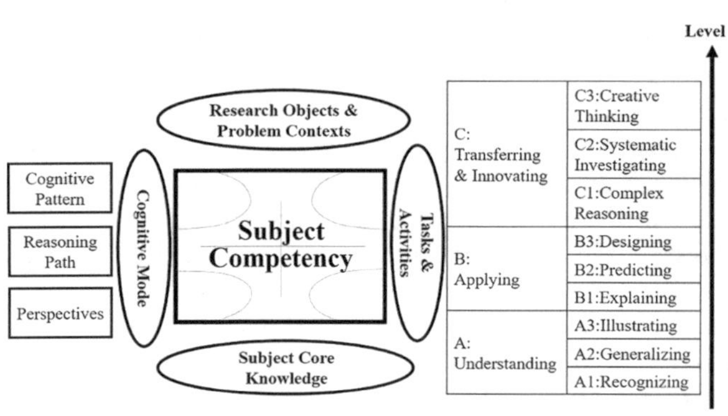

Figure 8.1 Subject Competency Framework (SCF) (Wang, 2016; Wang et al., 2022).

Table 8.2 The LP of the periodic law of elements theme

Progress variables			*Levels and performance of students*	
Research objects	*Perspective*	*Cognitive pattern*		
A group of elements	Similarity & Gradation (comparing)	Systematic "position-structure-property" cognitive mode	Level 6	Be able to design a substance's complex transformation path to solve real-life problems according to the periodic law of elements.
	Position (group) & Substance's property	atomic structure-element's property-substance's property; position-element's property-substance's property	Level 5	Be able to construct "position-structure-property" cognitive mode and predict and verify an unfamiliar element's property based on that of the familiar elements.
	Position (period) & Substance's property	atomic structure-element's property-substance's property; position-element's property-substance's property	Level 4	Be able to associate an element's property with a substance's property. Be able to explain and compare element similarity/gradation by designing an inquiry experiment with an appropriate substance.
	Position	position-element's property-substance's property	Level 3	Be able to illustrate and analyze a familiar element's position from the "position-structure-property" perspective.
2 elements	Atomic structure & Substance's property	atomic structure-element's property-substance's property	Level 2	Be able to associate atomic structure with an element's property and understand that the gain or loss of electrons is affected by the nucleus's attraction to its outermost electrons.
1 element	Element's property & Redox reaction	independent & simple correlation	Level 1	Be able to understand an element's property from a microscopic perspective. Be able to build a primary atomic structure perspective and recognize that a simple substance's oxidizability and an anion's reducibility are connected to an atom's gain or loss of electrons.

cognitive patterns as progress variables, constructing and demonstrating students' LP levels on this theme (Table 8.2, W. Y. Wang, 2018).

Concerning the progressive development of conceptual understanding in science, scholars have also explored the LPs of key practices in science. Scholars have identified key elements in scientific practices and proposed structured learning frameworks matching students' cognitive levels based on an epistemological approach to deconstructing key practices in science. The learning frameworks are then used as a basis to delineate the LP in terms of the perfection of structures and element levels (e.g., Yao & Y. Y. Guo, 2018b). Another progress variable used frequently is the "disciplinary competencies" system, which has been used to coordinate the LPs of various disciplines. This system has been empirically validated in studies describing levels of performance expectations, not only in the natural science fields (physics, chemistry, biology, geography) but also extended to research in mathematics, Chinese language, English language, and other social science disciplines (for a review, see Lei Wang, et al., 2018). The three core elements of disciplinary competencies ("learning and understanding," "applying and practicing," and "transfer and innovation"), which are a combination of learning activities and cognitive expectations, played the role of progress variables in several studies. The disciplinary competencies also provide a framework to integrate disciplinary knowledge and practice and provide progress variables for the design and development of the subsequent integrated LPs.

Research Paradigms of LPs

One of the first anchors for LP research is the identification of core ideas and key practices in science. Three large-scale international comparison studies were conducted independently in Europe, the USA, and China for science education in the early years of the 21st century. Based on a consensus on targeting core ideas and key practices of science, LP studies soon spread to other countries that shared a common academic discourse. With the knowledge gained worldwide, the research paradigm of LP continues to shift and evolve. In recent decades, the core research paradigm of LPs has been rapidly iterating, from LPs of disciplinary core ideas to LPs of scientific practices, as well as unidimensional LPs to integrated LPs, and from large-scale assessment-based LPs to instruction-based LPs (Jin, Mikeska, et al., 2019; Yao & Y. Y. Guo, 2018a).

The Assessment-based Approach

The assessment-based LP studies follow an evidence-driven research paradigm, which uses representative data from assessment to identify the progression levels (Figure 8.2). This paradigm is an important development of LP research from the conceptual trajectory research in the 1990s. The introduction of advanced psychometric techniques to education gradually made many research ideas of LPs possible. In addition, having a basis of reliable empirical data, which can be tested repeatedly, has led LP findings to be widely accepted. A typical assessment-based LP study starts by developing a hypothetical LP, which is delineated by the progress variable(s) mentioned above. Then, a measurement model that is suitable for measuring the construct in the LP is chosen, such as item response theory, latent category analysis, and rule space modeling (e.g., Yao & Y. Y. Guo, 2018a; Xin et al., 2015). In the early stage, the level division and adjustment heavily relied on expert experience, so the Angoff method and Ebel method were commonly used during this process. Currently, with the development of advanced psychometric techniques, the bookmarking method is becoming increasingly popular in LP studies.

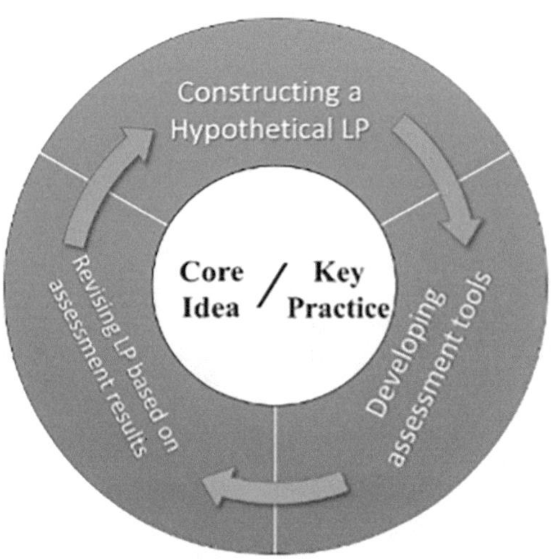

Figure 8.2 Research circle of assessment-based LPs.

The Instruction-based Approach

Scholars both in China and the USA are aware of the weakness of assessment-based LPs (e.g., Lehrer & Schauble, 2015; Yao & Y. Y. Guo, 2018a). As Lehrer and Schauble (2015, p. 434) note, "Although this approach to LPs may hold value for designers of curriculum and assessments, it has the disadvantage of reifying current educational practice." Therefore, scholars have sought to apply LPs in teaching and learning. In addition, researchers hope that the data collected in instruction can be used to refine LPs or sometimes even redesign LPs. This approach was named the instruction-based approach of LP development (Yao & Y. Y. Guo, 2018a). This approach accepts the "trading zone" view of LPs (Lehrer & Schauble, 2015). In this approach, university scholars and school teachers in China have traded their theoretical and practical wisdom and then developed the "learning progression-based design model" (the LPB design model, Figure 8.3) to use LPs in improving their instructional design and to collect student data and teacher feedback for refining LPs (Y. Y. Guo, et al., 2016).

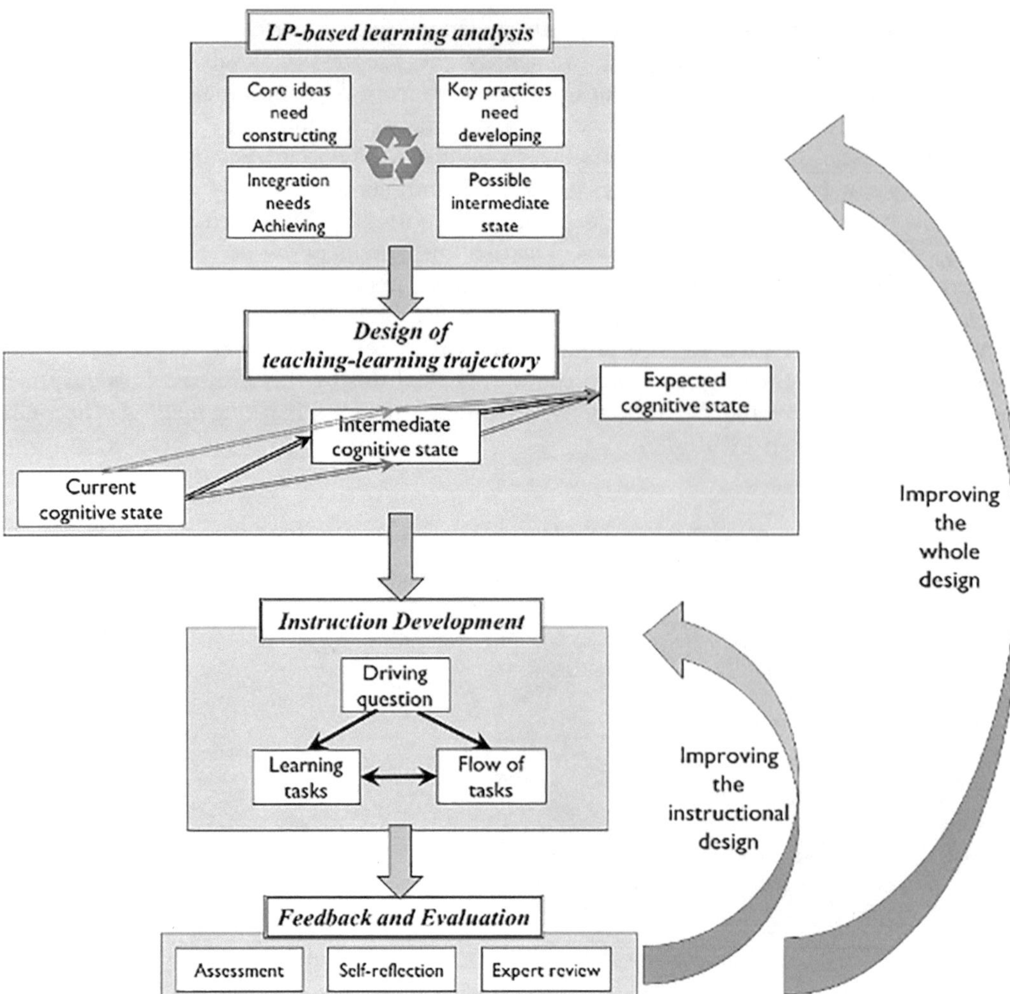

Figure 8.3 The LPB design model.

In this approach, the idea of "integrated learning progression" is vitally important (Lehrer & Schauble, 2012, 2015; Y. Y. Guo & Yao, 2016). An integrated learning progression is a model of cognitive development that incorporates both core scientific understanding and key scientific practices, in line with the contemporary core competence movement and more relevant to the actual needs in the science classroom. Returning to the nature of science education and scientific research, there should be neither knowledge understanding without practices nor practice development without knowledge. Therefore, the integrated LP is the ideal form of LP, which describes and summarizes the real cognitive development in the science context. The idea of "integrated learning progression" is also consistent with the above-mentioned "disciplinary competencies." A series of relevant design-based teaching improvement studies, named "High-End Lesson Study (HELS)," have been conducted combining the aims of improving local instruction quality and developing LPs with the instruction-based approach (Lei Wang et al., 2018; Lei Wang et al., 2022). The HELS is a series of quasi-experimental studies adopting mixed methods, including assessment as a certain scale, video studies, student interviews, etc., in an iterative circle (Figure 8.4).

Taking the study on the LP of the periodic law of elements as an example (W. Y. Wang, 2018), the theoretical model of LP integrated students' knowledge and modeling ability (see Table 8.3). After the instrument was developed based on the theoretical model, the test was piloted in several schools. The LP model was modified based on the collected data. To investigate the teaching conditions for achieving LPs and test the rationality of LPs, an instruction-based LP study was further carried out at the experimental school through the HELS process, as shown in Figure 8.4. Two rounds of study on the LP of the new teaching orientation and one round of study on the LP of the review class were carried out, while the

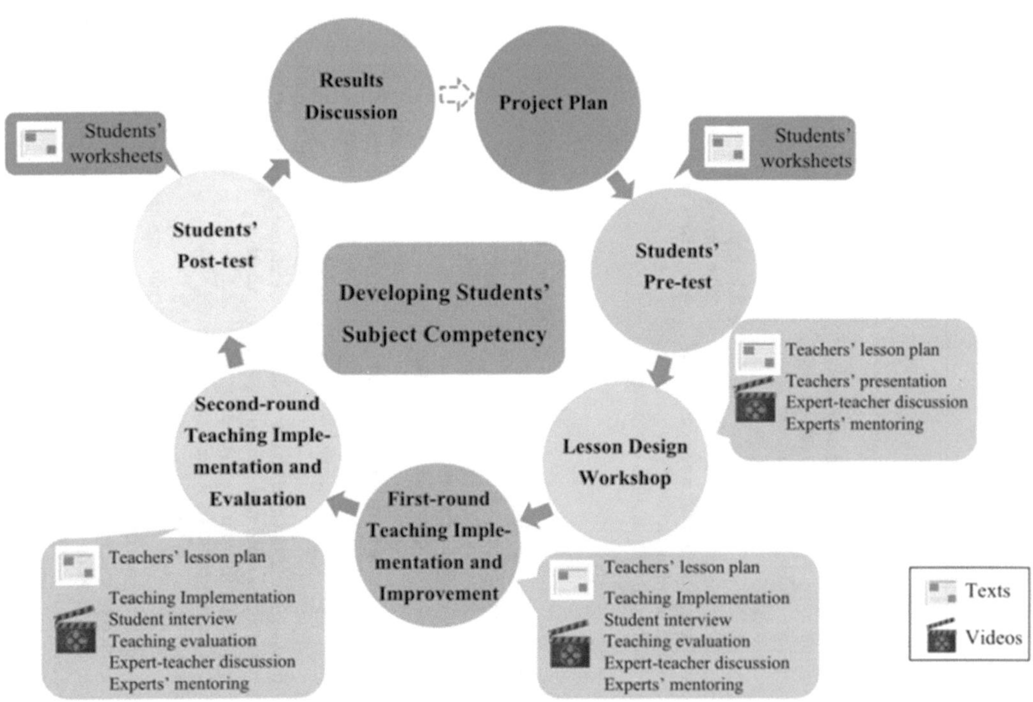

Figure 8.4 The process of HELS (Wang et al., 2022).

Table 8.3 Students' performance level of the periodic law of elements theme (W. Y. Wang, 2018)

Levels and performance of students			*Teaching variables*	
Integrated LPs levels	*Knowledge content*	*Model and modeling*	*Teaching sequence*	*Key teaching problems (KTP) and key learning tasks (KLT) examples*
Level 6 Be able to design a substance's complex transformation path to solve real-life problems according to the periodic law of elements.	Using position-structure-element's and substance's properties relationships	Level 4 Be able to conduct systematic analysis and solve complex problems in real situations based on models.	Lesson 6	KTP602: Systematically predicting the chemical properties of unfamiliar substances and explaining the basis.
Level 5 Be able to construct a "position-structure-property" cognitive model and predict and verify an unfamiliar element's property based on that of the familiar elements.	Adding position-structure-element's and substance's properties relationships	Level 3 Be able to integrate and develop models to describe causal relationships between elements. Be able to explain and predict actual phenomena. Be able to evaluate models and select models.	Lesson 5	KTP502: Designing experiments to verify the grading regularity of the properties of elements of the same group.
Level 4 Be able to associate an element's property with a substance's property. Be able to explain and compare element similarity/gradation by designing an inquiry experiment with an appropriate substance.	Adding the relationship between an element's properties and a substance's properties	Level 2 Be able to find model elements (multi-variable) for complex models and establish relationships between model elements	Lesson 4	KTP401: Predicting the grading regularity of an unfamiliar element's properties of the same period based on its atomic structure.
Level 3 Be able to illustrate and analyze a familiar element's position from the "position-structure-property" perspective.	Adding element position, position-Structure-element's property relationship		Lesson 3	KTP304: Illustrate the grading regularity of a familiar element's properties based on its atomic structure.
Level 2 Be able to associate atomic structure with an element's property and understand that the gain or loss of electrons is affected by the nucleus's attraction to its outermost electrons.	Relationship between atomic structure and an element's property	Level 1 Be able to find, discover and establish associations between elements of a simple model (single variable) based on empirical evidence (data or experimental facts) and revise them.	Lesson 2	KTP201: How do the factors of atomic structure affect the element's ability to gain or lose electrons?
Level 1 Be able to understand an element's property from a microscopic perspective. Be able to build a primary atomic structure perspective and recognize that a simple substance's oxidizability and an anion's reducibility are connected to an atom's gain or loss of electrons.	atomic structure		Lesson 1	KLT101: What factors of the atomic structure affect the element's ability to gain or lose electrons?

key teaching problems and key learning tasks to realize the LP were concluded. Students' data in the process were collected through paper-and-pencil evaluation of pre-test and post-test knowledge, performance evaluation of classroom observation, and interviews with students. Based on these data, the typical student's performance and the LP after HELS were described. Based on the teaching experiments and students' data, the LP was validated and supplemented. The teaching sequence was added to the LP model, and the key teaching problems, key learning tasks, and students' specific performance at each level and each lesson were extracted (see Table 8.3).

Representative LP Studies in China

After the clarification of the theoretical foundations, the progress variables, and the paradigms of LPs, we conducted a systematic literature review on LP studies in China. The database we used is the China National Knowledge Infrastructure (CNKI), which is the largest academic database in China. As education journals in China tend to publish short articles, almost all empirical LP studies, which are normally longer than 10 pages, are recorded among master's/doctoral dissertations. Considering that many theses of a few top universities are not included in CNKI, we also conducted additional searches on the theses of these universities using a search engine such as More Discovery. To provide a convincing and compatible overview of LP studies, we focused on the core inclusion criteria in a previous LP review: (1) a theoretical framework compatible with current research and (2) empirically supported validation (Jin, Mikeska, et al., 2019). With such conditions, we selected 126 empirically validated LPs developed by Chinese scholars (Figure 8.5), among which 99 LPs focus on disciplinary core ideas/concepts (DCIs/DCCs),[1] 17 LPs focus on scientific inquiry or scientific thinking (SI/ST), and 10 LPs integrate multiple dimensions of disciplinary core competences of science (ILP).

All dissertations or articles present research on student learning. For international comparison, we provide descriptive information following a framework in a previous review of LPs (Jin, Mikeska et al., 2019). As shown in Table 8.4, most studies focused on one or two grade bands. Most studies used K-12 students as the research sample, while only 4 studies included college students and 1 study included preschool children. Secondary school students received the most attention from scholars, as traditionally top normal universities focus more on secondary school education. Most large-scale studies relied on paper-and-pencil tests to

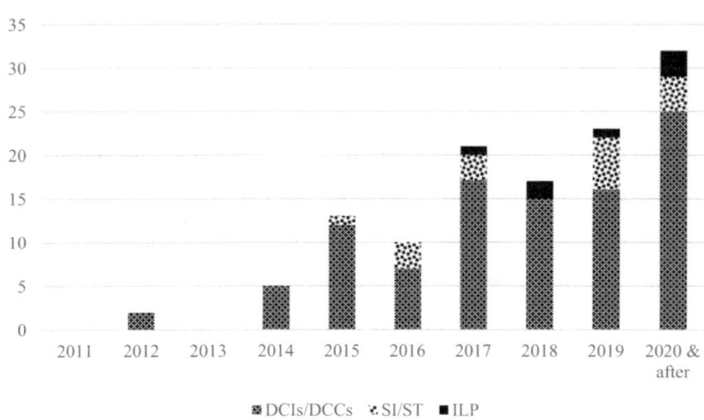

Figure 8.5 LP studies in China.

Table 8.4 Empirical studies that develop or validate LPs

Types of information	*Descriptive information and number of studies*				
Range of School levels	One grade band	Two grade bands	Three grade bands	Four grade bands	/
	91	28	7	/	/
Participants included in the study	Preschool children	Elementary students	Middle school students	High school students	College students
	1	11	39	108	4
Sample sizes[2]	Fewer than 100 participants	100–499 participants	500–999 participants	1,000–4,999 participants	5,000 and more than 5,000 participants
	10	69	23	18	3

collect data and quantitative methods such as item response theory (e.g., Rasch analysis), cognitive diagnosis (e.g., rule space model), and concentration analysis to analyze data.

LPs of Disciplinary Core Ideas/Concepts

In China, there is a tradition of discipline-based teaching in secondary schools, which means that physics, chemistry, biology, and geography are taught independently in most districts in grades 7–12. All of the above disciplines attach great importance to their disciplinary core ideas/concepts, whose scales are partly different from disciplinary core ideas (DCIs) in other countries. Some LP studies address disciplinary core ideas in curriculum standards (for example, "2. Energy" in physics, and "27. Ecosystems" in biology); some might feel these DCIs are too highly generalized and want to resolve them into subordinate concepts (such as "9. Chemical bonds" in chemistry, and "23. The life cycle of cells" in biology). Considering that building LPs of disciplinary core ideas/concepts is often taken as the first step in a series of LP studies, it is not surprising that more than 70% of LP studies in China involve LPs of disciplinary core ideas/concepts. Most of these studies focused on one core idea/concept, while a few studies covered more than one core idea/concept (Table 8.5).

In addition to the above studies, there are a large number of studies about the development of scientific concepts. Although many of these studies are titled "learning progression" (might double or triple above our statistics of LPs), they did not meet the criteria of true LP studies (Jin, Mikeska et al., 2019): they either lacked a robust theoretical framework or lacked a sound validation process. Therefore, these studies have not been included in our statistics. However, the prevalence of these studies shows that teachers in China truly appreciate the idea of LPs and are willing to share their experiences of students' conceptual development with the frameworks and opportunities offered by LP studies. The potential value and further application of these studies need further discussion.

LPs of Scientific Inquiry and Scientific Thinking

Science education in China attaches great value to both scientific inquiry and scientific thinking. In Table 8.6, we organized the inquiry and thinking according to Chinese national curriculum standards[3] (for a brief introduction to science curriculum standards in China, see Yao & Y. Y. Guo, 2018c). The 17 LPs in this area have covered five scientific practices in NGSS, with the other three absent ("Asking questions," "Analyzing and interpreting data,"

Table 8.5 LPs of disciplinary core ideas/concepts (selected)*

Disciplines	*Learning progressions (selected)*
Physics	1 Matter (Y. Yang, 2018)
	2 Energy (e.g., Yao & Y. Y. Guo, 2013; Wei, 2014; S. S. Han, 2021)
	3 Electromagnetic fields (Y. F. Zhang, 2016, 2020)
	4 Mechanics (L. X. Xie, 2015; Yao & C. S. He, 2020)
	5 Buoyancy (Y. Z. Gao, 2018)
Chemistry	6 Structure of matter (M. C. Huang, 2014)
	7 Energy (e.g., W. Z. Wang, 2015; Y. Zhou, 2016; L. Chen, 2020)
	8 Conservation in chemistry (Y. C. Wang, et al, 2020)
	9 Chemical bonds (Liang, 2019; Jing Wang, 2020)
	10 Interaction of particles (Tong, 2015; Zhao, 2017)
	11 Organic compounds (J. Q. Zhang, 2018)
	12 Inorganic compounds (Lan Wang, 2015)
	13 Chemical equilibrium (e.g., J. S. Huang, 2017; G. T. Wang, 2017; Jiang, 2019)
	14 Chemical reaction (Sun, 2015)
	15 Ionic reaction (J. C. Li, 2017; J. X. Yang, 2019; L. N. Li, 2020)
	16 Electrolyte solution (T. T. Xie, 2018; Qian, 2019)
	17 Redox reaction (K. X. Liu, 2012; H. Li, 2017)
	18 Electrochemistry (e.g., X. X. Ding, 2014; Y. P. Wu, 2017; Song, 2017)
Biology	19 Energy (H. H. Hu, 2019)
	20 Matter and energy (Y. X. He, 2020)
	21 The compositions of cells (Q. L. Li, 2020)
	22 The metabolism of cells (J. Z. He, 2017; C. L. Hu, 2020)
	23 The life cycle of cells (Y. N. Ding, 2020)
	24 Heredity (e.g., J. Gao & X. Y. Gao, 2015; Y. Y. Li & X. F. Li, 2018; Dong, 2017)
	25 Evolution (Xia, 2015; Y. Y. Chen, 2016; L. Liu, 2019)
	26 Stability and Regulation (Qin, 2017; Mao, 2020)
	27 Ecosystems (e.g., F. Gao, 2012; Du, 2018; Y. M. Liu, 2021)
Geography	28 Movement of Earth (X. M. Lin, 2018; X. L. Hu, 2019)
	29 Physical geography (J. Yang, 2015)
	30 Climate (Tan, 2019)
	31 Human–land relationship (M. Y. Wang, 2020)
	32 Area (P. X. Chen, 2020)
	33 Human activities (J. Yang, 2015)
	34 Resources (B. R. Chen & M. Wang, 2021)

Note
* Studies in this table are not listed in the reference list.

and "Obtaining, evaluating, and communicating information"). Four of these LPs focus on problem-solving (No. 3 and No. 4), systems thinking (No. 13), and scientific imagination (No. 14), which are not among the eight scientific practices in NGSS.

Integrated LPs and LPs of Subject Competencies

Initially, LP studies of scientific inquiry and scientific thinking were somewhat content-general (e.g., Yao & Y. Y. Guo, 2018b), which did not emphasize their specific combination with scientific content knowledge. In these studies, content knowledge was regarded as the context where students exhibited their inquiry or thinking ability.

Recently, an increasing number of Chinese scholars have sought to develop integrated LPs. For example, by combining previous LP frameworks in scientific explanation (Yao & Y. Y. Guo, 2018b) and disciplinary core ideas energy (Yao et al., 2017), knowledge

Table 8.6 LPs of scientific inquiry and scientific thinking*

Scientific inquiry or scientific thinking		*Learning progressions*
Scientific inquiry	1	An LP of physics experiments investigations (X. L. Wang, 2019)
	2	An integrated LP fusing matter and design investigation (H. Q. Wu & Y. Zhang, 2020)
	3	An LP of chemistry problem solving (W. Q. Hu, 2020)
	4	An LP of geography problem solving (X. B. Su et al., 2020
Scientific thinking	5	An LP of argumentation in physics (Y. Chen, 2016)
	6	An integrated LP fusing physics core ideas and argumentation (Mi, 2018)
	7	An LP of evidence-based reasoning (X. Guo, 2019; D. Zhang, 2019)
	8	An LP of scientific explanation (Yao & Y. Y. Guo, 2018b)
	9	An integrated LP fusing electromagnetic fields and scientific explanation (Yao, 2020)
	10	An LP of chemical modeling (Si, 2017; Shi, 2019; Y. M. Hu, 2019)
	11	An LP of modeling in physics (J. Zhang, 2014)
	12	An LP of stoichiometry (Huangfu, 2016)
	13	An LP of systems thinking in geography (Ren, 2020)
	14	An LP of scientific imagination (C. C. Wang, 2017)

Note

* Studies in this table are not listed in the reference list.

components are incorporated with scientific practices for each level of an integrated LP. The main issue in integrated LP research is validating this type of LP (identifying whether and how core ideas and key practices are progressing). The SCF model and SCF-based HELS (Lei Wang et al., 2022) are promising theoretical frameworks to solve this problem. Again, we take W. Y. Wang's LP study on the periodic law of elements as an example (W. Y. Wang, 2018). She integrated student performance on knowledge content and Model & Modeling ability (see Table 8.3). Three strategies are used to solve the validation problem. First, it is necessary to make theoretical presuppositions for the two dimensions, double-code the instrument tools, and evaluate students' performance on the two dimensions. Second, the teaching sequence should be added to the LP model, and the development levels of the two dimensions should be evaluated at each possible progression turning point. Third, all the data of the whole process of HELS should be collected, including quantitative data from the paper-and-pencil assessment of pre-test and post-test, as well as qualitative data from classroom observation and student interviews, to supplement and describe students' performance. Using these strategies, W. Y. Wang (2018) tested whether and how students have made progress in core ideas and key abilities at the same time, integrating the two dimensions. Only on this basis can we better validate whether and how core ideas and key abilities are progressing.

A few Chinese studies have developed LPs for other mental constructs in science education, such as disciplinary core competencies. In the newly revised science curriculum standards in China, each discipline proposed four or five disciplinary core competencies (for a review, see Yao & Y. Y. Guo, 2018a). Targeting curriculum–instruction–assessment coherence, several researchers have developed LPs to investigate student performance, such as C. W. Gao and M. Wang (2021) in geography education (geography practices ability).

Applications of LPs

Practicality is an important quality of LPs (Jin, Mikeska et al., 2019; Lehrer & Schauble, 2015; Yao & Y. Y. Guo, 2018a). Although there are a few controversies, at the macro level, the inferences of LP's description of students' levels have provided useful information for curriculum and assessment design. At the micro level, scholars have recently sought ways to use LPs to improve instruction.

Curriculum Standards Design Facilitated by LPs

From the macroscopic view, the design of science curriculum standards in China has been informed by the concepts and research results of LPs. At the primary school level, drawing from the results of international comparison research and LPs, the revision group redesigned the three learning stages (grades 1–2, grades 3–4; grades 5–6) of the *Primary School Science Curriculum Standards*. Stage expectations were proposed for each division of curriculum objectives: "scientific knowledge," "scientific inquiry," "scientific attitude," and "science, technology, society, and the environment." Taking "plan an investigation" in the dimension of scientific inquiry as an example (Table 8.7), in grades 1–2, standards only expect students to realize that scientific investigation needs a plan directed by their teacher. In grades 3–4, students' plan-making is still under the guidance of the teacher, but they are expected to make a simple investigation plan based on the knowledge they learned. In grades 5–6, students are expected to design simple investigation plans and experimental plans independently.

At the secondary school level, the newly revised science curriculum standards for senior high school proposed the "levels of subject core competencies," which are composed of 4 or 5 achievement levels for each dimension. The descriptions of learning performance at each level are based on a large-scale assessment, adopting an approach quite similar to the large-scale assessment-based approach of LPs, on a representative sample of Chinese high school students. Similar to the integrated LPs, integration of the learning performance in each dimension, which is called the "standards of academic quality," was proposed to guide the formative assessment in everyday instruction and qualification/graded examination. Now, the "standards of academic quality" in curriculum standards replaced the examination syllabus. Therefore, the curriculum design and assessment design, for the first time, achieved in-depth coherence and consistency in China (MOE, 2022).

Table 8.7 An example of the stage expectation in science curriculum standards for primary school

	Grades 1–2	*Grades 3–4*	*Grades 5–6*
Plan an investigation	Students are expected to … under the direction of teachers, realize that scientific investigation needs a plan.	under the guidance of teachers, make a simple investigation plan based on the knowledge they have learned.	make a relatively sound investigation plan based on the knowledge they have learned; design a single-variable experiment with the consciousness of variable control and demonstrate a preliminary ability in experimental design.

Applying LPs in Instruction Improvement

From the microscopic view, LPs should be useful for instruction improvement that helps students move toward the upper anchor of the LPs (Jin, van Rijn et al., 2019; Lehrer & Schauble, 2015; Yao & Y. Y. Guo, 2018a). Many groups, led by university scholars, have conducted experimental studies on LP-based instruction improvement. For example, at the primary school level, researchers have organized a series of case studies on the learning process of concepts such as "density" and "force and motion." LP application at the secondary school level is collaborating with an instruction-based approach to LP development. In this kind of action research, secondary school teachers not only engage in professional development but also act as research partners. Evidence from quasi-experimental studies (e.g., Yao et al., 2023) showed that teachers' use of LPs during their instructional design process had a positive impact on classroom instruction and students' performance. The LP-based professional development in the "Disciplinary Instruction-Improvement" project produced a significant influence on teachers. Many instructions won teaching awards at all levels. In addition, many LP-based instructional designs have provided excellent teaching cases for curriculum development and textbook revision (e.g., L. Wang & G. J. Chen, 2019).

Applying LPs in Assessment Development

Since its origins, research on LP has been closely linked to cognitive assessment in science education. On the one hand, constructing an LP requires information about student performance, which is obtained from educational measurements. On the other hand, assessment tools, which are developed with an LP research base, prove that LPs can provide high-quality, systematic information about students' academic progress. Assessment development in the 21st century has a basic assumption that all assessment systems should be designed based on students' cognitive models. As the influence of LP research continues to expand, many research institutions and scholars have used LPs as a basis model for assessment development and item design.

Taking chemistry as an example, Lei Wang and her team have made some achievements in applying LPs in assessment. Taking the Subject Competency Framework (SCF, Figure 8.1) and LPs as the theoretical framework, they developed a series of assessment instruments, conducted academic quality standards evaluations, and generated a series of evaluation reports. For example, Q. Wang (2021) applied LPs and SCF to evaluate the academic quality of inorganic subjects in senior high school. Moreover, the published assessment tools in LP studies and newly developed item pools with an LP foundation have been absorbed in online learning assistance systems such as "Smart Learning Partner" (http://slp.bnu.edu.cn/). LP-based assessment can also help improve the coherence between classroom assessments and large-scale assessments.

Discussion and Future Directions

The study of LPs essentially reflects the development of contemporary views of science education: a shift from the traditional segmented teaching of individual subject knowledge and individual inquiry skills to the systematic design of instructional sequences to help students achieve an integrated development of core ideas and key abilities, as well as scientific attitudes and responsibility or, more concisely, to develop students' core competencies in science. Comparing the studies of LPs in China and other countries, it is notable that two main differences can be found under the same research paradigm.

The first difference concerns the progress variables. While we share some of the progress variables (such as complexity and knowledge integration) with the English-speaking science education community, we also highlight some progress variables proposed by Chinese scholars. This chapter introduced a series of progress variables used in the LP studies in China, such as the construct model in Table 8.1 and Figure 8.1. These construct models for student mental development in science, despite their significant influence in the science education community whose working language is Chinese (the paper introducing the models in Table 8.1 and Figure 8.1 has been downloaded 15,000 times and cited more than 500 times in the last 5 years), are still not well known by the English-speaking science education community. Unfortunately, space limitation does not allow us to further illuminate these models, to explain the process of how we construct these models, or to provide a validity argument for these models in one chapter. However, we still want to present the academic and cultural context in which these models were developed.

Disciplinary education researchers (i.e., scholars specializing in physics education, chemistry education, biology education, geography education, etc.) have been leading the development of science education in China for a long time. Traditionally, these disciplinary education researchers focused on teaching, and learning in secondary school come from different faculties (e.g., college/departments of physics, chemistry, biology, and geography), which means that their colleagues are physicists, chemists, etc. Influenced by a domain-general research paradigm, psychological models, or pedagogical theories, the teachers first construct domain-specific theories for cognitive development in certain disciplines by combining those domain-general constructs with the practice of subject instruction. These multiple discipline-based theories of cognitive development would then draw on each other and, of course, engage in intense competition at times when the whole science education academic community communicates. Refined and tested theories will have an interdisciplinary impact on science education. This "general–specific–general" path of theoretical construction is the key to the continued evolution of science education in China.

The second difference is the credit given to quantitative validation. There are several ways of classifying LP studies (Duschl et al., 2011; Jin, Mikeska et al., 2019; Yao & Y. Y. Guo, 2018a). For example, according to the grain size of LPs, previous LP studies can be classified into two categories: "LP at the scale of instructional design" and "LP at the scale of curriculum planning." The LPs at the scale of instructional design are usually more specific LPs, and their topic scope is relatively small. In contrast, the LPs at the scale curriculum planning have a more general topic scale and normally cover several years of teaching and learning.

Two of the earliest LP studies published by Teachers College at Columbia University are typical examples of these two types of LPs (Corcoran et al., 2009). The first one, an LP of buoyancy, is an instructional design scale LP. The second one, an LP of atomic molecular theory, is an LP on the scale of curriculum planning. The key difference between the two types of LPs is not only in the scale of the topic but also in the methodology. Most LPs at the curriculum planning scale normally adopt an assessment-based approach. For a larger scale, the sensitivity of the measurement tools becomes less demanding. Therefore, both the measurement tool and the LP can be validated more easily. However, for the LP at the instructional design scale, it is difficult to quantitatively validate when the student's progression is small. Even though researchers have conducted teaching experiments and qualitative analysis for the development of this kind of LP, it is still not easy to achieve a certain level of adequate validation because the result is complex and can be explained differently as many influencing factors (e.g., curriculum, textbook, teacher, student, learning environment, etc.) have been included in teaching experiments. The impact of LPs on the success of a teaching experiment is only one of the many factors. As a result, a large proportion of the publications in international journals describe LPs at the scale of curriculum planning, while publication of an LP at the instructional design scale is relatively rare. Ironically, in regard to serving instructional practice, the LP at the instructional design scale has become more popular with teachers. Still using the previous example, the LP of buoyancy has been rapidly adopted in instruction in many countries, including China.

All LP researchers emphasize that LPs should eventually be used in instructional practice. Scholars both in China and the US have warned that too much focus on validation might hinder the progression of LP studies (e.g., Lehrer & Schauble, 2015; Yao & Y. Y. Guo, 2018a). We agree with the view that "LPs are just one of many different types of cognitive models. They by no means should be viewed as authoritative depictions of science learning" (Jin, van Rijn et al., 2019, p. 1326). Therefore, we conceptualize that an LP is a blueprint for describing the cognitive development of students in a particular domain under the influence of instruction (Yao & Y. Y. Guo, 2018a). Based on this conceptualization, the aim of LP research should be facilitating students' learning, the research content should be domain-specific mental constructs, and the cognitive development in this definition is instruction-assisted development. With this definition, we revisit the question of validation of LPs. LPs themselves are a blueprint that is difficult to validate directly, but the products that emerge from them (such as curriculum design, assessment tools, and instructional designs) and the effects they produce can be tested to some extent. In this view, the LP should not be viewed as a static hypothesis waiting for testing and corroboration but as a "trading zone" in which all practitioners of science education can share wisdom and be mutually enlightened.

Acknowledgment

This research was supported by the "International Joint Research Project" of the Faculty of Education, Beijing Normal University. The authors would like to thank Han Si-Si, Chen Xiao-Ge, and Zou Zi-Wei for their help in the literature review.

Notes

1 Some of the disciplinary core ideas printed in Chinese science curriculum standards share the same name as disciplinary core ideas and crosscutting concepts in NGSS but have different meanings from the latter. Therefore, we use the term "disciplinary core ideas/concepts (DCIs/DCCs)" when speaking of Chinese LPs of scientific knowledge. See section "LPs of Disciplinary Core Ideas/Concepts" for details of this issue.

2 The sum of this row does not equal the total number of Chinese empirical validated LPs selected in this review (126) because several of these studies did not report their sample size.

3 Scientific inquiry and scientific thinking are defined slightly differently in curriculum standards of each discipline. Taken together, scientific inquiry consists of raising questions, making plans, carrying out experiments and communication, while scientific thinking includes evidence-based reasoning, modeling, critical thinking or scientific argumentation, systems thinking and creative thinking.

References

Corcoran, T., Mosher, F. A., & Rogat, A. (2009). *Learning progressions in science: An evidence-based approach to reform (Report No. RR-63)*. Consortium for Policy Research in Education. https://repository.upenn.edu/cpre_researchreports/53/

Driver, R., Rushworth, P., Squires, A., & Wood-Robinson, V. (1994). *Making sense of secondary science: Research into children's ideas*. Routledge.

Duschl, R. A., Maeng, S., & Sezen, A. (2011). Learning progressions and teaching sequences: A review and analysis. *Studies in Science Education*, *47*(2), 123–182. 10.1080/03057267.2011.604476

Gao, C. W., & Wang, M. (2021). 高中生地理野外实践能力测评框架的构建与实证分析 [Construction and empirical analysis of evaluation framework of high school students' geography field practice ability]. 中国考试 *[Journal of China Examinations]*, (04), 54–59. 10.19360/j.cnki.11-3303/g4.2021.04.009.

Guo, Y. Y., & Yao, J. X. (2016). 基于核心素养学习进阶的科学教学设计 [Instructional design of science lessons based on the learning progression of key competencies]. 课程·教材·教法 *[Curriculum, Teaching Material and Method]*, *36*(11), 64–7010.19877/j.cnki.kcjcjf.2016.11.011.

Guo, Y. Y., Yao, J. X., & Zhang, J. (2013). 整合与发展——科学课程中概念体系的建构及其学习进阶 [Integration and development: The construction and learning progression of conceptual system in science curriculum]. 课程·教材·教法 *[Curriculum, Teaching Material & Method]*, *33*(2), 44–49. 10. 19877/j.cnki.kcjcjf.2013.02.010.

Harlen, W. (2010). *Principles and big ideas of science education*. Hertfordshire: Association for Science Education.

Jin, H., Mikeska, J. N., Hokayem, H., & Mavronikolas, E. (2019). Toward coherence in curriculum, instruction, and assessment: A review of learning progression literature. *Science Education*, *103*(5), 1206–1234. 10.1002/sce.21525

Lehrer, R., & Schauble, L. (2012). Seeding evolutionary thinking by engaging children in modeling its foundations. *Science Education*, *96*(4), 701–724. 10.1002/sce.20475

Lehrer, R., & Schauble, L. (2015). Learning progressions: The whole world is NOT a stage. *Science Education*, *99*(3), 432–437. 10.1002/sce.21168

Lin, C. D. (2016). *21*世纪学生发展核心素养研究 *[Research on core competencies for students' development in 21st century]*. Beijing Normal University Press.

Ministry of Education, P. R. China. (2022). 义务教育科学课程标准 *[Science curriculum standards for primary school]*. Beijing Normal University Press.

National Research Council. (2007). *Taking science to school: Learning and teaching science in grades K-8*. National Academies Press.

NGSS Lead States. (2013). *The next generation science standards: For States, By States*. National Academies Press.

Wang, Lei (2016). 学科能力构成及其表现研究—基于学习理解、应用实践与迁移创新导向的多维整合模型 [Exploring performance and intrinsic composition of disciplinary competency—Based on the multi-integrative model of learning-applying -innovating]. 教育研究 *[Educational Research]*, *37*(9), 83–92, doi:CNKI:SUN:JYYJ.0.2016-09-012.

Wang, L., & Chen, G. J. (2019). 普通高中化学教科书 *[Common high school chemistry textbooks]*. Shandong Science and Technology Press.

Wang, L., Zheng, G. M., Guo, Y. Y., Wang, Q., Cao, Y. M., & Wang, J. (2018). 基于核心素养的学科能力诊断评价和教学改进系统—九学科协同研究与实践 [Disciplinary competence diagnostic evaluation and teaching improvement system based on core competence—Nine disciplines collaborative research and practice].

Wang, Lei., Wang, Q., Kong, S., Hu, J., & Chen, X. (2022). Subject competency framework in fostering high-end lesson study—A case of teaching "properties of iron salts" unit in a senior high school. *International Journal for Lesson and Learning Studies*, *11*(2), 73–90. 10.1108/IJLLS-11-2021-0103

Wang, Q. (2021). 基于新课标的高中生必修无机物主题学业质量评价研究 [Research on the evaluation of academic quality of compulsory inorganic subjects for senior high school students based on the new curriculum standard] [Unpublished master's thesis]. Beijing Normal University.

Wang, W. Y. (2018). 基于学习进阶的元素周期律主题教与学研究 [The research for teaching and learning periodic law of elements—Based on learning progression] [Unpublished master's thesis]. Beijing Normal University.

Xin, T., Le, M. L., Guo, Y. F., & Jiang, Y. (2015). 学业质量标准的建立途径: 基于认知诊断的学习进阶方法 [The approach to establishing achievement standard: The learning progressions based on cognition diagnostic]. 教育学报 *[Journal of Educational Studies]*, *11*(05), 72–79. 10.14082/j.cnki. 1673-1298.2015.05.009

Yao, J. X., & Guo, Y. Y. (2014). 为学生认知发展建模：学习进阶研究回顾与展望 [Modeling students' cognitive development: A review of ten-year research on learning progression]. 教育学报 *[Journal of Educational Studies]*, *10*(5), 35–42. 10.14082/j.cnki.1673-1298.2014.05.005

Yao, J. X., & Guo, Y. Y. (2018a). 学习进阶：素养的凝练与范式的演变 [Learning progression: Focalizing on core competences and its paradigm reframing]. 教育科学 *[Education Science]*, *34*(4), 30–35. 10.3969/j.issn.1002-8064.2018.04.005

Yao, J. X., & Guo, Y. Y. (2018b). Validity evidence for a learning progression of scientific explanation. *Journal of Research in Science Teaching*, *55*(2), 299–317. 10.1002/tea.21420

Yao, J. X., & Guo, Y. Y. (2018c). Core competences and scientific literacy: The recent reform of the school science curriculum in China. *International Journal of Science Education*, *40*(15), 1913–1933. 10.1080/09500693.2018.1514544

Yao, J. X., Liu, Y. X., & Guo, Y. Y. (2023). Learning progression-based design: advancing the synergetic development of energy understanding and scientific explanation. *Instructional Science*, 51, 397–42110.1007/s11251-023-09620-0.

Yao, J. X., Guo, Y. Y., & Neumann, K. (2017). Refining a learning progression of energy. *International Journal of Science Education*, *39*(17), 2361–2381. 10.1080/09500693.2017.1381356

9

IMPLICATIONS OF MATHEMATICS LEARNING TRAJECTORIES FOR SCIENCE EDUCATION

Douglas H. Clements, Shannon Stark Guss, and Julie Sarama

Approaches to standards, curriculum development, and pedagogy in math and science are diverse; however, recent years have seen a growing movement to base each of these on *learning trajectories* or *learning progressions* (e.g., Clements & Sarama, 2014; National Research Council, 2007, Chapter 8; 2012, 2013). In this chapter, we compare the various conceptions and terms of this construct; present our definition; differentiate between our conception and others'; briefly review the current evidentiary base in our area, early childhood mathematics; and draw implications for future work on progressions in science.

Concepts and Definitions: A Need for Clarity

The Learning Trajectory Construct in Math

Learning trajectories are a tool supporting the development of a curriculum or a curriculum component. Simon stated that a "hypothetical learning trajectory" included "the learning goal, the learning activities, and the thinking and learning in which the students might engage" (1995, p. 133). Building on Simon's definition but emphasizing a cognitive science perspective and a base of empirical research, "we conceptualize learning trajectories as descriptions of children's thinking and learning in a specific mathematical domain, and a related, conjectured route through a set of instructional tasks designed to engender those mental processes or actions hypothesized to move children through a developmental progression of levels of thinking, created with the intent of supporting children's achievement of specific goals in that mathematical domain" (Clements & Sarama, 2004, p. 83). In other words, each learning trajectory has three parts: (a) a goal, (b) a developmental progression, and (c) instructional activities. To attain a specific mathematical competence in a given topic or domain (the goal), students learn each successive level (the developmental progression), aided by tasks (instructional activities) designed to build the mental actions-on-objects that enable thinking at each later level.

The name "learning trajectory" may indicate that it refers only to the developmental progression, but the term reflects its roots in a constructivist perspective (Simon, 1995). Although the name emphasizes learning over teaching, Simon's and Clements and Sarama's definitions involve instructional tasks and teaching strategies.

DOI: 10.4324/9781003170785-10

Both these aspects (developmental progressions of thinking and instructional sequences) serve the most important but often least discussed aspect of learning trajectories – the goal. Our learning trajectories base goals on both the expertise of mathematicians and research on students' thinking about and learning of mathematics (Clements, Sarama et al., 2004; Fuson, 2004; Sarama & Clements, 2009). This results in goals that are organized into the "big" or "focal" ideas of mathematics: overarching clusters of concepts and skills that are mathematically central and coherent, consistent with students' early intuitive thinking, and generative of future, more sophisticated thinking and learning (Clements, Sarama, et al., 2004; NCTM, 2006).

Once the mathematical goals are established, research is reviewed to determine if there is a natural developmental progression identified in theoretically- and empirically-grounded models of students' thinking, learning, and development (Carpenter & Moser, 1984; Griffin & Case, 1997, Sarama & Clements, 2009). The issue of what is meant by *natural* is sure to arise. These progressions play a unique role in children's cognition and learning because they are particularly consistent with the findings of cognitive and developmental science and children's intuitive knowledge and patterns of thinking and learning at various levels of development (at least in a particular culture but guided in all cultures by universal cognitive structures), with each level characterized by specific mental objects (e.g., concepts) and actions (processes). These actions-on-objects are children's primary way of operating on, knowing, and learning about the world (Sarama & Clements, 2009).

On this basis, researchers build a cognitive model of students' learning in a specific domain that is sufficiently explicit to describe the processes involved in the construction of the mathematical goal across several qualitatively distinct structural levels of increasing sophistication, complexity, abstraction, power, and generality (we will return to the methodologies used in a subsequent section). In math, we posited, and empirical evidence using multiple methodologies has confirmed, that these broad levels within a topic develop in a sequence. Finally, *interactions* between topics create a more complex network.

Levels are *not* stages (Lehrer & Schauble, 2015); instead, they are patterns of thinking that overlap (Sarama & Clements, 2009). That is, levels of thinking are theoretically nonrecurrent (Karmiloff-Smith, 1984); however, people can frequently "return" to earlier levels of geometric thinking in specific contexts. Therefore, we postulate the construct of *nongenetic levels* (Clements et al., 2001; Sarama & Clements, 2009). Nongenetic levels have two special characteristics. First, social influences and instruction determine progress through nongenetic levels more than age-linked development. (At this point, this only implies that progression does not occur by necessity with time but demands, in addition, instructional intervention. Certain levels may develop under maturational constraints; further research is needed on this issue.) Second, although each later nongenetic level builds on the knowledge that constitutes earlier-developing levels, its nongenetic nature does not preclude the instantiation and application of earlier levels in specific contexts (not necessarily limited to especially demanding or stressful contexts). There is a probability of evoking each level depending on intraindividual and contextual factors. However, this process is codetermined by conscious metacognitive control, and this control increases as one moves up through the levels. Therefore, people have increasing choices to override the default probabilities. Furthermore, using different levels is environmentally adaptive; thus, the adjective "later" should be understood as an increased level of abstraction and generality without the implication of either inherent superiority or the abandonment of earlier-developing levels as a consequence of the development of later levels of thinking. Nevertheless, the levels would constitute veridical qualitative changes in behavior, especially regarding the construction of mathematical schemes out of action (Sarama & Clements, 2009).

These concepts are part of our theoretical framework, a synthesis that we call *Hierarchic Interactionalism* (for a full explication, see Sarama & Clements, 2009). The term indicates the influence and interaction of global and local (domain-specific) cognitive levels and the interactions of innate competencies, internal resources, and experience (e.g., cultural tools and teaching). Foundations of mathematical ideas are represented first intuitively, then with language, then metacognitively–this last indicates that the child possesses an understanding of the topic and can access and operate on those understandings. The theory posits that developmental progressions are domain and topic-specific, that different developmental courses are possible (and that differences within and across individuals create variation, which is the wellspring of invention and development). These concepts and skills develop co-mutually, as expected probabilities in learning and development rather than linear stages that must be attained (Confrey, 2019). Although levels are purposefully ordered by increasing sophistication, a network of skills and concepts is also hypothesized and contributes to the design of instruction – see Table 9.2. Providing instructional experiences aligned with developmental progressions is usually more effective, efficient, and generative for the child than learning that does not follow these paths (Sarama & Clements, 2009).

This theoretical framework thus establishes the basis for the instructional tasks – the third and final component of a learning trajectory. Our work comprises key tasks designed to promote learning or assess learning (or *both* – curriculum-embedded assessments are often ideal) at a particular conceptual level or benchmark in the developmental progression. Extant research is used to identify tasks effective in promoting students' learning at each level by encouraging children to construct the concepts and skills that characterize the next level. We hypothesize the specific mental constructions – the *building* of mental actions-on-objects and patterns of thinking that constitute students' thinking at each level. We design tasks that include external objects and actions that mirror the hypothesized mathematical activity of students as closely as possible; for example, objects may be shapes or sticks, and actions might be creating, copying, uniting, dis-embedding, and hiding both individual units and composite units. Tasks require children to apply, externally and mentally, the actions and objects of the goal level of thinking (see Clements & Battista, 2000, for further description and examples of mathematical objects/concepts and mathematical actions/processes that operate on them). These tasks are sequenced, corresponding to the order of the developmental progressions to complete the hypothesized learning trajectory. Further, the tasks' characteristics and accompanying pedagogical interactions are explicitly linked to transitions *between* levels.

Instructional tasks that promote learning are chosen through formative assessment. Often embedded within the tasks that promote learning, formative assessment ties the instructional tasks with the developmental progression. Professional noticing of children's performance – the ease and confidence with which they engage in a task, the misconceptions or lack thereof, and their ability to think about their thinking (metacognition) – supports instructional decision-making about the extent to which students can engage at a later level of development. Alternatively, children's responses and behaviors may indicate the need to present material earlier in the developmental progression.

In addition to formative assessment, researchers call for using learning progressions in the design of summative assessments as an essential part of understanding children's learning (Shepard & Pellegrino, 2018). The development of the Research-based Early Mathematics Assessment incorporated at least two items for each developmental level (Clements et al., 2008). Although the correctness of tasks was designed to capture children's levels of thinking at a specific level, the strategies children use to come to that answer can be used to assess children's competence at several levels. For example, a correct score for a task in which children place enough objects to change the number in the set from six to ten can be completed in several ways. At earlier levels of development, children may count all the existing objects, then place the next four. Children may add on from six at the level the task was designed to assess. At a later level, children may make a verbal statement about a known combination ("six and four is ten."), then place the last four objects all at once. The benefits of using developmental progressions to develop assessments include support for understanding children's thinking and making decisions that follow the hypothesized learning trajectory. This provides qualitative interpretations about what students know and can do that align instruction to children's strengths and needs (Shepard & Pellegrino, 2018).

Learning Progressions in Science

The term most often encountered in science education is "learning progressions" (LPs). Is this the same concept? The precise answer depends on the source. However, progressing according to students' thinking appears to be a shared commitment:

> [W]e develop the idea of learning progressions as an approach to research synthesis that could serve as the basis for a dialogue that includes researchers, assessment developers, policymakers, and curriculum developers. Learning progressions describe the successively more sophisticated ways of thinking about a topic that can follow one another as children learn about and investigate a topic over a broad period (e.g., 6 to 8 years). They are crucially dependent on instructional practices if they are to occur. (National Research Council, 2007, p. 214)

The importance of the core component of LTs, developmental progressions of levels of thinking and learning, is at the heart of the shared conceptual framework. The first component of learning trajectories, the goal, with its conceptual and procedural structures, is implicit but explicit in the foundational works (e.g., National Research Council, 2007, Chapter 8; 2012, 2013), although they are treated as separate constructs: "Hence, core ideas and their related learning progressions are key organizing principles for the design of the framework (National Research Council, 2012, p. 26).

In addition to the idea of distinct levels of thinking and learning defined in mathematics, a framework for science incorporates dimensions of practices and crosscutting concepts, sometimes integrated and defined as progress variables (Corcoran et al., 2009; Kennedy et al., 2005). Some have said that LTs and LPs differ in granularity, with mathematics defining levels in small and specific terms and science defining levels in longer terms (Confrey, 2019). However, progress variables in science viewed as "representations of the knowledge, skills, and other competencies one wishes to increase through the learning activities associated with a curriculum," may be more akin to levels of developmental progressions in mathematics (Kennedy et al., 2005, p. 2). Progress variables can translate between assessment and curricular decisions, further likening them to developmental progression levels (Kennedy et al., 2005). The idea of progress variables in science progressions, including both understanding and practices, is also similar to math progressions, given that levels of progressions in math often incorporate behaviors or practices, such as verbal counting or object correspondence, with levels of thinking, such as knowing number words with or without the additional understanding that number words should correspond to objects being counted. The importance of connected dimensions within a progression is echoed in connections across progressions (see Table 9.2).

Further, several researchers have developed and evaluated quite specific learning progressions, often for older students. For example, learning progressions have been developed for concepts of electricity, the particle nature of matter, and energy (Anderson, 2008; Hadenfeldt et al., 2016; Jin & Anderson, 2012; Kaldaras et al., 2021; Liu, 2013; Merritt & Krajcik, 2009; Smith et al., 2006).

What is unclear in the notion of "progressions" is the inclusion of the third component of LTs, instructional activities, and strategies. Accepting that LPs are "crucially dependent" on them, the LP seems "outside" of instruction, thus equating LPs with LT's developmental progression. Indeed, another National Research Council report uses the specific phrase: "science education ... is built on the notion of learning as a developmental progression" (2012, pp. 10–11). However, that report later seems to include teaching within the definition: "Such progressions describe how students' understanding of the idea matures over time and the instructional supports and experiences that are needed for them to make progress" (National Research Council, 2012, p. 26).

Instructional practices that assess learning are more evident in research than practices that promote learning. Work has been done to align instructional experiences to hypothesized learning progressions (Kaldaras et al., 2021; Kennedy et al., 2005), and researchers call for extensions of LPs to be applied to curricular design (Duschl et al., 2011). Much of this work applies tasks associated with learning progressions in science to the development of assessments (Confrey et al., 2019; Kaldaras et al., 2021; Kaldaras et al., 2022; Merritt & Krajcik, 2009). The work on developing assessments uses developmental progressions that separate content and practices (Kennedy et al., 2005) and more complex frameworks that integrate core disciplinary ideas, crosscutting concepts, and practices (Kaldaras et al., 2021). Continued work in developing assessments will likely support the validation of learning progressions and further research and development of instructional experiences using these progressions.

Benefits of Relating LTs and LPs

Potential Contributions of Learning Trajectories Theory and Methods to Science Learning Progressions

Include Instruction as an Inextricable Component of LTs or LPs

Most science educators discuss instruction, but the inclusion or connections between developmental progressions and instruction vary widely. Although studying and applying either psychological developmental progressions or instructional sequences separately can be valid research goals, and studies of each can and should inform education, we believe the power and uniqueness of the learning trajectories construct stems from the inextricable interconnection between these two aspects.

Hierarchical Interactionalism posits that learning trajectories, as defined, is a particularly fecund instructional approach (Clements & Sarama, 2004). Based on the hypothesized, specific mental constructions (mental actions-on-objects) and patterns of thinking that constitute students' thinking at the levels of the developmental progression, curriculum developers design instructional tasks that include external objects and actions that mirror the hypothesized mathematical activity of children as closely as possible. These tasks are sequenced, with each corresponding to a level of developmental progressions, to complete the hypothesized learning trajectory. Such tasks will theoretically constitute a particularly efficacious educational program; however, there is no implication that the task sequence is the only path for learning and teaching, only that it is hypothesized to be one fruitful route. Empirical evaluations of that hypothesis have been positive (see the section "Developing and Evaluating LPs"). Tasks present a problem; people's actions and strategies to solve the problem are represented and discussed; reflection on whether the problem is solved, or partially solved, leads to new understandings (mental actions and objects, organized into strategies and structures) and actions. Specific learning trajectories are the main bridge that connects the "grand theory" of hierarchic interactionalism to particular theories and educational practice.

Further, such inextricable, detailed connections between the theory and empirical work on the developmental progressions and the instructional activities and strategies allow for researchers and other educators to *test the theory by testing the activity,* supporting curriculum development and evaluation as a scientific enterprise (Battista & Clements, 2000; Clements, 2007; Clements & Battista, 2000). An expectation that instruction elicits learning at a given level has made teaching experiments a common method for evaluating learning progressions. Instructional tasks designed to capture a specific level of thinking will produce evidence for or disconfirming evidence of *some* level of thinking, even if that is not the level for which the activity was intended to scaffold. The level of thinking observed or disconfirmed can be incorporated into the empirical data for the progression. Similarly, the activity can either be improved to better scaffold at the intended level or defined to elicit a different level of learning. Because children's learning and development are not tied to their immediate experiences, their location on a learning path can still be observed even when an activity is not optimally scaffolding their learning progress. Because of this, activities should be designed with embedded assessment – tasks that give teachers and students valuable information about whether a child needs to be challenged at earlier or later levels. The optimal activities include these tasks and suggestions on supporting learning at adjacent levels – testing these also provides evidence for the mechanisms through which teachers can build on children's current understanding.

Plan for an Efficacious Grain Size of the Developmental Progressions

Math and science LTs/LPs agree that they "extend over multiple years" (National Research Council, 2012, p. 26), thus constituting a major contribution to educational coherence (Clements & Sarama, 2021a). Many science LPs also specify grade-range changes over K-12. For example, "In grades K-2, we choose ideas about phenomena that students can directly experience and investigate," but in "grades 3–5, we include invisible but chiefly still macroscopic entities" (National Research Council, 2012, p. 33). In many cases, researchers have created detailed LPs (Hadenfeldt et al., 2016; Jin & Anderson, 2012; Kaldaras et al., 2021; Liu, 2013; Merritt & Krajcik, 2009; Smith et al., 2006).

Notably, the Framework (National Research Council, 2012) used the construct of "endpoints" mainly due to a lack of consistent and detailed research on the vast landscape of science topics, not because details were not wanted. Indeed, such details were eliminated due to feedback, and the Framework calls explicitly for research to help teachers and their students interact over the short term. However, they distinguish between such work in "units" versus the long term, which they identify as LPs.

In contrast to grade-band levels, math LTs have levels that are useful in daily teaching interactions as well as for planning and coordination across ages. Such detailed information is helpful for both curriculum development and teaching, especially the empirically validated strategy of formative assessment (Clements & Sarama, 2021a). Creating such information is a recommended direction for science LP research and development.

Ensure Levels are (Interim or Final) Goals

However, generating detailed development progressions may create *too many* levels, and science educators may benefit from one of our own mistakes. Initially, if children went through a level of thinking that would not be considered a goal but a transitional phase in figuring out mathematics and its application, we included that level in the developmental progression. For example, in students' development of concepts and strategies for comparing numbers, research shows a level at which they count to compare numerosities but not use the result of counting! That is, they may correctly count four large balls and six small balls but then–due to the distraction of the greater mass–declare that there are "more large balls" (Clements & Sarama, 2021a; Sarama & Clements, 2009). Of course, we wanted teachers to understand and notice this, but ... we did not anticipate that some teachers believed that they should teach children particular behaviors, such as ignoring the number and declaring there were more of the larger balls, which is never what we intended. Our solution to this dilemma was to place this thinking pattern in the level *just before,* as extant research notes that children might start to count but would not necessarily *use* that counting yet. Then the following level depicted children using counting accurately to compare the numbers into sets despite their physical appearance. In this way, we both inform teachers of exciting research on children's mathematical thinking but avoid misleading anyone into believing that those were goal behaviors to elicit.

Developing and Evaluating LPs

Consistent with work on science LP creation (Smith et al., 2006), research on math LTs can provide extensive models of the scientific and practical.[1] creation of learning trajectories such as using the Curriculum Research Framework (CRF, Clements, 2007). Many related examples are available and may be helpful (Baroody et al., under review; Barrett et al., 2017; Blanton et al., 2015; Clements & Sarama, 2004, 2021a; Clements, Sarama, Baroody et al., 2020; Clements et al., 2018; Clements, Wilson, et al., 2004; Confrey et al., 2014; Confrey et al., 2012; Confrey et al., 2019; Maloney & Confrey, 2010; Putri & Zulkardi, 2017; Sarama & Clements, 2009; Sarama et al., 2022; Wilson, 2009).

As just one example, Table 9.1 describes the categories and phases of the CRF, focusing on how they helped us create and evaluate learning trajectories in early math, including research and development questions asked, issues to consider, and research methods.

Resist Isolating LTs/LPs

Here is another mistake we made that others could avoid. When we first developed the learning trajectories, they were in a massive table with ages in rows and columns for every topic. The ages were, of course, not absolute. However, they served as a guide when we conducted our first synthesis of research from fields as diverse as mathematics education, early childhood education, developmental psychology, and cognitive science, and a broader diversity of methods, including single-age, cross-sectional, and longitudinal studies. This organizational structure allowed us to record and create *linkages between the various* topics (related to the construct of progress variables in science, Corcoran et al., 2009, p. 15). Unfortunately, when we published the learning trajectories in several different forms, every editor, without exception, separated all the learning trajectories into their sections. This isolation contributed to simplicity and printability. However, we did not realize until later what we had lost. People thought of the learning trajectories in isolation, and it was difficult

Table 9.1 Categories and phases of the Curriculum Research Framework (CRF) for LTs

Categories	Questions asked	Phases
A Priori Foundations. Extant research is reviewed in variants of the research-to-practice model, and implications for the nascent LT development effort are drawn.	What is already known that can be applied to the anticipated LTs? Phase 1 2 3	Established review procedures and content analyses (National Research Council, 2004) are employed to garner knowledge concerning the specific subject matter content, including the role it would play in students' development (*phase 1*), general issues concerning development (*phase 2*); and pedagogy, including the effectiveness of certain types of activities (*phase 3*).
Learning Trajectories. Activities are structured in accordance with empirically-based models of students' thinking and learning in each targeted subject-matter domain.	How might the LT be constructed to be consistent with evidence of students' thinking and learning (posited to have characteristics not arbitrary and not equally amenable to various instructional approaches/activities)? Phase 4	In *phase 4*, a developmental progression is created based on research on students' mathematical thinking and learning on the topic. A set of instructional activities and strategies is sequenced consistent with the cognitive actions-on-objects that define each level of that progression. Distinguishing phase 4 from 3 is not only the focus on the child's learning of a specific topic but also the iterative nature of its application. That is, especially when the research base is nascent for the topic, levels are often created and revised (or, not infrequently, created anew) dynamically, simultaneously with developing instructional activities using grounded theory methods, clinical interviews, teaching experiments, and design experiments.
Formative Evaluation. In these phases, empirical evidence is collected to evaluate and refine the LT, realized in some form.	How can market share for the LT be maximized? Phase 5	*Phase 5* focuses on marketability, using strategies such as gathering information about mandated educational objectives and surveys of consumers to ensure uptake within a target population.
	Is the LT usable by and effective with various student groups and teachers? How can it be improved in these areas or adapted to serve diverse situations and needs? Phase 6 7 8	Formative *phases 6 to 8* seek to understand the meaning that students and teachers give to the LT's objects and activities in progressively expanding social contexts; for example, the usability and effectiveness of specific components and characteristics of the LT as implemented by a teacher who is familiar with the materials with it in small groups (*phase 6*) and whole classes (*phase 7*) and, later, by a diverse group of teachers (*phase 8*). Methods include interpretive work using a mix of model testing and model generation strategies, including design experiments, micro-genetic, micro-ethnographic, and phenomenological approaches (*phase 6*), classroom-based teaching experiments, and ethnographic participant observation (*phase 7*), and these plus content analyses (*phase 8*). Finally, the LT is altered based on empirical results, focusing on expanding to include aspects of support for teachers.

Summative Evaluation. In these phases, rigorous, causal empirical evidence is collected to evaluate the LT.	What is the effectiveness (e.g., in affecting teaching practices and ultimately student learning) of the LT, now in complete form, as implemented in realistic contexts? Phase 9 10	*Phases 9 and 10* use randomized field trials and differ most markedly on the characteristic of scale. *Phase 9* may use 10 or 20 classrooms and "super-realization" support (Cronbach et al., 1980): a careful assessment of what the LT can accomplish in ideal circumstances. *Phase 10* examines the fidelity or enactment and sustainability of the LT, alone or embedded within a curriculum, when implemented on a large scale, and the critical contextual and implementation variables that influence its effectiveness. For both phases, experimental or carefully planned quasi-experimental designs, incorporating observational measures and surveys, help generate political and public support and their scientific advantages. Qualitative approaches continue to be necessary for dealing with the complexity and indeterminateness of educational activity.

then to remind them of all the interweaving conceptual support and even prerequisites that develop along each learning trajectory provided to the development of other topics. As a simple example, Table 9.2 shows how the close connections among four learning trajectories for topics in number are connected (this is a sample, the learning trajectories go further).

Contributions of Science Learning Progressions to Math Learning Trajectories

Although not the focus of this chapter, work on science LPs has ramifications for the creation, refinement, and evaluation of math LTs. We briefly mention two.

Include Boundaries

The National Research Council *Framework* has a recommendation we did not incorporate: "Standards should incorporate boundary statements" (2012, p. 301). This means that attention is given not just to what should be taught but what does *not* need to be taught, at least for children at a given level. The *Framework* mentions density as a topic often covered repeatedly in elementary school to little or no avail; in mathematics, probability may have a similar fate – as both concepts may be better addressed in secondary grades. Thus, the practical goals of developing and researching both learning trajectories and learning progressions should be to guide educators on meaningful, challenging, and achievable math and science learning objectives *focused on those* built on children's natural thinking and are generative of future learning. In math, we do not emphasize probability, for the early ideas children develop do not build during the early years.

Resist Isolating *Domains*

The *Framework* also calls for research "to explore how literacy, language skills, and mathematics intersect with learning in science across multiple school years. This research is important for understanding how the practices develop over time and how learning experiences in other subjects might leverage or be leveraged by learning in science" (National Research Council, 2012, p. 316). Although researchers in math LTs have done this type of work (Clements, Sarama, Layzer et al., 2020; Sarama et al., 2012), far more attention is needed. Similarly, science education's crosscutting concepts and other connections have gone beyond what math education has done with processes (PSSM) or practices (CCSS).

One caveat for *integrating* across domains: Although, given the connections, it is reasonable to claim that curricula and pedagogical approaches should fully integrate all aspects of science, math, and other domains, the history of such educational efforts and research evaluations raises concerns about such complete integration (Clements & Sarama, 2021b). For example, reviews of fully integrated curricula (e.g., all activities involve all domains) reveal little evidence that they are superior to traditional structures and that there are challenges in implementing such curricula (Czerniak et al., 1999; George, 1996; Preschool Curriculum Evaluation Research Consortium, 2008). Often, the mathematics content is weak. In contrast, we promote an *interdisciplinary approach* in which rich connections are made between domains. However, each retains its core conceptual, procedural, and knowledge foundation (epistemological) structures (Clements & Sarama, 2021b). Two or more domains would always (but *only*) be integrated when that combination is consistent and complementary with those structures for *each* domain (see Chapter 26 on mathematical thinking in science in this volume). Thus, children gain exposure to prerequisite mathematics

Table 9.2 Connections of the number and arithmetic learning trajectories (from Clements & Sarama, 2021a)

Broad levels	*Subitizing*	*Counting*	*Comparing, ordering*	*Adding and subtracting*
	Perceptual Subitizing	Counting with Cardinality	Matching and Comparison	Small #s, Direct Modeling
Visual / Perceptual Thinker	• Number Senser: Foundations • Very Small Number Recognizer • Maker of Small Collections • Small Collection Namer • Perceptual Subitizer to 4 • Perceptual Subitizer to 5	• Number Word Sayer: Foundations • Chanter • Reciter • Corresponder • Counter (Small Numbers) • Counter (10) • Producer (Small Numbers) • Counter and Producer (10+) • Counter Backward from 10	• Comparison Senser: Foundations • Many-to-One Corresponder • Object Corresponder • Perceptual Comparer • Early Comparer of Similar Items • Early Comparer of Dissimilar Items • Matching Comparer	• Arithmetic Senser: Foundations • Preverbal +/– • Small Number +/– • Find Result +/– • Make It N
	Conceptual Subitizing	Counting Strategies	Numerical Comparison	Flexible Strategies
Numerical/ Modeling Thinker	• Conceptual Subitizer to 5 • Conceptual Subitizer to 7 • Conceptual Subitizer to 10 • Conceptual Subitizer to 20	• Counter from N (N+1, N–1) • Counter On Using Patterns • Counter On Keeping Track • Counter of Quantitative Units/ Place Value	• Counting Comparer (Same Size) • Counting Comparer (5) • Mental Number Line to 5	• Find Change +/– • Counting Strategies +/– • Part–Whole +/– • Numbers in Numbers +/– • Deriver +/–:

skills in an appropriate sequence, and scientific inquiry is designed to promote a deep understanding of conceptual content and science processes. On the other hand, when connections *are* drawn between mathematics and science, they are genuine and detailed, with their impact undiluted by less fruitful attempts at integration.

The learning trajectories of early math were developed with an awareness of the interconnectedness of mathematical concepts and patterns of thinking. Clarifying the boundaries of topics was necessary for the precision of developing and studying the developmental progressions and for the practical use of clear instructional goals. The topics within which developmental progressions were developed needed strong evidence for their mathematical importance, acknowledging connections across topics – see Broad Levels in Table 9.2 for an illustration of these connections while being specific enough to focus instruction. Many activities can be fun for children and include mathematical ideas. However, the objectives and intentionality of instructional activities may be unclear without an overarching goal and understanding of how to reach that goal.

Specific Math LTs with Implications for Science LPs

In this section, we discuss mathematical topics that play a direct role in the science, technology, and engineering (STE) domains, demonstrating ways to bring either math or science to the foreground to progress in both fields of instruction (Silk et al., 2009). All mathematical topics are essential (e.g., palpably, numbers and arithmetic). However, here we focus on an example that may be less obvious on the surface and then turn to two topics wherein math and the STE domains deal with the same topics but in beneficial complementary ways. These are just examples. Remember the fundamental role of mathematics: As Galileo put it, "Mathematics is the alphabet with which God has written the universe."

Patterning and Algebraic Thinking

Patterning is an everyday activity in early childhood classrooms but is often done in simple ways, such as making red, green, red, green, red, green paper chains. Such activities can be a good start for many children, but they can disguise the depths of this topic and its relationships to science. Thus, *following the LT through all the levels* is critical in realizing the potential of mathematics for science.

First, in mathematics, even simple repeating patterns are more complex when done in depth. For example, not only should kids progress two different types of sequential patterns, such as AABC, AABC, but children should also be asked to identify the unit of the pattern and explicitly recognize that this sequential pattern is the repetition of that unit "AABC" in that example. Even such simple sequential patterns are ubiquitous in science, such as day and night or the four seasons.

Children can also analyze how that pattern is structured, 2 As, 1 B, 1 C. This leads to an advance from sequential repeating patterns to *growing patterns,* such as a stair made of connecting cubes, with each step containing one more cube than the previous one. More advanced growing patterns then follow, such as triangular numbers as shown in Figure 9.1a … up to the patterns in Pascal's triangle and the highly related Fibonacci sequence, golden rectangle, and golden spiral are ubiquitous in nature (Posamentier & Lehmann, 2007).

Further, following LTs develops an understanding of number sentences and algebraic thinking that supports the learning of science. Current curricula and teaching practices too often start and end with the equals sign as a "place to put the answer" (Kieran, 2018; Seo &

(a)

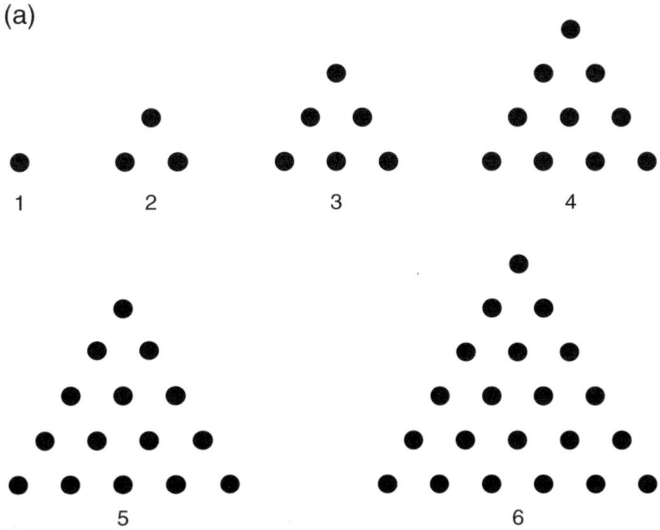

(b)

Triangle Number	Number of Circles
1	1
2	3
3	6
4	10
5	15

Figure 9.1 Triangular numbers illustrated in a visual pattern (a) that can be analyzed in a "T-table" (b).

Ginsburg, 2003; Warren, 2006). That is, children see 4 + 3 = 7 as 4 + 3 –> __. Therefore, some will fill in the blank in 4 + 3 = __ + 2 with 7 and queried about the "2," respond, "I don't know" or "It shouldn't be there" (Clements & Sarama, 2021a). Following LTs for Patterns, Structure, and Algebraic Thinking (Clements & Sarama, 2021a, LearningTrajectories.org) ensures that children understand equations in terms of equality and "balance" by (a) seeing diverse forms (e.g., 5 = 3 + 2), (b) figuring out that 5 makes 4 + 3 = __ + 2 true, and (c) discussing whether 58 + 37 = 37 + 58 is true or false …*without* calculating, and even whether generalizations such as a + b = b + a are always true (Franke et al., 2004).

Similarly, following this LT engages children in algebraic thinking in interpreting, explaining, and representing the patterns illustrated in the T-chart in Figure 9.1b. These foundations are instrumental to understanding and manipulating formulas *meaningfully* in later science instruction. Pattern recognition and generation are also critical in computer science (Relkin & Bers, 2020). That is recognizing a situation or structure as "seen previously" and eventually generalizable or repeatable to good effect in solving problems. In computational thinking, this develops into countable loops, recursion, and recognition and use of the cumulative effect.

Measurement

The concept of measurement in math focuses on assigning a number to a magnitude of some attribute (Clements & Sarama, 2021a). It can apply to such mathematical topics as length, area, volume, and angle size. Similarly, scientific efforts may focus on assigning a number to forces, temperature, the fecundity of living organisms, light, sound, or changing quantities over time. Measurable properties of objects, such as weight, length, area, and volume, are important for children's understanding of matter (Smith et al., 1985; Smith et al., 2006). Measurement is inextricably linked to the scientific process of observation. Thus, similar activities can be used to understand mathematical measurement or several scientific concepts. For example, most young children enjoy activities involving cars and ramps – a math activity might focus on the angle measure of the ramp. In contrast, a science activity might focus on the magnitude of force used to push a car down different sizes of ramps. Both types of measurement include foundational concepts of more or less, with children's developmental level indicating the precision they might measure (see an example at https://www.learningtrajectories.org/math-activities/exploring-ramps-stemie).

The assignment of a number to an observation – whether length or temperature – includes the underlying understanding that more or less of something exists. These foundations of measurement become increasingly complex as children experience comparisons of different magnitudes. Adults may be interested in immediately having children assign a number to their measurement (given that this is the ultimate goal of measurement). However, the developmental progressions of early math would indicate that children will first recognize that something is measurable, then make comparisons between concrete differences in magnitude, followed by (with adult help) the use of units to quantify measurement. The use of units for quantification occurs first (in geometric measurement) by filling the space with sufficient numbers of physical units, then iterating single units and using measurement tools that represent the repeated single units. Research shows that following this mathematics measurement LT leads to a robust understanding of measurement often not achieved using traditional approaches (Barrett et al., 2017; Sarama et al., 2022).

Even when children begin to use quantitative units, the study of developmental progressions of science will still need to attune to potential errors in children's thinking – such as the precise use of units, then the combination or fractional use of units. Across these steps, children's thinking should dominate the instruction rather than efforts to achieve adult levels of comprehension.

Classification and Data Analysis

Like measurement, data analysis is directly tied to the scientific observation process. Without intentional data collection through observation or other systematic means, there is no data to analyze. Because of the many intersections of math used in data, this may be an intimidating topic, but one that, again, can be guided by existing LTs in mathematics. Like early abilities to recognize the number of objects in a group without counting (subitizing) and comparing numbers, the early classification may be processed visually – with matching as a primary focus. In the same ways that young children notice similarities in geometric shapes, they may notice similarities in types of animals or plants. This first part of the progression – classification – builds towards quantifying categories. Once categories are quantified, they can be compared – so that classification leads to data analysis. Later in development, children may begin to notice patterns in data, such as the "S" shape that characterizes the rate of growth for many living organisms (Lehrer & Schauble, 2012). Developing skills using mathematical models can support both math and science learning.

An essential aspect of data analysis that transitions instructional experiences from being primarily math to primarily science is using data to answer a question or solve a problem. For example, in an activity about the weather, a mathematics-focused activity might focus on graphing temperatures over time and comparing changes in number – focusing on the benefit of such mathematical strategies rather than simply using procedures (Lehrer & Schauble, 2012). In contrast, the same instruction focused on science might graph temperature over time and reflect on how that relates to seasonal changes – with the specificity of discussions of temperature change over time led by children's understanding of weather and climate. In summary, using the mathematics LTs ensures (a) no children miss key levels of understanding, relegating them to rote learning and (b) children progress through the later levels of the LTs, ensuring they benefit from the full range of skills *and concepts* mathematics has to offer.

Conclusion

Comparing and synthesizing the various conceptions of learning trajectories and learning progressions can benefit research and development across domains. We argue for increased use of the learning trajectories construct in science education, considering its three components, models of creation and evaluation, and five potential contributions. We, in turn, are revising our conceptions based on work in science. Everyone wins. Across math and science research and development, there is agreement about the need to attend to the iterative development of big ideas carefully and how they are influenced by experience – increasingly articulated by developmental or learning progressions. These ideas may begin intuitively at the earliest levels, become explicitly expressed through exposure to language-defining concepts and mental structures, and eventually solidify in children's ability to think about their thinking. The instructional tasks adults design to promote and assess learning should be informed by hypothesized progressions, with the testing of these tasks continually benefitting the empirical body of knowledge about these hypotheses. Finally, the usefulness of these tasks must continually be grounded in the extent to which they support effective instructional decisions.

Thus, future research directions can focus on continuing to articulate, validate, and use developmental progressions to align curriculum and assessment and design more powerful instructional experiences for students. This includes experiences that help students achieve educational standards, connect concepts, and overcome barriers that may not be explicitly defined in performance expectations. Frameworks such as the Curriculum Research Framework have been helpful to our work in defining progressions and aligning useful curricula. Both fields must continue to rise to the challenge of incorporating multiple dimensions of concepts and practices while clarifying boundaries that retain the usefulness and effectiveness of the learning trajectories.

Acknowledgement

This chapter was supported by the Institute of Education Sciences, U.S. Department of Education through Grants R305A150243 and R305A220102; the National Science Foundation through Grant DRK-12 23006061908889; the Office of Special Education Programs (OSEP, U.S. Dept. of Education) through Federal Award H327G180006 and Subaward No. 5112267; and Heising-Simons Foundation Grants #2015–156 and #2022-3548 to the University of Denver. The opinions expressed are those of the authors and do not represent the views of the funding agencies. We have no known conflict of interest to disclose.

Note

1 That is, based on the wisdom of expert practice and other complementary types of research, including historical, aesthetic, literary, and narrative perspectives. Clements, D. H. (2007). Curriculum research: Toward a framework for 'research-based curricula'. *Journal for Research in Mathematics Education*, *38*(1), 35–70. https://doi.org/10.2307/30034927.

References

Anderson, C. W. (2008, May 22). *Conceptual and empirical validation of learning progressions*. Paper presented at the Meeting on Advancing Research on Adaptive Instruction and Formative Assessment, Philadelphia, PA.

Baroody, A. J., Eiland, M. D., Clements, D. H., & Sarama, J. in press Does a learning trajectory facilitate the learning of early cardinal-number concepts? *Journal of Mathematical Behavior*

Barrett, J. E., Clements, D. H., & Sarama, J. (2017). Children's measurement: A longitudinal study of children's knowledge and learning of length, area, and volume. *Journal for Research in Mathematics Education Monograph Series*, *16*, 262.

Battista, M. T., & Clements, D. H. (2000). Mathematics curriculum development as a scientific endeavor. In A. E. Kelly & R. A. Lesh (Eds.), *Handbook of research design in mathematics and science education* (pp. 737–760). Erlbaum.

Blanton, M., Brizuela, B. M., Gardiner, A. M., Sawrey, K., & Newman-Owens, A. (2015). A learning trajectory in 6-year-olds' thinking about generalizing functional relationships. *Journal for Research in Mathematics Education*, *46*, 511–558. 10.5951/jresematheduc.46.5.0511

Carpenter, T. P., & Moser, J. M. (1984). The acquisition of addition and subtraction concepts in grades one through three. *Journal for Research in Mathematics Education*, *15*(3), 179–202. 10.2307/748348

Clements, D. H. (2007). Curriculum research: Toward a framework for 'research-based curricula'. *Journal for Research in Mathematics Education*, *38*(1), 35–70. 10.2307/30034927

Clements, D. H., & Battista, M. T. (2000). Designing effective software. In A. E. Kelly & R. A. Lesh (Eds.), *Handbook of research design in mathematics and science education* (pp. 761–776). Erlbaum.

Clements, D. H., & Sarama, J. (2004). Learning trajectories in mathematics education. *Mathematical Thinking and Learning*, *6*, 81–89. 10.1207/s15327833mtl0602_1

Clements, D. H., & Sarama, J. (2014). Learning trajectories: Foundations for effective, research-based education. In A. P. Maloney, J. Confrey, & K. H. Nguyen (Eds.), *Learning over time: Learning trajectories in mathematics education* (pp. 1–30). Information Age Publishing.

Clements, D. H., & Sarama, J. (2021a). *Learning and teaching early math: The learning trajectories approach* (3rd ed.). Routledge. 10.4324/9781003083528

Clements, D. H., & Sarama, J. (2021b). STEM or STEAM or STREAM? Integrated or interdisciplinary? In C. Cohrssen & S. Garvis (Eds.), *Embedding STEAM in early childhood education and care* (pp. 261–275). Palgrave Macmillan. 10.1007/978-3-030-656224-9

Clements, D. H., Battista, M. T., & Sarama, J. (2001). *Logo and geometry* (Vol. 10). National Council of Teachers of Mathematics. 10.2307/749924

Clements, D. H., Sarama, J., & DiBiase, A.-M. (2004). *Engaging young children in mathematics: Standards for early childhood mathematics education*. Erlbaum.

Clements, D. H., Sarama, J., & Liu, X. (2008). Development of a measure of early mathematics achievement using the Rasch model: The research-based early maths assessment. *Educational Psychology*, *28*(4), 457–482. 10.1080/01443410701777272

Clements, D. H., Sarama, J., Baroody, A. J., & Joswick, C. (2020). Efficacy of a learning trajectory approach compared to a teach-to-target approach for addition and subtraction. *ZDM Mathematics Education*, *52*(4), 637–648. 10.1007/s11858-019-01122-z

Clements, D. H., Sarama, J., Layzer, C., Unlu, F., & Fesler, L. (2020). Effects on mathematics and executive function of a mathematics and play intervention versus mathematics alone. *Journal for Research in Mathematics Education*, *51*(3), 301–333. 10.5951/jresemtheduc-2019-0069

Clements, D. H., Sarama, J., Van Dine, D. W., Barrett, J. E., Cullen, C. J., Hudyma, A., Dolgin, R., Cullen, A. L., & Eames, C. L. (2018). Evaluation of three interventions teaching area measurement as spatial structuring to young children. *The Journal of Mathematical Behavior*, *50*(23), 23–41. 10.1016/j.jmathb.2017.12.004

Clements, D. H., Wilson, D. C., & Sarama, J. (2004). Young children's composition of geometric figures: A learning trajectory. *Mathematical Thinking and Learning*, 6, 163–184. 10.1207/s15327833mtl0602_1

Confrey, J. (2019). *A synthesis of research on learning trajectories/progressions in mathematics*. Organisation for Economic Co-operation and Development. http://www.oecd.org/education/2030/A-Synthesis-of-Research-on-Learning-Trajectories-Progressions-in-Mathematics.pdf

Confrey, J., Maloney, A. P., Nguyen, K. H., & Rupp, A. A. (2014). Equipartitioning: A foundation for rational number reasoning. Elucidation of a learning trajectory. In A. P. Maloney, J. Confrey, & K. H. Nguyen (Eds.), *Learning over time: Learning trajectories in mathematics education* (pp. 61–96). Information Age Publishing.

Confrey, J., Rupp, A. A., Maloney, A. P., & Nguyen, K. H. (2012). *Toward developing and representing a novel learning trajectory for equipartitioning: Literature review, diagnostic assessment design, and explanatory item response modeling* [not online].

Confrey, J., Toutkoushian, E., & Shah, M. (2019). A validation argument from soup to nuts: Assessing progress on learning trajectories for middle-school mathematics. *Applied Measurement in Education*, *32*(1), 23–42. 10.1080/08957347.2018.1544135

Corcoran, T. B., Mosher, F. A., & Rogat, A. (2009). *Learning progressions in science: An evidence-based approach to reform (Research Report No. 63) [not online]*. Corsortium for Policy Research in Education.

Cronbach, L. J., Ambron, S. R., Dornbusch, S. M., Hess, R. D., Hornik, R. C., Phillips, D. C., Walker, D. F., & Weiner, S. S. (Eds.). (1980). *Toward reform of program evaluation: Aims, methods, and institutional arrangements*. Jossey-Bass.

Czerniak, C. M., Weber, W. B., Jr., Sandmann, A., & Ahern, J. (1999). A literature review of science and mathematics integration. *School Science and Mathematics*, *99*(8), 421–430.

Duschl, R., Maeng, S., & Sezen, A. (2011). Learning progressions and teaching sequences: A review and analysis. *Studies in Science Education*, *47*(2), 123–182. 10.1080/03057267.2011.604476

Franke, M. L., Carpenter, T. P., Levi, L., & Zeringue, J. K. (2004). Algebra in elementary school: Developing relational thinking. *ZDM–The International Journal on Mathematics Education*, *36*.

Fuson, K. C. (2004). Pre-K to grade 2 goals and standards: Achieving 21st century mastery for all. In D. H. Clements, J. Sarama, & A.-M. DiBiase (Eds.), *Engaging young children in mathematics: Standards for early childhood mathematics education* (pp. 105–148). Erlbaum.

George, P. S. (1996). The integrated curriculum: A reality check. *Middle School Journal*, *28*(1), 12–19. https://www.jstor.org/stable/23024060

Griffin, S., & Case, R. (1997). Re-thinking the primary school math curriculum: An approach based on cognitive science. *Issues in Education*, *3*, 1–49.

Hadenfeldt, J. C., Neumann, K., Bernholt, S., Liu, X., & Parchmann, I. (2016). Students' progression in understanding the matter concept. *Journal of Research in Science Teaching*, *53*(5), 683–708. 10.1002/tea.21312

Jin, H., & Anderson, C. W. (2012). A learning progression for energy in socio-ecological systems. *Journal of Research in Science Teaching*, *49*(9), 1149–1180. 10.1002/tea.21051

Kaldaras, L., Akaeze, H., & Krajcik, J. S. (2021). Developing and validating next generation science standards-Aligned learning progression to track three-dimensional learning of electrical interactions in high school physical science. *Journal of Research in Science Teaching*, *58*(4), 589–618. 10.1002/tea.21672

Kaldaras, L., Yoshida, N. R., & Haudek, K. C. (2022). Rubric development for AI-enabled scoring of three-dimensional constructed-response assessment aligned to NGSS learning progression. *Frontiers in Education*, *7*. https://www.frontiersin.org/articles/10.3389/feduc.2022.983055

Karmiloff-Smith, A. (1984). Children's problem solving. In M. E. Lamb, A. L. Brown, & B. Rogoff (Eds.), *Advances in developmental psychology (Vol. 3)* (pp. 39–90). Erlbaum.

Kennedy, C., Brown, N., Draney, K., & Wilson, M. (2005, April). Using progress variables and embedded assessment to improve teaching and learning. Paper presented at the American Education Research Conference, Montreal, Canada.

Kieran, C. (2018). Seeking, using, and expressing structure in numbers and numerical operations: A fundamental path to developing early algebraic thinking. In C. Kieran (Ed.), *Teaching and learning algebraic thinking with 5- to 12-year-olds: The global evolution of an emerging field of research and practice* (pp. 79–105). Springer International Publishing. 10.1007/978-3-319-68351-5_4

Lehrer, R., & Schauble, L. (2012). Seeding evolutionary thinking by engaging children in modeling its foundations. *Science Education*, *96*(4), 701–724. 10.1002/sce.20475

Lehrer, R., & Schauble, L. (2015). Learning progressions: The whole world is NOT a stage. *Science Education*, *99*(3), 432–437. 10.1002/sce.21168

Liu, X. (2013). Difficulties of items related to energy and matter: Implications for learning progression in high school chemistry. *Emergent Topics on Chemistry Education, 14*(9), 1–7.

Maloney, A. P., & Confrey, J. (2010). *The construction, refinement, and early validation of the equipartitioning learning trajectory*. ICLS.

Merritt, J., & Krajcik, J. S. (2009). Developing a calibrated progress variable for the particle nature of matter. Paper presented at the Learning Progressions in Science Conference, Iowa City, IA.

National Research Council. (2004). *On evaluating curricular effectiveness: Judging the quality of K-12 mathematics evaluations*. Mathematical Sciences Education Board, Center for Education, Division of Behavioral and Social Sciences and Education, The National Academies Press. http://books.nap.edu/books/0309092426/html/

National Research Council. (2007). *Taking science to school: Learning and teaching sciences in grades K-8*. National Academies Press. https://www.nap.edu/catalog/11625/taking-science-to-school-learning-and-teaching-science-in-grades

National Research Council. (2012). *A framework for K-12 science education: Practices, crosscutting concepts, and core ideas*. National Academies Press.

National Research Council. (2013). *Next generation science standards: For states, by states*. The National Academies Press. 10.17226/18290

NCTM. (2006). *Curriculum focal points for prekindergarten through grade 8 mathematics: A quest for coherence*. National Council of Teachers of Mathematics.

Posamentier, A. S., & Lehmann, I. (2007). *The fabulous Fibonacci numbers*. Prometheus Books.

Preschool Curriculum Evaluation Research Consortium. (2008). *Effects of preschool curriculum programs on school readiness (NCER 2008–2009)*. Government Printing Office. http://ncer.ed.gov

Putri, R. I. I., & Zulkardi. (2017). Fraction in shot-put: A learning trajectory. AIP Conference Proceedings,

Relkin, E., & Bers, M. U. (2020). Designing an assessment of computational thinking abilities for young children. In L. Cohen & S. Waite-Stupiansky (Eds.), *STEM in early childhood education: How science, technology, engineering, and mathematics strengthen learning* (pp. 83–98). Routledge.

Sarama, J., & Clements, D. H. (2009). *Early childhood mathematics education research: Learning trajectories for young children*. Routledge. 10.4324/9780203883785

Sarama, J., Clements, D. H., Barrett, J. E., Cullen, C. J., & Hudyma, A. (2022). Length measurement in the early years: Teaching and learning with learning trajectories. *Mathematical Thinking and Learning, 24*(4), 267–290. 10.1080/10986065.2020.1858245

Sarama, J., Lange, A., Clements, D. H., & Wolfe, C. B. (2012). The impacts of an early mathematics curriculum on emerging literacy and language. *Early Childhood Research Quarterly, 27*(3), 489–502. 10.1016/j.ecresq.2011.12.002

Seo, K.-H., & Ginsburg, H. P. (2003). "You've got to carefully read the math sentence ...": Classroom context and children's interpretations of the equals sign. In A. J. Baroody & A. Dowker (Eds.), *The development of arithmetic concepts and skills: Recent research and theory* (pp. 161–187). Erlbaum.

Shepard, L. A., & Pellegrino, J. W. (2018). Classroom assessment principles to support learning and avoid the harms of testing. *Educational Measurement: Issues and Practice, 37*(1), 52–57.

Silk, E. M., Higashi, R., Shoop, R., & Schunn, C. D. (2009). Designing technology activities that teach mathematics. *Technology Teacher, 69*(4), 21–27.

Simon, M. A. (1995). Reconstructing mathematics pedagogy from a constructivist perspective. *Journal for Research in Mathematics Education, 26*(2), 114–145. 10.2307/749205

Smith, C. L., Carey, S., & Wiser, M. (1985). On differentiation: A case study of the development of the concepts of size, weight, and density. *Cognition, 21*, 177–237.

Smith, C. L., Wiser, M., Anderson, C. W., & Krajcik, J. S. (2006). Implications of research on children's learning for standards and assessment: A proposed learning progression for matter and the atomic-molecular theory. *Measurement: Interdisciplinary Research & Perspective, 14*(1&2), 1–98. 10.1080/15366367.2006.9678570

Warren, E. (2006). Exploring an understanding of equals as quantitative sameness with 5 year old students. In J. H. Woo, H. C. Lew, K. S. Park, & D. Y. Seo (Eds.), *Proceedings of the 31st Conference of the International Group for the Psychology in Mathematics Education* (Vol. 4, pp. 249–256). Charles University.

Wilson, P. H. (2009). *Teachers' uses of a learning trajectory for equipartitioning* [Dissertation, North Carolina State University]. Raleigh, NC.

10

PERSPECTIVES ON LEARNING PROGRESSION THEORIES AND METHODOLOGIES

Commentary for Section I

Christopher J. Harris

This handbook makes one thing abundantly clear – Learning progressions are among the strongest instantiations of models of how learning builds in a domain over time. They describe a progression of disciplinary knowing and learning that stretches from an early foundational base (often referred to as a lower boundary or lower anchor) upward, forming a pathway toward developing progressively more complex knowledge and reasoning that reaches a sophisticated level of performance or achievement (often referred to as an upper boundary or upper anchor). Most learning progressions focus on the development of discipline-specific knowledge and skills that go beyond the recall of factual knowledge or the procedural application of knowledge and toward applying knowledge in varying sophisticated ways. Between the lower and upper anchors, intermediary steps or levels are intended to describe a coherent interconnected pathway that maps out how students' knowledge and skills are likely to develop with the calibrated support of instruction. The research and development of learning progressions is driven by learning theory, grounded in prior research on student cognition and learning, and organized around disciplinary content and practices. As representations of theories of learning and how learning can advance with the support of instruction, learning progressions can be tested empirically to examine their validity. The central role of theory and the use of empirical evidence distinguishes learning progressions from the more familiar scope and sequence descriptions that ordinarily undergird curriculum and instruction.

Collectively, the chapters for Section I represent an intriguing set of perspectives on the development, evaluation, and validation of science learning progressions, raising a multitude of challenges and opportunities for the field. They cover a wide range of terrain, illustrating that in the relatively short time – less than two decades – since learning progressions emerged as an area of research and development, the work is expansive and now very much established as essential for increasing our understanding of how learning develops in domains over months and years. At the same time, the area is still emerging with new advances underway, critiques to be addressed, and much more development work to be done. The perspectives in Section I include examining the cognitive foundations of science learning progressions (Smith and Amin, Chapter 2), exploring the major critiques of learning progressions research (Jin and Krajcik, Chapter 3), considering issues around validity of learning progressions

DOI: 10.4324/9781003170785-11

(Kane, Chapter 4, and Kaladras and Krajcik, Chapter 5), describing the development of assessments based on a learning progression (Wilson, Yao, and Osborne, Chapter 6), reflecting on learning progression approaches used in different countries' educational systems (Neumann, Chapter 7, and Lei Wang and colleagues, Chapter 8), and comparing and synthesizing the various conceptions of learning trajectories in mathematics and learning progressions in science (Clements, Guss, and Sarama, Chapter 9). Each chapter contributes to the emergent thinking about where we are today with learning progression formulations and methodologies, where we are headed or could be headed next, and their important benefits to science education today and in the future.

In this commentary, I summarize the perspectives from each chapter and organize them around three themes related to the overarching topic of Section I. The first theme addresses the status of learning progression formulations and methodologies including some of the issues raised. The second theme focuses attention on the validity of learning progressions and the development of assessments. The third theme considers the value of comparing learning progression approaches. Chapters organized under the third theme encourage us to consider what we can learn from (1) comparing learning progression approaches across countries and (2) comparing learning progression efforts across disciplines. Drawing on insights gleaned from my reading of the collective set of Section I chapters, I close with challenges and opportunities for learning progression researchers and developers.

Status and Issues of Learning Progression Formulations and Methodologies

Smith and Amin, Chapter 2, launch the section with a delve into the early foundational learning that occurs well before children enter formal schooling and serves as the underpinning for science knowledge growth. Foundational learning refers to young learners' knowledge and reasoning resources that can serve as the productive starting point of a learning progression.

Smith and Amin illustrate how a systemic perspective, one that takes into account both intuitive and scientific conceptual systems, is needed to identify early and intermediate forms of understanding that might be causally connected to building learning progressions in science domains. They call to strengthen the bridge between foundational learning that sets the stage for domain-specific learning and the development of learning progressions for disciplinary core ideas and science practices. The high value in doing this bridging is that it stands to better characterize the student conceptual and reasoning resources of the lower anchor thereby providing information that will be essential for building the structure of a learning pathway toward an upper anchor that will work for a wide range of students. Moreover, these conceptual and reasoning resources in tandem with what is known about learning mechanisms and the target conceptions to be achieved in the domain, can strengthen the learning progression as a model for informing instructional design.

An intriguing idea proposed by Smith and Amin is a modeling approach to the design of learning progressions. They put forth that intermediate levels and upper anchors should be conceptualized as a progression of knowledge systems organized around increasingly powerful models. The idea is not new (see, for example, Rogat et al., 2011) but the case they make is compelling. They draw on prior research work from cognitive development, the cognitive and learning sciences, and science education research on student conceptions and conceptual change to illustrate how models can form a meaningful progression that can be evaluated and revised on an empirical basis, thereby reflecting the central scientific practice of modeling itself. As such, Smith and Amin view models as prime candidates for efficiently characterizing progressively more complex systems of knowledge within a domain.

Jin and Krajcik, Chapter 3, explore two major critiques of learning progressions research. One stems from the debate around the intermediary steps of learning progressions and whether progress in science learning from the lower to upper anchors is best represented as a sequence of a small number of achievement levels. Might it be more likely that between a lower and upper anchor there are many different steps and pathways for individual students to take? Closely related, the other critique originates from the debate about whether a single learning progression can adequately represent all learners in all contexts progressing reliably through a sequence as specified by the progression (see, for example, Hammer & Sikorski, 2015). Since learning is dynamic and context-dependent, can one learning progression really be useful for the broad range of learners we have in today's science classrooms?

According to Jin and Krajcik, perspectives on knowledge structure and knowledge development underly the debates. At the heart of the matter are perspectives on learning and conceptual change, namely a knowledge-in-pieces perspective developed by diSessa (1993; 2018) that students' knowledge can be represented as loosely connected fragments, known as "phenomenological primitives" or p-prims, and a naïve-theory perspective (Gopnik & Wellman, 1994) that learning is fundamentally theory like and viewed as a process of theory revision. Jin and Krajcik compare the two perspectives and point out that while the perspectives are quite different, both recognize the significance of students' prior knowledge and experiences as valuable resources for learning. Moreover, both perspectives have been applied in the development of learning progressions. They acknowledge, however, that the theory perspective has significantly influenced learning progressions research.

Regarding the first debate, Jin and Krajcik view this as fundamentally an issue relating to the nature of student knowledge. The knowledge-in-pieces perspective raises a critique of the value of learning progressions that have few steps or levels. The prevailing view is that it is most advantageous to identify universal learning pathways that represent a greater number of students rather than delineate many different levels. Jin and Krajcik acknowledge that it is of value to verify the right levels of achievement. Smith and Amin in Chapter 2 refer to this as a grain size issue and propose that the level of detail needed will depend on the intended use of the learning progression. Jin and Krajcik hold that if the goal is to build coherence across curriculum, instruction, and assessment, then utilizing validation techniques for confirming the appropriate levels will be paramount. The second debate is viewed by Jin and Krajcik as fundamentally an issue of knowledge development. The knowledge-in-pieces perspective raises the critique that it is unlikely that one progression can be applied to all students as they will have multiple paths for learning (i.e., one size does not fit all). Jin and Krajcik agree that it does not follow that all learners in all contexts will move reliably along a sequence as specified in a learning progression. But their reasoning, quite differently, is that learning progressions are not meant to portray individual differences in growth; rather, similar to points raised by Smith and Amin in Chapter 2, learning progressions are models of relatively efficient learning pathways toward progressively more complex knowledge growth or achievement. Research has also shown that there are commonalities in students' progress in domain-specific learning. Jin and Krajcik's final point is perhaps their strongest, that the most powerful benefits of learning progressions are not for individual tracking but for providing models that can be used to develop and align curriculum, instruction, and assessment in ways that promote students' productive progress in learning.

Validity of Learning Progressions and the Development of Assessments

In Chapter 4, Kane turns our attention to challenges in validating learning progressions. Validation, in the case of learning progressions, refers to examining whether the multiple claims regarding them are supported by adequate evidence. In its early development stages, validation of a learning progression is iterative and formative, with an emphasis on improving how well the learning progression functions (e.g., to what extent does the progression describe a pathway most students follow?); in its latter stages, the focus shifts to a more summative evaluation of how well the learning performance functions given its intended uses (e.g., does instruction based on the learning progression promote learning for students?). Kane puts forth that a learning progression can be evaluated in three ways: as a local theory of learning specific to some domain and grade level, in terms of its effectiveness in promoting student progress, and as an assessment program that yields results to be interpreted and used in instruction. Yet, Kane recognizes, too, that there are no clearly established standards to be relied on in validating learning performances. In response, the chapter proposes an iterative, argument-based approach to validation for them.

Kane does a deep dive into the basic structure of learning progressions and thoughtfully walks through each of the related ways to think about validation for them. Kane considers how the complex performances described in learning progressions can be evaluated, focusing attention for example on the progress variables that are to be used to indicate progress toward the target performances. Interestingly, the chapter also briefly considers implications for long-term learning progressions. Kane considers the issue of scaling up the basic assumptions of small and mid-size learning progressions to longer more comprehensive ones and speculates that their validation will require a more complex model, likely involving both summative as well as formative evaluations.

Kaladras and Krajcik, Chapter 5, discuss the validation of three-dimensional learning progressions that map the growth of students' integrated performance along the three dimensions of disciplinary core ideas, scientific and engineering practices, and crosscutting concepts as emphasized in the *Framework for K-12 Science Education* (NRC, 2012) and *Next Generation Science Standards* (NGSS Lead States, 2013). Three-dimensional learning refers to the process of applying the three dimensions to make sense of phenomena or solve problems. It is the means through which students are expected to build proficiency with the performance expectations of the NGSS over time. Kaladras and Krajcik contend that learning progressions should represent the latest models of knowledge-in-use wherein students use the three dimensions from the earliest grades upward. They point out that few such integrated learning progressions exist, yet developing these learning progressions is critical because the work is already underway for developing curriculum materials and classroom-based assessments that emphasize three-dimensional learning. Based on some of their own early efforts to develop three-dimensional learning progressions, Kaladras and Krajcik describe their explorations for designing sound approaches for determining validity. They describe how they have been addressing the issue of establishing the validity of the three-dimensional levels within a learning progression. What is especially clear from this chapter is that this represents an important area of future research and development.

In Chapter 6, Wilson, Yao, and Osborne illustrate the development and validation of assessments based on a learning progression for the scientific practice of argumentation. They describe how they investigate evidence for the validity of argumentation assessments. To carry this out, they apply exploratory factor analysis to examine the dimensionality of the assessments, and then apply item response modeling to examine whether the hypothesized order of waypoints of the learning progression is confirmed by the set of items used. Waypoints refer to the intermediate places on a progression that represent distinct ways of understanding on the way toward the upper anchor. Their assessment development work is a strong example of a rigorous and systematic effort to investigate the alignment of a learning progression with empirical data from an assessment. Noteworthy is that they found that the assessment provided evidence in support of the learning progression on scientific argumentation. Beneficially, they also share their challenges with developing items that elicit student performance indicative of waypoints on their learning progression, linking student performance on an item to a specific progression waypoint, and handling unexpected variables that may influence student success on items.

Comparing Learning Progression Approaches across Countries and Disciplines

We stand to learn a great deal by comparing and reflecting on learning progression approaches used in different countries' educational systems. In Chapter 7, Neumann introduces the German approach to learning progressions that are based on competency models of the knowledge, skills, and abilities that students are expected to develop as a function of science education. The models delineate levels of proficiency that serve as performance targets for assessments that make it possible to determine the extent to which students have mastered the competencies across the various science domains. Separate models have been developed for science core concepts and practices. According to Neumann, the development and validation of competency models is a growing area of research that is typically conducted at a large scale to obtain evidence that students indeed progress as hypothesized based on theoretical considerations. The German approach to learning progressions is characterized by careful delineation of how students develop competence in a domain into respective models; validation of these models in large-scale studies involving students from different grades and grade bands; and the use of specifically designed assessments and elaborated statistical procedures in the validation process. One concern raised by Neumann about the work is that there has been little evidence that the competency models have guided the development of improved curriculum or instruction. Neumann points out that the international approach to learning progressions, in contrast, has focused attention much more on using learning progressions as a basis for developing curriculum and informing instruction.

In Chapter 8, L. Wang, Yao, Q. Wang, Liu, Mou, J. Wang, and M. Wang report that Chinese scholars have conducted learning progression work for well over a decade. The central research and development focus of learning progressions has been evolving. The work has extended from learning progressions of disciplinary core ideas to the development of learning progressions of scientific practices, as well as from developing unidimensional learning progressions to integrated multidimensional ones that align more with the goals of contemporary science education. There has also been a shift in focus of applications of learning progressions from what Wang and colleagues refer to as an assessment-based learning progressions approach to more of an instruction-based approach that is carried out in partnership with teachers to inform the design of instruction, curricula, and professional learning. The assessment-based approach involves iterative cycles of constructing a learning progression, developing assessments, and using assessment data to confirm and validate the progression levels. Quite differently, the instruction-based approach turns to the instructional setting as the primary site for developing, using, and refining learning progressions while simultaneously improving the instructional design.

There is also much to learn from comparing the work of learning progressions and their various conceptions across disciplines. Co-authors Clements, Guss, and Sarama of Chapter 9 compare and synthesize learning progressions of science with the learning trajectories of mathematics. The learning trajectory construct in math has three parts – a specific mathematical competence goal, a developmental progression with levels describing the intermediary steps for building toward the goal, and instructional activities alongside teaching strategies that support learners in moving to later levels and ultimately achieving the competence goal. Clements, Guss, and Sarama emphasize that although levels are purposefully ordered by increasing sophistication, they are not linear stages that must be attained but rather "patterns of thinking" that represent concepts and skills that develop co-mutually and that are most productive for instruction. Math learning trajectories have levels that are intended to be useful in daily teaching interactions as well as for planning and coordination across ages and grades. Formative assessment ties the instructional activities with the developmental progression of the trajectory.

Their chapter explores how learning trajectories and learning progressions can benefit research and development across domains. They note that across math and science research and development, there is agreement about the need to attend carefully to the iterative development of big ideas and how they are influenced by instructional experience. There is also agreement that instructional tasks designed to promote and assess learning should be informed by hypothesized progressions, with the testing of these tasks continually benefitting the empirical body of knowledge about these hypotheses. Moreover, the usefulness of tasks must continually be grounded in the extent to which they support effective instructional decisions. Clements, Guss, and Sarama note that important points of distinction also exist, such as differences in the grain size of levels, and that some areas of learning trajectories development such as aligned instructional activities and strategies are far ahead of the efforts of those involved with science learning progressions. They also consider the productive intersections where potential contributions of learning trajectories theory and methods can benefit science learning progressions and vice versa. Inspirationally, their chapter directs both fields to learn from one another as we continue to develop, validate, and apply progressions and trajectories to align curriculum and assessment more strongly and design instructional experiences for the benefit of all students.

Closing: Challenges and Opportunities for Learning Progression Researchers and Developers

The chapters for Section I highlight that this is both an exciting and challenging time for learning progressions research and development. There have been impressive advances in the development, evaluation, and validation of science learning progressions. They have become an essential component of contemporary science education, providing powerful cognitive models grounded in evidence that can drive the development of curriculum, assessment, instruction, and professional learning aligned with one another. Yet, there are a multitude of challenges and opportunities for researchers and developers in the field. One significant challenge is incorporating the multiple dimensions of disciplinary core ideas, crosscutting concepts, and science and engineering practices into learning progressions and identifying and validating the performance levels. Taking a "knowledge-in-use" perspective more fully aligns with contemporary science education and the direction that more recent curriculum and assessment development has taken.

Another challenge is ramping up the work to cover the full range of science topics and domains. This represents a substantial undertaking that may be accelerated by mutually beneficial collaborations among researchers and science educators across educational systems internationally. Also, as science learning progressions continue to be mapped out and empirically validated, there is an opportunity for cross-pollination of learning across different disciplinary domains, principally mathematics but also others, that stands to enrich the work of all. Related, there has been increasing emphasis on integrated disciplinary learning of sciences, mathematics, engineering, and computer science and this raises discussion of progressions and trajectories that need to cut across established disciplinary boundaries. This represents a near-future direction for research. Finally, the advances in science learning progression research and development are changing how we think about the design and coordination of curriculum, instruction, and assessment. Learning progressions and the methodologies surrounding them hold promise for creating a more robust and coherent framework for science education spanning from the very youngest learners upward, supporting increasingly sophisticated learning performances and achievement.

References

diSessa, A. A. (1993). Toward an epistemology of physics. *Cognition and Instruction*, *10*(2-3), 105–225.

diSessa, A. A. (2018). A friendly introduction to "knowledge in pieces": Modeling types of knowledge and their roles in learning. In G. Kaiser , H. Forgasz , M. Graven , A. Kuzniak , E. Simmt , & B. Xu (Eds.), *Invited lectures from the 13th international congress on mathematical education* (pp. 65–84). Springer International Publishing

Gopnik, A., & Wellman, H. M. (1994). The theory theory. In L. A. Hirschfeld, & S. A. Gelman, (Eds.), *Mapping the mind: Domain specificity in cognition and culture* (pp. 257–293). New York: Cambridge University Press.

Hammer, D. & Sikorski, T. (2015). Implications of complexity for research on learning progressions. *Science Education*, *99* (*3*), 424–431.

National Research Council. (2012). *A framework for K-12 science education: Practices, crosscutting concepts, and core ideas*. Washington, DC: National Academies Press.

NGSS Lead States. (2013). *Next generation science standards: For states, by states*. Washington, DC: National Academies Press.

Rogat, A., Anderson, C., Foster, J., Goldberg, F., Hicks, J., Kanter, D., Krajcik, J., Lehrer, R., Reiser, B., & Wiser, M. (2011). *Developing learning progressions in support of the new science standards: A RAPID Workshop Series*. New York: CPRE. Teachers College, Columbia University.

SECTION II

Learning Progressions to Promote Student Learning

11
LEARNING PROGRESSIONS IN GENETICS

Ravit Golan Duncan and Moraima Castro-Faix

Introduction

In this chapter, we tell the evolution story of a genetics learning progression from its initial conception in 2009 through its revision over the course of multiple implementation studies. This particular progression is one of several in the domain of genetics and is one that has been studied by other researchers beyond those who initially developed it. The progression also informed the *Framework for K-12 Science Education* conceptualization of the *Heredity Disciplinary Core Idea* (LS3) and its development over grades and grade bands (National Research Council, 2012). We aim to highlight some of the challenges and lessons learned from our decade-long research agenda of refining a learning progression that can inform standards, curriculum, and assessment in the domain of genetics. We begin with a brief discussion of the motivation for this work and then describe and discuss three rounds of research on the progression. We conclude by discussing some of the remaining questions and challenges in our own work (in genetics education) and how these relate to broader issues relevant to progressions in science more broadly construed.

Take One: Initial Development of the Genetics Progression

Need for the Progression

Duncan, Rogat, and Yarden (2009) developed the initial progression based on an analysis of existing research in the domain including the developers' own work in genetics education. The motivation for this work stemmed from the stark contrast between scientific advancements in genetics on the one hand and genetics education on the other.

The late 19th century and early 20th century involved extensive developments in genetics research. Various new ideas and technologies, such as genetic screening, personalized medicine, genetically modified organisms, etc., were entering the public realm, and debates about the benefits, concerns, and ethics of these technologies were prevalent. At the time, and today, concepts in genetics were part of compulsory education in many countries yet contemporary research suggested that students' level of understanding of these concepts was sorely lacking (Duncan & Reiser, 2007; Kindfield, 1992; Lewis & Kattmann, 2004; Lewis & Wood-Robinson, 2000; Marbach-Ad & Stavy, 2000; Tsui & Treagust, 2003; Venville & Treagust, 1998).

DOI: 10.4324/9781003170785-13

The research literature documented multiple challenges in learning genetics. Some of these challenges were due to the nature of the domain, such as the complexity of genetic phenomena spanning multiple organization levels (genes, proteins, cells, organs, expressed traits) and the unfamiliarity of the molecular entities and mechanisms. Other challenges were due to how concepts were taught (Kurth & Roseman, 2001). For example, the instructional focus on molecular details (e.g., detailed structure of DNA and RNA) at the expense of helping students understand the "big picture" (nature of the genetic information). Often most of genetics instruction is relegated to high school during which these challenging ideas (with all their details) are crammed into an already over-crowded biology curriculum. Yet these ideas are complex and require recurrent engagement with them that expands and deepens students' understandings. Duncan and colleagues (2009) concluded that "Given the complexity of the domain of modern genetics and the inadequacy of current instructional methods and materials, it seems that a more coherent and extensive approach to genetics learning is needed." (pp. 656). Learning progressions provide such an approach as they involve multiple ideas, revised in growing sophistication over time.

Initial Progression

The initial genetics progression was based on a framework of genetic literacy developed by Stewart, Cartier, and Passmore (2005). This framework posited that genetic literacy entailed the understanding of, and fluent movement between, three models in genetics: a) the inheritance model, which explains patterns of correlation between genotype (genes) and phenotype (expressed trait); b) the meiotic model, which explains how genes are transferred from one generation to the next via sex cells; and c) the molecular model, which explains how the genetic information encoded in genes brings about the physical manifestation of the trait across organization levels--cells, tissues, organs, etc. (See Figure 11.1).

Duncan et al. (2009), through an extensive review of existing research on students' understandings of genetics, inductively identified eight constructs (ideas) that map onto these three models and the relationships between them and are encompassed under two questions: 1) How do genes influence how we, and other organisms, look and function? And 2) Why do we vary in how we, and other organisms, look and function? The progression described the development of student reasoning for each of these eight constructs across three levels of sophistication that loosely mapped onto late elementary Grades (4–5) middle school (grades 6–8), and high school (grades 9–10). Appendix A describes this initial progression.

Challenges

While the initial progression was developed based on the research available at the time, several aspects of the progression were more tenuous than others due to gaps in that body of research. First, there was little in the literature to guide any assertions about dependencies between constructs. That is, whether some level of sophistication needs to be attained on some constructs before progress can be made on other constructs. Thus, the Duncan et al. (2009) progression did not include any conjectures about such relationships and dependencies. Second, the mapping of levels onto grade bands was also largely conjectural. It was unclear what would be possible for younger students to understand with carefully designed instruction. Third, there was simply not enough information about some of the constructs to ascertain what kinds of ideas students hold about them and how these might change over time. For example, there was not a lot of understanding about how students reason about gene expression— the idea that genes can be turned on and off based on internal and external stimuli (e.g., tomato ripening occurs when

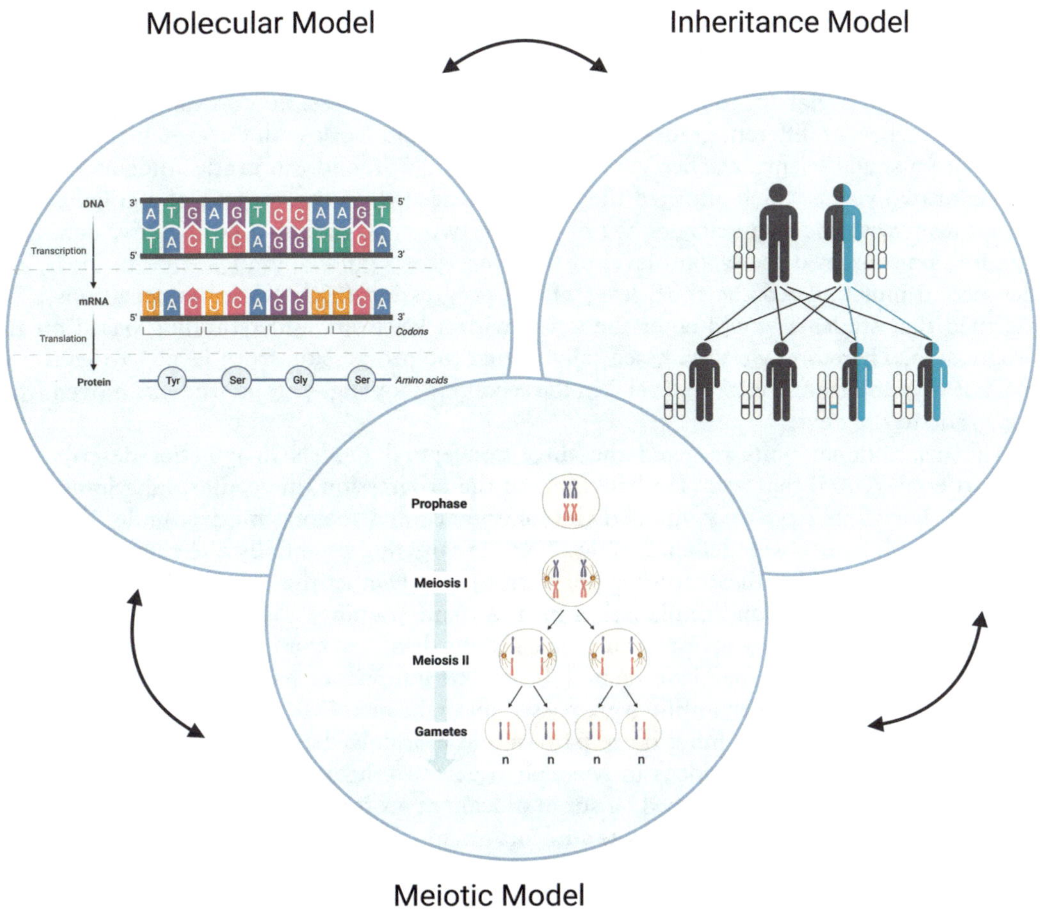

Figure 11.1 Model of genetics literacy (based on Stewart et al., 2005).

exposure to the plant hormone ethylene results in turning on multiple genes involved in various processes of ripening). Thus, the ideas in the proposed construct were largely conjectural. Fourth and final, the genetics progression did not explicitly focus on scientific practices. There was little information to inform how genetics ideas (concepts) and practices interface and how progression along dimensions of practice (e.g., scientific modeling) might afford or constrain progression along the constructs. As we later discuss, the progression informed the development of a curricular unit that engaged students in scientific practices, in particular modeling and argumentation. These practices were a good fit for the content constructs as students were developing explanatory models that reflected the three organizing models of genetics literacy (Stewart et al., 2005) that undergird the progression.

Thus, the initial progression was a hypothetical learning progression, developed inductively, that needed to be tested and refined empirically (Cobb & Bowers, 1999; Stevens et al., 2010). Our next phase of the research involved a longitudinal intervention study in which we developed instructional units based on the initial progression and studied their implementation over two consecutive years with middle school students. We aimed to begin developing a validity argument for the progression through this study.

Take Two: Longitudinal Study of the Genetics Progression

The most obvious research design for a progression is a longitudinal study. However, such studies are not trivial to conduct given attrition and the possible confounding effects of different teachers at different grades. We therefore chose to work with two small K-8 schools, each with a single science teacher who taught the 6th, 7th, and 8th grade students over the course of two years, which afforded the continuity needed. In the context of a collaborative design team with the two teachers, we developed two units over two consecutive summers. The first unit targeted the second level of the progression (middle grades expectations), and the second unit targeted the third level of the progression (high school expectations). We assumed that students would enter the study with a level one understanding based on the progression. Therefore, we were essentially testing the progression from level 1 to level 2 in the first year and from level 2 to level 3 in the second (following) year as students moved from one grade to the next.

The instructional units targeted the three conceptual models in genetics described by Stewart et al. (2005) that were the backbone of the progression (molecular, inheritance, and meiotic). For simplicity, we combined the inheritance and meiotic models under the more familiar umbrella of "Mendelian genetics." While targeting essentially the same constructs, the two units, differed in the anchoring phenomena and whether the unit involved engineering design or investigation and explanation as the main framing. The first unit began with molecular genetics ideas by investigating three genetic disorders from the symptom level to the molecular level. Students then investigated the genetic inheritance patterns for these disorders and how genes from one generation were passed on to the next (Mendelian genetics). The core framing was explaining the inheritance pattern and molecular basis of the genetic disorders. The second unit began with ideas in Mendelian genetics through a design-based project in which students attempted to "breed" a strain of leaner pigs for healthier bacon. The unit then addressed ideas in molecular genetics as students attempted to genetically engineer pigs with a healthier fat content. The core framing for this second unit was thus engineering design. Moreover, the sequence of teaching Mendelian and molecular genetics concepts was reversed in the two units in relation to each other. This was not intentional in the sense that we did not aim to reverse the order on purpose. Instead, the order in each unit was an artifact of the design given the phenomenon and framing.

Both science teachers implemented the units in their respective schools over a period of five weeks in each year (i.e., five weeks per unit). However, the fidelity of implementation in one of the schools was inadequate, and thus, only data from the other school (n = ~60) was analyzed. The focus data for this study included pre-post written assessments and interviews. The written assessment included 35 questions, the majority multiple choice. Of these, 20 items targeted Mendelian genetics constructs, and 15 questions targeted molecular genetics constructs. The items were coded based on the levels of the progression. The semi-structured interview protocol consisted of three tasks that, taken together, also prompted students' understanding of the three models and were coded similarly based on the progression levels.

Lessons Learned

Our findings from this longitudinal study were reported in multiple publications (Duncan, Castro-Faix, & Choi, 2016; Duncan et al., 2011; Freidenreich, Duncan, & Shea, 2011; Shea & Duncan, 2013). Several interesting insights emerged.

In our analysis of the first-year data, we found that overall, students made the most gains in learning the molecular model (effect size 1.54), followed by the meiotic model (effect size 1.17) and then the inheritance model (effect size 0.82). Accordingly, the study supported the notion that younger students (middle school) can engage with the molecular model at a reasonable level of sophistication. This is important to note in light of critiques of other researchers, and our concerns, that molecular entities are unfamiliar (and unobservable) to students and therefore reasoning about them would require sophisticated interdisciplinary knowledge (e.g., chemical structures and interactions) that is beyond the capacity of younger students (Duschl, Maeng, & Sezen, 2011). It seems that students were able to reason about molecular entities in productive ways without deep knowledge of chemistry.

However, there were aspects of their understanding that, while improved, remained problematic. For example, even after instruction about a third of the students did not conceive of plants and bacteria as having genes (inheritance model). Thus, their notion of the universality of genes (an understanding embodied in construct A) was not at the expected level based on the progression. In fact, understanding that all organisms have genes is a level 1 understanding of construct A, meaning that a third of the students did not shift beyond that level. There were other ideas that students struggled to grasp and for which they made less progress: (1) students still had difficulties predicting the genetic context of sex cells from individuals with different genotypes (meiotic model), (2) students' repertoire of protein functions was somewhat limited (molecular model), and (3) students' understanding of the connections between the three genetic models was not robust (Freidenreich et al., 2011).

Using this data set, we were also interested in determining whether learning any of the genetics models first provided any benefit to learning the other models. As noted earlier, the research literature did not provide any information about whether progress in understanding one genetic model constrains or affords learning of the other models. This question was pertinent because there were other competing progressions in genetics that took opposing stances on this issue. Dougherty (2009) suggested that the preferred sequence would be to teach the inheritance model first but for more complex traits (multifactorial traits), followed by the molecular model, and, only later, the inheritance model. Roseman, Caldwell, Gogos, and Kurth (2006) in their progression proposed that students learn the molecular model first as it supports an understanding of the biological mechanisms that underly the inheritance patterns of the meiotic and inheritance models (combined in our study as the Mendelian model). In our work, we were positioned to tackle this question, at least partially. Our analysis showed that students' knowledge of both models develops simultaneously and at a similar rate over time. There were also modest correlations between students' abilities in molecular genetics at the end of year one or the beginning of year two and their final abilities in Mendelian genetics (end of year two). We interpreted these findings as suggesting that students' knowledge of molecular genetics may provide a bootstrapping effect for their learning of Mendelian genetics (Duncan et al., 2016).

The last finding from the longitudinal study we wish to review relates to using data to revise and refine the progression (Shea & Duncan, 2013). As noted earlier, students' understanding of the kinds of functions proteins have, and their role in genetic phenomena was somewhat limited. This was not a particularly novel finding, we had seen it in earlier work (Duncan & Reiser, 2007; Duncan, 2007). We also knew that students did not always conceive of genes as containing information, i.e., coding for proteins (Duncan & Reiser, 2007; Venville & Treagust, 1998; Lewis & Kattmann, 2004). The idea that genetic information contains universal instructions that specify protein structure was captured in construct B of the progression, and the idea that proteins have a central role in the functioning of all living organisms and are the mechanism that connects genes and traits was captured on construct C of the progression (Duncan et al., 2009). Both constructs had three levels. As part of our analyses of the longitudinal data, we set out to carefully examine these constructs and the relationship between them (Shea & Duncan, 2013). As a result, we added several new levels to each construct. These included a level 0, with ideas that were of lesser sophistication than the initial lower anchor (level 1) of the constructs. Some of the existing levels were further split into additional levels to provide more nuance to students' ideas. Overall, the revised levels were of a finer grain size compared to the original progression.

The issue of grain size of a progression is manifested in the number of levels (and to some extent constructs) in a progression. In the case of the genetics progression, we initially opted for a coarse grain size of three levels that each spanned multiple grades. The revision of the constructs, based on observed students' ideas from the longitudinal implementation study, led to a finer grain size with substantially more levels. How does one go about making decisions regarding when it makes sense and is productive to combine, split, or remove levels? A core heuristic that guided our decision-making was "that grain size is determined in terms of cognitive and instructional productivity (e.g., add a level if it highlights an important and productive conceptual shift that can be promoted through instruction). Thus, levels should correspond to important conceptual steppingstones (Wiser et al., 2009) rather than reflect an exhaustive array of incremental understandings." (Shea & Duncan, 2013, pp. 25). We have since used this heuristic to guide the progression revision of other constructs (Castro-Faix, Duncan, & Choi, 2021).

Challenges

While the longitudinal study provided valuable insights into the genetics progression, this work posed several challenges. First, the items in the assessments were developed to ascertain students' understandings of the three genetic models (molecular, meiotic, and inheritance) more generally, they were not specifically developed per construct. Recall that the eight constructs mapped onto the three models and the relationships between them, thus there were multiple constructs per model. When trying to revise the constructs it became apparent that the assessment items were not equally distributed per construct, nor did they provide adequate information about specific levels. It was therefore impossible to develop a validity argument for all constructs and all levels. A much more systematic approach was needed to do this work.

Second, and relatedly, it also became clear that the specific context of assessment questions mattered. Students appeared to be reasoning at higher levels based on the interview questions which allowed for more follow-up. They also seemed to reason better about genetic phenomena in humans than in other organisms; perhaps due to having more familiarity with the general biology of humans compared to other organisms. Given that the assessment tasks did not systematically vary contextual factors, it was impossible to make any claims about how such factors impact learning and how such differences can be captured in the progression itself.

Third, the lack of more substantial gains in the second year was difficult to interpret. One could assert that the shift from level 2 to level 3 understandings of constructs was too difficult for middle schoolers and consequently developmentally inappropriate. Therefore, the progression, as a model of learning, needs to be modified to reflect this constraint. However, one could also question the role of the instructional unit. As we noted earlier, the two units were rather different in framing, anchoring phenomenon, context (humans versus animals), etc. It may be that the second instructional unit was simply not that good at supporting learning. Perhaps, the engineering design task was overly complex and this distracted from the targeted ideas (Kanter, Schwille, MacKenzie & Reiser, 2003). There was some anecdotal evidence that students did find the tasks in the second unit more challenging. This dilemma is not unique to our particular study. If we are to assume that movement along a progression is not developmentally inevitable and requires targeted instruction (Duncan & Rivet, 2018) then there will always be a question about the efficacy of the specific instantiation of instruction and whether a lack of learning is a result of developmental constraints (i.e., the problem is with the hypothesized progression) or due to ineffective instruction (Duncan & Gotwals, 2015).

Our next phase of research, therefore, aimed to address some of these problems through more systematic development of assessments and instruction, as well as making some judicial decisions about which issues to tackle and on which to defer. The next phase was also poised to address two key questions in the field of learning progressions—to what extent are the learning paths supported by a progression developmentally constrained? And what role does instruction play in supporting or hindering such progress?

Take Three: Implementation Study of Alternative Instructional Sequences

Learning progressions take a developmental perspective and describe a hypothetical learning path through a domain. The theoretical assumption underlying these progressions is that there are developmental constraints and affordances that drive the learning route. "Developmental" constraints in this sense are not necessarily related to age; rather they reflect the conceptual constraints that may exist in a learning path because the acquisition of new knowledge depends on existing knowledge. Instruction plays a key role in helping students move through this path as progress is not developmentally inevitable. This common conceptualization of progressions raises the question of how strong the constraints of the learning process are. It may be that constraints are relatively loose and many different learning paths are possible that start from the same lower anchor and lead to the same upper anchor. Theories of learning that take a more situated, knowledge-in-pieces, and dynamic ecologies perspective (diSessa, 1988, 2014; Lave & Wenger, 1991; Scherr, 2008) would predict such a scenario of a multitude of paths.

If the developmental constraints are strong then learning progressions research can help us identify such constraints and thus inform standards, assessment, and instruction. However, if there is a large number of possible paths it is not clear whether hypothetical progressions can contribute to the development of a single set of standards (across grades), appropriate assessments, and effective instructional approaches. The question that arises is therefore: how strong are the proposed developmental constraints and are limited well-defined paths likely? A study of two progressions in the same domain can provide some of the evidence needed to address these questions. On the one hand, such a study may show that students' actual learning paths are similar regardless of the instructional conditions. That is, the development of certain ideas necessarily precedes the development of others regardless of which ideas were presented first instructionally. However, it is expected that more students would achieve higher levels of sophistication in the instructional condition that reflects the progression that is more closely aligned with the actual learning path. Such outcomes will provide some support for the notion that the conceptual constraints are strong and that in this domain there may be a progression that appropriately models learning paths. On the other hand, learning paths may look entirely different under the two instructional conditions suggesting that the developmental constraints on learning are weak (at least in this domain) and that multiple learning paths are both viable and likely. Such evidence will undermine, at least to some extent, the theoretical premise that there are substantive and critical developmental constraints and affordances that can be captured in a progression.

As noted earlier, there are alternative progressions in the domain of genetics and largely these vary in terms of whether they propose learning the molecular model first or the inheritance model first. If, in fact, the instructional order matters because the developmental constraints make it difficult to understand one model before learning the other, then students learning the pre-requisite model first will do better overall (i.e., more of them will reach higher levels of sophistication compared to students in the alternative order). In the third phase of the genetics progression research, we set out to test this hypothesis. The implementation study involved developing two parallel and highly comparable instructional units in molecular and classical genetics (combined inheritance and meiotic models) and using them in two instructional conditions: molecular first (MC) or classical first (CM). We worked with five biology teachers and their multiple classes (sections) of high school biology (11th graders) in a suburban district (n = 285). For each teacher, half the classes experienced the MC condition and the other half the CM condition. Classes were not randomly assorted given logistical challenges, but we did our best to ensure that classes across both conditions were comparable in terms of pretest scores and other related variables, such as students' chemistry scores from the prior year. We conducted two rounds of implementation for two consecutive years with the teachers (there were some changes to staffing in the school and we did have some new teachers in the second round).

The instructional units were developed using the model-based inquiry (MBI) approach and engaged students in the development, evaluation, and revision of models of genetics phenomena such as the inheritance pattern of hair texture and the molecular basis of genetic resistance to HIV (Castro-Faix, Duncan, & Choi, 2021; Duncan, Choi, Castro-Faix, & Cavera, 2017). The two units were comparable in terms of the MBI approach, the number and type of modeling opportunities, and the focus on human traits. Each unit took about 4 weeks to teach and after both units were completed (in either order) students engaged with a week-long bridge module that was aimed at connecting across models by having students develop a general explanation of the molecular basis of various inheritance patterns. Over the summer after the first implementation, the units were revised slightly by the collaborative team of researchers and teachers based on our collective experiences of the enactments.

In addition to developing units, we also developed a set of written assessments. This time around we were much more systematic and used evidence-centered design to carefully develop a sufficient number of items per construct and level (Mislevy, Steinberg & Almond, 2003; Mislevy, Almond & Lukas, 2004). We chose to focus the instruction and assessments on only five constructs (A, B, C for the molecular unit, E and F for the classical unit) in order to keep the assessment development task and instructional time manageable, and because these constructs embodied ideas that were already central in the high school genetics curriculum and were familiar to the teachers. We also restricted the contextual factors in items to be about human genetics in almost all items (except those for construct A which were about the universality of the genetic code). Given limited assessment time (tests could not take more than one class period) and ease of scoring, we elected to develop Ordered Multiple Choice (OMC) items for our assessments. In these multiple-choice items, each response option maps onto one level of understanding based on a learning progression (Briggs et al., 2006). The scoring of OMC items allows for partial credit to all item responses, rather than binary right or wrong scoring—thus, more information about students' level of understanding is provided through these items. We developed a pool of 85 OMC items through an iterative process that involved three pilot tests. Item fit statistics were obtained for each item, and based on the

Construct F-3 Item- *Inheritpattern*

F.3. Relationship between molecular and Classical models	**L0**- Dominant allele overpowers recessive allele. No protein-based mechanism.	**L1**-Proteins bring about the phenotype. Lack of a protein can result in a disorder.	**L2**-Dominant allele creates a protein that can compensate for missing protein of recessive allele. Does not imply understanding of mechanism for dominance. May assume no normal protein made in heterozygote dominant.	**L3**-Dominant traits can have several molecular mechanisms: dosage, interference, continuous signaling. Definitions of dominant and recessive alleles are categorical.

ALP is a genetic disease. People with ALP disease make a mutated protein that causes the cells to constantly divide. In normal people, cells only divide when they receive a hormone signal. Based on this information, ALP disease is likely to be:

A. It is recessive- because the protein made from the normal gene can compensate for the malfunctioning protein made from the mutated gene. [L2]
B. It is recessive- because the malfunctioning protein made from the mutated gene is canceling the normal functioning of the protein made from the normal gene. [L1]
C. It is dominant- because the malfunctioning protein made from the mutated gene is canceling the normal functioning of the protein made from the normal gene. [L3]
D. It is dominant- because the protein made from the normal gene is compensating for the malfunctioning of the protein made from the mutated gene. [L1]

Figure 11.2 Ordered multiple choice item example.

items' statistics and functioning, we revised items for subsequent pilots or removed them from consideration; for more on the item evaluation and refinement process see Choi and Duncan (2016). We then selected the best-performing 56 OMC items from the pool to include in the final assessment forms. There were about 9–13 items per construct (for the five constructs). Most of the items included response options that mapped onto two to four levels of a particular construct (we added a level zero based on data from the prior longitudinal study). These items were incorporated into two parallel forms of the written test each of which contained 44 OMC items with 32 of those being linking items (same in both forms of the test). Figure 11.2 provides an example of an item we termed *inheritpattern* aligned to construct F-3. Note that the original construct F was divided into three sub-constructs for this phase of the research and F-3 relates specifically to the relationship between the classical and molecular models during this phase of the research. This item related to three levels of F-3 with two response options at level 1, one response option at level 2, and one response option at level 3 (most sophisticated).

This construct was one of the more challenging for students (see below), especially in terms of relating the molecular and classical models as is needed to reach level 3 on this item.

Lessons Learned

Our analyses focused on the second year of implementation since the overall gains in that year were larger. This is likely because the teachers were familiar with the units and had more ownership over them given their involvement in the revisions of the unit during the intervening summer. Several interesting findings emerged and we discuss them briefly, for more information see Castro-Faix et al. (2021) and Duncan et al. (2017). Overall, we found that students made significant gains across all constructs, often moving up about one level of understanding along the progression. The greatest gains, but also spread, were for construct B—the idea that genes have information for proteins. Many of the students in the study held level 1 understandings at the beginning of the unit and most ended up with a level 2 understanding, with a substantial minority ending with level 3 understandings. Conversely, construct F- patterns of inheritance showed the smallest gains. The upper levels of construct F entail connecting inheritance patterns to their underlying molecular mechanisms; essentially linking the molecular and inheritance models. This linking was primarily taught through the week-long bridge module. We believe that the instruction was insufficient to help students develop a robust understanding of the relationships between the models such that they could achieve level 3 understandings for this construct.

Regarding our question about the impact of alternative learning progressions, we found no statistically significant differences in learning under the two instructional conditions (CM or MC), suggesting that the cognitive constraints on learning in genetics may not be very strong. However, the MC condition gains were slightly higher across all constructs. While not statistically significant, it was surprising that the trend was for all constructs. This trend also contradicted a reasonable assumption that students may show larger gains for the constructs that they learned last and closer to the administration of the test (i.e., the classical constructs E and F for condition MC, and vice versa). The across-the-board slightly higher gains for the molecular-first condition suggest that there may be some learning differences between the two conditions that are not adequately captured by our study design or instrument.

In addition to administering the OMC written assessments, we also conducted interviews with a subset of students in each class (n = 60) with tasks adapted from, and revised based on, the interviews from the longitudinal study. Interview data allowed us to revise and refine the remaining constructs A, E, and F (Castro-Faix et al., 2021) using the same heuristics developed by Shea and Duncan (2013) when revising constructs B and C. Of note is that these revisions led to the splitting of construct E into two distinct, but related, constructs E1 and E2. The original construct included ideas about both the passage of genetic information from one generation to the next and the role that sex cells play in that passage. It appeared from the data that these two sub-ideas developed in parallel and distinctly from each other for the middle levels of the progression. However, the higher levels of the E construct were integrated and students were able to develop mechanistic explanations using sex cells as the mechanism for physical passage of genetic information. That is, the progression for construct E started off as two subconstructs that were distinct and then merged at the upper levels. This was a novel move for us in terms of the nature of the revisions, and we therefore expanded Shea and Duncan's (2013) heuristics to include guidance for when to split constructs. Namely, suppose it seems that the students' reasoning about a construct involves two distinct ideas and that students can be at different levels of sophistication on each of these sub-ideas. In that case, it may be that there is more than one latent variable that needs to be unpacked or described further. Such a splitting also provides centrality, consistency, and unity to the main idea in each construct.

The interview data was also used to tease out relationships and dependencies between constructs using a Guttman analysis (Castro-Faix et al., 2021). For example, we found that understandings of constructs E1 and E2 provided somewhat of a bootstrapping effect for construct F (which as noted earlier is a difficult construct for students to master). It seems that students need to know about the role of sex cells in meiosis and the inheritance of alleles as half a set (construct E) before they can understand more difficult patterns of inheritance (construct F-Level 4). Similarly, our analyses also suggest that there may be some influence of construct F on learning construct E. That is, the understanding that there are "hidden alleles" (construct F) seems to support a better understanding of both meiosis and the physical passage of genes (constructs E1 and E2). These relationships between constructs help to flesh out the complexity and add nuance to depictions of how students' learning unfolds across multiple constructs.

Challenges

Like the longitudinal study, our two-year (replicated) implementation study provided valuable information but was not without its challenges, we expand on two of them.

The development of assessments for this study was much more systematic and allowed us to make claims about student reasoning at different levels for each construct. Yet, this development process was arduous and not trivial (Choi & Duncan, 2016). The various item fit analyses provided information about where items were falling short and hinted at what needs to be fixed, but these analyses do not prescribe fixes, and any change to an item to address one problem can "break" it in another sense and result in other problems. Often, while we knew the nature of the problem with an item, the revisions were an art of guesswork as to how changing particular words might impact student interpretation of the item prompt or response options. Moreover, once the progression is revised and there are additional (or different) levels and possibly constructs, any set of OMC items needs to be revised to capture these changes. In our own work, we have yet to develop a new set of items that reflects the revisions to the constructs based on our prior research. Thus, the development of assessment items, validation of the interpretations, and use of the assessment results are an ongoing effort and we caution researchers to build in sufficient time (and relevant expertise) to conduct this work.

Second, our study showed that high school students were able to move along the progression as anticipated over the course of a 9–10-week instructional intervention. The study took place in a district that sequences the science courses from physics (9th grade), to chemistry (10th grade) to biology (11th grade and the focus grade for this study). Earlier we noted that in our longitudinal study we speculated that middle school students struggled to progress to the highest level on the molecular constructs (which were targeted for high school) possibly due to a lack of relevant chemistry or physics knowledge. The implementation study showed that high school students can reason at these higher levels, as anticipated, but did not address the question of whether this was, in fact, dependent on pre-requisite knowledge in other domains, such as chemistry. We remain with this question at this time and hope to more fully explore it in future work. However, we can provide a few speculations regarding how chemistry knowledge may intersect with the genetics progressions. Understanding construct A, the universality of the genetic materials, likely depends on some grasp of the physical nature of the genetic material. That is, the idea that genes are sequences of nucleotides and that DNA is a long molecule that looks and functions in the same way across organisms. Understanding the function of proteins, construct C, also entails understanding their physical structure and some chemical properties. In our curriculum we introduced students to only several chemical properties, namely hydrophobic/hydrophilic, having a positive or negative charge, and disulfide bonds between cysteine amino acids. Familiarity with these properties allowed students to conceptualize some aspects of protein structure such as transmembrane domains, bends and grooves in proteins, and repelling or attracting other molecules. Students with deeper chemistry knowledge would be able to reason about additional properties and their impact on protein structure and function, thus expanding their understanding of construct C. At this time, we can only speculate about the value added by chemistry knowledge or any other domain of knowledge. Studying interactions between progressions in

Table 11.1 Summary of progression development

Design of instruction and assessment	*Methods of research*	*Findings*	*Revisions*
Phase I: Initial Development of LP			
NA	Inductive analysis of prior literature in genetics education and the empirical research therein.	Developed a progression with eight constructs that map onto three genetic models: inheritance, meiotic, and molecular.	NA
Phase II: Longitudinal study			
Developed two five-week units informed by the progression to study movement from level 2 to level 3 of the progression. Designed written assessments and interview tasks that mapped onto the three genetic models.	The two units were implemented sequentially over two years by the same teacher in a mixed 6–8th grade class (n = ~60)	Gains across all three models with the largest gains for the molecular model and smallest gains for the inheritance model. No clear developmental dependencies between models.	Added level zero to constructs to reflect understandings that are precursors to level. Revised constructs B and C- added levels to capture nuance in the data.
Phase III: Implementation study of Alternative Instructional Sequences			
Developed two four-week modules corresponding to constructs molecular and classical genetics (combined inheritance and meiotic models). Developed a bank of OMC items corresponding to constructs A, B, C, E, F, as well as interview tasks.	Five teachers implemented the modules in alternating order (molecular and classical or vice versa) across multiple sections of their 11th-grade biology classes (n = 285).	Gains of approximate one level across constructs with largest gains for construct B and least gains for construct F. Small but consistent trends in larger gains for students in the molecular-first condition suggesting weak bootstrapping effect of molecular constructs (A, B & C) for classical constructs (E & F).	Revised constructs A, E, and F. Split E into two sub-constructs that merge at the upper levels of this construct. Added a level to each of the constructs A and F to capture nuance in students' reasoning at the lower end of the progression.

the same domain, let alone other domains, is complex and a heavy empirical lift. To date, we are not aware of any other work that has explored how progressions in different science domains impact each other.

In summary, our work over the past decade has provided an illustrative example of the successes, tensions, and challenges involved in learning progressions work. Table 11.1 summarizes the stages of development of the progression.

While we have made progress on some of the questions that drove our research agenda for the genetics progressions, other questions remain elusive. A key message from this story is that learning progressions work requires time, is iterative in nature, and in some sense never really done. To keep this chapter focused we have elected to describe only our own work on the genetics LP. However, we would be remiss if we did not mention that while we were working on the progression a parallel group was also developing instructional materials and assessments informed by the progression and engaging in empirical studies at the high school and college levels to further test and revise the progression (Todd & Kenyon, 2015; Todd & Romine, 2017; Todd, Romine & Cook-Whitt, 2017). Findings from these lines of research supported the progression and also further defined the stepping-stone ideas for its constructs (Todd & Kenyon, 2015). Since the development of the initial progression, other research on genetics learning and teaching has brought to light some important implications and ramifications that were not taken into account in our own and earlier research. We discuss these next and their implications for further work on the genetics progression and genetics instruction writ large.

New Research with Critical Implications

Over the past decade and especially the last five years, there have been growing concerns about how genetics instruction, in particular a focus on Mendelian genetics, contributes to deterministic and essentialist views of genetics (Dar-Nimrod & Heine, 2011; Dar-Nimrod et al., 2021; Donovan, 2014, 2016, 2017; Donovan et al., 2020). This was not fully on our radar in 2009 but has become more salient and problematic. Specifically, work by Donovan, 2014, 2016, and 2017 has shown that emphasizing single gene disorders and the association between some of these genetic disorders and races (e.g., noting that Sickle Cell Anemia is common in African-Americans) can reinforce students' beliefs that there are racial differences in other complex traits such as intelligence and athletic ability and that these are genetic and therefore unchangeable (Donovan, 2016). Several researchers have explored possible solutions to this problem (Donovan et al., 2019; Donovan et al., 2020; Jamieson & Radick, 2017) and have found that learning about human genetic variation, emphasizing the role of the environment and learning about ancestry, can decrease essentialist and deterministic thinking.

The findings from this body of work are particularly important given the increasing attention in the field to issues of equity and justice. Understanding the fallacies and dangers of genetic determinism and essentialism is critical for genetic literacy. It now behooves us, and the field of genetics education, to carefully examine and reconsider the progressions we have and re-conceptualize what it would mean to foster genetic literacy that is anti-racist. A revised progression needs, at a minimum, to deepen students' understanding of the role of the environment in genetic phenomena by emphasizing the constructs associated with gene expression and gene regulation (D and H) and enhancing the focus on multifactorial traits.

However, countering essentialist and deterministic views of genetics, while absolutely necessary, will likely be insufficient as students may replace them with cultural and moral essentialism and the "othering" of marginalized groups as less culturally and morally sophisticated. We believe that anti-racist genetics instruction needs to intentionally address the socio-historical-political context in which gene-environment interactions take place. Understanding that genetic variations in traits are (often) largely a result of different environments (rather than genes) still leaves the environment as a benign and neutral space. But environments are not neutral, they are political and shaped by historical context (e.g., redlining policies in the US that led to the current and ongoing segregation). Environments are also not benign but rather can be incredibly harmful especially the environments experienced by marginalized communities. Our teaching of genetics needs to develop socio-political genetic awareness and understandings that: a) almost all traits are multifactorial – many genes interact with the environment, b) the environment in gene-environment interactions are non-neutral and non-benign, c) these environments interact with genomes that are more similar than they are different across populations, and d) racial categories are neither genetic nor biological but these categories have profound and often devastating impacts on all aspects of the health and well-being of racially marginalized groups.

A Few Concluding Considerations

In this chapter, we have attempted to describe the evolution of our genetics progression over the course of a decade of research about it in classroom settings. In this section, we discuss a few remaining key challenges and future work regarding the genetics progression as well as raise a few issues in learning progressions research that are more broadly relevant.

The discussion in the prior section regarding developing students' critical and socio-political genetics awareness raises the question of the aims and focus of genetics education; how do we anticipate that students might use their genetics literacy (however we construe it) in their everyday lives now and in the future? This question has implications for genetics learning but is also endemic to any other progression and its upper anchor— the anticipated most sophisticated level of understanding.

In the case of genetics, what qualifies as "most sophisticated" depends on the situation. At times, using simple Mendelian genetics may be sufficient (e.g., to explain the likelihood of having a child with Cystic Fibrosis given heterozygous parents); in other situations, it may not (e.g., explaining the risk of developing breast cancer if one carries the *BRCA1* gene). Even reasoning that seems to be epistemically inappropriate, for example, ignoring anomalous data, is appropriate under some conditions, and scientists employ such reasoning strategies themselves (Chinn & Brewer, 1993). For many progressions, including the genetics progressions, definitions of the upper anchors stem from normative (aligned with scientific cannon) expectations that are often informed by experts' analyses of the domain and what students should know about it. However, Sikorski (2019) has argued that we may need to adopt a more expansive definition of the upper anchor that is much more sensitive to the situational appropriateness of reasoning. That is, what counts as sophisticated reasoning at the highest level depends on the situation, and there may be a plurality of upper anchors that are appropriate for different situations.

We can think of the upper anchor in terms of becoming a competent outsider (Feinstein, 2011)—someone who can make appropriate judgments about the credibility of claims based on cues, such as source credibility (e.g., professional reputation, publication venue) and degree of consensus in the field, even when they lack a deep understanding of the discipline and its methods of inquiry. Positioning genetic literacy as being a competent outsider expands what counts as "most sophisticated." That is, the upper anchor should include knowledge needed by the competent outsider to solve a genetic issue or problem relevant to the situation they are dealing with and given the cultural and political context in which they operate. For example, one might expect a competent outsider to be able to make sense of claims related to genetic issues that appear in the news. An attempt to identify these types of understanding was carried out by Shea (2015), who analyzed genetics-related articles in the New York Times science section. She found that those articles often entailed understanding molecular genetics ideas that corresponded to the highest levels of the Duncan et al. (2009) progression, but only a lower-level understanding of Mendelian genetics ideas. Moreover, Shea (2015) found that some articles entailed an understanding of conserved genetic sequences, and comparisons of sequences across species, which went beyond the upper-level expectations of the genetics learning progression as conceptualized by Duncan et al. (2009). It is therefore clear that the upper anchor expectations of progressions, including ours, do not always overlap with what may actually be needed to be a genetically-literate competent outsider. Moreover, competency also depends on historical and cultural relevance. Our argument for a critical socio-political genetic awareness expands what it means to be a competent outsider to include understanding the history and ramifications of genetic concepts and research findings. The upper anchor should reflect the understandings and views we would like to espouse in a more politically aware, democratic, and justice-seeking public. In addition, the upper anchor/s should reflect the plurality of situations in which such understandings and views are likely to be used by competent outsiders.

The consideration of context matters in another sense as well as "diagnosing" students' levels of reasoning through various forms of assessments (written, interview, etc.) In our work, we have found that students' level of reasoning as inferred from interview tasks was higher than what they exhibited on the written OMC assessments. Moreover, even within a single assessment, students do not reason consistently (i.e., at a single level) across items for the same construct. This apparent inconsistency is known as the "messy middle" phenomenon (Gotwals & Songer, 2010; Steedle & Shavelson, 2009). The messy middle can partly be attributed to noise in the data, but it may not always be a simple case of measurement error. It is also likely that some contexts are, in fact, more difficult for students to reason about and represent possible (sub) levels of difficulty. For example, students may find it more difficult to reason about the function of proteins in relation to biological phenomena (or organisms) they are less familiar with. Developing a robust understanding of proteins' role in genetic mechanisms (construct C) entails learning about a wide variety of protein functions in different organisms (Haskel-Ittah, Duncan, Vázquez-Ben, & Yarden, 2019). This level of nuance is not fully captured by the genetic progression (or the assessments). To some extent, this is also an issue of grain size and while students may appear inconsistent reasoners (due to a not-yet robust understanding) at one grain size, they may be much more consistent at a coarser grain size. This suggests that we may need to develop progressions at different grain sizes to support different purposes associated with teaching and learning. A grain size useful for large-scale assessment may be too coarse for supporting day-to-day classroom instruction (Duncan & Rivet, 2018). In our work, we have been able to revise the progression in ways that provide a finer grain size (more levels) than can be useful for planning and implementing instruction. However, we have yet to develop assessments to reflect these revisions partly because it is difficult to develop OMC items that span more than 3–4 levels. More levels (or more constructs) require more items, thus there is also a practical constraint to maintaining greater nuance.

The last issue we wish to consider in closing is that of instructional dependency and the relationship between the progression, science standards in the U.S., and available instructional materials. The current science standards in the U.S. are fairly well-aligned with the Duncan et al. (2009) progression in that expectations across grade bands reflect the progress along the progression constructs (NGSS, 2013). Over the past decade, there have been tremendous and productive efforts to develop high-quality NGSS-aligned instructional materials for middle (OpenSciEd at openscied.org) and high school science (inquiryHub at colorado.edu/program/inquiryhub) that are open source and thus freely available. The genetics unit for the 8th grade (OpenSciEd) and the high school unit (inquiryHub) both address many (but not all) of the constructs of the progression. Neither unit addresses construct D (differentiation and regulation of gene expression), and while both units address some aspects of construct G (changes to the genetic information) they do not deeply engage with the evolutionary aspects of that construct (conserved genetic sequences and differences across species). The middle school unit explicitly addresses the role of environment in trait variation and students are expected to develop multifactorial models of inheritance (construct H). However, the high school unit focuses more on Mendelian inheritance patterns and their molecular basis and less on the role of the environment. However, the high school unit does address some of the historical-social-political aspects of genetics research and technologies in ways that can support the development of critical socio-political genetic awareness of the sort we have advocated. This is accomplished through the exploration of past and present attempts to change human traits, the socio-political assumptions underlying these technologies, their outcomes, and repercussions (e.g., human eugenics, gene therapy and testing, CRISPR-cas9 system). Engaging with the political is an emboldened approach taken by inquiryHub and one that is much needed.

In summary, we hope that our account of the development and evolution of the genetics progression provides some insights about the obstacles, and possible solutions, that dot the path of learning progressions research. While this path, like learning progressions themselves, is not linear nor straightforward, to transverse it has been a productive journey for us and one that has yielded important new knowledge to inform standards, curriculum, and assessment in the domain. We look forward to the next chapter in this saga.

References

Briggs, D. C., Alonzo, A. C., Schwab, C., & Wilson, M. (2006). Developmental assessment with ordered multiple-choice items. *Educational Assessment*, 11(1), 33–63. doi: 10.1207/s15326977ea1101_2

Castro-Faix, M., Duncan, R. G., & Choi, J. (2021). Data-driven refinements of a genetics learning progression. *Journal of Research in Science Teaching*, 58(1), 3–39. doi:/10.1002/tea.21631

Chinn, C. A., & Brewer, W. F. (1993). The role of anomalous data in knowledge acquisition: A theoretical framework and implications for science instruction. *Review of educational research*, 63(1), 1–49. doi:/10.3102/00346543063001001

Choi, J., Duncan, R. G. (2016 April). *Validating assessments for a genetics learning progression*. Paper presented at the Annual Meeting of the American Educational Research Association, Washington, DC.

Cobb, P. & Bowers, P. (1999). Cognitive and situated learning perspectives in theory and practice. *Educational Researcher*, 28(2), 4–15. doi:/10.3102/0013189X028002004

Dar-Nimrod, I., & Heine, S. J. (2011). Genetic essentialism: On the deceptive determinism of DNA. *Psychological bulletin*, 137(5), 800. doi: 10.1037/a0021860

Dar-Nimrod, I., Kuntzman, R., MacNevin, G., Lynch, K., Woods, M., & Morandini, J. (2021). Genetic essentialism: The mediating role of essentialist biases on the relationship between genetic knowledge and the interpretations of genetic information. *European Journal of Medical Genetics*, 64(1), 104119. doi:/10.1016/j.ejmg.2020.104119

Donovan, B. M. (2014). Playing with fire? The impact of the hidden curriculum in school genetics on essentialist conceptions of race. *Journal of Research in Science Teaching*, 51(4), 462–496. doi:/10.1002/tea.21138

Donovan, B. M. (2016). Framing the genetics curriculum for social justice: An experimental exploration of how the biology curriculum influences beliefs about racial difference. *Science Education*, 100(3), 586–616. doi:/10.1002/sce.21221

Donovan, B. M. (2017). Learned inequality: Racial labels in the biology curriculum can affect the development of racial prejudice. *Journal of Research in Science Teaching*, 54(3), 379–411. doi:/10.1002/tea.21370

Donovan, B. M., Semmens, R., Keck, P., Brimhall, E., Busch, K. C., Weindling, M., ...Salazar, B. (2019). Toward a more humane genetics education: Learning about the social and quantitative complexities of human genetic variation research could reduce racial bias in adolescent and adult populations. *Science Education*, 103(3), 529–560. doi:10.1002/sce.21506

Donovan, B. M., Weindling, M. & Lee, D. (2020). From basic to humane genomics literacy. *Science & Education*, 29, 1479–1511 (2020). doi:10.1007/s11191-020-00171-1

di Sessa, A. A. (1988). Knowledge in pieces. In G. Forman & P. B. Pufall (Eds.), *Constructivism in the computer age* (pp. 49–70). Lawrence Erlbaum Associates, Inc.

diSessa, A. A. (2014). A history of conceptual change research: Threads and fault lines. comparisons. *Science Education*, 99(3), 410–416. doi:/10.1017/CBO9781139519526.007

Dougherty, M. J. (2009). Closing the gap: Inverting the genetics curriculum to ensure an informed public. *The American Journal of Human Genetics*, 85(1), 6–12. doi:10.1016/j.ajhg.2009.05.010

Duncan, R. G., Castro-Faix, M., & Choi, J. (2016). Informing a learning progression in genetics: Which should be taught first, Mendelian inheritance or the central dogma of molecular biology? *International Journal of Science and Mathematics Education*, 14(3), 445–472. doi:/10.1007/s10763-014-9568-3

Duncan, R. G., Choi, J., Castro-Faix, M., & Cavera, V. L. (2017). A study of two instructional sequences informed by alternative learning progressions in genetics. *Science & Education*, 26(10), 1115–1141. doi:10.1007/s11191-017-9932-0

Duncan, R. G., Freidenreich H. B., Chinn C. A., & Bausch, A. (2011). Promoting middle school students' understanding of molecular genetics. *Research in Science Education*, 41(2), 147–167. doi: 10.1007/s11165-009-9150-0

Duncan, R. G., & Gotwals, A. W. (2015). A tale of two progressions: On the benefits of careful comparisons. *Science Education*, 99, 410–416. doi: 10.1002/sce.21167

Duncan, R. G., & Reiser, B. J. (2007). Reasoning across ontologically distinct levels: Students' understandings of molecular genetics. *Journal of Research in Science Teaching*, 44, 938–959. doi:10.1002/tea.20186

Duncan, R. G. & Rivet, A. E. (2018). Learning progressions. In F. Fisher, C. E. Hmelo-Silver, & S. R. Goldman (Eds.), *International Handbook of the Learning Sciences* (pp. 422–432). New York: Routledge. doi:/10.4324/9781315617572-41

Duncan, R. G., Rogat, A., & Yarden, A. (2009). A learning progression for deepening students' understanding of modern genetics across the 5th-12th grades. *Special Issue on Learning Progressions for the Journal of Research in Science Teaching*, 46(6), 644–674. doi:/10.1002/tea.20312

Duschl, R., Maeng, S., & Sezen, A. (2011). Learning progressions and teaching sequences: A review and analysis. *Studies in Science Education*, 47(2), 123–182. doi: 10.1080/03057267.2011.604476

Feinstein, N. (2011). Salvaging science literacy. *Science Education*, (95), 168–185. 10.1002/sce.20414

Freidenreich, H. B., Duncan, R. G., & Shea, N. (2011). Exploring middle school students' understanding of three conceptual models in genetics. *International Journal of Science Education*, 33(17), 2323–2349. doi: 10.1080/09500693.2010.536997

Gotwals, A. W., & Songer, N. B. (2010). Reasoning up and down a food chain: Using an assessment framework to investigate students' middle knowledge. *Science Education*, 94(2), 259–281. doi: 10.1002/sce.20368

Haskel-Ittah, M., Duncan, R. G., Vázquez-Ben, L., & Yarden, A. (2019). Reasoning about genetic mechanisms: Affordances and constraints for learning. *Journal of Research in Science Teaching*, 57(3), 342–367. doi:10.1002/tea.21595

Jamieson, A., & Radick, G. (2017). Genetic Determinism in the Genetics Curriculum. *Science & Education*, 26(10), 1261–1290. doi:10.1007/s11191-017-9900-8

Kanter, D. E., Schwille, K., MacKenzie, S., & Reiser, B. J. (2003). Curriculum design strategies support content learning in task-structured science curricula. Paper presented at the Annual Meeting of the *National Association for Research in Science Teaching*. Philadelphia, PA.

Kindfield, A. C. H. (1992). Teaching genetics: Recommendations and research. In M. U. Smith, & P. E. Simmons (Eds.), *Teaching genetics: Recommendations and research proceedings of a national conference* (pp. 39–43). Cambridge, MA: National Science Foundation.

Kurth, L. A., & Roseman, J. (2001, April). Findings from the high school biology curriculum study: Molecular basis of heredity. Paper presented at the Annual Meeting of the *National Association for Research in Science Teaching*, St. Louis, MO.

Lave, J., & Wenger, E. (1991). *Situated learning: Legitimate peripheral participation*. Cambridge university press.

Lewis, J., & Kattmann, U. (2004). Traits, genes, particles, and information: Re-visiting students' understandings of genetics. *International Journal of Science Education*, 26(2), 195–206. doi:10.1080/0950069032000072782

Lewis, J., & Wood-Robinson, C. (2000). Genes, chromosomes, cell division and inheritance - do students see any relationship? *International Journal of Science Education*, 22(2), 177–195. doi:10.1080/095006900289949

Marbach-Ad, G., & Stavy, R. (2000). Students' cellular and molecular explanations of genetics phenomena. *Journal of Biological Education*, 34(4), 200–205. doi:10.1080/00219266.2000.9655718

Mislevy, R. J., Almond, R. G., & Lukas, J. F. (2004). *A brief introduction to evidence-centered design (CSE Technical Report 632)*. Los Angeles: National Center for Research on Evaluation, Standards, and Student Testing (CRESST), Center for the Study of Evaluation, UCLA. [Retrieved June 6, 2005 from http://www.cse.ucla.edu/reports/r632.pdf]

Mislevy, R. J., Steinberg, L. S., & Almond, R. G. (2003). Focus article: On the structure of educational assessments. *Measurement: Interdisciplinary Research & Perspective*, 1(1), 3–62. doi:10.1207/s15366359mea0101_02

Next Generation Science Standards [NGSS] Lead States. (2013). *Next generation science standards: For states, by states*. National Academies Press. doi: /10.17226/18290=

National Research Council. (2012). *A framework for K-12 science education: Practices, crosscutting concepts, and core ideas*. Washington, DC: The National Academies Press. 10.17226/13165.

Roseman, J., Caldwell, A., Gogos, A., & Kurth, L. A. (2006). Mapping a coherent learning progression for the molecular basis of heredity. Paper presented at the Annual Meeting of the *National Association of Research in Science Teaching*, San Francisco, CA.

Scherr, R. E. (2008). Gesture analysis for physics education researchers. *Physical Review Special Topics - Physics Education Research*, 4(1). doi:10.1103/physrevstper.4.010101.

Shea, N. A. (2015). Examining the nexus of science communication and science education: A content analysis of genetics news articles. *Journal of Research in Science Teaching*, 52(3), 397–409. doi:10.1002/tea.21193

Shea, N. A., & Duncan, R. G. (2013). From theory to data: The process of refining learning progressions. *Journal of the Learning Sciences*, 22(1), 7–32. doi: /10.1080/10508406.2012.691924

Sikorski, T. R. (2019). Context-dependent "upper anchors" for learning progressions. *Science & Education*, 28(8), 957–981. doi:/10.1007/s11191-019-00074-w

Steedle, J. T., & Shavelson, R. J. (2009). Supporting valid interpretations of learning progression level diagnoses. *Journal of Research in Science Teaching*, 46(6), 699–715. doi:/10.1002/tea.20308

Stevens, S. Y., Delago, C. & Krajcik, J. S. (2010). Developing a hypothetical multi-dimensional learning progression for the nature of matter. *Journal of Research in Science Teaching*, 47(6), 687–715. doi:/10.1002/tea.20324

Stewart, J., Cartier, J. L., & Passmore, P. M. (2005). Developing understanding through model-based inquiry. In M. S. Donovan, & J. D. Bransford (Eds.), *How students learn* (pp. 515–565). Washington, D.C.: National Research Council. doi:/10.1111/j.1949-8594.2009.tb17870.x

Todd, A., & Kenyon, L. (2015). Empirical refinements of a molecular genetics learning progression: The molecular constructs. *Journal of Research in Science Teaching*, 53(9), 1385–1418. doi:/10.1002/tea.21262

Todd, A., & Romine, W. L. (2017). Empirical validation of a modern genetics progression web for college biology students. *International Journal of Science Education*, 39(4), 488–505. doi:/10.1080/09500693.2017.1296207

Todd, A., Romine, W. L., & Cook Whitt, K. (2017). Development and validation of the learning progression–based assessment of modern genetics in a high school context. *Science Education*, 101(1), 32–65. doi:/10.1002/sce.21252

Tsui, C.-Y., & Treagust, D. F. (2003). Genetics reasoning with multiple external representations. *Research in Science Education*, 33, 111–135. 10.1023/A:1023685706290

Venville, G. J., & Treagust, D. F. (1998). Exploring conceptual change in genetics using a multidimensional interpretive framework. *Journal of Research in Science Teaching*, 35, 1031–1055. doi:/10.1002/(SICI)1098-2736(199811)35:9<1031::AID-TEA5>3.0.CO;2-E

Wiser, M., Smith, C. L., Asbell-Clarke, J., & Doubler, S. (2009, April). Developing and refining a learning progression for matter: The inquiry project: Grades 3–5. In CS (Chair), Developing and refining a learning progression for matter from pre-K to grade. Symposium conducted at the meeting of the *American Educational Research Association*, San Diego, CA.

Appendix A The Genetics Learning Progressions (Duncan, Rogat, & Yarden, 2009)

Big idea: 1) How do genes influence how we, and other organisms, look and function?	*Level 1: (5th–6th Grade)*	*Level 2: (7th to 8th Grade)*	*Level 3: (9th to 10th Grade)*
A **Universality of the genetic material. The genetic information is hierarchically organized**	All the cells in an organism have genetic information.	Genes are found in chromosomes. In most sexually reproducing organisms, all cells have the same set of chromosomes except for sex cells.	Chromosomes are made of DNA molecules. Genes are sequences of nucleotides in the DNA.
B **The genetic instructions have information for the structure and function of proteins**	Genes are instructions that help organisms develop and function.	Genes are instructions for molecules that carry out important functions in the organism.	The genetic code is translated into amino acids to make proteins.
C **Proteins have important functions in living organisms and they connect genes and traits**	Cells carry out many functions. Organisms are composed of multiple levels of organization that interact with one another (cells, organelles, etc.) Changes in one level may lead to changes in another.	Changes to the gene can lead to changes in the protein. Proteins have certain properties that afford their function and they do the work of the cell.	The three-dimensional shape of proteins is determined by their sequence of amino acids. This can determine how the protein functions. Mutations can affect the structure and function of proteins.
D **All cells have the same genes but they express different genes.**	Different cells have different structure and functions	Different kinds of cells have different types of proteins that carry out their basic and unique functions.	All the cells in an organism have the same genetic material. However, it is expressed differently depending on the cell.

Big idea: 2) Why do we vary in how we, and other organisms, look and function?	*Level 1: (5th–6th Grade)*	*Level 2: (7th to 8th Grade)*	*Level 3: (9th to 10th Grade)*
E **Reproduction is the transfer of information from one generation to the next.**	In sexual reproduction, each parent contributes half of the genetic material. All organisms reproduce and transfer their genetic information.	Sex cells are created through the process of meiosis which randomly assorts the chromosomes into the gametes. This process divides the genetic material in half.	During meiosis, the chromosomes swap sections and create new combinations of genes in a chromosome leading to genetic variation. The replication of DNA is regulated to prevent errors.
F **There are patterns of correlation between genes and traits.**	For each trait, there are several variations. Different organisms	Alleles are versions of a gene. Each individual has two versions of each allele. The	The difference between each allele is the nucleotide sequence that results in different

Big idea: 2) Why do we vary in how we, and other organisms, look and function?	*Level 1: (5th–6th Grade)*	*Level 2: (7th to 8th Grade)*	*Level 3: (9th to 10th Grade)*
	may have different versions of a trait.	chromosomes carry one version of the gene and there are correlations between the alleles and the trait.	or missing proteins leading to a phenotype. The dominant and recessive patterns of inheritance can be explained at the molecular level as a consequence of the function of the proteins made by each allele.
G **Changes to the genetic information lead to changes in the phenotype and lead to genetic variation.**	There is phenotypic variation in a population because the genetic information is different.	Genetic information can change and the effects on the organism depend on the changes. There are chromosomal differences between the sexes.	Genetic variation is caused by mutation, some sequences vary while others are conserved. The DNA sequence varies between individuals in a population allowing us to differentiate between them.
H **The environment can lead to changes in the way genes are expressed.**	The environment influences our traits. Related organisms may end up behaving or looking differently due to environmental effects.	The environment can lead to changes at the protein level that can influence cell function	The environment can alter gene expression or cause mutations that affect the expression of proteins.

12

DEVELOPING THREE-DIMENSIONAL LEARNING PROGRESSIONS OF ENERGY, INTERACTION, AND MATTER AT MIDDLE SCHOOL LEVEL

A Design-Based Research

Peng He, Namsoo Shin, and Joseph Krajcik

Introduction

The complex real-world problems facing humans—today and in the future—challenge the science, education, and policy communities to improve science education worldwide by shifting the goals of science education to knowledge-in-use (i.e., usable knowledge) rather than just knowing fragmented pieces of knowledge (National Research Council [NRC], 2007). That is, through learning science, all students should have the ability to make sense of compelling and challenging phenomena, solve complex problems, make informed decisions related to their daily lives, know how to learn more when needed, and consider pursuing science-related careers (NRC, 2012; National Academies of Sciences, Engineering, and Medicine [NASEM], 2019). The complex nature of phenomena and problems requires students to have the ability to selectively incorporate ideas from a range of knowledge domains and science practices and apply them to new contexts to solve problems. In other words, a single core idea is not sufficient. Instead, students must synthesize ideas (e.g., matter, energy, interaction) across discipline and science practices to apply them to new contexts. For students to reach knowledge-in-use learning goals (NRC, 2000), educators must consider what students should know and be able to do and what paths they can take to achieve each goal. Furthermore, students' development of knowledge-in-use proficiency takes time (He et al., 2023a; Li et al., 2023; Krajcik et al., 2023); it depends on experiences and education (NRC, 2007), where individuals have opportunities to use their knowledge to explain phenomena or solve problems (NRC, 2012). To this end, three-dimensional (3D) learning—incorporating disciplinary core ideas (DCIs), science and engineering practices (SEPs), and crosscutting concepts (CCCs)—has been suggested as a potential mechanism to achieve student knowledge-in-use learning goals (Krajcik et al., 2023; NRC, 2012). With 3D learning, students can apply various ideas by engaging in multiple SEPs to make sense of compelling phenomena or solve complex problems.

DOI: 10.4324/9781003170785-14

In terms of guiding students' knowledge-in-use development, learning progressions (LPs) have received wide-scale attention in depicting pathways that students travel as they progress toward proficiency (NRC, 2007; Wilson & Berenthal, 2006). LPs offer one way to align assessment, instruction, and learning experiences to support learners in developing usable knowledge (He et al., 2023a, 2023b). As such, students can gain opportunities to develop their understanding of essential content (DCIs and CCCs) and practices (SEPs) (NRC, 2007; Smith et al., 2006). However, the question remains about developing three-dimensional learning progressions (3DLPs) to support all students' sustainable knowledge-in-use development. To our knowledge, little evidence exists about whether 3DLP can provide opportunities for a wide variety of students to access and sustain science learning, regardless of their ethnicity, gender, SES, and cultural and educational background. Using a design-based research approach, this study explores developing a 3DLP of matter, interaction, and energy at the middle school level and offers essential design principles for developing 3DLPs.

Addressing the Challenges with Relevant Literature

3D Learning Progression for Integrating Scientific Ideas with Scientific Practices

Building upon decades of research on how students learn, *A Framework for K–12 Science Education* (hereafter, *Framework*) and the NGSS (NASEM, 2019; NRC, 2007, 2012; NGSS Lead States, 2013) emphasize a developmental approach to learning science. LPs provide a way to depict students' developmental process toward a deep understanding of scientific knowledge (Corcoran et al., 2009; NRC, 2007; Smith et al., 2006). The *Framework* describes the theoretical basis of 3D learning across grades. However, science education and the learning science research communities lack design principles for developing more fine-grained, evidence-based 3DLPs that integrate DCIs, SEPs, and CCCs. In the past two decades, most LP studies have focused on student development of science content (e.g., Breslyn et al., 2016) or scientific practices (e.g., Osborne et al., 2016). For instance, several LP studies investigate students' understandings of energy across a wide period (e.g., Herrmann & DeBoer, 2018; Liu & McKeough, 2005; Neumann et al., 2013) and different disciplines (e.g., Lee & Liu, 2010; Park & Liu, 2016). Those energy LPs focus on measuring content knowledge rather than mapping out the progression of students' science proficiency using energy to make sense of phenomena or solve problems. In addition, most LP studies of Matter depict students' progression of the content knowledge of matter (e.g., Morell et al., 2017; Smith et al., 2006; Stevens et al., 2010) without integrating SEPs. To date, Kaldaras and colleagues (2021) have responded to this challenge by developing a 3DLP for *Matter and its Interaction* in high physical science and formulated some principles for constructing 3DLPs. Thus, we need additional LP research that considers integrating scientific knowledge (DCIs, CCCs, and SEPs) to achieve learning goals.

Another challenge in developing 3DLPs is that making sense of sophisticated scientific phenomena requires multiple DCIs, SEPs, and CCCs. For instance, various DCIs (e.g., matter, energy, and interaction) intersect to explain phenomena or solve problems such as water evaporation and corrosion of metals. A 3DLP needs to provide the potential learning paths to show how students' 3D knowledge—the integration of multiple DCIs, SEPs, and CCCs—develops over time and becomes more sophisticated; this latter allows learners to use their knowledge in new and challenging situations. However, few LP studies address overlapping ideas and practices (i.e., multiple constructs) necessary to understand sophisticated scientific phenomena. The complex and practical nature of 3D learning makes it challenging to develop 3DLPs that deal with various constructs in each dimension (He et al., 2020; Shin et al., 2021). This is because 3DLPs integrate all three dimensions with multiple constructs into one sequence, which brings variation and uncertainty to 3DLP development. Therefore, how does one select SEPs and CCCs for multiple DCIs? How does one articulate the cognitive levels of different combinations of those three dimensions? Addressing these challenges would map students' evidence-based learning paths toward a deeper understanding of sophisticated, compelling phenomena and complex problems.

3D Learning Progression Offer Students Opportunity to Learn

The NRC's (2012) *Framework* recognizes that equity in science education requires all students to have opportunities to learn science by using SEPs, DCIs, and CCCs to make sense of phenomena. To support equitable student learning over time (NASEM, 2019), a well-designed 3DLP may provide a promising venue for a wide range of learners, regardless of ethnicity, gender, SES, and cultural and educational background. Although most existing LPs (e.g., Hadenfeldt et al., 2016) are developed based on empirical evidence, they provide students with a unified and fixed learning path, which may fail to depict the development of a wide variety of learners. Students with diverse learning backgrounds may follow different paths in their development. A unified and fixed learning path benefits some learners, but inclusive access for diverse learners requires multiple pathways. These issues highlight the vital need for researchers to develop integrated 3DLPs to prompt a wide variety of learners to build a deeper understanding of sophisticated DCIs integrated with SEPs and CCCs. Because of experiences and educational backgrounds, students develop an understanding of science in different ways and at different rates (Krajcik & Shin, 2023). 3DLPs should provide inclusive learning opportunities for all students based on their prior learning experiences, culture, and proficiency status. For a given grade band (e.g., 6–8), the NGSS has anchored the starting and ending points of a 3DLP using performance expectations. However, there is much flexibility for intermediate levels between the upper and lower levels, which provides equitable access for students from diverse backgrounds to build toward the upper levels. This gap in the existing LP literature leaves a question about developing 3DLPs that address a wide range of learners from diverse backgrounds. This gap creates the need for a 3DLP with multiple flexible pathways based on educational, social, and cultural backgrounds to be developed.

3D Learning Progression for Guiding Curriculum, Instruction, and Assessment

LPs play a vital role in providing a guide for designing and aligning curriculum materials, instruction, and assessment in a coherent educational system (Fortus & Krajcik, 2012; He et al., 2023b; Jin et al., 2019). Coherence is essential to a high-quality educational system, ensuring student learning goals are achieved effectively (He et al., 2023a; Li et al., 2023; Wilson & Berenthal, 2006). Research proves that LP-based curricula, instruction, and assessment significantly improve student science learning (e.g., Anderson et al., 2018; He et al., 2023a; Shin et al., 2019). Teachers and students participating in these research projects benefited from the research-based LP-aligned curriculum, instruction, and assessment with professional learning support. However, with one learning path, most LPs were developed for a particular research-based learning environment (Plummer & Krajcik, 2010; Plummer & Maynard, 2014; Songer et al., 2009). Such LPs might not be helpful to other teachers when they adapt and revise their existing materials. Because LPs use what is known about student learning and from a developmental perspective, many teachers would need professional learning to use them effectively. There is a need to use 3DLPs to adapt curriculum materials, organize teaching sequences, and provide learning support to students. Knowing whether and how teachers can apply the research-based 3DLP as a guide to coherently and efficiently adapt their curriculum, instruction, and assessment would benefit stakeholders (e.g., policymakers, curriculum and assessment developers, teachers, and students).

Methodology

Design-Based Research

To address the above-mentioned challenges, this study explores a design-based research approach (Barab & Squire, 2004; Brown, 1992; Collins et al., 2004) to develop a 3DLP that incorporates multiple DCIs (matter, interaction, and energy), CCCs, and SEPs and establishes essential design principles for designing such 3DLP. Based on Reeves's (2006) design research process, we articulated a design process for developing 3DLPs with four stages (see Figure 12.1): defining problems of LPs from literature and classroom practices (Stage 1), designing principles and process for developing 3DLPs (Stage 2), receiving feedback and iteratively revisions of 3DLPs (Stage 3), and reflecting the existing work and informing next round of developing and implementing 3DLPs (Stage 4). Our design research process uses an iterative refinement cycle that allows the designers to revise the 3DLPs and design principles. This study describes two phases of developing the 3DLP based on the four-stage process.

Systematic Design Approach

The development of the 3DLP was an iterative process by modifying Harris and colleagues' (2019) design approach, which emphasizes learning goal alignment (Krajcik et al., 2008) and the logic of evidentiary reasoning (Mislevy & Haertel, 2006). Figure 12.2 presents our process

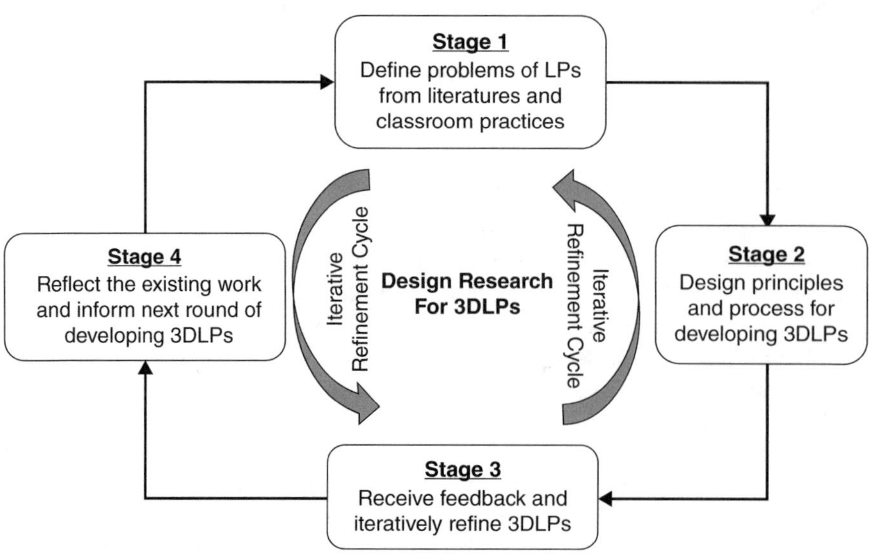

Figure 12.1 Design research for developing 3DLPs.

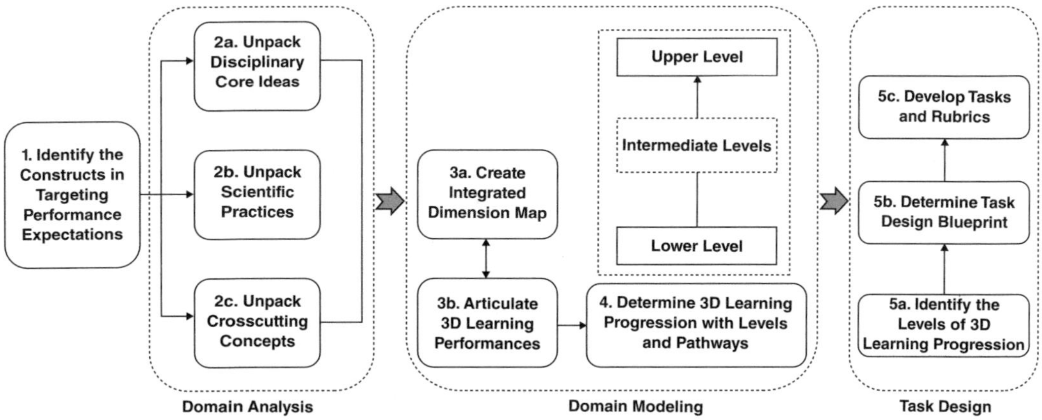

Figure 12.2 The systematic design process of developing 3DLPs.

consisting of five steps: 1) identifying the constructs in targeting NGSS performance expectations, 2) unpacking DCIs, SEPs, and CCCs, 3) articulating 3D learning performances, 4) determining 3D learning progression with levels and pathways, and 5) developing tasks and rubrics. The first phase describes unpacking each dimension of the performance expectations. The second phase focuses on developing the 3DLP. Regarding the 3DLP levels, we specify the lower boundary (prior knowledge) and upper boundary (grade band expectations) from the NGSS performance expectations and identify student challenges and potential phenomena. Next, we describe possible learning pathways between the starting and ending points to describe student progression toward knowledge-in-use proficiency. Then, we create claims for the 3DLP and specify the evidence for students' achievement at each level. Next, the final step is to develop the 3D assessment tasks to match the claims and evidence of the 3DLP.

Two-Round Design-Based Research Process

This section describes two rounds in developing the 3DLP. We unpack our design-based research with four stages in each round, including defining problems, design guidelines and products, feedback from researchers and teachers, and reflection from this round of research.

First Round 3DLP Development

Stage 1. Defining Problems

We describe the practical problems based on the gaps in the literature and the needs of teachers to teach NGSS-aligned classrooms. Most existing studies focus on developing LP with one dimension, either one DCI or one SEP, which may not be beneficial for mapping out student proficiency for 3D science learning. Moreover, the existing studies mainly mapped out student progress at the end of grade bands (Grades 2, 5, 8, and 12), which provides boundaries for student science learning but cannot be used to support the design of a learning system within grade bands (e.g., middle school level). Thus, small grain-size 3DLPs within a specific grade band need development to map student learning progress and support teachers in developing coherent and efficient learning systems.

Stage 2.1. Design Guidelines

Based on existing LP studies (e.g., Duschl, 2019; Hadenfeldt et al., 2016; Jin et al., 2019; Kaldaras et al., 2021), we articulate seven design guidelines for developing 3DLPs (see Table 12.1). We used these design guidelines to develop the initial 3DLP and associate assessment tasks. Moreover, the version of the design guidelines is modified iteratively based on multiple data sources (e.g., feedback from science education experts and experienced teachers and student performances on assessment tasks). Each design guideline is described in Stage 3 based on the feedback from science education researchers.

Stage 2.2. Design Products

Because of students' knowledge-in-use of energy, the cognitive complexity of knowledge (Neumann et al., 2013) was identified as an effective way to help students make sense of phenomena (Harris et al., 2019). We defined levels of an initial 3DLP by the complexity of the relations of energy ideas. We specified the three-dimensional components (DCI, SEP, and CCC) of a learning performance statement at each level in the 3DLP (see *Online Resource 12.1*). Based on our unpacking of performance expectations, five SEPs, and five CCCs were selected. Two constructs of the 3DLP— the manifestations of Energy and Energy Transfer —were identified. The manifestations of energy are kinetic and potential energy at the macroscopic and microscopic levels. The categories of energy transfer can be recognized respectively at the macroscopic and microscopic levels.

In this round, we constructed students' knowledge-in-use of energy with macro and micro levels from a system of one single object to a system of multi-component particles (see the *Online Resource 12.1.*). At the macro level, the lower anchor (Level Ia, Ib) of the 3DLP was defined as students' making sense of phenomena at the macroscopic level related to a system of a moving object (related to kinetic energy, Level Ia) and a system of two interacting objects (related to potential energy, Level Ib). At Level II, students use energy transfer at the macro level in the context of science and engineering practices. At the low anchor of the micro level (Level III), students use the idea of the motion of particles (related to the average kinetic energy) and the interaction of molecules (related to energy in the chemical process) to explain phenomena in scientific activities. Finally, in the upper anchor of this 3DLP, students make sense of phenomena related to energy transfer at the microscopic level (Level IV).

We then identified twenty-two assessment tasks associated with the 3DLP, including rubrics. Two to five assessment tasks were assigned to each level of the 3DLP. Tasks are technology-based, made available online, anchored in a phenomenon, and contextualized with brief scenarios (Harris et al., 2019; He et al., 2023c). We provide an example of assessment tasks and associate rubrics that align with a learning performance statement at Level IV (see the *Jennifer's teapot* task in the *Online Resource 12.2*). At Level IV, the learning performance claims that "students develop a model that includes a particle view of matter to predict how the average kinetic energy and the temperature of a substance change when thermal energy is transferred from or to a sample." This learning performance integrates the core idea of "the average kinetic energy and the temperature of a substance change when thermal energy is transferred from or to a sample" and the crosscutting concept of "cause and effect" (CaE) with the scientific practice of *"develop a model."* Prompts (i.e., draw a model) align with the evidence statements in the design pattern (step 5 in Figure 12.2). We designed the rubric that is aligned with the evidence statements. Each evidence statement served as the highest level in the rubric to judge student proficiency (see the *Online Resource 12.2*).

Table 12.1 The design guidelines for developing 3DLPs

Aspects	*Theoretical design guideline*	*Modified design guideline (1st round: adding researcher feedback)*	*Modified design guideline (2nd round: adding teacher feedback)*
1. Scope	1.1. The 3DLP scope depicts how student science proficiency develops and how to support moving students from one level to the next (Duschl, 2019; Jin, 2019; NRC, 2012).	1.1. The 3DLP maps how student science proficiency develops and how to support moving students from one level to the next (Duschl, 2019; Jin, 2019; NRC, 2012). 1.2. The 3DLP focuses on one type of science proficiency, with the same SEP, CCC, and DCI in each level of the 3DLP. *[Refer to researcher feedback]*	1.1. The 3DLP should map how student knowledge-in-use proficiency develops and how to support moving students from one level to the next (Duschl, 2019; Jin et al., 2019; NRC, 2012). 1.2. The 3DLP focuses on knowledge-in-use proficiency with the same SEPs, CCCs, and DCIs at each level. (Kaldaras et al., 2021). *[Refer to researcher feedback]*
2. Constructs	2.1. The 3DLP constructs are identified based on core ideas in a discipline or across different disciplines.	2.1. The 3DLP constructs are identified based on core ideas in a discipline or across different disciplines. 2.2. The 3DLP constructs are elaborated and aligned with the vision of the NRC's (2012) *Framework* and the NGSS. *[Refer to researcher feedback]*	2.1. The 3DLP constructs are identified based on core ideas in a discipline or across different disciplines. 2.2. The 3DLP constructs are elaborated and aligned with the vision of the NRC's (2012) *Framework* and the NGSS. *[Refer to researcher feedback]*
3. Levels and (Pathways)	3.1. The 3DLP levels are determined based on theoretical perspectives, such as knowledge enrichment, theory transformation, or knowledge integration (Jin et al., 2019).	3.1. The 3DLP levels are determined based on theoretical perspectives, such as knowledge enrichment, theory transformation, or knowledge integration (Jin et al., 2019) 3.2. One level of the 3DLP includes multiple energy constructs. *[Refer to researcher feedback]*	3.1. The 3DLP levels should be crafted based on theoretical perspectives, such as knowledge enrichment, theory transformation, and knowledge integration (Jin et al., 2019). 3.2. The 3DLP should craft the intertwined levels of multiple related constructs as students learn in a complex situation. *[Refer to researcher and teacher feedback]* 3.3. Multiple pathways of intertwined levels are optimal for fostering the development of a wide range of students in terms of ethnicity, gender, SES, and cultural and educational background. *[Refer to teacher feedback]*
4. Performance Statements	4.1. The 3DLP statements articulate science proficiency through performance learning goals integrating	4.1. The 3DLP statements articulate science proficiency through performance learning goals integrating DCIs, CCCs, and SEPs (NRC, 2013; Krajcik et al., 2008).	4.1. The 3DLP level statements should articulate student knowledge-in-use proficiency through integrating DCIs, CCCs, and SEPs (NGSS Lead States, 2013).

(*Continued*)

Table 12.1 (Continued)

Aspects	*Theoretical design guideline*	*Modified design guideline (1st round: adding researcher feedback)*	*Modified design guideline (2nd round: adding teacher feedback)*
	DCIs, CCCs, and SEPs (NGSS Lead States, 2013; Krajcik et al., 2008).	4.2. The 3DLP level statements are articulated using the elaborations of the NGSS PEs but with teachable and assessable fine-grained 3D learning goals. (Harris et al., 2019). *[Refer to researcher feedback]*	4.2. The 3DLP level statements are articulated using the elaborations of the NGSS PEs but with teachable and assessable fine-grained 3D learning goals. (Harris et al., 2019). *[Refer to researcher feedback]*
5. Assessments	5.1. A coherent sequence of three-dimensional assessments measures student proficiency aligned with 3DLP levels (Kaldaras et al., 2021). 5.2. Three-dimensional assessments provide complex and compelling phenomena that engage students to elicit science proficiencies aligned with 3DLP levels (Harris et al., 2019) 5.3. Each three-dimensional assessment provides information about science proficiency for at least one level of a specific construct.	5.1. A coherent sequence of three-dimensional assessments measures student proficiency aligned with 3DLP levels (Kaldaras et al., 2021). 5.2. Three-dimensional assessments provide complex and compelling phenomena that engage students to elicit science proficiencies aligned with 3DLP levels (Harris et al., 2019). 5.3. Each three-dimensional assessment provides information about science proficiency for at least one level of a specific construct. 5.4. One 3D assessment provides information to place students at different levels of LP depending on their responses and the complexity of a real-world phenomenon. *[Refer to researcher feedback]*	5.1. The 3D assessments deploy sequentially to measure student proficiency aligned with the levels of the 3DLP (Kaldaras et al., 2021). 5.2. Each 3D assessment should provide complex and compelling phenomena that engage a wide range of students to perform their knowledge-in-use proficiency. 5.3. The 3D assessments should provide information to place students at different levels of the 3DLP, depending on their responses and the complexity of the real-world phenomenon. *[Refer to researcher feedback]* 5.4. A set of 3D assessments are needed to cover all levels of the 3DLP. *[Refer to researcher and teacher feedback]* 5.5. Clear prompts and suitable scaffoldings should be designed to address the levels in different constructs. *[Refer to teacher feedback]*

Stage 3. Feedback from Science Education Researchers

In this stage, we recruited an external review panel of three science education researchers to provide feedback on our Round 1 version of 3DLP and the associated assessment tasks selected from the Next Generation of Science Assessment project (NGSA, 2023).[1] The three experts (Experts A, B, and C) have expertise in NGSS teaching, curriculum, assessment development, and learning progression. These experts were selected because they were not part of our study and had research and practice experience related to our 3DLP development. In their feedback, they connected our design products to their research experience and shared their comments and suggestions. Therefore, the feedback from the three experts provides external validity for our 3DLP. They offered feedback based upon two prompts: 1) the appropriateness of the initial 3DLP (the scope, performance statements, constructs, and levels); and 2) the appropriateness of the associated assessment tasks for the 3DLP. We summarized the three experts' feedback based on the five aspects (see Table 12.1).

The scope of 3DLP. We specify multiple types of SEPs and CCCs for each level in the 3DLP, including "analyzing and interpreting data to find patterns ...", "develop a model to explain ..." and "construct a scientific explanation about ..." Expert A suggested, "pick one SEP and CCC for each of your performance expectations. That way, it is a separate LP for each NGSS performance expectation, and you have the same SEP and CCC associated with a given NGSS performance expectation, so it is easier to write out learning progression levels." Expert B suggested that "There are at least three levels of progression, one for each dimension of 3D learning: one for the DCI, one for the practice, and one for the CCC. ... It's good that you address the same practices in each level." Based on their feedback, we decided that the 3DLPs should be developed by identifying the targeted DCIs that incorporate the same SEP and CCC (see design guideline 1.2 in Table 12.1).

The statement of 3DLP. We articulated student science proficiency using grain-size learning performances at each level to maintain the 3-dimensions. Expert A commented, "I like that the LP levels are expressed as performance expectations because it makes it easier to see how students are using their knowledge related to energy." Expert C also expressed, "In general, the learning performances seem appropriately aligned to the performance expectations from which they were derived." We took the messages to use small grain-size 3D learning performance goals to state student achievement on each level of the 3DLPs (see design guideline 4.2 in Table 12.1).

The constructs of 3DLP. We defined the constructs of the 3DLP, including kinetic energy, potential energy, energy transfer, and thermal energy. Expert B referred to his ideas on defining constructs of 3DLP based on some existing LPs in published journal articles and suggested that the constructs of our 3DLP should align with the prior empirical findings. Expert C critiqued that "While the levels seem arranged in order of sophistication, there is a massive leap in difficulty from Level 2 to Level 3. In Levels 1 and 2, students are to reason about macroscopic phenomena in purely macroscopic terms. For Level 3, students must reason about phenomena in terms of atomic/molecular behavior, which is notoriously counterintuitive." According to their feedback, we realized that students could not make sense of real-world phenomena with only one DCI (e.g., energy) but should include some related DCIs, such as force interaction and the particle nature of matter (see design guideline 2.1 in Table 12.1).

The levels of 3DLP. We determined the 3DLP levels based on the complexity of the system from a single object and two interacting objects (at the macroscopic level) to multiple particles (at the microscopic level) with the notion of knowledge integration. Within a system (e.g., kinetic energy), we created several levels based on students' deeper understanding (with the idea of knowledge enrichment and theory transformation). Expert A shared her research experience on creating a learning progression in an NGSS-aligned curriculum: "We first start at the macro level talking about how energy of macroscopic objects changes when they interact, and what are some indicators of energy transfer "(light, sound, etc.). We later transition to microscopic level and talk about how energy changes can be described at the atomic level." Expert C suggested that "Weaving together discussion of potential energy, kinetic energy, and total system energy would seem the most promising path forward. This would require building toward multiple PEs simultaneously." We took those messages to determine the levels of 3DLP based on multiple intertwined constructs for mapping out students‘ authentic learning progress (see design guideline 3.2 in Table 12.1).

The assessments of 3DLP. We developed sequenced 3D assessments with learning performances in each level of the 3DLP. 3D assessment allows students to perform science proficiency by engaging in a complex and compelling phenomenon. Each assessment addresses a single construct, such as thermal energy (see the *Online Resource 12.2*). Expert B pointed out, "In the real world, phenomena are not aligned with a single PE; they are related to several PEs. By assessing students on single PE, we are creating artificial situations that may be psychometrically easier to grapple with but less predictive of students' abilities at making sense of real-world issues." Following his comments, he further suggested that "having a single interesting phenomenon and asking students several questions about the same phenomenon, each one addressing different PEs." We accepted his suggestion that develops a 3D assessment of multiple constructs to make sense of a complex and compelling phenomenon (see design guideline 5.4 in Table 12.1).

Stage 4. Reflection from This Round of Research

Three key takeaways from the researchers' feedback on developing the 3DLP and associated assessment tasks at the middle school level include: (a) the primary constructs of the 3DLP can simultaneously support student development of DCIs (e.g., Kinetic energy and potential energy) with a SEP (e.g., Develop models) and CCC (e.g., Cause and Effect); (b) fine grain-size 3D integrated learning performances can be used to guide the 3DLP for instructional and assessment purposes; and (c) the progression of energy ideas is intertwined with other concurrent ideas (Force and Motion) for figuring out phenomena. Accordingly, we added five more design guidelines for developing the next round of 3DLP after analyzing data from round 1 (see Table 12.1).

Second Round 3DLP Development

Stage 1. Defining Problems

Based on the key takeaways from the Round 1 research, we revised the initial 3DLP and developed the second version in the second round. First, we articulated a series of fine-grain-sized learning performances to map out the levels of the 3DLP. Next, three compelling and complex real-world scenarios, assessment tasks, and exemplar responses were designed to align with the 3DLP to assess students' 3D proficiency of energy, matter, and interaction with the SEP of Developing and Using Models and the CCC of Cause and Effect.

Stage 2. Design Process and Products

NGSS performance expectations (PEs) related to energy and the associated ideas of interaction and matter at the middle school level were unpacked to articulate a set of learning performances, which are 3D learning goals that stem from the PE of NGSS but are smaller in scope. We developed a coherent set of learning performances, assessment tasks, and rubrics. Previous LP studies on energy (e.g., Herrmann & DeBoer, 2018) and the knowledge-in-use perspective (Harris et al., 2019) were employed to map out the LPs for middle school student science proficiency of energy using the learning performances (see Table 12.2).

Using the NGSA design process (Harris et al., 2019), we designed three assessment tasks to track students' energy, matter, and interaction proficiency at the middle school level. Assessment tasks and associated rubrics were developed by integrating the targeting DCIs with SEP of modeling and CCC of cause and effect. We developed 3D assessments that include a complex and compelling phenomenon that addresses student proficiency in multiple constructs. We also created student response examples (rubrics) to identify multiple constructs' levels (He et al., 2023c). Here, we provided an example of assessment tasks to assess students' proficiency in kinetic energy, potential energy, and associated ideas of force and motion (see the *Online Resource 12.3*) and an example of a student's model and explanation at Level 2 proficiency (see the *Online Resource 12.4*).

Stage 3. Feedback from Experienced Science Teachers

Four experienced science teachers with NGSS teaching experience at the middle school provided feedback on the second round of 3DLP and assessment tasks. We prompted them to provide feedback on the appropriateness of the 3DLP and the associated assessment tasks.

Overall, experienced teachers agreed that the designed scenario could fit well with the 3DLP, and its levels could be used to describe their students' understanding of the phenomena. For instance, one teacher commented that "The LP nicely progresses from Level 1 to Level 4 to show a progression of what student understanding could look like. Each level gets progressively more sophisticated." They suggested that one complex 3D assessment cannot fully cover all levels in different related constructs. Based on the specific feedback, we added a design guideline for developing 3DLP (see it in Table 12.1). Experienced teachers also commented that the prompts and exemplar models might be challenging for their students. They suggested separating the prompts into three parts, and the exemplar models should be pictorial representations of the scenario. Some minor comments, such as the appropriateness of the video itself and some DCI aspects, need reorganization.

Based on the feedback from teachers, we made minor adjustments to the 3DLP. Initially, we used the following holistic prompt: "Construct a model to explain why the basketball rolling down the slide can move, push the Yoga ball forward, and both of them stop." According to teachers' feedback, we changed this holistic prompt into three separate prompts (see Figure 12.1). In the examples of models, we provided three kinds of student models (Model 1-fully abstract, Model 2- partial abstract, and Model 3-pictorial drawing). One teacher commented, "In the context of being scaffolded, students could get close to exemplar 1 and 2; Exemplar 3 is the model that most students would draw like."Another teacher stated, "all three models are good. I like Model 3 because it has students explaining what they have drawn - step-by-step." However, we decided to keep the drawn abstract models because of the requirement of developing models at the middle school level. We believe students should obtain their modeling proficiency as an abstract model and not draw a replica of the

Table 12.2 The four levels of 3D learning performances in 2nd round 3DLP

Level	*Disciplinary core ideas (DCIs) at each level*	*3D learning performance*
Level 1	Force and motion When objects collide, the forces exert each other, *the strength of the forces is equal, but the direction is opposite. (MS-PS2-1)	Students construct models to show that when a moving object collides with another object, it exerts a force on that object.
Level 2	Force and motion • When objects collide, the forces exert with each other, and the strength of the forces are equal, but the direction is opposite. (MS-PS2-1) Kinetic Energy • Motion energy is properly called kinetic energy. (MS-PS3-1) Energy Transfer (not including the DCI) • The motion of an object changes and the kinetic energy of the object is transferred to some other changes in energy (such as potential energy, temperature, or kinetic energy) (MS-PS3-5).	Students construct models to explain that the contact force exerted on one object to another will change the object's motion upon collision and that the kinetic energy of the moving object was transferred to another object involved in the collision.
Level 3	Force and motion • When objects collide, the forces exert each other, *the strength of the forces is equal, but the direction is opposite. (MS-PS2-1) • When two objects interact at a distance, each can exert a force on the other. The strength of the forces is equal, but the direction of the forces is opposite. (MS-PS2-1) • Gravitational forces are always attractive. (MS-PS2-4) Kinetic Energy • Motion energy is properly called kinetic energy. (MS-PS3-1) Energy Transfer (not including the DCI) The motion of an object changes, and the object's kinetic energy is transferred to other changes in energy (such as potential energy, temperature, or kinetic energy) (MS-PS3-5).	Students construct models to explain that the contact and non-contact forces changed the object's motion, and the object's kinetic energy was transferred to another object due to the collision.
Level 4	Force and motion • When objects collide, the forces exert with each other, and the strength of the forces are equal, but the direction is opposite. (MS-PS2-1) • When two objects interact at a distance, each can exert a force on the other. The strength of the forces is equal, but the direction of the forces is opposite. (MS-PS2-1) • Gravitational forces are always attractive. (MS-PS2-4) • A system of objects exists in a field that causes the objects to exert forces at a distance (electrical, magnetic, and gravitational). (MS-PS2-5)Kinetic Energy • Motion energy is properly called kinetic energy. (MS-PS3-1) Potential Energy • PE1: A system of objects may contain stored energy (potential energy), depending on their relative positions. (MS-PS3-2) Energy Transfer (not including the DCI) • The motion of an object changes, and the kinetic energy of the object is transferred to some other changes in energy (such as potential energy, temperature, or kinetic energy) (MS-PS3-5). When objects interact at a distance, each one exerts a force on the other that transfers potential energy to change the kinetic energy of objects. (MS-PS3-2)	Students construct models to explain that the contact and non-contact forces changed the object's motion, and the potential energy of the object and the Earth was transferred to the object's kinetic energy and transferred to another object due to the collision.

Note: the gray color aspects of ideas were omitted in the 3DLP.

phenomena (He et al., 2023a; Li et al., 2021; Schwarz et al., 2009). According to this specific teacher feedback, we added a design guideline of "5.5 Clear prompts and suitable scaffoldings should be designed to address the levels in different constructs" (see it in Table 12.1). In addition, one teacher mentioned that "I just wonder if you have considered what would happen if students were taught in a different order. Is there flexibility in the levels? By the end of middle school, the levels will work. However, if you test 6th or 7th graders you might get a lot of variation depending on what the unit progression is." We carefully considered her comment and realized that multiple pathways of intertwined levels might depict students' learning progression depending on their diverse backgrounds, such as ethnicity, gender, SES, and cultural and educational backgrounds (see design guideline 3.3 in Table 12.1).

Stage 4. Reflection from This Round of Research

The three key takeaways from the four experienced teachers' feedback on developing the 3DLP and associated assessment tasks at the middle school level include: (a) students with diverse backgrounds may have multiple pathways to achieve the upper level; (b) a set of 3D assessment tasks need to be carefully designed to cover all levels of 3DLP; and (c) clear prompts and suitable scaffoldings should be employed to address the levels in multiple constructs. Those key takeaways were added to the design guideline for developing the next round of 3DLP (see Table 12.2).

3DLP Development Move Forward

Taking advantage of two rounds of preliminary research, we have received funding to continue our research that aims to continually develop and test the 3DLP and use it to guide teachers' adaptation of local learning materials to support students' knowledge-in-use development. To achieve our goal, we take the invaluable messages from the above explorations to tailor the scope of the under-developed 3DLP that depicts students' understanding of two fundamental questions in physical science: What is everything made of? And why do things happen? (NRC, 2012). These two questions are related to a wide range of natural and human-created phenomena that occur in our everyday lives, such as water evaporation and the corrosion of metals. To understand such phenomena, one must ultimately consider the structure of matter at atomic and subatomic scales and physical and chemical interactions in terms of forces among objects (particles), related energy transfers, and their consequences (NRC, 2012). The three DCIs in physical science have explanatory powers to explain how and why certain phenomena occur. They can explain and predict the physical and chemical subprocesses within most systems at the macroscopic or microscopic levels. Developing a sophisticated understanding of these ideas requires a developmental approach across years of education. We developed a 3DLP that depicts and supports middle school students' knowledge-in-use. Students who become proficient can employ multiple DCIs (matter, interaction, and energy) with SEPs and CCCs to make sense of phenomena and solve problems. With respect to the core ideas of matter, interaction, and energy, a broad range of NGSS PEs is targeted: Matter and its Interaction (MS-PS1-1 to MSPS1-6) and Energy (MS-PS3-1 to MS-PS3-5). Although many SEPs and CCCs are involved in these performance expectations, we focus on two SEPs (*Developing and Using Models* and *Constructing Explanations*) and two CCCs (*System and System Models* and *Cause and Effect*) essential to developing knowledge-in-use proficiency. We designed three phases for developing and testing the 3DLP in Figure 12.3.

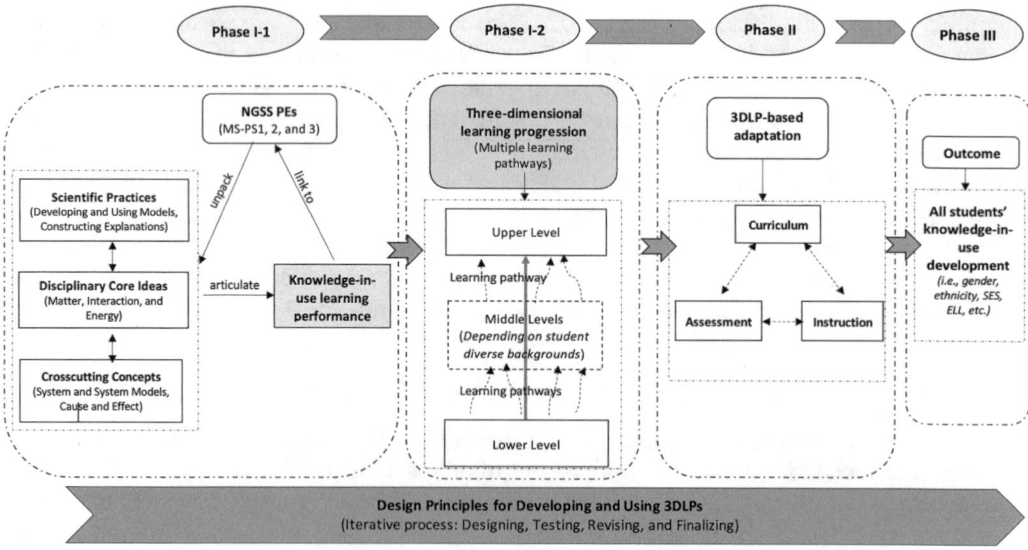

Figure 12.3 Three phases for developing and testing a 3DLP.

Our project employs a design-based research approach to design, revise, and finalize the LP iteratively over three years (Brown, 1992). We collaborate with teachers from three school districts. Two districts are in major Midwest cities with low SES and underrepresented student populations (i.e., mainly black students). The third school district has a predominantly middle-class white population. These specific contexts will provide information across different school cultures. In Phase 1, we developed the 3DLP using a construct-centered design (CCD, Shin et al., 2010) aligned with the lower boundary (prior grade band expectations) and upper boundary (grade band expectations) from the NGSS. Our work with experienced teachers articulates intermediate learning pathways based on previous literature and teachers' instructional experiences. The experienced teachers used their experiences and perspectives to compare and contrast adaptations of the 3DLP. We, therefore, have an initial LP and associated classroom-embedded assessment tasks. We further revise the LP by conducting semi-structured interviews with middle school students from diverse backgrounds by race (e.g., Asian, Black, Latina/o, white), levels of SES (Low and Middle), and educational experiences (high, middle, and low performances). Analyzing students' interviews helps refine the intermediate learning pathways of students with diverse backgrounds. Teachers also helped us identify the main intermediate learning pathways based on their students' performances (see Phase 1–2). We then have the second version of our 3DLP. We sent the second version to three other teachers and external experts to receive feedback. The participating teachers will receive professional learning to adapt their curriculum materials using the assessment tasks and instructional strategies based on the 3DLP (see Phase 2). In Phase 3, we will collect cross-sectional data on students' performance on the LP-aligned assessment tasks in Grades 6, 7, and 8 over two years. Students' responses to the classroom assessments are analyzed to identify the intermediate learning pathways. In addition, teacher and student interviews, classroom observations, and teacher and student artifacts are collected and used to refine the levels of 3DLP. The cross-sectional data are analyzed to investigate students' 3D learning along the 3DLP. The follow-up analysis explores learning growths by group (e.g., race, gender, school SES, etc). The findings inform the final revision of the 3DLP. The 3DLPs with various learning pathways provide equitable learning opportunities for a wide range of students, including low SES, Black and White, and middle SES populations.

Concluding Remarks

This chapter delves into the intricate process of design-based research, focusing on the development of 3DLPs for multiple DCIs, including energy, matter, and interaction with the integration of various SEPs and CCCs to make sense of real-world phenomena. The discussion encompasses a two-phase design process and the current state of research. Throughout our project, we have made significant strides in developing the 3DLP, establishing concrete design guidelines (Table 12.1) and a systematic design process (Figure 12.2). Additionally, we created and refined early-stage 3DLP products with ongoing development and testing. Despite the progress we have made, we acknowledge the persistent challenges that arise in the development and application of 3DLPs, such as defining and specifying levels along 3DLP. The intricate nature of science education calls for continuous research and critical examination of these challenges. We highlight notable hurdles and offer relevant suggestions for future research. By addressing these challenges, we not only shed light on the obstacles faced in this field but also provide a compass to guide future research efforts.

Articulating the levels of learning performances for 3DLPs during the developmental phase presents a challenging task that requires capturing students' evolving knowledge. Our study adopted three theoretical perspectives as design principles, including knowledge enrichment, theory transformation, and knowledge integration (Jin et al., 2019). These principles serve as guiding tenets in the development of 3DLPs. Furthermore, it is crucial to meticulously articulate all levels of 3DLPs using fine-grained 3D learning performance objectives (Harris et al., 2019). Although we have outlined the levels of 3DLPs (Table 12.2), we acknowledge the need for a comprehensive discussion on the detailed process of articulating these levels, which was not explicitly addressed in our research. In mapping out the levels of our 3DLP, we only considered the levels of DCIs but ignored the levels of SEPs and CCCs in the two-round design process. While our approach maintains the salient features of three-dimensional learning performances in our 3DLP, we recognize the need to further explore the levels of SEPs and CCCs. Previous studies have also provided valuable insights into learning progressions of scientific practices, including modeling (Schwarz et al., 2009), argumentation (Osborne et al., 2016), explanation (Songer et al., 2009), and crosscutting concepts such as size and scale (Delgado, 2009). However, articulating learning performance levels for 3DLPs becomes increasingly complex due to integrating DCIs, SEPs, and CCCs with different levels. In light of this, we urge the field to intensify empirical research efforts to illuminate methodologies to address the challenge. These explorations will significantly contribute to articulating learning performance levels for 3DLPs, enabling the creation of students' knowledge-in-use proficiency.

Exploring multiple learning pathways for students with diverse backgrounds remains an underexplored area in current research. As students' understanding develops in unique ways, influenced by their prior experiences and backgrounds (Krajcik & Shin, 2023; NRC, 2017), we recognize the importance of incorporating multiple learning pathways within our 3DLP. We aim to capture the diverse trajectories of students from lower to upper levels, acknowledging that these pathways may vary. However, our study is in the process of identifying the main learning pathways at the intermediate levels that delineate distinct roadmaps for students' progression from lower to upper levels. Additionally, we need to investigate whether these learning pathways differ among various demographic groups, such as those based on ethnicity or language backgrounds. Therefore, researchers must decide to dedicate their efforts to developing innovative methodologies that uncover and illuminate the intricate dynamics of learning pathways, paving the way for more effective and inclusive educational practices. We can create a more inclusive and effective learning environment by exploring multiple learning pathways and identifying the factors that influence students' trajectories.

Supporting students' development over time poses an ongoing challenge in science education. 3DLPs can serve as valuable tools for guiding the coherence of curriculum, instruction, assessment, and professional learning to ensure effective learning systems that align with science standards (Fortus & Krajcik, 2012; Jin et al., 2019). Although students have the opportunity to learn science in such a learning system, students still need immediate feedback and instructional support to move from lower to upper levels aligned with a 3DLP. How can we effectively develop and deliver timely feedback and instructional support tailored to individual needs? Moreover, how can we support the development of all students, regardless of their backgrounds, as they progress over time? Innovative technologies, such as artificial intelligence (AI), hold great promise in achieving these goals. Leveraging innovative technologies like AI opens up possibilities for creating personalized learning experiences that adapt to individual needs and provide immediate guidance. This fosters continuous development, enabling students to reach their full potential. Achieving this requires interdisciplinary collaboration, and pooling the expertise of science educators, equity and social justice advocates, and computer scientists. The potential impact of our collective efforts in this research agenda is substantial, as we can overcome the challenges associated with supporting students' development over time.

This chapter introduces our design-based research that sheds light on the fundamental design principles and systematic processes involved in the development of 3DLPs. While we have made notable progress in our endeavors, numerous challenges persist, necessitating further exploration by researchers in the field. Through our design-based research, we have uncovered invaluable insights into the framework of 3DLP development. By identifying design principles and establishing a systematic process, we have laid a solid foundation for advancing science education. However, our journey is far from complete, as various challenges pose significant obstacles. As researchers, we recognize the importance of continuous exploration and innovation. Development and implementation of effective 3DLPs require ongoing dedication to addressing the complexities and nuances inherent in science education. By persistently delving into these challenges, we can unlock new perspectives and approaches that will shape the future of designing and implementing 3DLPs to support student knowledge-in-use development.

Acknowledgments

This study is supported by the National Science Foundation (Grant Number DRL-2201068). Any opinions, findings, conclusions, or recommendations expressed in this material are those of the authors and do not necessarily reflect the views of the National Science Foundation.

Note

1 See exemplar assessment tasks from the NGSA website: http://nextgenscienceassessment.org/

References

Anderson, C. W., de los Santos, E. X., Bodbyl, S., Covitt, B. A., Edwards, K. D., Hancock, I. J. B., & Welch, M. M. (2018). Designing educational systems to support enactment of the Next Generation Science Standards. *Journal of Research in Science Teaching*, *55*(7), 1026–1052. 10.1002/tea.21484

Barab, S., & Squire, K. (2004). Design-based research: Putting a stake in the ground. *The Journal of the Learning Sciences*, *13*(1), 1–14. 10.1207/s15327809jls1301_1

Breslyn, W., McGinnis, J. R., McDonald, R. C., & Hestness, E. (2016). Developing a learning progression for sea level rise, a major impact of climate change. *Journal of Research in Science Teaching*, *53*(10), 1471–1499. 10.1002/tea.21333

Brown, A. L. (1992). Design experiments: Theoretical and methodological challenges in creating complex interventions in classroom settings. *The Journal of the Learning Sciences*, *2*(2), 141–178. /10.1207/s15327809jls0202_2

Collins, A., Joseph, D., & Bielaczyc, K. (2004). Design research: Theoretical and methodological issues. *The Journal of the Learning Sciences*, *13*(1), 15–42. 10.1207/s15327809jls1301_2

Corcoran, T., Mosher, F. A., & Rogat, A. D. (2009). *Learning progressions in science: An evidence-based approach to reform (RR-63)*. New York: Consortium for Policy Research in Education (CPRE), Teachers College, Columbia University.

Delgado, C. (2009). Development of a research-based learning progression for middle school through undergraduate students' conceptual understanding of size and scale. University of Michigan.

Duschl, R. A. (2019). Learning progressions: framing and designing coherent sequences for STEM education. *Disciplinary and Interdisciplinary Science Education Research*, *1*(1), 1–10. 10.1186/s43031-019-0005-x

Fortus, D., & Krajcik, J. (2012). Curriculum coherence and learning progressions. In: B. Fraser, K. Tobin., C. McRobbie (Eds.), *Second international handbook of science education* (pp. 783–798). Springer International Handbooks of Education, vol 24. Springer, Dordrecht. 10.1007/978-1-4020-9041-7_52

Hadenfeldt, J. C., Neumann, K., Bernholt, S., Liu, X., & Parchmann, I. (2016). Students' progression in understanding the matter concept. *Journal of Research in Science Teaching*, *53*(5), 683–708. 10.1002/tea.21312

Harris, C. J., Krajcik, J. S., Pellegrino, J. W., & DeBarger, A. H. (2019). Designing Knowledge-In-Use Assessments to Promote Deeper Learning. *Educational Measurement: Issues and Practice*, *38*(2), 53–67. 10.1111/emip.12253

He, P., Chen, I.-C., Touitou, I., Bartz, K., Schneider, B., & Krajcik, J. (2023a). Predicting student science achievement using post-unit assessment performances in a coherent high school chemistry project-based learning system. *Journal of Research in Science Teaching*,*60*(4), 724–760. 10.1002/tea.21815

He, P, Krajcik, J, & Schneider, B (2023b). Transforming standards into classrooms for knowledge-in-use: an effective and coherent project-based learning system. *Disciplinary and Interdisciplinary Science Education Research*, *5*(1),22.10.1186/s43031-023-00088-z.

He, P., Zhai, X., Shin, N., Krajcik, J. (2023c). Applying Rasch measurement to assess knowledge-in-use in science education. In: Liu, X., Boone, W. J. (eds.), *Advances in Applications of Rasch Measurement in Science Education*. Contemporary Trends and Issues in Science Education, vol 57. Springer, Cham. 10.1007/978-3-031-28776-3_13.

He, P., Shin, N., Li, T. & Krajcik, J. (2020, March 15-18). *Developing an integrated learning progression and assessments to measure middle school student proficiency of energy*. Paper presented at the annual meeting of the NARST-A Worldwide Organization for Improving Science Teaching and Learning through Research, Portland, OR. (Conference canceled).

Herrmann-Abell, C. F., & DeBoer, G. E. (2018). Investigating a learning progression for energy ideas from upper elementary through high school. *Journal of Research in Science Teaching*, *55*(1), 68–93. 10.1002/tea.21411

Jin, H., van Rijn, P., Moore, J. C., Bauer, M. I., Pressler, Y., & Yestness, N. (2019). A validation framework for science learning progression research. *International Journal of Science Education*, *41*(10), 1324–1346. 10.1080/09500693.2019.1606471

Kaldaras, L., Akaeze, H., & Krajcik, J. (2021). Developing and validating Next Generation Science Standards-aligned learning progression to track three-dimensional learning of electrical interactions in high school physical science. *Journal of Research in Science Teaching*, *58*(4), 589–618. 10.1002/tea.21672

Krajcik, J., McNeill, K. L., & Reiser, B. J. (2008). Learning-goals-driven design model: Developing curriculum materials that align with national standards and incorporate project-based pedagogy. *Science education*, *92*(1), 1–32. 10.1002/sce.20240

Krajcik, J., Schneider, B., Miller, E., Chen, I.-C., Bradford, L., Baker, Q., Bartz, K., Miller, C., Li, T., Codere, S., & Peek-Brown, D. (2023). Assessing the Effect of Project-Based Learning on Science Learning in Elementary Schools. *American Education Research Journal*, *60*(1), 70–102. 10.3102/00028312221129247

Krajcik, J., & Shin, N. (2023). Student conceptions, conceptual change, and learning progressions. In S. K. Abell & N. G. Lederman (Eds.), *Handbook of research on science education* (pp. 121–157). Routledge.

Lee, H. S., & Liu, O. L. (2010). Assessing learning progression of energy concepts across middle school grades: The knowledge integration perspective. *Science Education*, *94*(4), 665–688. 10.1002/sce.20382

Li, T., Chen, I., Miller, E., Miller, C., Schneider, B., & Krajcik, J. (2024). The relationships between elementary students' knowledge-in-use performance and their science achievement.*61*(2),358–418. *Journal of Research in Science Teaching*. 10.1002/tea.21900

Li, T., Miller, E., Chen, I. C., Bartz, K., Codere, S., & Krajcik, J. (2021). The relationship between teacher's support of literacy development and elementary students' modelling proficiency in project-based learning science classrooms. *Education 3-13: International Journal of Primary, Elementary and Early Years Education*. *49*(3), 302–316. 10.1080/03004279.2020.1854959.

Liu, X., & McKeough, A. (2005). Developmental growth in students' concept of energy: Analysis of selected items from the TIMSS database. *Journal of Research in Science Teaching*, *42*(5), 493–517. 10.1002/tea.20060

Mislevy, R., & Haertel, G. (2006). Implications of evidence-centered design for educational testing. *Educational Measurement: Issues and Practice*, *25*(4), 6–20. 10.1111/j.1745-3992.2006.00075.x

Morell, L., Collier, T., Black, P., & Wilson, M. (2017). A construct-modeling approach to develop a learning progression of how students understand the structure of matter. *Journal of Research in Science Teaching*, *54*(8), 1024–1048. 10.1002/tea.21397

National Academies of Sciences, Engineering, and Medicine. (2019). *Science and engineering for grades 6-12: Investigation and design at the center*. The National Academies Press, Washington, DC. 10.17226/25216

National Research Council. (2000). *How people learn: Brain, mind, experience, and school: Expanded Edition*. The National Academies Press, Washington, DC. 10.17226/9853

National Research Council. (2007). *Taking science to school: Learning and teaching science in grades K–8*. The National Academies Press, Washington, DC. 10.17226/11625

National Research Council. (2012). *A framework for K-12 science education: Practices, crosscutting concepts, and core ideas*. The National Academies Press, Washington, DC. 10.17226/13165

Neumann, K., Viering, T., Boone, W. J., & Fischer, H. E. (2013). Towards a learning progression of energy. *Journal of Research in Science Teaching*, *50*(2), 162–188. 10.1002/tea.21061

NGSA. (2023). *Next generation science assessment*. https://ngss-assessment.portal.concord.org

NGSS Lead States. (2013). *Next generation science standards: For states, by states*. The National Academies Press, Washington, DC. 10.17226/18290

Osborne, J., Henderson, B., MacPherson, A., Szu, E., Wild, A., & Yao, S.-Y. (2016). The development and validation of a learning progression for argumentation. *Journal of Research in Science Teaching*, *53*(6), 821–846. 10.1002/tea.21316

Park, M., & Liu, X. (2016). Assessing understanding of the energy concept in different science disciplines. *Science Education*, *100*(3), 483–516. 10.1002/sce.21211

Plummer, J. D., & Krajcik, J. (2010). Building a learning progression for celestial motion: Elementary levels from an Earth- based perspective. *Journal of Research in Science Teaching*, *47*(7), 768–787. 10.1002/tea.20355

Plummer, J. D., & Maynard, L. (2014). Building a learning progression for celestial motion: An exploration of students' reasoning about the seasons. *Journal of Research in Science Teaching*, *51*(7), 902–929. 10.1002/tea.21151

Reeves, T. C. (2006). Design research from a technology perspective. In J. van den Akker (Ed.), *Design methodology and developmental research in education and training* (pp. 64–78). Kluwer, The Netherlands.

Schwarz, C. V., Reiser, B. J., Davis, E. A., Kenyon, L., Achér, A., Fortus, D., ... & Krajcik, J. (2009). Developing a learning progression for scientific modeling: Making scientific modeling accessible and meaningful for learners. *Journal of Research in Science Teaching*, *46*(6), 632–654. 10.1002/tea.20311

Shin, N., Choi, S., Stevens, S. Y., & Krajcik, J. S. (2019). The impact of using coherent curriculum on students' understanding of core ideas in chemistry. *International Journal of Mathematics and Science Education*, *17* (2), 295–315. 10.1007/s10763-017-9861-z

Shin, N., Stevens, S. Y., & Krajcik, J. (2010). Tracking student learning over time using construct-centred design. In S. Rodrigues (Ed.), *Using Analytical Frameworks for Classroom Research* (pp. 56–76). Routledge.

Shin, N., He, P., Li, T., & Krajcik, J. (2021, April 7-10). *A three-dimensional integrated learning progression and aligned assessments to monitor middle school student proficiency of energy, modeling and cause and effect.* Paper presented at the annual meeting of the NARST-A Worldwide Organization for Improving Science Teaching and Learning through Research, Virtual Conference.

Smith, C. L., Wiser, M., Anderson, C. W., & Krajcik, J. (2006). Implications of research on children's learning for standards and assessment: A proposed learning progression for matter and the atomic molecular theory. *Measurement: Interdisciplinary Research and Perspectives*, *14*(1-2), 1–98. 10.1080/15366367.2006.9678570

Songer, N. B., Kelcey, B., & Gotwals, A. W. (2009). How and when does complex reasoning occur? Empirically driven development of a learning progression focused on complex reasoning about biodiversity. *Journal of Research in Science Teaching*, *46*(6), 610–631. 10.1002/tea.20313

Stevens, S. Y., Delgado, C., & Krajcik, J. S. (2010). Developing a hypothetical multi-dimensional learning progression for the nature of matter. *Journal of Research in Science Teaching*, *47*(6), 687–715. 10.1002/tea.20324

Wilson, M. R., & Berenthal, M. W. (2006). *Systems for state science assessment.* National Academies Press, Washington, DC. 10.17226/11312

Appendix

Online Resource 12.1

The 1st Round Three-Dimensional Learning Progression

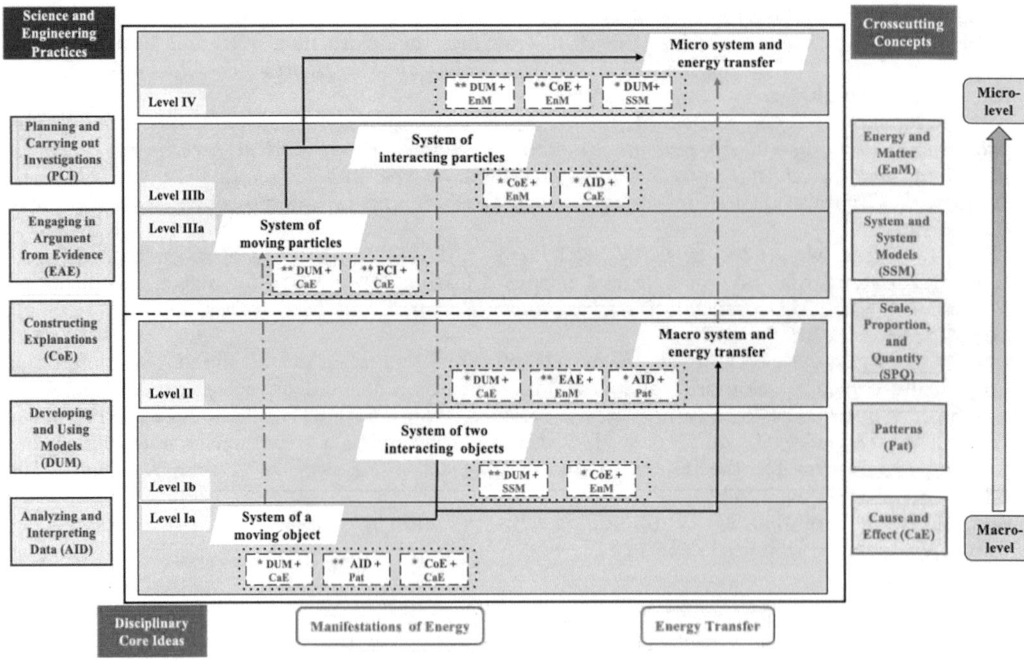

Online Resource 12.2

An Example of Assessment Tasks and Rubrics

Jennifer's teapot (ID#057.02-e05)

Jennifer heats water in a teapot. While she does this, thermal energy is transferred to the water.

Question #1

Draw a model that shows what will happen to the molecules in the water and the temperature of the water **before** and **after** heating. Your model should include pictures and a key.

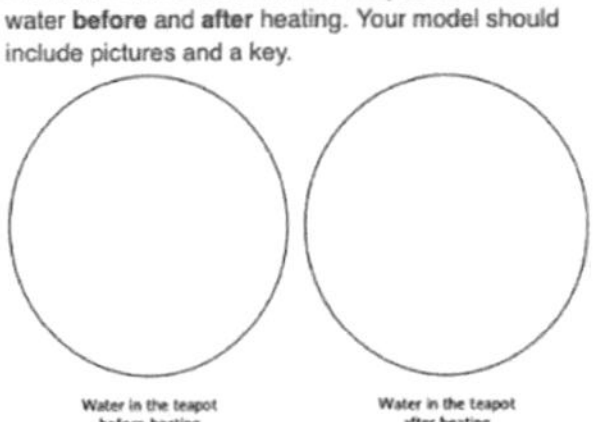

Explain how your model shows that transferring thermal energy to the water changes the movement of the water molecules and the temperature of the water.

Make drawing

The RUBRIC of ASSESSMENT TASK (Jennifer's teapot, ID#057.02-e05)

Rubric Component (Evidence Statement)	Proficiency Level		
	Proficient	Developing	Beginning
1. Model shows causal relationship between particle speed and transfer of thermal energy.	Particles are indicated as moving faster (slower) AND having higher (lower) temperature after the transfer of thermal energy to (away from) the sample (2)	Particles are indicated as moving faster (slower) OR having higher (lower) temperature after the transfer of thermal energy to (away from) the sample (1)	Model shows no clear relationship between thermal energy transfer and particle speed or temperature (0)
2. Legend shows how the identity and speed of particle motion should be represented in the model.	Legend shows how the identity of particles (e.g. a circle) OR particle speed (e.g. an arrow) should be represented in the model (1)	Legend shows how the identity of particles (e.g. a circle) OR particle speed (e.g. an arrow) should be represented in the model (1)	Legend shows NEITHER how the identity of particles (e.g. a circle) NOR particle speed (e.g. an arrow) should be represented in the model (0)
3. Statement that transferring thermal energy to (away from) a substance increases (decreases) particle speed in the substance.	Student states that that transferring thermal energy to (away from) a substance increases (decreases) particle speed in the substance (1)	N/A	Student does not correctly relate thermal energy transfer to particle speed (0)
4. Statement that temperature is the average kinetic energy (motion) of particles.	Student states that temperature is the average kinetic energy (motion) of particles (1)	N/A	Student does not state that temperature is the average kinetic energy (motion) of particles (0)

Online Resource 12.3
An Assessment Task for the 2nd Round 3DLP

(A video is embedded in the task)

Scenario: Jane is playing with the basketball on a slide. When she releases the ball from the top of the slide, she finds it rolls down and continues to move forward on the ground. Then, the moving basketball collides with the orange ball on the ground. After a collision, the basketball moves slower, but the orange ball starts to move. Finally, both the basketball and the Yoga ball stop.

Prompt 1: Construct a model to explain why the basketball rolling down the slide can continue to move on the ground.

Prompt 2: Construct a model to explain why the basketball pushes the orange ball forward.

Prompt 3: Construct a model to explain why both the basketball and the orange balls stop at the end.

Online Resource 12.4

Construct models:

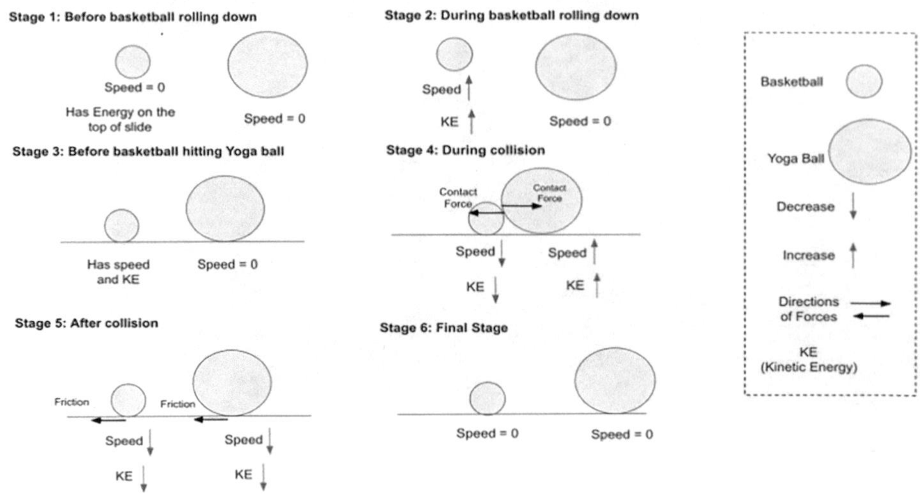

Explanation using the model above: The basketball has energy at the top of the slide. The basketball rolling down the slide can move on the ground. So, the basketball has kinetic energy. When the basketball hits the orange ball, it exerts a force on the orange ball. So, some of the kinetic energy of the basketball is transferred to the kinetic energy of the orange ball. Thus, the orange ball has kinetic energy so that it can move a distance. There are friction forces between the basketball and the ground and the orange ball and the ground that make both of them stop, and the kinetic energy of the basketball and the orange ball is transferred to other types of energy.

13

RETHINKING LEARNING PROGRESSIONS FOR ENERGY

David Fortus and Jeffrey Nordine

The Importance and Challenges of Learning about Energy

People living in the 21st century face a wide range of energy-related personal and social issues: deciding between purchasing a hybrid, electric, or turbo-charged car, eating an omnivorous or an herbivorous diet, supporting or opposing local and national policies regarding the production and storage of energy, or taking steps to mitigate extreme weather events. To reach well-informed opinions and decisions on these issues and others, it is essential to have knowledge of energy and be able to use this knowledge to make sense of problems and naturally occurring events, both qualitatively and quantitatively (National Research Council, 2012). For example, although it may seem intuitive that installing solar cells will reduce the amount of energy required from fossil fuels, the true challenge in replacing fossil fuels with solar cells is related to the rate of energy transfer. The peak rate of energy transfer from solar cells occurs around noon, while peak demand for energy transfer to households occurs in the evening, and this mismatch in peak energy transfer may be addressed through energy storage technologies (e.g., batteries), again making issues of rate of energy transfer as well as total storage capacity critical for ultimate success. Quantitative models of the rate of energy transfer between interconnected systems are needed to evaluate the opportunities and challenges associated with solar cells; the same is true for understanding climate change, the impact of dietary choices, and so on (Smil, 2005, 2006).

Using energy to make sense of phenomena and problems requires more than knowledge about the energy concept itself. Rather, students need to be able to use energy ideas to make sense of phenomena and problems (Harris et al., 2019; Pellegrino & Hilton, 2012). When students effectively use their science knowledge, they can activate science ideas in concert with other conceptual tools in the context of real-world scenarios to predict and explain the behavior of phenomena and solve problems. The *Framework for K-12 Science Education* (National Research Council, 2012) provides a valuable vision for how students should be able to use science ideas, known as three-dimensional learning. In three-dimensional learning, students integrate disciplinary core ideas with science and engineering practices and cross-cutting concepts[1] to investigate, interpret, and explain novel problems and naturally occurring events (i.e., phenomena).

DOI: 10.4324/9781003170785-15

Several studies have shown that by the end of K-12 schooling, most students are not able to use energy ideas to make sense of a wide range of phenomena (Herrmann-Abell & DeBoer, 2017; Liu & McKeough, 2005; Neumann et al., 2013). It is important to note that such studies have been conducted within the scope of "typical" energy instruction (e.g., not tied to a specific intervention). Researchers have questioned whether the conceptual foundations of traditional approaches to energy instruction (typically based on the identification of different forms of energy and the transformations occurring among them) may unwittingly be contributing to students' difficulties in developing and using energy ideas (Brewe, 2011; Ellse, 1988; Nordine et al., 2011; Swackhamer, 2005).

To overcome these challenges, coherent instruction that follows a learning progression is needed. In this chapter, we describe how learning progressions are developed, the traditional energy learning progression, and the concerns with this learning progression. We then describe a non-traditional approach to energy instruction that was inspired by the US *Framework for K-12 Science Education* (National Research Council, 2012) and shows promise for supporting students' ability to develop and use energy ideas, discussing its implications for supporting students' learning about energy over time and the possible revisions needed to the existing energy learning progression.

Learning Progressions

Learning progressions are not only descriptions of increasingly sophisticated levels of understanding of core concepts, such as energy, but also descriptions of typical instruction which would foster a more sophisticated level of understanding and how students' level of understanding may be assessed (Krajcik et al., 2012).

The development of a learning progression typically starts by defining upper and lower anchors, and then a description of intermediate levels of understanding (Alonzo & Wenk Gotwals, 2012). The upper anchor is traditionally informed by policy documents that state what students are expected to know (these expectations may be realistic or not, as we will see regarding a learning progression for energy), while the lower anchor is usually based on research describing children's naïve understandings of the core concept. The intermediate steps draw upon research on students' learning of the concept or practice (e.g., Schwarz et al., 2009) and analysis of the domain (e.g., Hadenfeldt et al., 2016).

There are two prominent approaches to constructing a learning progression in the existing literature: instructional and cross-sectional. The instructional approach begins with an intervention designed to foster students' progression in learning about the key concept and then tracking learning outcomes to determine whether students indeed progress as hypothesized. This is then followed by a modification of the intervention, the development of an additional instructional intervention intended to take students to the next level, or the development of an instructional intervention intended to support students' construction of the prior knowledge upon which the original intervention was built. For example, Nordine and colleagues (2011) developed a forms-based approach to energy instruction for 8th-grade students that highlighted the importance of indicators and factors in helping students identify which forms of energy were relevant in phenomena and how the magnitude of the energy manifested in each of these forms changed during the phenomenon. The approach was implemented in an 8-week-long unit and enacted in a middle school. Using pre/post assessments and interviews, they studied how students' knowledge of energy increased following the enactment of the unit and compared these students' knowledge with that of 9th-grade students who had participated in the intervention unit the previous year as well as

10th- and 11th-grade students at the same school who had studied energy with the same teachers using a more traditional approach to energy instruction. The findings of this study informed a learning progression of energy by indicating how, given that students have reached a predefined level of prior knowledge of energy (i.e., at the beginning of the intervention in 8th grade), they can be helped to progress from that prior level to a higher level of knowledge (i.e., at the end of the intervention in 8th grade). This new higher level of knowledge could then become the new level of prior knowledge to serve as the basis for an additional intervention that could raise students' knowledge of energy even higher (i.e., after their subsequent 9th-grade science class).

The instructional approach to constructing a learning progression is a longitudinal approach, in which groups of students are followed for the duration of interventions, usually relatively short periods of time, each group of students studied independently and then all the results combined to define A) increasingly sophisticated levels of understanding of the core concept, B) descriptions of instructional approaches which support students in moving from one level to the next, and C) ways in which students' knowledge can be assessed at each level.

The cross-sectional approach starts with the development of an instrument meant to investigate students' progression from the lower through the upper anchors and assumes that all participants have learned from curricula that are similar to each other in their approaches to fostering learning of the key concept (e.g., Herrmann-Abell & DeBoer, 2017; Neumann et al., 2013) or that the learning of this key concept is the outcome of natural cognitive development (e.g., Liu & McKeough, 2005; Shavelson & Kurpius, 2012).

Both approaches enrich our understanding of learning progressions, with cross-sectional studies providing a picture of large-scale patterns and instructional providing insights into how student learning develops along the large-scale learning progression or investigating the influence of instruction in providing alternative pathways within the learning progressions identified by cross-sectional studies.

Traditional Energy Instruction Underlying the Traditional Energy Learning Progression

Student difficulties with the energy concept and in particular with energy conservation have been studied for decades (for an overview, see Doménech et al., 2007), and much of this research has been conducted in the context of traditional energy instruction, which is the basis for the traditional energy learning progression. This learning progression (Chen et al., 2014; Neumann et al., 2013; Nordine et al., 2011) – see Figure 13.1 – shows that starting from nascent ideas developed through everyday experiences (e.g., that energy is somehow required to do things), students begin to know about different forms of energy (e.g., kinetic, gravitational, electric). As this knowledge grows, they begin to construct knowledge about energy transformations (conversions between forms) and transfer (exchanges between systems), and they do so before they have full knowledge of energy forms. Students then begin to develop an understanding of energy degradation (transformations into thermal energy) and dissipation (energy spreading to the surroundings as heat); this too occurs before a full understanding of transformations and transfer is reached. Energy conservation is the upper anchor of the learning progression, and students begin to develop ideas about energy conservation in parallel with the former stage(s), but a robust understanding of energy conservation seems to develop last. Further, research on this learning progression has indicated that few students reach a robust understanding of conservation by the end of high school (Herrmann-Abell & DeBoer, 2017; Neumann et al., 2013).

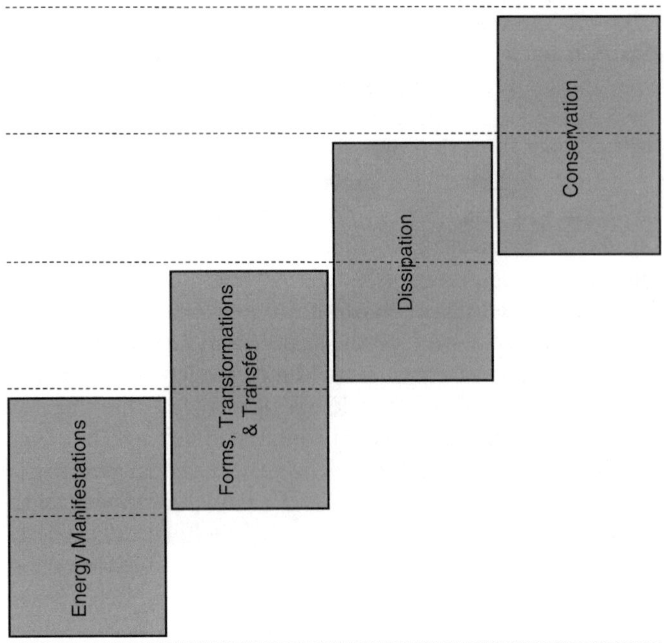

Figure 13.1 The traditional energy learning progression (e.g. Nordine et al., 2011; Neumann et al., 2013; see Chen et al., 2014).

Energy has traditionally been introduced and defined in the context of mechanics as the ability to do work (e.g., Halliday et al., 2010; Serway & Jewett, 2018). This approach has been criticized as limiting students' understanding of energy to the context of mechanics (Duit, 2014; Papadouris & Constantinou, 2011), and scholars have also noted that it may also impede students' understanding of energy conservation (Bryce & MacMillan, 2009). Instead of introducing energy in relation to mechanical work, many curricula present energy as a quantity that manifests itself in different forms and then proceed to discuss energy transformations and transfer (Kaper & Goedhart, 2002). Approaches that include the idea that energy exists in various forms are common in secondary schools (e.g., Hewitt, 2009; Kolodner et al., 2003; Krajcik et al., 2012). We refer to any approach in which students analyze phenomena by identifying different forms of energy as a "forms-based" approach, though such approaches typically emphasize both energy transformation and transfer. While forms-based approaches successfully support student learning of certain aspects of energy, many students still struggle with developing and using ideas about energy conservation (Lee & Liu, 2010; Neumann et al., 2013). Forms-based approaches have been criticized for emphasizing labeling instead of promoting a deeper understanding (Swackhamer, 2005), for implying to many students that each form of energy is conserved independently (e.g., that kinetic energy is conserved) (Nordine et al., 2011), and may actually impede future learning about energy (Bransford & Schwartz, 1999; Kaper & Goedhart, 2002). Research also indicates that many students have difficulties in correctly identifying the relevant system(s) under study, particularly when performing quantitative energy analysis (Lindsey et al., 2012; Van Huis & van den Berg, 1993).

In a forms-based perspective, forms of energy that arise due to interaction at a distance are classified as potential energy (e.g., gravitational energy, chemical energy). Students who learn in these approaches often struggle with the idea of potential energy, often thinking that potential energy is not actually energy, but rather the potential to have energy (Cooper & Klymkowsky, 2013; Lindsey et al., 2012). These students also commonly associate potential energy with single objects (e.g., the potential energy of a ball thrown into the air) rather than a system of objects interacting at a distance (e.g., the ball-Earth system) (Doménech et al., 2007; Jewett Jr, 2008; Summers, 1983). From a quantitative forms-based perspective, calculating values for potential energy raises difficulties, since this always involves defining some configuration of a system that corresponds to zero potential energy. Thus, the potential energy of the same system may be assigned different values of potential energy for different choices of a zero point, and the potential energy of a system may even be (and commonly is) assigned negative values. Conceptualizing and using the idea of potential energy is challenging for many students who learn about energy from a forms-based perspective.

While many researchers have questioned whether the conceptual foundations of traditional approaches to energy instruction may be responsible for students' difficulties (Brewe, 2011; Ellse, 1988; Nordine et al., 2011; Swackhamer, 2005), we are unaware of any studies – until recently – that have empirically investigated the effects of approaches to energy instruction that do not rely on the idea of energy forms. Therefore little has been known about whether forms-based instruction contributes to student difficulties in learning and using energy ideas and whether students may be better supported in developing and using energy ideas through approaches that do not require energy forms (Nordine et al., 2018).

The Systems-Transfer Approach to Energy Instruction

In an effort to more fully understand how students can be better supported in developing and using energy ideas, Nordine and colleagues (2018) engaged in a rethinking of the approaches that guided much, if not most energy instruction in the past few decades. They developed an approach to middle school physical science energy instruction, called the systems-transfer approach (STA), that reflects a fundamental departure from traditional forms-based approaches. The STA explains the phenomenon by focusing on the processes that increase or decrease the energy of interacting systems as energy is transferred between the systems. This perspective stands in stark contrast with the notion, common in forms-based approaches, that the energy of a system can/should be calculated without any changes occurring in a system (e.g., calculating the kinetic energy of a car moving at a certain speed or the gravitational potential energy of a boulder on a cliff). In contrast to forms-based approaches in which energy ideas may be applied even when no changes are occurring, energy analysis within the STA only makes sense in the context of active or ongoing changes. The STA does not require energy forms or energy transformations, and it emphasizes the connections between energy, systems, and fields that mediate interaction at a distance (Swackhamer, 2005). Such departures from traditional instruction align with the recommendations for middle school energy instruction outlined in the US *Framework for K-12 Science Education* (National Research Council, 2012), which notes in particular that the idea that energy exists in different forms is misleading (p. 122), stresses that energy conservation is fundamentally connected to the idea of systems (pp. 120–121), and that middle school students should begin to connect ideas about energy to ideas about the fields that mediate interaction-at-a-distance (p. 123).

Fortus and colleagues (2019) compared the learning of students who learned about energy using a respected curriculum building off a traditional forms-based approach with that of students who learned from the STA and found that students who learned from the STA

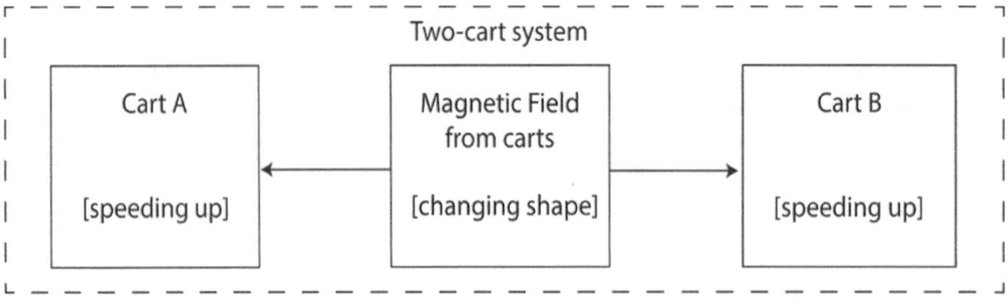

Figure 13.2 ETM for two repelling magnetic carts.

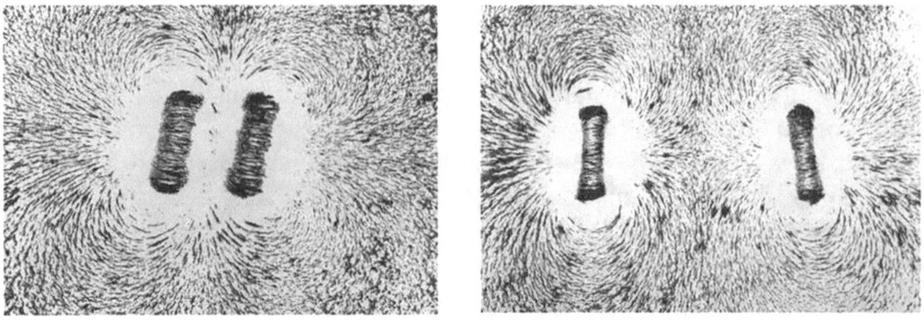

Figure 13.3 As the distance between the magnets changes, energy is transferred to/from the field, changing the shape of the field.

outperformed their peers in every aspect they considered, regardless of their prior knowledge of energy.

The systems-transfer perspective is represented by an Energy Transfer Model, shown in Figure 13.2, for the phenomenon of two magnetic carts repelling each other. The two magnetic-cart system is comprised of three subsystems: Cart A, Cart B, and the magnetic field from the carts. When the carts are held close together and then released, they move away from each other, each undergoing an energy-increasing process (speeding up). In this process, energy is transferred to them by the magnetic field, which undergoes an energy-decreasing process, corresponding to a change in shape in the magnetic field as the distance changes between the magnets (see Figure 13.3).

The Role of Fields

Note that the carts' magnetic field is treated as a subsystem (every rectangle is a system) of the two magnetic-cart system. The explicit inclusion of fields is required in the systems-transfer approach. This manner of interpreting this phenomenon is quite different from a more traditional approach that relies on energy forms, which would say that the explanation of this phenomenon involves the transformation of magnetic potential energy into kinetic energy. The systems-transfer approach includes a more explicit focus on modeling the systems and subsystems involved, identifying the changes happening in each of those systems as energy is transferred, and treats energy as a unitary quantity that must always be associated with an object or a field.

A key aspect of the systems-transfer approach is that it makes the fields concept necessary in the context of energy instruction. This stands in contrast to most traditional forms-based approaches; despite the fact that every type of potential energy (i.e., gravitational energy, electrical energy) involves objects that interact at a distance via fields, the role of fields is routinely ignored in traditional forms-based energy instruction. However, the idea that energy can be stored and transferred via fields is foundational to the formulation of the energy concept itself (Coopersmith, 2015). In the repelling magnets phenomenon discussed previously and represented in Figures 13.2 and 13.3, the field from the interacting magnets is necessary to identify the system that transferred energy to the carts as their speed increased. In fact, in the systems-transfer unit investigated by Fortus and colleagues (2019), the repelling magnets phenomenon is used to motivate a need to know about fields; by constructing an initial Energy Transfer Model of the repelling magnets phenomenon, students quickly recognize that there is a missing system that energy is transferred *from* as the speed of the magnets increases.

When building on fields, potential energy is as actual as kinetic energy – it is simply energy manifested in fields rather than in objects. While fields are abstract entities known to challenge students thinking (Bradamante & Viennot, 2007), traditionally, fields have been introduced as mathematical constructs rather than as physical entities (Swackhamer, 2005). In the STA, fields are presented as actual physical entities rather than mathematical abstractions. Fields provide a "place" for energy when it is not manifested in objects, something that has typically been a source of confusion for students (Lindsey et al., 2012; Summers, 1983). Prior research identified energy conservation as the most challenging idea about energy for students (Herrmann-Abell & DeBoer, 2017; Liu & McKeough, 2005; Neumann et al., 2013), but Fortus and colleagues' (2019) work indicates that introducing fields may actually help students develop an understanding of the transfer of energy within a system and energy conservation (Kubsch et al., 2021). From the systems-transfer perspective, energy must always be in some physical system – whether an object or a field – it does not exist independently and cannot simply pop into or out of existence as different forms.

As mentioned earlier, Fortus and colleagues (2019) investigated the extent to which students developed usable knowledge of energy by comparing two groups of matched middle school students, one which learned according to the systems-transfer approach, the other which learned by a more traditional forms-based approach. In every aspect, the knowledge developed by the systems-transfer students was superior. The systems-transfer students outperformed the students in the other group on 9 out of 10 items assessing three-dimensional learning of energy. A propensity-score-matched t-test based on post-test results indicated the clear superiority of systems-transfer scores, with a between-groups effect size of .63 ($t = 4.79$, $df = 230$, $p < .001$). Conceptual networks (Kubsch et al., 2020) showed that the systems-transfer students developed more parsimonious knowledge of energy, organized primarily around the idea of transfer without any reference to transformation, while the other students' knowledge of energy drew heavily upon forms, transformation, and transfer (Fortus et al., 2019).

Empirical results from the STA strongly suggest that middle school students do not need the idea of energy forms or transformation to begin constructing and effectively using energy knowledge, and this finding opens the possibility of identifying promising energy learning progressions that do not require these ideas but may more strongly involve connections to others, such as fields and systems thinking.

One or More Energy Learning Progressions?

An important question for energy learning progression research is the extent to which alternative pathways to developing and using energy knowledge are viable or even more effective than the traditional energy learning progression. Alternative pathways may exist

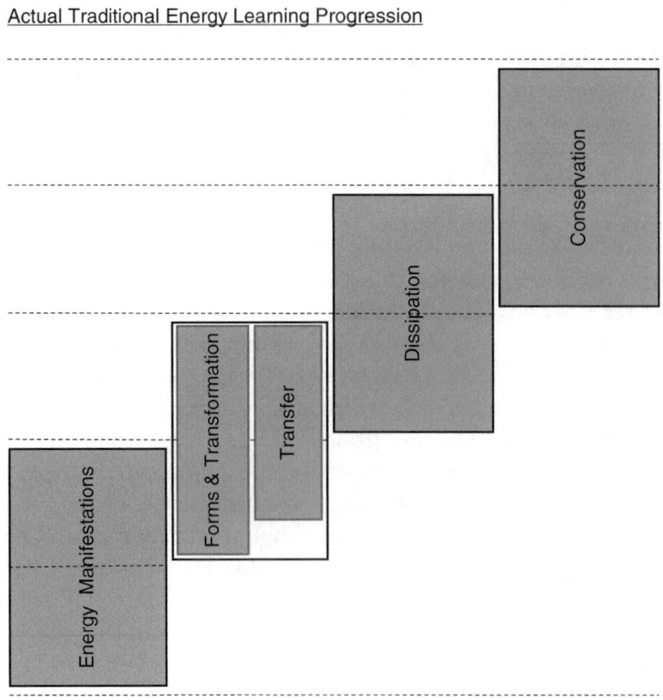

Figure 13.4 Actual traditional energy learning progression.

within the traditional learning progression or there may be alternatives to the traditional learning progression. Returning to the systems-transfer approach, we see that this approach may provide an alternative pathway. It aligns with the traditional learning progression (Figure 13.1) at its start – students begin from nascent ideas developed through everyday experiences, learning that energy is manifested in many ways, such as when objects move, change their temperatures, or when electricity flows. But then it bifurcates away from the traditional learning progression, which indicates that students begin to learn about energy forms and energy transfer in parallel. In fact, in traditional energy instruction, the concepts of energy forms and transformations are introduced first, and then, shortly after, energy transfer is added, so actually the second step in the traditional energy learning progression is made of two mostly overlapping but slightly shifted sub-steps, energy forms and transformations, then energy transfer – see Figure 13.4. However, in the STA students do not learn about forms at all and in fact, the study by Fortus and colleagues (2019) showed that the conceptual models of students who learned about energy in the STA do not even include the concepts of energy forms or transformations. This indicates that it may be possible to bypass the sub-step of energy forms and transformations in the traditional energy learning progression and move directly from nascent everyday ideas to energy transfer – see Figure 13.5.

An Energy Learning Progression Based upon a Systems-Transfer Perspective

When energy forms and transformations are not introduced, students begin to develop more formal ideas about energy. They first learn to interpret phenomena in terms of energy moving from one object to another. For example, students may consider how shining a flashlight on a

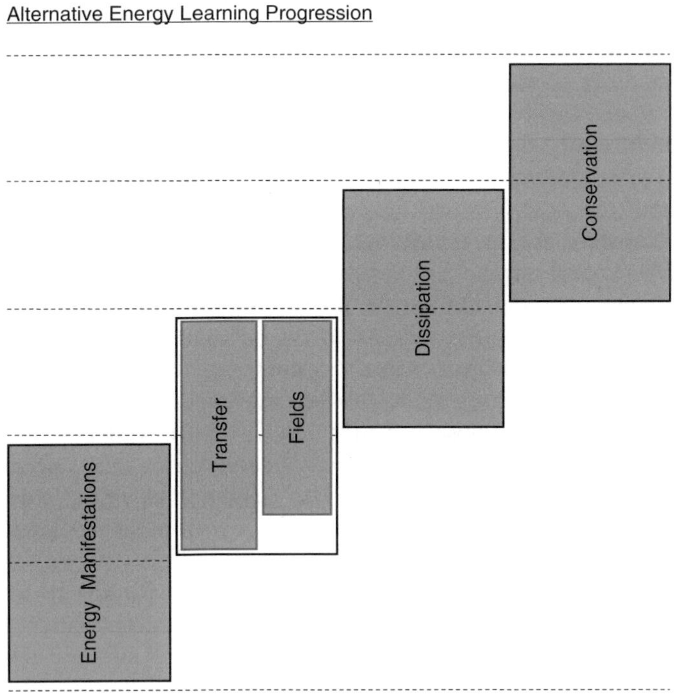

Figure 13.5 Alternative energy learning progression.

Crooke's Radiometer can cause the radiometer to turn. Building on their everyday ideas about energy, students readily associate motion with energy – and more motion with more energy. As such, students readily accept that energy has been somehow "provided" to the radiometer as it begins to spin. Additionally, and likely due to their everyday experiences with light bulbs and/or sunlight, students readily associate light with energy. Thus, students' nascent ideas gleaned from everyday experiences prime them for more formal thinking about the radiometer phenomenon in terms of energy transferred from the flashlight and to the radiometer. Likewise, students readily associate energy with heating/hot things, priming them to use the same energy concepts to make sense of new heating-related phenomena in the same energy transfer terms. A major benefit of students' tendency to associate speed and temperature with energy is that the energy processes associated with changes in speed/temperature give clear evidence that energy was either transferred to or from an object – if an object either speeds up or heats up, then energy was clearly transferred to that object from another system or subsystem. This is different than the conclusion that would have been reached in traditional forms-based energy instruction. There, students would have learned that kinetic energy or thermal energy was transferred, which requires the students to first develop a rudimentary understanding of the concepts of kinetic and thermal energy. At the beginning of a learning progression that does not include forms or transformation, students' nascent ideas about energy are sufficient to help interpret a wide range of phenomena in terms of the systems involved, the energy-related processes occurring in those systems, and the resulting direction of energy transfer between the systems.

To begin developing the next level of understanding in such a learning progression, students need additional ideas about fields as they begin to explore phenomena that require fields to make a full accounting of the energy transfers involved (e.g., the repelling magnets phenomenon discussed above, a ball falling from a table) – see Figure 13.5. For example, students must learn that fields have a particular structure in space and that when the configuration of objects that interact via fields changes, the structure of the field changes as well. Such structure changes can be easily visualized using iron filings or ferrofluids in the case of magnetic interactions (see Figure 13.2). The change in the structure of a field as energy is transferred to/from it is important because it constitutes the energy change process that the field undergoes during energy transfers. At the middle school level, the structure of the field should not be described in mathematical terms. Without mathematical analysis to determine whether changes in field structure are associated with a positive or negative change in the energy of the field, middle school students can simply observe the behavior of the objects that interact via the field. If one or more of the objects speed up, energy is transferred from the field; if the object(s) slow down, energy is transferred to the field that mediates the interaction at a distance. In our empirical work, we have observed that middle school students are well-equipped to interpret macroscopic magnetic and gravitational phenomena in terms of energy transfers between objects and fields, but that microscopic phenomena (e.g., chemical interactions in which particles that interact via fields rearrange) are best left to later (see Kubsch et al., 2021).

The progression from tracking energy transfers between interacting systems to understanding energy dissipation is relatively straightforward. Students clearly notice from everyday macroscopic phenomena (such as a pendulum swinging or a bowl of soup cooling down) that in many phenomena, there is an increasing amount of energy "missing" from the system of interest. From the systems-transfer perspective, this indicates that there are previously undetected energy transfers to a system or subsystem that interacts with the system of interest – this new system is usually the surrounding environment. A thermal camera can be particularly helpful for detecting the relatively small, short-lived changes in temperature that constitute the energy change processes that occur in the environment as energy is transferred to it (Nordine et al., in press). Thus, developing and using ideas about energy dissipation to explain phenomena are layered on top of a set of previously existing ideas in which students interpret phenomena by identifying interacting systems, the energy change processes they undergo when phenomena occur, and the associated direction of energy transfer between those systems. In contrast, when students learn about energy from a forms-based perspective, the idea that energy is transferred to the surroundings must typically be introduced for the first time, and these new dissipation ideas must be especially attached to a particular form of energy: thermal energy. Thus, building new ideas about energy dissipation in the context of a forms-based learning progression requires that students learn about energy transfer (if they haven't already), apply ideas about systems (e.g., energy leaving a system of interest) which are typically not foregrounded in earlier stages of a forms-based learning progression (because they are not typically needed until now), and attach special behavior to a particular form of energy (thermal energy). It is of course possible, even likely, that students in a forms-based approach do not explicitly learn about energy transfer, as dissipation could simply be described as thermal energy in the surroundings. In this case, energy transfer and systems ideas are not explicitly foregrounded and not well-developed, potentially contributing to student difficulty using systems in the context of energy (Lindsey et al., 2012). A learning progression that does not require energy forms and transformations but does require more explicit connections to systems provides a more straightforward path to understanding energy dissipation that is more solidly based upon lower anchor understanding.

Building ideas about energy conservation is similarly more straightforward in a learning progression that does not include forms or transformation, and conservation ideas may also be built by more explicitly tapping into students' everyday ideas. For example, students commonly develop a quasi-material conception of energy (Brewe, 2011; Duit, 1987; Watts, 1983) in which they discuss energy in substance-like terms. While it has been established that energy is not a substance (Coopersmith, 2015), several physics educators point to the benefits of such a quasi-material conception (Brewe, 2011; Duit, 1984; Swackhamer, 2005), and at least one empirical study has failed to detect any deleterious effects that arise from a quasi-material conception of energy (Nordine et al., 2011). From a systems-transfer perspective, a quasi-material conception of energy has some clear benefits: students readily accept that energy must be "in" some physical system and that energy (much like material objects) does not simply appear or disappear. Accordingly, energy that is "lost" is not gone, but simply in some system where it is difficult to detect. When students conceptualize energy as always associated with some physical system, they can begin to develop intuitive notions of conservation that align with their notions of object permanence that they have been developing since childhood. In contrast, a forms-based learning progression is fraught with nuanced ideas about the nature, location, and permanence of potential energy. Accordingly, learning about energy conservation in a forms-based approach is more abstract, as energy is discussed in terms of being "in" one form or another (e.g., kinetic or gravitational) rather than "in" one physical system or another (e.g., a cart or a magnetic field). An empirical study of 8th-grade students in a forms-based approach found that many students continued to struggle with qualitative energy conservation (i.e., the idea that energy is not created/destroyed) (Nordine et al., 2011); conversely, 7th-grade students learning about energy using a system-transfer approach found that students successfully progressed toward a qualitative understanding of energy conservation (Kubsch et al., 2021).

Building and Using Energy Ideas across Disciplines

The energy learning progressions literature tends to be heavily physics-centric, often failing to account for how energy is represented and discussed in other science and engineering disciplines. One potential benefit of a learning progression based upon the systems-transfer perspective is its alignment with non-physics disciplines, which tend to heavily emphasize transfers between systems rather than transformations between energy forms. In its description of energy as a crosscutting concept, the US *Framework for K-12 Science Education* stresses that conservation is best interpreted in terms of energy transfers between systems, stating, "One of the great achievements of science is the recognition that, in any system, certain conserved quantities can change only through transfers into or out of the system" (National Research Council, 2012, p. 94). Indeed, food webs representing ecosystems, Sankey diagrams representing the energy usage of societies, and Earth energy budget representations all represent energy transfers between interacting systems. When students build energy ideas in physical science that more readily align with how energy is discussed and represented in other disciplines, they are in a better position to use the same ideas across contexts, setting the stage for deeper and more transferrable learning (Schwartz & Goldstone, 2016).

Conclusion

Energy learning progressions have been the target of over 15 years of theoretical and empirical work, and much of this work has demonstrated remarkable alignment. Yet, existing empirical studies have been conducted using cross-sectional designs and without reference to specific instructional approaches. Recent work to develop a systems-transfer approach to middle school energy instruction suggests that an alternative learning progression may be particularly effective in supporting students in learning about energy dissipation and conservation without the need to develop ideas about energy forms or transformation.

While a system-transfer learning progression aligns with and leverages many of students' intuitive ideas about energy, studies on the systems-transfer approach (Fortus et al., 2019; Kubsch et al., 2021) indicate that the learning of energy cannot be the outcome of natural cognitive maturation, but rather that it is dependent on the instructional approach used to learn about energy. Different instructional approaches can lead to different trajectories in a learning progression. This is an important conclusion, not only for energy learning progressions but possibly, and perhaps even likely, for learning progressions of other key concepts or practices.

This conclusion indicates that it may be critical for future investigations to account for how learning progressions may bifurcate or perhaps combine as learners develop; for example, although it appears that the systems-transfer approach bypasses forms and transformation, as long as high school energy instruction remains unchanged, students who learn by this approach will eventually encounter forms and transformation ideas. Learning progressions research may help illuminate whether and how these learners successfully navigate this transition. Learning progressions research should also investigate how learning about energy in the systems-transfer approach supports students in applying their ideas about energy to disciplines other than physics and whether it leads learners to perceive energy as a single entity, identical for physics, chemistry, life science, and earth science.

Note

1 Disciplinary core ideas are ideas, specified by the *Framework* National Research Council. (2012). *A framework for K-12 science education: Practices, crosscutting concepts, and core ideas*. National Academies Press., are organizing principles for a scientific discipline and keys for understanding, investigating, and solving more complex ideas and problems in that discipline. Science and engineering practices are the practices that scientists and engineers engage in when making sense of phenomena and solving problems. Crosscutting concepts are a set of conceptual tools, ideas, and heuristics that span disciplinary boundaries and are essential for explaining phenomena (in science) and defining solutions (in engineering).

References

Alonzo, A. C., & Wenk Gotwals, A. (Eds.). (2012). *Learning progressions in science: Current challenges and future directions*. Sense Publishers.

Bradamante, F., & Viennot, L. (2007). Mapping gravitational and magnetic fields with children 9–11: Relevance, difficulties and prospects. *International Journal of Science Education*, *29*(3), 349–372.

Bransford, J. D., & Schwartz, D. L. (1999). Rethinking transfer: A simple proposal with multiple implications. *Review of Research in Education*, *24*(1), 61–100.

Brewe, E. (2011). Energy as a substance like quantity that flows: Theoretical considerations and pedagogical consequences. *Physical Review Special Topics - Physics Education Research*, *020106*(7), 1–14.

Bryce, T. G. K., & MacMillan, K. (2009). Momentum and kinetic energy: Confusable concepts in secondary school physics. *Journal of Research in Science Teaching*, *46*(7), 739–761. 10.1002/tea.20274

Chen, R. F., Eisenkraft, A., Fortus, D., Krajcik, J. S., Neumann, K., Nordine, J. C., & Scheff, A. (Eds.). (2014). *Teaching and learning of energy in K – 12 education*. Springer.

Cooper, M. M., & Klymkowsky, M. W. (2013). The trouble with chemical energy: Why understanding bond energies requires an interdisciplinary systems approach. *CBE—Life Sciences Education*, *12*(2), 306–312.

Coopersmith, J. (2015). *Energy, the subtle concept: The discovery of Feynman's blocks from Leibniz to Einstein*. Oxford University Press.

Doménech, J. L., Gil-Pérez, D., Gras-Martí, A., Guisasola, J., Martínez-Torregrosa, J., Salinas, J., Trumper, R., Valdés, P., & Vilches, A. (2007). Teaching of energy issues: A debate proposal for a global reorientation. *Science & Education*, *16*(1), 43–64.

Duit, R. (1984). Learning the energy concept in schools - Empirical results from The Phillippines and West Germany. *Physics Education*, *19*, 59–66.

Duit, R. (1987). Should energy be illustrated as something quasi-material? *International Journal of Science Education*, *9*(2), 139–145.

Duit, R. (2014). Teaching and learning the physics energy concept. In R. F. Chen, A. Eisenkraft, D. Fortus, J. S. Krajcik, K. Neumann, J. C. Nordine, & A. Scheff (Eds.), *Teaching and learning of energy in K – 12 Education* (pp. 67–85). Springer.

Ellse, M. (1988). Transferring, not transforming energy. *School Science Review*, *69*, 427–437.

Fortus, D., Kubsch, M., Bielik, T., Krajcik, J. S., Lehavi, Y., Neumann, K., Nordine, J. C., Opitz, S. T., & Touitou, I. (2019). Systems, transfer, and fields: Evaluating a new approach to energy instruction. *Journal of Research in Science Teaching*, *56*(10), 1341–1361. 10.1002/tea.21556

Hadenfeldt, J. C., Neumann, K., Bernholt, S., Liu, X., & Parchmann, I. (2016). Students' progression in understanding the matter concept. *Journal of Research in Science Teaching*, *53*(5), 683–708.

Halliday, D., Resnick, R., & Walker, J. (2010). *Fundamentals of physics, Chapters 33-37*. John Wiley & Sons.

Harris, C. J., Krajcik, J. S., Pellegrino, J. W., & DeBarger, A. H. (2019). Designing knowledge-in-use assessments to promote deeper learning. *Educational Measurement: Issues and Practice*, *2*, 53–67.

Herrmann-Abell, C. F., & DeBoer, G. E. (2017). Investigating a learning progression for energy ideas from upper elementary through high school. *Journal of Research in Science Teaching*, *55*(1), 68–93. 10.1002/tea.21411

Hewitt, P. G. (2009). *Conceptual physics: The high school program*. Prentice-Hall.

Jewett Jr, J. W. (2008). Energy and the confused student II: Systems. *The Physics Teacher*, *46*(2), 81–86.

Kaper, W. H., & Goedhart, M. J. (2002). 'Forms of Energy', an intermediary language on the road to thermodynamics? Part I. *International Journal of Science Education*, *24*(1), 81–95.

Kolodner, J. L., Camp, P. J., Crismond, D., Fasse, B., Gray, J. T., Holbrook, J., Puntambekar, S., & Ryan, M. (2003). Problem-based learning meets case-based reasoning in the middle-school science classroom: Putting learning by design(tm) into practice. *Journal of the Learning Sciences*, *12*(4), 495–547. 10.1207/s15327809jls1204_2

Krajcik, J. S., Reiser, B. J., Sutherland, L. M., & Fortus, D. (2012). *IQWST: Investigating and questioning our world through science and technology, (Middle School Science Curriculum Materials)*. Activate Science.

Krajcik, J. S., Sutherland, L. M., Drago, K., & Merritt, J. D. (2012). The promise and value of learning progression research. In S. Bernholdt, K. Neumann, & P. Nentwig (Eds.), *Making it tangible: Learning outcomes in science education* (pp. 261–283). Waxmann.

Kubsch, M., Opitz, S., Nordine, J., Neumann, K., Fortus, D., & Krajcik, J. (2021). Exploring a pathway towards energy conservation through emphasizing the connections between energy, systems, and fields. *Disciplinary and Interdisciplinary Science Education Research*, *3*(1), 2. 10.1186/s43031-020-00030-7

Kubsch, M., Touitou, I., Nordine, J. C., Fortus, D., Neumann, K., & Krajcik, J. S. (2020). Transferring knowledge in a knowledge-in-use task: Investigating the role of knowledge organization. *Education Sciences*, *10*(20). 10.3390/educsci10010020

Lee, H.-S., & Liu, O. L. (2010). Assessing learning progression of energy concepts across middle school grades: The knowledge integration perspective. *Science Education*, *94*(4), 665–688.

Lindsey, B. A., Heron, P. R. L., & Shaffer, P. S. (2012). Student understanding of energy: Difficulties related to systems. *American Journal of Physics*, *80*(2), 154–163.

Liu, X., & McKeough, A. (2005). Developmental growth in students' concept of energy: Analysis of selected items from the TIMSS database. *Journal of Research in Science Teaching*, *42*(5), 493–517.

National Research Council. (2012). *A framework for K-12 science education: Practices, crosscutting concepts, and core ideas*. National Academies Press.

Neumann, K., Viering, T., Boone, W. J., & Fischer, H. E. (2013). Towards a learning progression of energy. *Journal of Research in Science Teaching*, *50*(2), 162–188. 10.1002/tea.21061

Nordine, J., Weßnigk, S., & Greinert, L. (in press). Using thermal cameras in secondary physics to support learning about energy. In H. J. F. Jeppsson, & K. Schönborn (Eds.), *Thermal cameras in science education*. Springer Nature.

Nordine, J. C., Fortus, D., Lehavi, Y., Neumann, K., & Krajcik, J. S. (2018). Modelling energy transfers between systems to support energy knowledge in use. *Studies in Science Education*, *54*(2), 177–206. 10.1080/03057267.2018.1598048

Nordine, J. C., Krajcik, J. S., & Fortus, D. (2011). Transforming energy instruction in middle school to support integrated understanding and future learning. *Science Education*, *95*(4), 670–699.

Papadouris, N., & Constantinou, C. P. (2011). A philosophically informed teaching proposal on the topic of energy for students aged 11–14. *Science & Education*, *20*(10), 961–979.

Pellegrino, J. W., & Hilton, M. L. (2012). Committee on defining deeper learning and 21st century skills. *Center for Education*.

Schwartz, D. L., & Goldstone, R. L. (2016). Learning as coordination: Cognitive psychology and education. In *Handbook of educational psychology* (pp. 75–89). Routledge.

Schwarz, C. V., Reiser, B. J., Fortus, D., Davis, E. A., Kenyon, L., & Shwartz, Y. (2009). Developing a learning progression of scientific modeling: Making scientific modeling accessible and meaningful for learners. *Journal of Research in Science Teaching*, *46*(6), 632–655.

Serway, R. A., & Jewett, J. W. (2018). *Physics for scientists and engineers with modern physics*. Cengage Learning.

Shavelson, R. J., & Kurpius, A. (2012). Reflections on learning progressions. In *Learning progressions in science* (pp. 13–26). Brill Sense.

Smil, V. (2005). *Energy at the crossroads: global perspectives and uncertainties*. MIT Press.

Smil, V. (2006). *Transforming the twentieth century: technical innovations and their consequences* (Vol. 2). Oxford University Press.

Summers, M. K. (1983). Teaching heat – an analysis of misconceptions. *School Science Review*, *64*, 670–676.

Swackhamer, G. (2005). *Cognitive Resources for Understanding Energy*. Retrieved May 1 from http://modeling.asu.edu/modeling/RESOURCES_10-03.pdf

Van Huis, C., & van den Berg, E. (1993). Teaching energy: a systems approach. *Physics Education*, *28*(3), 146.

Watts, M. (1983). Some alternative views of energy. *Physics Education*, *18*, 213–217.

14

GEOLOGY AND EARTH SYSTEMS SCIENCES LEARNING PROGRESSIONS

Richard A. Duschl, Kathryn M. Bateman, Seungho Maeng, and Scott McDonald

Introduction

Plate tectonic theory launched a new view of the Earth which, over the span of 50 years, led to dynamic advancements in geosciences and geographical sciences research programs. This evolution, in turn, impacted our conceptualization of Earth Systems Science. The dynamic interactions among the Earth's biosphere, hydrosphere, atmosphere, cryosphere, and lithosphere informed the identification and characterizations of the challenges humans, and other life forms, encounter on Earth; e.g., understanding, and adapting to complex Earth systems. The National Research Council's Board on Earth Sciences and Resources formed a Committee on Strategic Directions for the Geographical Sciences in the Next Decade. The report – *Understanding the Changing Planet: Strategic Directions for the Geographical Sciences* (NRC, 2010) – is organized around a set of Strategic Research Questions regarding a set of complex adaptive systems:

> How are humans changing the physical environment of the Earth's surface?
> How can we best preserve biological diversity and protect endangered ecosystems?
> How are climate and other environmental changes affecting the vulnerabilities of coupled Human-Environment systems?

"Many of the central challenges of the 21st century are tied to changes to the spatial organizations and character of the landscapes and environments of Earth's surface as populations move, natural resources are depleted, and climate shifts" (NRC, 2010, p. 1). The NRC report also makes it very clear that technological developments and new priorities have inspired the rapid growth of geographical sciences over the past two decades. The result is a broader set of researchers (e.g., economists, biologists, epidemiologists, sociologists, climate scientists, geologists, etc.) now engaging in transdisciplinary geosciences research. A foundational model for the Earth as a dynamic system emerges from the ongoing development and enrichment of Plate Tectonic theory.

Learning Progressions that work in the geosciences must take this transdisciplinary complexity into account, though one individual LP cannot consider it all. In this paper, we will discuss two learning progressions approaches to the complexity of geoscience learning.

DOI: 10.4324/9781003170785-16

The first describes an LP that adopts a foundational plate system perspective for a dynamic Earth model and takes a sociocultural perspective on the classroom implementation of the LP. The second describes an LP that adopts a geocognitive perspective and focuses on the assessment of the temporal and spatial relationships of a dynamic Earth model of Plate Tectonics. The final section "Trends in Earth Science Education" examines interdisciplinary and transdisciplinary approaches within Earth Systems Science (ESS). ESS provides, in turn, new frameworks for examining and understanding how Complex Adaptive Systems created by human populations are impacting planet Earth. We argue that building from Plate Tectonic models and mechanisms, additional design-based research is needed to develop geo/geographical science learning progressions within other ESS and Complex Adaptive Systems contexts.

Plate Tectonics – The Driving Phenomenon of Earth Science

The Theory of Plate Tectonics (PT), developed between 1950 and 1970, is a relatively new scientific science discovery. PT theory emerged from centuries of charting coastlines and mapping the location and physical features of the Earth's major continents and oceans. The accounts of Plate Tectonics typically found in Earth Science textbooks begin with the legacy of Alfred Wegener's Continental Drift Theory which emerged from the recognition of the fit between continental coastlines. But up until the 1960s there was no mechanism to account for the movement of continents. The pedagogical problem in Earth Science is that without a mechanism story, elementary-level school science activities/lessons, like those that have students cutting out the continents as jigsaw puzzle pieces and assembling them into Gondwana Land and Pangaea arrangements, can lead to serious and pernicious misunderstandings among learners. Such a historical, descriptive, and pragmatic orientation to Earth Science continues to dominate curriculum materials designs of K-12 earth science programs and courses of study. This is also a significant problematic issue given research on learning. Stressing descriptive "What we know" learning goals without also including systems thinking that incorporates higher level "How and Why we know" reasoning learning goals, negates opportunities for developing epistemic knowledge practices that can be used when examining, building, and refining frameworks for Earth Systems and Earth Models (Duschl et al., 2021; Lehrer & Schauble, 2004; Miller et al., 2018; Stroupe et al., 2019).

Advances in technology, especially computing power for simulations, data analytics, artificial intelligence, visualizations, and sensors in scientific understanding coupled with a growing global sense of responsibility have shifted both the focus and the methods of the geo/geographical sciences. The trends for several decades have been toward a systems analysis of the Earth (Mayer, 2002). With respect to inquiry methods, the trend is toward model-based and systems-thinking science education. Models are increasingly being used for exploring and explaining the complex dynamics and structures of the Earth's surface. Systems thinking targets issues of and questions about the habitability and sustainability of Planet Earth. Both are paramount for developing better predictive models that can, in turn, be used to shape policy and enhance the management of resources. As such, learning progressions targeting models and modeling practices and Earth System Science need to be a strong thread in the geoscience and geographical science curricula (Mayer, 2002; Oreskes, 2003). Plate Tectonic Theory is an excellent context for introducing students to and engaging them in the study and reasoning of patterns and phenomena that occur in ESS.

The appearance of "puzzle pieces" that fit together, dates to the 16th century map makers (Oreskes, 2003). But explanations of why and how continents move did not arrive until

400 years later. An edited volume on the emergence of the Plate Tectonics by Naomi Oreskes (2003), delves into the post-World War II thinking regarding the modern theory development of the Plate Model of the Earth. Oreskes raises an interesting philosophical thesis that Plate Tectonics, unlike other major 17th, 18th, and 19th-century scientific theories, is the first theory *not* born out of a conceptual framework. Rather, Plate Tectonic theory emerged from the accumulation, assimilation, and representation of massive amounts of geophysical data collected by an array of new tools and technologies, e.g., sonar, radar, magnetometers, and telecommunications. Effective Learning Progressions for teaching and learning about the development of the Theory of Plate Tectonics are critically important. We present two examples below.

This historical development of the Theory of Plate Tectonics offers valuable insights that informed our efforts in developing LPs. To help the reader begin thinking about the design of LPs in geology and the relationships to ESS, we consider two different learning progressions for teaching/learning Plate Tectonics. LPs in Earth Science have challenged the step-wise learning pathways proposed by many standards and policy documents. Here, we present two learning progressions that took different theoretical stances on learning. The first, from McDonald and colleagues (2019), approached student understanding of Plate Tectonics from a sociocultural perspective, considering the classroom cultures and learning practices in their work. The second, from Maeng (2020) uses cognitive frameworks to better understand the reasoning practices of learners around Plate Tectonics. This resulted in variations in progress variables between the two LPs, however, both offer insightful pedagogical information that helps inform the teaching and learning of Earth Science in K-12 contexts.

Both LPs are empirically grounded approaches to understanding the pathways to sophisticated explanations of a phenomenon (Corcoran et al., 2009), Plate Tectonics. These pathways begin as theoretical ideas about how learning can be empirically tested, both in classrooms and laboratory settings (Wilson, 2009). Along these pathways, the researcher is exploring what students or learners understand about phenomena early on as compared to an endpoint (Corcoran et al., 2009). This endpoint, determined by the researchers, is known as the *upper anchor*, bounding the expectations for student understanding. The LP researcher should also define a *lower anchor* to identify the students' expected prior knowledge (Duncan & Hmelo-Silver, 2009; Duschl et al., 2011). The development of movements between upper and lower anchors is further broken down into strands of interconnected information called *progress variables,* which serve as a framework for assessment of the upper anchor (Corcoran et al., 2009; Wilson, 2009). Finally, the LP researcher assumes that movement along the path from the lower to the upper anchor is non-linear and requires mediation. Movement along the learning progression is not an inherent or natural process that will develop in a learner without guidance and instructional support (Duncan & Hmelo-Silver, 2009).

A Sociocultural and Implementation Approach to Plate Tectonics LPs

McDonald and colleagues (McDonald et al., 2019) approached the development of a learning progression around Plate Tectonics as an evolutionary learning progression. That is, the learning progression is developed to "refine and define the developmental pathway(s) through identification of mid-levels or steppingstones that are then used to bolster meaning making and reasoning employing crafted instructional interventions" (Duschl et al., 2011, p. 157). Creating a hypothetical learning progression (Corcoran et al., 2009) involves drawing on a variety of K-12 science education documents including the Earth Science Literacy Initiative

(ESLI, 2010); the NGSS (NGSS Lead States, 2013), Benchmarks for Science Literacy (American Association for the Advancement of Science; AAAS, 2001) and the Framework for K-12 Science Education (NRC, 2012). From these documents, the team developed an upper anchor to measure the level of sophistication of student explanations of the following:

> Volcanoes, earthquakes, and mountains are the surface manifestations of large-scale movement of both solid and near-solid earth materials. Such movement can be explained as a natural consequence of the initial formation of the planet, its subsequent differentiation into chemically distinct units (crust, mantle, core), and the continual transfer of heat from the interior of the planet to the surface.

The upper anchor was then divided into three progress variables – Plate Movement, Mechanism of Plate Motion, and Plate System (See Figure 14.1). Next, a series of interviews and classroom observations helped to document students' progressive understanding of Plate

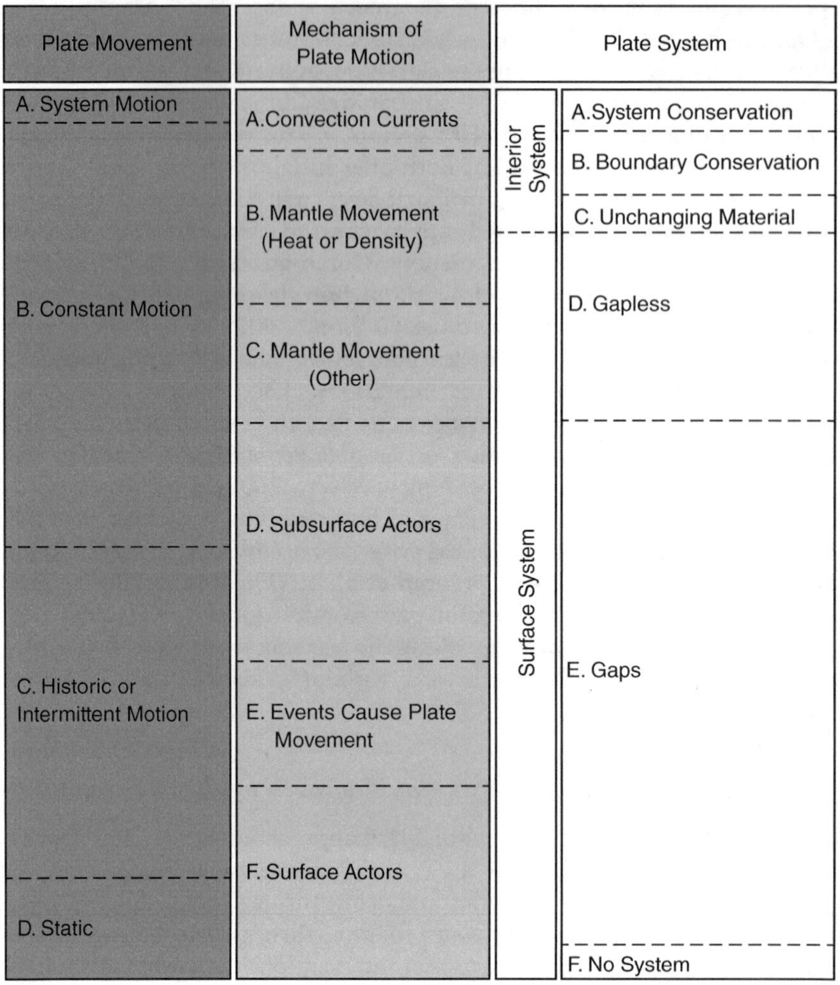

Figure 14.1 Learning progression for plate tectonics (McDonald et al., 2019).

Tectonics. Within the progress variable *Plate Movement,* they found four stages of students' motion representations: static, historic, constant (no mechanism), and systemic (mechanism). In the "messy middle," students frequently expressed the idea that plates only move intermittently or historically, or that plates move constantly, but independent of any system.

For the progress variable, *Mechanism of Plate Motion,* students' understanding of how the plates moved was assessed. For coding on this progress variable, students at the lowest level attributed motion to surface factors such as wind and water. Next higher levels of attribution selected events like earthquakes, volcanic eruptions, and other subsurface processes but did not make connections to the Earth's Mantle layer. At the next higher level of interior motions, students connected plate movement with Mantle movement and Mantle movement related to heat or density. The most sophisticated explanation, connected plate motion to the dynamics of convection currents in the Upper Mantle.

The third progress variable, *Plate System,* examined students' representations of the interconnected system of plates. Here, the progress variable was split into two contexts – Surface Systems and Internal Systems. At the lower anchor levels, students did not express any system of plates but rather saw plate interactions as having gaps between plates, bumping into one another, or referring to something "having to be" between the plates so they can move. At the "messy middle" Surface System level, plates did not have gaps but also did not make connections with anything below the surface. Within the "messy middle" Interior System section, students only see subsurface processes connecting with surface processes. For example, understanding that subducted materials move down into the Earth but without describing/understanding the changes that occur. At the next higher level, Boundary Conservation, students can explain that material was conserved since material subducted was also creating new crustal material at divergent plate boundaries. However, here students retained a localized, often single-plate explanation of the process. Lastly, a small number of students were able to make System Conservation connections across the interactions of Major and Minor global plate systems.

Learning about Teaching from Learning Progressions Research in Earth Science

To better understand the reasoning behind student explanations of Plate Tectonics, McDonald and colleagues also examined classroom instruction. Findings from student interviews revealed that although curriculum and standards only asked for students to explain patterns in surface-level phenomena, students were able to achieve a system-level understanding of plate tectonics (McDonald et al., 2019). Middle school students were indeed capable of comprehending plate tectonics as a system.

Two teachers participated in interviews and observations – Brendan and Talia. Generally, these teachers shared similar educational backgrounds (preparation in both geoscience and education) and seven and ten years of teaching experience respectively. They taught in the same geographical region of the State. Differences between the two teachers were due to school district characteristics. Brendan taught in an affluent suburban school PT Middle School (PTMS) with low poverty and predominantly white students. Talia, on the other hand, taught a more diverse student population, in an urban school W Middle School (WMS) with 67% of students receiving free and reduced lunch.

Regarding pedagogical approaches, Brendan implemented "Ambitious Science Teaching" (Windschitl et al., 2018) pedagogy to prompt and explore the question, "Why does [our area] look the way that it does?" Students worked through investigations, data reading, and mapping. Classroom norms included frequent discussion of the activities and construction of explanations of phenomena. Overall, data showed that Brendan created a student-centered, constructivist classroom at PTMS that linked to greater learning achievement in our post-instruction interviews.

At WMS, Talia structured tasks such that students worked on their own or as a whole class. Activities followed more traditional teaching norms and included reading from the textbook, structured step-by-step hands-on activities, answering questions with closed answers, and initiate, response, evaluate (IRE) discourse patterns (Lemke, 1990). This positioned the teacher and the textbook as the authorities on science knowledge. Interspersed between the textbook reading were small group activities around the *Discovering Plate Boundaries* data visualizations (Sawyer et al., 2005), as well as hands-on activities such as creating a convection current in water with food dye. After completing demonstrations or hands-on activities, students then completed lab worksheets independently. Talia's instruction was a more teacher-centered model with a focus on transmitting the normative explanation. Though many of the instructional activities were superficially similar in the two cases, students in PTMS described a more sophisticated understanding of plate tectonics more often than their peers at WMS in the post-unit interviews. Only students at PTMS were coded as describing each progress variable at the uppermost level, demonstrating an understanding of the upper anchor.

In addition to pedagogical approaches, curricular sequencing, such as the order of introduction of progress variables also appears to influence students' understanding of the upper anchor. When explanatory science concepts were not developed before students were introduced to historical evidence, students were more likely to misattribute evidence as causes. At WMS, after Talia introduced the *Discovering Plate Boundaries* maps, instruction focused on the historical development of plate tectonics as a theory, starting with Alfred Wegener, and included the pieces of evidence he used to develop his theory. In post-unit interviews, students at WMS were less likely to describe an upper-level explanation of the progress variable *Plate System*. The students recalled the independent, discrete pieces of knowledge related to the development of the theory but did not tie them together to develop a more coherent system-level understanding.

The connections and throughlines in the curricular and pedagogical approaches taken by the two teachers resulted in differential movements along the learning progression. Talia's instruction started and ended with the *Discovering Plate Boundaries* data visualizations as an attempt to frame the set of activities, but instruction around the progress variable *Mechanism of Plate Movement* was disconnected from the data visualizations which were designed to support students' understanding of the progress variable *Plate System*. Brendan's use of a focal phenomenon and driving question helped students at PTMS assemble the discrepant pieces of evidence from activities related to each of the three progress variables over the course of the unit of instruction into a more sophisticated and systems-focused explanation. This difference could be linked with the students' post-instructional interview responses, in which PTMS students moved closer to the upper anchor of the LP than their peers at WMS, especially in the progress variable *Plate System*. Reaching the upper anchor, a dynamic system-level understanding of Plate Tectonics, was best explained by the instruction that tied back to explaining a phenomenon and answering a driving question that provided students epistemic agency (Stroupe, 2014) within instructional activities. Brendan's approach of engagement in science and engineering practices and crosscutting concepts while discussing Plate Tectonics supported students' understanding of the phenomenon. These findings are a sound starting point to further develop Earth Science learning progressions that take a systems approach.

A Geocognition and Assessment Approach to Plate Tectonics LPs

The Geocognition Learning Progression framework from Maeng (2020), adopts a different set of progress variables for PT. Geoscience learning contexts, like ESS contexts, have unique challenges regarding the reasoning practices and meaning-making needed to engage with core concepts and mechanisms, as in PT. The focus on reasoning practices such as spatial reasoning, systems thinking, and deep-time/deep-space phenomena, as well as a strong emphasis on modeling as a core investigatory and knowledge-building process, provides a different set of considerations for learning progressions researchers in an Earth Systems Science domain. Kastens et al. (2009) articulated such reasoning practices as the nature of geoscience thinking. Maeng (2020) in a study of secondary-level students, re-identified such reasoning practices as geocognition. That is, as a set of cognitive processes necessary to engage with geoscience models, mechanisms, and theories.

Maeng's (2020) approach to developing PT learning progressions was based on a psychometric investigation of students' reasoning practices in geocognition. Ordered multiple-choice assessment items were used, in which the item options addressed progressively more sophisticated levels of reasoning practices. Maeng conducted a Rasch analysis to generate estimations of student proficiency and item difficulty on the same scale. According to Rasch's analysis estimation, Table 14.1 presents the hypothetical geocognition LPs which reveals how students' reasoning practices in geocognition improve progressively across four levels. The progress variables for this learning progression framework represent three kinds of reasoning practices: spatial reasoning, temporal reasoning, and systems thinking.

Table 14.1 A hypothetical learning progression of geocognition for understanding plate tectonics (revised from Maeng, 2020)

Level	*Progress variables*		
	Spatial reasoning	*Temporal reasoning*	*Systems thinking*
4 **upper anchor**	Able to identify the types of plate boundaries and plate movement based on a three-dimensional understanding of geological terrains by synthesizing map-view and cross-sectional data.	Able to understand geological time scale, identify temporal order of geological events, and infer the speed of plate movement.	Able to synthesize the information on volcanoes, earthquakes, and geological terrains, and construct explanations on plate movements.
3	Able to identify the type of plate boundaries by synthesizing map-view and cross-sectional data.	Able to recognize the sequential order of geological events and infer the direction of plate movement based on geological time data.	Able to synthesize the information on volcanoes, earthquakes, and geological terrains, and relate them with plate boundaries.
2	Able to match map-view data one-to-one with its cross-sectional view image and apply them to conceiving geological terrains.	Able to recognize sequential order from geological data (e.g., age of rocks) and understand geological time scale.	Able to integrate different geological information and make meaning of the data.
1 **Lower anchor**	Able to perceive the reliefs of oceanic floors and continents.	Able to recognize the temporal differences of given geological data.	Able to identify different geological data separately.

Assessments with and from Learning Progressions in Plate Tectonics

Maeng's (2020) geocognition LP emphasizes how students employ spatial reasoning, temporal reasoning, and systems thinking to understand plate boundaries and movements. The LP focused more on how students think about visual data on plate tectonics than on what they know during their learning process. In contrast, McDonald et al.'s (2019) LP emphasizes students' systematic understanding of plate movement, plate motion mechanism, and plate system conservation. Compared with Maeng's (2020) study, McDonald et al.'s (2019) LP focused more on what students know than how they engage with PT cognitive reasoning. Despite the different approaches, results from the two LP frameworks reported a strong use of systems thinking and a fully systemic understanding of plate movement and plate system as the upper anchor of the LPs. Therefore, reaching the highest level of the two LPs, students use a higher level of systems thinking to synthesize complex information about earthquakes, volcanoes, and movements including mantle convection and the interaction of global plate systems with system conservation. Given this end goal, however, additional information is needed for more rigorous assessments and measures of complex reasoning practices to better understand the connections between the assessment and the learning of systematic mechanisms of plate movement and the Earth's internal plate system. Considering the disciplinary uniqueness of geoscience and geographical science data, research on learning progressions in geoscience requires novel designs for the development of appropriate assessments. Such assessments are needed to guide the evaluation and description of (1) students' geoscientific reasoning practices (e.g., spatial reasoning/systems thinking) and (2) students' conceptual learning and understanding within geosciences LPs. Reflecting on the data-driven theory model for PT (Oreskes, 2003) as the context for developing the theoretical components of Plate Tectonics, an important factor for designing and guiding PT LPs research is the use of appropriate geoscientific data on plate movements and its impact on surface phenomena.

Technological support on data visualization in geoscience and geography provides various applications, simulations, and visualizing tools. While only focusing on surface phenomena of the Earth, geological terrains, earthquakes, and volcanoes, Maeng (2020) implemented data visualization on plate movement with GeoMapApp (Ryan et al., 2009) and provided hypothetical descriptions of students' progressive use of geoscience reasoning practices. For more concrete visual data about the Earth's interior, see Epler-Ruths et al. (2020) who used Seismic Explorer to examine students' spatial skills and their understanding of plate tectonics. Examples of data visualization employing GeoMapApp or Seismic Explorer to examine students' understanding of geoscientific reasoning practices appear in the figures as follows.

When students learn about plate movement and its related geoscientific phenomena, they need to interpret and analyze geoscience and geographic map data using geoscientific reasoning practices. Spatial reasoning in geoscience includes recognizing a shape or pattern from complicated map data, ascribing geoscientific meaning to the shape or pattern, and then synthesizing two-dimensional observations into a three-dimensional image (Kastens & Ishikawa, 2006). Figures 14.2 and 14.3 represent an exemplary case using spatial reasoning for understanding plate movement. Here, spatial reasoning focuses on analyzing and interpreting map-view data along with cross-sectional view data of geological terrains. In this case, the data is taken at the convergent boundary between the Nazca plate and the South American plate. In Figure 14.2, the reliefs of landforms and sea-floor basins are shown with differently graded colors. Figure 14.2 also provides the distribution of earthquakes and volcanoes. Earthquakes are shown with different sizes and colored circles which represent the magnitudes and focal depths of earthquakes, respectively. Figure 14.3 is a cross-sectional

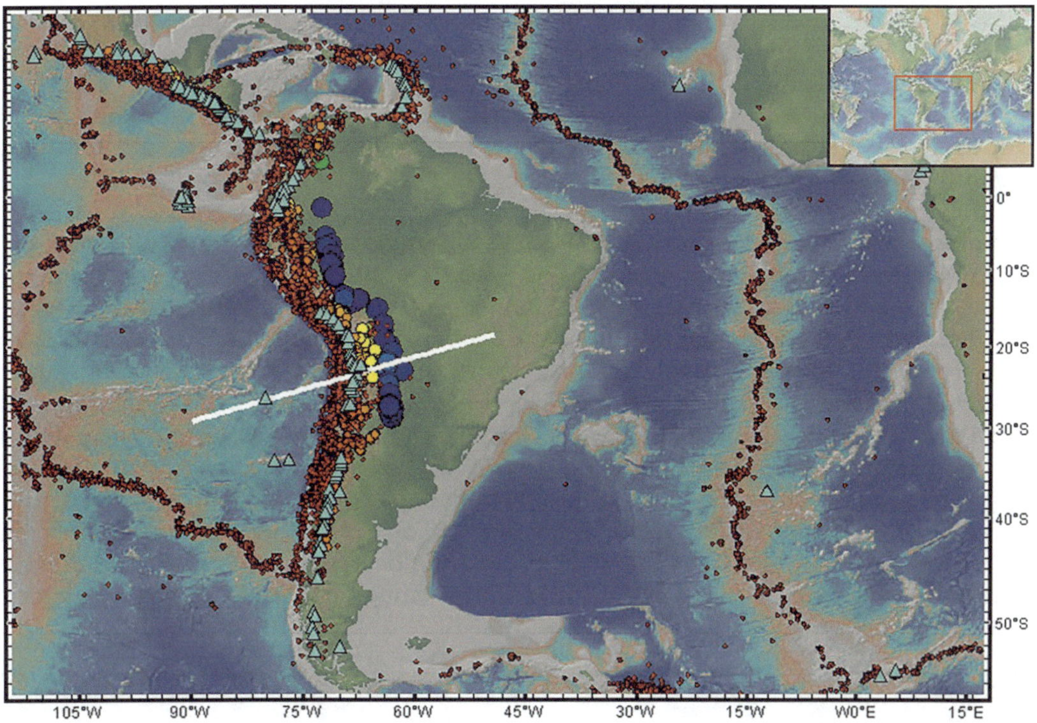

Figure 14.2 The locations of earthquakes and volcanoes along the South American-Nazca plate boundary, depicted on a topographic map (captured from GeoMapApp), Triangles represent volcano locations. Circles indicate earthquakes with red being the shallowest, increasing the depth of focus (hypocenter) along with orange, yellow, green, blue, and purple. The size of the circle indicates the strength of the earthquakes, with larger circles indicating stronger earthquakes.

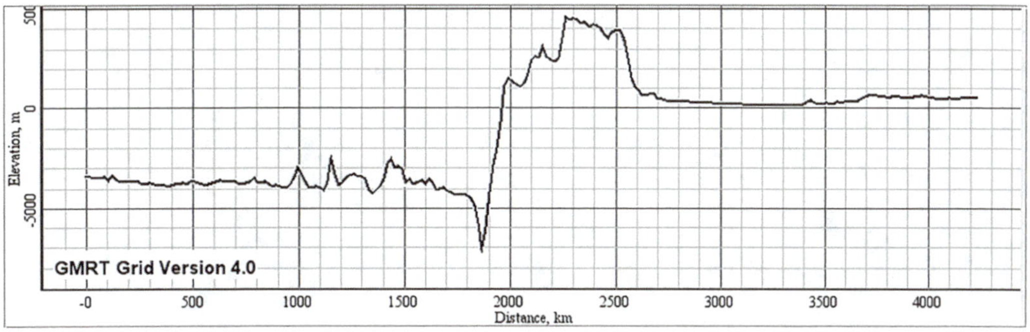

Figure 14.3 A cross-sectional profile of the area along the white line of Figure 14.2 (captured from GeoMapApp).

profile of the area cut along the white line in Figure 14.2. Both figures are obtained by manipulating GeoMapApp.

When students observe map data in Figure 14.2 and think about characteristic distributions of earthquakes and volcanoes along with specific terrains, they need to perceive the reliefs of landforms and sea-floor configurations from the different color gradations depicted on the topographic map. Students use spatial reasoning to perceive geomorphological patterns in the eastern Pacific Ocean and South America from the map (Figure 14.2), and then mentally create a cross-sectional image based on their perception of geomorphological patterns along the white line and compare their mental image with the cross-section diagram drawn from GeoMapApp (Figure 14.3).

Conceiving plate movement requires learners' perception of the vast scale of deep-time spanning such geological events, as well as temporal arrangements of those events in a logically ordered way based on relative as opposed to absolute dating principles (Dodick & Orion, 2006). In Figure 14.4, color gradients depict the age distribution of the bedrock (basalt) in the Atlantic Ocean in millions of years (MY). Figure 14.5 provides a cross-sectional profile of the bedrock age distribution of the area along the white line in Figure 14.4. Here, too, both figures are obtained by manipulating GeoMapApp. Students face a geocognition challenge when they (1) observe and interpret geoscientific map data as shown in Figures 14.4 and 14.5, then (2) compare the bedrock ages at each point along the white line in Figure 14.4, and (3) infer the direction of plate movement around the Atlantic Ocean ridge based on the symmetrical distribution of bedrock age. The challenge is to coordinate and grasp the vast time scale of geoscientific events and arrange the order of bedrock age distribution according to the relative dating of different geological times in each bedrock point.

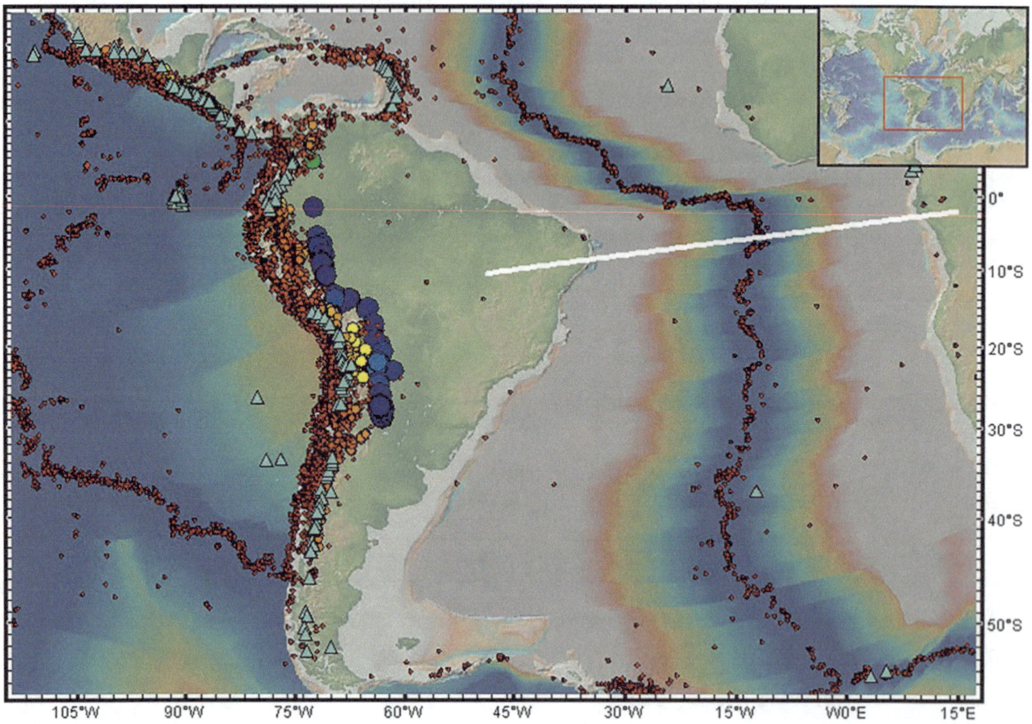

Figure 14.4 Bedrock age distribution in the Atlantic Ocean with earthquakes (captured from GeoMapApp). In the area of the Atlantic Ocean colors represent bedrock age with blue being the youngest, increasing in age with green, yellow, pink, and light pink. This map also uses the same representation of earthquakes and volcanoes as Figure 14.2.

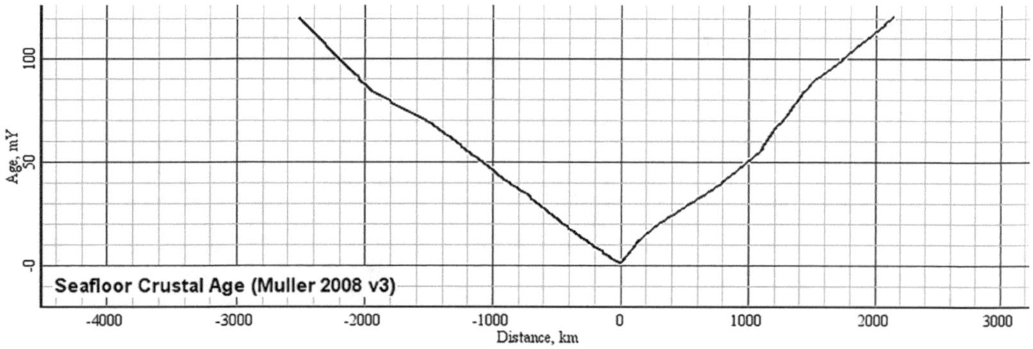

Figure 14.5 Cross-sectional profile of the bedrock age distribution along the white line in Figure 14.4 (captured from GeoMapApp).

While spatial and temporal reasoning within geological time contexts can be addressed through Figures 14.2 to 14.5 above, systems thinking also needs to be considered for students to understand plate movement based on synthesizing the geoscience data and data structures in those same figures. To infer the oblique subducting movement of the Nazca plate below the South American plate from Figures 14.2 and 14.3, and to construe diverging movement of the South American plate and the African plate from Figures 14.4 and 14.5, students need to engage in an information synthesis of data of five elements: geological terrains, locations of earthquakes and volcanoes, variations of focal depths of earthquakes, age distributions of bedrocks in the ocean basins, and long period of bedrock age distributions. Students' next step is to use the synthesized information, relate it to plate boundaries, and then construct an explanatory model "big picture" of plate movement near the western part of South America and the Atlantic Ocean.

Seismic Explorer tools provide overlays for placing earthquakes, volcanoes, and tectonic plates data onto an Earth map. Seismic Explorer also provides a three-dimensional piece of the Earth cut out of the map to show the distribution of earthquake focus points below the surface. Figure 14.6 is an example of graphic data obtained from Seismic Explorer, here showing the distribution of earthquakes and volcanoes near South America, Africa, and the Atlantic Ocean. When observing the figure, students see the location of earthquakes and volcanoes, the array of earthquakes, and must use spatial reasoning to comprehend the height and depth of land and ocean basins based on the reliefs of color images. Viewing Figure 14.7, in which the focal points of earthquakes are depicted three-dimensionally along with the sliced white box out of Figure 14.6 between the points P1 and P2, students can perceive three-dimensional spatial information and envisage various images of the depth and distribution of earthquakes by scrolling the diagram in every direction and comparing the top-view and side-views of this box area. It is another kind of geocognition involving spatial reasoning and mental rotation, which can be achieved and improved using Seismic Explorer. Synthesizing the graphic data, students use systems thinking to compare the patterns of earthquake and volcano distributions with patterns of geomorphological features, and then construct an explanation or graphical representation for plate movement at the boundary between the Nazca Plate and the South American Plate.

Regarding the hypothetical LPs for plate tectonics, both Maeng (2020) and Epler-Ruths et al. (2020) recognize the importance of working and reasoning with graphic data. In their respective studies, visual data obtained from GeoMapApp and Seismic Explorer were helpful for students to construct explanatory accounts of how surface phenomena are related to

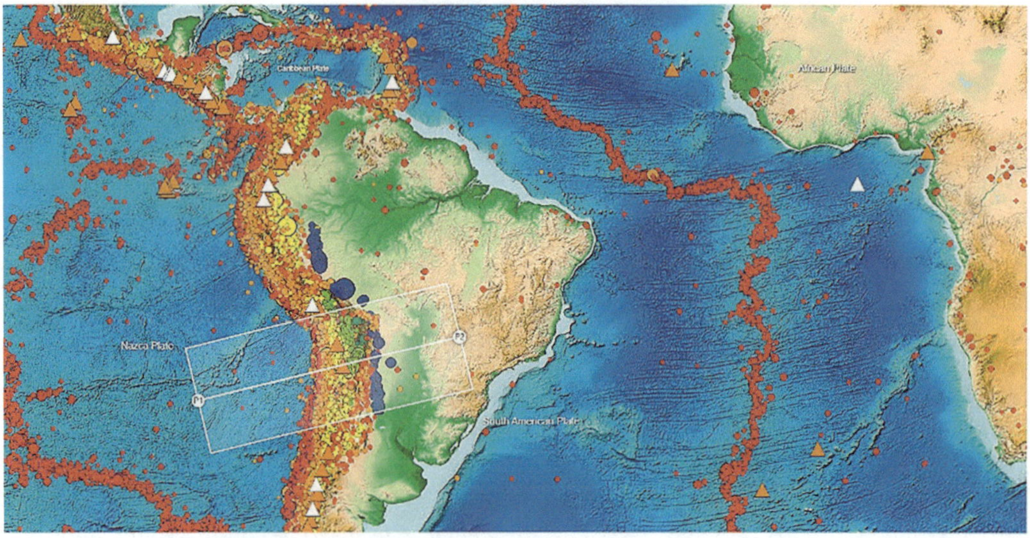

Figure 14.6 An example of graphic data including earthquakes and volcanoes obtained from Seismic Explorer. Triangles represent volcano locations. The color of the circles indicates the focus depth of the earthquakes, and the size of the circle indicates the strength of the earthquake, following the same representation as Figure 14.2.

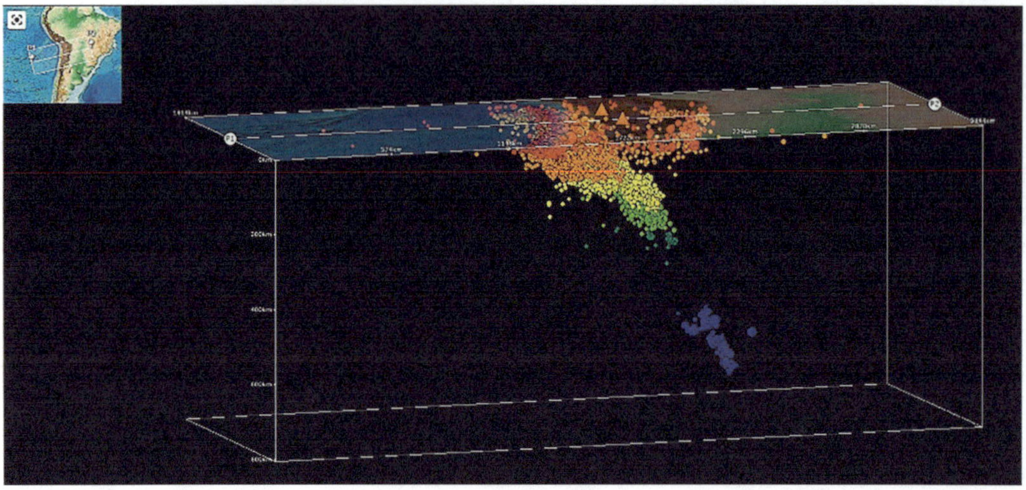

Figure 14.7 An example of a 3D cross-sectional piece along the box out of Figure 14.3 (obtained from Seismic Explorer). Triangles represent volcano locations. The color of the circles indicates the focus depth of the earthquakes and the size of the circle indicates the strength of the earthquake, following the same representation as Figure 14.2.

below-surface dynamics of plate movements. For both research studies, the data visualization was transformed into data narratives (Radinsky, 2020), from which students then described and interpreted the data structures that informed their ways of reasoning. Therefore, one can infer that the exchange or mobility of students' narratives of graphic data on plate movement and its surface phenomena can be used as evidence regarding learners' understanding of mechanics and Earth systems that drive Plate Tectonic movements. Thus, the students' narratives and reasoning can be used to establish progress variables and upper anchors of hypothetical Plate Tectonic LPs. This process is guided by and related to the data-driven theory-building cases on Plate Tectonics by geoscience experts and scholars of the history of geosciences (e.g., Oreskes, 2003).

The geosciences and geographical sciences in school science education programs continue to take a back seat to the other science disciplines (King et al., 2021; Mayer, 2002; Orion, 2019). Science education in the years surrounding the development of the Theory of Plate Tectonics, favored the biological and physical sciences (Dodick & Orion, 2003). With the onset of numerous global crises that are nested in ESS contexts (e.g., climate change, pandemics, rising sea levels, melting glaciers/sea ice, deforestation), new frameworks and trajectories are needed to support reasoning in complex adaptive systems. Learners' engagement with the models, mechanisms, and databases of PT Theory as shown above has the potential to positively influence the development of systems thinking and reasoning.

Trends in Earth Science Education

We argue that taking an interdisciplinary or transdisciplinary approach with both natural and social science perspectives helps students to better understand Earth Science and Earth System Science (ESS). Over the last 50+ years, changes in geosciences and geographical sciences have been dramatic. The reconceptualization of Earth Science as ESS, as noted above, is a relatively recent development and coincides with the development of technologies. For example, the access to new inquiry methods and new forms of data and data representations that Geographical Information Systems (GIS) and Earth satellites provide is contributing new evidence to deepen our understanding of the planet's geophysics and the mechanics of Earth systems. Early on, the pursuit of geographical and geoscience knowledge in the 19th and 20th centuries was more closely aligned with Earth's surface geomorphic features used to address problems of transportation, commerce, and industry.

The core knowledge and core practices in Earth System Science and Geography curricula provide complementary and unique affordances for the coupled design of science, engineering, and social science education learning progressions. The geo/geographical sciences offer up natural, social, environmental, and historical contexts that can facilitate the development of spatial reasoning, data-driven theory development, model-based reasoning, and mechanistic causal frameworks that, in turn, can facilitate comprehension of deep-time and deep-space contexts. The *Learning to Think Spatially* NRC (2006) synthesis research study report is a valuable resource for tackling the current focus on Place-based and Phenomenon-Based orientations for STEM education. Herein, we describe three major trends in Earth Science education that warrant further investigation by the field. First, we discuss the use of crosscutting concepts in Earth Sciences and their implications for refocusing on Earth Sciences from a complex adaptive systems approach. Next, we discuss the implications for learning progression research in designing for future Earth Science learning. Finally, we discuss the use of GIS as a tool to support students' Earth Science learning.

Earth Sciences Education and Crosscutting Concepts

Current views in science education tout the importance of leveraging transdisciplinary approaches to science education, and this is especially true for geosciences. Moving forward, the development of learning progressions in the geo/geographical sciences should consider the value of transdisciplinary formats to solve complex, systemic problems grounded in locally relevant phenomena.

Another helpful report for aiding the design of ESS and Geo/Geographical Sciences learning environments is the *Geography Road Map Report* (Bednarz et al., 2013). The *Road Map* report identifies several ways the geographical sciences may make unique contributions to K-16 education. One aspect, outlined above, focuses on the tools and technologies embedded in various GIS platforms. Another aspect is the affordances the study of geographically based complex adaptive systems can provide in developing system thinking and conceptual reasoning. Yet another is how the geo/geographical sciences can affect STEM education goals in the realm of scientific observations, measurements, and representations. The Geo/Geographical sciences offer up compelling contexts for engagement with complex adaptive systems, core disciplinary concepts, and investigative and epistemic practices – modeling, argumentation, and explanation.

Complex adaptive systems that are embedded within ESS, for example, represent what some geographers see as crosscutting concepts (CcC) that unite the perspectives of different subfields of geography (Massey, 1999). Among the seven CcC found in the *Framework* (NRC, 2012), four are particularly suited to highlight geo/geographic practices:

Patterns

Observed patterns of forms and events guide organization and classification, and they prompt questions about relationships and the factors that influence them.

Scale, Proportion, and Quantity

In considering phenomena, it is critical to recognize what is relevant at different measures of size, time, and energy and to recognize how changes in scale, proportion, or quantity affect a system's structure or performance.

Systems and System Models

Defining the system under study – specifying its boundaries and making explicit a model of that system – provides tools for understanding and testing ideas that are applicable throughout science and engineering.

Energy and Matter: Flows, Cycles, and Conservation

Tracking fluxes of energy and matter into, out of, and within systems helps one understand the systems' possibilities and limitations.

In geography there are likely many such similar concepts; however, with respect to Geo/Geographical sciences, examining ***complex adaptive systems*** provides affordances for understanding and modeling complexity which are central to understanding human-environment interactions and how these are shaping Earth's surface.

Implications for Earth Science Learning Progressions

For developing PT learning progressions, our focus is on a set of conceptual understandings and cognitive/epistemic practices that (1) will lead to more robust understandings of Geo/

Geographical phenomena and (2) address different types and mechanisms of complex adaptive systems within ESS. Effective LP contexts need to be *motivating*, to encourage sustained engagement with a particular phenomenon (Blumenfeld et al., 1991). LP contexts are intended to be *generative*, in that by engaging with them, learners can develop an understanding of complex systems that prepare them for future problem solving (Bransford & Schwartz, 1999). Finally, the LP contexts are intended to be *representative* of the broad range of phenomena that the Geo/Geographical sciences and Earth Systems Sciences (ESS) examine.

Learning progressions can be designed to focus on the study of ESS that target the three levels of human agency – Individual Agents, Community Agents, and Earth System Agents. The context is to engage learners in the development of Human and Planetary Histories (HPH). ESS contexts for K-5 might target the HPH of local places, e.g., school yards, neighborhoods, and communities. Grades 6–8 might examine ESS contexts and HPH on broader scales, e.g., State, Regional, and National contexts such as water and transportation infrastructures, agricultural practices, as well as shoreline development and erosion. ESS contexts for upper secondary grades 9–12, and post-secondary STEM education courses, might take up global Earth System and HPH contexts, e.g., oceans, urban and rural communities, tropical/temperate rainforests, indigenous people, continents, habitats, polar and equatorial zones.

Motivating geoscience contexts might include providing K-8 students with local, community, or regionally grounded phenomena that directly impact their daily lives. For example, students living along the San Andres fault in California may be more motivated to understand the interactions of plates in a system as they have experienced earthquakes that cause damage to their local community. Watersheds of major rivers and flooding are another viable context for examining ESS. An equally important way to motivate students is to bring their culture and funds of knowledge to the classroom as valuable assets. Megan Bang, Ananda Marin, and colleagues have worked with Indigenous peoples to rethink what science education might look like in partnership with local communities (Bang et al., 2014) as a means of dismantling the settled expectations of nature-culture relations (Bang & Marin, 2015). Specifically, within the geosciences, Riggs (2005) determined successful Indigenous geoscience courses were place-based and problem-based and emphasized outdoor contexts. Motivation may involve more than just content but also instructional practices.

Young people are likely to need supports to engage with ESS complex adaptive systems in the form of rich tasks, good instructional materials, and effective teaching. It is much easier for young learners to apprehend structures than the relationships among structures, functions, and behaviors in systems (Hmelo-Silver et al., 2007). Further, students often reason about complex systems as if the systems operated through central control rather than self-organization, and they have difficulty understanding how small actions can have large effects on systems (Jacobson, 2001). It is especially difficult for students to be able to think in terms of different levels of systems, an important ability for being able to reason about emergent phenomena (Wilensky & Resnick, 1999). The same sort of difficulties exist with students thinking across scales (Cheek et al., 2017) and spatial contexts (Uttal & Cohen, 2012). A promising but still developing line of research is exploring how providing students with direct interactions with complex systems and opportunities to model systems can help develop student understanding (see Yoon et al., 2018 for a review of this literature).

Geographic Information Systems (GIS) Contexts for Earth System Science

Learning with GIS and other modeling tools is critically important for ESS LPs. As a learning resource, GIS helps students appreciate maps as representations of data and as tools for constructing geographic accounts of phenomena (Lund & Sinton, 2007). Because GIS allows students to manipulate data and visualizations, they gain a potentially deeper insight into geographic phenomena (Medyckyj-Scott, 1994). Despite their central importance, the educational value of GIS for teachers and students has not been the subject of much rigorous research (cf., Kerski, 2003). GIS tools are centrally important to Civil and Environmental Engineering domains. In addition to GIS tools, agent-based modeling tools deserve more research in education because they are especially designed to help students gain a better understanding of complex adaptive systems (Levy & Wilensky, 2005). At the same time, the ubiquity and importance of tools like GIS for doing geography makes the question of *whether* such tools make a difference less relevant for advancing understanding of teaching and learning. What is most needed is research on *how* and *under what conditions* such tools can be effective (Means & Penuel, 2005).

Spatial concepts and analysis are central to learning in ESS courses, but also in other sciences, mathematics, social studies, and history. ESS courses and modules often incorporate activities that require students to interpret and use maps and models of the Earth's surface (Rapp et al., 2007). Students in history classes can use GIS tools to learn to use maps as a resource for historical reasoning (Knowles, 2002; Radinsky, 2008). Students who engaged in geoscience learning contexts that incorporated spatial skills such as map reading often found this impacted their learning outside of the geoscience course (Gold et al., 2018). Studying learning when engaging with maps and GIS tools across different content areas can help researchers develop an understanding of how learning of spatial concepts and analysis skills varies by disciplinary domains and contexts.

Learning that involves working with data visualizations and engaging in data analytics are other core practices used when reasoning, modeling, and problem solving in ESS contexts. Computer-based modeling tools allow students to see and interact with data on large spatial and temporal scales usually unachievable in observations. Though GIS is one such tool, others, such as Seismic Explorer (the Concord Consortium, seismic-explorer.concord.org) enable students to manipulate data and engage in argumentation practices when scaffolded with appropriate instruction (McDonald et al., 2020). The Seismic Explorer tool allows students to manipulate representations of data from earthquakes and volcanic eruptions onto topographic map views and three-dimensional, subsurface views of Earth (Pallant et al., 2020). Another visualizing tool, GeoMapApp (www.geomapapp.org; Ryan et al., 2009) supports constructing graphic data for teaching and learning geoscience topics. GeoMapApp provides direct access to the Global Multi-Resolution Topography (www.gmrt.org). GeoMapApp also allows students to depict data visualizing maps to stack various geological and geophysical information such as earthquakes, volcanoes, sediment thicknesses, or the age of bedrocks onto the GMRT maps. Such tool use not only supports students in examining real-world data and engaging in the epistemic practices of professional scientists but also helps refine and set out learning progression pathways that support students' epistemic agency.

Earth Science Education: A Transdisciplinary Approach to Systems Thinking

The emergence and development of the Theory of Plate Tectonics is a story of international teams of scientists discovering, measuring, and coordinating a set of physical, chemical, biological, geologic, oceanographic, and atmospheric patterns that occurred over hundreds of millions of years on the Earth. The recent 20th-century development of new technologies and tools made it possible to observe and measure unforeseen Earth dynamics and patterns. The depth-of-focus earthquake dynamics along deep-sea trenches, linear locations of earthquakes along mid-ocean ridges, and active volcanoes of the Pacific Ocean Ring-of-Fire are explained by Plate Tectonics.

The development and design of learning progressions and teaching sequences are needed to address the challenging transdisciplinary contexts of developing and teaching Earth Science curricula that target the development of systems thinking, model-based reasoning, and mechanistic thinking. The Earth Sciences and ESS when considered through the pluralistic lenses across the biosphere, hydrosphere, atmosphere, lithosphere, cryosphere, and the human "technosphere" presents accessible and fertile contexts for situating complex reasoning and students' epistemic agency.

We have argued that Plate Tectonics is a viable context for K-12 students to initiate engagements in ESS. The model-based reasoning and the explanations of mechanisms needed to describe and explain Plate Tectonic geological dynamics, provide a platform for which other Earth Systems (e.g., Climate Risk Management, Coastal Sea-level Rise, Flooding/ Droughts) can become a focus for engaging in transdisciplinary approaches to science education. Advances in technology, especially computing power for simulations, data analytics, artificial intelligence, visualizations, and sensors, and in scientific understanding coupled with a growing global sense of responsibility have shifted both the focus and the methods of the Geo/Geographical sciences. The trends for several decades have been toward a systems analysis of the Earth (Mayer, 2002). With respect to inquiry methods, the trend is toward model-based and systems-thinking science education. Models are increasingly being used for exploring and explaining the complex dynamics and structures of the Earth's surface. Systems thinking targets issues of and questions about the habitability and sustainability of Planet Earth. Both are paramount for developing better predictive models that can, in turn, be used to shape policy and enhance the management of resources. As such, learning progressions targeting models and modeling practices in ESS need to be a strong thread in the geoscience and geographical science curricula (Mayer, 2002; Oreskes, 2003). Plate Tectonic Theory is an excellent visual and spatial context for engaging students to initiate the study of patterns and phenomena associated with ESS.

Anthropocene Epoch

Recent scholarly developments, based on the impacts human civilizations are having on the planet, have researchers in both the sciences and the humanities discussing and advocating for a new geological epoch to be added onto the Geologic Time Scale. Verburgt and de Waal (2022) state "the Anthropocene problematizes history as a concept that encompasses past, present, and future in a temporal whole" (p. 367). They go on to state that "the Anthropocene has started to live 'two lives': as a singular scientific or geological notion; and as a plural humanities – and social sciences – driven concept" (p. 371). The upshot is that geology is not being viewed separately from ESS following on from the "Great Acceleration" of Plate Tectonic tools, technology, and thinking. An Anthropocene Working Group (AWG) was set up in the 1950s by the International Commission on Stratigraphy (ICS) to examine the emergence of the "techno-sphere," e.g., technologies developed to monitor and measure human global activities. "[T]he Anthropocene is exploding the spatial and temporal boundaries ... forcing [the] study of technological, societal, and economic developments that eventually enabled humans to change the Earth System" (Verburgt & de Waal, 2022, p. 375).

Simon and Thomas (2022) assert that the Anthropocene "arises not only from geology but also from Earth System Science (ESS)" and together "incorporate both physical systems and human systems in an integrated view of our planet" (p. 396). Human agency then has three forms: Individual, Collective, and Earth System according to Simon and Thomas. The challenge they state is to understand how "human activities coalesced in the mid-twentieth century to become an Earth System agent impacting the Earth" (p. 405). We agree and argue that it is time for the infusion of transdisciplinary "systems thinking" curricula into Middle, High School, and post-secondary education to develop the three forms of human agency among youth.

The emergent Anthropocene perspective also stresses the need for interdisciplinary collaboration. Coen and Jonsson (2022) in their article "Between History and Earth System Science" advocate for an ESS approach that incorporates natural and social scientific data to study global change. Through the lens of geo-epistemology (Rankin, 2016), ESS would be characterized by traits, such as, (i) reliance on remote sensing, (ii) use of computer modeling, (iii) exclusion of humanity from Earth System, and (iv) the assumption that international governance needs to address global-scale environmental problems (Coen & Jonsson, 2022, p. 408). With regards to the importance of pluralism in arriving at scientific consensus, Coen and Jonsson state, "[c]limate change is not one phenomenon but many, carrying different meanings for different people and other living things. Grappling with the ethical as well as epistemic challenges of climate change demands not one way of knowing but many" (p. 421).

The coalition of geoscience and geographic sciences fits both the Anthropocene and science education transdisciplinary agendas well, particularly with respect to a focus on "Complex Adaptive Systems." Many important geographic questions focus on complex phenomena. Studying human-environment interactions, climate change migration and mitigation, and geopolitical conflicts all require Geo/Geographical scientists to consider many different components, actors, and interactions. It is essential for students to engage in discussions about understanding and addressing the impacts of large, rapid changes to the Earth's surface along with developing capacities to model complexity (NRC, 2010).

Complex Adaptive Systems

Complex adaptive systems, not unlike ESS, have several features in common that distinguish them from simpler systems (Holland, 2006; Johnson, 2002; Sawyer, 2005). First, they have large numbers of components. Studying the human-environment interactions that shape Earth's surface requires Earth Scientists and Social Scientists to consider human political, social, cultural, and economic activities, as well as natural processes such as climate, weathering, and erosion. These components and processes also interact in complex adaptive systems and enable researchers and students to adapt and learn while examining them. For example, expanded road capacity via increasing the number of highway lanes leads to increased traffic flows, as people adapt their patterns of activity to the increase in access to different places (Sheppard, 1995). Further, as flows change in a region, students could study their own community's behavior to emergent properties of the system (e.g., increased traffic) by changing modes of transportation, altering work schedules, and the like.

Many concepts in human and physical geography resonate with core ideas about complex adaptive systems (O'Sullivan, 2004). For example, the idea that spatial configurations and patterns emerge from local interactions underlies geographical analyses of a wide range of phenomena, from the formation of industrial clusters in cities (Brenner, 2004; Malmberg et al., 1996) to the spread of infectious diseases (Haggett, 1994a, 1994b; Root et al., 2009). For example, Steven Johnson's 2006 novel *The Ghost Map* provides opportunities to engage students in primary literacy skills, map reading, epidemiology, civil engineering, modeling, and computational reasoning (see Jungck, 2012 for examples). Engaging in complex adaptive systems grounded in real-world problems provides opportunities for students to not only learn content but also the interconnectedness across different contents and contexts. This movement towards an interconnected, systems approach to contextualized science is also presented as a goal for the 2024 PISA Strategic Vision and Direction (OECD, 2020).

Furthermore, many emergent patterns in physical systems that shape Earth's surface are self-organizing; that is, they emerge without any central agent or authority. An example is networks of streams, drainage basins, and deltas, which all have characteristic patterns that are the result of multiple interacting processes (Phillips, 1995). Feedback loops, another core idea in complexity theory, figure in geographers' models that show how residential segregation can emerge, even when individuals prefer to live in integrated neighborhoods (Clark, 2008). These arenas too can be seen as transdisciplinary adventures. Residential segregation can be tied to differential environmental injustices and a lack of educational opportunities (Bateman et al., 2021).

Earth System Science

To date, Learning Progressions on Plate Tectonics have provided a solid start to understanding how students learn within Earth System Science contexts. We are maintaining that Plate Tectonics is a sound platform to build from and foster students' engagements with other Earth Systems. Additional research is needed though on ESS and Geo/Geographical science learning progressions found in other Earth Science domains and within contexts of the Anthropocene Epoch complex adaptive systems contexts.

The progressive approach we are recommending is one that recognizes that ESS and complex adaptive systems function within inherently transdisciplinary real-world contexts, i.e., involving human agency. Efforts to develop a transdisciplinary approach to Earth Science education require a rethinking of several facets of the broader educational system. Although the NGSS and other modern standards documents incorporate Earth Science concepts and interweave systems thinking and modeling practices, such as the GEODE curriculum used in Epler-Ruths et al. (2020), the presence of Earth Sciences and Geography in K-12 classrooms remains minimal, especially for secondary students. Asking for new courses to be woven into the school day, already packed with other equally valuable content, is unlikely to be well received (Johnson, 2013). Rather, the idea of taking an inter/transdisciplinary approach through the lenses of Anthropocene dynamism – melding natural and social sciences with relevant historical contexts of human agency at individual, community, and Earth System levels – would enable Geographic and Geosciences to merge with other academic domains to better facilitate authentic discussions regarding Earth System Science.

Reimagining curricular sequencing as transdisciplinary work, however, requires a shared vision from stakeholders, inclusive of teachers, industry, and community members, not only policymakers or administrators (Johnson, 2012). Research in integrated STEM program implementation can provide researchers and practitioners with starting points to build effective, efficient transdisciplinary programs that support the inclusion of three-dimensional Earth Science in K-12 schooling (see NRC, 2011). A reimagining of the curricular sequencing in ways that move away from "STEM enhanced" designs, in which courses are added to supplement the traditional biology-chemistry-physics sequence, and towards an integrated, transdisciplinary approach to STEM curricula could better infuse science and engineering practices, technology, and computational thinking (Johnson, 2013). Bringing in traditionally non-STEM disciplines such as geography, the arts, and literacy helps us transcend disciplines and move toward real-world problems (Gess, 2017).

References

American Association for the Advancement of Science. (2001). *Designs for science literacy*. Oxford, UK: Oxford University Press.

Bang, M., Curley, L., Kessel, A., Marin, A., Suzumori III, E. S., & Stracke, G. (2014). Muskrat theories, tobacco in the streets, and living Chicago as Indigenous land. *Environmental Education Research*, *20*(1), 37–55.

Bang, M., & Marin, A. (2015). Nature–culture constructs in science learning: Human/non-human agency and intentionality. *Journal of Research in Science Teaching*, *52*(4), 530–544.

Bateman, K. M., Steele, D., & Sexton, C. M. (2021). Sustainability science education: our animalistic response-ability. *Cultural Studies of Science Education*, *16*(3), 841–855.

Bednarz, S. W., Heffron, S., & Huynh, N. T. (Eds.). (2013). *A road map for 21st century geography education: Geography education research (A report from the Geography Education Research Committee of the Road Map for 21st Century Geography Education Project)*. Washington, DC: Association of American Geographers.

Blumenfeld, P. C., Soloway, E., Marx, R. W., Krajcik, J. S., Guzdial, M., & Palincsar, A. (1991). Motivating project-based learning: Sustaining the doing, supporting the learning. *Educational psychologist*, *26*(3-4), 369–398.

Brenner, N. (2004). *New state spaces: Urban governance and the rescaling of statehood*. Oxford, UK: Oxford University Press.

Bransford, J. D., & Schwartz, D. L. (1999). Rethinking transfer: A simple proposal with multiple implications. *Review of Research in Education*, *24*(1), 61–100.

Cheek, K. A., LaDue, N. D., & Shipley, T. F. (2017). Learning about spatial and temporal scale: Current research, psychological processes, and classroom implications. *Journal of Geoscience Education*, *65*(4), 455–472.

Clark, W. A. V. (2008). Geography, space, and science: Perspective from studies of migration and geographical sorting. *Geographical Analysis*, *40*(3), 258–275.

Coen, D. R., & Jonsson, F. A. (2022). Between history and Earth system science. *ISIS*, 113(2), 407–416.

Corcoran, T., Mosher, F. A., & Rogat, A. (2009). *Learning progressions in science: An evidence-based approach to reform, CPRE Research Report #RR-63*.

Dodick, J., & Orion, N. (2006). Building on understanding of geologic time: A cognitive synthesis of the "macro" and "micro" scales of time. In C. A. Manduca & D. W. Mogk (Eds.), *Earth and Mind: How geologists think and learn about the Earth* (pp. 77–93). Boulder: Geological Society of America.

Dodick, J., & Orion, N. (2003). Geology as a historical science: Its perception within science and the education system. *Science & Education*, *12*(2), 197–211.

Duncan, R. G., & Hmelo-Silver, C. (2009). Learning progressions: Aligning curriculum, instruction, and assessment. *Journal of Research in Science Teaching*, *46*, 606–609.

Duschl, R., Avraamidou, L., & Azevedo, N. H. (2021). Data-texts in the sciences. *Science & Education*, *30*(5), 1159–1181.

Duschl, R., Maeng, S., & Sezen, A. (2011). Learning progressions and teaching sequences: A review and analysis. *Studies in Science Education*, *47*(2), 123–182.

Earth Science Literacy Initative (ESLI). (2010). Earth science literacy principles. Retrieved from http://www.earthscienceliteracy.org/

Epler-Ruths, C. M., McDonald, S., Pallant, A., & Lee, H.-S. (2020). Focus on the notice: Evidence of spatial skills' effect on middle school learning from a computer simulation. *Cognitive Research: Principles and Implications*, *5*, 61.

Gess, A. H. (2017). Steam education: separating fact from fiction. *Technology and Engineering Teacher*, *77*(3), 39–41.

Gold, A. U., Pendergast, P. M., Ormand, C. J., Budd, D. A., Stempien, J. A., Mueller, K. J., & Kravitz, K. A. (2018). Spatial skills in undergraduate students—Influence of gender, motivation, academic training, and childhood play. *Geosphere*, *14*(2), 668–683. 10.1130/GES01494.1

Haggett, P. (1994a). Geographical aspects of the emergence of infectious diseases. *Geografiska Annaler: Series B, Human Geography*, *76*(2), 91–104.

Haggett, P. (1994b). Prediction and predictability in geographical systems. *Transactions of the Institute of British Geographers*, *19*(1), 6–20.

Hmelo-Silver, C. E., Marathe, S., & Liu, L. (2007). Fish swim, rocks sit, and lungs breathe: Expert-novice understanding of complex systems. *The Journal of the Learning Sciences*, *16*(3), 307–331.

Holland, J. H. (2006). Studying complex adaptive systems. *Journal of Systems Science and Complexity*, *19*(1), 1–8.

Jacobson, M. J. (2001). Problem solving, cognition, and complex systems: Differences between experts and novices. *Complexity*, *6*(2), 1–9.

Johnson, C. C. (2012). Implementation of STEM education policy: Challenges, progress, and lessons learned. *School Science and Mathematics*, 112(1), 45–55.

Johnson, C. C. (2013). Conceptualizing integrated STEM education. *School Science and Mathematics*, *113*(8), 367–368.

Johnson, S. (2002). *Emergence: The connected lives of ants, brains, cities and software*. New York: Scribner.

Johnson, S. (2006). *The ghost map: The story of London's most terrifying epidemic--and how it changed science, cities, and the modern world*. New York: Penguin.

Jungck, J. R. (2012). Incorporating quantitative reasoning in common core courses: mathematics for the ghost map. *Numeracy*, *5*(1), 7.

Kastens, K. A., & Ishikawa, T. (2006). Spatial thinking in the geosciences and cognitive sciences: A cross disciplinary look at the intersection of the two fields. In C. A. Manduca & D. W. Mogk (Eds.), *Earth and mind: How geologists think and learn about the Earth*. Special Paper 413 presented at the meeting of the Geological Society of America, Denver, CO (pp. 53–76).

Kastens, K. A., Manduca, C. A., Cervato, C., Frodeman, R., Goodwin, C., Liben, L. S., Mogk, D. W., Spangler, T. C., Stillings, N. A., & Titus, S. (2009). How geoscientists think and learn. *Eos, Transactions American Geophysical Union*, *90*(31), 265–266.

Kerski, J. J. (2003). The implementation and effectiveness of geographic information systems technology and methods in secondary education. *Journal of Geography*, *102*(3), 128–137.

King, C., Gorfinkiel, D., & Frick, M. (2021). International comparisons of school-level geoscience education – the UNESCO/IGEO expert opinion survey. *International Journal of Science Education*, *43*(1), 56–78.

Knowles, A. K. (Ed.). (2002). *Past time, past place: GIS for history*. Redlands, CA: ESRI Press.

Lehrer, R., & Schauble, L. (2004). Modeling natural variation through distribution. *American Educational Research Journal*, 41(3), 635–679.

Lemke, J. L. (1990). *Talking science: Language, learning, and values*. Norwood, NJ: Ablex Publishing Corporation, 355 Chestnut Street, Norwood, NJ 07648 (hardback: ISBN-0-89391-565-3; paperback: ISBN-0-89391-566-1).

Levy, S. T., & Wilensky, U. (2005). *Student patterns in exploring NetLogo models embedded in the Connected Chemistry curriculum*. Paper presented at the Annual Meeting of the American Educational Research Association. Montreal, CA.

Lund, J. J., & Sinton, D. S. (2007). Critical and creative visual thinking. *Understanding place: GIS mapping across the curriculum*, 1–18.

Maeng, S. (2020). Investigating a learning progression for reasoning practices of geocognition using GeoMapApp-Based Assessment. *Asia-Pacific Science Education*, 6(2), 584–614.

Mayer, V., (Ed.) (2002). *Global science literacy*. Dordrecht, NL: Kluwer Academic Publishers.

Malmberg, A., Sölvell, Ö., & Zander, I. (1996). Spatial clustering, local accumulation of knowledge and firm competitiveness. *Geografiska Annaler: Series B, Human Geography*, *78*(2), 85–97.

Massey, D. (1999). Space-time, 'science' and the relationship between physical and human geography. *Transactions of the Institute of British Geographers*, *24*(3), 261–276.

McDonald, S., Bateman, K., Gall, H., Tanis-Ozcelik, A., Webb, A., & Furman, T. (2019). Mapping the increasing sophistication of students' understandings of plate tectonics: A learning progressions approach. *Journal of Geoscience Education*, *67*(1), 83–96.

McDonald, S., Wray, K., McCausland, J., Bateman, K., Pallant, A., & Lee, H.-S. (2020). Taking up the mantle of knowing: Supporting student engagement in progressive scientific discourse in geoscience. In M. Gresalfi & I. S. Horn (Eds.), *The interdisciplinarity of the learning sciences, 14th international conference of the learning sciences (ICLS) 2020*, Vol. 1 (pp. 565–568). Nashville, TN: International Society of the Learning Sciences.

Medyckyj-Scott, D. (1994). Visualization and human-computer interaction in GIS. In H. M. Hearnshaw & D. Unwin (Eds.), *Visualization in geographical information systems* (pp. 200–211). New York, NY: Wiley.

Means, B., & Penuel, W. R. (2005). Research to support scaling up technology-based educational innovations. In C. Dede, J. P. Honan, & L. C. Peters (Eds.), *Scaling up success: Lessons from technology-based educational improvement* (pp. 176–197). San Francisco, CA: Jossey-Bass.

Miller, E., Manz, E., Russ, R., Stroupe, D., & Berland, L. (2018). Addressing the epistemic elephant in the room: Epistemic agency and the Next Generation Science Standards. *Journal of Research in Science Teaching*, *55*(7), 1053–1075.

National Research Council. (2006). *Learning to think spatially: GIS as a support system in the K-12 curriculum*. Washington, DC: The National Academies Press.

National Research Council (2010). *Understanding the changing planet: Strategic directions for the geographical sciences*. Washington, DC: The National Academies Press.

National Research Council. (2011). *Successful K-12 STEM education: Identifying effective approaches in science, technology, engineering, and mathematics*. Washington, DC: National Academies Press.

National Research Council. (2012). *A framework for K-12 science education: Practices, crosscutting concepts, and core ideas*. Washington, DC: The National Academies Press.

NGSS Lead States. (2013). *Next generation science standards: For states, by states*. Washington, DC: The National Academies Press.

OECD. (2020). *PISA 2024 strategic vision and direction for science: A vision for what young people should know about science and be able to do with science in the future*. Paris: OECD Publishing.

Oreskes, N. (Ed.) (2003). *Plate Tectonics: The insider's history of the modern theory of the Earth*, with Homer Le Grand. Boulder, CO: Westview Press.

Orion, N. (2019). The future challenge of Earth science education research. *Disciplinary and Interdisciplinary Science Education Research*, *1*, 1–8.

O'Sullivan, D. (2004). Complexity science and human geography. *Transactions of the Institute of British Geographers*, *29*(3), 282–295.

Pallant, A., McDonald, S., & Lee, H.-S. (2020). Shifting plates, shifting minds: Plate tectonics models designed for classrooms. *The Earth Scientist*, *36*(1), 40–46.

Phillips, J. D. (1995). Self-organization and landscape evolution, *Progress in Physical Geography*, *19*(3), 309–321.

Radinsky, J. (2008). A GIS learning environment to teach historical understanding. In A. J. Milson & M. Alibrandi (Eds.), *Digital geography: Geospatial technologies in the social studies classroom* (pp. 99–117). New York, NY: IAP.

Radinsky, J. (2020). Mobilities of data narratives. *Cognition and Instruction*, *38*(3), 374–406.

Rankin, W. (2016). *After the map: Cartography, navigation, and the transformation of territory in the twentieth century*. Chicago, IL: University of Chicago Press.

Rapp, D. N., Culpepper, S. A., Kirkby, K., & Morin, P. (2007). Fostering students' comprehension of topographic maps. *Journal of Geoscience Education*, *55*(1), 5–16.

Riggs, E. M. (2005). Field-based education and indigenous knowledge: Essential components of geoscience education for Native American communities. *Science Education*, *89*(2), 296–313.

Root, E. D., Meyer, R. E., & Emch, M. E. (2009). Evidence of localized clustering of gastroschisis births in North Carolina, 1999–2004. *Social Science & Medicine*, *68*(8), 1361–1367.

Ryan, W. B. F., Carbotte, S. M., Coplan, J. O., O'Hara, S., Melkonian, A., Arko, R., Weissel, R. A., Ferrini, V., Goodwillie, A., Nitsche, F., Bonczkowski, J., & Zemsky, R. (2009). Global multi-resolution topography synthesis. *Geochemistry, Geophysics, Geosystems*, *10*(3), 1–9.

Sawyer, D. S., Henning, A. T., Shipp, S., & Dunbar, R. W. (2005). A data rich exercise for discovering plate boundary processes. *Journal of Geoscience Education*, *53*(1), 65–74.

Sawyer, R. K. (2005). *Social emergence: Societies as complex systems*. Cambridge, UK: Cambridge University Press.

Sheppard, E. (1995). Modeling and predicting aggregate flows. In S. Hanson (Ed.), *The geography of urban transportation* (pp. 100–128). New York: Guilford Press.

Simon, Z. B., & Thomas, J. A. (2022). Earth system science, Anthropocene historiography, and three forms of human agency. *ISIS*, *113*(2), 396–406.

Stroupe, D. (2014). Examining classroom science practice communities: How teachers and students negotiate epistemic agency and learn science-as-practice. *Science Education*, *98*(3). 10.1002/sce.21112

Stroupe, D., Michaels, S., & Moon, J. (2019). Epistemic tools in science education. *Science Education*, *103*(4), 948–951.

Uttal, D. H., & Cohen, C. A. (2012). Spatial thinking and STEM education: When, why, and how? In *Psychology of learning and motivation* (Vol. 57, pp. 147–181). San Diego, CA: Academic Press.

Verburgt, L. M., & de Waal, E. (2022). Introduction: Rethinking history of science in the Anthropocene. *ISIS*, *113*(2), 366–376.

Wilensky, U., & Resnick, M. (1999). Thinking in levels: A dynamic systems approach to making sense of the world. *Journal of Science Education and Technology*, *8*(1), 3–19.

Wilson, M. (2009). Measuring progressions: Assessment structures underlying a learning progression. *Journal of Research in Science Teaching*, *46*(6), 716–730.

Windschitl, M., Thompson, J., & Braaten, M. (2018). *Ambitious science teaching*. Boston, MA: Harvard Education Press.

Yoon, S. A., Goh, S. E., & Park, M. (2018). Teaching and learning about complex systems in K–12 science education: A review of empirical studies 1995–2015. *Review of Educational Research*, *88*(2), 285–325.

15

SUPPORTING CURRICULUM DEVELOPMENT WITH A LEARNING PROGRESSION FOR MATTER-TRACING INVESTIGATIONS

Emily E. Scott, Kirsten D. Edwards, and Charles (Andy) W. Anderson

Introduction

This chapter and Chapter 16 of this volume both describe learning progressions developed by *Carbon: Transformations in Matter and Energy* (*Carbon TIME*: https://carbontime.create4stem.msu.edu/), a research-practice partnership for middle and high school classrooms focusing on carbon cycling at multiple scales. This project is consistent with the *Framework for K-12 Science Education* (NRC, 2012) and the *Next Generation Science Standards* (NGSS Lead States, 2013), including the framework of three-dimensional science learning.

The learning progression described in this chapter is deeply intertwined with curriculum development. We start with a discussion of how we defined the goals and key elements for our learning progression work. We then discuss the interactive development of teaching units and the learning progression, paying specific attention to how students evaluate claims (i.e., hypotheses), interpret mass change data, and generate questions during investigations of combustion and living systems. We conclude with a discussion of the implications of this work for curriculum development and learning progression research.

We begin by considering a basic question: *what do learning progressions need to do*, especially in supporting curriculum development? We propose an answer based on Chapter 8 of *Taking Science to School* (NRC, 2007), one of the original publications about learning progressions. Learning progressions need to describe a sequence of students' sensemaking performances about scientifically relevant phenomena that "bridge the gap" between valued upper-anchor performances and observed lower-anchor performances, which reveal the ways that learners make sense of phenomena before instruction. For the purposes of curriculum development, we believe that the upper anchor performance, and all the steps in between, describe ways that students make sense of phenomena in an inherently three-dimensional way (i.e., using science practices, disciplinary core ideas, and crosscutting concepts).

In this chapter, we describe the development of a learning progression for students' investigations of carbon-transforming processes in four parts:

 DOI: 10.4324/9781003170785-17

A The upper anchor: we began with our decisions about phenomena and sensemaking performances that we value as important goals for science education—what we label as environmental science literacy with respect to carbon-transforming processes.
B Prior research on students' lower-anchor performances: we briefly review prior research on how students trace matter through physical and chemical changes.
C Our research and curriculum development: we devote most of the chapter to the iterative process, involving both research and curriculum development, through which we describe a progression of students' sensemaking performances that bridge the gap between the lower and upper anchors, as well as how we used that research to design scaffolds for students questions and investigations in *Carbon TIME* units.
D Evidence for the effectiveness of the curriculum. We provide evidence from student assessments for the effectiveness of the revised units.

Defining the Upper Anchor (Environmental Science Literacy)

We begin with our value-based choices about learning goals—the upper anchor for our learning progression work. Our definition of the upper anchor includes (1) a *cluster of related phenomena* that play important roles in students' lives and (2) *valued three-dimensional performances* that students will need as informed citizens.

Choosing a Cluster of Phenomena around Carbon-Transforming Processes

The world provides an infinity of phenomena, but K-12 students have time to study only a few. The environmental science literacy project began in 2004 with a basic question: *which phenomena are so important that all students need to study them?* We chose to focus on phenomena that students needed to understand as informed citizens in our democracy, particularly phenomena associated with socioscientific issues, including climate change.

We wanted students to understand how climate change is driven by imbalances in global carbon cycles leading to the accumulation of CO_2 and other greenhouse gases in the atmosphere. But global carbon cycles are consequences of phenomena that take place at macroscopic scales (such as burning fuels, growth and movement of animals and plants, and decay) which scientists explain using atomic-molecular processes such as combustion, photosynthesis, and cellular respiration. So, our LP focuses on explaining this cluster of phenomena and the relationships among them across multiple scales.

These phenomena fall into two groups, with one group including macroscopic-scale phenomena and the other group including ecosystem- and global-scale phenomena. We have published articles focusing on students' sensemaking about global systems (Covitt et al., 2020; Johnson & Anderson, 2017; Miller & Anderson, 2017). Our focus in this volume is on how students learn about tracing matter and energy through macroscopic systems. The phenomena we observe at the macroscopic scale, including flames, plant and animal growth and movement, and decay, can be explained with models of chemical change at the atomic-molecular scale, and they contribute to carbon cycling and energy flow at ecosystem and global scales, as shown in Table 15.1.

Defining Environmental Science Literacy as Three-Dimensional Performances

The *Carbon TIME* project originated with a general definition of environmental science literacy, which we defined as "preparing students to use scientific knowledge and practices in

Table 15.1 Phenomena included in macroscopic-scale carbon learning progressions

Macroscopic (observable) phenomenon	*Atomic-molecular process*	*Impact on the global carbon cycle*
Flames	Combustion: (Organic materials with CH) + $O_2 \rightarrow CO_2 + H_2O$	Adds C to the atmosphere as CO_2
Animal and plant metabolism/ movement; decomposition	Cellular respiration: $C_6H_{12}O_6 + 6O_2 \rightarrow 6CO_2 + 6H_2O$	Adds C to the atmosphere as CO_2
Plant growth	Photosynthesis: $6CO_2 + 6H_2O \rightarrow C_6H_{12}O_6 + 6O_2$	Removes C from the atmosphere into plant biomass
Animals eating, decomposition	Digestion: large organic molecules broken into smaller organic molecules	No direct effects on C in the atmosphere
Plant, animal, fungus growth	Biosynthesis: small organic molecules linked together in large organic molecules	No direct effects on C in the atmosphere

their decisions about environmental issues" (Anderson et al., 2018, p. 1031). This requires (a) a three-dimensional understanding of environmental systems and how they are affected by human technologies and (b) what Bransford and Schwartz (1999) described as *preparation for future learning:* the ability to learn about and respond to new issues or new data as they arise. The upper anchor of our LP involves three-dimensional performances around a set of science and engineering practices:

- *Asking questions:* preparation for future learning includes the ability to ask productive questions about systems and phenomena that we encounter. For *Carbon TIME* units, we were particularly interested in helping students to ask questions that involved *tracing matter and energy through systems.* Such questions are productive because tracing matter and energy is a key scientific strategy for making sense of phenomena, and is essential for making sense of global carbon cycling and its effects on climate change.
- *Conducting and critiquing investigations:* preparation for future learning also involves conducting investigations of carbon-transforming phenomena and critiquing investigations and arguments from evidence by other people, including investigations and data in media reports. To do this successfully, students must recognize how the crosscutting concept of matter and energy conservation makes it possible to use data about mass changes and the chemical composition of materials to trace matter through systems.
- *Explanation and prediction:* students (a) explain how matter—particularly carbon—and energy move and are changed in flames, animals, plants, and decomposers and (b) predict how changes in matter and energy will affect systems. This requires that students integrate core ideas about carbon-transforming processes with crosscutting concepts when developing their explanations and predictions. These performances have been the focus of much of our earlier learning progression research (e.g., Jin & Anderson, 2012; Mohan, Chen, & Anderson, 2009).

In this chapter, we focus on the development of students' proficiencies in asking questions and in conducting and critiquing investigations. A learning progression focused on explanation is the focus of Chapter 16 of this volume.

Evidence about Lower-Anchor Performances

We now shift from our goals for upper-anchor performances to empirical evidence about lower-anchor performances: the interests, understandings, and practices of middle and high school students. What questions do students ask about flames, animals, plants, and decomposers? How do they conduct investigations and interpret data about these systems? In particular, we are interested in questions and investigations that involve *tracing matter* and energy through these systems.

Developmentally, we see tracing matter as something that young children are naturally curious about, as indicated by their steady improvement on Piagetian conservation tasks (Driver, 1985; Elkind, 1961; Piaget, 1951; see also Anderson, Nordine, & Welch, 2021; Smith et al., 2006). Even without formal instruction, children learn to trace matter through simple physical changes, such as pouring liquids from one container to another and changes of state. For these simple physical phenomena, children learn that the amount of matter does not change, so they are curious about where the matter goes and they investigate how it changes.

But when it comes to more complex systems like flames, animals, plants, and decomposers, learners often shift to foci other than tracing matter (Mohan, Chen, & Anderson, 2009). For these systems, it's not possible to visually trace matter as it changes state or chemical composition. Furthermore, these phenomena provide many other things to be curious about—development, nutrition, energy, health, etc.—important and worthwhile questions to investigate—but they do not lead toward the kinds of matter-tracing performances that we described for our upper anchor above.

To achieve our goal of environmental science literacy we needed a learning progression—and a curriculum—that enabled students to ask, and answer through investigations, questions about how matter and energy in general and carbon-containing materials in particular are transformed by flames, animals, plants, and decomposers. We now turn our attention to the interactive development of the curriculum and the learning progression.

Bridging the Gap through Curriculum Development

Our early learning progression work (e.g., Mohan, Chen, & Anderson, 2009; Jin & Anderson, 2012) provided the foundation for an iterative, multi-year process where we both improved our learning progressions and designed the six *Carbon TIME* units focusing on different sub-clusters of phenomena. Four of those units focus on macroscopic-scale phenomena: *Systems and Scale* (focusing on combustion), *Plants, Animals,* and *Decomposers.* The other two units, *Ecosystems* and *Human Energy Systems*, focus on energy flow and carbon cycling at ecosystem and global scales. Our focus in this chapter is on the macroscopic-scale units and their associated learning progressions.

Unit Instructional Models

We organized each unit around a storyline that engages students in the three groups of science practices described above (i.e., questioning, investigations, and explanation practices). Our definition of environmental science literacy entails scaffolding students' proficiency in all three groups of practices:

1 Students as questioners: each unit begins with a phenomenon and driving question. For example, in the *Animals* unit, the teacher shows a time-lapse video of a child growing over

multiple years, and students consider the driving question, "How does a child grow, move, and function?" A key outcome of this consensus-seeking discussion is a set of student questions about the phenomenon that they will answer during the unit, including questions that involve tracing matter and energy (e.g., How does the child use food to grow?). These questions prepare students for future learning.

2 Students as investigators: in each unit, the initial discussion leads first to a lesson providing students with foundational knowledge (e.g., cells and food molecules), then to an investigation in which students trace matter and energy through systems. During the investigations, students (a) plan an investigation, (b) collect, analyze, and interpret data (e.g., observing mealworms eating potatoes and noting changes in energy, mass, and the concentration of CO_2 in the air using Bromothymol blue or BTB as an indicator), and (c) develop arguments from evidence about how matter and energy are moving and changing. Students reach evidence-based conclusions and record unanswered questions to be addressed later in the unit.
3 Students as explainers: during the explanations part of the unit (discussed in Chapter 16, this volume) students use scientific models to explain and predict observations of phenomena. Each unit concludes with a series of activities where students use molecular models to model the underlying chemical changes (as shown in Table 15.1), and then develop and discuss rigorous scientific explanations that answer the driving question. They then use what they have learned to explain other related phenomena (e.g., how dolphins, mealworms, and salmon move and grow).

We supported this development work by creating and revising two different learning progressions. In this chapter, we focus on students' questioning and investigation practices. In Chapter 16 we discuss the learning progression focused on students' explanation practices. We note that these learning progressions are closely related because they involve the same cluster of phenomena, and because students' performances across different practices use many of the same crosscutting concepts and disciplinary core ideas.

Results of Pilot Work

In our pilot work, we saw that students generated many different questions in the initial phases of the units. Many questions involved the role of different forms of matter (such as food, fuel, and air) and energy in phenomena such as burning ethanol, plant growth, or children's growth and movement. However, questions explicitly about tracing matter and energy (e.g., What happens to food in our bodies after we eat it? Where in our bodies does the CO_2 that we breathe out come from?) were relatively rare.

The investigations in the *Carbon TIME* units extended over several class periods (or several weeks for investigations of plant growth and decay) as students generated hypotheses and questions, carried out the investigations, analyzed data, constructed evidence-based arguments, and identified unanswered questions.

Although students and teachers enjoyed the investigations and students clearly learned from them, we noted that students offered differing interpretations of mass change data. For example, many students noticed *similarities* between mass change data for plants and animals: growing radish plants gained mass while the growing medium (paper towels or hydroponic beads) lost some mass; similarly, growing mealworms gained mass while the potatoes that they ate lost mass. These students often did not pay careful attention to *differences* in the mass change data: the growing plants gained more mass than their growth media lost, while the growing mealworms gained less mass than the potatoes lost.

Since questions about tracing mass are important for students' sensemaking about the phenomena (Where did the extra mass that the plants gained come from? Where did the extra mass that the potatoes lost go?), we chose to focus on how students made sense of mass change data in our learning progression research.

Bridging the Gap through Learning Progression Research

We could see that our early versions of the *Carbon TIME* units were not bridging the gap between our lower and upper anchors for many students. Even after completing the unit investigations, some students were not recognizing matter-tracing questions as interesting and important. Similarly, many were not noticing key patterns in mass change data and interpreting those patterns as evidence of transformation and movement of matter (e.g., Jin et al., 2017). This led us to a deeper investigation of students' questions and approaches to interpreting mass change data.

Asking Students to Evaluate a Hypothetical Investigation

To examine the ways students engaged in sensemaking around the *Carbon TIME* investigations, we developed a sequence of interview tasks that walked 25 volunteer middle school, high school, undergraduate, and graduate students through several hypothetical investigations about changes in the mass of a system: animal growth, combustion, decomposition, or plant growth. These hypothetical investigations were similar to the ones in the *Carbon TIME* units but focused on different phenomena and were simpler in design. Each hypothetical investigation included three interview tasks:

Task 1: Evaluating Claims

During the first task in the sequence, we asked students to evaluate three claims that were based on the claims that students made during *Carbon TIME* investigations in units and our previous research (Jin & Anderson, 2012; Mohan, Chen, & Anderson, 2009). The full set of claims for the four different investigations is in Table 15.2. The claims for each system suggested three different types of explanations:

- *Matter-energy conversion:* the first claim explained how a carbon-transforming process worked by interconverting matter and energy, such as suggesting animals lose mass by "burning" it off or that plants grow by converting sunlight into biomass. This claim can be invalidated without the need for an investigation if students understand how to apply the crosscutting concept of matter and energy conservation.
- *Movement and transformation of observable matter:* the second claim explained how a carbon-transforming process worked by describing carbon movement through observable pathways involving only solids and liquids, like animals excreting waste products or a plant growing by taking up soil nutrients. This claim aligns with student ideas that are based on what they can see, like when matter changes phases. At this stage of the interview tasks, this claim is plausible and consistent with matter and energy conservation but must be verified with data to confirm its appropriateness.

Table 15.2 Competing claims participants were presented with during interviews for each task

	Animals task	*Combustion task*	*Decomp task*	*Plant task*
Question	What happens to the mass of food not used in animal growth?	Why does a match lose mass when it burns?	Why does a dead tree covered in shelf fungus lose mass as it decays?	Where does the mass of a plant come from?
Claim 1: Matter/energy conversion	Daryll claims: "The five-year-old's body turned the mass of the food into energy in order to grow."	Li claims: "The molecules in the match were converted into heat and light energy that was given off by the match."	Keller claims: "The change in mass of a decaying tree covered in shelf fungus is the result of the fungus changing the tree's atoms into energy for the fungus to use to grow."	Oscar claims: "I think most of the plant's weight comes from the sunlight."
Claim 2: Observable matter-tracing pathway	Bai claims: "The five-year-old's body got rid of most of the extra mass of the food as waste, like poop and pee."	Lupe claims, "The extra mass from the match turned into ashes that might have fallen on the floor."	Latisha claims: "When a decaying tree covered in shelf fungus loses mass, most of the change in mass is due to atoms moving from the tree to the soil."	Mike claims: "A growing plant gains most of its weight from materials that came from the soil."
Claim 3: Atomic-molecular matter-tracing pathway	Marisol claims: "The five-year-old breathed out most of the extra mass of the food as gases, like CO_2."	Tyrone claims: "The decrease in the mass was caused by the movement of atoms from the match to the air as CO_2."	Song claims: "When a decaying tree covered in shelf fungus loses mass, most of the change in mass is due to atoms moving from the tree to the fungus to the air."	Lucia claims: "A plant gains most of its weight from gasses in the air."

- *Chemical change involving gases:* the third claim explained how a carbon-transforming process worked by describing carbon movement through unobservable pathways that involve gases and chemical changes explained at the atomic-molecular scale. This claim aligns with scientific explanations for the question being investigated. However, at this stage of the interview tasks, students cannot conclusively decide between this claim and the previous one without additional evidence.

Task 2: Analyzing Mass Change Data

The second task in the interview sequence described a hypothetical investigation that generated data about changes in mass. Each data set revealed a discrepancy between the amount of solid and liquid mass entering and leaving an organism and the mass change of that organism. For example, in the investigation of animal growth, the change in the average mass of five crickets is less than the average mass of food eaten, even after accounting for matter lost via waste (Table 15.3). This mass discrepancy is important because it indicates that matter was lost (or gained, in the case of plants) to a matter pool not measured or observed in the investigation.

After students explained what patterns they saw in the data table from the investigation, we asked them to revisit the three claims from the first interview task to explain whether the data supported their initial ideas about which claims were plausible.

Table 15.3 Data table from the animal growth task provided during interviews

Sample	*Change in mass of cricket (g)*	*Change in mass of food (g)*	*Mass of waste (g)*
1	0.2	2.0	0.4
2	0.2	2.1	0.5
3	0.3	2.3	0.5
4	0.1	1.9	0.4
5	0.4	2.3	0.7
Average	**0.3**	**2.1**	**0.5**

Task 3: Identifying Unanswered Questions

In the last task of the interview sequence, we asked students to identify any unanswered questions that were not addressed by the investigation. This question was designed to evaluate students' preparation for future learning (Bransford & Schwartz, 1999) by revealing which aspects of the investigation they viewed as incomplete or inconclusive. When students identified questions they would like to further explore, we also prompted them to brainstorm about an investigation that could address their questions. This allowed us to further consider if and how they considered questions and approaches to tracing matter and energy.

Characterizing Student Thinking in a Learning Progression Framework

Our analysis of transcripts of the student interviews enabled us to develop a learning progression framework that described students' sensemaking performances during the three interview tasks. Our learning progression framework consists of three progress variables that describe how students: PV1) evaluate claims about carbon-transforming processes, PV2) analyze mass change data about carbon-transforming processes, and PV3) identify unanswered questions not addressed by the data (Table 15.4). These progress variables

Table 15.4 Learning progression levels for three progress variables

Level	*PV1: Evaluating claims explaining mass changes*	*PV2: Analyzing mass change data*	*PV3: Asking questions after investigations*
Upper Anchor	Students evaluate claims based on whether they conserve matter (i.e., apply matter conservation principle) including unobservable matter pools (i.e., gases).	Students sum mass data across different carbon pools and identify a mass discrepancy, which they attribute to an unmeasured gaseous matter pool.	Students propose questions that identify limitations in the data, focus on matter tracing, and are constrained by principles.
Intermediate levels	Students evaluate claims based on whether they conserve matter (i.e., apply matter conservation principle) OR identify unobservable matter pools (i.e., gases)	Students sum mass data across different carbon pools and identify a mass discrepancy, which they attribute to carbon-energy interconversions or a non-gaseous matter pool.	Students propose questions that partially address limitations in the data; may not use principles to constrain questions.
Lower Anchor	Students evaluate claims based on whether they describe something an organism does or needs to function properly.	Students use the data to describe how individual organisms are behaving or to describe general increases or decreases of observable matter pools.	Students identify aspects of the system they are curious about independent of the data, or they critique the experimental design.

describe patterns in student responses that we observed across all four interview hypothetical investigations—combustion, animal growth, plant growth, and decay. In general, Table 15.4 describes a progression from:

- *An Organisms-as-actors* approach to investigations, in which students focus on investigations as a way to understand what animals, plants, or decomposers need and how they use it, to
- A *matter-tracing* approach to investigations, in which students interpret mass discrepancies as evidence of matter movement among carbon pools.

Student Exemplars

In this section, we describe three students' interview responses from the Animal Growth Investigation Task to illustrate the broad reasoning patterns we identified at the different levels of our progress variables. In the Animal Growth Investigation Task, students:

- Considered hypothetical claims about how an animal grows (Table 15.2), made by:
  - Daryll claiming matter-energy conversion ("The five-year-old's body turned the mass of the food into energy in order to grow"),
  - Marisol claiming matter movement including gases ("The five-year-old breathed out most of the extra mass of the food as gases, like CO_2"), and
  - Bai claiming matter movement among solid and liquid matter pools ("The five-year-old's body got rid of most of the extra mass of the food as waste, like poop and pee") then
- analyzed a data table showing food consumption and mass changes for five crickets (Table 15.3), and
- identified unanswered questions and suggested possible follow-up investigations.

Lowest Level: Investigations as a Way to Study Organisms as Actors

At the lowest level, students framed the inquiry scenario as an investigation into what organisms do (i.e., *What do animals do with the food they eat?*) rather than one of matter movement and change. This is consistent with a view of animals as actors who use food for purposes including growth and movement.

When **Jeremiah**, a middle school student in a general science class, was asked to evaluate the three claims (PV1), his reasoning focused on different ways that animals used food as they grew.

- He rejected Marisol's claim, saying, *"Marisol's completely wrong because I really don't think that food and mass like that could change into a gas straight up … As far as I know there's not a process in the body that can do that."*
- Jeremiah agreed with Daryll's claim because it described a resource necessary for growth: *"I agree with Daryll fully. There's a lot of energy needed to grow, I'm sure there's lots of food is changed to that energy to grow …"*
- Jeremiah also agreed with Bai because he recognized waste production as part of how animals function, saying, *"I guess, I agree with Bai too because a lot of waste that is made is obviously gotten rid of … only the waste was changed into waste. Not all the good stuff …"*

Jeremiah's focus on what organisms need to grow led him to interpret the data from the investigation about cricket growth (PV2) by looking for patterns in how much food was consumed, and how much waste was produced, by individual crickets, saying: *"These two [crickets 3 and 5] gained 2.3 [grams] and they had a very similar change in mass, and this one [cricket 4] ate the lowest amount of food ..."* Because Jeremiah focused on how individual crickets behaved when analyzing the data, he did not notice any new information that challenged his initial ideas about the three claims:

- *"I don't think Marisol's still right ... because then they would not have that ... big of a change in mass because they're breathing out their nutrients which would be terrible."*
- *"I still think that Daryll is very definitely right ... most of the food that they had ate did transfer into the change of the mass ..."*
- He agreed with Bai's claim because *"there was definitely waste produced."*

When asked what questions he had about the inquiry investigation, Jeremiah identified questions aimed at how crickets used food to grow, move, and stay healthy (PV3):

Jeremiah: *I'd test how much activity they're getting in their little box because definitely activity does affect the mass of the cricket ...*

Interviewer: *Would that help you figure out–would that give you more evidence to support Daryll's claim or Bai's claim maybe or?*

Jeremiah: *I feel it would be a little bit more on Daryll's claim because muscle mass is very important when you're growing to stay strong enough to hold your body up ... What if they're active, like being more active and their energy is being used for more activity than growing, they don't grow as much or they grow more ...*

Other students reasoning at this level identified unanswered questions about the experimental set-up that they would change or improve upon if the experiment were to be run again (e.g., increase the number of crickets being measured) or expressed an interest in learning more about the habitats where an organism was found. In all of these cases, students said if they had more information about the organism, they would be better able to critique the claims. Helping students recognize that mass changes are connected to matter movement during chemical processes will be an important next step for students' reasoning at this level in understanding the purpose of mass change investigations.

Intermediate Level: Investigations as a Way to Trace Matter through Systems with Uncertainty around Matter Conservation Principles or Unobservable Matter Pools

At the intermediate level, student accounts framed the inquiry scenario as providing data useful for matter tracing but struggled to apply the matter conservation principle or discounted unobservable matter pools. These uncertainties led students to agree with some of the invalid claims.

Amanda, an undergraduate student studying forensic chemistry, reasoned in ways that were consistent with our intermediate levels. When Amanda considered the three claims (PV1), she recognized that they described chemical processes but struggled to apply principles of matter and energy conservation.

- She thought Marisol's claim had a *"little tiny of a merit"* because she knew that CO_2 was involved in breathing but said it was unconnected to food: *"I guess I agree with the fact that we do breathe out carbon dioxide but I would say that it doesn't … I guess the reason for the extra mass."*

- Amanda thought Daryll's claim that food was turned into energy was most valid:

Amanda: *I would say that I agree with that [Daryll's claim].*
Interviewer: *So the body turns … the mass, it went into energy for growth?*
Amanda: *Yeah, I mean with the caveat that not all of it goes into growth but some of it does, you know, turn into energy to grow.*

- For Bai, she agreed with the claim but disagreed with the idea that "most" of the extra mass went to waste: *"I wouldn't say most but I would say that it's–definitely some of the extra mass goes there."* This detail suggests she was not only attentive to the nature of the pathways described in each of the claims, but also recognized that considering quantities of matter movement was important.

Despite Amanda's tendency to conflate matter and energy, she demonstrated a strong commitment to tracing and conserving matter in the way she analyzed the data (PV2), which allowed her to identify the mass discrepancy in the data: *"First off I would say one would assume that the mass of the cricket and the mass of the waste would be equal to the mass of the change in food. So that's a fairly big difference. So that's kind of important."* She then used this finding to reevaluate her initial thoughts about Daryll's and Bai's claims, saying:

Amanda: *I mean … these two would have to be equal to that which–*
Interviewer: *So in order for Daryll to be true and Bai to be true the change in mass of the cricket and the mass of the waste would need to equal–?*
Amanda: *Yeah, which it doesn't so obviously there's something else going on that is not accounted for.*

When asked how this pattern relates to Marisol's claim of matter leaving as CO_2, she said: "*Obviously there's something else going on that isn't accounted for … I wouldn't have expected that to make such a big difference but it could be the CO_2 kind of stuff.*" Amanda acknowledged that Marisol's claim was valid and aligned with the data, even though it went against her expectations. This indicated she used evidence-based reasoning to understand how a system works, in line with a scientific approach to investigations. However, her continued uncertainty around energy in living systems muddled how she used this evidence to evaluate Darryl's claim: *"The mass–if it were turning into energy … I would say that definitely there should be some relationship between the energy that [it] uses to turn into growing and the actual change in mass of the cricket."*

When asked what questions she still had after the investigation (PV3), Amanda continued to focus on matter-tracing concerns, noting, *"There's obviously a change in mass and it's not abundantly clear where that mass goes."* She proposed doing more testing to determine if the missing mass was going into a gas by using a *"closed system so you can really contain everything and account for everything."* Included in her rationale for using a closed system was the idea that *"there's no energy lost … so you can see where the energy went, I guess,"* showing her continued uncertainty in discriminating between matter and energy in this scenario.

While students who reason at the intermediate level still have some uncertainty in applying scientific principles to living systems, or connecting these principles with data, we suggest that these students are prepared for future learning because they recognize the importance of matter tracing and matter conservation. Amanda exemplified this through her quantitative reasoning strategies and reliance on evidence to inform her ideas about a system, even when they were contrary to her initial expectations.

Students reasoning at this level also identified unanswered questions that would generate more data to evaluate the proposed claims rather than general questions about topics they were curious about, suggesting they understood that there were specific types of questions that would advance their understanding of matter and energy in relation to the phenomenon.

Highest Level: Investigations as a Way to Trace Matter through Chemical Processes in a Manner Consistent with Scientific Principles

At the highest level, students used matter conservation principles to evaluate claims and analyze data. This kind of reasoning aligns with formal scientific thinking and is a benchmark described in NGSS (NGSS Lead States, 2013). Students frame animal growth as a balance of matter moving into and out of organisms through materials that are both observable (e.g., feces) and unobservable (e.g., CO_2).

Melissa, an undergraduate student majoring in environmental science, demonstrated reasoning consistent with our highest level. When Melissa critiqued the three claims (PV1), she quickly recognized the validity of Marisol's claim: "*So right away, I agree with Marisol based on my explanation.*" She also brought this understanding when she considered Bai's claim, distinguishing between waste at a macroscopic scale (i.e., Bai's claim) and waste at a microscopic scale, noting: "*Our body's not growing at like this huge macro scale. It's growing at a very microscopic, on a cell to cell basis ...*" Finally, by using matter and energy conservation principles, she was able to reject Daryll's claim: "*Daryll's claim seems again to be this turning into or conversion of mass and energy.*"

When Melissa analyzed the data (PV2), she quickly recognized there was a pool of matter not accounted for in the data:

Melissa: *So the change in the mass of the food is not equal to the mass of waste. So there's, quote, unquote, missing mass there because you would assume that if there was no change in the mass of the cricket, that [the change in the mass of the food and waste] would be equal. So there's a non-equal. So the cricket must be growing in some way. But the cricket does not make up for the amount that's missing. So it has to be going somewhere else.*

This recognition that there was matter not accounted for in the data helped Melissa discriminate between the two matter-tracing claims. She said:

Melissa: *My thought is that it's going out into the air as CO_2 ... So that would be agreeing with Marisol. Because there's this missing mass that's not accounted for by the growth of the cricket or the amount of waste. So this is disproving Bai's claim. Because conservation of mass states that you can neither create nor destroy it. So it has to be going somewhere.*

Melissa also recognized that the data did not provide the kind of evidence needed to assess Daryll's claim. However, she was able to draw on her knowledge of matter and energy conservation to invalidate Daryll's claim.

When asked what unanswered questions she had from the investigation (PV3), Melissa recognized that more investigation was needed to confirm CO_2 as the likely missing matter pool. She framed her question as if she were a teacher instructing a science lesson, saying, "*You talk about this missing mass, like there's mass not accounted for. So where could it be going?*" She then suggested leading a discussion with students that brought in processes that generated CO_2.

Significance of Findings

These findings gave us a deeper understanding of how and why students were struggling to interpret the results of their investigations in the *Carbon TIME* units and suggested strategies for improvement. We knew that most students were beginning the units with understandings that resembled Jeremiah's, so we needed to design opportunities to learn that would enable students to pose questions about tracing matter and energy, to find patterns indicating movements of matter in mass change data, and to use those patterns as a basis for arguments from evidence about how flames and organisms transform matter and energy.

Using Learning Progression Research to Support Curriculum Development

We used the results from our learning progression research to return to curriculum development. Specifically, we drew on the learning progression research to design and refine scaffolds for the investigations in each unit. Scaffolds are intended to support students in doing tasks that they are not yet able to do on their own (i.e., a student's zone of proximal development or ZPD; Vygotsky, 1978).

In the case of the *Carbon TIME* units, the learning progression research showed that the ZPD was not the same for all students; Lower-level students like Jeremiah had different needs from mid-level students like Amanda. Therefore, we designed scaffolds that would be meaningful for students at all levels and enable students at the lower levels of the learning progression to move toward the higher levels. Our curriculum design work was iterative across multiple years. As part of the iterative revisions, we used the learning progression research to revise three key scaffolds—what we call process tools—that support students' work as questioners and investigators in each unit. (For a discussion of scaffolding students' work as explainers see Chapter 16 of this volume.)

Each process tool served as the focus for a lesson organized around a discourse routine (see Appendix A) in which (a) students wrote their ideas privately, then shared them in a class discussion, (b) the teacher led a consensus-seeking discussion identifying both ideas that students agreed on and unanswered questions, and (c) students revised their writing based on how their ideas had changed after the discussion. As we revised the process tools based on the learning progression research, we concurrently revised the accompanying discourse routines. We discuss one general scaffold (The Three Questions) and three process tools that scaffold students' questioning and investigating practices below.

Foundation for the Process Tools: The Three Questions

Each tool was designed around the Three Questions (see Chapter 16, this volume,) which scaffold all three core practices—questioning, investigating, and explaining. The Three Questions focus on (1) matter movement, (2) matter change, and (3) energy change.

The learning progression research helped us think more clearly about the functions of the Three Questions in the early parts of each unit (students as questioners and investigators). In particular, we recognized that teachers needed to help students like Jeremiah recognize that matter-tracing questions were important and that the students could conduct investigations to answer them. We also worked with teachers to emphasize how the third column of the Three Questions tool: Evidence We Can Observe, could be helpful to mid-level students like Amanda, including the connection between mass change and matter movement: "A change in mass shows that molecules are moving."

Students as Questioners: The Expressing Ideas and Questions Tool

Each unit starts with a phenomenon and a driving question (e.g., "How does a child grow, move, and function?" for the *Animals* unit). The Expressing Ideas and Questions process tool (See Figure B1 of Appendix B) provides students with space to generate and share their own initial explanations of and questions about the phenomenon. The claims evaluated by the interviewed students during the first learning progression task (Table 15.2) were based on ideas that students expressed using this tool.

During the discourse routine, the teacher strategically draws attention to students' initial explanations and questions that are related to matter and energy tracing (for example, questions about how our bodies use food to move and grow). The teacher also helps the students recognize gaps, inconsistencies, or differences among students in their initial explanations of the phenomenon. This helps lower-level students like Jeremiah recognize matter- and energy-tracing questions as important and as questions that they can discuss, investigate, and eventually answer.

In units after the first unit, the teacher can also ask students to consider their initial thinking in light of the matter and energy conservation principles in the Three Questions. This press to check initial explanations against the matter and energy conservation principles provides students with a first opportunity to evaluate claims in this way (as was also required in Task 1 of the interview). This helps mid-level students like Amanda use the conservation principles as heuristics to prepare for future learning.

The learning progression research helped us to recognize that the most important outcome of the consensus-seeking discussions was not agreement on ideas, but shared questions—particularly questions that focus on tracing matter and energy. This led us to change the name of the tool—from Expressing Ideas to Expressing Ideas and Questions—and to emphasize shared questions as a key outcome of the class discussion in our teacher-facing materials.

Many teachers use a Driving Question Board to record key ideas and questions and return to them later in the unit (Johnson, 2019). The teacher facilitation throughout the Students as Questioners portion of the unit is key in building student interest and investment in matter and energy-tracing questions that they will follow throughout the unit, preparing them for future learning about the phenomena that are the focus of the unit.

Starting Investigations: The Predictions and Planning Tool

The learning progression research led to changes in how students start unit investigations. As described above, these investigations focused on tracing matter by using BTB to detect changes in CO_2 concentrations and digital balances to detect mass changes in the systems that they were investigating—in the case of the *Animals* unit, mealworms eating potatoes. In the early versions of the units, students completed the Predictions Tool, which asked them to predict what would happen to mass and CO_2 concentrations and explain the reasons for their predictions. We found that although the students liked observing and measuring the mealworms, they often failed to see what they were doing as a matter-tracing investigation. They often made correct predictions for incorrect reasons (e.g., predicting that the food would lose mass because the mealworms "ate it up").

Based on the learning progression research, we developed a revised version of what we now call the Predictions and Planning Tool. Students now start the investigations with (a) the questions that they generated about tracing matter and energy as the mealworms ate, grew, and moved, and (b) a discussion of how different laboratory tools could be helpful: the digital scales could detect matter movement and BTB could detect chemical changes. They then write and discuss how they could use the laboratory tools and methods to trace matter and identify how matter and energy change during the investigation. Students answer questions such as, "How will you detect and record matter changes?" and, "What data will you collect and how will the data help answer the question?" They are also again prompted to check their predictions against the principles of matter and energy conservation.

In the consensus-seeking discussion that follows, students compare their predictions and their plans with each other, as well as anticipate what different patterns in the data would tell them about how the matter was moving and changing. These discussions prime students to look for patterns in the data that provide evidence relevant to mass change questions. This helps lower-level students like Jeremiah anticipate how the data they collect will be connected to the questions that the class generated in the previous discussion. It helps mid-level students like Amanda see how their ideas about matter movement and change lead to testable predictions about mass changes and CO_2 concentrations.

Concluding Investigations: The Evidence-Based Arguments Tool

Following the investigation, the Evidence-Based Arguments Tool builds on the planning and predictions work by prompting students to consider how their investigation evidence specifically (a) provides evidence to support conclusions about each of the Three Questions (matter movement, matter change, and energy change) and (b) generates additional questions to be answered. We discuss each purpose below.

Reaching Warranted Conclusions

The Evidence-based Arguments Tool (See Figure B2 of Appendix B) coupled with small and whole group discourse helps students to focus on the mass discrepancies in the data. In the case of the *Animals* investigation, for example, the whole system—potato, mealworms, and mealworm waste—loses mass while the BTB shows that CO_2 is going into the air. This is especially important for lower-level students like Jeremiah who may notice broad trends (like the mealworms gaining weight while the food loses weight) without noticing other patterns that indicate chemical changes involving gases). In the consensus-seeking discussion, the teacher and students work toward an evidence-based conclusion—that some of the matter in the potato is going through the mealworms and coming out in CO_2.

Identifying Unanswered Questions

As in the interview scenarios, the data from the investigation is insufficient evidence to construct a full explanation of the phenomenon. The Evidence-Based Arguments Tool also serves a role here in scaffolding students to identify questions that remain unanswered about what happened to matter and energy during their investigation. Key unanswered questions at this point are related to where the "missing" mass went and where the CO_2 in the system came from.

From the learning progression research, we knew these types of unanswered questions are very different from the questions students proposed at the lower level of the learning progression—related to critiques of the investigation or tangential curiosities. These questions, though, are essential to preparation for future learning, where citizens need to ask themselves, and others, how investigations are incomplete and what conclusions are not warranted.

The consensus-seeking discussion guides students toward unanswered questions "zooming in" to make sense of what is happening in animals at the cellular and atomic-molecular scale. This step is beyond what they can collect direct data on using tools and observations in the classroom. Therefore, the students explore these questions *as explainers* by using hands-on atomic-molecular models to model carbon-transforming processes. These models are then used by students as the basis for writing model-based explanations of the phenomena (discussed in more detail in Chapter 16).

Summary

Together the scaffolds help students to ask matter-tracing questions about the initial phenomenon; to participate in planning investigations that will collect that evidence; to interpret their results and develop evidence-based arguments; and to generate additional questions about transformations of matter and energy that they will answer later in the unit. Teachers can also use the process tools and class discussions as formative assessments. Teachers can interpret students' writing and talk and can use probing and follow-up questions during whole class or small group discourse in places where students may need additional support.

Results of Written Assessments

We assessed the effectiveness of our revised units by creating written assessment items that mirrored the interview tasks and administering them to middle and high school students who were participating in the *Carbon TIME* research project. Like the interview tasks, the written assessments ask students to evaluate competing claims about how animals, plants, and decomposers moved and transformed matter and explain their reasoning, both before and after seeing the results of investigations similar to those presented in Table 15.2 above. We worked with Jay Thomas of ACT to develop and validate automated scoring protocols for students' constructed responses. The assessments are available on the *Carbon TIME* website.

Our combined data include about 50,000 pre-tests and post-tests for the *Animals, Plants,* and *Decomposers* units during the 2017–18 and 2018–19 school years from a diverse sample of middle and high school classrooms; almost all of the students were taking middle school science classes or high school beginning biology classes. The results showed consistent patterns across both years and all three units, so we show coding results for the *Animals* pre-tests and post-tests for the 2018–19 school year (Figure 15.1). (The codes for the Asking Questions progress variable were not sufficiently reliable, so Figure 15.1 shows results for the Evaluating Claims and Analyzing Mass Change Data progress variables.)

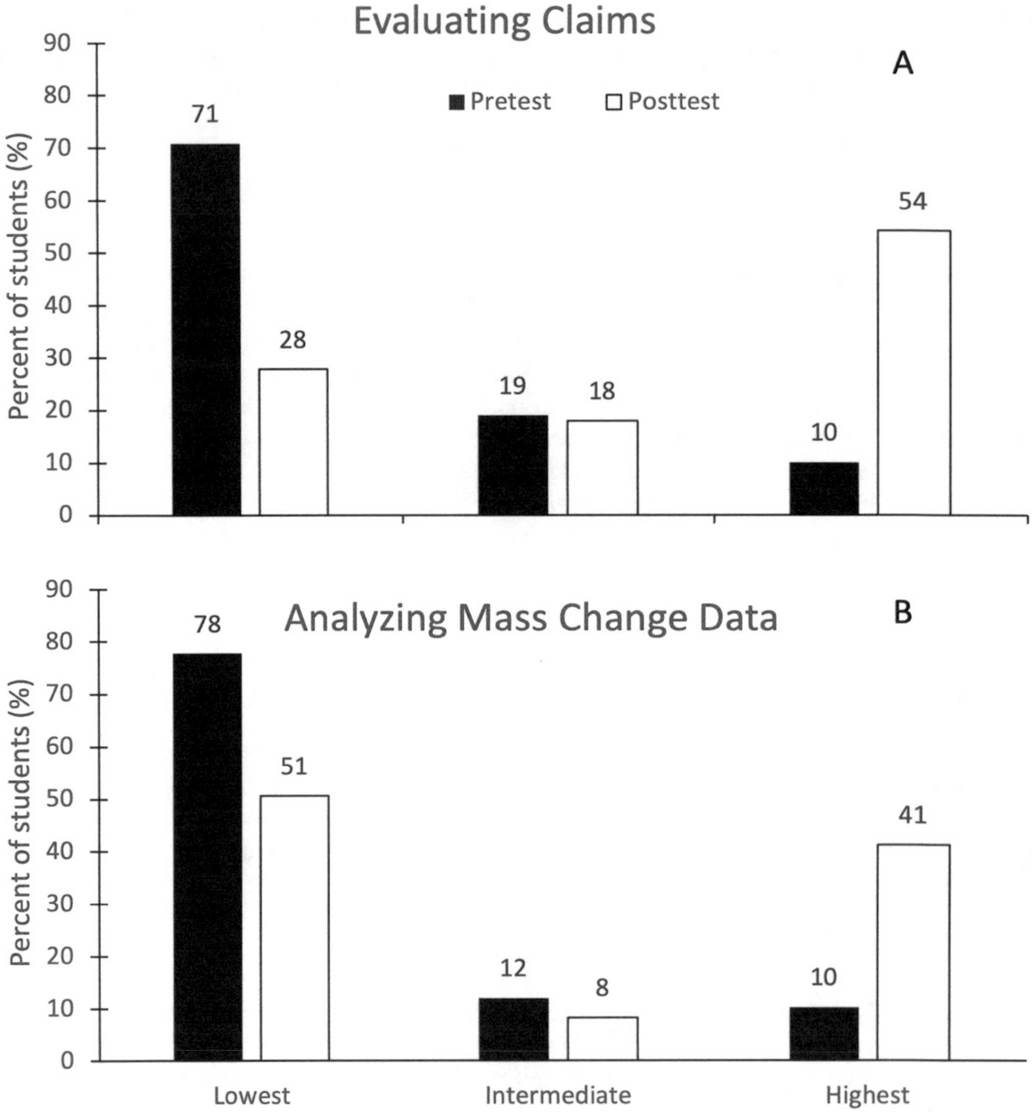

Figure 15.1 Comparing pre-test and post-test results for Animals investigation questions for the 2018–19 school year (Pre-test N = 4,835, post-test N = 4,745). Results are shown for the Evaluating Claims (A) and Analyzing Mass Change Data (B) progress variables.

We see these results as encouraging while showing that we still have work to do. The pre-post differences are statistically highly significant. Using a paired T-test, for Evaluating Clams, $t = 58.63$, $p < 0.0001$; for Analyzing Mass Change Data, $t = 35.38$, $p < 0.0001$. A Wilcoxon rank sum test treating the data as ordinal also shows highly significant results. In order for students to learn effectively from the investigations, we feel that students need to achieve at least Level 2 in the learning progression (the level attained by Amanda above). On the post-tests, about 72% of students had attained this level for Evaluating Claims and 49% for Analyzing Mass Change Data. This represents substantial progress from the students' pre-test performances but still significant numbers of students who did not identify key patterns in the data and use those patterns to construct arguments from evidence.

Conclusion

In this chapter, we described the development of a learning progression that focuses on how students critique claims, interpret data, and ask questions about mass changes during carbon-transforming processes. We believe that this is a strategic focus, for two reasons.

First, this learning progression describes students' approaches to truly important phenomena, practices, crosscutting concepts, and core ideas. These particular phenomena are the ones that drive carbon cycling on local and global scales, so their impact on students' lives will be profound. Matter tracing is also truly a Big Idea, central to our understanding of chemical change, living systems, and Earth systems. Students who are prepared to act as informed citizens as we confront climate change-related issues must be able to understand and critique matter-tracing investigations at local and global scales.

Second, we believe that this work illustrates some key principles for the development of learning progressions that support three-dimensional curriculum development. If we want teaching to engage students in making sense of phenomena, then all levels of our learning progressions need to describe students' sensemaking. True sensemaking performances are inherently three-dimensional, requiring students to use SEPs, DCIs, and CCCs in combination. Thus practices such as analyzing data or evaluating hypotheses are fundamentally different for different phenomena, since they require students to use different DCIs and CCCs. This means that as we define our upper anchor goals and analyze lower-level student performances, we must choose phenomena and three-dimensional performances that we truly value. We hope that these findings can also be useful to other science educators engaged in assessment and instructional design.

Acknowledgments

This research is supported in part by a grant from the National Science Foundation: Sustaining Responsive and Rigorous Teaching Based on *Carbon TIME* (NSF 1440988). Any opinions, findings, and conclusions or recommendations expressed in this material are those of the author(s) and do not necessarily reflect the views of the National Science Foundation.

We wish to acknowledge contributions to this research from Qinyun Lin, who conducted statistical analyses; Jay Thomas, who led in the development of machine scoring of student assessments; Jennifer Doherty, who developed the initial version of the interview protocol; and Sarah Bodbyl, Pingping Zhao, and Christa Haverly, who conducted interviews.

Please visit our website: https://carbontime.create4stem.msu.edu/

References

Anderson, C. W., de los Santos, E. X., Bodbyl Roels, S., Covitt, B., Edwards, K. D., Hancock, J. B., Lin, Q., Morrison Thomas, C., Penuel, W., & Welch, M. (2018). Designing educational systems to support enactment of the *Next Generation Science Standards*. *Journal of Research in Science Teaching*, *55*(7), 1026–1052.

Anderson, C. W., Nordine, J., & Welch, M. (2021). Energy and matter: Flows, cycles, and conservation. In J. Nordine, & O. Lee (Eds.), *Crosscutting concepts: Strengthening science and engineering learning* (pp. 165–194). Arlington, VA: NSTA Press.

Bransford, J. D., & Schwartz, D. L. (1999). Chapter 3: Rethinking transfer: A simple proposal with multiple implications. *Review of Research in Education*, *24*(1), 61–100.

Covitt, B. A., Parker, J. M., Kohn, C., Lee, M., Lin, Q., & Anderson, C. W. (2020). Understanding and responding to challenges students face when engaging in carbon cycle pool-and-flux reasoning. *The Journal of Environmental Education*, *52*(2), 98–117.

Driver, R. (1985). *Children's ideas in science*. London: McGraw-Hill Education (UK).

Elkind, D. (1961). Children's discovery of the conservation of mass, weight, and volume: Piaget replication study II. *The Journal of Genetic Psychology*, *98* (2), 219–227.

Jin, H., & Anderson, C. W. (2012). A learning progression for energy in socio-ecological systems. *Journal of Research in Science Teaching*, *49*(9), 1149–1180.

Jin, H., Johnson, M. E., Shin, H.-J., & Anderson, C. W. (2017). Promoting student progression in science classrooms: A video study. *Journal of Research in Science Teaching*, *54*, 852–883.

Johnson, W. (2019). Using a Driving Question Board educator resource. *Carbon TIME Project*: Michigan State University. https://carbontime.create4stem.msu.edu/sites/default/files/library/recurring_features_strategies/Driving_Question_Board_Educator_Resource.pdf

Johnson, W. R., & Anderson, C. W. (2017). Unpacking the climate change performance expectations in the *Next Generation Science Standards*. In D. Shepardson, A. Roychoudhury, & A. Hirsch (Eds.), *Teaching and learning about climate change: A framework for educators* (pp. 106–119). New York: Routledge.

Miller, H. K., & Anderson, C. W. (2017). Using NGSS crosscutting concepts as a tool for climate change and citizenship education. In D. Shepardson, A. Roychoudhury, & A. Hirsch (Eds.), *Teaching and learning about climate change: A framework for educators* (pp. 181–193). New York: Routledge.

Mohan, L., Chen, J., & Anderson, C. W. (2009). Developing a multi-year learning progression for carbon cycling in socio-ecological systems. *Journal of Research in Science Teaching*, *46*(6), 675–698.

National Research Council. (2007). *Taking science to school: Learning and teaching science in grades K-8*. Washington, DC: The National Academies Press.

National Research Council. (2012). *A framework for K-12 science education: Practices, crosscutting concepts, and core ideas*. Washington, DC: The National Academies Press.

NGSS Lead States. (2013). *Next generation science standards: For states, by states*. Washington, DC: The National Academies Press. https://www.nextgenscience.org/

Piaget, J. (1951). *The child's conception of the world*. London: Routledge.

Smith, C., Wiser, M., Anderson, C. W., & Krajcik, J. (2006). Implications of research on children's learning for standards and assessment: A proposed learning progression for matter and the atomic-molecular theory. *Measurement: Interdisciplinary Research and Perspectives*, *4*(1-2), 1–98.

Vygotsky, L. S. (1978). *Mind in society*. Cambridge, MA: Harvard University Press.

Appendix A: Carbon TIME Discourse Routine

The ***Carbon TIME*** **classroom discourse routine** is an intentional sequence of private and public talk and writing surrounding each Process Tool in a unit. The routine establishes consistent times and places for students to think and write individually, to share and edit ideas with others, and for the class to listen and comment together.

The classroom discourse routine serves several purposes.

1 It provides necessary opportunities for students to understand and clarify their own thinking and questions, which is requisite for conceptual learning.
2 The routine allows *the classroom community and the teacher* to know about and understand students' different and similar ideas, and for the teacher to make responsive instructional decisions.
3 Also, consistently foregrounding and returning to student ideas is important for providing students with agency in the classroom – to support their ownership, motivation, and curiosity aimed at *figuring it out*.

Engaging in classroom discourse guided by this discourse routine scaffolds students in developing scientific explanations that answer the driving question for each unit.

The *Carbon TIME* classroom discourse routine involves 5 steps.

1 Introduction: The teacher establishes the purpose of the Process Tool, making connections to the unit's driving question and supporting students in drawing on prior knowledge and experiences, as well as preceding activities and discussions.
2 Private thinking and writing: Students use the Process Tool individually to think about, draw, and write their ideas, questions, conclusions, and explanations.
3 Partner or small group work: Students share and compare ideas in pairs or small groups.
4 Sharing ideas in whole-class discussion: As a whole group, the class engages in discussions that elicit, clarify, and compare ideas and questions.
5 Consensus-seeking discussion accompanied by public writing: As a whole group, the class comes to a consensus (agreement) about some things, documenting these publicly. The consensus-seeking purpose is *different*, depending on the Process Tool, as shown below.

Consensus-Seeking Purposes of Different Process Tools

Expressing Ideas Tool	*Predictions Tool*	*Evidence-Based Arguments Tool*	*Explanations Tool*
Students are …			
Questioners	*Investigators*	*Investigators*	*Explainers*
Class seeks to reach consensus regarding …			
• Similarities and differences among students' ideas • Points of disagreement • Wondering questions	• Similarities and differences among students' ideas • Points of disagreement • Wondering questions	• Relevant patterns in data • Warranted conclusions • Unanswered questions	• Coherent explanations that answer the Three Questions while following the rules in ways consistent with evidence

Appendix B: Carbon TIME Sample Process Tools

1.2 Expressing Ideas and Questions Tool: How do *you* think a child grows, moves, and functions?

Draw labeled arrows to show *your ideas* about what might be moving into, out of, or through the child. Think about what materials are moving and how they might be moving.

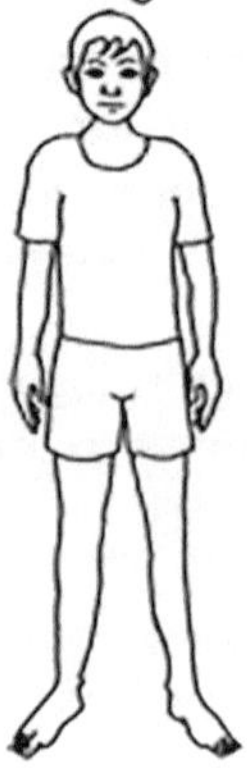

What goes in: List all the things you can think of that go into the child.	**What comes out:** List all the things you can think of that come out of the child.
What happens to the food the child eats?	**What happens to the air the child breathes?**

Your Explanation: Write your explanation for how a child grows, moves, and functions. Think about what materials and processes you listed in your drawing above.

____

____

____

____

Your Questions: What questions do you currently have about how a child grows?

____

Figure B1 Expressing ideas and questions tool from the animals unit.

3.3 Evidence-Based Arguments Tool: What Happens When A Mealworm Eats?

Think about your mealworm investigations. Complete this tool with patterns from class evidence, conclusions, and unanswered questions.

	Class Evidence What patterns did we find in our class evidence about each of the Three Questions?	Conclusions What can we conclude about each of the Three Questions using this evidence?	Unanswered Questions What do we still need to know in order to answer each of the Three Questions?
The Matter Movement Question			
The Matter Change Question			
The Energy Change Question			

Figure B2 Evidence-based arguments tool from the animals unit.

16

USING A LEARNING PROGRESSION TO ASSESS AND SCAFFOLD STUDENTS' EXPLANATIONS OF CARBON-TRANSFORMING PROCESSES

Hannah K. Miller, Wendy R. Johnson, Allison L. Freed, Jennifer H. Doherty, and Charles (Andy) W. Anderson

Introduction

This chapter and Chapter 15 of this volume describe a pair of learning progressions developed by *Carbon: Transformations in Matter and Energy* (*Carbon TIME*: https://carbontime.create4stem.msu.edu/), a research-practice partnership for middle- and high-school classrooms focusing on carbon cycling at multiple scales. The development of both learning progressions was deeply intertwined with curriculum development, including the instructional model for the six units developed by the project, all building toward a goal of *environmental science literacy*. Each unit scaffolds student work in three different roles: questioner, investigator, and explainer. Chapter 15 describes how we developed and used a learning progression focused on students' questioning and investigating practices. In this chapter, we focus on the third role: students as explainers.

When we began the project, we had already developed a learning progression focused on students' explanations of carbon-transforming processes, which we have described in previous publications (Jin & Anderson, 2012; Mohan et al., 2009). In this chapter, we explain how we extended and modified this learning progression framework to make it more useful for assessing and scaffolding students' explanation performances. We also describe how we used this modified framework to revise and improve our unit materials. We address these issues in five parts:

A *Prior work on the explanation learning progression.* We describe key characteristics of the initial explanation learning progression, as well as its relationship to other learning progressions that we used to support curriculum and assessment development for the *Carbon TIME* project.

B *Theoretical foundation.* We discuss ideas from the literature that play an essential role in this learning progression, including the distinction between scientific principles and scientific facts and the important role of epistemic heuristics.

DOI: 10.4324/9781003170785-18

C *Explanation practices in the Carbon TIME instructional model.* We describe the role of explanation practices in the instructional model that guided the development of *Carbon TIME* units, and the limitations of the prior learning progression for supporting teachers who needed to scaffold and assess their students' explanations.

D *Developing an improved explanation learning progression framework.* We describe how we analyzed transcripts of students' explanations to develop an improved framework that identified specific progress variables—characteristics of students' explanations that teachers and curriculum developers could use to scaffold and assess students' explanations.

E *Using the learning progression to improve the curriculum.* We describe how we used the modified learning progression to improve the teaching materials and share evidence for their effectiveness.

Prior Work on the Explanation Learning Progression

In this chapter, as in Chapter 15, we suggest that learning progressions supporting curriculum development need to include:

- A goal, or *upper anchor,* that identifies a cluster of related phenomena that play important roles in students' lives and indicates valued three-dimensional performances that students need as informed citizens.
- A *learning progression framework* describing levels of performance: a sequence of students' sensemaking performances related to the phenomena that "bridge the gap" between valued upper-anchor performances and observed lower-anchor performances—the ways that learners make sense of phenomena before instruction; and
- *Assessment resources* that elicit student responses and use rubrics to identify their learning progression levels.

One of the starting points for the *Carbon TIME* project was a learning progression that included all three of these components. During the project, we also developed two other learning progressions, one focusing on students' questioning and investigation practices (discussed in Chapter 15 of this volume) and one focused on students' reasoning about carbon cycling and energy flow in ecosystems and Earth systems (Covitt et al., 2020). We discuss each of the three components below.

Upper Anchor: Three-Dimensional Performances and Cluster of Phenomena

The *Carbon TIME* project is organized around the general goal of environmental science literacy, which we defined as "preparing students to use scientific knowledge and practices in their decisions about environmental issues" (Anderson et al., 2018, p. 1031). This requires (a) a three-dimensional understanding of environmental systems and how they are affected by human technologies and (b) what Bransford and Schwartz (1999) described as *preparation for future learning:* the ability to learn about and respond to new issues or new data as they arise. Figure 16.1 provides an overview of the upper anchor and framework; the top three rows (i.e., role in the carbon cycle, chemical changes, macroscopic phenomena) address different aspects of the upper anchor.

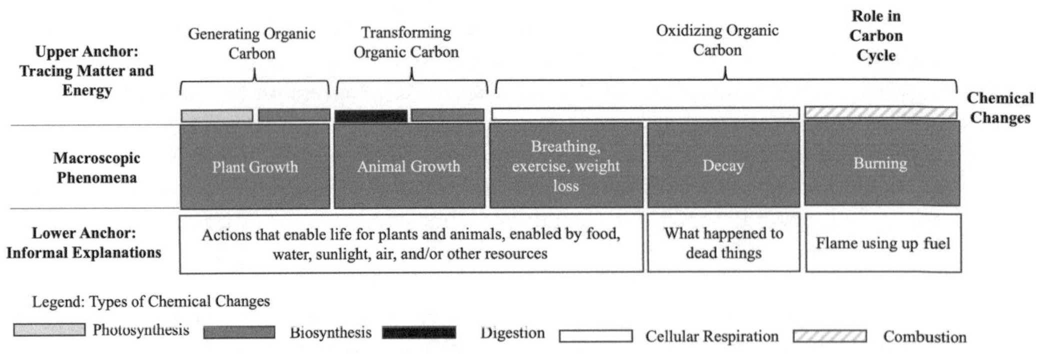

Figure 16.1 Contrasting ways of grouping carbon-transforming processes.

CLUSTER OF RELATED PHENOMENA THAT INVOLVE CARBON-TRANSFORMING PROCESSES

Our learning progressions focus on phenomena related to global carbon cycling and its effects on climate change. This entails using pool-and-flux models to reason about carbon cycling at a global scale (see Covitt et al., 2020) as well as understanding how global carbon cycling is related to macroscopic phenomena that we experience every day. These phenomena, all involving carbon-transforming processes, are listed in the third row of Figure 16.1 (and discussed in more detail in Chapter 15). They include burning organic materials and phenomena related to metabolic processes in animals, plants, and decomposers, such as growth, movement, eating, breathing, and decay.

Our goals for scientific sensemaking about these systems include all three dimensions of the *Framework for K-12 Science Education* (NRC, 2012) and the *Next Generation Science Standards* (NGSS Lead States, 2013). Scientific sensemaking requires performances that combine science and engineering practices (SEPs), disciplinary core ideas (DCIs), and crosscutting concepts (CCCs).

SCIENCE AND ENGINEERING PRACTICES (SEPs)

We are interested in using scientific causal models to explain and predict phenomena (Braaten & Windschitl, 2011). In particular, the causal model involved in this learning progression framework focuses on *tracing matter and energy through systems*. The upper-anchor explanations use this model to account for how matter—particularly carbon—and energy move and are changed in flames, animals, plants, and decomposers, relating those changes to observable changes in systems.

DISCIPLINARY CORE IDEAS (DCIs)

The second row of Figure 16.1 (i.e., chemical changes) points to key DCIs needed to trace matter and energy through the processes that we have identified. These processes include chemical changes that play key metabolic roles in all organisms, so explaining these processes involves core ideas from physical science (PS1: Matter and Its Interactions, PS3: Energy) and life sciences (LS1: From Molecules to Organisms: Structures and Processes). The top row shows how all these processes are connected through carbon cycling. Within and across these processes, carbon transforms between organic and inorganic forms, while energy flows with degradation.

CROSSCUTTING CONCEPTS (CCCs)

We discuss the role of CCCs in much more detail below. Here we note that some CCCs also play a key role in explaining these phenomena: most obviously Energy and Matter, but also Scale and Systems and System Models.

Learning Progression Framework: From Force-Dynamic to Matter- and Energy-Tracing Explanations

The bottom row of Figure 16.1 (i.e., lower anchor: informal explanations) summarizes lower-anchor explanations of carbon-transforming processes. Most middle- and high-school students initially provide what Pinker (2007) and Talmy (1988) describe as force-dynamic explanations—characterizing plants and animals as actors who need resources such as sunlight, food, and water to sustain their lives—in contrast with mechanistic explanations that use atomic-molecular models to explain phenomena in terms of chemical change.

As Figure 16.1 shows, these scientific and force-dynamic explanations lead to different ways of seeing phenomena as alike and different. In force-dynamic terms, plant and animal growth and movement are all closely related phenomena involving living actors and their needs, while decay and combustion are quite different because they consume dead things. In scientific explanations, on the other hand, photosynthesis is unique in its ability to create organic carbon compounds from inorganic reactants, while animal movement, decay, and combustion are all similar in that they involve the oxidation of organic carbon compounds.

Based on interviews and written responses from elementary, middle, and high school students, we developed a learning progression framework that included four levels of achievement as shown in Table 16.1 (based on Mohan et al., 2009; Jin & Anderson, 2012). For the *Carbon TIME* project, we found that the distinctions between Level 1 and Level 2—force-dynamic explanations with different levels of detail—were not very useful for analyzing responses of middle- and high-school students, so we relied on a framework that distinguished Levels 2, 3, and 4.

Assessment Resources: Interviews, Written Assessments, and Rubrics for Learning Progression Levels

We also developed a variety of assessment tools used for both research and classroom teaching. These included interview protocols and written assessments. All of these tools were

Table 16.1 Levels of explanations in the learning progression framework

Level 4: Coherent Scientific Explanations. Students successfully apply fundamental principles such as conservation of matter and energy to phenomena at multiple scales in space and time (generally consistent with Next Generation Science Standards).
Level 3: Incomplete or Confused Scientific Explanations. Students show awareness of important scientific principles and models at smaller and larger scales, but they have difficulty connecting accounts at different scales and applying principles consistently.
Level 2: Elaborated Force-Dynamic Explanations. Students' explanations continue to focus on actors, enablers, and natural tendencies of inanimate materials, but they add detail and complexity, especially at larger and smaller scales.
Level 1: Simple Force-Dynamic Explanations. Students' explanations focus on actors, enablers, and natural tendencies of inanimate materials, using relatively short time frames and macroscopic scale phenomena.

built around transfer tasks: they elicited students' explanations of phenomena that were different from the phenomena that the students had previously studied. They were all accompanied by rubrics that enabled teachers and researchers to identify the LP Levels of student responses.

Theoretical Foundation for Learning Progressions

In this section, we focus on two characteristics of the explanation learning progression that are especially important when we use it for assessment and curriculum development. First, the upper anchor includes both explanation performances and *preparation for future learning.* Second, CCCs play a key role in upper-anchor performances. In this section, we discuss both of these characteristics in the context of related literature.

Preparing Students for Future Learning

Bransford and Schwartz (1999) situate their ideas about preparation for future learning in the literature on *transfer:* the ability to ask questions and figure out answers in new situations. For this reason, our learning progression research analyzes student explanations of phenomena that they have not previously studied. We note two aspects of student performances that are especially useful for explaining new phenomena: distinguishing principles from facts and using epistemic heuristics.

DISTINGUISHING SCIENTIFIC PRINCIPLES FROM SCIENTIFIC FACTS

Scientific explanations in most disciplines rely on causal models for explanations (Braaten & Windschitl, 2011)—models that are constrained by principles such as the energy and matter conservation laws. Principles are particularly important in knowledge transfer situations, where students are figuring out explanations of phenomena that they have not previously explained.

Our previous research found that some students treated key scientific principles as isolated "facts" that could be used or ignored as needed when explaining phenomena (Dauer et al., 2015; Miller et al., 2013). The students incorporated new facts into their explanations without recognizing how these facts relate to the governing principles at work in all chemical changes. Other researchers have also documented how students can treat general principles as context-specific facts (Collins et al., 1991; Furtak et al., 2014; Lee & Liu, 2009; Modell, 2000; Stevens et al., 2009).

USING EPISTEMIC HEURISTICS

Krist et al. (2019) describe "epistemic heuristics" that students use in constructing mechanistic explanations. In contrast with principles—rules that constrain models and explanations—heuristics are generative: they suggest questions and hypotheses leading toward scientific explanations. We suggest that epistemic heuristics play a key role in preparation for future learning by suggesting important questions or generating hypotheses.

Figure 16.1 illustrates how different learning progression levels can be associated with different epistemic heuristics. Lower-level explanations rely on epistemic heuristics associated with force-dynamic reasoning; they suggest questions and hypotheses that identify actors (such as animals and plants), their purposes (to survive and grow), and enablers they need to achieve those purposes (such as food, water, and air). In contrast, epistemic heuristics that focus on tracing matter and energy through systems can help students see how what they have learned about one process can help them generate explanations for other processes—for example, the similarities among combustion, animal movement, and decay. In each case, the epistemic heuristics identify goals for sensemaking, suggest analogies and relevant evidence, and help people decide when they have figured out a satisfactory explanation.

The Important Role of Crosscutting Concepts

We noted above that CCCs played an important role in defining the upper anchor and framework for our learning progressions. In the context of this research, we focus specifically on students' use of the three CCCs: (a) *Energy and matter: flows, cycles, and conservation;* (b) *Scale, proportion, and quantity*; and (c) *Systems and system models.* In particular, we suggest that CCCs can function as principles and epistemic heuristics.

CROSSCUTTING CONCEPTS AS PRINCIPLES

Some CCCs, especially Energy and Matter, are based on principles (conservation laws) that define what kinds of explanations are acceptable. For example, "I got a good night's sleep, so I have lots of energy today" is an acceptable statement in colloquial discourse. However, interpreting this statement as "Sleep is a source of energy for humans" would generally not be accepted by scientists, because, according to the principle of energy conservation, energy is not generated by sleeping. As we have argued elsewhere (Anderson et al., 2021), using the law of energy conservation to constrain explanations plays a key role in students' progress toward upper-anchor explanations of phenomena.

Thus using the laws of conservation of matter and energy to constrain accounts of carbon-transforming processes is a powerful tool for interpreting global processes (e.g., carbon cycling and climate change) as well as everyday events (e.g., decay, plant growth) (Hartley et al., 2011; Rice et al., 2014). Like other principles, this CCC often operates in the background, providing rules that constrain scientific explanations without conscious attention from participants who have mastered the principles.

CCCs AS EMBODYING EPISTEMIC HEURISTICS

CCCs can also play an important role as epistemic heuristics for figuring out explanations of new phenomena. We discussed above how the Energy and Matter CCC invokes the laws of conservation to constrain scientific explanations, ruling out possible explanations that do not follow these principles. This CCC can also be generative, functioning as a powerful epistemic heuristic that suggests useful questions about how a system conserves and transforms matter and energy. Because of the conservation laws, questions like "*Where does the matter come from?*" or "*Where does the energy go?*" can lead toward upper-anchor scientific explanations.

Following (a) the second and third CCCs, (b) Scale, proportion, and quantity, and (c) Systems and system models, we identify a hierarchy of systems at four different scales:

1 Large scale (carbon pools and fluxes, like the biomass, fossil fuel, or atmosphere pools);
2 Macroscopic scale (organisms and materials, like plants, animals, or air);
3 Microscopic/cellular scale (cells and cell components);
4 Atomic-molecular scale (atoms and molecules).

Krist et al. (2019) demonstrate how "connecting scales" functions as an epistemic heuristic for generating mechanistic explanations: we often explain macroscopic scale phenomena (such as the growth and movement of plants and animals) through the structure and function of their microscopic and atomic-molecular scale subsystems.

In summary, our work on the explanation learning progression included a focus on preparation for future learning—how students learned to suggest scientific explanations for new phenomena that they had not previously studied. We have suggested that in these situations, principles (that constrain acceptable explanations) and heuristics (that generate worthwhile questions, goals, and hypotheses) are especially important, and that CCCs can be useful as both principles and heuristics.

Explanation Practices in the Carbon TIME Instructional Model

As we explained in Chapter 15, *Carbon TIME* units are organized around an instructional model that scaffolds student practices in three roles that together encompass almost all of the science and engineering practices. Chapter 15 focuses on the first two roles: questioner and investigator. In this chapter, we focus on the third role: explainer. When students explain carbon-transforming processes, three SEPs play a prominent role: using models, constructing explanations, and obtaining, evaluating, and communicating information.

When we analyzed the instructional practices of teachers using *Carbon TIME,* we found that the teachers with higher student learning gains effectively combined divergent and convergent instructional practices (Covitt et al., 2024). The divergent instructional practices engaged students in scientific sensemaking, supporting them as they expressed their ideas, formulated questions, and designed investigations. The convergent practices relied on ideas associated with cognitive apprenticeship (Collins et al., 1989; Lave & Wenger, 1991), coaching students as they achieved proficiency in scientific practices.

Our learning progression research had the potential to support teachers' convergent instructional practices in two ways (a) guiding formative assessments that enabled teachers to analyze and respond to student performances, and (b) scaffolding students' explanations of phenomena. Below we discuss both the potential and the shortcomings of our initial LP for each purpose.

Using the Learning Progressions for Formative Assessment

Our work builds on a body of research that suggests how learning progressions might be used as tools for assessment in science education. Furtak et al. (2014), for example, suggest that using learning progressions as tools for formative assessment of student learning may help teachers move beyond interpreting student ideas as "right and wrong." Alonzo (2018) similarly reviewed potential uses of learning progressions as tools for formative assessment and suggested that learning progressions may be used as tools for framing and eliciting student thinking and then interpreting and responding to those ideas.

While the potential of learning progressions for formative assessment is clear, using learning progressions for this purpose has proved challenging. Researchers have pointed out ways that teachers' practices and ideas about teaching and learning can affect their ability to use learning-progression-based information (Covitt et al., 2018; Furtak et al., 2014; Gotwals, 2018). Our work with teachers alerts us to another issue having to do with the nature of learning progression frameworks and assessments. Gotwals (2018) summarizes the issue in this way:

> For example, a LP with larger scope and grain size may be more appropriate to inform K–12 standards because standards need to consider students' understandings from when they enter school to when they leave school. However, to be useful for formative assessment purposes, a LP with a smaller scope and grain size, which provides nuances in the shifts in student thinking about specific content, would be more likely to support inferences that teachers could use to support student learning. (Gotwals, 2018, p. 158)

Our experience generally confirmed Gotwals' hypothesis. During the course of instruction, teachers need specifics. "Here is one specific thing to notice and comment on" is more helpful to a teacher than "Student A is at Level 2." Thus we needed to develop tools for fine-grained analysis of students' explanations and to identify specific "next steps" that students can make to improve their explanations.

Using the Learning Progression to Scaffold Students' Explanations

We also wanted to use our learning progressions to design *process tools* that provided scaffolding or temporary support for students' initial explanations, so that teachers could later withdraw the scaffolds as students gained proficiency. We focused on designing four kinds of Explanation Tools, focusing on the four kinds of chemical changes on the second row of Figure 16.1 above: photosynthesis, biosynthesis, cellular respiration, and combustion. Students can use these atomic-molecular scale processes to construct explanations tracing matter and energy through the macroscopic processes in the third row: plant and animal growth, metabolism, and movement, decay, and burning of organic materials.

We found, though, that designing Explanation Tools required more specific analyses of students' performances than our initial learning progression framework provided. Good scaffolds help students to break large complex tasks (like explaining plant growth) into simpler, more manageable tasks, and they guide students in terms that they can understand, even before they are proficient in the higher LP levels. So again, we needed more precise analyses of students' performances to guide our design of Explanation Tools.

Developing an Improved Learning Progression Framework

With these needs in mind, we began the project with the learning progression framework described in Tables 16.1 and 16.2 above, distinguishing three levels of proficiency as students progressed from force-dynamic explanations (Level 2) to model-based explanations (Level 4) that traced matter and energy. In order to develop a learning progression framework that had more of the details we needed for instructional scaffolding and formative assessment, we analyzed interviews conducted by 19 volunteer teachers during the 2012–2013 school year. The teachers interviewed 25 secondary students (21 high-school students and 4 middle-school students) before they began teaching units, and again after three (or more) units were complete.

These 50 interview transcripts were valuable to us because they recorded students talking in their own words about carbon-transforming phenomena. Several interview tasks involved explaining phenomena such as tree growth, a girl jumping, a match burning, a tree decaying, and functions of plant parts. These tasks started with general prompts which students could choose to answer in colloquial or scientific language. Follow-up probes encouraged students to provide more detailed scientific explanations. Other tasks encouraged students to explain relationships among phenomena and ideas, including card-sorting exercises that asked students to group event or materials cards and then explain their reasoning behind their grouped cards.

Table 16.2 A framework for analyzing explanations of carbon-transforming processes in secondary science students

LP levels	*Context-specific knowledge*	*Orientation towards principles of matter and energy*	*Systems & system models; scale*	*Precision in matter and energy words*
Level 4	**Advanced:** Context-specific descriptions of chemical change. This level is consistent with NGSS performance expectations.	**Principle-Oriented:** Using the principles of energy and matter conservation to constrain explanations of familiar and unfamiliar natural phenomena and trace matter and energy separately through carbon-transforming processes.	**All Scales:** Explanations that connect macroscopic scale phenomena with cellular scale and atomic-molecular scale mechanisms.	**More Precise:** Clear distinctions between matter words (*matter, materials, atoms, element, nutrient*, and *molecules*) and energy words (*energy, forms of energy*, & *energy transformation*).
Level 3	**Intermediate:** Confused or incomplete explanations lacking some essential context-specific knowledge.		**Limited Atomic-Molecular Scale:** Explaining atomic-molecular scale processes in *some* contexts, with varying degrees of detail regarding the roles of matter/energy.	
		Fact-Oriented: Describing matter/energy conservation as "facts to be memorized" rather than "reasoning principles that apply to all contexts," thus applying these principles *inconsistently* across contexts and sometimes conflating matter and energy.	**Macroscopic Scale Only:** Explanations either stay within the macroscopic scale or use the vocabulary of the atomic-molecular scale, but in ways that ascribe macroscopic properties to atomic-molecular objects.	**Less Precise:** Conflating the meanings between matter words (*matter, materials, atoms, element, nutrient*, and *molecules*) and energy words (*energy, forms of energy*, and *energy transformation*).
Level 2	**Beginning:** Focusing on actors, enablers, and natural properties of inanimate materials.	**Force-Dynamic:** Characterizing energy as a "life force" and matter as resources for life, alertness, growth, or survival.		

Four of the five co-authors worked through an iterative process of developing a revised learning progression framework and using it to code the transcripts. We were especially interested in identifying *progress variables* describing specific aspects of students' responses that changed at each level. We needed these progress variables to make the learning progression more useful for scaffolding and formative assessment.

Overall, two major findings emerged from this analysis and our goal to develop an LP that would be a stronger tool for instruction and assessment. First, our extended discussions among coders enabled us to develop a revised learning progression framework for interpreting and assessing student explanations. This framework is organized around four progress variables—the columns in Table 16.2. We discuss levels of each progress variable below with examples of student responses at each level. Second, we were able to look at patterns across the 50 pre- and post-interviews to identify how students' explanations changed after studying the *Carbon TIME* pilot units.

Revised Framework with Four Progress Variables

Overall, our analysis identified four progress variables that describe how students' reasoning changed as they advanced toward three-dimensional scientific explanations. Table 16.2 compares levels of achievement for each of these progress variables.

PROGRESS VARIABLE 1: CONTEXT-SPECIFIC KNOWLEDGE

Our first progress variable, Context-Specific Knowledge, refers to content knowledge associated with the Disciplinary Core Ideas from the NRC Framework and NGSS (NGSS Lead States, 2013; NRC, 2012). The phenomena we focused on in the interviews include growth, functioning, movement, death, and decay of living things; and combustion. Scientific explanations of these phenomena require context-specific knowledge about the reactants and products of chemical processes: combustion, photosynthesis, biosynthesis, digestion, and cellular respiration, as well as knowledge of the structure and function of living systems at the cellular and macroscopic scales.

Logan,[1] who began the year with *Force-Dynamic* context-specific knowledge, described tree growth in terms of what a tree needs to "help it live." In the two excerpts from Logan's pre-interview below, Logan explained how a tree uses environmental resources to grow.

INTERVIEWER: How does the tree use soil to grow?
LOGAN: The tree uses the soil to grow because it takes nutrients from the soil and it takes those and it helps it to live.

. . . .

INTERVIEWER: And will the tree do different things with the different kinds of energy?
LOGAN: Use it to grow.
INTERVIEWER: Why does a tree need energy?
LOGAN: Because [of] how it grows.

Scientists would describe these environmental resources as forms of matter (e.g., water, air, soil) and energy (e.g., sunlight). In this pre-interview, however, Logan described what the tree needs to stay alive and grow, rather than attempting context-specific descriptions of how trees transform matter and energy to build biomass. Narrative accounts focusing on organisms' needs and purposes like this were coded *Force-Dynamic* for this first progress variable, which signifies less sophisticated content knowledge.

Compare this to Logan's post-interview, in which he provided a much more sophisticated explanation for tree growth.

INTERVIEWER: How does the tree use air to grow?
LOGAN: The tree takes in the CO_2 and it uses it with water for a process called photosynthesis and sunlight triggers this. Basically, what it does is it transforms those items into food for itself with the sunlight going into chemical energy.
INTERVIEWER: And how does the tree use water to grow?
LOGAN: The tree needs water for photosynthesis and so it takes it in and uses it for that process.
INTERVIEWER: How does the tree use soil nutrients to grow?
LOGAN: It needs some of the nitrogen inside the soil nutrients for its biomass.
INTERVIEWER: And how does the tree use sunlight to grow?
LOGAN: The sunlight triggers the photosynthesis process …
INTERVIEWER: All right. Does the tree do anything with the air that surrounds it?
LOGAN: It takes in the CO_2 and puts out oxygen.

In this post-interview, Logan's accounts of the processes and materials involved in tree growth are what we would expect in an advanced explanation. He identified inputs and outputs of photosynthesis for both matter and energy, as well as mentioning the role of soil nutrients in biosynthesis. Because of these context-specific details included in his account, this particular segment was coded *Advanced* for the context-specific knowledge progress variable (Table 16.2).

However, it is important to note that context-specific knowledge does not transfer to other processes. Although Logan offered a thorough explanation for how a tree used matter and energy for growth in the excerpt above, other exchanges in his post-interview revealed that, at the end of the year, he lacked a complete explanation for how cellular respiration occurs in multiple contexts. In the following segment from his post-interview, Logan was asked to sort cards into groups based on how the phenomena on the cards change the air and to then explain the reasoning behind his grouping strategy. Here is his explanation of why he separated the tree decaying card from other cards involving oxidation of organic carbon (girl jumping, the car running, flame burning, little girl growing, the tree growing).

LOGAN: Because the flame would take in gas. And then, like respiration for [girl growing and tree growing], I guess that's what I would call that group.
INTERVIEWER: Do you think the tree decaying is doing any sort of respiration process?
LOGAN: Well, it might. I don't think it's so, at least if it's anything, it's not as much as [the items in the second group].
INTERVIEWER: And by respiration, can you explain to me what you mean by that?
LOGAN: Like the exchange of gases.

This exchange shows that Logan's context-specific knowledge did not extend to explaining the role of cellular respiration by decomposers in the decay process, partly because he described respiration as like breathing at the macroscopic scale: "the exchange of gases." This was a common pattern, where students used different levels of context-specific knowledge in their explanations of different phenomena.

PROGRESS VARIABLE 2: USE OF PRINCIPLES OF MATTER AND ENERGY

Our second progress variable, *Use of Principles of Matter and Energy*, refers to the principles most closely associated with the CCC "Matter and energy: flows, cycles, and conservation." We call these "principles" because they represent rules and laws that apply across contexts. Conservation of matter is a powerful principle because it tells us that regardless of how atoms are arranged and rearranged, they cannot be created or destroyed in physical or chemical changes. Similarly, conservation of energy is a principle saying that explanations must trace energy through processes, rather than suggesting that energy can be created or "used up."

Kayla began the year with reasoning that we coded *Force-dynamic* with respect to the second progress variable (Table 16.2), meaning that she focused on the purposes and results of processes in living systems rather than on transformations of matter and energy. Kayla's pre-interview contained a variety of excerpts in which she associated energy with life and living things. For example, in this excerpt, the teacher asked her to describe how some items (dead wood, living branches, and a cow) are different with respect to energy.

INTERVIEWER: Can you think of a way that these are different in terms of energy?
KAYLA: Okay. The dead wood is not alive. The living branches are alive. The muscles in the cow are still alive. So the dead wood is dead. The living branch is alive and also the cow is alive.

In Kayla's post-interview, her explanations showed a shift toward treating matter and energy as entities that could be traced through systems, but her explanations were not always constrained by the principles of matter and energy conservation. For example, in the following excerpt, her teacher asked her if matter was ever "used up" in an ecosphere, which is an enclosed glass container with a few living organisms and water:

INTERVIEWER: Do you think that matter changes from one form to another in the ecosphere?
KAYLA: Yes, matter can recycle through an ecosystem.
INTERVIEWER: Does matter get used up when it changes from one form to another?
KAYLA: I wouldn't say it gets used up. I would just say it gets changed into a different type of matter until that matter is not needed in that plant. So like if you get food you don't use the feces that come out.

However, later in the same interview, she contradicted this commitment to conservation when her teacher asked about energy conservation in an ecosphere:

INTERVIEWER: Do you think that energy that's going into the ecosphere changes from one form to another?
KAYLA: Yes. I think it goes in. As for like instance, like the light go into the algae, [and] then it goes into heat energy and then goes through a cycle going through matter using different chemical energies and then going back out into different parts of the environment as heat energy.
INTERVIEWER: So does the energy get used up when it changes form?
KAYLA: Yes.

Sometimes Kayla traced energy and matter through systems in a way we associate with *Principle-Oriented* use of conservation laws of matter and energy. However, when the teacher asked if the energy could be used up, Kayla broke the rule of energy conservation by confirming that energy could be used up.

This inconsistent application of the principles of matter and energy was common in other students' transcripts as well. For example, another student, Xavier, stated in his pre-interview that "matter cannot be created or destroyed," but earlier in the same transcript contradicted this by suggesting that as a tree grows it "would be creating matter."

We argue that Kyla and Xavier interpreted the conservation laws as context-specific facts instead of principles that constrain all natural systems. Because of this inconsistency, Kayla and Xavier's post-interview transcripts received a *Fact-oriented* code for the use of principles of matter and energy progress variable.

PROGRESS VARIABLE 3: SCALE

Coding for this progress variable revealed three levels: *Macroscopic Scale Only* (less sophisticated), *Limited Atomic-Molecular Scale* (mid-level), and *All Scales* (most sophisticated). *Macroscopic Scale Only* explanations either (a) omitted the atomic-molecular scale entirely, or (b) used the language of the atomic-molecular scale in ways that ascribed macroscopic properties to atoms or molecules. For example, Daniel's explanation of tree growth was coded *Macroscopic Scale Only*.

INTERVIEWER: Is there a connection between exchanging gases and growing for the tree? So you just said it exchanges [gases]. It brings in the carbon dioxide and then it releases the oxygen. So that's the gas exchange we're talking about.

DANIEL: Okay.

INTERVIEWER: Does that gas exchange do anything to help with growing for the tree? Is there a connection there?

DANIEL: I think there could be. I guess like the carbon dioxide … I don't know that much about trees, but it could be something that affects the growth, just because …

INTERVIEWER: Go ahead and think about it, and tell me how would you describe that possibly affecting the growth. So specifically, the carbon dioxide that you just mentioned.

DANIEL: That's tricky. I don't know. Maybe one was of it allows oxygen and carbon dioxide to move through the tree. And the more it breathes in, the more space it needs to clean it or whatever it does, and let it out.

While using the terms "oxygen and carbon dioxide" repeatedly throughout his interview, he ascribed only macroscopic movements and properties (e.g., states of matter, temperature, color) to these gases and other materials. In this case, the gases take up space and need to be "cleaned," but do not interact within systems at an atomic-molecular scale. These types of accounts often contained false indicators of sophistication, because they included names of molecules but not reasoning at the atomic-molecular scale.

Explanations were coded *Limited Atomic-Molecular* described matter or energy interaction at the atomic-molecular scale in some (but not all) contexts, with some errors in their reasoning. Students who were able to connect spatial scales consistently in their explanations were coded *All Scales*. These students (a) began to tell a story of chemical change as atoms being rearranged to make new molecules, (b) identified that energy is stored in bonds or molecules, and (c) characterized energy as the movement of molecules.

For example, Patrick described matter at the atomic-molecular scale by telling a story of chemical change as atoms being rearranged to make new molecules: "Like when photosynthesis, obviously the energy source is the sun and then it's used to split the carbon dioxide and oxygen apart. So the carbon helps build the tree and the oxygen is used as a waste product." This signified the use of scale as an epistemic heuristic to explain mechanisms of tree growth at an atomic-molecular scale.

PROGRESS VARIABLE 4: PRECISION IN MATTER AND ENERGY WORDS

Our final progress variable, precision in matter and energy words, refers to how clearly students distinguish between different words used in scientific and familiar language to refer to forms of matter and energy. These included:

- Words that were already familiar to most students, but were used with more precise and restricted meanings in scientific contexts, such as *energy, matter, food, sugar, nutrient, mineral,* and others.
- Words that specifically described matter at the atomic-molecular scale, such as *atom, molecule,* and the names of specific atoms and molecules.
- Words that could be used to describe both macroscopic materials and atoms or molecules, such as *oxygen, carbon dioxide, sugar, glucose, water, fat, protein, carbohydrate,* and others.

This progress variable is divided into just two levels (Table 16.2). *Less Precise* explanations conflate the meanings within and among matter words (matter, materials, atoms, element, nutrient, and molecules) and energy words (words related to energy, forms of energy, and energy transformation) and among words referring to matter and energy at different scales. *More precise* explanations make consistently clear distinctions within and among matter and energy words and among systems at different scales.

For example, Isabella used Less Precise language to describe the differences among three organic materials.

INTERVIEWER: Okay. Can you think of ways [living branches, muscles in a cow, and dead wood] are different with respect to energy?

ISABELLA: Sure. The living branches are absorbing energy from the air, whereas the muscles in the cow are just moving. It doesn't absorb any air from the atmosphere. It gets it from these branches or you make these branches grass or something like that … . And then the dead wood is basically decomposed wood that it was given back that energy that it had back to the air.

In this excerpt, Isabella recognized that the exchange of matter and energy between organisms and their environments was important, but she lacked the precise language needed to clearly distinguish between matter and energy and explain how they were transformed as they moved within the systems in question.

In contrast, Victoria was asked to sort materials into two groups. In the excerpts below, Victoria's teacher asked her to explain why she sorted these specific items into the same group:

INTERVIEWER: So you said that living branches, gasoline, dead wood, and cows are alike?

VICTORIA: Uh-hmmm.

INTERVIEWER: Can you explain to me why you put those in the alike pile?

VICTORIA:	... Because they're all organic and they have C to C and C to H bonds which store a lot of energy.
	....
INTERVIEWER:	So you say that sand and glass and carbon dioxide and water and soil have the least amount of stored energy in them?
VICTORIA:	Uh-hmmm.
INTERVIEWER:	Can you tell me why?
VICTORIA:	... like the water, it has hydrogen atoms in it, but it doesn't have carbon atoms to make a high-energy bond. And the same with the soil and the same with the carbon, glass, and sand.

In these explanations, Victoria described how atoms bond to each other, and identified the specific bonds needed to construct molecules that "store energy," demonstrating a command of precise language needed to describe what makes the items "organic." Due to Victoria's precise use of words related to atoms, molecules, and matter, Victoria's transcript was coded *More Sophisticated* for Precision in Matter and Energy Words. The linguistic features of her explanation demonstrated a consistent use of key terms related to matter and energy needed to form a precise scientific explanation.

While we generally don't advocate for nit-picking about definitions of specific vocabulary words, in this case, the principles behind the other progress variables rely on precise word usage. In everyday language, we can and should continue to say, "I have lots of energy today because I got a good night's sleep." But the principle of energy conservation requires a much more precise and restricted definition of "energy." Similarly, it is important to recognize that molecules, but not atoms, can be destroyed or created in chemical changes.

RELATIONSHIPS AMONG PROGRESS VARIABLES

These examples show how all four progress variables play essential and interdependent roles in scientific explanations that trace matter and energy through carbon-transforming processes. For example, an explanation of tree growth that applies a more sophisticated epistemic heuristic for the conservation of matter and energy would need to include the following:

1 Context-specific information about trees' structure and function. For example, trees have dual circulation systems, carrying water and minerals from the roots to all parts of the tree, and sugar from the leaves to all parts of the tree;
2 Conservation of matter and energy. The explanation is constrained by this principle, separately tracing matter from carbon dioxide and water to glucose and oxygen, then to other molecules that constitute the tree, and tracing energy from sunlight to energy stored in chemical bonds.
3 Connecting scales. This heuristic prompts students to connect macroscopic phenomena to atomic-molecular models. In colloquial discourse, it seems reasonable the trees would be able to convert one colorless, odorless gas—carbon dioxide—into another colorless, odorless gas—oxygen. On the other hand, it seems absurd to claim that the mass of a mature tree comes mostly from carbon dioxide. Only by connecting macroscopic phenomena with atomic-molecular models can students explain how the latter claim is supported by scientific models and evidence.
4 Precision in language use. For example, explanations must distinguish between "carbon" the element and "carbon dioxide" the gas, or between "sunlight" as a form of energy and "glucose" as a form of matter that has available chemical energy.

In combination, these four progress variables also explain how students can be prepared for future learning—another key goal for our project. In particular, only the first progress variable, Context-specific Knowledge, relies on specific facts about different phenomena. The other three progress variables each invoke principles or heuristics that students can use to generate questions and hypotheses and to constrain the explanations that they generate for new phenomena.

Evidence about Students' Learning from Pilot Units

Overall, patterns across our 50 interviews were consistent with patterns we saw in earlier work: almost all students improved from pre-interviews to post-interviews, but only a few advanced to the highest level (consistent with *NGSS* performance expectations) for all four progress variables. One question this raises for us is: *why is that last step so difficult?*

Our reading of patterns across the interviews suggests two answers to this question. One focuses on the complexity of the goal performances: our revised learning progression framework (Table 16.2), unpacks many aspects of students' reasoning that were sometimes mentioned in our earlier framework (Table 16.1) but not specifically analyzed. The post-interview results suggest that the scaffolds and assessment rubrics in the pilot units—based on the Table 16.1 framework—were sufficient to help most students attempt explanations that focused on matter and energy (i.e., Level 3 reasoning) but inadequate to help them achieve full Level 4 proficiency.

Our second answer focuses on the distinction between context-specific *facts* and more generally applicable *principles*. In the post-interviews, 24 of the 25 students referred to the facts of matter and energy conservation, but only 8 students applied them consistently as principles. The principles can be stated simply; most students learned to state them and sometimes use them appropriately. What they failed to do is develop proficiency in applying the conservation laws to explanations of every phenomenon.

These results suggest that Level 4 discourse—constructing explanations for specific scientific phenomena that use the principles of matter and energy—is a challenging goal for secondary students. Most students needed better scaffolding and more precise feedback on their explanation performances than teachers using *Carbon TIME* pilot units were able to provide.

Using the Learning Progression to Improve the Curriculum

Our revised LP framework (Table 16.2) represents a significant shift in perspective from the LP framework that we had at the beginning of the project (Table 16.1). The original framework focused on *levels* of student explanations (rows of Table 16.1), while in the revised framework the emphasis shifted to *progress variables* (columns of Table 16.2). The four progress variables identified in these analyses enabled us to improve scaffolding and assessment for explanations in the curriculum in several ways. (All unit materials are available on the *Carbon TIME* website: https://carbontime.create4stem.msu.edu/),

- *Designing frameworks and Explanation Tools.* We have organized units around frameworks and tools that explicitly target these progress variables. We discuss an example—the Photosynthesis Explanation Tool—below.
- *Developing classroom discourse routines.* We worked to develop classroom discourse routines that use sequenced divergent and convergent instructional practices to scaffold students' engagement with phenomena (described in the *Carbon TIME Discourse Routine* educator resource on the project website).

- *Formative and summative assessment.* We developed tools for formative assessment that reveal how successfully students are engaging each progress variable, as well as grading tools that suggest ways to give students credit for elements of their explanations that match each progress variable.
- *Professional development.* We developed activities and materials that engage teachers in interpreting student texts and scaffolding students' talk and writing around these progress variables (available in the Professional Development Resources in the *Carbon TIME* Library on the website).

An Example: The Photosynthesis Explanation Tool

The Photosynthesis Explanation Tool (Figure 16.2) is one example of how we used the LP framework and progress variables to design scaffolds for students' explanations. As discussed in Chapter 15, students use the Explanation Tools after they have completed the questioning

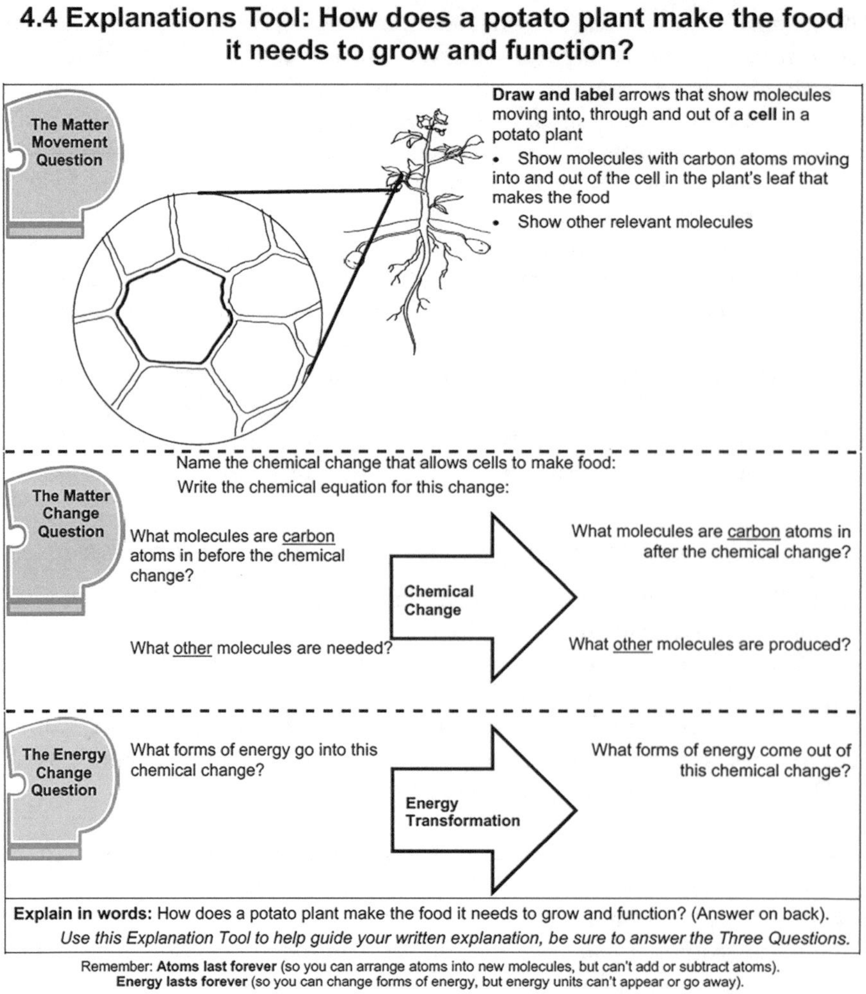

Figure 16.2 Photosynthesis explanations tool.

and investigating parts of the unit, learning from their plant growth investigations that plants absorb carbon dioxide as they grow, and that growing plants gain more dry mass than the soil loses. During the Explaining part of the unit students study how plant growth and functioning involve a three-step process (see Figure 16.1 above), with each step having its own Explanation Tool:

1 Photosynthesis: plant leaf cells make glucose from carbon dioxide and water, with oxygen gas as a waste product.
2 Biosynthesis: the glucose is transported to other parts of the plant, where cells make other organic molecules from glucose and soil minerals.
3 Cellular respiration: all plant cells get energy by oxidizing glucose and using the energy released by the chemical change.

Before they use the Photosynthesis Explanation Tool, students have also developed context-specific knowledge by using molecular models to make carbon dioxide and water molecules and then rearranging the atoms into glucose and oxygen molecules. The tool is designed for a lesson organized around the *Carbon TIME* discourse routine: students (a) discuss the goal of the lesson and how it fits into the unit storyline, (b) complete the process tool individually, (c) compare and revise their explanations in small groups, and (d) participate in a class consensus-seeking discussion about the key elements of a good explanation of photosynthesis.

The goal performance for the explanation tool is a good answer to the last question near the bottom of the tool: *how does a potato plant make the food it needs to grow and function?* This question requires a paragraph-long explanation that addresses all four progress variables. The first part of the Explanation Tool consists of simpler questions that provide scaffolds for the students' final paragraphs. All Explanation Tools are organized around the Three Questions, a general scaffold saying that matter- and energy-tracing explanations need to answer the Matter Movement Question, the Matter Change Question, and the Energy Change Question. The contents and organization of the tool address each of the four progress variables, as described below.

PROGRESS VARIABLE 1: CONTEXT-SPECIFIC KNOWLEDGE

The students' prior experiences, including investigations, molecular modeling, and presentations, provide access to the context-specific information they need to explain photosynthesis, including information about the structure and function of plants and plant cells and the reactants and products of the photosynthesis chemical change. So their challenge at this point is to use that information as they construct a coherent explanation—constrained by principles and using heuristics—of the chemical changes in context. The Explanation Tool prompts them to remember and use this knowledge.

PROGRESS VARIABLE 2: USE OF PRINCIPLES OF MATTER AND ENERGY

Both the reminders of the conservation laws at the bottom of the tool and the structure of the tool itself help the students construct an explanation that is constrained by the conservation principles. Each section of the tool, for example, requires students to trace matter and energy by identifying reactants and products while accounting for all of the atoms and types of energy involved.

PROGRESS VARIABLE 3: SCALE

Like other CCCs, spatial scale often operates implicitly when we speak. Teachers, scientists, and curriculum materials commonly switch from one scale to another without explicit verbal signposts or markers to indicate (a) what spatial scale they were speaking about currently or (b) when they were transitioning from one scale to another. This tool and other scaffolds use the metaphor of "zooming in" from the macroscopic to atomic-molecular scale to help students work through how visible phenomena at the macroscopic scale can be explained by invisible mechanisms at the cellular and atomic-molecular scale.

PROGRESS VARIABLE 4: PRECISION IN MATTER AND ENERGY WORDS

To trace matter through the process of photosynthesis, learners must not only recognize and be able to verbally express the difference between atoms (e.g., carbon, oxygen, hydrogen), molecules (e.g., oxygen, carbon dioxide, glucose), and materials (e.g., air, wood, leaf), but they must also use these terms to express their knowledge of what is happening to the matter and energy when a potato plant grows. The structure of the three questions and the specific prompts enable students to make careful distinctions between matter and energy, between atoms and molecules, and among system models at different scales.

SUPPORTING FORMATIVE ASSESSMENT AND GRADING

As we noted above, the levels of our initial LP framework (Table 16.1) afforded teachers insight into their students' reasoning but not specific guidance about how they could respond to students' lower-level explanations or grade their students' work. The progress variables enabled us to improve guidance for teachers about assessing and grading the program's process tools and tests. Our revised assessing and grading rubrics suggested ways to recognize specific problems and provide students with feedback, as well as to grade students' work using rubrics based on the progress variables. We were also able to improve teachers' professional learning by helping them use the progress variables to recognize specific strengths and weaknesses in students' talk and writing.

Evidence about the Effectiveness of Changes

While the pilot *Carbon TIME* units enabled most students to make progress, very few achieved our goals for all four progress variables. We believe that the changes we made in the units and assessments improved the units substantially. But what does the evidence about student learning show? In this section, we discuss evidence about how teachers used the revised units and the effects on student learning.

Evidence for Overall Effectiveness

Our quantitative analysis (Lin et al., 2022) reports results from a data set that included 59,654 student assessments collected in classrooms of 133 middle- and high-school teachers over a four-year period. Two key findings are important for this article.

1 *Overall success: Carbon TIME* resources improved students' overall proficiency in practices related to carbon cycling compared to other curricula used by the same teachers, and students with lower pre-test scores showed higher learning gains.

2 *Variation among classrooms*: Demographic factors historically associated with inequities in educational systems (percent free and reduced lunch, and percent marginalized students of color) were also associated with different student learning gains, but these accounted for much less of the variance than differences among teachers.

DIFFERENCES AMONG TEACHERS

We also used these analyses to construct *value-added measures of learning gains for individual teachers* that control for pre-test scores and school demographic factors: although the measures do not support precise ranking of individual teachers, they can enable examination of patterns of variation between classrooms with higher and lower learning gains.

We found that our revised *Carbon TIME* resources enabled virtually all the participating teachers to engage their students in active learning and improve the quality of their students' explanations, but there were important differences in the expertise and commitments of individual teachers. Teachers with lower learning gains used *Carbon TIME* resources for what Hess and Azuma (1991) refer to as "quick and snappy" teaching, relying on hands-on activities, novel facts, and a fast pace to keep students engaged. Teachers with higher learning gains made much more effective use of the improved resources for scaffolding and formative assessment, using apprenticeship teaching strategies to develop proficiency around all four progress variables in their explanations (Covitt et al., 2024).

Conclusion

The main story we have to tell in this chapter is about how we created a more useful version of our learning progression for explanations that trace matter and energy through carbon-transforming processes, shifting from a framework that emphasized learning progression levels (rows of Table 16.1) to a framework that emphasizes progress variables (columns of Table 16.2). The new framework affords us deeper insights into the key components of three-dimensional scientific explanations and into how students develop proficiency in each of those components.

Traditional teaching prioritized context-specific knowledge (DCIs) and positioned teachers as graders of students' answers. This framing causes teachers to focus their attention on right versus wrong answers rather than on students' reasoning processes. Our work demonstrates that it is the CCCs and SEPs that prepare students for future learning. Teachers who used the CCCs as principles and heuristic tools better-prepared students for explaining new phenomena.

The revised framework and its progress variables are also important because they afford insights into a key aspect of environmental science literacy: preparation for future learning. One of the four progress variables—context-specific knowledge—requires learning new facts about each new phenomenon that students encounter; students can prepare for future learning by knowing questions that they can ask about these facts. The other three progress variables each describe CCC-based principles or epistemic heuristics that students can use for all matter- and energy-tracing explanations of these phenomena. Thus, students who are proficient in these progress variables are well prepared to generate model-based explanations of new phenomena, preparing them to be productive citizens in a world where they need to understand how carbon-transforming processes will shape their lives.

We have also described how we used the progress variables from this learning progression to improve scaffolding and assessment resources for student explanations in *Carbon TIME* units, as well as evidence that these improvements helped to increase student learning in virtually all participating classrooms. The most significant improvements, though, came in the classrooms of teachers who shared a commitment to environmental science literacy for all students and expertise in apprenticeship teaching strategies. These findings help us to understand the challenges that lie ahead.

Acknowledgement

This research was supported in part by grants from the National Science Foundation: Enacting Responsive and Rigorous Teaching Based on Carbon TIME (DRL 1020187) and Sustaining Responsive and Rigorous Teaching Based on *Carbon TIME* (NSF 1440988). Any opinions, findings, and conclusions or recommendations expressed in this material are those of the author(s) and do not necessarily reflect the views of the National Science Foundation.

Please visit our website: https://carbontime.create4stem.msu.edu/

Note

1 All student names are pseudonyms.

References

Alonzo, A. C. (2018). An argument for formative assessment with science learning progressions. *Applied Measurement in Education*, *31*(2), 104–112. doi:10.1080/08957347.2017.1408630

Anderson, C. W., de los Santos, E. X., Bodbyl Roels, S., Covitt, B., Edwards, K. D., Hancock, J. B., Lin, Q., Morrison Thomas, C., Penuel, W., & Welch, M. (2018). Designing educational systems to support enactment of the *Next Generation Science Standards*. *Journal of Research in Science Teaching*, *55*(7), 1026–1052.

Anderson, C. W., Nordine, J., & Welch, M. (2021). Energy and matter: Flows, cycles, and conservation. In J. Nordine & O. Lee (Eds.), *Crosscutting concepts: Strengthening science and engineering learning* (pp. 165–194). Arlington, VA: NSTA Press.

Braaten, M., & Windschitl, M. (2011). Working toward a stronger conceptualization of scientific explanation for science education. *Science Education*, *95*(4), 639–669. doi: 10.1002/sce.20449

Bransford, J. D., & Schwartz, D. L. (1999). Rethinking transfer: A simple proposal with multiple implications. *Review of Research in Education*, *24*(1), 61–100.

Collins, A., Brown, J. S., & Holum, A. (1991). Cognitive apprenticeship: making thinking visible. *American Educator*, *6*(11), 38–46.

Collins, A. B., Brown, J. S., & Newman, S. E. (1989). Cognitive apprenticeship: Teaching the craft of reading, writing and mathematics. In L. B. Resnick (Ed.), *Knowing, learning and instruction: Essays in honor of Robert Glaser* (pp. 453–494). Hillsdale: Lawrence Erlbaum Associates, Inc.

Covitt, B. A., de los Santos, E. X., Lin, Q., Morrison Thomas, C., & Anderson, C. W. (2024). Instructional practices in secondary science: How teachers achieve local and standards-based success. *Journal of Research in Science Teaching*, *61*(1), 170–202.

Covitt, B. A., Parker, J. M., Kohn, C., Lee, M., Lin, Q., & Anderson, C. W. (2020). Understanding and responding to challenges students face when engaging in carbon cycle pool-and-flux reasoning. *The Journal of Environmental Education*, *52*(2), 98–117.

Covitt, B. A., Gunckel, K. L., Caplan, B., & Syswerda, S. (2018). Teachers' use of learning progression-based formative assessment in water instruction. *Applied Measurement in Education*, *31*(2), 128–142. doi: 10.1080/08957347.2017.1408627

Dauer, J. M., Miller, H. K., & Anderson, C. W. (2015). Conservation of energy: An analytical tool for student accounts of carbon-transforming processes. In R. Chen, A. Eisenkraft, D. Fortus, J. Krajcik, K. Neumann, & A. Scheff (Eds.), *Teaching and learning of energy in K-12 education* (pp. 47–61). Switzerland: Springer International.

Furtak, E. M., Morrison, D. E. B., & Kroog, H. (2014). Investigating the link between learning progressions and classroom assessment. *Science Education*, *98*(4), 640–673. doi: 10.1002/sce.21122

Gotwals, A. M. (2018). Where are we now? Learning progressions and formative assessment. *Applied Measurement in Education*, *31*(2), 157–164. doi: 10.1080/08957347.2017.1408626

Hartley, L. M., Wilke, B. J., Schramm, J. W., D'Avanzo, C., & Anderson, C. W. (2011). College students' understanding of the carbon cycle: Contrasting principle-based and informal reasoning. *BioScience*, *61*(1), 65–75. doi: 10.1525/bio.2011.61.1.12

Hess, R. D., & Azuma, H. (1991). Cultural support for schooling: Contrasts between Japan and the United States. *Educational researcher*, *20*(9), 2–9.

Jin, H., & Anderson, C. W. (2012). A learning progression for energy in socio-ecological systems. *Journal of Research in Science Teaching*, *49*(9), 1149–1180. doi: 10.1002/tea.21051

Krist, C., Schwarz, C. V., & Reiser, B. J. (2019). Identifying essential epistemic heuristics for guiding mechanistic reasoning in science learning. *Journal of the Learning Sciences*, *28*(2), 160–205.

Lave, J., & Wenger, E. (1991). *Situated learning: Legitimate peripheral participation.* Chicago: Cambridge University Press.

Lee, H.-S., & Liu, O. L. (2009). Assessing learning progression of energy concepts across middle school grades: The knowledge integration perspective. *Science Education*, *94*(4), 665–688. doi: 10.1002/sce.20382

Lin, Q., Frank, K. A., Bathia, S., Thomas, J., Draney, K., & Anderson, C. W. (2022). Factors affecting students' learning from a design-based implementation research project in diverse education systems. *Journal of Research in Science Teaching*, *59*(5), 808–840.

Miller, H., Webster, A., Dauer, J., Anderson, C. W. (2013). Alternative learning trajectories toward Understanding matter and energy in socio-ecological systems. Presented at the Annual Meeting of the National Association for Research in Science Teaching, Rio Grande, Puerto Rico, April 6-9, 2013.

Modell, H. I. (2000). How to help students understand physiology? Emphasize general models. *Advances in Physiology Education*, *23*, 101–107.

Mohan, L., Chen, J., & Anderson, C. W. (2009). Developing a multi-year learning progression for carbon cycling in socio-ecological systems. *Journal of Research in Science Teaching*, *46*(6), 675–698. doi: 10.1002/tea.20314

NGSS Lead States. (2013). *Next generation science standards: For states, by states.* Washington, DC: The National Academies Press.

National Research Council. (2012). *A framework for K-12 science education: Practices, cross cutting concepts, and core ideas.* Washington, DC: National Academies Press.

Pinker, S. (2007). *The stuff of thought.* New York: Penguin.

Rice, J., Doherty, J. H., & Anderson, C. W. (2014). Principles, first and foremost: a tool for understanding biological processes. *Journal of College Science Teaching*, *43*(3), 74–82.

Stevens, S. Y., Delgado, C., & Krajcik, J. S. (2009). Developing a hypothetical multi-dimensional learning progression for the nature of matter. *Journal of Research in Science Teaching*, *47*(6), 687–715. doi: 10.1002/tea.20324

Talmy, L. (1988). Force dynamics in language and cognition. *Cognitive Science*, *12*(1), 49–100.

17

REFLECTION ON THE LEARNING PROGRESSION APPROACH TO PROMOTING STUDENT LEARNING

Commentary for Section II

Xiufeng Liu

As one of the foundations for the *Next Generation Science Standards* (NGSS Lead States, 2013), learning progressions (LPs) have been driving reform efforts in curriculum, instruction, and assessment in the United States (U.S.) over the past decade (Krajcik & Shin, 2023). The six chapters in this section represent LP development efforts to promote student learning in major K-12 disciplinary domains, such as biology, chemistry, earth science, environmental science, and physics. In this chapter, I will first reflect on the origins of LPs; I will then summarize the nature and characteristics of LPs put forward in the literature. Based on the nature and characteristics of LPs, I will propose a framework of review criteria for LPs, and using this framework, I will comment on specific LPs reported in Section II. I will conclude this chapter by identifying a few areas that should be attended to in future research and development of LPs.

Drivers for Learning Progressions

LPs are a result of at least three related driving forces. The first is productive research on students' conceptions and conceptual change in science (e.g., Chi et al., 1994; Pintrich et al., 1993; Posner et al., 1982; Tyson et al., 1997; Wandersee et al., 1994). Here, I highlight alternative conceptions and conceptual change research, although this area is closely related to research in other areas including didactics, teaching experiments, theory of mind, and metacognitive development (Duschl et al., 2011). Alternative conceptions research began in the late 1970s (e.g., Driver, 1983; Driver & Easley, 1978). As our knowledge about students' conceptions of almost every major concept in science accumulated, researchers began to make sense of conceptual trajectories, or students' concept development over the years, and to call for designing curriculum and instruction based on these conceptual trajectories (e.g., Driver et al., 1994a, 1994b). New theories to explain how students' conceptions of science develop over time were also in demand. For example, Liu and his colleagues applied neo-Piagetian theories to examine how students from elementary to high school might develop their concepts of matter and energy (Liu & Collard, 2005; Liu & Lesniak, 2005, 2006; Liu & McKeough, 2005).

DOI: 10.4324/9781003170785-19

The *Atlas of Science Literacy* (AAAS, 2007) maps pathways of students' science concept development from K-12 in all domains of science from an expert perspective. However, these early attempts to map students' science concept development over the years were detached from curriculum and instruction; at most, they reflected typical or status quo science teaching and learning that was considered unsatisfactory. Also, these initial maps were disconnected from "doing science" or engaging in science practices. Effective curriculum and instruction facilitated students' long-term concept development theories and strategies were desired. LPs fulfill this desire because, in addition to mapping out students' long-term concept development, they also integrate with science and engineering practices (NRC, 2007), and later with crosscutting concepts as well (NGSS Lead States, 2013; NRC, 2012). LPs intend to promote students' abilities to apply their knowledge and understanding to solve real-world problems; they value knowledge-in-use (Krajcik & Shin, 2023).

Another driving force for LPs is the wave of curriculum comparison studies in the early 2000s. The 1996 Third International Math and Science Study (TIMSS) revealed that U.S. students in 4th, 8th, and 12th grades hovered only around average performance among participating countries, which sent a shock wave in academic, social, and political communities. Searches for factors that might have contributed to American students' mediocre performance followed. One identified factor was the U.S. science curriculum, which was described as "a mile wide and an inch deep" (Schmidt et al., 2002, p. 122). Compared to their counterparts abroad, U.S. science teachers cover a variety of subjects every school year, and, as a result, students seldom get to explore key topics in any depth. In a follow-up study, Schmidt and collaborators (2005) compared the science content coverage in different grades in the U.S. *National Science Education Standards* (NRC, 1996), as well as the science standards in 12 U.S. states, to the content topics covered in each grade of 6 TIMSS highest-achieving countries. They found that in the U.S., topics entered and lingered, so each grade typically devoted instructional attention to substantially more topics, as compared with top-performing countries, where new science topics were gradually introduced, part of instruction for a few grades, and then left the curriculum as separate topics. Researchers subsequently called for coherent, "less is more" curricula across grade levels.

To facilitate the development of coherent curricula, Project 2061 released a resource, *Designs for Science Literacy* (AAAS, 2001). *Designs* defines different types of curricular coherence. The logical coherence is a sequence of topics that is logically reasonable and, if appropriate, represents the disciplinary content's hierarchical nature from which the subject matter derives, such as from qualitative to quantitative, descriptive to predictive, observational to experimental, macroscopic to microscopic, experimental to computer modeling, and simple to complex. The developmental coherence is a sequence of topics that is consistent with the psychological sequence of students' content learning, such as from prior knowledge/alternative conceptions to scientific conceptions; sensorimotor to pre-operational to concrete operational to formal operational; type-of characteristics to causal to logico-hypothetico relations; inner speech to outer speech; and number sense to relational to spatial to statistical reasoning. Literacy coherence is a sequence of topics that is consistent with the stated vision of scientific literacy (e.g., *Benchmarks of Scientific Literacy*) in terms of content specification and grade level specification. Finally, thematic coherence is a sequence of topics that is consistent with the given set of unifying themes, such as systems, models, constancy and change, and scale. LPs respond to the above call for curriculum coherence across grade levels; it goes further to include coherence among curriculum, instruction, and assessment (see below).

A third driving force for LPs is the call for systems of assessment. As a result of the 2002 *No Child Left Behind Act* in the U.S., U.S. states must develop and implement challenging standards and assess students' annual progress in achieving the standards from grades 3 through 8. To guide the states in developing standards-based assessments, the National Research Council Committee developed a report, *Systems for State Science Assessment* (NRC, 2006). The report called for states to develop systems of assessment that are horizontally coherent (among assessment, instruction, and curriculum), developmentally coherent (from grade to grade), and vertically coherent (among classroom assessment, school/district assessment, state assessment, and national assessment such as NAEP). To achieve the above coherences, *Systems* introduced the idea of LPs for the first time as descriptions of successively more sophisticated ways of thinking (or moving towards more expert understanding) about an idea. Sample LPs in physical science and biology were commissioned (Catley et al., 2005; Smith et al., 2006). LPs are intended to achieve the coherences called for in the academic, policy, and practitioner communities.

In summary, LPs are a result of advances in research on science teaching and learning, and curriculum and assessment; they are intended to be a lever for promoting in-depth student learning and coherence among curriculum, instruction, and assessment. LPs take a systems approach by simultaneously considering major factors affecting students' learning and conceptual change, such as students' prior knowledge, curriculum coherence, meaningful and engaging contexts, development over time, and self-reflection (Krajcik & Shin, 2023). LPs offer a viable solution to the long-sought goal of achieving coherence among curriculum, instruction, and assessment.

Nature and Characteristics of LPs

The *Framework* (NRC, 2012) defines LPs as a roadmap: "If mastery of a core idea in a science discipline is the ultimate educational destination, then well-designed learning progressions provide a map of the routes that can be taken to reach that destination" (NRC, 2012, p. 26). There are other categorizations of LPs in the literature. LPs can be viewed as necessary "strategies for formulating and developing environments of learning that align curriculum, instruction and assessment" and "conjectural or hypothetical model pathways of learning over periods of time that have been empirically validated" (Duschl et al., 2011, p. 124). Along the line of LPs as conceptual models, Lehrer and Schauble (2015) describe LPs as "models of how student learning develops over an extended period of time—ideally, several years" (p. 433). As models, LPs are developed for specific purposes and can explain and predict how student learning develops over time. Lehrer and Schauble (2015) particularly emphasize that LPs aim for transformative changes in teaching and learning instead of conserving conventional or status quo teaching and learning. From this perspective, LPs are aspirational and progressive. In fact, instructional support as part of LPs is explicitly stated in the *Framework*: "Such progressions describe both how students' understanding of the idea matures over time and the instructional supports and experiences that are needed for them to make progress" (NRC, 2012, p. 26).

Duschl and colleagues (2011) differentiate two types of LPs: validation LPs and evolutional LPs. Validation LPs promote standards-based progressions with instruction as interventions; they are theory-driven, with fixed progress variable steps and targets. Validation LPs adopt a misconception-based "fix it" view of conceptual change with a domain-general orientation to topic selection. On the other hand, evolutionary LPs are based on sequencing teaching experiments across multiple grades, with instruction refining progressions; they are evidence-driven, with flexible progress variable steps and targets. Evolutionary LPs adopt an intuition-based "work with it" view of conceptual change, with a domain-specific orientation to topic selection.

Given the abovementioned conceptual nature of LPs, researchers have identified important characteristics of LPs. Based on the literature, Krajcik and Shin (2023) summarize the following essential features, which may be considered characteristics of LPs: (a) LPs characterize learning in terms of sets of related big ideas to develop more sophisticated understandings to explain phenomena and solve problems; (b) LPs integrate sets of disciplinary core ideas (DCIs) and crosscutting concepts (CCCs) with science and engineering practices (SEPs) so that learners can apply them when necessary; (c) LPs delineate progress levels to guide curricular and instructional development to support an individual in constructing deeper understanding following various paths; (d) LPs introduce DCIs and CCCs to learners from early childhood through college for developing knowledge-in-use of science ideas over time with various experiences; and (e) LPs describe critical conceptual change as a developmental process with qualitatively different levels of understanding.

Criteria for Reviewing and Comparing LPs

Previous systematic reviews of LPs have found that there is no standard or commonly adopted convention in reporting LPs (Duschl et al., 2011; Jin et al., 2019). Despite the diversity in describing and reporting LPs, there is a common understanding that LPs should include the following essential components: (a) upper and lower anchors, (b) developmental and comprehensive levels, (c) description of progress across levels, and (d) aligned curriculum, instruction, assessment, and professional learning (Krajcik & Shin, 2023). Based on the literature, Krajcik and Shin further propose that LPs should be based on research, three-dimensional integration of DCIs, SEPs, and CCCs, and include multiple paths and assessments to monitor growth over time.

Duschl et al. (2011) propose that an analysis of LP research studies must address two questions: (a) How well developed is the identification of the foundational knowledge that facilitates and advances pathways of reasoning and understanding? and (b) How thorough is the description of the teacher-mediated learning pathways? These two questions pertain to Krajcik and Shin's components a, b, and c, and partially d. Jin and collaborators (2019) conceptualize the above review criteria as developmental coherence, horizontal coherence, and vertical coherence. Developmental coherence refers to the degree to which an LP is consistent with current understandings of how students learn science. They specifically emphasize that current understandings of how students learn science include the necessity for integration of knowledge and practices, domain-specific reasoning, and conceptual change in the forms of knowledge enrichment, transformation, and integration. Horizontal coherence refers to the alignment among LP-facilitated instruction, curriculum, and assessment. Finally, vertical coherence refers to the alignment of assessments at different levels, such as classroom and large-scale state assessments. Together, the three coherences fulfill the two promises of LPs to promote student knowledge-in-use and alignment among curriculum, instruction, and assessment.

The abovementioned review criteria of LPs focus on LPs as products. Given that LPs are models that continue to evolve, there is also a need for a set of review criteria related to the development and validation of LPs. Krajcik and Shin (2023) propose a design-based research approach with design-test-revise cycles for developing, testing, and studying three-dimensional LPs and associated curriculum materials and assessments. Design-based research requires collaboration among researchers, classroom teachers, and assessment experts. Initial LPs may be proposed as conjectures or hypotheses; they are then tested in classroom settings. Based on testing, researchers revise the initial LPs and instructional materials and suggest teacher practices. The LPs will be tested again, and the revision and testing will continue iteratively until there is sufficient evidence of students reaching the specified levels of performance in LPs.

For both conceptions of LPs as products and processes, LPs as research must demonstrate internal validity, external validity, and ecological validity (Campbell & Stanley, 1968; Bracht & Glass, 1968). Internal validity is the degree to which the design, data collection, and analysis support the claims made about the LPs—how trustworthy is the development and validation of LPs? External validity is the degree to which the claims about the LPs can be generalized beyond the sample and time; ecological validity is the degree to which the LPs can be implemented in real classroom settings and conditions.

Reviews of LPs in Section II

Chapter 12 by Duncan and colleagues reports on a genetics LP that has gone through multiple revisions over the past decade. The development of the initial LP was motivated by the complexity of the domain of modern genetics, students' challenges, and the need for a coherent and extensive approach to learning genetics (Duncan et al., 2009). The initial LP was based on a genetics literacy framework (Stewart et al., 2005) that consists of understanding and fluent movement among three models, the inheritance model about the correlation between genotype and phenotype, the meiotic model about gene transfer from one generation to another, and the molecular model about the relationship between genetic information encoded in genes and manifestations of traits across levels of organization (i.e., cells, tissues, organs, organ systems, and multicellular organisms). Eight constructs related to these models were identified based on the literature, and the initial LP was described as students' development of reasoning along the eight constructs across three grade bands: 4–5, 6–8, and 9–10. However, this initial LP was hypothetical because it was not empirically tested, nor was it associated with any curriculum and instruction. Most importantly, the initial LP neither integrated science and engineering practices, nor presented the three-dimensional learning envisioned in the *NGSS*.

To validate the initial LP, researchers and teachers co-developed two, five-week units for grades six through eight. Teachers taught the two units over two years; pre- and post-test and interview data were collected. Results showed that students made the most gains in learning the molecular model, followed by the meiotic model, followed by the inheritance model. Student knowledge of both inheritance and meiotic models developed simultaneously and at a similar rate, suggesting that students' knowledge of molecular genetics may have provided a bootstrapping for their learning of Mendelian (i.e., inheritance and meiotic) models. However, about a third of students did not develop the notion of the universality of genes, such as recognizing that plants and bacteria have genes. They also had difficulties in predicting the genetic content of sex cells from individuals with different genotypes, understanding protein functions, and drawing connections among the three genetic models. The data also revealed limitations with the LP. Specifically, the assessment for the LP lacked sufficient coverage of the eight constructs and all of the developmental levels. The variation of the contexts in the assessment tasks was limited; students did not demonstrate substantial learning gains in year two from Levels 2 to 3, suggesting either the developed curriculum and instruction might not be effective or there was a flaw in the LP for defining Levels 2 and 3. These findings informed the revision of the initial LP, with more levels reflecting important conceptual stepping stones.

The next round of revision and validation, consisting of an implementation study, intended to address the previous issues. Specifically, the implementation study intended to answer the question: is it possible to have multiple pathways from the same lower anchor to the same upper anchor? To answer this question, the researchers tested two parallel and highly comparable instructional units in molecular and classical genetics (combined inheritance and meiotic models) and used them in two instructional conditions: molecular first or classical first. Five high school biology teachers who taught multiple sections of students were involved in the study. Each unit took about four weeks to teach, and after both units were completed (in either order), students engaged in a week-long bridging module that was aimed at making connections across models by having students develop a general explanation of the molecular basis of various inheritance patterns. In addition, the written assessment was systematically developed to ensure its construct validity. Results showed that, once again, students made significant gains across all constructs in both instructional conditions from Levels 1 to 2, and still only a small number of students ended with Level 3. Further, there were no statistically significant differences in learning gains between the two instructional conditions, although, across the board, there was a slightly higher gain for the molecular first condition. Interview data also suggested that there might be some between-construct influence. These findings suggested a need to further revise the constructs and the aligned assessment items, i.e., the LP.

By applying the review criteria for the genetic LP, we can see that the genetic LP has clearly defined upper and lower anchors, developmental levels, progress variables, and aligned curriculum, instruction, and assessment. There was evidence for developmental coherence with a focus on the "work with it" orientation for conceptual development and three-dimensional integration, as well as horizontal coherence by simultaneously developing and testing curriculum, instruction, and assessment materials. The development and validation process utilized design-based research approaches through multiple iterations of revision and development. The LP integrated DCIs, SEPs, and CCCs and had multiple pathways. There was evidence for the LP's internal validity. However, it is unclear how the LP might promote vertical coherence by aligning classroom assessments with external, such as standards-based state assessments. Further, although the LP has gone through multiple rounds of iteration, the sample sizes of teachers and students involved in the validation appeared to be small and homogeneous (e.g., high schools with the physics-chemistry-biology course sequence); its external validity and ecological validity remain to be established.

Chapter 13 by He and colleagues reports the development and validation of an LP on energy, interaction, and matter for the middle school grades. The LP integrates multiple DCIs, SEPs, and CCCs to develop useable student knowledge, or knowledge-in-use, by explicitly attending to students' diversity by creating multiple pathways for students of diverse backgrounds. Two rounds of design-based research were completed to develop and refine the LP. During the first round, a hypothetical LP was proposed based on the research literature and *NGSS*. This hypothetical LP describes students' learning performance development through four levels on two progress variables: manifestations of energy and transfer of energy through integrating various DCIs, SEPs, and CCCs. Twenty-three technology-based assessment tasks, including rubrics, were then developed based on the hypothetical LP. An invited external review panel of science teachers reviewed the hypothetical LP and the associated assessment. The teachers agreed that the two progress variables of the LP can simultaneously support student development of DCIs with SEPs and CCCs; that the fine-grained, three-dimensional learning performances could guide three-dimensional instruction and assessment; and that the energy DCI seamlessly connects to other DCIs, such as force and motion, in making sense of everyday phenomena. Teachers' feedback resulted in adding five more design guidelines for developing the next version of the LP.

The second round of design-based research repeated the first-round stages by defining the problem (i.e., designing compelling and complex, real-world scenarios), designing processes and products (i.e., a coherent set of learning performances, assessment tasks, and rubrics), and seeking feedback from experienced teachers. The teachers in general agreed that the designed scenario could fit well with the revised LP, and its levels could be used to describe their students' understanding of the phenomena. Students of diverse backgrounds could follow multiple pathways to achieve an upper level. The teachers also suggested that assessment tasks needed to be carefully designed to cover all levels of the LP, and clear prompts and suitable scaffoldings were necessary to address the levels in multiple constructs. Based on this feedback, the design guidelines were updated again to guide the next round of LP revision.

The next rounds of design-based research will involve teachers from multiple school districts representing diverse student backgrounds. Teachers will receive professional development (PD) on the LP, assessment tasks, rubrics, and how to adapt their own curriculum and instruction according to the LP and assessment. Cross-sectional data on the LP-aligned assessment task performance at grades six, seven, and eight over two years will be collected. Student responses to the classroom assessment will be analyzed to identify the intermediate learning pathways. Data collection will include teacher and student interviews, classroom observations, and teacher and student artifacts that will be analyzed to inform the refinement of the LP levels. The development and validation of the LP are still ongoing.

Applying the review criteria of LPs, we can claim that the energy, interaction, and matter LP for middle school grades has clearly defined upper and lower anchors, developmental levels, progress variables, and aligned formative and summative assessments. It is design-based and three-dimensionally integrated, includes assessment, and attends to multiple learning pathways. The LP promotes developmental coherence through knowledge enrichment, theory transformation, and knowledge integration; it also promotes horizontal coherence by providing teachers with professional development to adapt teachers' curriculum, instruction, and assessment according to the LP. Because the LP has not been empirically validated in real classrooms, and data on student learning remains to be collected and analyzed, it is premature to claim the internal, external, and ecological validities for the LP. Moreover, no information is available on how the LP may promote vertical coherence by aligning the classroom assessment with the external standardized assessment.

Chapter 14 by Fortus and Nordine discusses progress and challenges related to LPs on energy. The importance of developing student knowledge—particularly useable knowledge—on energy is indubitable as it is both a DCI and CCC in the *NGSS*. It is expected that by the end of high school, students should be able to develop an understanding of energy conservation both qualitatively and quantitatively. However, previous research has consistently shown that this goal is eluded by a majority of students (Herrmann-Abell & DeBoer, 2017; Liu & McKeough, 2005; Neumann et al., 2013). As Fortus and Nordine pointed out, the unsatisfactory progress toward energy conservation may be associated with typical instruction that starts with identifying energy forms and transformations among forms, and the validation of the LP was based on cross-sectional surveys. LPs provide an ideal opportunity for new approaches to teaching energy, which is what Fortus and Nordine have done.

To develop a new LP that promotes more effective student progression toward energy conservation, they first conducted a series of teaching experiments to investigate whether or not form-based instruction indeed inhibited student learning progress. Specifically, they developed an approach to middle school physical science energy instruction, called the systems-transfer approach (STA), consistent with the *Framework* (NRC, 2012). STA focuses on modeling the systems and subsystems involved, identifying the changes happening in each of those systems as energy is transferred, and treating energy as a unitary quantity that must always be associated with an object or a field that mediates interaction at a distance. STA introduces field as a physical entity (rather than a quantity) in instruction. Fortus and colleagues found that in every aspect, the students' knowledge developed by the systems-transfer instruction was superior to that by energy-form instruction. Conceptual network analysis showed that the systems-transfer students developed more parsimonious knowledge of energy, organized primarily around the idea of transfer without any reference to transformation. Seventh-grade students learning about energy using a system-transfer approach could successfully progress toward a qualitative understanding of energy conservation. The above research findings allowed the researchers to propose an alternative LP that replaces energy forms and transformation with energy transfer and fields. The two LPs, the form-based and STA-based, overlap in their lower anchor (Level 1) and upper anchor (Level 4) as well as degradation (Level 3); the only difference is at Level 2. Thus, the two LPs may still be considered as one learning progression with two pathways.

Although the Fortus and Nordine chapter does not report the development and validation of a complete LP that includes curriculum, instruction, and assessment, they propose a viable alternative pathway to the conventional LP in the literature. Their modified LP reaffirms the purpose of LPs to promote transformative instruction; it also highlights the importance of considering multiple pathways in an LP. The development and validation of LPs is an ongoing process. What is important is how the LP may demonstrate not only internal validity (how effective the LP promotes student conceptual change) but also external validity (how generalizable the evidence is for all students) and ecological validity (how likely the LP can be implemented in everyday classrooms); evidence for the above validities is needed.

Chapter 15 by Duschl et al. describes two LPs on geology and earth systems science learning. The first LP adopts a foundational plate system perspective for a dynamic earth model and takes a sociocultural perspective in classroom implementation. McDonald and colleagues (2019) drew from a variety of policy documents on geology and earth science education to define the upper anchor around the use of systems thinking and a full systemic understanding of plate movement and plate systems. The upper anchor was then divided into three progress variables: plate movement, mechanisms of plate motion, and plate systems. They then conducted interviews with students and classroom observations to define the lower anchor and identify students' progressive understandings of each of the progress variables. These findings resulted in a hypothetical LP on plate tectonics. The second LP adopts a geocognitive perspective and focuses on the assessment of temporal and spatial relationships of a dynamic earth model of plate tectonics (Maeng, 2020). It adopts a different set of progress variables for plate tectonics with a focus on reasoning practices and modeling: spatial reasoning, temporal reasoning, and systems thinking. To develop a hypothetical LP, Maeng adopted the same upper anchor as in the first LP, i.e., the use of systems thinking and a full systemic understanding of plate movement and plate systems. Maeng then developed a set of ordered multiple-choice items with choices representing four possible levels of geoscience thinking. Rasch's analysis of student response data was conducted to validate the appropriateness of the three progress variables and the four developmental levels.

Both LPs developed assessments and measures of complex reasoning practices about the systematic mechanisms of plate movement and the Earth's internal plate systems. Considering the disciplinary uniqueness of geoscience and geographical science data, assessment and instruction took advantage of technological support, e.g., GeoMapApp and Seismic Explorer, through simulations and data visualization. Using the technological tools, students were able to construct explanatory accounts of how surface phenomena are related to below-surface dynamics of plate movements and transform the data visualization into data narratives. Both LPs aim to support student reasoning in complex adaptive systems through engagements with the models, mechanisms, and databases of plate tectonics theory.

Applying the review criteria to the two geology and earth science LPs, we see that both have clearly defined upper and lower anchors, developmental levels, progress variables, and aligned curriculum, instruction, and assessment. Both are research-based, three-dimensionally integrated, and include assessment. Because both LPs have the same upper anchor, the two LPs may be considered as one LP with different pathways. Since no data are presented in this chapter, the internal validity, external validity, and ecological validity of the LPs are unclear.

Chapter 16 by Scott and colleagues describes the development and validation of an LP on middle and high school environmental literacy. The LP is a result of the project *Carbon: Transformations in Matter and Energy (Carbon TIME)*. To develop the LP, the researchers started with identifying desirable student performances at the end of grade 12: investigating macroscopic phenomena related to carbon-transforming processes including flames; animal, plant, and fungus growth; and animal and plant metabolism, movement, and decomposition. They then operationalized student performance expectations based on the *NGSS* performance expectations around three science and engineering practices: asking questions, conducting and critiquing investigations, and explaining and predicting. For the lower anchor, researchers consulted earlier research on students' conceptions of matter and conservation of matter (e.g., Driver, 1985; Elkind, 1961; Piaget, 1951); they also observed student-generated questions about complex and everyday phenomena, such as flames, animals, plants, and decomposers. They noticed that students tended not to focus on matter tracing; instead, they attended to other aspects such as development, nutrition, energy, health, and so on. As a result of this early exploration, the researchers developed six curriculum units focusing on different phenomena, such as combustion, animals, plants, and decomposers to bridge the gap between the upper and lower anchors.

Initial piloting of the curriculum units found that students did not make desirable progress from the lower anchor to the upper anchor. The researchers then conducted a series of interviews with middle school, high school, undergraduate, and graduate students on the hypothetical investigations in the initial curriculum units. Student responses related to making claims, analyzing data, and identifying unanswered questions were analyzed and categorized, and an intermediate performance level between the lower and upper anchors was proposed. A hypothetical LP was proposed to consist of three progress variables: evaluating claims and explaining mass changes, analyzing mass change data, and asking questions after investigations. It also included three performance levels from organism-as-actors approach to investigations to a matter-tracing approach to investigations.

The hypothetical LP was then used to guide the revision of the curriculum units, specifically to design and refine scaffolds (i.e., process tools) for the investigations in each unit. Particularly, since interview studies found that the Zone of Proximal Development (ZPD) (Vygotsky, 1978) was different for different students, process tools must be meaningful for students at all levels. Each process tool served as the focus for a lesson organized around a discourse routine: students writing their ideas individually and then sharing with the class, teachers leading consensus-seeking discussions, and students revising their ideas. Teachers can also use the process tools and class discussions as formative assessments; they can use probing and follow-up questions during whole class or small group discourse in places where students may need additional support. A multi-year testing and revision of the process tools and discourse routines occurred. The foundations for all process tools are three questions about matter movement, matter change, and energy change.

Pre-tests and post-tests involving 50,000 diverse middle and high school students from multiple states taking biology courses were used to evaluate the effectiveness of the LP-facilitated curriculum and instruction. If a Level 2 performance was considered acceptable, about 72% of students had attained this level for evaluating claims and 49% for analyzing mass change data. The researchers acknowledged that although the achievement represents substantial progress, a significant number of students did not identify key patterns in the data, nor did they use those patterns to construct arguments from evidence.

Applying the review criteria of LPs to the *Carbon TIME LP* summarized above, we can see that the LP included upper and lower anchors, developmental levels, progress variables bridging the developmental levels, and aligned curriculum, instruction, and formative assessment. The LP is based on design research, three-dimensionally integrated, and includes formative assessment. However, it is unclear how multiple pathways are developed, although the researchers stated that students were diverse in their ZPDs. The developmental coherence can be considered achieved through a conceptual change approach of "working with it" and the integration of DCIs and SEPs; the LP promotes transformative learning toward environmental literacy. Horizontal coherence is also achieved by developing LP-aligned formative and summative assessments as part of curriculum material development. However, the achievement of vertical assessment is unknown, as there is no mention of how the LP is aligned with external assessment, particularly state standards-based assessments. Finally, from a validity perspective, the LP has strong internal validity and external validity based on extensive validation and data; nevertheless, ecological validity is unclear because no information is available about teacher professional development and the required resources to implement the LP in actual classrooms.

Chapter 16 by Miller et al. reports the development and validation of an LP for the same goal of environmental literacy as reported in Chapter 15. In order to achieve the environmental literacy outlined by *Carbon TIME*, students must develop competencies as questioners, investigators, and explainers. While Chapter 15 focuses on the LP for the first two competencies, Chapter 16 focuses on the LP for the third competency. Building on a previous version of LPs on students' explanation of carbon-transforming processes, Miller et al. undertook a systematic process of revising the previous LP and developing and testing curriculum and assessment resources aligned with the revised LP. The initial LP had four levels of progression from force-dynamic explanations (characterizing plants and animals as actors who need resources, such as sunlight, food, and water to sustain their lives) to mechanistic explanations, i.e., using atomic-molecular models involving change in matter and energy. Written assessments and rubrics for the levels were also developed.

Revision of the initial LP was based on analysis of interviews with teachers and students. Four closely related progress variables—context-specific knowledge; use of principles of matter and energy; scale; and precision in matter and energy words—were identified with examples of student responses at each level. Pre- and post-instructional interviews with students showed that almost all students improved their explanation performance, but only a few advanced to the highest level (Level 4 of the *NGSS* performance expectations) on all four progress variables, suggesting that Level 4—constructing explanations for specific scientific phenomena that use the principles of matter and energy—was challenging for secondary students, and better scaffolding and more precise feedback on their explanation performances than that provided in the initial *Carbon TIME* pilot units were needed.

The revised LP with four progress variables enabled researchers to improve scaffolding and assessment for explanations in the curriculum units. Specifically, the researchers developed design frameworks and explanation tools for each progress variable to organize curriculum units, classroom discourse routines, formative and summative assessments, and teacher professional development. Results from close to 60,000 students from the classrooms of 133 middle and high school teachers over four years showed that students significantly improved their proficiency in explanation performance related to carbon cycling after learning the units, and this learning gain was significantly higher than that based on other curricula used by the teachers. Further, there was a large, educationally significant difference in pre-post learning gains among teachers. Specifically, there were higher learning gains associated with teachers effectively using the curriculum resources for scaffolding and formative assessment, as well as apprenticeship instructional strategies to develop proficiency around all four progress variables in their explanations. School demographic factors (e.g., percent of students receiving free and reduced-price lunch and percent of marginalized students of color) accounted for much less variance than teachers.

Applying the review criteria to the LP in Chapter 16, we see that, as with the LP reported in Chapter 15, the explanation LP included upper and lower anchors, developmental levels, progress variables bridging the developmental levels, and aligned curriculum, instruction, and assessment. The LP is based on design research, three-dimensionally integrated, and includes both formative and summative assessments. The developmental coherence is achieved through a conceptual change approach of "working with it" and the integration of DCIs, SEPs, and CCCs; the LPs promote transformative learning toward environmental literacy. Horizontal coherence is also achieved by developing LP-aligned formative and summative assessments as part of the curriculum material development. However, the achievement of vertical assessment is unknown as there is no mention of how the LP is aligned with external assessment, particularly the state standards-based assessments. Also, it is unclear how multiple pathways are taken into consideration, as the progression among the four progress variables still appears to be linear. Finally, from a validity perspective, the LP has strong internal validity and external validity based on extensive validation processes, although ecological validity is unclear because no information is available regarding teacher professional development and the resources needed to implement the LP.

Conclusion

Based on the above reviews, all six LPs are exemplary in that they engage design-based research; are three-dimensionally integrated; include upper and lower anchors, developmental levels, and progress variables; and align curriculum, instruction, and assessment. They all promote ambitious and transformative instruction, developmental coherence, and horizontal coherence. However, the above LPs differ in the extent to which they address internal validity, external validity, and ecological validity, and no LP explicitly considers vertical coherence among classroom formative assessment, summative assessment, and external standards-based assessments, such as state tests. Particularly, achieving external validity and ecological validity requires testing an LP with larger and more diverse student samples and in diverse classroom settings. Teacher knowledge and school resources needed to implement an LP also need to be explicitly considered and empirically studied.

The promises of LPs to facilitate ambitious, transformative instruction and to promote developmental, horizontal, and vertical coherence are highly desirable, yet challenging goals. Future research on LPs should make explicit how the curriculum and instruction aligned with a particular LP is ambitious and transformative and what evidence is available to support the claim. Learning is complex and idiosyncratic (Hammer & Sikorski, 2015); no LP is likely to account for all variations of the process. Thus, what is necessary and important is for LPs to explicitly state the types of students and conditions the LP is developed for. Future LP research needs to consider how different LPs and different pathways of the same LP can be developed for different students in different learning conditions. Diversifying LPs and LP pathways is necessary to achieve equity and inclusion in science teaching and learning.

The LPs reported in this section are still limited to a few grade levels; larger grade span LPs, such as from elementary through high school, are needed for external, standards-based assessments. These LPs, with larger grade spans, should coordinate with smaller grade span LPs, such as those in this section, that are specifically for classroom instruction. As more learning progressions are developed in the future, coordination among different LPs for the same subject at the same grade levels, such as between a force and motion LP and an energy LP in high school physics, will also become important for coherent student learning.

Statistician, George Box, once famously said "all models are wrong, but some are useful." LPs as conceptual models cannot and should not be expected to explain and predict all learning trajectories. Thus, future LPs should always be evaluated based on how they promote effective teaching and learning and coherence among curriculum, instruction, and assessment.

References

American Association for the Advancement of Science (AAAS) (1993). *Benchmarks for science literacy.* New York: Oxford University Press.

American Association for the Advancement of Science (AAAS) (2001). *Designs for science literacy*. New York: Oxford University Press.

American Association for the Advancement of Science (AAAS) (2007). *Atlas of science literacy.* Washington, DC: AAAS Project: the Author.

Bracht, G. H., & Glass, G. V. (1968). The external validity of experiments. *American Education Research Journal*, *5*, 437–474.

Campbell, D. T., & Stanley, J. C. (1968). *Experimental and quasi-experimental designs for research.* New York: Houghton Mifflin Company.

Catley, K., Lehrer, R., & Reiser, B. (2005). *Tracing a prospective learning progression for developing understanding of evolution.* Paper Commissioned by the National Academies Committee on Test Design for K-12 Science Achievement, 67.

Chi, M. T. H., Slotta, J. D., & deLeeuw, N. (1994). From things to processes: A theory of conceptual change for learning science concepts. *Learning and Instruction*, *4*, 27–43.

Driver, R. (1983). *The pupil as scientist?* London: Milton Keynes.

Driver, R. (1985). *Children's ideas in science*. London: McGraw-Hill Education (UK).

Driver, R., & Easley, J. (1978). Pupils and paradigms: A review of the literature related to concept development in adolescent science students. *Studies in Science Education*, *5*, 61–84.

Driver, R., Leach, J., Scott, P., & Wood-Robinson, C. (1994a). Young people's understanding of science concepts: Implications of cross-age studies for curriculum planning. *Studies in Science Education*, *24*, 75–100.

Driver, R., Squires, A., Rushworth, P., & Wood-Robinson, V. (1994b). *Making sense of secondary science: Research into children's ideas*. London: Routledge.

Duncan, R. G., Rogat, A., & Yarden, A. (2009). A learning progression for deepening students' understanding of modern genetics across the 5th-12th grades. *Journal of Research in Science Teaching*, *46*(6), 644–674. 10.1002/tea.20312

Duschl, R. A., Maeng, S., & Sezen, A. (2011). Learning progressions and teaching sequences: A review and analysis. *Studies in Science Education*, *47*, 123–182. 10.1080/03057267.2011.604476

Elkind, D. (1961). Children's discovery of the conservation of mass, weight, and volume: Piaget replication study II. *Journal of Genetic Psychology*, *98*(2), 219–227.

Hammer, D., & Sikorski, T.-R. (2015). Implications of complexity for research on learning progressions. *Science Education*, *99*, 424–431. 10.1002/sce.21165

Herrmann-Abell, C. F., & DeBoer, G. E. (2017). Investigating a learning progression for energy ideas from upper elementary through high school. *Journal of Research in Science Teaching*, *55*(1), 68–93. 10.1002/tea.21411

Jin, H., Mikeska, J. N., & Hokayem, H. (2019). Toward coherence in science, instruction, and assessment: A review of learning progression literature. *Science Education*, *103*, 1206–1234. 10. 1002/sce.21525

Krajcik, J., & Shin, N. (2023). Student conceptions, conceptual change, and learning progressions. In N. G. Lederman, D. L. Zeidler, & J. S. Lederman (Eds.), *Handbook of research on science education: Vol. III* (Chapter 5, pp. 121–157). New York: Routledge.

Lehrer, R., & Schauble, L. (2015). Learning progressions: The whole world is NOT a stage. *Science Education*, *99*, 432–437. 10.1002/sce.21168

Liu, X., & Collard, S. (2005). Using Rasch model to validate stages of understanding the energy concept. *Journal of Applied Measurement*, *6*(2), 224–241.

Liu, X., & Lesniak, K. (2005). Students' progression of understanding the matter concept from elementary to high school. *Science Education*, *89*(3), 433–450.

Liu, X., & Lesniak, K. (2006). Progression in children's understanding of the matter concept from elementary to high school. *Journal of Research in Science Teaching*, *43*(3), 320–347.

Liu, X., & McKeough, A. (2005). Developmental growth in students' concept of energy: Analysis of selected items from the TIMSS database. *Journal of Research in Science Teaching*, *42*(5), 493–517.

Maeng, S. (2020). Investigating a learning progression for reasoning practices of geocognition using GeoMapApp-Based Assessment. *Asia-Pacific Science Education*, *6*(2), 584–614.

McDonald, S., Bateman, K., Gall, H., Tanis-Ozcelik, A., Webb, A., & Furman, T. (2019). Mapping the increasing sophistication of students' understandings of plate tectonics: A learning progressions approach. *Journal of Geoscience Education*, *67*(1), 83–96.

National Research Council. (1996). *National science education standards*. Washington, DC: National Academies Press.

National Research Council. (1999). *How people learn: Bridging research and practice*. Washington, DC: National Academies Press.

National Research Council. (2006). *Systems for state science assessment*. Washington, DC: National Academies Press.

National Research Council. (2007). *Taking science to school: Learning and teaching science in grades K-8*. Washington, DC: National Academies Press.

National Research Council. (2012). *A framework for K-12 science education: Practices, crosscutting concepts, and core ideas*. Washington, DC: National Academies Press.

Neumann, K., Viering, T., Boone, W. J., & Fischer, H. E. (2013). Towards a learning progression of energy. *Journal of Research in Science Teaching*, *50*(2), 162–188. 10.1002/tea.21061

NGSS Lead States. (2013). *Next generation science standards: For states, by states*. Washington, DC: National Academies Press.

Piaget, J. (1951). *The child's conception of the world*. London: Routledge.

Pintrich, P. R., Marx, R. W., & Boyle, R. A. (1993). Beyond cold conceptual change: The role of motivational beliefs and classroom contextual factors in the process of conceptual change. *Review of Educational Research*, *63*(2), 167–200.

Posner, G., Strike, K., Hewson, P., & Gertzog, W. (1982). Accommodation of a scientific conception: Toward a theory of conceptual change. *Science Education*, *66*(2), 211–227.

Schmidt, W. H., Curtis, C. M., Raizen, S. A., etc. (2002). *A splintered vision: An investigation of U.S. science and mathematics education*. Dordrecht: Springer. 10.1007/0-306-47209-0.

Schmidt, W. H., Wang, H., & McKnight, C. C. (2005). Curriculum coherence: An examination of US mathematics and science content standards from an international perspective. *Journal of Curriculum Studies*, *37*(5), 525–559.

Smith, C. L., Wiser, M., Anderson, C. W., & Krajcik, J. S. (2006). Implications of research on children's learning for standards and assessment: A proposed learning progression for matter and the atomic-molecular theory. *Measurement: Interdisciplinary Research and Perspectives*, *4*(1–2), 1–98. 10.1080/15366367.2006.9678570

Stewart, J., Cartier, J. L., & Passmore, P. M. (2005). Developing understanding through model-based inquiry. In M. S. Donovan, & J. D. Bransford (Eds.), *How students learn* (pp. 515–565). Washington, DC: National Research Council. 10.1111/j.19498594.2009.tb17870.x

Tyson, L. M., Venville, G. J., Harrison, A. G., & Treagust, D. F. (1997). A multidimensional framework for interpreting conceptual change events in classroom. *Science Education*, *81*, 387–404.

Vygotsky, L. S. (1978). *Mind in society: The development of higher psychological processes*. Cambridge, MA: Harvard University Press.

Wandersee, J. H., Mintzes, J., & Novak, J. (1994). Research on alternative conceptions in science. In D. Gabel (Ed.), *Handbook of research on science teaching and learning*. New York: Macmillan.

SECTION III

Curriculum, Instruction, and Teacher Learning

18

KEY COMPONENTS OF LEARNING PROGRESSION-BASED EDUCATIVE CURRICULUM MATERIALS DESIGNED TO SUPPORT TEACHERS AND THEIR DIVERSE STUDENTS

Kristin L. Gunckel, Beth A. Covitt, and Julie A. Bianchini

Since learning progressions were first widely introduced to the field of science education in *Systems for State Science Assessments* (National Research Council, 2006) and became a foundation for *A Framework for K-12 Science Education* (National Research Council, 2012) and the Next Generation Science Standards (NGSS Lead States, 2013), many have argued that learning progressions have the potential to bring coherence to curriculum, instruction, and assessment (Corcoran et al., 2009; Duschl et al., 2011; Jin et al., 2019). The proposed argument is that once patterns in how students' ideas change over the course of several grades are identified, learning progressions can be used as frameworks to develop curricula, instruction, and assessments that are aligned with each other, meet students where they are, and support them in reaching higher levels of performance in science.

Curriculum materials, which we define in this chapter as the suite of resources necessary to guide teachers in their instructional and assessment practices and students in their learning (Ball & Cohen, 1996; Brown, 2009), can potentially play an important role in facilitating this coherence by bringing learning progression frameworks into classrooms. Curriculum materials are enacted by teachers, and the extent to which these materials impact student learning depends on teachers' understanding of the learning progressions upon which the curriculum materials are based (Jin et al., 2019). Evidence suggests, however, that teachers do not always find the learning progression frameworks to be easy tools to draw upon for informing their curricular, instructional, and assessment practices (Covitt et al., 2018; Furtak, 2012; Gunckel et al., 2018; Jin et al., 2019). For example, teachers may find learning progression frameworks to be too abstract and/or articulated at a grain size too coarse to inform their instruction (Alonzo & Elby, 2019; Alonzo et al., 2022). In addition, teachers may use learning progressions in ways that are different from the originally envisioned uses of learning progressions—for example, viewing them as checklists of ideas students might hold for the purposes of identifying and fixing their students' misconceptions (Alonzo et al., 2022; Furtak, 2012). Alternatively, some teachers find learning progressions most useful as starting

DOI: 10.4324/9781003170785-21

points for testing their own insights into student thinking (Alonzo & Elby, 2019). Research also suggests that teachers are often better at making inferences about their students' levels of understanding than they are at responding to those levels of understanding with appropriate next steps for instruction (Furtak & Tayne, 2019; Heritage et al., 2009; Schneider & Gowan, 2013). These findings about teachers' experiences using learning progressions highlight both the need and opportunity for comprehensive learning progression-based curriculum materials that attend to both teachers' and students' needs by guiding teachers in using learning progressions-based assessment strategies, instructional activities, and pedagogical approaches to support diverse students in achieving rigorous learning goals.

In this chapter, we describe key components of comprehensive learning progression-based curriculum materials designed to bring coherence at the classroom level to curriculum, instruction, and assessment for diverse learners. These components include educative features designed to support teacher learning and practice (Davis et al., 2014; Davis & Krajcik, 2004) as well as student learning. Along with other strategies, such as professional development, the inclusion of educative features in learning progression-based curriculum materials has the potential to help teachers develop greater capacity over time for enacting learning progression-informed instruction for their diverse students (Anderson et al., 2018; Gunckel et al., 2018). Key components of learning progression-based curriculum materials include (1) formative assessments with educative features to support teachers' use of learning progressions to interpret student thinking, (2) instructional sequences and pedagogical tools for responding to student thinking in ways that support students in achieving higher levels of performance on a learning progression, and (3) language supports for multilingual students to be able to engage and excel in learning progression-based instructional activities. To illustrate these components, we draw on examples from three related curriculum materials development projects based on a learning progression for water in environmental systems.

- School Water Pathways—Middle and high school curriculum materials designed to engage students in tracing the flow of water along multiple pathways and through multiple processes into, through, and away from their schoolyards (Caplan et al., 2013).
- Reasoning Tools for Understanding Water Systems—A suite of pedagogical tools designed for middle school teachers to integrate into their existing instruction to support students in making progress toward the next level of achievement in the learning progression (Covitt et al., 2018).
- Comp Hydro—Curriculum materials designed to build high school students' capacity to use computational models to make sense of the groundwater system (Gunckel et al., 2022).

Below we first describe the learning progression for water in environmental systems and then discuss how this learning progression informed the design of each key component and related educative features included in these curriculum materials.

Overview of the Learning Progression for Water in Environmental Systems

The Water Systems Learning Progression (WSLP) describes how students learn to provide accounts of water as it moves through connected environmental systems, such as the atmosphere, surface, and groundwater system, and the human-designed elements of those systems. Upper anchor performances on the learning progression are model-based explanations and predictions that trace water through environmental systems at multiple scales (from atomic-molecular through the landscape) and adhere to scientific principles that include the forces that drive (i.e., gravity, pressure) and constrain (i.e., topography, permeability) water movement. This definition of the upper anchor aligns with Next Generation Science Standards performance expectations that require students to develop and use models as well as numerous crosscutting concepts (e.g., energy and matter; systems and system models; scale, proportion, and quantity; structure and function; stability and change) to explain the cycling of water through Earth's systems (e.g., see NGSS Performance Expectations MS-ESS2-4 and HS-ESS2-2). Progress from one level of performance to the next requires a change in the ways that students talk about where and how water moves (Gunckel et al., 2012). This framing of what changes is based on the perspective that knowledge and practices are situated in and shaped by ways of talking, thinking, and acting within a community (Gee, 1991; Wenger, 1998).

Students' ways of using language to make sense of phenomena change as they access the discourse practices of new communities. Initially, students are embedded in home communities that employ force-dynamic ways of explaining events, based on the vernacular grammatical structure of languages (Pinker, 2007; Talmy, 1988). At levels 1–2,[1] students explain the movement of water in terms of actors with abilities to move water, such as clouds that suck water up into the sky or the ground that soaks water up like a sponge. Later, when they go to school, students begin to develop level 3 school science stories for tracing water and naming and ordering processes such as evaporation or precipitation, but they do not yet explain why or how water moves through systems. It is not until students gain access to more formal scientific discourse that they provide level 4 model-based accounts to explain how and why water moves (Braaten & Windschitl, 2011; Lehrer & Schauble, 2006). Level 4 accounts about water in environmental systems include the forces that move water (e.g., gravity, pressure) and the constraints (e.g., topography, permeability) on where water goes. Level 4 accounts also include hidden and invisible pathways, such as underground or through the atmosphere as water vapor. Table 18.1

Table 18.1 Water systems learning progression (WSLP)

Level of performance	*Description of student accounts*	*Example of student discourse*
Level 4 Model-based accounts	Includes forces that move water and constraints that limit where water goes. Includes hidden processes and pathways (e.g., underground, atmospheric) at multiple scales.	Where does water in a puddle go? "Depending on the topography, the water will eventually make it into the groundwater system. Gravity pulls the water down until it hits an impermeable layer, where it will then collect." How does water get into a river? "A) Runoff from either side. B) Rain directly on the river. C) Filled by the water table, through the permeable layer."
Level 3 School science stories	Names processes (e.g., evaporation, condensation, melting). Puts events in order.	Where does water in a puddle go? "It evaporated into the air, where it condensed [sic] in the clouds, and precipitated in another area." How does water get into a river? "Water gets into a river by rain, or by the water melting off a mountain, and traveling by stream until it gets to a river."
Level 1–2 Force-dynamic accounts	Identifies actors and enablers that move water.	Where does water in a puddle go? "It dried up." How does water get into a river? "It could get into a river by being rained into it."

Table 18.2 Strengths and challenges at each level

Level of performance	*Description of student accounts*	*Key strengths of students to build on*	*Key challenges students experience*
Level 3 School science stories	Names processes (e.g., evaporation, condensation, melting). Puts events in order.	Names processes and traces water through connected systems; connects two-dimensional representations (i.e., maps, cross-sections) to the physical world.	Using principles to account for how water moves.
Level 1–2 Force-dynamic accounts	Identifies actors and enablers that move water.	Traces water along one pathway; recognizes that there are mechanisms that move water; connects three-dimensional models to the physical world.	Tracing water along multiple potential pathways; identifying processes that move water (e.g., evaporation, condensation, precipitation, etc.); connecting representations (e.g., map features) to the physical world.

provides an overview of the levels of performance in the WSLP and examples of student discourse at each level.

The utility of the learning progression for designing curriculum materials is that the levels of performance identify important accomplishments that students have achieved in learning that can provide a foundation for progressing to the next level as well as key challenges that students may encounter in the process. For example, level 1–2 force-dynamic accounts tend to trace water along one pathway, such as a circular pathway reflected in the classic water cycle diagram (i.e., from the ocean to the clouds to the mountains and back to the ocean). These accounts include mechanisms that move water, but these mechanisms place agency in entities such as clouds, the ground, or the water itself. The next step for students who provide level 1–2 accounts is to be able to identify, describe, and use processes that move water, such as evaporation, condensation, precipitation, infiltration, and runoff, to trace water along multiple potential pathways. The processes that trace water through invisible phases (e.g., evaporation and condensation) or hidden locations (e.g., groundwater) are the most challenging for students to learn to explain.

Accounts of water at level 3 (school science stories) make connections between three-dimensional representations and the physical world and trace water through connected systems (e.g., atmosphere to surface water to groundwater). The next step for students who provide level 3 accounts is to use model-based principles to trace water through systems. This includes learning to use two-dimensional representations, such as maps and cross-sections, to trace water and to identify driving forces and constraints to explain and predict where water flows. Table 18.2 provides an overview of some key accomplishments and challenges for each level.

Learning Progression-Based Formative Assessment Embedded into Curriculum Materials

Effective uses of learning progressions and learning progression-based curriculum materials require teachers to formatively assess how student performances align with the learning progression and then respond with appropriate instruction that supports learners in making progress toward the next level of performance (Covitt et al., 2018; Gunckel et al., 2018). One approach to formative assessment involves embedding assessment into instructional activities. Embedded assessments are "opportunities to assess student progress and performance [that] are integrated into the instructional materials and are virtually indistinguishable from day-to-day classroom activities" (Wilson & Sloane, 2000, p. 182). Embedding assessments directly into curricular activities can emphasize the match between instruction and assessment and provide multiple time points for teachers (and students) to check on students' ideas and learning over the course of a unit.

Through our collaborations with teachers, we have found that learning progression-based embedded assessments are more useful for teachers when they are relatively brief and possible to integrate within a portion of a class period and, similarly, when they provide conceptually clear ways for teachers to review students' responses without needing to devote a large investment in time. Learning progression-based assessments that elicit students' thinking can help teachers gain a window into students' ideas that are richer than just employing a binary—*did or did not get it*—frame (Alonzo et al., 2022; Covitt et al., 2018).

One example of a formative assessment with educative features that we developed in collaboration with teachers is the School Map Assessment (Figure 18.1), which is embedded within the *School Water Pathways* curriculum materials (Caplan et al., 2013). The School Map Assessment was designed to help teachers identify ways that students experience difficulty mentally toggling between a two-dimensional representation of a landscape on a map and a three-dimensional conception of the shape of the landscape. While some maps (e.g., contour maps) are specially designed to convey topography, most maps that include water features have some information about topography implicitly embedded. For example, a water feature on a map will generally be at a relatively lower elevation than the land immediately surrounding it. This model-based principle is not always obvious to students (Gunckel et al., 2012).

The School Map Assessment provides teachers with evidence of how students are making spatial and conceptual inferences from maps and then provides teachers with ideas for instructional responses based on students' answers to the assessment questions. The assessment includes two questions, both of which use the form of a forced-choice combined with a constructed response. We have found questions that ask for a forced-choice response along with an explanatory constructed response to be productive for eliciting student ideas relevant to multiple dimensions of three-dimensional science learning (e.g., a disciplinary core idea related to principles governing water movement along with science and engineering practices such as constructing an explanation and analyzing and interpreting data).

For example, the first question in the School Map Assessment asks students to choose from multiple options representing the shape of the land and then explain their choice. The forced-choice component makes the question more accessible to students' reasoning in varying ways (i.e., students do not need to figure out from scratch what is meant by "a line representing the shape of the land across the distance from Point X to Point Y"—a task that could be quite daunting for students reasoning at level 1–2 and having difficulty connecting two-dimensional representations to the physical world). Then, the query for a constructed response elicits a fuller performance that can shed light on students' growing knowledge and practice in multiple dimensions. Considering questions 1 and 2 in the School Map Assessment together, students' choices and explanations illuminate how students are analyzing and interpreting the abstracted data in the map to explain (science and engineering practice) the direction of surface water flow governed by drivers and constraints (disciplinary core idea).

School Map

Below is a map of a school campus.

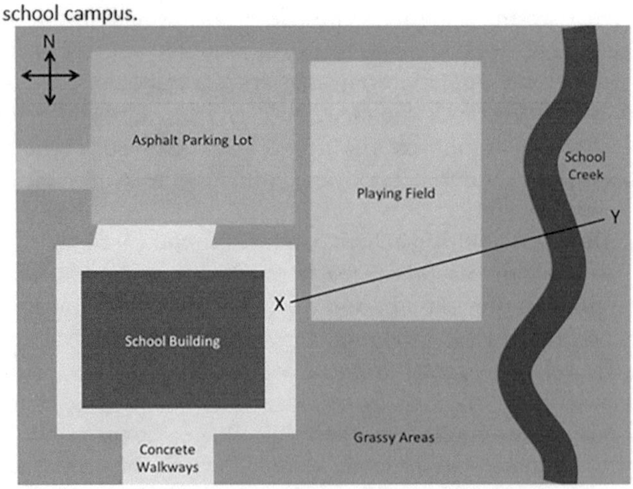

1. If you were looking from the side instead of from above, what would the shape (height) of the land be like across the distance from Point X to Point Y? (Circle the answer you think is the best.)

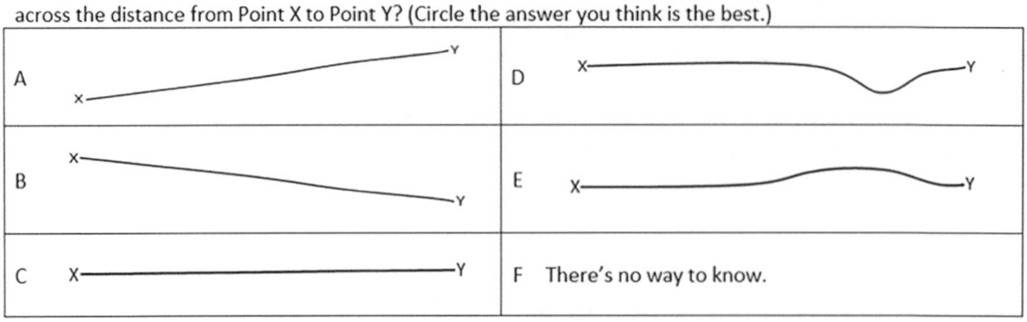

Explain your reasons for your answer.

____

____

2. Circle which direction you think School Creek is flowing: North South You can't tell from the map

Explain how you know.

____

____

Figure 18.1 School map assessment.

The teacher materials accompanying the School Map Assessment include (1) the purpose of the assessment, (2) a target performance description, (3) suggestions for administration, (4) a description of responses at different learning progression levels, and (5) suggestions for instruction based on student responses. We provide a description of elements four and five below to illustrate how the assessment and associated materials serve as educative resources designed to make learning progressions useful from a teacher's perspective (e.g., specific regarding students' ideas and classroom implementation implications).

Element 4. Descriptions of Common Responses Reflecting Knowledge and Practice at Each Performance Level

The assessment materials include examples of the kinds of responses teachers might expect to see from students. Table 18.3 shows examples of responses characteristic of levels 2 through 4 reasoning on the WSLP. Note that consistent with the idea that a learning progression approach to formative assessment is more about students' ideas and less about whether a student is "right or wrong," some of the example level 3 responses are not, in fact, "wrong." While some choices are "correct," the explanations do not address scientific principles underlying how and why water moves (see level 4 example for contrast—in which example response explains that water flows down due to gravity and that the direction of water flow with respect to the map depends on relative elevations in different locations). This distinction between levels 3 and 4 is consistent with what Braaten and Windschitl (2011) have identified as the importance of pressing for student explanations.

The amount of time and effort that teachers put into reviewing embedded assessment responses can vary. For those teachers with limited time, just reviewing the distribution of forced-choice responses can provide helpful insights into their students' reasoning. For example, if a student responds to the first question by selecting option A (i.e., the line slopes upward from X to Y), they may be reasoning with a literal reference to what is shown on the map rather than making inferences about topography by interpreting various map elements and symbols. To the extent that teachers can delve deeper into reviewing constructed responses and/or eliciting students' ideas during classroom discussion, they can gain additional insights into student thinking. For example, without providing an opportunity for students to provide explanations for question 2, teachers would not be able to tell if students who chose "N" for creek direction did so because they thought that the compass direction with the arrow was showing the direction of the creek, or because they thought that all creeks just flow north, or because they had some other reason for choosing north. The embedded assessments are designed to provide insights into student thinking rather than serving as grading tools, and the teachers we have worked with have generally been familiar with this purpose, given their experiences with formative assessment practices (Covitt et al., 2018).

Element 5. Suggestions for Instructional Responses for Students Performing at Each Level

Because research has found that teachers are often better at interpreting student assessment responses than knowing what to do next (Heritage et al., 2009; Schneider & Gowan, 2013), the teacher materials also include suggestions for instruction to help students move toward capacity for more formal knowledge and practice (Shepard et al., 2018). For example, if teachers find that many students provide level 1–2 responses to the School Map Assessment, the accompanying materials suggest that teachers engage students in firsthand activities to build conceptual connections between two-dimensional map representations and three-dimensional landscapes. Example activities described in the materials include taking a map of the area to a creek, river, or pond close to their school and comparing the height of the land around the water to the height of land under the water. Teachers can ask students to describe what they notice (i.e., that land will always be lower where there is water), explain why this is the case, and connect their explanations to inferences about the land height that can be made from reading their maps.

Similarly, the educative information in the curriculum materials suggests to teachers that students responding at level 3 might be ready for more complex activities, such as making

Table 18.3 Educative features of formative assessment materials connecting student responses to the learning progression

Level of performance	*If you were looking from the side, what would the shape (height) of the land be like from X to Y?*	*Which direction is the creek flowing?*	*Response descriptions (consistent with strengths and challenges in Table 18.2)*
Level 4 Model-based accounts	**D.** () Playing fields are usually pretty level and in D, the land starts out level from X. A creek has to be lower than the land around it because water flows down due to gravity. In option D, the land dips down where the creek is.	**You can't tell from the map.** Sometimes you can tell the direction of water from a map, but not here. There are no tributaries, topo lines, or other clues to suggest elevation and direction.	Responses include inferences connecting 2-dimensional maps to 3-dimensional landscapes. Because the driving force of gravity moves surface water to places of lower elevation, a river will be lower in elevation than the land on either side of the river. On the map, there are no clues provided that would enable someone to figure out the relative elevation along the length of School Creek that is shown.
Level 3 School science stories	**B.** () The land goes down to the creek. OR **D.** (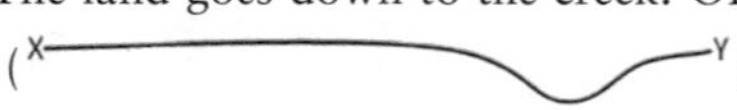) The creek is lower.	**You can't tell from the map.** The map doesn't show which way the creek is going.	Responses show understanding that the map is a representation of a 3-dimensional landscape and make some inferences about the landscape from the map. However, responses generally do not use drivers and constraining factors to explain what is happening. Thus, they sometimes reflect mistakes in interpretations and inferences.
Level 2 Force-dynamic accounts	**A.** The line slants up from point X to point Y like line does on the map. Land must go up like the line goes up. OR **E.** The land is bumpy near the creek. OR **F.** (There's no way to know) The map doesn't show the shape of the land.	North. I can tell because the compass is pointing North.	Responses begin to connect maps to landscapes. However, students have trouble connecting details of map representations to 3-D landscapes. While they realize the map represents a place, their responses may just describe what is on the map without making connections to things like topography.

Note. This example separates levels 1 and 2. For the purposes of this chapter, our discussions of these two levels have been combined.

inferences about the direction of water flow on maps that show connected tributaries, rivers, and larger bodies of water, including lakes or oceans. The materials also suggest that students responding at level 3 could be ready to start working with different types of representations, such as topographic maps, and with making inferences about elevation and direction of water flow based on map contours. Another suggestion informed by learning progression research is that teachers could ask students to reason in the opposite direction, for example, by creating a contour map for a physical model of a landscape (e.g., a tarp or crumpled paper model of a watershed).

We have found that creating learning progression-based embedded assessments with associated teacher materials can help transform learning progressions from research tools that teachers may view as too abstract or coarse-grained into usable and useful instructional tools that have clear connections to activities in classrooms and ideas that students express (Carpenter et al., 2020; Gunckel et al., 2018). Furthermore, the development of educative features in learning progression-informed materials also highlights the collaborative nature of learning progression research and using learning progressions to create curriculum materials. Our work with teachers and schools contributed to our understanding of students' ideas and ways of reasoning (informed by extensive discussions with teachers and observations in classrooms) and to the design and refinement of curriculum materials (informed by iterative, critical feedback from teachers who implemented the learning progression-based units designed by our projects).

Learning Progression-Aligned Instructional Sequences and Pedagogical Tools

Comprehensive learning progression-based curriculum materials can support teachers in responding to students' level of performance with appropriate next steps for instruction. In this section, we describe two ways learning progression-based curriculum materials can help teachers support students in making progress toward more rigorous learning outcomes.

Using Learning Progressions to Build Stepping Stones Toward Rigorous Learning Goals

One of the first tasks of writing curriculum materials is defining the learning goals that the instructional activities should address (Kesidou & Roseman, 2002; Krajcik et al., 2008). Learning progressions can be a useful resource for this task because they help identify the intermediate steps, sometimes referred to as stepping stones, needed to reach a desired upper anchor learning performance (Campbell et al., 2016; Wiser et al., 2012). Instructional activities can then be designed and sequenced to target these intermediate steps and build on students' growing proficiencies (Wiser et al., 2012).

In the Comp Hydro curriculum project, one of our learning goals was for students to use model-based principles to account for the shape of groundwater contamination plumes and make predictions about future directions and rates of flow of contamination, especially given a variety of remediation scenarios. We expected students to be able to use computational models based on maps and cross-sections to explain and predict the flow of water (and contamination) through aquifers. Our previous research developing the WSLP showed that one challenge students experience in developing model-based accounts of water moving through environmental systems is in making the connection between three-dimensional representations (such as three-dimensional physical models) and two-dimensional representations (such as maps and cross-sections). (See Table 18.2.) For example, students performing at levels 1–2 recognize that three-dimensional models are small-scale representations of the physical world. Therefore, we know that while students performing at levels 1–2 would be able to recognize the water table when using a physical model of a groundwater system, they would need support to infer the shape of the water table when using computer models of groundwater flow. We identified interpreting the three-dimensional shape of a water table based on maps and cross-sections as a stepping stone that our curriculum materials would need to include. This goal offered an intermediate step between performances many students had achieved and the goal of using maps and cross-sections to trace contaminated water through aquifers.

In our resulting curriculum materials, we included a sequence of instructional activities that integrated multiple representations of water tables, defined as the surface of the saturated zone in an aquifer. We began with a schoolyard-sized model of a network of piezometers (i.e., wells used to measure elevations of the water table). Each "well" was a section of vertically oriented PVC pipe filled with a different volume of water. The top of the water surface in each pipe represented the surface of the water table in that location. Students measured the level of water in each pipe, marked the water level on the outside of the pipe, and then used string to connect the pipes in such a way as to show the elevation of the water table, noting how the elevation sloped toward the surface water features such as rivers and lakes. This activity built on students' recognition that there is water underground and helped them see the top of an aquifer as a three-dimensional, sloped surface (Figure 18.2).

To support students in connecting the three-dimensional representation of the water table to the two-dimensional representation of a water table on a contour map, we created a

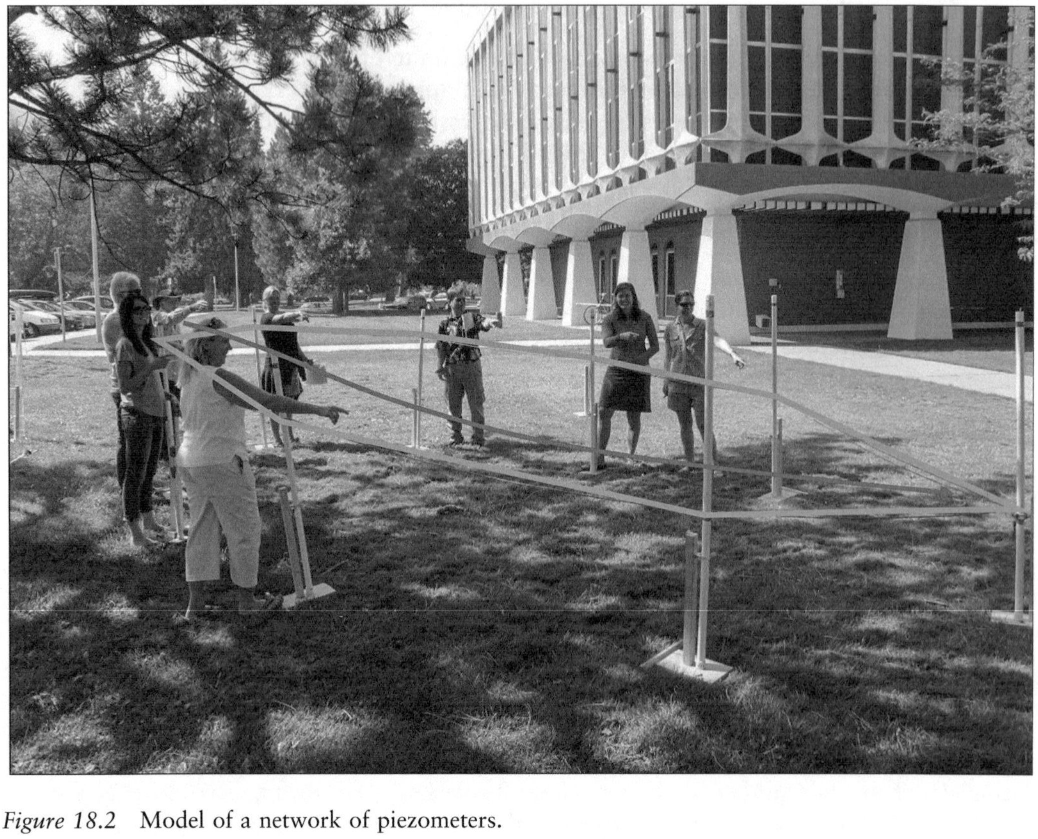

Figure 18.2 Model of a network of piezometers.

Figure 18.3 Tabletop model of a network of piezometers.

tabletop-sized version of the piezometer network that used painted dowels to represent the wells. Students again used string to connect the tops of the water table marked on each dowel to create a three-dimensional representation of the water table. This time, students used rulers to measure the height of the string at different locations and used those measurements to create a two-dimensional contour map of the water table elevations (Figure 18.3). This activity built on students' ability at levels 1–2 to connect a three-dimensional representation to the physical world and pushed students to translate those three-dimensional representations into more abstract two-dimensional representations. At this point, students were ready to work with computational models of groundwater contamination plumes.

The learning progression helped us sequence these activities to build on students' developing proficiencies in interpreting the water table as depicted on contour maps and cross-section representations of water systems, a step necessary to construct level 4 accounts of water moving through groundwater systems. The curriculum materials included explanations to the teachers about how this sequence fit the learning progression and how each activity supported students in making the steps to the next level of performance on the learning progression.

Using Learning Progressions to Provide Multiple Entry Points to Instructional Sequences

Typically, students do not move through learning progressions at the same rate or even along the same pathway (National Research Council, 2007). Students in each grade or classroom will likely not all be at the same level of performance at the beginning of a new unit of instruction. For example, when learning to trace water through systems, students performing at levels 1–2 recognize that water flows through systems, but they typically recognize only a

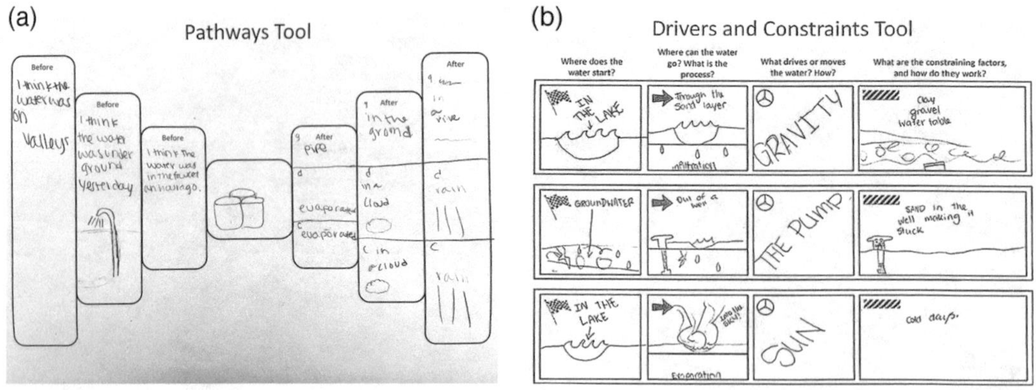

Figure 18.4 Pedagogical reasoning tools.

single pathway at a time. A stepping stone goal for them to make progress along the learning progression is to learn to trace water through multiple pathways. At the same time, students who perform at level 3 already recognize that there are multiple pathways and are learning to explain how and why water flows along particular pathways, a level 4 performance (e.g., why water on a sidewalk will be more likely to evaporate and water on a grassy sports field or a sandy playground will be more likely to infiltrate). Therefore, curriculum designers need to provide multiple entry points so that students performing at different levels can productively engage in instructional activities.

In the Tools for Reasoning about Water project, we used the WSLP to design *pedagogical tools for reasoning* that would scaffold students performing at different levels in tracing water through various environmental systems. We designed these tools for reasoning, which took the form of graphic organizers, so that teachers could integrate them into their existing instructional sequences in a variety of ways and support students performing at different levels on the WSLP to make progress toward the next level. The first tool, called the Pathways Tool (Figure 18.4a), supports students performing at levels 1–2 in working toward level 3 performances. On this tool, students begin in the center by identifying a familiar local water source, such as a nearby body of water or even water in a glass, bathtub, or swimming pool, and then trace both backward (i.e., where the water came from) and forwards (i.e., where the water goes next). The boxes become progressively larger in each direction to encourage students to consider multiple potential locations for water at each point along the pathway. The second tool, called the Drivers and Constraints Tool (Figure 18.4b), supports students performing at level 3 in working toward level 4 learning performances. This tool prompts students to identify likely pathways for water but focuses students' attention on the scientific principles that govern where water moves and why.

Neither of these tools functions as an instructional activity by itself. Educative features in the curriculum materials provide teachers with instructions on how to integrate the tools into other instructional activities, such as stream table activities or permeability activities, to engage students in reasoning about the flow of water. Thus, the tools provide teachers with flexibility to serve students working at different points along the learning progression. In our classroom observations, teachers used the tools in a variety of ways, sometimes having groups of students use different tools and sometimes having all students use one tool at a time.

As models of learning, learning progressions can provide insight into challenges that students may experience while working toward achieving rigorous learning goals as well as strengths that students bring to learning more complex disciplinary core ideas. This, in turn, can inform the development of instructional sequences and pedagogical tools that help teachers respond to students' level of thinking and support students in reaching a higher level of performance.

Integrating Learning Progression-Based Reasoning Tools with Instructional Language Routines to Support Multilingual Students

A third component of learning progression-based educative curriculum materials that can help teachers both use new lenses to interpret their diverse students' thinking and engage their students in making progress on a learning progression is instructional language routines. Instructional language routines come from mathematics education and are structured ways for students to regularly and productively engage with concepts and practices (Kelemanik et al., 2016; Zwiers et al., 2017). Instructional language routines help frame progress across levels of performance in learning progressions by emphasizing that it is not mastering a new vocabulary that is important when learning science, but rather engaging in a new discourse employing disciplinary core ideas, science and engineering practices, and crosscutting concepts. Rigorous, language-rich opportunities are particularly important for multilingual learners, who develop their English language proficiency in interaction with their science understanding (Lee et al., 2019). Below, we discuss two examples of instructional language routines designed to support multilingual learners. In our examples, the routines are implemented in conjunction with the School Water Pathways curriculum materials (see also Roberts & Bianchini, 2019).

The need to intentionally integrate instructional language routines for multilingual learners in learning progression-based curriculum materials emerged from our research with teachers and students from diverse geographic and demographic contexts (Carpenter et al., 2020). We found teachers' efforts to engage their students in reasoning and sensemaking on concepts central to the WSLP were uneven. In some cases, teachers engaged their students in generative discussions of ideas, practices, and concepts to help move students' thinking forward on the learning progression; in other cases, teachers provided only a cursory opportunity to discuss water pathways. Merging the use of pedagogical tools for reasoning (e.g., the Pathways Tool), discussed in the section above, with instructional language routines that promote student language development and sensemaking helps to better scaffold the learning of disciplinary thinking and discourse by all students, particularly multilingual learners. The language routines can be considered an educative feature, providing teachers guidance in how to productively implement reasoning tools with their students.

To elaborate, pedagogical tools for reasoning, such as those described in the previous section, are graphic organizers designed to scaffold students performing at different levels on the WSLP as they trace water through various environmental systems. Instructional language routines are structured routines to scaffold multilingual learners' efforts to engage with text and their peers in rigorous science learning. Instructional language routines support students to (a) participate in sensemaking by organizing tasks and amplifying language (i.e., in contrast to simplifying, or watering down, disciplinary language, anticipating where students might need support and providing multiple ways) to access rigorous content and engage in reasoning; (b) cultivate conversations by creating opportunities and supports for constructive discussions in pairs, groups, and as a whole class; (c) optimize output by requiring students to describe their thinking to others orally, visually, and/or in writing; and (d) maximize linguistic and cognitive meta-awareness by forging connections across core concepts, reasoning, and language (Kelemanik et al., 2016; Zwiers et al., 2017). Examples of instructional language routines that can be included in educative curriculum materials include Co-Craft Questions & Problems; Three Reads; Stronger & Clearer; and Clarify, Critique & Correct (Roberts et al., 2021). We note for readers that the term *correct* in the title of the last instructional routine should be understood as short-hand for elaborate, revise, and/or refine; as stated in a previous section, learning progression research encourages both teachers and students to move away from notions of correct and incorrect conceptions (Furtak, 2012).

These language routines are written into educative curriculum materials as supports for both teachers and their students—to provide teachers concrete ways to help their students use reasoning tools to make progress along the learning progression. As an example, in the School Water Pathways curriculum, we integrated the use of the Pathways Tool for reasoning (Figure 18.4a) with the instructional language routine Stronger & Clearer to help move students' thinking, particularly multilingual learners' thinking, from levels 1–2 to level 3 on the WSLP. Stronger & Clearer is a routine that provides multilingual learners a structured and interactive opportunity to revise and refine both their ideas and their verbal and written output through individual work and peer discussions (Zwiers et al., 2017). As a graphic organizer, the Pathways Tool represents a type of multilingual learner instructional support that encourages language production. Pairing the Stronger & Clearer routine with the Pathways Tool offers a structured, interactive, and iterative opportunity for students to revise and refine their ideas about where water comes from and where it goes—both verbally and in writing. As such, multilingual learners' own languages and linguistic practices as well as their efforts to comprehend and produce language are supported so that they engage in productive sensemaking (Meier et al., 2020).

In this combined reasoning tool and language routine, multilingual learners begin with an individual *prewrite*, filling in the various boxes in the Pathways Tool to trace where water in a local context they are familiar with (e.g., a glass, bathtub, swimming pool) comes from and where it goes. They then engage in *think time* to decide and rehearse what they will say about the water pathways to their peers. Next, in a *pair share*, two peers take turns presenting and listening to ideas, respectively. Each asks clarifying questions when needed, especially related to evidence and reasoning. Multilingual learners *switch partners* one to three more times, continuing to present their ideas, listen to their peers, and ask clarifying questions. In these pair shares, students are learning to talk about water in ways that help them trace water along multiple pathways to work toward level 3 performances. Finally, they complete an individual *post-write*, where they return to their seats and revise their Pathways Tool to more thoroughly trace the multiple pathways water takes from their target context – to solidify the thinking and talking they practiced and refined in their pair shares. In completing this combined reasoning tool and language routine, multilingual learners participate in sensemaking; discuss constructively in pairs; optimize output by having to describe their thinking to others orally and in writing; and forge connections across core concepts, reasoning, and language.

As a second example, we integrated the use of the Drivers and Constraints Tool (Figure 18.4b) with the instructional language routine Clarify, Critique & Correct to scaffold multilingual learners in moving from level 3 to level 4 on the WSLP. Clarify, Critique, & Correct is a language routine that gives multilingual learners a piece of work that is not their own to analyze, reflect on, and develop. The intent is to prompt students to reflect on an incomplete written argument, explanation, or solution, and then to improve upon the written work by clarifying thinking, offering critique, and revising the initial response. As with the Pathways Tool, the Drivers and Constraints Tool is already in the form of a graphic organizer; it includes visuals with the text as well. The Clarify, Critique & Correct routine gives students an opportunity to reflect on, analyze, and revise a response to this tool that is not their own—a response either written by a previous student or drafted by the teacher—as a stepping stone to making progress on the learning progression.

In this combined reasoning tool and language routine, working in pairs or as a group, multilingual learners examine a partially completed Drivers and Constraints graphic organizer and work to improve the response using three steps. Students first *clarify* the reasoning of the initial response. They are given sentence starters to help frame their discussion if needed: I notice … . I wonder how they decided …? Next, students critique the response, using their own understanding of the scientific principles that govern where water moves and why to analyze the response. For this step, in addition to using sentence starters, they are explicitly asked to identify at least one strength and one suggestion for improvement. Third, in *correct*, students work collaboratively to write an improved response, keeping those parts of the initial response that are strong, revising those parts that are vague or confusing, and adding details where helpful. The routine concludes with pairs or groups sharing their improved Drivers and Constraints Tool responses and making additional refinements as needed during the ensuing whole group discussion. By engaging in this combined reasoning tool and language routine, students are given structured practice in scientific discourse through thinking, talking, and writing about the various drivers of water movement.

As these two examples illustrate, by pairing research on effective instruction for multilingual learners with learning progressions research, curriculum materials can help teachers support all their students in developing both fluency in disciplinary language and deep conceptual understanding aligned with a learning progression.

Conclusion

Learning progression-based curriculum materials are, in many ways, the linchpin for realizing the coherence across curriculum, instruction, and assessment that learning progressions promise to offer. At a minimum, learning progression-based curriculum materials can provide instructional sequences that are responsive to students' ways of making sense of the world and supportive of students in developing ways of thinking and talking that align with scientific models of phenomena. However, comprehensive learning progression-based curriculum materials can also provide crucial scaffolds for supporting teachers' interpretations of students' sensemaking and productive responses to students' ideas during instruction. Comprehensive learning progression-based curriculum materials that incorporate formative assessments, pedagogical reasoning tools, instructional language routines, and educative features for teachers have the potential to support teachers in using learning progressions in classrooms and support diverse students in achieving high-level performances on learning goals.

We have found that designing these components of comprehensive learning progression-based curriculum materials benefits from close collaborations with teachers, schools, and school districts (Anderson et al., 2018). Teachers and districts can make significant contributions to the development of these materials through collaboration on identifying key areas of focus, collecting student assessment data, informing the design and testing of instructional materials, and providing feedback. These partnerships enable researchers and curriculum developers to use learning progressions to design curriculum materials that are both responsive to student sensemaking and supportive of teacher practice, resulting in curriculum materials that can increase coherence across assessment, curriculum, and instruction in the classroom.

Note

1 For the purposes of this chapter, levels 1 and 2 have been combined to level 1–2.

References

Alonzo, A. C., & Elby, A. (2019). Beyond empirical adequacy: Learning progressions as models and their value for teachers. *Cognition and Instruction*, *37*(1), 1–37. 10.1080/07370008.2018.1539735

Alonzo, A. C., Wooten, M. M., & Christensen, J. (2022). Learning progressions as a simplified model: Examining teachers' reported uses to inform classroom assessment practices. *Science Education*, *106*(4), 852–889.

Anderson, C. W., de los Santos, E. X., Bodbyl, S., Covitt, B. A., Edwards, K. D., Hancock, J. B., Lin, Q., Morrison Thomas, C., Penuel, W. R., & Welch, M. M. (2018). Designing educational systems to support enactment of the Next Generation Science Standards. *Journal of Research in Science Teaching*, *55*(7), 1026–1052. 10.1002/tea.21484

Ball, D. L., & Cohen, D. (1996). Reform by the book: What is - or might be- the role of curriculum materials in teacher learning and instructional reform? *Educational Researcher*, *25*(9), 6–8, 14.

Braaten, M., & Windschitl, M. (2011). Working toward a stronger conceptualization of scientific explanation for science education. *Science Education*, *95*(4), 639–669. 10.1002/sce.20449

Brown, M. W. (2009). The teacher-tool relationship: Theorizing the design and use of curriculum materials. In J. T. Remillard, B. A. Herbal-Eisenmann, & G. M. Lloyd (Eds.), *Mathematics teachers at work: Connecting curriculum materials and classroom instruction* (pp. 17–36). Routledge.

Campbell, T., Schwarz, C., & Windschitl, M. (2016). What we call misconceptions may be necessary stepping-stones toward making sense of the world. *Science and Children*, *53*(7), 28.

Caplan, B., Gunckel, K. L., Warnock, A., & Cano, A. (2013). Investigating water pathways in schoolyards. *Green Teacher*, *98*(Winter), 28–33.

Carpenter, S. L., Kim, J., Nilsen, K., Irish, T., Bianchini, J. A., & Berkowitz, A. R. (2020). Secondary science teachers' use of discourse moves to work with student ideas in classroom discussions. *International Journal of Science Education*, *42*(15), 2513–2533. 10.1080/09500693.2020.1820620

Corcoran, T. B., Mosher, F. A., & Rogat, A. (2009). *Learning progressions in science: An evidence-based approach to reform CPRE Research Report #RR-63*.

Covitt, B. A., Gunckel, K. L., Caplan, B., & Syswerda, S. (2018). Teachers' use of learning progression-based formative assessment in water instruction. *Applied Measurement in Education*, *31*(2), 128–142. 10.1080/08957347.2017.1408627

Davis, E., Palincsar, A. S., Arias, A. M., Bismack, A. S., Marulis, L., & Iwashyna, S. (2014). Designing educative curriculum materials: A theoretically and empirically driven process. *Harvard Educational Review*, *84*(1), 24–52. 10.17763/haer.84.1.g48488u230616264

Davis, E. A., & Krajcik, J. (2004). Designing educative curriculum materials to support teacher learning. *Educational Researcher*, *34*(3), 3–14.

Duschl, R., Maeng, S., & Sezen, A. (2011). Learning progressions and teaching sequences: A review and analysis. *Studies in Science Education*, *47*(2), 123–182. 10.1080/03057267.2011.604476

Furtak, E. M. (2012). Linking a learning progression for natural selection to teachers' enactment of formative assessment. *Journal of Research in Science Teaching*, *49*(9), 1181–1210. 10.1002/tea.21054

Furtak, E. M., & Tayne, K. (2019). Affordances and constraints of learning progression designs in supporting formative assessment. In *Bridging research and practice in science education* (pp. 241–256). Springer.

Gee, J. P. (1991). What is literacy? In C. Mitchell, & K. Weiler (Eds.), *Rewriting literacy: culture and the discourse of the other* (pp. 3–11). Bergin & Garvey.

Gunckel, K. L., Covitt, B. A., Berkowitz, A. R., Caplan, B., & Moore, J. C. (2022). Computational thinking for using models of water flow in environmental systems: Intertwining three dimensions in a learning progression. *Journal of Research in Science Teaching* (59), 1159–1203. 10.1002/tea.21755

Gunckel, K. L., Covitt, B. A., & Salinas, I. (2018). Learning progressions as tools for supporting teacher content knowledge and pedagogical content knowledge about water in environmental systems. *Journal of Research in Science Teaching*, *55*(9), 1237–1362. 10.1002/tea.21454

Gunckel, K. L., Covitt, B. A., Salinas, I., & Anderson, C. W. (2012). A learning progression for water in socio-ecological systems. *Journal of Research in Science Teaching*, *49*(7), 843–868. 10.1002/tea.21024

Heritage, M., Kim, J., Vendlinski, T., & Herman, J. (2009). From evidence to action: A seamless process in formative assessment? *Educational Measurement: Issues and Practice*, *28*(3), 24–31. 10.1111/j.1745-3992.2009.00151.x

Jin, H., Mikeska, J. N., Hokayem, H., & Mavronikolas, E. (2019). Toward coherence in curriculum, instruction, and assessment: A review of learning progression literature. *Science Education*, *103*(5), 1206–1234. 10.1002/sce.21525

Kelemanik, G., Lucenta, A., & Creighton, S. J. (2016). *Routines for reasoning*. Heinemann.

Kesidou, S., & Roseman, J. E. (2002). How well do middle school science programs measure up? Findings from Project 2061's curriculum review. *Journal of Research in Science Teaching*, *39*, 522–549. 10.1002/tea.10035

Krajcik, J., McNeill, K. L., & Reiser, B. J. (2008). Learning-goals-driven design model: Developing curriculum materials that align with national standards and incorporate project-based pedagogy. *Science Education*, *92*(1), 1–32. 10.1002/sce.20240

Lee, O., Llosa, L., Grapin, S., Haas, A., & Goggins, M. (2019). Science and language integration with English learners: A conceptual framework guiding instructional materials development. *Science Education*, *103*(2), 317–337. 10.1002/sce.21498

Lehrer, R., & Schauble, L. (2006). Cultivating model-based reasoning in science education. In R. K. Sawyer (Ed.), *Cambridge handbook of the learning sciences* (pp. 371–388). Cambridge University Press.

Meier, V., Aminger, W., McLean, M., Carpenter, S. L., Moon, S., Hough, S., & Bianchini, J. A. (2020). Preservice secondary science teachers' understanding of academic language: Moving beyond "just the vocabulary". *Science Education*, *104*(2), 222–251. 10.1002/sce.21560

National Research Council. (2006). Systems for state science assessment. In M. Wilson, & M. W. Bertenthal (Eds.), *Committee on test design for K-12 science achievement*. National Academies Press.

National Research Council. (2007). Taking science to school: Learning and teaching science in grades K-8. In R. A. Duschl, H. A. Schweingruber, & A. W. Shouse (Eds.), *Committee on science learning kindergarten through eighth grade*. National Academies Press.

National Research Council. (2012). *A framework for K-12 science education: Practices, crosscutting concepts, and core ideas*. National Academies Press.

NGSS Lead States. (2013). *Next generation science standards: For states, by states*. National Academies Press. www.nextgenscience.org

Pinker, S. (2007). *The stuff of thought: Language as a window into human nature*. Penguin Group.

Roberts, S. A., & Bianchini, J. A. (2019). Scaffolding preservice science teacher learning of effective English learner instruction: A principle-based lesson cycle. *Innovations in Science Teacher Education*, *4*(3). Retrieved from https://innovations.theaste.org/scaffolding-preservice-science-teacher-learning-of-effective-english-learner-instruction-a-principle-based-lesson-cycle/

Roberts, S. A., Nguyen, T., Dexter Torti, C., Mansour, R., & Bianchini, J. A. (2021). *Using mathematical language routines to engage high school students in mathematical disciplinary practices*. AERA Annual Meeting, Online.

Schneider, M. C., & Gowan, P. (2013). Investigating teachers' skills in interpreting evidence of student learning. *Applied Measurement in Education*, *26*(3), 191–204. 10.1080/08957347.2013.793185

Shepard, L. A., Penuel, W. R., & Pellegrino, J. W. (2018). Using learning and motivation theories to coherently link formative assessment, grading practices, and large-scale assessment. *Educational Measurement: Issues and Practice*, *37*(1), 21–34. 10.1111/emip.12189

Talmy, L. (1988). Force dynamics in language and cognition. *Cognitive Science*, *12*(1), 49–100.

Wenger, E. (1998). *Communities of practice: Learning, meaning, and identity*. Cambridge University Press.

Wilson, M., & Sloane, K. (2000). From principles to practice: An embedded assessment system. *Applied Measurement in Education*, *13*, 181–208. 10.1207/S15324818AME1302_4

Wiser, M., Smith, C. L., & Doubler, S. (2012). Learning progressions as tools for curriculum development: Lessons from the Inquiry Project. In *Learning progressions in science* (pp. 357–403). Brill Sense.

Zwiers, J., Dieckmann, J., Rutherford-Quach, S., Daro, V., Skarin, R., Weiss, S., & Malamut, J. (2017). *Principles for the design of mathematics curricula: Promoting language and content development*. Stanford University.

19

CROSSCUTTING CONCEPTS AND LEARNING PROGRESSIONS

Jeffrey Nordine and Okhee Lee

Doing science and engineering involves using a set of conceptual tools to make sense of phenomena and solve problems in the natural and designed world. In the US, *A Framework for K-12 Science Education* (National Research Council [NRC], 2012) and the *Next Generation Science Standards* (NGSS Lead States, 2013a) identify three categories of conceptual tools that are essential to doing science and engineering: science and engineering practices (SEPs), disciplinary core ideas (DCIs), and crosscutting concepts (CCCs). The *Framework* refers to SEPs, DCIs, and CCCs as three "dimensions" of science learning. SEPs consist of a set of practices (e.g., asking questions, using models) in which scientists and engineers engage. DCIs specify central and powerful ideas from science and engineering disciplines (e.g., natural selection, structure and properties of matter, engineering design process). CCCs consist of a set of concepts "that unify the study of science and engineering through their common application across fields" (NRC, 2012, p. 2), these are as follows:

1 Patterns
2 Cause and effect: mechanism and explanation
3 Scale, proportion, and quantity
4 Systems and system models
5 Energy and matter: flows, cycles, and conservation
6 Structure and function
7 Stability and change.

While CCCs are essential to doing science and engineering, the role they have played in science instruction has historically been implicit rather than explicit. The ideas contained within the CCCs appeared in previous standards documents (American Association for the Advancement of Science [AAAS], 1993; NRC, 1996), but their role within previous documents was peripheral – acknowledged as common across science disciplines but not explicitly included in learning standards. In contrast, the *Framework* and NGSS explicitly include CCCs in each of the NGSS performance expectations (which specify how students should be able to use their science learning at various grade levels or bands). The *Framework* and NGSS recognize the need for CCCs to be intentionally and explicitly included in science instruction.

DOI: 10.4324/9781003170785-22

Like DCIs and SEPs, CCCs are complex and nuanced ideas that require extended time to develop; yet, in contrast to DCIs and SEPs, little has been done to investigate how students develop an understanding of CCCs over time. The *Framework* authors acknowledged this paucity of research: "The research base on learning and teaching the crosscutting concepts is limited. For this reason, the progressions we describe should be treated as hypotheses that require further empirical investigation" (NRC, 2012, p. 84). Despite decades of recognition that CCCs are an indispensable part of doing science and engineering, the *Framework* authors were unable to draw upon a sufficient base of empirical research to outline evidence-based learning progressions for CCCs. And despite the call for additional empirical research in this area, little additional work has been done since the release of the *Framework* (Fick & Arias, 2022).

Although little empirical research has been done to construct learning progressions for CCCs, researchers and practitioners have made strides in further clarifying the role of CCCs in science learning and teaching more broadly. In our previous work (Nordine & Lee, 2021), we argued that CCCs play three primary roles in science teaching and learning: (a) integrating with SEPs and DCIs to make sense of phenomena and solve problems, (b) providing a set of complementary conceptual lenses on phenomena, and (c) connecting across contexts and disciplines. Further, CCCs can uniquely connect to students' existing science ideas and funds of knowledge, providing learners with multiple points of entry to sensemaking and serving to broaden access and inclusion in the science classroom.

Despite more than a decade since the release of the *Framework*, learning progressions for CCCs in K-12 science education remain poorly understood. In this chapter, we consider the important role of CCCs in science learning, examine the current research evidence on learning progressions for CCCs, and offer guidance for future research on CCC learning progressions.

CCCs and Their Role in Science Learning

Recognizing and Situating CCCs in Science Learning in Earlier Efforts

Prior to their inclusion in the NGSS, CCCs were included in two major standards documents in the US: *Benchmarks for Science Literacy* (AAAS, 1993) and *National Science Education Standards* (*NSES*; NRC, 1996), albeit named differently. The *Benchmarks* included four "common themes": (a) systems, (b) models, (c) constancy and change, and (d) scale. The *NSES* identified five "unifying concepts and processes": (a) systems, order, and organization; (b) evidence, models, and explanation; (c) constancy, change, and measurement; (d) evolution and equilibrium; and (e) form and function. Both standards documents noted the widespread use of these ideas across science disciplines and recognized their role in supporting scientific thinking. The *Benchmarks* authors described the common themes as "really ways of thinking rather than theories or discoveries" (AAAS, 2009), while the *NSES* asserted that the unifying concepts and processes "provide students with productive and insightful ways of thinking" (NRC, 1996, p. 115).

Both the *Benchmarks* and the *NSES* described precursor CCC ideas as "ways of thinking" that were productive across science disciplines. Whether an intentional decision or not, the *Framework* does not describe CCCs as "ways of thinking;" instead, it stresses the "explanatory value" of CCCs across science and engineering disciplines. When describing their potential value for students, the *Framework* asserts that they "help provide students with an organizational framework for connecting knowledge from the various disciplines into a coherent and scientifically based view of the world" (NRC, 2012, p. 83).

Since the release of the *Framework*, some researchers have refuted the assertion that CCCs bring greater coherence to science learning, arguing that there is little to no theoretical or empirical basis for this assertion (Osborne et al., 2018). Others have noted that the term "crosscutting" contributes to confusion about the role of CCCs in science learning, as it implies that their primary value is in supporting cross-disciplinary connections (Cooper, 2020). CCCs can look quite different when used in different disciplines (Cooper & Klymkowsky, 2013; Nordine, 2016), thus raising legitimate questions about how scientists apply CCCs and whether they do so in a similar manner across disciplines (Osborne et al., 2018).

Though there has been some debate about whether CCCs bring coherence across science disciplines, there is broad agreement that the CCCs are a set of conceptual tools used by scientists from all disciplines (albeit often quite differently) as they make sense of phenomena and solve problems (Fick et al., 2019; Quinn, 2021). Since the publication of the *Framework*, several groups of scholars have offered clarifications regarding the role of CCCs in science learning using both theoretical and practical perspectives.

Clarifying the Role of CCCs in Science Learning

One of the first efforts to clarify the intended role of CCCs in science learning was made by Rivet et al. (2016), who analyzed the *Framework*, NGSS appendices, and other NGSS supporting documents to identify metaphors used within these documents to articulate the theoretical role of CCCs in science learning. They identified four major metaphors: (a) lenses, (b) bridges, (c) tools, and (d) rules of the game. The lenses metaphor refers to using CCCs to think about phenomena and problems – just as different optical lenses emphasize some parts of an object and de-emphasize others, different CCCs can focus student attention on different features of phenomena and problems. The bridges metaphor refers to the potential of CCCs to support students in seeing connections across contexts and disciplines. The tools metaphor refers to the potential for CCCs to help students strengthen their explanations of phenomena and their solutions to problems. Finally, the rules of the game metaphor refer to the role of CCCs in supporting students' understanding of epistemologies and the shared language of science.

In attempting to clarify the intended role of CCCs in science instruction, Krajcik and Reiser (2021) identified and elaborated four primary roles: (a) guiding students' engagement in SEPs, (b) applying ideas across contexts, (c) problematizing and deepening student explanations, and (d) providing explicit tools for sensemaking. CCCs may guide engagement in SEPs by playing a focusing role in which students use CCCs to consider key aspects of phenomena and problems as they engage in SEPs. CCCs may also guide engagement in SEPs by playing an evaluative role in which students reflect on the adequacy of the products they create through engagement in SEPs. CCCs may help students apply ideas across contexts by supporting them in activating relevant prior knowledge as they encounter new phenomena. CCCs may support students to problematize and deepen explanations as they consider whether and how explanations of phenomena are adequate and provide conceptual tools for revising and refining explanations to more fully account for how or why phenomena occur. Finally, CCCs may serve as explicit tools for sensemaking when students learn to use CCCs to guide their questioning, direct model construction, and evaluate explanations of phenomena. Krajcik and Reiser stress that CCCs are not ends in and of themselves but rather tools to be used in conjunction with SEPs and DCIs to make sense of phenomena and solve problems.

In a comprehensive review of the peer-reviewed science education research literature published between 2012 and 2019, Fick and Arias (2022) identified five themes regarding the importance of CCCs for science learning: (a) providing opportunities to learn, (b) drawing upon funds of knowledge, (c) connecting to big ideas, (d) linking to practice, and (e) connecting across topics and disciplines. CCCs play a role in supporting opportunities for science learning through their inclusion in curriculum, instruction, and assessment. Though Fick and Arias did not identify consensus with respect to how CCCs support science learning opportunities, they found that many articles stressed the importance of explicitly including CCCs in science learning. As supports for drawing upon students' funds of knowledge, CCCs may play a powerful role in connecting students' everyday experiences with science learning opportunities in school. CCCs connect to the big ideas of science (e.g., DCIs) and may play an important role in deepening students' understanding and ability to use science ideas. Likewise, CCCs are strongly connected to SEPs – Fick and Arias refer to "intrinsic links" between specific SEPs and CCCs, since CCCs are required to fully engage in SEPs. Finally, CCCs have the potential to provide common language, analogies, representations, and models to consistently frame learning contexts, which supports students in connecting their science learning across topics and disciplines.

To summarize, looking across efforts to clarify the role of CCCs in science learning, we identify three areas of consensus. First, CCCs are a set of conceptual tools that are indispensable for doing science and are essential for using DCIs and SEPs to make sense of phenomena and solve problems. Second, CCCs function as complementary lenses on phenomena. Using CCCs as lenses can focus students' attention on salient features of phenomena and problems and set the stage for productive sensemaking and problem-solving. Using multiple CCCs to view the same phenomenon and problem can deepen understanding. Third, CCCs span contexts and disciplines. Using these tools consistently across a range of contexts can help learners notice deep structure across phenomena and problems (Schwartz & Goldstone, 2016).

Broadening Access to Science

Broadening access to science is a central theme of the *Framework* and the NGSS, hence "all standards, all students" (NGSS Lead States, 2013b). Traditionally, it has been assumed that students from nondominant backgrounds bring limited prior canonical science knowledge. However, as the NGSS emphasizes "all standards, all students," it is imperative that science be made real and relevant to all students. A small body of literature illustrates how CCCs can play a role in broadening access to science learning (Goggins et al., 2019, 2021; Nordine & Lee, 2023). Currently, this literature serves more as conjectures than a set of results.

All students bring to the science classroom their funds of knowledge from their homes and communities. In their everyday lives, students use various resources to make sense of the world. In the NGSS classroom, students use CCCs to make sense of phenomena and design solutions to problems as scientists and engineers do in their work. CCCs can be thought of as resources that students use in their everyday lives to make sense of the world (e.g., What are the lengths of shadows over the course of a day? Why do the lengths of shadows change over the course of a day?) and bring to the science classroom to help make sense of phenomena (e.g., I want to find lengths of shadows over the course of a day; I want to figure out why the lengths of shadows change over the course of a day). This perspective on *CCCs as resources* that all students bring to the science classroom to make sense of phenomena is particularly important for students who do not see science as relevant to their lives or future careers. This perspective on CCCs creates relevance by capitalizing on students' funds of knowledge from their homes and communities (Fick & Arias, 2022; González et al., 2005). This perspective also promotes the participation of all students by offering multiple points of entry and concrete conceptual tools for engagement in the science classroom. When CCCs are thought of as resources that students have and can use to make sense of the world, teachers demonstrate value for students' cultural and linguistic resources. This perspective on CCCs as resources is a shift from a traditional perspective on CCCs as academically demanding to students, especially those from nondominant backgrounds who were perceived to bring limited prior canonical science knowledge.

If students are expected to use their understanding of CCCs to explain and predict phenomena and design solutions to problems, students need to engage with compelling phenomena and problems involving the use of CCCs. They are motivated to figure out phenomena that are real and relevant to them, their families, and their communities. In figuring out the phenomena, students bring their funds of knowledge from their homes and communities. In particular, they can leverage CCCs as resources in the context of local phenomena and problems rooted in their experiences, so that local phenomena and CCCs are mutually supportive for making their' ideas visible (Lee, 2020).

Initially, students' understanding of CCCs is often intuitive based on their experiences in their homes and communities (Fick et al., 2019; Goggins et al., 2019, 2021). Given that students come to school with intuitive ideas about CCCs, science instruction should leverage these intuitive ideas. Through instruction, CCCs need to be made explicit so students can use CCCs to make sense of phenomena and problems (Fick et al., 2017). As students develop a deeper understanding of CCCs, they use CCCs more intentionally to examine the same phenomenon or a new phenomenon. And whenever students make use of CCCs to make sense of phenomena and problems, they further develop their understanding of CCCs.

Some CCCs are more intuitive to students than others (Fick et al., 2019; Nordine & Lee, 2023). For example, humans innately recognize patterns in stimuli in their surroundings, such as in language acquisition, facial recognition, and music appreciation. On the other hand, students' ability to use energy and matter conservation principles develops only through extended effortful practice. To promote all CCCs as resources, designing instruction around students' use of more intuitive CCCs at the beginning of the school year lays the foundation for students to better use all CCCs over time. By explicitly naming the CCCs students use intuitively at the beginning of the year, students are made aware of the resources they bring to the classroom and see themselves doing the work of scientists and engineers. In turn, students develop more intentional use of CCCs over time as they make sense of different disciplines and different phenomena (Goggins et al., 2019, 2021).

To summarize, building on CCCs as resources during instruction can make students' intuitive ideas about CCCs visible and can guide students in using CCCs intentionally over the course of instruction. This perspective on CCCs, which builds on students' funds of knowledge, is powerful for all students, and especially powerful for students from nondominant backgrounds. This perspective calls for a shift in instruction from a deficit perspective (i.e., students from underserved groups come to the science classroom with limited sense-making resources) to an asset perspective (i.e., students from underserved groups come to the science classroom with rich sense-making resources). By explicitly building on students' intuitive ideas, CCCs can act to both broaden participation and strengthen science learning for all students.

Theoretical Efforts to Define CCC Learning Progressions

The *Framework* authors recognized that the progressions they outlined for the CCCs constituted hypotheses in need of further research. Like all well-formed hypotheses, these progressions were not simply guesses but rather rested on a foundation of existing knowledge about how students learn.

The Common Themes that appear within the *Benchmarks* were first identified as important ideas across all science disciplines in *Science for All Americans* (AAAS, 1989), though this book focused on describing the role that the Common Themes play in science without describing how students might build understanding over time. A major contribution of the *Benchmarks* was to elaborate a possible sequence for how students might learn more sophisticated ideas about the Common Themes throughout Grades K-12. Drawing upon existing research into student ideas about science (e.g., Driver et al., 1985), the *Benchmarks* described what students should know about each Common Theme across Grade Bands K-2, 3–5, 6–8, and 9–12. Importantly, much of the research available at this time consisted of studies focusing on student conceptions about specific science topics. Notably, very little of the research that undergirded the *Benchmarks* progressions for the Common Themes employed cross-sectional or longitudinal designs to explore how students build ideas over time.

In contrast to the Common Themes in the *Benchmarks*, the Unifying Concepts and Processes in the *NSES* opted not to specify how learners might develop a more sophisticated understanding over time: "Each of the concepts and processes of this standard has a continuum of complexity that lends itself to the K-4, 5–8, and 9–12 grade level clusters used in the other content standards. In this standard, however, the boundaries of disciplines and grade-level divisions are not distinct – teachers should develop students' understanding continuously across grades K-12" (National Research Council, 1996, pp. 115–116).

In outlining hypothetical learning progressions for the CCCs, the *Framework,* and NGSS build on the efforts of the *Benchmarks* and *NSES* by recognizing that each CCC is comprised of complex sets of ideas that must be learned over extended periods of time. The *Framework* and NGSS extended the *Benchmarks* and *NSES* by elaborating how learners might use CCCs along with SEPs and DCIs to make sense of phenomena and solve problems. Much of this elaboration is presented in NGSS Appendix G, which elaborates on grade band endpoints for each CCC across Grades K-12 and provides examples of NGSS performance expectations to illustrate how students might use their understanding of CCCs to engage in three-dimensional learning. The hypothetical learning progressions outlined in the *Framework* and NGSS Appendix G are not explicitly linked to research on student learning, making it difficult to ascertain the extent to which these progressions are based upon synthesis across empirical research or best guesses of how students might build more complex ideas over time. In either case, the *Framework* is clear that CCC progressions are best regarded as hypotheses in need of empirical work.

There is little debate that the ideas captured under Common Themes, Unifying Concepts and Processes, and Crosscutting Concepts are essential parts of science, but little is known about how students should build their understanding and ability to use these ideas over time. The *Framework* and NGSS build upon the *Benchmarks* and NSES in two important ways: (1) recognizing that student learning about CCCs is most powerful and useful when it is integrated with DCIs and SEPs, and (2) hypothesizing a progression of how students might develop CCC ideas throughout K-12 as they integrate CCCs with DCIs and SEPs to make sense of phenomena and solve problems. A clear implication of the *Framework* and NGSS is that any work to further develop the hypothetical learning progressions proposed within these documents should focus not only on students' knowledge of the CCCs themselves but their increasing sophistication in using CCCs in conjunction with SEPs and DCIs to make sense of phenomena and solve problems. That is, CCC learning progressions should focus on students' knowledge-in-use (Pellegrino & Hilton, 2012) about CCCs rather than simply their knowledge of CCCs.

Empirical Research to Define CCC Learning Progressions

Since the release of the *Framework*, little work has been done to further define and empirically investigate how students build and use CCCs over time. In their literature review, Fick and Arias (2022) identified only five peer-reviewed publications addressing CCC learning progressions in any explicit way, and they noted that these five papers primarily focused on CCCs as one aspect of developing a learning progression for some other science topic (e.g., evolution) or SEP (e.g., planning and carrying out investigations).

Although there is not yet a robust base of empirical research into how students develop an increasingly sophisticated understanding of CCCs throughout K-12, several ideas related to CCCs have been actively researched using a learning progression focus. Since the release of the *Framework*, the most active area of empirical research using a CCC frame has been related to the CCC "matter and energy: flows, cycles, and conservation" (Fick & Arias, 2022). Below, we discuss empirical research related to this CCC to illustrate how existing research might inform CCC learning progressions and to identify gaps that could be addressed by future CCC learning progressions research.

The Case of Matter and Energy

The energy concept is central to all sciences, as is an understanding of the nature of matter, but neither matter nor energy is CCCs on their own. Rather, the crosscutting idea is that certain quantities are conserved (Quinn, 2021). Although matter and energy are not the only conserved quantities (other examples include momentum and charge), matter and energy are particularly useful in making sense of a wide range of phenomena and problems that students are likely to encounter in their lives.

Learning progressions have been extensively researched for the matter concept (He et al., this volume) and the energy concept (Fortus & Nordine, this volume). Much of this research has focused on how students build an understanding of each concept separately. To systematically investigate and identify levels of student learning about matter and energy, researchers have identified key aspects of each concept. In a review of research into student learning about matter, Hadenfeldt et al. (2014) specified four aspects of learning about the matter concept: (a) structure and composition, (b) physical properties and change, (c) chemical properties and change, and (d) conservation. These four aspects have been empirically shown to be distinct but interrelated aspects of a single overarching concept (Hadenfeldt et al., 2016). Researchers have similarly identified four aspects of learning about energy: (a) transformation, (b) transfer, (c) degradation, and (d) conservation. Like the aspects of the matter concept, these aspects of the energy concept have been empirically shown to be distinct yet interrelated aspects of a single overarching concept (Herrmann-Abell & DeBoer, 2018; Neumann et al., 2013).

Existing research into matter and energy learning progressions converges in finding that an understanding of conservation takes the most time to develop. Although the conservation principle for both matter and energy is stated relatively simply – matter and energy are never created nor destroyed – researchers have found that students' ability to restate the conservation principle is not a good indicator of whether they will adhere to it when making sense of phenomena (Chabalengula & Mumba, 2012). Developing the ability to apply conservation principles to make sense of phenomena and problems takes many years of effortful engagement guided by instruction. While previous research has found that an understanding of conservation is typically the last to develop as students learn about matter and energy, little is known about how students develop an increasingly sophisticated understanding of conservation over time.

In describing the conservation of energy and matter as a crosscutting idea, the *Framework* argues, "The ability to examine, characterize, and model the transfers and cycles of matter and energy is a tool that students can use across virtually all areas of science and engineering" (National Research Council, 2012, p. 95). More generally, we have argued that developing a more sophisticated understanding of CCCs results in students' ability to: (a) integrate between DCIs and SEPs to make sense of phenomena and solve problems, (b) use CCCs as conceptual lenses on phenomena, and (c) connect across contexts and disciplines (Nordine & Lee, 2021). While research has illuminated relatively little about how students develop their understanding of conservation over time, some studies have shed light on how students may begin to develop their ability to use their understanding of matter and energy to integrate DCIs and SEPs, to view phenomena through this these conceptual lenses, and to connect across disciplines.

Integrating CCCs with DCIs and SEPs

To integrate conservation of matter and energy with DCIs and SEPs, students must learn to explicitly activate conservation principles as they engage in SEPs to explain phenomena and solve problems. Like many other CCCs, conservation principles are often implicit rather than explicit – for example, as students learn to recognize the various manifestations of energy (e.g., sound, light, motion) in phenomena, the transfer of energy from one system/subsystem to another is often omitted from energy analysis (Nordine et al., 2018). There has been some promising work in better understanding how students explicitly use the conservation of matter and energy to learn within disciplinary contexts.

Jin and Anderson (2012) constructed an initial learning progression describing how students use energy in the context of carbon-transforming processes. They defined two progress variables – association (students' ideas about how processes, materials, or forms of energy are alike or different) and tracing (students' ways of identifying what endures and what changes during an event) – to characterize students' increasing ability to construct explanatory and predictive accounts of carbon-transforming processes. In discussing their results, they argued that "a learning progression focusing on energy can never be just about energy because energy is not a typical scientific concept. It is a foundation for an approach to analysis that uses matter and energy conservation to connect and constrain accounts of processes and systems in the material world" (Jin & Anderson, 2012, p. 1175). Rather than focusing on how students learn the energy concept itself, Jin and Anderson emphasized students' ability to use conservation ideas in conjunction with their knowledge of carbon-transforming processes to construct explanations and predictions.

Scott et al. (2023) have investigated how students use the idea of mass balance to reason about matter flows in biological systems. Through a series of student interviews and analysis of student written responses, they identified two progress variables to describe how students use mass balance: (1) identifying/using matter fluxes to construct explanations and (2) using mathematical thinking to explain impacts on material amounts. The researchers used a three-dimensional science learning perspective to analyze how well students were able to integrate mathematical thinking, understanding of the nature of matter, and matter conservation to make sense of biological phenomena. They argue that these progress variables can be used to understand how students progress in their ability to track mass fluxes to interpret phenomena from a range of contexts and disciplines.

Using Conservation as a Conceptual Lens on Phenomena

Using conservation as a lens on phenomena involves using conservation laws to set bounds on the behavior of systems to generate new insights and/or questions about how and why the phenomenon occurs or the system behaves as it does. Students must develop some nascent ideas about how matter and energy are conserved as phenomena occur. At a minimum, students must develop a sense of the "qualitative" conservation of energy and matter, which is to say that energy and matter have the quality that they are never created nor destroyed. This is distinct from developing an understanding of the "quantitative" conservation of energy and matter, which means that the quantity of energy and matter in a system can only change by the net amount transferred to and from it. Both ways of understanding conservation are useful for generating insights into the behavior of phenomena, but simply knowing *that* energy and matter are conserved is not enough, as students commonly agree with statements of conservation principles but violate these principles when asked to qualitatively explain phenomena (Chabalengula et al., 2012) and struggle to use conservation principles to quantitatively explain phenomena (Lindsey et al., 2012).

When students use conservation ideas to generate insights about phenomena, they trace matter/energy flows such that they realize something about the phenomena that they hadn't noticed before or recognize a gap in their existing knowledge. To determine students' ability to use conservation as a generative lens on phenomena, researchers need to go beyond tracking students' ability to construct explanations or solve problems and focus instead on how students activate and use conservation ideas as they encounter novel phenomena in which new learning is required. Very little research has shed light on how students use conservation ideas as a lens for generating insights about phenomena. One example is Kubsch et al. (2020), who studied how constructing an energy representation in which students qualitatively track energy transfers between interacting systems supported students in speculating about unseen changes occurring in observable systems and recognizing when unaccounted-for systems must play a role in phenomena (even if students were puzzled about what those systems were).

Connecting across Contexts and Disciplines

Using conservation to connect across disciplines involves interpreting phenomena and problems from different disciplinary contexts using the same interpretive heuristics. The bulk of existing work investigating how students learn and use energy/matter ideas consistently across disciplines has focused on energy rather than matter.

Some researchers have used a theoretical perspective to dispute the idea that energy is a useful concept for connecting disciplines, noting that the concept is represented and used quite differently in different disciplines (Osborne et al., 2018). Yet, empirical research suggests that students do begin to develop a cross-disciplinary understanding of energy. Opitz et al. (2019) assessed Grade 6–10 students' understanding of energy across biology, chemistry, and physics and found no evidence of discipline-specific differentiation in their understanding. Similarly, Park and Liu (2021) assessed university students' understanding of energy across environmental science, biology, chemistry, and physics and found that item difficulty was not significantly different across disciplines. Furthermore, they noted that item difficulty levels varied less for items focused on energy degradation/conservation ideas than items focused on energy source/form/transfer ideas, suggesting that energy degradation/conservation plays a potentially more powerful role in connecting disciplines.

Researchers have also collected evidence that students successfully build energy ideas across disciplines during instruction. Fortus et al. (2015) tracked student assessment performance on energy items across instructional units that were part of a comprehensive curriculum designed to be coherent across units. They found that students successfully used their energy ideas learned in one disciplinary unit to learn effectively in other disciplinary units. Similarly, Podschuweit and Bernholt (2018) explored whether learning about energy across physics, biology, and chemistry contexts supported students' ability to transfer their energy learning to new contexts compared to students who learned about energy in a predominantly physics context. They found that students who learned across contexts were more successful in using their existing energy knowledge to make sense of new situations across disciplinary contexts.

Despite arguments that energy is applied in fundamentally different ways across disciplines, the empirical research suggests that students develop energy ideas that do span disciplines, that instruction plays a significant role in supporting students' ability to use energy ideas across disciplines, and that conservation is a particularly important idea for connecting across contexts and disciplines.

Summary

The research provides some evidence that students can productively use their understanding of matter and energy in ways that align with the three primary roles of CCCs in science learning: integrating with SEPs and DCIs, using CCCs as conceptual lenses on phenomena, and connecting across contexts and disciplines. Yet, much of this research has focused on students learning about the concepts of matter or energy independently rather than the overarching conservation principle that lies at the heart of the CCC. As a result, even though "energy and matter: flows, cycles, and conservation" is the most researched CCC (Fick & Arias, 2022), little remains known about how students build an increasingly sophisticated understanding of and ability to use conservation principles over time.

A Path Forward for Research into CCC Learning Progressions

CCCs are a powerful set of conceptual tools used across science and engineering disciplines, yet they have historically played an implicit rather than an explicit role in science and engineering learning and instruction. The inclusion of CCCs within the *Framework* and the acknowledgment by the authors that a robust research base for CCCs did not exist represented a call to action for the research community to investigate how students learn to use CCCs. More than a decade later, much work remains, but existing research has clarified the contours of the path forward.

Focusing on the Role of CCCs in Science and Engineering Learning

Theoretical and empirical work has clarified three primary roles of CCCs in strengthening science and engineering learning and instruction: (a) integrating with DCIs and SEPs, (b) serving as a set of conceptual lenses on phenomena, and (c) connecting across contexts and disciplines. Future research to develop and validate CCC learning progressions should attend to these roles, discussed next.

First, attending to the integration of CCCs with DCIs and SEPs as students make sense of phenomena means that CCC learning progressions research should go beyond focusing on how students build knowledge of CCCs alone and instead emphasize how students use CCCs in conjunction with DCIs and SEPs as they make sense of phenomena. Focusing on how students use CCCs in this way aligns with approaches to assessing three-dimensional science (NRC, 2014) and knowledge-in-use (Harris et al., 2019).

Second, when learning progressions research attends to understanding how students use CCCs as conceptual lenses on phenomena, it provides insight into students' ability to use CCCs to generate new insights about how and why phenomena occur. To focus on how students use CCCs to generate new insights, assessments must be embedded in contexts in which new learning is required. Such assessments may be designed to, for example, elicit and categorize student questions about unfamiliar phenomena and problems (Bransford & Schwartz, 1999), track student learning in subsequent instruction (e.g., Fortus et al., 2015; Nordine et al., 2011), or monitor students as they engage in a learning activity (Schneider & Blikstein, 2018).

Finally, attending to how students use CCCs to connect across contexts and disciplines involves focusing on students' ability to productively apply the same CCCs across a range of phenomena and problems. Research on how to develop this ability requires going beyond looking for correlations in correct assessment responses across contexts and disciplines, and instead focusing on student thinking as they engage in sensemaking and problem-solving. Think-alouds, student interviews, and real-time observations of student work may be particularly useful in understanding whether and how students activate and use the same CCCs across phenomena and problems.

Leveraging CCCs to Broaden Access to Science

All learning progressions should include a focus on how students use their prior knowledge and experiences to engage in new learning, and this focus is especially important for research on CCC learning progressions. CCCs have broad applicability in students' lives outside of school, and as such, CCCs can serve as resources that reciprocally bridge students' funds of knowledge with the science being learned in the science classroom. To describe how students use CCCs to make sense of phenomena or solve problems, CCC learning progressions research must be embedded in learning environments in which students have opportunities to engage with phenomena and problems that are authentic and relevant to them, their families, and their communities. In this way, researchers may better understand how students from diverse backgrounds activate their funds of knowledge to develop a more sophisticated understanding and ability to use CCCs over time. This research allows for multiple points of entry for students to make sense of phenomena and problems and provides opportunities for researchers to identify multiple pathways with all students.

Embedding Research in Practice

Learning progressions are crucially dependent upon instructional practices (NRC, 2007), and many existing learning progressions have been developed and empirically validated using design-based research that relies on iterative and collaborative teaching experiments (Duncan & Rivet, 2018). In the case of CCC learning progressions, teaching experiments have been initiated by practicing classroom teachers in three-dimensional NGSS-aligned classrooms without the benefit of a robust research base to guide their instructional practices. Thus, there already exists a practical body of knowledge within the teaching community about how students might build and use CCCs over time (Miller & Shelton, 2021). Researchers should seek to learn from practitioners' experiences and expertise. Design-based research and research-practice partnerships may be particularly effective approaches for researchers to collaborate with practitioners in developing and validating CCC learning progressions.

In closing, the *Framework* authors readily acknowledged the paucity of research on CCCs and described CCC learning progressions as hypotheses in need of empirical research. Since the release of the *Framework*, the pace of activity in CCC learning progressions research has been slow. Though some promising research to inform the development of CCC learning progressions has begun (e.g., Gunckel et al., 2022; Scott et al., 2023; Yoon et al., 2019), this area of research is still nascent. The continued development and validation of CCC learning progressions research will require methodological approaches that reflect CCCs' unique roles in science and engineering learning and instruction, broaden access to science learning with all students by leveraging their everyday lives and funds of knowledge, and prioritize collaboration between researchers and practitioners with complementary areas of expertise.

References

American Association for the Advancement of Science. (1989). *Science for all Americans*. Oxford University Press.

American Association for the Advancement of Science. (1993). *Benchmarks for science literacy*. Oxford University Press.

American Association for the Advancement of Science. (2009). *Benchmarks for science literacy online*. http://www.project2061.org/publications/bsl/online/index.php

Bransford, J. D., & Schwartz, D. L. (1999). Rethinking transfer: A simple proposal with multiple implications. *Review of Research in Education*, *24*, 61–100.

Chabalengula, V. M., & Mumba, F. (2012). Understanding energy conservation: Intersection between biological and everyday life contexts. In A. Z. Ahmed (Ed.), *Energy conservation* (pp. 73–92). 10.5772/51872

Chabalengula, V. M., Sanders, M., & Mumba, F. (2012). Diagnosing students' understanding of energy and its related concepts in biological context. *International Journal of Science and Mathematics Education*, *10*(2), 241–266. 10.1007/s10763-011-9291-2

Cooper, M. M. (2020). The crosscutting concepts: Critical component or "third wheel" of three-dimensional learning? *Journal of Chemical Education*, *97*(4), 903–909. 10.1021/acs.jchemed.9b01134

Cooper, M. M., & Klymkowsky, M. W. (2013). The trouble with chemical energy: Why understanding bond energies requires an interdisciplinary systems approach. *CBE—Life Sciences Education*, *12*(2), 306–312. 10.1187/cbe.12-10-0170

Driver, R., Guesne, E., & Tiberghien, A. (1985). *Children's ideas in science*. Open University Press.

Duncan, R. G., & Rivet, A. E. (2018). Learning progressions. In F. Fischer, C. E. Hmelo-Silver, S. R. Goldman, & P. Reimann (Eds.), *International handbook of the learning sciences* (pp. 422–432). Routledge.

Duschl, R. A., Schweingruber, H. A., & Shouse, A. W. (Eds.). (2007). *Taking science to school: Learning and teaching science in Grades K-8*. National Academies Press.

Fick, S. J., & Arias, A. M. (2022). A review of literature that uses the lens of the next generation science crosscutting concepts: 2012–2019. *Journal of Research in Science Teaching*, *59*(6), 883–929. 10.1002/tea.21747

Fick, S. J., Baek, J., & Arias, A. M. (2017). Unit planning using the crosscutting concepts. *Science Scope*, *40*(9), 40–45.

Fick, S. J., Nordine, J., & McElhaney, K. W. (Eds.). (2019). *Proceedings of the summit for examining the potential for crosscutting concepts to support three-dimensional learning*. University of Virginia. http://curry.virginia.edu/CCC-Summit

Fortus, D., Sutherland Adams, L. M., Krajcik, J., & Reiser, B. (2015). Assessing the role of curriculum coherence in student learning about energy. *Journal of Research in Science Teaching*, *52*(10), 1408–1425. 10.1002/tea.21261

Goggins, M., Haas, A., Grapin, S., Januszyk, R., Llosa, L., & Lee, O. (2021). Broadening access to science: Crosscutting concepts as resources in the NGSS classroom. In J. Nordine & O. Lee (Eds.), *Crosscutting concepts: Strengthening science and engineering learning* (pp. 43–59). National Science Teaching Association.

Goggins, M., Haas, A., Grapin, S., Llosa, L., & Lee, O. (2019). Integrating crosscutting concepts into science instruction. *Science & Children*, *57*(1), 56–61.

González, N., Moll, L. C., & Amanti, C. (2005). *Funds of knowledge theorizing practice in households, communities, and classrooms*. Erlbaum.

Gunckel, K. L., Covitt, B. A., Berkowitz, A. R., Caplan, B., & Moore, J. C. (2022). Computational thinking for using models of water flow in environmental systems: Intertwining three dimensions in a learning progression. *Journal of Research in Science Teaching*, *59*(7), 1169–1203. 10.1002/tea.21755

Hadenfeldt, J. C., Liu, X., & Neumann, K. (2014). Framing students' progression in understanding matter: A review of previous research. *Studies in Science Education*, *50*(2), 181–208. 10.1080/03057267.2014.945829

Hadenfeldt, J. C., Neumann, K., Bernholt, S., Liu, X., & Parchmann, I. (2016). Students' progression in understanding the matter concept. *Journal of Research in Science Teaching*, *53*(5), 683–708. 10.1002/tea.21312

Harris, C. J., Krajcik, J. S., Pellegrino, J. W., & DeBarger, A. H. (2019). Designing knowledge-in-use assessments to promote deeper learning. *Educational Measurement: Issues and Practice*, *38*(2), 53–67. 10.1111/emip.12253

Herrmann-Abell, C. F., & DeBoer, G. E. (2018). Investigating a learning progression for energy ideas from upper elementary through high school. *Journal of Research in Science Teaching*, *55*(1), 68–93. 10.1002/tea.21411

Jin, H., & Anderson, C. W. (2012). A learning progression for energy in socio-ecological systems. *Journal of Research in Science Teaching*, *49*(9), 1149–1180. 10.1002/tea.21051

Krajcik, J., & Reiser, B. J. (2021). How crosscutting concepts, disciplinary core ideas, and science and engineering practices work together in the classroom. In J. Nordine & O. Lee (Eds.), *Crosscutting concepts: Strengthening science and engineering learning* (pp. 19–42). National Science Teaching Association.

Kubsch, M., Nordine, J., Fortus, D., Krajcik, J., & Neumann, K. (2020). Supporting students in using energy ideas to interpret phenomena: The role of an energy representation. *International Journal of Science and Mathematics Education*, *18*(8), 1635–1654. 10.1007/s10763-019-10035-y

Lee, O. (2020). *Making everyday phenomena phenomenal: Using phenomena to promote equity in science instruction*. https://www.nsta.org/science-and-children/science-and-children-septemberoctober-2020/making-everyday-phenomena

Lindsey, B. A., Heron, P. R. L., & Shaffer, P. S. (2012). Student understanding of energy: Difficulties related to systems. *American Journal of Physics*, *80*(2), 154–163. 10.1119/1.3660661

Miller, E. C., & Shelton, T. (2021). The role of crosscutting concepts in teacher sensemaking and empowerment. In J. Nordine & O. Lee (Eds.), *Crosscutting concepts: Strengthening science and engineering learning* (pp. 357–376). National Science Teaching Association.

National Research Council. (1996). *National science education standards*. National Academies Press. 10.17226/4962

National Research Council. (2007). *Taking Science to School: Learning and Teaching Science in Grades K-8* (R. Duschl, H. Schweingruber, & A. Shouse, Eds.). The National Academies Press. https://doi.org/10.17226/11625

National Research Council. (2012). *A framework for K-12 science education: Practices, crosscutting concepts, and core ideas*. National Academies Press.

National Research Council. (2014). *Developing assessments for the next generation science standards*. National Academies Press.

Neumann, K., Viering, T., Boone, W. J., & Fischer, H. E. (2013). Towards a learning progression of energy. *Journal of Research in Science Teaching*, *50*(2), 162–188. 10.1002/tea.21061

NGSS Lead States. (2013a). *Next generation science standards*. https://www.nextgenscience.org

NGSS Lead States. (2013b). *NGSS appendix D-"All standards, all students": Making the Next Generation Science Standards accessible to all students*. https://www.nextgenscience.org

Nordine, J. (2016). Why is energy important? In J. Nordine (Ed.), *Teaching energy across the sciences, K-12* (pp. 3–16). National Science Teaching Association.

Nordine, J., Fortus, D., Lehavi, Y., Neumann, K., & Krajcik, J. (2018). Modelling energy transfers between systems to support energy knowledge in use. *Studies in Science Education*, *54*(2), 177–206. 10.1080/03057267.2018.1598048

Nordine, J., Krajcik, J., & Fortus, D. (2011). Transforming energy instruction in middle school to support integrated understanding and future learning. *Science Education*, *95*(4), 670–699. 10.1002/sce.20423

Nordine, J., & Lee, O. (Eds.). (2021). *Crosscutting concepts: Strengthening science and engineering learning*. National Science Teaching Association.

Nordine, J., & Lee, O. (2023). On the nature and utility of crosscutting concepts. *Education Sciences*, *13*, 640. 10.3390/educsci13070640

Opitz, S. T., Neumann, K., Bernholt, S., & Harms, U. (2019). Students' energy understanding across biology, chemistry, and physics contexts. *Research in Science Education*, *49*(2), 521–541. 10.1007/s11165-017-9632-4

Osborne, J., Rafanelli, S., & Kind, P. (2018). Toward a more coherent model for science education than the crosscutting concepts of the Next Generation Science Standards: The affordances of styles of reasoning. *Journal of Research in Science Teaching*, *55*(7),962–981. 10.1002/tea.21460

Park, M., & Liu, X. (2021). An investigation of item difficulties in energy aspects across biology, chemistry, environmental science, and physics. *Research in Science Education*, *51*(S1), 43–60. 10.1007/s11165-019-9819-y

Pellegrino, J. W., & Hilton, M. (Eds.). (2012). *Education for life and work: Developing transferable knowledge and skills in the 21st century*. National Academies Press. 10.17226/13398

Podschuweit, S., & Bernholt, S. (2018). Composition-effects of context-based learning opportunities on students' understanding of energy. *Research in Science Education*, *48*(4), 717–752. 10.1007/s11165-016-9585-z

Quinn, H. (2021). The role of crosscutting concepts in three-dimensional science learning. In J. Nordine & O. Lee (Eds.), *Crosscutting concepts: Strengthening science and engineering learning* (pp. xi–xix). National Science Teaching Association.

Rivet, A. E., Weiser, G., Lyu, X., Li, Y., & Rojas-Perilla, D. (2016). What are crosscutting concepts in science? Four metaphorical perspectives. In C. K. Looi, J. L. Polman, U. Cress, & P. Reimann (Eds.), *Transforming learning, empowering learners: The International Conference of the Learning Sciences* (pp. 970–973). International Society of the Learning Sciences. https://repository.isls.org/handle/1/356

Schneider, B., & Blikstein, P. (2018). Tangible user interfaces and contrasting cases as a preparation for future learning. *Journal of Science Education and Technology*, *27*(4), 369–384. 10.1007/s10956-018-9730-8

Schwartz, D. L., & Goldstone, R. (2016). Learning as coordination: Cognitive psychology and education. In L. Corno & E. M. Anderman (Eds.), *Handbook of educational psychology* (3rd ed., pp. 61–75). Routledge.

Scott, E. E., Cerchiara, J., McFarland, J. L., Wenderoth, M. P., & Doherty, J. H. (2023). How students reason about matter flows and accumulations in complex biological phenomena: An emerging learning progression for mass balance. *Journal of Research in Science Teaching*, *60*(1), 63–99. 10.1002/tea.21791

Yoon, S., Goh, S.-E., & Yang, Z. (2019). Toward a learning progression of complex systems understanding. *Complicity: An International Journal of Complexity and Education*, *16*(1). 10.29173/cmplct29340

20

SCIENCE TEACHER EDUCATORS AND SCIENCE TEACHER LEARNING PROGRESSIONS

Resource, Roadmap, and Representation

Julie A. Luft, Yuxi Huang, and Harleen Singh

Science teacher educators are focused on the learning of teachers. To support science teachers' learning, they work in different realms with teachers. Some science teacher educators primarily serve as instructors of science teachers. In this role, they design the instruction in a class, sequence the learning experiences within a course, or plan programs of study. Within each of these areas, there is a consideration for the teachers' attainment of specific standards. These standards often encompass content and curricular knowledge, an understanding of learners, an ability to plan and enact sound science instruction, an understanding of equity, and ways to be a professional (e.g., see Luft et al., 2015; National Academies of Science, Engineering, and Medicine [NASEM], 2015).

Other science teacher educators function as researchers and are interested in a myriad of areas that describe and explore teacher learning. Their examinations can consist of, for example, the use of different professional learning interventions to enhance learning, the changing nature of teacher knowledge, or the role of practice-based communities in supporting learning (Erduran & Guilfoyle, 2022). Examining these different areas involves using a variety of research methods and theoretical frames (see Luft & Jones, 2022).

Depending upon the requirements of the science teacher educator's professional position, there may be an opportunity to focus on both instruction and research. These hybrid positions often have science teacher educators embedding their research in their work with teachers. This hybrid position allows science teacher educators to build their own understanding of teachers, as well as contribute to the knowledge of teacher learning in the community.

Science teacher educators also work within the continuum of teacher professional learning. Within this continuum are multiple opportunities for teachers to learn. For example, field experiences are central to the preparation of preservice science teachers. These experiences occur in different settings, have different durations, and are structured to build beliefs, self-efficacy, and identity or to cultivate the knowledge and practices essential to teaching science (Stroupe, 2022). Induction teachers, however, need access to different supports to help them learn the rules, policies, and expectations within their instructional context (Navy et al., 2022). These supports can consist of colleagues or various forms of professional programming. Experienced science teachers often build their teaching ability in more nuanced ways, which may involve learning how to be leaders, how to use new technologies in their

DOI: 10.4324/9781003170785-23

disciplines, or how to build their capacity for equitable science instruction (e.g., Borgerding et al., 2013; Braaten & Sheth, 2017; Whitworth et al., 2022).

In this chapter, we recognize that science teacher educators work in different realms of science teacher education. We belong to the realm that views teachers as assets (we have a lot to learn from teachers), and intertwines working with teachers and examining their learning. This unique space allows us to adopt different positions to explore science teacher learning. For instance, we can focus on how teachers learn over time, within a classroom setting, or within the context of specific school policies. Regardless of the position we adopt, we see teachers as learning progressively. This chapter examines how researchers contribute to our understanding of the progressive learning of science teachers. Through the descriptions of progressive science teacher learning, science teacher educators can be guided in their work with science teachers.

To prepare for this chapter, we searched for articles about science teachers' learning progressions to determine what learning progressions existed and how learning progressions have been used to support science teacher learning. Knowing about the continuous learning of science teachers can be useful to science teacher educators. Prior to describing the review process and results, we suggest a definition of science teacher learning progressions. Then we share our focused review of learning progression research in the area of science teachers, which highlighted how (a) learning progressions can be a *resource* for science educators and science teachers, (b) learning progressions can serve as a *roadmap* toward a specific outcome or goal, or (c) learning progressions can empirically *represent* teacher learning. These different portrayals of the use of learning progressions will be elucidated in this chapter. A general overview of these areas can be found in Table 20.1.

Table 20.1 Learning progressions for teachers as a resource, roadmap, or representation

	Resource	*Roadmap*	*Representation*
Focus	Supporting the learning of science teachers	Guiding the professional growth of teachers	Descriptions of science teacher learning
Elaboration of use in science teacher education	Progression is used by science teacher educators to help teachers understand the thinking of their students, or to design instructional programming for teachers. For instance, science teacher educators could design a program with learning progressions in mind, or they can support teachers in learning about student ideas.	Progression is used by both science teacher educators and science teachers to articulate the knowledge, skills, and practices that are associated with different professional roles or for teacher development. For instance, a science teacher educator could design programming that supports teachers in becoming department heads. Science teachers could also use the progression to advance their knowledge and skills to become mentors.	Progression is developed empirically by science teacher education researchers to depict the complexity of science teacher learning, and to guide further examinations of learning progressions for science teachers. Science teacher education researchers, for instance, study and develop a progression about how science teachers use modeling in the life sciences, or how science teachers acquire equitable practices.

Learning Progressions in Science Teacher Education

Learning progressions typically describe how students increase the complexity of their understanding of disciplinary ideas. These learning progressions have three major areas which consist of (a) a lower anchor that depicts base knowledge, (b) intermediate levels that progressively reveal more sophisticated ways of understanding, and (c) an upper anchor that represents the target learning outcome (Duncan & Hmelo-Silver, 2009). Learning progressions are valuable in the classroom because they can guide the development of curriculum, serve as a way to assess student learning, and assist teachers in their classroom instruction (Jin et al., 2019).

Science teachers stand to benefit from the development of learning progressions. Jin et al. (2019) noted two important benefits for teachers: (a) a better understanding of how students learn an idea, and (b) the building of a repertoire of ways to respond to the varied ideas of students. When these benefits are evident among teachers, they can then appraise student knowledge, determining how instructional events should be sequenced to support student learning or selecting appropriate assessments to monitor student learning. In the class, this can result in the differentiation of instruction (see Tobin & Tippett, 2014) and the linking of the science and engineering practices (NGSS Lead States, 2013). With the expanded knowledge and practices that come from using student learning progressions, teachers can engage in more consequential instruction that supports the learning of science among all students.

Science teacher educators can also benefit from learning progressions associated with teachers. With teacher-based learning progressions, science teacher educators can make informed decisions when they work with their teachers. These progressions would be similar to the structure (i.e., high anchor or low anchor) and knowledge focus of student-based learning progressions. However, they would be different in the focal areas, which could include instructional practices, beliefs, attitudes, or skills. They would also take into account the ongoing learning of teachers throughout their careers and the different contextual settings in which teachers work. With a different audience and different focal learning areas, teacher-based learning progressions for science teacher educators would be defined as follows:

> Focused descriptions about science teacher learning that align with specific practices, skills, or cognitive areas found in science teacher standards, guidelines, or other guiding documents. These descriptions have progressive steps that consist of a low anchor and intermediate steps that build consequentially toward a specific high anchor. The anchors of teacher-based learning progressions are subject to change and reside within the continuum of teacher learning and the varied contextual nature of teaching science.

There are limited examples of science teacher learning progressions. Schneider and Plasman (2011) provided one example. Their review examined 25 years of research on science teachers' pedagogical content knowledge (PCK) to describe the development and potential sequences of science teacher PCK. In their review, they concluded that science teachers' PCK increased in sophistication over time, but there were alternative knowledge pathways along the way. The result of their literature review provided science teacher educators with a description of PCK development at different times in a teacher's career (e.g., preservice, induction, professional development). For science teacher educators, Schneider and Plasman provided insights into the educational opportunities teachers should have to build their PCK throughout their careers.

Learning progressions associated with science teachers are important for science teacher educators. When science teacher educators draw upon teacher learning progressions in their work with teachers, the learning progressions have the potential to be a resource, a roadmap, or a representation. As a resource, learning progressions are important in supporting the learning of teachers. As a roadmap, learning progressions can be used to target or guide teacher learning towards a specific outcome. As a representation, they can be empirical descriptions of what learning should look like among science teachers. These different orientations toward learning progressions are described in the following sections.

Learning Progressions as Resources, Roadmaps, or Representations

This chapter is an outgrowth of our interest in learning progressions in science teacher education. We specifically wanted to know what science teacher learning progressions existed and how they were used in science teacher education. To address these questions, we engaged in a review process focused on the what and how of science teacher learning progressions. This review began by searching for articles focused on existing science teacher learning progressions. This search, which focused on the question, "What are the existing science teacher learning progressions?" was conducted through EBSCO, which is a database of educational research. The search terms "learning progression" and "science teacher" were used in this first search of abstracts of peer-reviewed papers from 2006 to 2021. Of the identified articles, most came from the later years (post-2010). These articles were likely a result of both increased external funding and an increased interest of researchers in this area. The initial search identified over 50 articles. These articles were reviewed to determine if the articles focused on science teachers and illustrated a progression of science teacher learning. After this review, eight articles were retained.

Several of the articles from this first search did not focus on science teacher learning progressions. Some of the articles had terms within phrases that corresponded to science education. For example, "physical" was a common term in the phrase "physical education teacher." This is a plausible search result because physical science is an instructional area for science teachers. Other articles were focused on general learning progressions, which were not specifically about science teachers.

During the review of the first set of articles, answers to the second question began to emerge about how the learning progressions were and could be used by science teacher educators. The different approaches to science teacher learning progressions pointed to their use as resources, roadmaps, and representations. This became the organizing framework for the chapter and necessitated another search for other learning progression papers to expand upon our emerging ideas.

The next step in locating learning progression papers involved both a Google Scholar and snowball search. We looked for articles in Google Scholar that captured the diverse ways the learning progressions were used as resources, roadmaps, or representations. We used specific terms that were aligned with our classification scheme. For instance, "learning progressions to support preservice teacher learning" was used to locate articles that could enhance our discussion of the learning progressions as a resource. If the article was not accessible through Google Scholar, we turned to EBSCO. When the article was in EBSCO, we examined the other suggested articles. Finally, we looked at the references of the articles we located to see if these additional articles added clarity to our discussion. Two more articles were added to our collection, and they also helped expand our ideas about the use of science teacher learning progressions by science teacher educators.

This review of the literature was neither comprehensive nor exhaustive. Instead, this process was meant to determine what science teacher learning progressions existed in publications and to conceptualize how learning progressions can be and are used by science teacher educators. In addition, research on science teacher learning progressions is an emerging area, as most reviews of learning progressions pertain to students (e.g., Duschl et al., 2011). Through this review, we hoped to depict the current and potential use of science teacher learning progressions in science teacher education, which is seen through the learning progressions being resources, roadmaps, or representations.

Learning Progressions as a Resource

When learning progressions are a resource, they are important in supporting the learning of teachers. An important way that a learning progression can be a resource to science teacher educators pertains to the design of a specific learning event, which could be daily instruction, a course, or a program of study. As a guide, learning progressions can suggest specific learning targets for teachers. Learning progressions can also be an important resource for science teacher educators in their assessment of the learning of science teachers. By using the anchors and intermediate steps in learning progressions, science teacher educators can evaluate what science teachers know and determine the next instructional steps.

In a straightforward example, Scherz et al. (2011) used a learning progression as a resource for the assessment of the professional growth of science teachers. The assessments were associated with continuous professional development (CPD) programs that supported the science teachers' development toward accomplished practice in the area of life science instruction. Embedded within the assessments were the categories of perception of instruction, model of instruction, using learning materials, assessments, influence on the school system, and evidence of preparation. These categories had competencies that ranged from low to high. Scherz et al. documented teacher learning through the assessments, which allowed them to monitor the process of teacher learning and the product of teachers' learning. The researchers concluded that the ongoing evaluation of professional growth allowed for adjustments and the tailoring of the CPD programs for the participating teachers.

Slough and Rupley (2010) also illustrated how learning progressions can be used to design learning events. They, however, used student learning progressions to support the design of a program for elementary teachers. In their article, learning progressions were the backbone for the design of the instructional program for elementary science teachers that integrated literacy and science activities. They demonstrated how to select a point in a student learning progression and then place informational text and activities in the learning progression. The purposeful selection of text and activities, along with prompts for teachers to use, enabled teachers to guide the learning of students toward the learning progression anchors and steps. This use of learning progressions is akin to the idea of an educative curriculum (Davis & Krajcik, 2005), which supports the learning of the teachers and the students.

Learning progressions of students can also be used as a resource to support progressive science teacher learning. This resource is not focused on the design of learning environments and the assessment of learning. Instead, student learning progressions are used to guide the learning and work of science teachers. Science teacher educators can use this approach within the different structured learning events for teachers. For instance, teacher professional learning communities can study student learning to create student learning progressions. Studying student learning is a recognized way to support science teacher learning (NASEM,

2015). When these types of teacher communities are supported by science teacher educators, the progressive nature of science teacher learning can be guided and documented.

When teachers use learning progressions to understand the thinking of their students, there is a clear advantage for the teachers. However, it matters how learning progressions are used with teachers. For instance, Furtak and Heredia (2014) were interested in how teachers developed and utilized learning progressions. They studied two schools from one district in the United States that were developing and using learning progressions differently. One school was responsible for developing and using a learning progression focused on their students. The development of the learning progression among the teachers resulted in their coordination of instruction and contemplation about how to best monitor the learning of their students. The teachers at the other school were provided with a student learning progression, and they struggled to align the learning progression to their classroom work. Furtak and Heredia suggested that the progression was difficult to use because the teachers had not crafted the learning progression for their students and context. The researchers concluded that teacher learning can be improved when teachers develop their own student learning progressions.

Similarly, von Aufschnaiter and Alonzo (2018) investigated how learning progressions helped science teachers interpret their students' thinking. Their study involved preservice teachers who were asked by science teacher educators to examine video clips of students' ideas in the content area of force and matter. The learning progressions were used as a tool to help the preservice teachers focus more specifically on the ideas of the students. In their analysis of the noticing of the preservice teachers, they reported that the learning progressions supported the preservice teachers in moving beyond their general interpretations of student thinking and enabled them to focus on more specific aspects of student thinking in force and matter.

There are different ways in which learning progressions can be resources in the learning and instruction of teachers. In these examples, learning progressions were resources that supported the learning of the teachers. One type of resource guided the design of the teachers' learning environments through a description of teacher learning and aligned assessments. The learning environment was structured with attention to the learning progression. An enhancement in this use of the learning progression involves a corresponding assessment of the teachers, which can be used to modify the professional learning environment. For science teacher educators who use learning progressions in this manner, the amount of time they have to develop the program and assessments will certainly be a constraint.

Another way that learning progressions can be a resource is when teachers examine a student learning progression in an area in which they will teach or that corresponds to their students. When used as a resource to understand student thinking, science teachers begin to recognize student ideas, which influences how they structure their learning environments. Teacher learning is still the object of this type of use of a learning progression, but there is a direct focus on a teacher's knowledge of student ideas. This approach can be expanded to address how teachers make instructional decisions to support student learning, which contributes to a teacher's knowledge about students and instructional practices. In this setting, science teacher educators need to be familiar with student learning progressions that will be utilized by teachers. These progressions will be used to drive science teacher thinking about students, which is often a goal for science teacher educators. When science teacher educators use learning progressions in this way, they need familiarity with both the different learning progressions and the school context (e.g., teachers, students, and environment).

Learning Progressions as a Roadmap

When learning progressions are a roadmap, they suggest how teachers progressively build their knowledge and practices in different educational roles. The essential element of a learning progression as a roadmap is that it offers direction to a science teacher for growth, development, or advancement within the continuum of a career. This unique form of learning progression can be developed by science teacher educators, science education researchers, or science teachers. When using learning progressions as a roadmap, science teacher educators and science teachers work together to curate the learning experiences of science teachers that take them from one position in their career to another.

One way that a learning progression can be a roadmap pertains to descriptions of how a teacher can develop over time. Discussions of a teacher's PCK as a new, developing, or experienced teacher can be used as a roadmap. Schneider and Plasman (2011), for instance, reviewed over 90 research papers on the PCK of science teachers throughout their careers. They sought to understand how science teachers' PCK changed in terms of orientations, student ideas, instructional strategies, curriculum, and assessment. After examining these different papers, they were also able to suggest experiences that were important in advancing science teachers' PCK. Science teacher educators would certainly have an important role in curating the learning experiences of science teachers who are working toward a specific educational level, which could be a novice, an accomplished teacher, or a master teacher, for instance.

Roadmaps can be useful in helping teachers advance in one area of their practice. For instance, Schneider (2015) described an early career teacher learning progression that spanned preservice education to the first years in the classroom. In this progression, she highlighted a cycle of instruction that included planning, enacting, and reflection. She connected this cycle to areas in PCK that included orientations to teaching science, science curriculum, science teaching, student thinking about instruction, and instructional strategies. In this depiction of early career learning, she highlighted the planning knowledge of a teacher in the areas of orientation and curriculum. This chapter serves as a roadmap for preservice teachers in building their planning knowledge to become more accomplished early career science teachers. For science teacher educators, the specific recommendations are important transitional markers.

In a different form of learning progression in science teacher education, Luft et al. (2019) wrote about how teachers could progress professionally. They envisioned the process of teacher learning occurring over the span of a career but toward different professional points. These trajectories were based on empirical work but focused on the knowledge, practices, and attributes that science teachers should hold in moving toward the different positions. In the article, Luft et al. suggested different learning points as a science teacher progresses toward an administrative position such as district science coordinator or department head. Underlying this type of progress is transformative learning theory, which recognizes that learners are transformed as they learn and reframe their understandings of their current position (Mezirow, 1997, 2012).

For science teacher educators, these types of trajectories are important in planning a pathway and noting the progress of a teacher toward a specific outcome. By knowing a trajectory, science teacher educators can suggest different learning experiences that would support the development or advancement of a teacher. These types of trajectories also afford opportunities for new pathways. For instance, there may be specific knowledge bases that are shared by both master teachers and department heads, which can branch off into curriculum directors or professional development program leaders. These different pathways provide new opportunities for the growth and transformation of teachers.

When learning progressions become roadmaps for science teachers and science teacher educators, they provide a direction for the purposeful development of a teacher. Science teachers can use them to plan a course for learning or envision opportunities for their own advancement. Science teacher educators can share these different road maps with science teachers so that they can contemplate the different educational opportunities available to them. They can also help science teachers plan their course of advancement. Roadmaps are empirically based pathways that depict the continuum of teacher growth, development, and advancement. These roadmaps exist within the terrain of science education, which has many waypoints.

Learning Progressions as Representation

Within the learning progression literature, the research pertaining to learning progressions as representations is by far the most abundant. This research is often carried out by science teacher educators or science education researchers and is focused on empirical descriptions of the learning of science teachers. These descriptions can have utility for those who work with science teachers, but representations were not constructed with this outcome in mind. Roadmaps and resources, in contrast, connect learning progressions to specific outcomes for science teachers within the domain of science education.

Within this area, learning progressions as representations for science teacher learning are often discussed in one of two ways. First, studies in this area articulate the learning of science teachers through empirical investigations. These studies reveal how science teachers progress in their understanding or use of the knowledge associated with the teaching of science. Second, there are empirical studies that explore the context in which learning progressions can support the learning of science teachers. These empirical depictions reveal how science teachers progress. Some of these studies also attend to the external factors that can influence teachers.

Studies that depict how science teachers progress in the development of their instructional knowledge are important for science teacher educators, allowing them to contemplate how to create a learning environment that ensures the teacher progresses to a more complex understanding of the area of instruction. For instance, Hamed et al. (2020) analyzed the data from over 340 preservice primary teachers to examine how they progressed in their understanding of teaching science through inquiry. As do many studies of science teachers and learning progressions, this study was situated in PCK but focused on inquiry instruction. They embarked on this study both to describe and document the progression of their students who were teachers. By studying their preservice primary teachers, they found that most of them attained an intermediate level of understanding. The pathways in which the preservice teachers engaged to attain this level, however, were not always similar.

Like the previous study, Adadan and Oner (2014) investigated two preservice chemistry teachers' PCK in terms of behaviors of gases during a teaching method course. Researchers measured the preservice chemistry teachers' PCK twice during the semester and then conducted two interviews with each participant before and after the semester. These data were the basis of their analysis. Like Hamed et al. (2020), they found that both preservice chemistry teachers improved their PCK representations in different areas. However, the two teachers engaged in different patterns regarding their PCK representation.

In the second type of representation, there are studies about how science teachers use learning progressions. These studies have considered what teachers do, as well as the contextual features that support their use of learning progressions. Science teacher educators benefit from these studies in their descriptions of teacher learning. With better descriptions of the ways teachers learn, science teacher educators can, for instance, better respond to the teachers they work with or use these descriptions as resources (which were described earlier).

An example of this type of representation can be found in Covitt et al. (2018). In this study, the teachers' use of a learning progression was documented and became a learning progression. Covitt et al. were focused on how teachers used a learning progression for student ideas about hydrologic systems. They collected the science teachers' written responses, and then used item response theory to assess their proficiency in setting goals, interpreting student ideas, and responding to ideas with instruction. After analyzing the data, the researchers created a three-level framework (not aligned, didactic, and ambitious) to examine science teachers' practices. Their results showed most science teachers at the didactic or direct instruction level, which involves an emphasis on the student being right or wrong.

Another example of this type of learning progression can be found in Gunckel et al. (2018). They focused on how learning progression-based instructional materials could support teachers in developing more sophisticated content knowledge and PCK. To investigate the potential of learning progressions, Gunckel et al. had teachers participate in a professional development program, in which they learned about water in environmental systems and were introduced to the instructional materials. As the teachers used the instructional materials over a year, they completed written assessments regarding their knowledge. The researchers found that teachers who used the instructional materials had a greater change in their content knowledge and PCK (although modest) compared to the teachers who did not use the materials.

Learning progressions as representations are important to science teacher educators. They are empirical studies into science teacher learning that explore the progressive learning of a teacher or the ways teachers learn from working with learning progressions. This type of information is useful to science teacher educators in that it provides evidence about the complexity of science teacher learning. Knowing about the learning of science teachers enables science teacher educators to, for instance, provide better instructional guidance to support the learning of science teachers or develop more effective curricular materials. Ultimately, the empirical basis of science teacher learning can highlight additional areas in need of study among science teacher education researchers and ensure that science teacher educators make more evidence-based decisions in their work with teachers.

Learning Progressions in Science Teacher Education

In science teacher education, learning progressions can have different purposes. In this chapter, we propose three different purposes. A learning progression can support the work of teaching (resource), it can guide teachers in their development or advancement (roadmap), or it can be the object of study that provides insight into the ways teachers learn or the ways their learning is impacted by a learning progression (representation). These different purposes provide different forms of information to science teacher educators who work with teachers.

These different forms of learning progressions can exist independently or be intertwined with each other. For instance, representations could be resources supporting science teacher learning, or a science teacher educator could form a roadmap about the development of secondary science teachers over time through the process of representation. The isolated or intertwined nature of learning progressions in science teacher education is important to acknowledge and is useful in helping science teacher educators consider the different ways they work with teachers.

Understanding how teachers learn over time is important, which has been a focal point for many educational researchers (e.g., Avalos, 2011; Berliner, 2001; Sprott, 2019). Much of the early work in teacher learning was focused on the development of teachers in different phases of their careers. These studies laid important groundwork in the field and guided teacher educators in their work with teachers. More recent studies of teacher learning have been focused on how teachers learn in different school settings, disciplines, working communities, and informal settings, among other areas (see Russ et al., 2016). These studies provide more insights into the complexity of teacher learning.

In science teacher education, learning progressions focused on science teachers can offer insights into the ways teachers advance, develop, strengthen, or expand their science instruction, understanding of science teaching, or knowledge to teach science, along with other areas. This type of information has direct implications for the design of learning environments or different types of teacher programming. For instance, science teacher educators can put scaffolds in place to support the learning of science teachers when they know how early career secondary science teachers build their conceptions of inquiry or scientific practices. These scaffolds can consist of science mentors (see Bradbury, 2010), educative curriculum (see Davis et al., 2017), or professional learning communities (see Dogan et al., 2016).

Science teacher learning progressions ultimately reveal the dimensionality of teacher learning by looking at different areas of science teacher learning and by providing descriptive accounts of teacher learning. Science learning progressions have the important role of elucidating the dimensionality of science teacher learning. For science teacher educators, this dimensionality can enrich our understanding and advance our theoretical orientations to teacher learning, and can influence the work of science teacher educators with science teachers. In addition, with advanced theoretical orientations, science teacher education researchers will have insights into the identified gaps in theories, will be able to combine theories that capture emerging qualities of science teacher learning, or will garner new theoretical insights.

Moving Forward

In looking toward the future, science teacher educators should continue to study and report on the learning of their teachers. Three areas of study should be pursued initially. To begin with, understanding the learning of science teachers in how they utilize and learn from different instructional practices is important. Many learning progression studies focus on the ways science teachers use formative assessments or inquiry in the classroom. There is room to look at different instructional practices, the disciplinary nature of teaching science, the contextual influence on the teaching of science, and the ways science teachers learn to teach in equitable and consequential ways.

There is also a need to broaden the frame of a learning progression with teachers. Understanding how teachers build their knowledge and practices over time and position is of essential use to the field of science teacher education. The existing descriptions are useful, but following teachers over periods of time and in different positions is beneficial to the field; for instance, understanding how a preservice teacher builds complexity in using scientific practices or inquiry in their early years of teaching, or studying how science teachers develop into department or district leaders. This is not a normal format for a learning progression but is essential to understanding how to better support teacher learning.

Finally, there is a need to consider how to construct learning progressions from existing empirical work, and this knowledge should be accessible to both the teacher and a science teacher educator. Many authors have discussed the different ways in which learning progressions are developed. Science teacher learning progressions can and will look different, and suggestions on how to create these progressions are important.

References

Adadan, E., & Oner, D. (2014). Exploring the progression in preservice chemistry teachers' pedagogical content knowledge representations: The case of "behavior of gases". *Research in Science Education*, *44*(6), 829–858.

Avalos, B. (2011). Teacher professional development in teaching and teacher education over ten years. *Teaching and Teacher Education*, *27*(1), 10–20.

Berliner, D. C. (2001). Learning about and learning from expert teachers. *International Journal of Educational Research*, *35*(5), 463–482.

Borgerding, L. A., Sadler, T. D., & Koroly, M. J. (2013). Teachers' concerns about biotechnology education. *Journal of Science Education and Technology*, *22*(2), 133–147. 10.1007/s10956-012-9382-z

Bradbury, L. U. (2010). Educative mentoring: Promoting reform-based science teaching through mentoring relationships. *Science Education*, *94*(6), 1049–1071.

Braaten, M., & Sheth, M. (2017). Tensions teaching science for equity: Lessons learned from the case of Ms. Dawson. *Science Education*, *101*(1), 134–164.

Covitt, B. A., Gunckel, K. L., Caplan, B., & Syswerda, S. (2018). Teachers' use of learning progression-based formative assessment in water instruction. *Applied Measurement in Education*, *31*(2), 128–142.

Davis, E. A., & Krajcik, J. S. (2005). Designing educative curriculum materials to promote teacher learning. *Educational Researcher*, *34*(3), 3–14.

Davis, E. A., Palincsar, A. S., Smith, P. S., Arias, A. M., & Kademian, S. M. (2017). Educative curriculum materials: Uptake, impact, and implications for research and design. *Educational Researcher*, *46*(6), 293–304.

Dogan, S., Pringle, R., & Mesa, J. (2016). The impacts of professional learning communities on science teachers' knowledge, practice and student learning: A review. *Professional Development in Education*, *42*(4), 569–588.

Duncan, R. G., & Hmelo-Silver, C. E. (2009). Learning progressions: Aligning curriculum, instruction, and assessment. *Journal of Research in Science Teaching*, *46*(6), 606–609.

Duschl, R., Maeng, S., & Sezen, A. (2011). Learning progressions and teaching sequences: A review and analysis. *Studies in Science Education*, *47*(2), 123–182.

Erduran, S., & Guilfoyle L. (2022). The importance of research in science teacher education. In J. A. Luft & M. G. Jones (Eds.), *Handbook of research on science teacher education* (pp. 5–15). Taylor & Francis.

Furtak, E. M., & Heredia, S. C. (2014). Exploring the influence of learning progressions in two teacher communities. *Journal of Research in Science Teaching*, *51*(8), 982–1020.

Gunckel, K. L., Covitt, B. A., & Salinas, I. (2018). Learning progressions as tools for supporting teacher content knowledge and pedagogical content knowledge about water in environmental systems. *Journal of Research in Science Teaching*, *55*(9), 1339–1362.

Hamed, S., Ezquerra, Á., Porlán, R., & Rivero, A. (2020). Exploring pre-service primary teachers' progression towards inquiry-based science learning. *Educational Research*, *62*(3), 357–374.

Jin, H., Mikeska, J. N., Hokayem, H., & Mavronikolas, E. (2019). Toward coherence in curriculum, instruction, and assessment: A review of learning progression literature. *Science Education*, *103*(5), 1206–1234.

Luft, J. A., & Jones, M. G. (2022). *Handbook of research on science teacher education*. Taylor & Francis.

Luft, J. A., Dubois, S., Nixon, R., & Campbell, B. (2015). Supporting newly hired teachers of science: Attaining professional teaching standards. *Studies in Science Education*, *51*(1), 1–48.

Luft, J. A., Whitworth, B., Berry, A., Navy, S., & Kind, V. (2019). Trajectories of science teacher learning: Charting the course for teachers, educators, researchers, and policy makers. *Journal of Science Teacher Education*, *30*(1), 63–79.

Mezirow, J. (1997). Transformative learning: Theory to practice. *New Directions for Adult and Continuing Education*, *74*, 5–12.

Mezirow, J. (2012). Learning to think like an adult: Core concepts of transformation theory. In E. W. Taylor & P. Cranton (Eds.), *The handbook of transformative learning: Theory, research, and practice* (pp. 73–98). Jossey Bass.

National Academies of Sciences, Engineering, and Medicine. (2015). *Science teachers learning: Enhancing opportunities, creating supportive contexts*. Committee on Strengthening Science Education through a Teacher Learning Continuum. Board on Science Education and Teacher Advisory Council, Division of Behavioral and Social Science and Education. The National Academies Press.

Navy, S., Luft, J. A., & Msimanga, A. (2022). The learning opportunities of newly hired science teachers. In J. A. Luft & M. G. Jones (Eds.), *Handbook of research on science teacher education* (pp. 239–250). Taylor & Francis.

NGSS Lead States. (2013). *Next generation science standards: For states, by states*. The National Academies Press.

Russ, R. S., Sherin, B. L., & Sherin, M. G. (2016). What constitutes teacher learning. In D. H. Gitomer & C. A. Bell (Eds.), *Handbook of research on teaching* (5th ed.; pp. 391–438). American Educational Research Association.

Scherz, Z., Bialer, L., & Eylon, B. S. (2011). Towards accomplished practice in learning skills for science (LSS): The synergy between design and evaluation methodology in a reflective CPD programme. *Research in Science & Technological Education*, *29*(1), 49–69.

Schneider, R. M. (2015). Pedagogical content knowledge reconsidered: A teacher educator's perspective. In A. Berry, P. Friedrichsen, & J. Loughran (Eds.), *Re-examining pedagogical content knowledge in science education* (pp. 172–187). Routledge.

Schneider, R. M., & Plasman, K. (2011). Science teacher learning progressions: A review of science teachers' pedagogical content knowledge development. *Review of Educational Research*, *81*(4), 530–565.

Slough, S. W., & Rupley, W. H. (2010). Re-creating a recipe for science instructional programs: Adding learning progressions, scaffolding, and a dash of reading variety. *School Science and Mathematics*, *110*(7), 352–362.

Snoek, M., Dengerink, J., & de Wit, B. (2019). Reframing the teacher profession as a dynamic multifaceted profession: A wider perspective on teacher quality and teacher competence frameworks. *European Journal of Education*, *54*(3), 413–425.

Sprott, R. A. (2019). Factors that foster and deter advanced teachers' professional development. *Teaching and Teacher Education*, *77*, 321–331.

Stroupe, D. (2022). Understanding the role of field experiences in preservice science teacher preparation. In J. A. Luft & M. G. Jones (Eds.), *Handbook of research on science teacher education*. Taylor & Francis.

Tobin, R., & Tippett, C. D. (2014). Possibilities and potential barriers: Learning to plan for differentiated instruction in elementary science. *International Journal of Science and Mathematics Education*, *12*(2), 423–443.

von Aufschnaiter, C., & Alonzo, A. C. (2018). Foundations of formative assessment: Introducing a learning progression to guide preservice physics teachers' video-based interpretation of student thinking. *Applied Measurement in Education*, *31*(2), 113–127.

Whitworth, B. A., Wenner, J. A., & Turbin, D. (2022). Science teacher leadership: The current landscape and paths forward. In J. A. Luft & M. G. Jones (Eds.), *Handbook of research on science teacher education*. Taylor & Francis.

21

LEARNING PROGRESSIONS AS SUPPORT FOR SCIENCE TEACHERS' FORMATIVE ASSESSMENT PRACTICES

Alicia C. Alonzo

Introduction

In recent years, there has been significant interest in the use of learning progressions (LPs) to support teachers' formative assessment practices (e.g., Black et al., 2011). However, empirical evidence to date (e.g., Furtak et al., 2014; Gunckel et al., 2018) has been mixed. When teachers do not use LPs in ways researchers envision (e.g., Furtak & Tayne, 2019), this raises questions about the viability of researchers' proposals for how LPs might support teachers' formative assessment practices. In this chapter, I put these proposals in dialog with empirical evidence of teachers' use of LPs. I highlight ways that teachers have/have not taken up LPs amidst the complexity of classroom practices, with particular attention to a key feature of LPs: levels.

Formative Assessment

Formative assessment (FA) is "part of everyday practice by students, teachers, and peers that seeks, reflects upon, and responds to information ... in ways that enhance learning" (Klenowski, 2009, p. 264). Thus, for teachers, FA entails the practices of eliciting, interpreting, and responding to evidence of students' understandings[1] (e.g., Ruiz-Primo & Furtak, 2006). These three practices rest on a fourth practice involved in FA: "planning, developing, and communicating clear learning goals" (Gotwals, 2018, p. 159). While FA requires effective tools and strategies for eliciting evidence of student thinking (Bennett, 2011), it is not a type of assessment; instead, it refers to the use of assessment tools and strategies within this larger set of practices (e.g., McManus, 2008).

Effective FA requires what Minstrell et al. (2011) refer to as a "learning and learner stance" (p. 2). This stance includes an orientation towards making sense of, rather than solely evaluating, students' understandings (e.g., Covitt et al., 2018). The learning and learner stance is central to "continuously working on and with students' emerging science ideas to move student thinking forward" (Carpenter et al. 2020, pp. 2513–2514). Teachers' responses to student understandings may take the form of feedback to students, attempts to learn more about students' understandings, and/or adjustments to instruction (e.g., Andersson & Palm, 2017). Regardless of the form, effective FA responses require "build[ing] upon student thinking as it develops from naïve to more sophisticated" (Minstrell et al., 2011, p. 3).

 DOI: 10.4324/9781003170785-24

Challenging Aspects of FA Practices

Despite widespread interest in FA (Shepard, 2009), these practices are notoriously challenging (e.g., Gotwals & Birmingham, 2016; Stahnke et al., 2016). Researchers have posited that one source of teachers' difficulties may be their framing of student understandings (e.g., Levin et al., 2009). Teachers often frame student understanding dichotomously ("gets it"/"doesn't get it"; Otero, 2006). If they do attend to students' understandings, teachers may do so to identify and eliminate "misconceptions" (e.g., Larkin, 2012) and/or they may struggle to identify which understandings can be leveraged to support student learning (i.e., changes in thinking over time).

Dichotomous framing may lead teachers to focus on *eliciting* only information required to make such evaluative judgments—e.g., vocabulary (Otero, 2006) or facts—using "'can you guess what I (the teacher) am thinking' questions" (Gotwals & Birmingham, 2016, p. 373) that can be easily interpreted as right or wrong. Consistent with a dichotomous framing, teachers often make holistic ("right"/"wrong") *interpretations* of student understandings—even if they have elicited more nuanced information (e.g., Gotwals & Birmingham, 2016). While teachers may be "able to identify incorrect content in students' responses," "interpreting students' intuitive ideas" is more challenging (Jin et al., 2015, p. 1292).

Responding to interpretations of student thinking could involve planning instruction that addresses students' learning needs and/or providing feedback to advance student understanding. Both are challenging: Heritage et al. (2009) reported that planning instruction is more difficult than interpreting student thinking, while Schneider and Gowan (2013) found that providing feedback was more difficult than both interpreting student thinking and planning instructional responses to that thinking. The responding part of FA is often omitted, even when teachers elicit and interpret student understandings (e.g., Ruiz-Primo & Furtak, 2006). When teachers do provide feedback, it tends not to be in a form that supports students' learning (Kluger & DeNisi, 1996). As Shepard (2009) reminds us, "formative information is of little use if teachers don't know what to do" with it (p. 37).

It can be particularly difficult to know what to do when interpretations only indicate whether students "get it," so teachers' instructional responses may entail simply providing students with the correct answer or re-teaching the content when students "don't get it" (Stahnke et al., 2016). When teachers interpret students' understandings as misconceptions, they may try to replace (Alonzo et al., 2022) or "squash" (Furtak, 2012, p. 1195) these understandings, rather than using them as building blocks for learning.

In response to these challenges, researchers recognize that teachers need support and have identified an array of knowledge and skills required to engage in FA practices fully and effectively (e.g., Bennett, 2011; Heritage et al., 2009). Pedagogical content knowledge (PCK; Shulman, 1987) figures prominently in these recommendations. Tools and professional development have been recommended (e.g., Bennett, 2011) to support teachers' FA practices and the knowledge that underlies them. As far back as 1998, LPs have been recommended as one such support (Black & Wiliam, 1998).

Features of LPs Related to FA

LPs are models (e.g., Lehrer & Schauble, 2015), representing "the successively more sophisticated ways of thinking about a topic that can follow one another as children learn" (NRC, 2007, p. 219).[2] Most LPs include an upper anchor that articulates targeted understandings, a lower anchor that details resources for learning (i.e., prior understandings and experiences) students are likely to have at the beginning of the progression, and intermediate levels that describe incremental steps through which students might—with appropriate learning experiences—move from the lower to the upper anchor (e.g., Duschl et al., 2011). LPs are "more than an ordered set of ideas, curriculum pieces, or instructional events"; they "should characterize benchmarks of conceptual change" (Wilson & Lehrer, 2021, p. 2).

Like models in the natural sciences, LPs are both a) empirically and theoretically grounded and b) simplifications of complex phenomena. Simplification is inherent in, and a key characteristic of, all models (Bailer-Jones, 2003). For LPs, aggregating across many students to create LP levels is both an affordance, allowing for the identification of broad patterns in students' thinking and learning, and a constraint, in that a particular student may not fit those patterns. Thus, LPs both "illuminate" and "eliminate" information (Lehrer & Schauble, 2015, p. 731). In the sub-sections below, I highlight features of LPs with particular relevance to FA, including critical consideration of the use of LP levels (as both an affordance and a constraint) of this construct.

Scope and Grain Size

There is wide variation in the scope, or breadth, and corresponding grain size of LPs (e.g., Duschl et al., 2011). For example, LPs in the environmental literacy project (e.g., Mohan et al., 2009) have a broad scope (grades 4–12) and levels with a broad grain size, representing major shifts in students' reasoning (e.g., from "reasoning about actors and enablers" to "reasoning about 'stuff'"; Jin et al., 2013, pp. 1676–1677). In contrast, other LPs have a much smaller scope, describing learning during a single instructional unit—e.g., Yin et al.'s (2014) LP for sinking and floating, Alonzo and Steedle's (2009) LP for force and motion. Levels in these LPs have a smaller grain size, representing more modest shifts in understanding—e.g., from considering one factor (mass or volume) to considering two factors (mass and volume) to explain sinking and floating. Although it is possible for an LP to have a broad scope and small grain size, this is unusual. Most LPs in science seem to have 4–6 levels. With the same number of levels, an LP that is broad in scope will have a larger grain size than an LP that is narrow in scope.

Researchers argue that LPs that are "zoomed in" (i.e., have a smaller scope and smaller grain size) are likely more useful for FA practices (e.g., Furtak & Tayne, 2019; Gotwals, 2018). For FA, teachers need enough detail about students' developing understandings to make day-to-day instructional decisions. LPs with smaller grain sizes make smaller distinctions between student thinking at different levels and, thus, provide more detailed information about students' current status and learning needs. For example, Smith et al. (2006) proposed a broad LP (grades K-8) for matter and atomic theory, with each level describing learning appropriate for students in a three-grade span (K-2, 3–5, 6–8). While invaluable for laying out standards, this LP does not help a 4th-grade teacher to know how her students might be progressing with respect to broad goals for the 3–5 grade band (e.g., students should learn that "weight is an additive property of objects that can be measured," p. 14). Wiser et al. (2012) expanded the broad progression to describe learning for students in grades 3–5. Here, the teacher can see that students progress by first recognizing that solid objects have weight that is measured with a scale (rather than felt by hefting); then by recognizing that granular materials and liquids also have weight; and finally recognizing that "any piece of material, however small, visible or not, has weight" (p. 368). A teacher could find out what kinds of objects students think do and do not have weight and use that information, along with these finer-grained levels, to identify appropriate instruction for those students.

Identification of Typical Student Understandings

By describing how students might think about a topic with varying degrees of sophistication, LPs identify a wide variety of understandings that students might hold as they learn. This is similar to the "misconceptions" research of the 1980s in that, rather than considering only whether a student is "right" or "wrong," there is a focus on students' "wrong" (or less sophisticated) understandings. For example, students at Level 3 of an LP for carbon cycling in socio-ecological systems (Mohan et al., 2009), tend to "incorrectly convert matter to energy" (p. 689), and an intermediate level in Furtak's (2012) LP for natural selection focuses on "need-based change," in which students believe that evolutionary "changes arise from deliberate acts by the parent organism or the species as a whole" (Furtak & Heredia, 2014, p. 995). However, in contrast to misconceptions research, LPs treat these "errors" or "wrong" understandings as important foundations for further learning. For example, converting matter to energy in Level 3 of Mohan et al.'s (2009) LP supports students in conserving matter and/or energy and, thus, lays a foundation for conserving matter and energy, as distinct concepts, in Level 4. LPs reflect what is known (by researchers and/or teachers) about the nature of students' understandings before they are "right" and how those understandings might be leveraged to support further learning.

LP Stance Toward Learning

By definition, LPs represent students' ways of thinking, rather than disciplinary logic (e.g., Shea & Duncan, 2013), and encourage attention to "growth over status" (Briggs & Furtak, 2019, p. 146)—i.e., attention to learners and learning. In addition, by portraying intermediate steps between initial and targeted understandings, most LPs treat understandings at higher LP levels as revisions of those at lower levels (e.g., Shea & Duncan, 2013), often with the assumption that understandings build on one another (e.g., Schramm et al., 2018). One example is the transition from Level 3 to Level 4 in Mohan et al.'s (2009) LP described above. Another example is Wiser et al.'s (2012) construction of an LP for matter using "stepping stones" (p. 367), such as the compositional model of material, which "represents materials as continuous and homogenous" (p. 374). This model is limited in its accuracy and its exclusion of gases. However, it supports students' development of a "beginning particulate model" (p. 370) that includes gases as matter.

Representation of Student Thinking Using Levels

A key characteristic of LPs is the way that understandings are organized to describe student thinking at different levels of sophistication. Not only are student understandings ordered hierarchically, but they are organized into levels, which are more than a collection of understandings of similar difficulty. LP levels represent patterns in students' thinking, such as understandings that seem to reflect the same underlying perspective (e.g., Wiser et al., 2012), and thus that students tend to hold at the same time (e.g., Alonzo & Steedle, 2009).

A key affordance of LPs is the identification of these patterns and the accompanying assumption (whether implicit or explicit) that the patterns (levels) can be used to characterize individual students' thinking (e.g., Plummer & Maynard, 2014). Although some researchers have begun to describe LP levels not as coherent stages but as different ways of thinking that can be deployed in different contexts (e.g., Sikorski, 2019), any work involving assessment with respect to LP levels (e.g., Jin, van Rijn et al., 2019) requires that levels can be used to reliably characterize individual students' thinking—i.e., that there is consistent evidence of an individual student's reasoning at a given LP level (Bernholt and Sevian, 2018).[3]

However, there is extensive evidence suggesting that individual students both do not display consistent reasoning (e.g., Nehm & Schonfeld, 2008) and do not hold understandings at a given LP level as a coherent whole but rather may express understandings at different levels at the same time (e.g., Hokayem & Gotwals, 2016).[4] As such, LP levels, depicting aggregate patterns, represent a simplification (Alonzo & Elby, 2019),[5] as compared to the complexity of how individual students may be thinking in a particular moment.

Representation of Student Learning Using Levels

In addition to describing students' thinking at a given point in time through descriptions in levels, the hierarchical ordering of levels represents a pathway that students might follow as they learn. As such, LPs serve as a "map" for students' progress (e.g., Black et al., 2011).

In order to portray such a map, LPs typically aggregate evidence from many students to depict a single, linear pathway (e.g., Duncan & Gotwals, 2015).[6] Indeed, some similarity across students is fundamental to the LP construct, which assumes that "although some variability exists ... many students have similar patterns of development, learning, and thinking that can be generalized across the population" (Ketterlin-Geller et al., 2019, p. 2). Aggregate patterns in learning are essential for constructing standards and curricula that allow students to build understandings over time.

However, aggregate patterns cannot capture the complexity of learning for every student. This is because students build understanding on individual resources that they bring to the science classroom and develop within particular contexts and through relationships between teachers and students (e.g., Engelhard & Sullivan, 2011). As a result, a single pathway is unlikely to capture all students' learning, and pathways are "rarely replicated in individual students' context-dependent and non-linear learning trajectories" (Pierson et al., 2019, p. 834). This is not because researchers have done a bad job. It is an inherent feature of LPs: To represent aggregate patterns, LPs necessarily cannot capture all of the variation in students' thinking and learning.

LPs as Support for Teachers' FA Practices

By laying out the landscape that students traverse in coming to a full understanding of a topic, LPs may serve as a "road map," providing a sense for the typical terrain of students' understandings, as well as a path through that terrain and "steps along the route" (Black et al., 2011, p. 75). As such, LPs provide a framework that may guide stances towards student thinking and learning (e.g., Furtak, 2012); inform the setting of learning goals (e.g., Covitt et al., 2018); support eliciting, interpreting, and responding to evidence of students' understanding (e.g., Black et al., 2011); and foster teacher learning supportive of FA practices (e.g., Alonzo & Elby, 2019). A basic model for LP use goes something like this: a teacher elicits evidence of students' understanding relative to an LP, uses the results to diagnose students' LP levels, and then responds based on the gap between students' current level and the next one.

Different LP uses have different reliance on the features of LPs described in the previous section. Some uses rely on LP levels and the important information they provide. Other uses rely primarily on the identification of student understandings and stance toward learning represented in an LP, without tapping into the affordances of LP levels. For an extended example of LP use that, for the most part, represents the latter, please see Gotwals et al. (2021).

Where LP levels play a more prominent role, the ways they represent student thinking and learning become relevant. The basic model of LP use for FA requires reliable assessment of student understandings. Without this, diagnoses become uncertain and reliant on characteristics of the assessment task (e.g., Bernholt and Sevian, 2018). If we cannot assume that the understandings at a given level "hang together"—such that students holding one understanding at a given LP level are likely to hold other understandings at that level—teachers might provide LP-based instruction that is appropriate for "some students identified at a given level but not other students with a similar level diagnosis" (Shavelson & Kurpius, 2012, p. 15). Similarly, if the pathway laid out by an LP does not capture the way an individual student learns, the potential exists to misinterpret their progress (e.g., Shavelson & Kurpius, 2012). These issues can be somewhat ameliorated by the opportunities that FA provides to revise faulty interpretations through ongoing interactions with students (Black et al., 2011; Briggs & Furtak, 2019).

In the following sub-sections, I present proposed uses of LPs to support FA. These uses rely on the LP stance toward learning: They emphasize the learner's perspective (rather than disciplinary logic) and describe how learning occurs by building on students' current understandings. However, the proposed uses vary in whether and how they use LP levels as representations of student thinking and/or learning. To support the interpretation of the empirical evidence considered in the next section, I attend to this variation in the sub-sections below.

Guiding Stance Towards Student Thinking and Learning

Whether their levels are emphasized or not, LPs may offer affordances for shifting perspectives about student thinking and learning. LPs focus attention on students' understandings, portraying less sophisticated understandings in terms of progress towards—rather than only as deficient in comparison to—targeted understandings (Alonzo & von Aufschnaiter, 2018). Including students' understandings that are "wrong" but typical—and often even productive and necessary—may focus teachers' attention on the full range of understandings students are likely to exhibit, rather than dismissing all understandings not deemed "correct" (e.g., Furtak, 2012). As such, LPs may help teachers move from a "gets it/doesn't get it" perspective towards a more nuanced view of student thinking (e.g., Briggs & Furtak, 2019), which might support a "move away from a stance of 'diagnose and fix' toward an LP stance of 'identify and use'" (Gotwals, 2018, p. 162).

Setting Learning Goals

LP levels can be used to identify appropriate sub-goals (e.g., Heritage, 2007) by breaking larger instructional goals (i.e., the upper anchor) into smaller pieces (i.e., intermediate levels). In this way, LP levels themselves might be used directly as learning goals. However, LP levels might also more indirectly "influence the types of learning goals that teachers may develop" (Gotwals, 2018, p. 160). For example, LPs aligned with the three-dimensional *Next Generation Science Standards* (NGSS Lead States, 2013) could be used to support teachers' move towards developing their own, more three-dimensional learning goals (e.g., Covitt et al. 2018).

Eliciting Evidence of Student Understanding

Black et al. (2011) argue that the value of an FA task "rests mainly on its potential to provoke a range of responses" (p. 74). By illuminating understandings of varying sophistication, LPs may be used to support such provocation. By describing a range of ways students may engage with the content, LPs allow assessment developers to anticipate, and thus plan to elicit, a range of potential responses (e.g., Furtak et al., 2014). This may occur through a) an assessment task that provides opportunities for students to express a range of understandings, "shifting attention from a mostly disciplinary focus on strategies or representations to a focus on bringing forth students' informal and previous instructional experiences" (Sztajn et al., 2012, p. 150), or b) an assessment task that elicits understandings indicative of level(s) of an LP. The latter might be accomplished by linking the task to LP levels, e.g., ordered multiple choice (OMC; Briggs et al., 2006).

While researchers may provide assessment tasks for use with a particular LP (e.g., Yin et al., 2014), LPs may also support teachers in creating or adapting assessment tasks themselves (e.g., Furtak, Heredia, & Circi, 2018). LPs focus attention on possible student responses (Alonzo et al., 2022), may suggest varying types of scaffolding that might be provided to students during an assessment task (Furtak et al., 2014; Songer et al., 2009) and help to ensure that students with varying proficiency can engage with the task (Sztajn et al., 2012). In addition, by identifying critical shifts in understanding, LPs may support teachers in determining when to create an FA opportunity, e.g., introduce an assessment task (Yin et al., 2014) or ask a question.

Interpreting Evidence of Student Understanding

LP-based assessment tasks may provide more diagnostic information than is available with traditional item formats (Alonzo et al., 2022; Briggs et al., 2006), whether the tasks are designed to yield an LP-level diagnosis or to identify specific understandings a student might hold. In particular, LP-based assessment tasks give teachers information about a student's understandings, rather than just whether they responded correctly or incorrectly.

A diagnosis in terms of LP levels provides a) a description of student understanding from the LP level and b) a sense of students' progress with respect to a larger continuum (Black et al., 2011). Multiple items, potentially representing multiple item types, can be connected through an LP, since—unlike "partial credit" or "full credit"—LP levels have the same meaning across items.

Level-based interpretations may also support inquiries into students' responses. For example, because LP diagnoses are often messy, reflecting the inconsistencies and context-dependence of students' responses, teachers may investigate "how consistent students are in using ideas across task types and contexts" (Gotwals, 2018, p. 161), as well as how consistent students' performances are with respect to LP levels that contain multiple components. In doing so, teachers may seek to understand patterns in students' performance (Alonzo & Elby, 2019; Furtak et al., 2014), focusing particular attention on patterns that do not fit neatly with the LP levels. As Hammer and Sikorski (2015) argue, "idiosyncrasies ... are essential to student learning" and should not be treated—as is often the case when aggregating responses (for an individual student or a whole group of students)—as "conceptually insignificant noise" (p. 425). Teachers' inquiries may lead them to "qualitative insights about student understandings and misconceptions" (Shepard, 2008, p. 296). These insights emerge from using the inherent discrepancy between LP levels and the complexity of students' thinking and learning as a resource, rather than a potential flaw of the LP (Alonzo & Elby, 2019).

In contrast to level-based diagnoses and investigations, teachers may use understandings represented in an LP to interpret students' understandings and performances without attention to LP levels. For example, teachers may use understandings from an LP to categorize students' understandings and identify ones to highlight (e.g., Furtak & Heredia, 2014). This may happen in formal assessment events but also "on-the-fly" in classroom discussions or observations of students' performances. Awareness of significant student understandings that may serve as "stepping stones" (e.g., Wiser et al., 2012, p. 367) may be especially significant for using an LP to identify understandings that could be important to build upon (e.g., Furtak, 2012).

Teachers' interpretations of student understandings, whether during instruction or when reviewing performance on assessment tasks, may be extended to those that do not appear on the LP at all. LPs may serve as an interpretive framework (Yin et al., 2014) for making sense of student contributions. As such, understandings on an LP may serve "as the starting point to understand the nature of student thinking" (Furtak, Heredia, & Circi, 2018, p. 144).

Responding to Evidence of Student Understanding

In some cases, LP developers provide suggested instructional responses with the LP (e.g., Briggs & Furtak, 2019), which teachers could adapt for their own students. In other cases, teachers need to develop their own responses, using the LP as a framework (Shepard, 2009). Either way, an LP can help to identify the gap between current and targeted understandings and, based on that gap, what students would need to do to close it. As such, LPs may "suggest instructional paths forward" (Furtak et al., 2014, p. 642) or targeted feedback to support students in revising their understandings (Furtak & Heredia, 2014). Because differences between levels suggest what students need to learn (e.g., Yin et al., 2014), LPs can divide a pathway to targeted understanding into smaller moves between levels (Alonzo, 2011).

Incremental level-by-level moves might support students' progress toward targeted understandings. However, researchers caution that "simply because teachers are provided information about where students are located on an LP does not mean that the next steps in instruction are clear" (Furtak et al., 2013, p. 7). Wilson and Lehrer (2021) suggest a strategy that could be used to respond to evidence of student understanding relative to any LP: "juxtaposing student solutions at adjacent levels of the construct … and inviting whole-class reflection … to help students to extend their reasoning to higher levels" (p. 12). Additional strategies, both general (applicable across LPs) and specific (applicable only to a given LP), are needed to support teachers (and other instructional designers) to craft responses that can move students from one LP level to the next.

The proposal to support student learning through level-by-level shifts relies heavily on LP levels as reflective of both student thinking and student learning. A teacher needs to assume that she can trust both the level diagnosis and the pathway forward from that level to construct a response. However, an LP might also open up possibilities for teachers to consider how to respond to individual understandings represented in the LP without regard to their level. While not taking advantage of the affordance of an LP pathway, this approach focuses attention on student learning as represented in the LP and may reflect part of the LP stance: that students build new understandings on the foundation of older ones.

Teacher Learning

In addition to their direct use to set learning goals and elicit, interpret, and respond to evidence of student understanding, LPs may also support teacher learning about student thinking and learning. This support may occur through the content and pedagogical content knowledge embedded in LPs (e.g., Alonzo, 2011) and the potential for teachers to develop this knowledge (Gunckel et al., 2018, p. 1340; Sztajn et al., 2012) as they engage with an LP. These benefits may be realized, in particular, when educative LP-based curricula are provided (Gunckel et al., 2018).

LPs may also provide a conceptual framework that seeds teachers' generation of new knowledge. In this view, teachers do not just learn as they receive LPs from researchers but rather as they make sense of them (Sztajn et al., 2012) in relation to their own classroom practice. As models, LPs may, like models in science, be used to "notice and characterize phenomena that don't fit the model" and serve as a conceptual framework to ask and investigate questions (Alonzo & Elby, 2019, p. 6). The investigations suggested above as part of teachers' interpretation practice are one example. Alonzo and Elby (2019) even posit that the benefit of using LPs may come, not from teachers' direct use of LPs, but rather from the development of PCK and the use of this knowledge to inform their practice more generally.

Use of LPs in Professional Development

Researchers do not assume that LPs can, by themselves, offer envisioned support for teachers' FA practices. In addition to the LP-based materials, such as assessment tasks (e.g., Alonzo et al., 2022; Briggs & Furtak, 2019) and educative curricula (e.g., Carpenter et al., 2020) that are sometimes provided, most LP research includes attention to teacher professional development (PD). This is often intensive, involving sustained interactions with researchers—often over multiple years, in summer workshops (e.g., Covitt et al., 2018), meetings and other support during the school year (e.g., Furtak et al., 2014; Heredia, 2020), or a combination of the two (e.g., Alonzo et al., 2022).

A common feature of PD to support teachers' LP use is practice assigning student understandings to levels of an LP. This is true both of research studies that envision FA practices as largely dependent on LP levels (e.g., Covitt et al., 2018) and those that, while perhaps including a role for LP levels, assume more expansive uses of LPs (e.g., Alonzo et al., 2022; Briggs & Furtak, 2019). This practice supports teachers in developing baseline understandings of an LP, as well as opportunities to offer critiques and suggestions for contextual adaptations. Familiarity with LP levels supports uses of the LP for FA—even uses that do not directly depend on LP levels.

LPs may be used as a "jumping off point" for teachers' sensemaking, such as using an LP to understand patterns in score reports (e.g., Furtak et al., 2014) or as a model for teachers as they invent their own categories for analyzing student work (Furtak & Tayne, 2019). Alonzo et al. (2022) focus teachers' attention on evidence of students' reasoning not predicted by LPs and support teachers in using LPs as lenses to identify what is ambiguous about students' reasoning and, thus, what follow-up questions might help to clarify. Familiarity with LP levels scaffolds engagement with understandings that don't conform to the LP level descriptions and supports teachers in moving beyond strictly level-based FA practices. This is consistent with Briggs and Furtak's (2019) approach of engaging teachers in assigning LP scores to student responses *and* further analyzing those responses to qualitatively identify strengths and weaknesses.

Empirical Evidence for LPs as Support for Teachers' FA Practices

Empirical evidence regarding proposals for teachers' use of LPs is mixed (e.g., Covitt et al., 2018; Furtak et al., 2014) and somewhat sparse. In addition to my work (e.g., Alonzo & Elby, 2019; Alonzo et al., 2022), there are two major lines of research in this area: one undertaken by Erin Furtak and colleagues (e.g., Furtak et al., 2014; Heredia, 2020) and the other undertaken in a large environmental literacy project (e.g., Carpenter et al., 2020; Gunckel et al., 2018). Across projects, researchers have documented multiple ways that teachers used LPs as part of their FA practices. These vary widely within a study and even within teachers, such that the same study might yield disparate evidence of LP-supported FA.

In this section, I start by considering LP use in terms of the LP stance toward learning. Then, I describe how teachers used LPs as supports for FA practices both directly (through setting learning goals and eliciting, interpreting, and responding to evidence of student learning) and indirectly (through teacher learning). I divide evidence of teachers' FA practices into that showing the use of LP levels as described in the basic model of LP-based FA and alternatives that rely on other features of LPs.

LP Stance toward Thinking and Learning

Researchers (e.g., Alonzo et al., 2022; Furtak et al., 2014; Gunckel et al., 2018) have documented teachers' LP use that seems inconsistent with the LP stance described above but consistent with typical FA practice. Across studies, teachers interpreted students' understandings in terms of correctness, using comparison to the highest ("correct") level as "evidence of what students do not know" (Gunckel et al., 2018, p. 1354). Interpretations in terms of misconceptions were accompanied by perceptions of LPs as "catalogs of misconceptions to be 'squashed'" (Furtak, 2012, p. 1181). With these interpretations, teachers responded by providing correct content (e.g., Gunckel et al., 2018) and attempting to remove or replace students' understandings (e.g., Alonzo et al., 2022). Gunckel et al. (2018) reported that, even though teachers demonstrated growth in PCK (i.e., awareness of student understandings), their proposed responses to student understandings continued to emphasize providing accurate content to correct those understandings.

Furtak et al. (2014) provide some explanation for these results by documenting how teachers who frame student understandings in ways that are inconsistent with the LP construct have difficulty engaging with an LP. Because these framings are likewise challenging for FA, it may be particularly difficult for teachers with these views to use the LP to support FA.

However, across multiple research studies, LP use appears to support teachers in moving beyond a "gets it"/"doesn't get it" perspective (e.g., Furtak, 2012) to focus more on "students' understandings and challenges" (Covitt et al., 2018, p. 138). Furtak and colleagues (2014) report that an LP influenced teachers' framing and a shift "away from talking about student ideas [only as] 'wrong' toward using the language in the LP to identify and discuss different student ideas" (Furtak et al., 2012, pp. 425–426). Once they began to attend to students' understandings, teachers sometimes expressed new understandings of student thinking and learning—e.g., recognition that telling students the answer does not change their thinking (Alonzo et al., 2022).

In addition, as proposed, there is evidence of LPs supporting teachers' awareness of specific understandings students could be expected to have. Teachers in Alonzo et al. (2022) directly reported increased awareness of student understandings, and Furtak (2012) saw evidence of this in teachers' identification of student understandings during assessment conversations. In addition, a broadened perspective on student understandings may be particularly useful for teachers with weak FA practices (Furtak et al., 2015). Although less common, there is also some evidence of teachers moving away from a "fix it" to a "work with it" perspective on the role of students' current understandings in their further learning (Alonzo et al., 2022).

Setting Learning Goals

Covitt, Gunckel, and colleagues investigated the practice of setting learning goals using teachers' responses to a written assessment that focused on LP-based FA practices. Although

Gunckel et al. (2018) found that teachers' learning goals improved from "a mix of disconnected facts" (p. 1352) to content aligned with an LP, teachers did not seem to use the LP to inform the nature of their expectations for students' learning. In both this study and Covitt et al.'s (2018) study, "few teachers went beyond the didactic practice of setting 'naming' type goals to ask students to perform more ambitious tasks" (p. 138). Exhibiting a similar focus on content, teachers in Furtak and Tayne's (2019) study used LPs to identify content from their existing curriculum that should be highlighted in instruction. While these uses both preceded teachers' engagement with their students' understandings, teachers in Zhai et al.'s (2018) study made substantial adjustments to their learning goals based on an LP and results from an associated diagnostic pre-test. In all three cases, teachers used descriptions in the LP to identify goals for students' learning, but these were based on teaching correct content, without attention to the level-based steps laid out in the LPs.

Eliciting Evidence of Student Understanding

Stemming from their new/increased awareness of student understandings, teachers selected, adapted, and/or developed assessment tasks to elicit these understandings.

Elicitations Using LP Levels

Teachers often used level-based assessment tasks (e.g., OMC items or open-ended questions designed to elicit understandings at multiple levels of an LP) from the respective LP project (Alonzo et al., 2022; Covitt et al., 2018; Furtak & Heredia, 2014). As described above, these tasks are often designed to elicit a wide range of student understandings, and task developers use LP levels to support that goal. For example, while a question like "What happens to an apple when it is eaten?" allows for sophisticated explanations (top level of the LP for carbon cycling; Mohan et al., 2009), students could also respond with Level 1 reasoning (e.g., "The apple is made into little pieces," p. 685), Level 2 reasoning (e.g., The apple "gets digested and it stores energy for him," p. 686), or Level 3 reasoning (e.g., The apple "is turned into glucose and used as energy," p. 688). When student responses to an assessment task are explicitly linked to levels of an LP (e.g., through options of an OMC item or a scoring rubric for an open-ended task), students' responses yield evidence that could be used to interpret students' LP levels. However, teachers may or may not use the tasks for this purpose (e.g., Alonzo et al., 2022).

Elicitations Using Other LP features

Teachers often used projects' level-based tasks to focus on student understandings, rather than their levels. They also adapted the assessment tasks to meet classroom needs in ways that did not necessarily rely on LP levels.

In Alonzo et al. (2022), teachers reported using items provided by the research team to assess specific understandings. They selected assessment tasks from those provided by the project to elicit evidence of specific understandings from LPs, rather than evidence of students' LP levels: The LP served as a "menu" (p. 6), rather than an ordered presentation of understandings. Teachers in Furtak and colleagues' PD programs developed and/or adapted assessment tasks aligned to an LP (e.g., Furtak, Heredia, & Circi, 2018; Furtak & Heredia, 2014). Similar to Alonzo et al. (2022), the tasks tended to focus on eliciting specific understandings from the LP.

Shifts in teachers' practice provide further evidence of how LPs may impact teachers' eliciting. After project participation, Furtak et al. (2015) found that teachers developed higher quality LP-based FA tasks, and Furtak et al. (2018) found an increase in the rate of teachers' questions "that surfaced student thinking" and "a decrease in the number of talk turns that contained no questions" during FA conversations (p. 279). Teachers in Alonzo et al. (2022) reported that increased awareness of the general nature of students' thinking and rationales for common student understandings led them to adapt existing assessment practices (in general and beyond the topics presented in LPs) to focus more on understanding students' reasoning.

Interpreting Evidence of Student Understanding

Once evidence of students' understandings had been elicited, teachers interpreted this evidence to make inferences about students' understandings.

Interpretations Using LP Levels

In evidence to date, and consistent with concerns outlined above about the coherence of student thinking relative to its portrayal in LP levels, teachers seem to interpret specific student understandings, more than students themselves, with respect to LP levels (e.g., Alonzo et al., 2022; Covitt et al., 2018).

Most level-based interpretations reported to date are based on researcher-generated score reports. Teachers in Furtak and Tayne's (2019) study used score reports to identify topics for which student understanding was lower and, thus, where they needed to spend more instructional time. However, Furtak et al.'s (2014) initial plans for teachers to "make valid inferences about what students know" (p. 656) based on an LP and LP-based score reports were challenged by multiple forms of complexity. They found multiple ways that student performances are messier (i.e., inconsistent, less coherent) than suggested by simple LP level diagnoses—e.g., students responding differently to items assessing the same content, students receiving the same LP level diagnosis despite different patterns of performance relative to the LP.

Interpretations Using Other LP Features

Across studies, there is evidence of teachers using LPs "as frameworks for describing, articulating, and naming common student ideas" (Furtak & Tayne, 2018, p. 249) and, thus, supporting interpretations beyond the correctness of students' responses (e.g., Covitt et al., 2018). LPs seem to support teachers in making valid inferences about students' understandings (e.g., Furtak, 2012), perhaps even providing "some mitigation against speculative or unwarranted interpretations" (von Aufschnaiter & Alonzo, 2018, p. 124). Such interpretations extend beyond the LP itself, allowing teachers to make sense of understandings not included in the LP. For example, a teacher in Alonzo and Elby's study used students' responses to LP-based assessment items to draw a conclusion about his students' understandings with respect to gravity, a concept not included in the LP. Teachers in Furtak and Heredia (2014) "uncovered a new student idea" (p. 998) not included in their LP and related it to an existing LP level.

Using LPs, teachers focused on both strengths and weaknesses in student understanding (e.g., Furtak et al., 2014) and engaged in "richer and more in-depth" consideration of student understandings (Furtak et al. 2015, p. 12). These benefits have been observed in teachers' analysis of written (e.g., Gunckel et al., 2018) and video (e.g., von Aufschnaiter & Alonzo, 2018) evidence of student understanding, as well as in teachers' conversations with each other (e.g., Furtak & Tayne, 2018) and "on-the-fly" during assessment conversations (e.g., Furtak, 2012). For example, a student in Furtak's (2012) study explained an evolutionary change as:

> ... the pronghorn antelope have to adapt to the change in speed to be able to outrun [the cheetahs]. So somewhere in the genes there was a mutation where the offspring have like longer legs so they can outrun the cheetah. (p. 1200)

It would be easy for a teacher to focus on the student's understanding that there was a gene mutation that caused a change in the antelope and assume understanding (or "correctness"). However, recognizing that students may hold needs-based views of evolutionary change, the teacher pressed the student to explain what she meant by adapt: Did running away from the cheetahs cause antelope to become faster (needs-based view)?

In attending to student understandings, teachers often treated the LP as a checklist (e.g., Alonzo et al., 2022), without attention to the associated LP levels. Indeed, teachers' interpretations sometimes entailed explicit pushback against levels in the LP construct. Teachers in Furtak and Tayne's (2019) study invented their own categories for analyzing students' performances, rather than relying on LP levels. While using LP-based score reports, teachers in Alonzo and Elby's study (2019) often challenged LP levels and their use as a pathway for student learning. The teachers offered interpretations of student thinking that rejected the assumption of coherence within a given level. Instead, they invented a range of finer-grained interpretations of student understandings to account for the complexity they had observed in students' thinking—e.g., differentiating between items set on Earth vs. in outer space or noting that students applied Level 4 reasoning for "maintaining motion" but not for "speeding up motion."

Responding to Evidence of Student Understanding

Interpretations based on evidence of students' understandings sometimes informed instructional responses designed to enhance those understandings; however, this FA practice was reported less frequently than the others.

Responses Using LP Levels

As a general response, teachers in Alonzo et al. (2022) and Kim and Lehrer (2015, as cited in Wilson & Lehrer, 2021, p. 3) used LP levels to juxtapose student ideas during what teachers called "formative assessment conversations." These conversations were designed to support learning from peers with ideas at different LP levels but did not have an explicit goal of level-by-level progress for individual students. Similarly, there is some evidence from the environmental literacy project (e.g., Covitt et al., 2018) of teachers using students' LP levels to respond with appropriate instruction. However, consistent with curriculum materials provided as part of this project, these responses were not designed to support incremental development of students' understanding (i.e., promoting students' progress through an LP). Rather, teachers provided instruction aimed at moving students from their current level to the top (Level 4) of the LP.

There is limited evidence of teachers' responding to students' understanding to support incremental development from one LP level to the next. As suggested by the basic model of LP-based FA, teachers in Zhai et al.'s (2018) study made on-the-fly adjustments designed to move students from one LP level to the next. Furtak and Tayne (2019) observed that "in-depth conversations around next steps for instruction" based on students' performance (relative to LPs) on FA tasks were "unfortunately rare" (p. 251), despite facilitator encouragement, but that teachers did discuss instructional next steps when working with researcher-provided score reports. Similarly, although teachers in Alonzo et al. (2022) considered how planned instruction might be interpreted by students at different LP levels and described level-by-level progress as a goal of LP-based FA, there were very few examples of teachers proposing or taking this kind of incremental approach. One exception was a teacher who used the LP to *preemptively* plan instruction, assuming his students would hold ideas in the LP. He started with familiar language sufficient for ideas at lower LP levels, gradually introducing and switching to the language required for more sophisticated understanding at higher levels (Alonzo et al., 2022).

Responses Using Other LP Features

Although limited, there is some evidence of teachers drawing on an LP, but not necessarily LP-level-based interpretations of student understanding, to propose and/or enact responses. Indeed, Alonzo and Elby (2019) found that teachers' proposals for instructional responses were more specific and actionable when based on interpretations of students' performance that were finer-grained (e.g., specific instruction to address a particular student idea) than LP-level diagnoses (e.g., vague plans to address a level).

Consistent with "checklist" interpretations, teachers' responses (both proposed and enacted) often focused on addressing specific student understandings (e.g., Alonzo et al., 2022). These responses did not require teachers' attention to LPs as pathways for student learning, in that teachers often addressed student understandings by devising strategies to move students directly from their current level of understanding to the desired one (i.e., top level of the LP).

Initially, these strategies focused on getting rid of ideas from the LP (e.g., teachers in Furtak's 2012 study identifying students' ideas from the LP that were "No no's," p. 1195). Over time and with PD, some teachers became more responsive to students' understanding. Although not common in either study, Alonzo et al. (2022) and Covitt et al. (2018) both reported teachers proposing or enacting strategies to support students in revising their ideas and practices (i.e., building on current understanding to develop new ones) rather than simply trying to remove or replace those understandings. Similarly, Furtak et al. (2018) observed teachers providing less evaluative feedback and providing more "pushing feedback" (p. 272) when using LPs.

Teacher Learning

While a small, but growing, number of studies have explored how teachers' practices change over time with LP use and accompanying LP-based PD, only a few have directly examined corresponding growth in teachers' knowledge. Using a pre/post measure, Gunckel et al. (2018) reported moderate to large growth in teachers' content knowledge (CK) and PCK, respectively, for teachers who participated in LP-based PD and subsequently used LP-based curriculum materials. Notably, they observed no change in CK and only a small change in PCK for teachers who only attended the PD. However, they note that teachers' performance left room for improvement: Even at post-test, teachers' CK did not reach the highest level of the LP, and their PCK remained consistent with traditional school science instruction.

Beyond knowledge directly contained in an LP, teachers in Alonzo and Elby's (2019) study generated new understandings of student thinking based on their interactions with LP-based score reports. In addition, teachers expressed evidence-based skepticism towards the LP model and an inquiring stance toward student thinking. In doing so, they "not only drew specific inferences about the particular students represented in the score report, but also used the score report to generate and test [more general] hypotheses about students' thinking" (p. 23). Much of the teachers' skepticism and subsequent investigation of student thinking stemmed from their observations of inconsistencies between students' response patterns and LP-based expectations. While these investigations reflected concerns about the LP levels and their use to represent a pathway for learning, they also demonstrate how levels have both constraints (in terms of describing individual students' thinking and learning) and affordances (in terms of guiding investigation into students' thinking and learning).

Discussion

In this chapter, I have considered proposals for and empirical evidence of teachers' use of LPs to support FA practices, considering stances towards learning (as represented in LPs and in teachers' practice) and differentiating between a) LP uses that rely on the grouping of understandings into levels and the ordering of levels hierarchically and b) LP uses that do not rely on these features of LP levels. Below, I offer some reflections on the significance of the evidence to date, propose additional considerations that should be further explored to more fully understand teachers' use of LPs, and identify implications of that use for researchers' original proposals for LPs' utility.

Continued Engagement with Teachers' Views of Student Thinking and Learning

As Furtak et al.'s (2014) work has demonstrated, there is a bidirectional relationship between teachers' perspectives on students' thinking and learning and their engagement with LPs. While LPs may help to shift teachers' perspectives towards more asset-oriented, constructivist views of student thinking and learning, deficit perspectives and assumptions that emphasize eliminating students' "wrong" understandings may persist because, as Gunckel et al. (2018) argue, these beliefs are deeply embedded in our school systems. Even multi-year PD does not seem sufficient to fully reshape these perspectives, which are still observed even as teachers take up LP-based FA practices. For example, teachers in Alonzo et al. (2022) sometimes tried to remove student understandings and other times tried to leverage those understandings to support further learning. This suggests that teachers' newly adopted constructivist perspectives were not strongly held and, thus, may not be sustained beyond project participation. Thus, while we continue to consider tools, such as educative curricula, to support temporary shifts in teachers' stances toward student thinking and learning, broader structural changes may be needed. One potentially important target for structural change is summative assessments and their alignment with LPs and with teachers' goals for FA (Briggs & Furtak, 2019).

Interpreting Teachers' Uses of LP Levels

As discussed above, there is somewhat limited evidence of teachers using level-based FA practices, and some researchers have documented challenges to doing so. There is a tension between learning from teachers what they find useful about LPs (and adjusting models for LP use accordingly) and supporting teachers in using affordances of the LP construct. When asked to use LPs in any way that made sense to them, teachers in Alonzo et al. (2022) used FA practices that rely on LP levels much less frequently than those relying on other LP features. Their use of LPs was more consistent with Sikorski's (2019) call for LPs to be treated "as toolboxes or repertories of ways of thinking, rather than a stack-ranked list of ways of thinking" (p. 974). In addition, teachers' use of LP levels and pathways sometimes seemed performative, in that teachers seemed to be "playing along" with what they interpreted researchers wanted them to do while expressing doubts about these approaches.

Did teachers make these choices because it was more familiar to engage in FA without levels such that further PD might support the use of levels? Or did teachers make these choices because they recognize the simplifications inherent in the LP levels and do not find them as useful amidst the complexity of a class full of individual learners whose thinking and learning may not fit neatly into those levels? Because teachers' use of LPs is consistent with ways that LP levels simplify thinking and learning (to present an aggregate description), teachers' practices may serve as a guide to envisioning new ways of using LPs, as well to rethinking existing assumptions about how these tools might be useful in the classroom. Continued efforts—and partnerships with teachers—are required to explore the extent to which level-based FA practices are useful to teachers, beyond meeting expectations of participation in a research project.

As part of such partnerships, shared language is important. In the work of Alonzo and colleagues (Alonzo & Elby, 2019; Alonzo et al., 2022), teachers used language that suggested a levels-based approach to using LPs for FA, while viewing LPs in ways that challenged the LP construct's notion of levels as coherent sets of understandings that evolve through a pathway defined by those levels. For example, teachers in both studies viewed LP levels as comprising sets of understandings that students may or may not hold, rather than considering understandings in an LP level to have conceptual coherence.

Continued Attention to Diversity among Students

Research is needed to explore the extent to which LP levels (groups of student understandings organized hierarchically) simplify the phenomena of students' thinking and learning and what implications this might have for FA. Following Pierson et al.'s (2019) claim about the rarity of alignment between individual students' learning and aggregate descriptions in LPs, we might ask: How much do individual students' learning pathways differ from that depicted in an LP? How much variation exists for students who reach the upper anchor of the LP? (Do students' pathways vary but only some are successful?) What groups of students have learning pathways that do/do not align with an LP? There are equity implications if students' ways of thinking and learning are not represented in an LP (Alonzo, 2023).

Tackling these issues from a different point of view, Duncan and Gotwals (2015) call for an investigation of the conceptual constraints inherent in LPs. Loose constraints suggest that "many different learning paths are possible that start from the same place and lead to the same outcome" (p. 411). This would indicate more flexible LPs, with multiple ways of defining and ordering levels and/or new conceptualizations of levels themselves. In contrast, strong constraints suggest that only a small number of pathways (or even a single pathway) optimize learning. This would suggest that a less flexible LP (i.e., one with a single, clear pathway through set levels) would be more appropriate to ensure that students have opportunities to learn valued ideas and practices. Two studies heeding Duncan and Gotwals's (2015) call did not find strong constraints (Castro-Faix et al., 2019; Lombard et al., 2018), i.e., students exhibited "diverse and often surprising" (Lombard et al., 2018, p. 101) pathways. This reaffirms concerns about ways that students' thinking and learning might differ from that reflected in an LP and invites attention to the ways that diverse pathways play out in classroom assessment practices.

Concerns about whose thinking and learning is and is not represented in an LP are not confined to uses that rely on LP levels. Although LPs can increase teachers' awareness of student understandings (e.g., Furtak, 2012), if teachers attend only to these understandings (e.g., von Aufschnaiter & Alonzo, 2018), other understandings may be marginalized. Although a few studies are beginning to engage teachers with the diversity of student thinking and learning (e.g., supporting attention to understandings beyond those in an LP), at the grain size necessary for FA, there is likely to be significant variation in students' thinking and learning. Attention both to how LPs may be designed and/or interpreted to be more flexible (e.g., Sikorski, 2019) and how teachers might be supported in using these tools in more expansive ways is needed.

Reframing Roles

Is there a role for LP levels and pathways, given their broad characterization of students' thinking and learning and, thus, potential limitations in terms of applicability for any given student? It seems as if the answer is yes, in at least two ways. First, Furtak and colleagues argue that—even when not used directly by teachers, LPs (with their levels and pathways) can serve as an important foundation for PD (Furtak & Tayne, 2018) and/or for the development of resources that teachers might use more directly (Furtak & Tayne, 2019). The common practice of engaging teachers in analyzing student work using LP levels appears to have a number of benefits even if this is not the way that teachers eventually engage with evidence of their students' understanding. Second, many of the uses of LPs that rely on other features stem from level-based practices. Teachers in some studies generated new interpretations of student understandings, based on (level-based) OMC items and score reports, and level-based discrepancies between students' responses and LP expectations prompted teachers to inquire and generate new understandings (Alonzo & Elby, 2019; Alonzo et al., 2022; Furtak et al., 2014). Therefore, rather than rejecting uses of LPs that rely on LP levels, I argue that these provide an important foundation for PD and for uses that rely on other features of LPs.

By acknowledging that LP levels necessarily simplify the complex phenomena of students' thinking and learning and inviting teachers to investigate evidence of the complexity of student understandings, PD providers can support the skepticism and inquiring stance required for teachers to generate new knowledge. Inviting teachers to inquire into their students' understandings, rather than treating LP levels as necessarily applicable to all students in all contexts, suggests a different role for teachers with respect to LPs and PD. If LPs are framed as "generative launching pads for teachers' learning about student thinking" (Alonzo & Elby, 2019, p. 2), PD becomes about supporting teachers as knowledge creators.

Other Features

In this chapter, I have focused on levels as a key feature of LPs that may impact teachers' use of LPs for FA. However, these are not the only relevant determinants. First, as researchers acknowledge (e.g., Furtak et al., 2014), the LP used in a given study may influence the results—e.g., Furtak (2012)'s LP has a much smaller grain size than LPs in the environmental literacy project (e.g., Mohan et al., 2009). Furtak and Tayne's (2019) work to investigate teachers' use of multiple LPs is an important step towards understanding the affordances and constraints of different LP features for teachers' FA practices.

Second, PD and other supports are likely to have a significant impact on teachers' subsequent use of LPs as part of FA practices. Gunckel et al.'s (2018) comparison of teacher learning in PD vs. PD + curriculum enactment is crucial for understanding how different forms of PD might impact teachers' use of LPs, but comparisons across different PD models are also important. This is especially true since different groups (with different LPs) provide different resources and PD opportunities with different emphases. Detailed attention may be needed, as there could be significant variation in activities with the same description.

Third, researchers have documented the importance of alignment between teachers' current environments and the LP and its embedded stance towards students' thinking and learning. Prevailing discourse and narratives (e.g., Covitt et al., 2018), resources (Furtak & Tayne, 2019), and expectations (Heredia, 2020) may all impact teachers' use of an LP and associated materials.

Conclusion

Levels are a key feature of LPs, critical to many applications (such as curriculum and standards), in which aggregate understandings of students' thinking and learning are essential. For applications such as FA, in which variations among students may be important, levels have affordances and constraints. Further work and partnerships with teachers are required to develop more robust models for LP-supported FA that account for the complexity and messiness of both classrooms and the student thinking and learning LPs are designed to influence.

Acknowledgments

This chapter is based upon work supported by the US National Science Foundation under Grant No. DRL-1253036. Any opinions, findings, and conclusions or recommendations expressed in this chapter are those of the author and do not necessarily reflect the views of the National Science Foundation.

This chapter draws on research conducted in collaboration with Andrew Elby, Julie Christensen, and Michelle Wooten and reported in Alonzo and Elby (2019) and Alonzo et al. (2022).

Notes

1 I use "understandings" throughout the chapter to refer to both ideas and practices and "thinking" to refer to students' reasoning with and about understandings.
2 See Alonzo, 2018, for discussion of the differences between science-specific and more subject-general conceptualizations of LPs.
3 While attempting to characterize a common interpretation of the LP construct, I acknowledge that this is not universal. For example, Metz et al. (2019) explicitly state that "no one-to-one correspondence of child and learning progression level exist, even at one point in time" (p. 567).
4 Further research is required to investigate the role of LP grain size in the consistency of students' thinking with respect to an LP. Jin, Mikeska et al. (2019) argue that the apparent consistency of students' reasoning is simply a matter of grain size, and Jin et al. (2013) found evidence that "students' reasoning is relatively coherent and consistent" (p. 1687); however, this coherence and consistency was observed with respect to an LP with a large grain size, which may have less utility for FA.

5 LP levels—and the practice of interpreting student ideas and/or students themselves in terms of their relative distance from the upper anchor—have also been critiqued as promoting a deficit view of students and their ideas (Pierson & Clark, 2019).
6 A key exception is Sevian and Talanquer's (2014) use of a potential energy surface to represent the complex learning landscape of an LP. In this representation, regions of relative stability are identified but not arranged in a sequence or pathway.

References

Alonzo, A. C. (2011). Learning progressions that support formative assessment practices. *Measurement: Interdisciplinary Research & Perspective*, 9(2-3), 124–129. https://doi.org/10.1080/15366367.2011.599629

Alonzo, A. C. (2018). An argument for formative assessment with science learning progressions. *Applied Measurement in Education*, *31*(2), 104–112. https://doi.org/10.1080/08957347.2017.1408630

Alonzo, A. C. (2023). Learning progressions: Affordances, critiques, and pathways forward. In R. Tierney, F. Rizvi, K. Ercikan, & G. Smith (Eds.) *International encyclopedia of education* (4th ed.). Elsevier. 10.1016/B978-0-12-818630-5.14064-3

Alonzo, A. C., & Elby, A. (2019). Beyond empirical adequacy: Learning progressions as models and their value for teachers. *Cognition and Instruction*, *37*(1), 1–37. 10.1080/07370008.2018.1539735

Alonzo, A. C., & Steedle, J. T. (2009). Developing and assessing a force and motion learning progression. *Science Education*, *93*(3), 389–421. 10.1002/sce.20303

Alonzo, A. C., & von Aufschnaiter, C. (2018). Moving beyond misconceptions: Learning progressions as a lens for seeing progress in student thinking. *The Physics Teacher*, *56*, 470–473.

Alonzo, A. C., Wooten, M. W., & Christensen, J. A. (2022). Learning progressions as a simplified model: Examining teachers' reported uses to inform classroom assessment practices. *Science Education*, *106*(4), 852–889. 10.1002/sce.21713

Andersson, C., & Palm, T. (2017). The impact of formative assessment on student achievement: A study of the effects of changes to classroom practice after a comprehensive professional development programme. *Learning and Instruction*, *49*, 92–102. 10.1016/j.learninstruc.2016.12.006

Bailer-Jones, D. M. (2003). When scientific models represent. *International Studies in the Philosophy of Science*, *17*(1), 59–74. 10.1080/02698590305238

Bennett, R. E. (2011). Formative assessment: A critical review. *Assessment in Education: Principles, Policy & Practice*, *18*(1), 5–25. 10.1080/0969594X.2010.513678

Bernholt, S., & Sevian, H. (2018). Learning progressions and teaching sequences—Old wine in new skins? *Chemistry Education Research and Practice*, *19*(4), 998-997. 10.1039/C8RP90009D

Black, P., & Wiliam, D. (1998). Assessment and classroom learning. *Assessment in Education: Principles, Policy and Practice*, *5*(1), 7–74. 10.1080/0969595980050102

Black, P., Wilson, M., & Yao, S.-Y. (2011). Road maps for learning: A guide to the navigation of learning progressions. *Measurement: Interdisciplinary Research and Perspectives*, 9(2-3), 71–123. 10.1080/15366367.2011.591654

Briggs, D. C., Alonzo, A. C., Schwab, C., & Wilson, M. (2006). Diagnostic assessment with ordered multiple-choice items. *Educational Assessment*, *11*(1), 33–63.

Briggs, D. C., & Furtak, E. M. (2019) Learning progressions and embedded assessment. In S. M. Brookhart & J. H. McMillan (Eds.), *Classroom assessment and educational measurement* (pp. 146–169). Routledge. 10.4324/9780429507533-9

Carpenter, S. L., Kim, J., Nilsen, K., Irish, T., Bianchini, J. A., & Berkowitz, A. R. (2020). Secondary science teachers' use of discourse moves to work with student ideas in classroom discussions. *International Journal of Science Education*, *42*(15), 2513–2533. 10.1080/09500693.2020.1820620

Castro-Faix, M., Todd, A., Romine, W., & Duncan, R. G. (2019). Do alternative instructional approaches result in different learning progressions? In J. Kay & R. Luckin (Eds.), *Rethinking learning in the digital age: Making the learning sciences count, 13th International Conference of the Learning Sciences (ICLS) 2018*, Volume 2 (pp. 808–815). International Society of the Learning Sciences.

Covitt, B., Gunckel, K. L., Syswerda, S., & Caplan, B. (2018). Teachers' use of learning progression-based formative assessment in water instruction. *Applied Measurement in Education*, *31*(2), 128–142. 10.1080/08957347.2017.1408627

Duncan, R. G., & Gotwals, A. W. (2015). A tale of two progressions: On the benefits of careful comparisons. *Science Education*, *99*(3), 410–416. 10.1002/sce.21167

Duschl, R. A., Maeng, S., & Sezen, A. (2011). Learning progressions and teaching sequences: a review and analysis. *Studies in Science Education*, *47*(2), 123–182. 10.1080/03057267.2011.604476

Engelhard, G., Jr., & Sullivan, R. K. (2011). An ecological perspective on learning progressions as road maps for learning. *Measurement: Interdisciplinary Research & Perspectives*, *9*(2-3), 138–145. 10.1080/15366367.2011.603615

Furtak, E. M. (2012). Linking a learning progression for natural selection to teachers' enactment of formative assessment. *Journal of Research in Science Teaching*, *49*(9), 1181–1210. 10.1002/tea.v49.9

Furtak, E. M., & Heredia, S. C. (2014). Exploring the influence of learning progressions in two teacher communities. *Journal of Research in Science Teaching*, *51*(8), 982–1020. 10.1002/tea.21156

Furtak, E. M., Heredia, S. C., & Circi, R. K. (2018). Exploring alignment among learning progressions, teacher-designed formative assessment tasks, and student growth: Results of a four-year study. *Applied Measurement in Education*, *31*(2), 143–156. 10.1080/08957347.2017.1408624

Furtak, E. M., Kiemer, K., Swanson, R., de León, V., & Circi, R. K. (2015, April). Learning progressions, formative assessment, and professional development: Results of a longitudinal study. *Paper presented at the NARST annual international conference*, Chicago, IL.

Furtak, E. M., Morrison, D., & Kroog, H. (2014). Investigating the link between learning progressions and classroom assessment. *Science Education*, *98*(4), 640–673. 10.1002/sce.21122

Furtak, E. M., Morrison, D., & Iverson, H. (2013, April). Challenges in developing classroom assessments linked to multidimensional learning progressions. *Paper presented at the annual meeting of the National Association of Research in Science Teaching*, Puerto Rico.

Furtak, E. M., & Tayne, K. (2018). Mobilizing learning progressions for teacher use: Examining the utility of outside learning progressions in task co-design. In J. Kay & R. Luckin (Eds.), *Rethinking learning in the digital age: Making the learning sciences count, 13th International Conference of the Learning Sciences (ICLS) 2018*, Volume 1 (pp. 520–527). International Society of the Learning Sciences.

Furtak, E. M., & Tayne, K. (2019). Affordances and constraints of learning progression designs in supporting formative assessment. In E. McLoughlin, O. Finlayson, S. Erduran, & P. Childs (Eds.), *Bridging research and practice in science education: Contributions from science education research* (Vol. 6) (pp. 241–256). Springer. 10.1007%2F978-3-030-17219-0_15

Furtak, E. M. Thompson, J. Braaten M. , & Windschitl, M. (2012). Learning progressions to support ambitious teaching practices. In Alonzo, A. C. & Gotwals, A. W. (Eds.), *Learning progressions in science: Current challenges and future directions*. Rotterdam, The Netherlands: Sense Publishers. (pp. 405–433).

Gotwals, A. W. (2018). Where are we now? Learning progressions and formative assessment. *Applied Measurement in Education*, *31*(2), 157–164. 10.1080/08957347.2017.1408626

Gotwals, A. W., Alonzo, A. C., Kintz, T., & Roeber, E. (2021). *A guide to the use of formative assessment practice in science*. Guidebook commissioned by the Michigan Department of Education (Formative Assessment for Michigan Educators Initiative). https://famemichigan.org/wp-content/uploads/2022/05/FAMEGuide_Science.pdf

Gotwals, A. W., & Birmingham, D. (2016). Eliciting, identifying, interpreting, and responding to students' ideas: Teacher candidates' growth in formative assessment practices. *Research in Science Education*, *46*(3), 365–388. 10.1007/s11165-015-9461-2

Gunckel, K. L., Covitt, B. A., & Salinas, I. (2018). Learning progressions as tools for supporting teacher content knowledge and pedagogical content knowledge about water in environmental systems. *Journal of Research in Science Teaching*, *55*(9), 1339–1361. 10.1002/tea.21454

Hammer, D., & Sikorski, T.-R. (2015). Implications of complexity for research on learning progressions. *Science Education*, *99*(3), 424–431. 10.1002/sce.21165

Heredia, S. C. (2020). Exploring the role of coherence in science teachers' sensemaking of science-specific formative assessment in professional development. *Science Education*, *104*(3), 581– 604. 10.1002/sce.21561

Heritage, M. (2007). Formative assessment: What do teachers need to know and do? *Phi Delta Kappan*, *89*(2), 140–145. 10.1177/003172170708900210

Heritage, M., Kim, J., Vendlinski, T., & Herman, J. (2009). From evidence to action: A seamless process in formative assessment? *Educational Measurement: Issues and Practice*, *28*(3), 24–31. 10.1111/j.1745-3992.2009.00151.x

Hokayem, H., & Gotwals, A. W. (2016). Early elementary students' understanding of complex ecosystems: A learning progression approach. *Journal of Research in Science Teaching*, *53*(10), 1524–1545. 10.1002/tea.21336

Jin, H., Mikeska, J. N., Hokayem, H., & Mavronikolas, E. (2019). Toward coherence in curriculum, instruction, and assessment: A review of learning progression literature. *Science Education*, *103*(5), 1206–1234. 10.1002/sce.21525

Jin, H., Shin, H., Johnson, M. E., Kim, J., & Anderson, C. W. (2015). Developing learning progression-based teacher knowledge measures. *Journal of Research in Science Teaching*, *52*(9), 1269–1295. 10.1002/tea.21243g

Jin, H., van Rijn, P., Moore, J. C., Bauer, M. I., Pressler, Y., & Yestness, N. (2019). A validation framework for science learning progression research. *International Journal of Science Education*, *41*(10), 1324–1346. 10.1080/09500693.2019.1606471

Jin, H., Zhan, L., & Anderson, C. W. (2013). Developing a fine-grained learning progression framework for carbon-transforming processes. *International Journal of Science Education*, *35*(10), 1663–1697. 10.1080/09500693.2013.782453

Ketterlin-Geller, L. R., Shivraj, P., Basaraba, D., & Yovanoff, P. (2019). Considerations for using mathematical learning progressions to design diagnostic assessments. *Measurement: Interdisciplinary Research and Perspectives*, *17*(1), 1–22. 10.1080/15366367.2018.1479087

Klenowski, V. (2009). Assessment for learning revisited: An Asia-Pacific perspective. *Assessment in Education: Principles, Policy and Practice*, *16*(3), 263–268. 10.1080/09695940903319646

Kluger, A. N., & DeNisi, A. (1996). The effect of feedback interventions on performance: A historical review, a meta-analysis, and a preliminary feedback intervention theory. *Psychological Bulletin*, 119(2), 254–284. 10.1037/0033-2909.119.2.254

Larkin, D. (2012). Misconceptions about "misconceptions": Preservice secondary science teachers' views on the value and role of student ideas. *Science Education*, *96*(5), 927–959. 10.1002/sce.21022

Lehrer, R., & Schauble, L. (2015). Learning progressions: The whole world is NOT a stage. *Science Education*, *99*(3), 432–437. 10.1002/sce.21168

Levin, D. M., Hammer, D., & Coffey, J. E. (2009). Novice teachers' attention to student thinking. *Journal of Teacher Education*, *60*(2), 142–154. 10.1177/0022487108330245

Lombard, F., Merminod, M., Widmer, V., & Schneider, D. K. (2018). A method to reveal fine-grained and diverse conceptual progressions during learning. *Journal of Biological Education*, *52*(1), 101–112. 10.1080/00219266.2017.1405534

McManus, S. (2008). *Attributes of effective formative assessment*. Washington, DC: Council of Chief State School Officers. https://ccsso.org/resource-library/attributes-effective-formative-assessment

Metz, K. E., Cardace, A., Berson, E., Ly, U., Wong, N., & Wilson, M. (2019). Primary grade children's capacity to understand microevolution: The power of leveraging their fruitful intuitions and engagement in scientific practices. *Journal of the Learning Sciences*, *28*(4-5), 556–615. 10.1080/10508406.2019.1667806

Minstrell, J., Anderson, R., & Li, M. (2011, May). Building on learner thinking: A framework for assessment in instruction. Commissioned paper for the Committee on Highly Successful STEM Schools or Programs for K–12 STEM Education, National Academy of Sciences. https://sites.nationalacademies.org/cs/groups/dbassesite/documents/webpage/dbasse_072645.pdf

Mohan, L., Chen, J., & Anderson, C. W. (2009). Developing a multi-year learning progression for carbon cycling in socio-ecological systems. *Journal of Research in Science Teaching*, *46*(6), 675–689. 10.1002/tea.20314

National Research Council. (2007). *Taking science to school: Learning and teaching science in grades K–8*. Washington, DC: The National Academies Press. https://doi.org/10.17226/11625

Nehm, R. H., & Schonfeld, I. S. (2008). Measuring knowledge of natural selection: A comparison of the CINS, an open-response instrument, and an oral interview. *Journal of Research in Science Education*, *45*(10), 1131–1160. 10.1002/tea.20251

NGSS Lead States. (2013). *Next generation science standards: For states, by states*. Washington, DC: The National Academies Press. 10.17226/18290

Otero, V. K. (2006). Moving beyond the "get it or don't" conception of formative assessment. *Journal of Teacher Education*, *57*(3), 247–255. 10.1177/0022487105285963

Pierson, A. E., & Clark, D. E. (2019). Sedimentation of modeling practices: Dimensions of co-operative action at a classroom scale. *Science & Education*, *28*(8), 897–925. 10.1007/s11191-019-00050-4

Pierson, A., Clark, D., & Kelly, G. (Eds.) (2019). Scientific practices, epistemic aims, and learning progressions. *Science & Education*, *28*(8), 833–841. 10.1007/s11191-019-00070-0

Plummer, J. D., & Maynard, L. (2014). Building a learning progression for celestial motion: An exploration of students' reasoning about the seasons. *Journal of Research in Science Teaching*, *51*(7), 902–929. 10.1002/tea.21151

Ruiz-Primo, M. A., & Furtak, E. M. (2006). Informal formative assessment and scientific inquiry: Exploring teachers' practices and student learning. *Educational Assessment*, *11*(3-4), 205–235. 10.1080/10627197.2006.9652991

Schneider, M. C., & Gowan, P. (2013). Investigating teachers' skills in interpreting evidence of student learning. *Applied Measurement in Education*, *26*(3), 191–204. 10.1080/08957347.2013.793185

Schramm, J. W., Jin, H., Keeling, E. G., Johnson, M., & Shin, H. J. (2018). Improved student reasoning about carbon-transforming processes through inquiry-based learning activities derived from an empirically validated learning progression. *Research in Science Education*, *48*(5), 887–911. 10.1007/s11165-016-9584-0

Sevian, H., & Talanquer, V. (2014). Rethinking chemistry: A learning progression on chemical thinking. *Chemistry Education Research and Practice*, *15*(1), 10–23. 10.1039/C3RP00111C

Shavelson, R. J., & Kurpius, A. (2012). Reflections on learning progressions. In A. C. Alonzo & A. W. Gotwals (Eds.), *Learning progressions in science: Current challenges and future directions* (pp. 13–26). Rotterdam, The Netherlands: Sense Publishers. 10.1007/978-94-6091-824-7_2

Shea, N., & Duncan, R. G. (2013). From theory to data: The process of refining learning progressions. *Journal of the Learning Sciences*, *22*(1), 7–32. 10.1080/10508406.2012.691924

Shepard, L. A. (2008). Formative assessment: Caveat emptor. In C. A. Dwyer (Ed.), *The future of assessment: Shaping teaching and learning* (pp. 279–303). Lawrence Erlbaum Associates.

Shepard, L. A. (2009). Commentary: Evaluating the validity of formative and interim assessment. *Educational Measurement: Issues and Practice*, *28*(3), 32–37. 10.1111/j.1745-3992.2009.00152.x

Shulman, L. S. (1987). Knowledge and teaching: Foundations of the new reform. *Harvard Educational Review*, *57*(1), 1–22.

Sikorski, T.-R. (2019). Context-dependent "upper anchors" for learning progressions. *Science & Education*, *28*(8), 957–981. 10.1007/s11191-019-00074-w

Smith, C. L., Wiser, M., Anderson, C. W., & Krajcik, J. (2006). Implications of research on children's learning for standards and assessment: A proposed learning progression for matter and the atomic-molecular theory. *Measurement: Interdisciplinary Research & Perspective*, *4*(1-2), 1–98. 10.1080/15366367.2006.9678570

Songer, N. B., Kelcey, B., & Gotwals, A. W. (2009). How and when does complex reasoning occur? Empirically driven development of a learning progression focused on complex reasoning about biodiversity. *Journal of Research in Science Teaching*, *46*(6), 610–631. https://doi.org/10.1002/tea.20313

Stahnke, R., Schueler, S., & Roesken-Winter, B. (2016). Teachers' perception, interpretation, and decision-making: A systematic review of empirical mathematics education research. *ZDM*, *48*, 1–27. 10.1007/s11858-016-0775-y

Sztajn, P., Confrey, J., Wilson, P. H., & Edgington, C. (2012). Learning trajectory based instruction: Toward a theory of teaching. *Educational Researcher*, *41*(5), 147–156. 10.3102/0013189X12442801

von Aufschnaiter, C., & Alonzo, A. C. (2018). Foundations of formative assessment: Introducing a learning progression to guide pre-service physics teachers' video-based interpretation of student thinking. *Applied Measurement in Education*, *31*(2), 113–127. 10.1080/08957347.2017.1408629

Wilson, M., & Lehrer, R. (2021). Improving learning: Using a learning progression to coordinate instruction and assessment. *Frontiers in Education*, *6*(654212). 10.3389/feduc.2021.654212

Wiser, M., Smith, C. L., & Doubler, S. (2012). Learning progressions as tools for curriculum development. In A. C. Alonzo & A. W. Gotwals (Eds.), *Learning progressions in science: Current challenges and future directions* (pp. 359–403). Sense Publishers. 10.1007/978-94-6091-824-7_16

Yin, Y., Tomita, M. K., & Shavelson, R. J. (2014). Using formal embedded formative assessments aligned with a short-term learning progression to promote conceptual change and achievement in science. *International Journal of Science Education*, *36*(4), 531–552. 10.1080/09500693.2013.787556

Zhai, X., Li, M., & Guo, Y. (2018). Teachers' use of learning progression-based formative assessment to inform teachers' instructional adjustments: A case study of two physics teachers' instruction. *International Journal of Science Education*, *40*(15), 1832–1856. 10.1080/09500693.2018.1512772

22

USING LEARNING PROGRESSIONS IN PROFESSIONAL DEVELOPMENT PROGRAMS

Erin Marie Furtak, Kate Henson, and Jason Y. Buell

Introduction

International efforts in science education are seeking to reframe science learning as more than just a catalog of facts, but rather to engage students through accessible, everyday scenarios that give them opportunities not just to apply what they know but also to learn about the practices of scientists or science competencies (Organisation for Economic Cooperation and Development [OECD], 2009; National Research Council [NRC], 2012; Ropohl et al., 2018). In the US, the vision for science education reconceives classroom learning as opportunities for students to participate in three-dimensional learning experiences that combine the use of science practices, crosscutting concepts, and disciplinary core ideas (NGSS Lead States, 2013) to make sense of phenomena or solve ill-structured problems. Learning experiences are then deeply embedded in using phenomena that are interesting and relevant to students' daily lives (National Academies of Science, Engineering, and Medicine [NASEM], 2019).

These reforms are richly informed by previous waves of inquiry-based teaching and learning (Furtak & Penuel, 2019) as well as sociocultural turns in science education (e.g., Ford & Forman, 2006) and science studies (e.g., Latour & Woolgar, 1979; Pickering, 1995). At the same time, they raise expectations for the ways teachers design and facilitate learning environments that reposition student thinking at the center of classroom activity (NASEM, 2018), and place high demands on teachers to enact tasks in ways that will keep – rather than lower – the cognitive demand of tasks that organize students' and teachers' activity (Kang et al., 2016).

Learning progressions have been a key element of this most recent wave of science education reforms, providing foundational structures for the design of curriculum, assessment, and professional learning (Duschl, Maeng & Sezen, 2011; NRC, 2012). As described in the preceding chapters, learning progressions help to make explicit the ways in which learning experiences are designed to support advances in student learning over time, as well as to provide interpretive frameworks for student responses to assessment.

Researchers have also examined the ways that learning progressions can serve as scaffolds for teachers' classroom practice; that is, they can serve to support teachers in planning for and interpreting student work, as well as interpreting student ideas on the fly. In this chapter, we describe a partnership project in which we used a set of learning progressions as the

 DOI: 10.4324/9781003170785-25

foundation of a long-term, school-based professional learning experience for high school science teachers. We highlight the different ways the unique features of the learning progression supported teachers as they engaged in professional learning for formative assessment planning, design, and enactment across high school content areas.

Theoretical Framework

Penuel & Shepard (2016) associated learning progressions with socio-cognitive assessment interventions that support classroom learning. These interventions feature resources that support teachers in understanding the development of students' knowledge and engagement in practice over time. As such, learning progressions can chart goals from lower- to upper-anchors for student learning over spans of time (e.g., Corcoran, Mosher & Rogat, 2009), and thus serve as orienting frameworks for systems of assessment (Shepard, Penuel, & Pellegrino, 2018).

Learning progressions can also be viewed as resources that can organize particular types of activity (Kang & Furtak, 2021). From this perspective, we can view classrooms as activity systems comprised of participants occupying different roles, with multiple lived histories and working within a particular local organizational context toward a particular outcome, and supported by tools and routines (Engeström, 2001; Henson, 2019; Moss, 2008; Wertsch, 1998). We can envision a classroom activity system comprised of teachers and students engaged in a common activity to investigate and learn more about the practices and disciplinary core ideas of science. Teachers and students all have their own prior experiences and identities that inform their participation in the learning environment, which is also informed by the institutional context of the school (and likely also the district and state policy environments). In addition, teachers' and students' participation in learning can be organized and supported by tools and routines, such as tasks that encourage students to share their ideas about a particular phenomenon linked to a science learning goal, or a routine in which students share what they notice and wonder about an anchoring phenomenon.

Within this framework, we can also view learning progressions as tools that can orient teachers as they navigate student engagement in complex learning environments in which students are engaged in three-dimensional science learning; that is, science learning in which students learn disciplinary core ideas by engaging in science practices and applying crosscutting concepts (NRC, 2012). By building on what students know, and mapping how student ideas develop within a domain, learning progressions can make a formerly invisible landscape more visible for teachers, and create space for teachers to anticipate the kinds of responses and feedback they may want to provide to students. For example, many three-dimensional unit storylines begin with an anchoring phenomenon and invite students to generate ideas and questions that could support later inquiry. In these kinds of learning environments, learning progressions can help teachers anticipate the kinds of ideas and questions students might share, and predict how these ideas are related to each other based on empirical evidence.

Three-Dimensional Science Learning: Foundations for US Science Standards

The *Framework for K-12 Science Learning* (NRC, 2012) has served as the foundational document for US science standards over the past 10 years. A major advance of these standards has been their structure that combines multiple elements of knowing and doing science, often referred to as the three dimensions. First, the *Framework* describes the disciplinary core ideas, or the big, orienting ideas of science, within different domains of life, physical, and earth and space science. These ideas move away from granular science facts and toward organizing concepts within these domains, like evolution in the life sciences, kinetic molecular theory in physical sciences, and plate tectonics in earth and space sciences.

The second dimension of the *Framework* is comprised of science and engineering practices, or the disciplinary practices in which scientists and engineers go about their work. These overlap with ways of conceptualizing science inquiry in previous waves of science education reform (e.g., Schwab, 1962), and draw upon what has been called the "practice turn" in sociocultural theory (Ford & Forman, 2006). Science and engineering practices include developing and using models, constructing explanations, and engaging in arguments from evidence.

The third and newest element of the *Framework* is crosscutting concepts or themes that cut across the science disciplines. A way of thinking about these concepts is to envision the kinds of questions scientists might ask themselves when they encounter a novel scenario, such as looking for patterns, examining scale and proportion, or considering systems that might be involved (Furtak et al., 2021).

The innovation of the *Next Generation Science Standards* (NGSS Lead States, 2013), the foundation for standards in most US states, is that each standard is comprised of a performance expectation that includes all three of the dimensions. For example, HS-LS-1-7 states that students will "Use a model to illustrate that cellular respiration is a chemical process whereby the bonds of food molecules and oxygen molecules are broken and the bonds in new compounds are formed resulting in a net transfer of energy." This standard combines the science and engineering practice of *developing and using models* with the disciplinary core idea of *cellular respiration* and the crosscutting concept of *flow of energy and matter*.

Formative Assessment

Students' three-dimensional science learning in the US has been accompanied by calls for supporting student learning through ongoing formative assessment (NRC, 2014). Formative assessment has been defined as the process of creating space for students to share their developing ideas while learning is still in progress to provide instructional support to help them advance in their learning (e.g., Shepard & Penuel, 2018). It has gained attention in international educational reforms because of its promise to support all learners and create opportunities to provide additional support to those who need it (Black & Wiliam, 1998). More specifically, formative assessment consists of (1) clarifying learning goals within a broader progression of learning; (2) eliciting and analyzing evidence of student thinking; (3) engaging in self-assessment and peer feedback; (4) providing actionable feedback; and (5) using evidence and feedback to move learning forward by adjusting learning strategies, goals or next instructional steps (Council of Chief School Science Officers, 2018, p. 2–3).

The phrase "formative assessment" has been used to refer to the tasks that teachers use to organize classroom activity, the practices in which students and teachers participate around these tasks, as well as informally while listening and responding to student ideas (Bennett, 2011; Furtak, Heredia & Morrison, 2019). By creating opportunities for students and teachers to surface and discuss ideas while instruction is in progress, formative assessment creates space for teachers to connect students' ideas and experiences with science and to adapt and adjust learning experiences to center students' interests and needs.

Formative assessment *tasks* are designed to consist of multiple components that create space for students to share their thinking through open-ended questions, drawing models, writing explanations, and other features (Kang et al., 2014; NRC, 2014; Fine & Furtak, 2020). Formative assessment *practices* are the actions teachers and students engage in to ask each other questions, push each other's thinking, and better understand each other's ideas (Cowie et al., 2011; Dini et al., 2020). While formative assessment remains a key mechanism for teachers enacting Framework-aligned instructional experiences in their classrooms, it is at the same time a complex practice that involves drawing out student ideas and experiences, interpreting them relative to scientific models and theories, and then using that information to determine subsequent instructional moves, and to provide feedback to students (e.g., Dini et al., 2020). Navigating the complexity of student ideas can be challenging, and as such, tools or frameworks such as learning progressions have been used to help teachers navigate and make inferences about student thinking. They may also be able to scaffold the enactment of effective formative assessment practice to support science learning (Alonzo & Elby, 2019; Furtak, 2012).

Returning to the theoretical framing described above, formative assessment tasks can serve as tools that help to organize teachers and students around multiple elements of formative assessment. Engaging teachers in professional learning around additional tools to support these activities – including learning progressions – can help to bring into alignment science learning experiences.

Learning Progressions as Tools to Support Teacher Professional Learning: The Formative Assessment Design Cycle

Scholars of learning progressions have identified several uses for learning progressions as supports for teachers' classroom assessment and formative assessment (Alonzo et al., 2022; Gunckel et al., 2018; Furtak, 2012). These uses include planning learning goals and where to embed assessments, designing tasks to draw out student thinking, and using the learning progression to interpret student thinking and plan the next steps for instruction.

Over the course of several interconnected studies, we have developed a set of routines for teachers' co-design of formative assessment tasks centered on learning progressions which we call the Formative Assessment Design Cycle (Furtak & Heredia, 2014; Furtak, Morrison & Kroog, 2014). These routines build on other studies of teachers' iterative collaboration and reflection around co-planned tasks in professional learning communities (e.g., Borko et al., 2008; Horn & Little, 2010) and align with other efforts conducted in parallel to create spaces for teacher learning about ambitious science teaching reforms (e.g., Thompson, Hagenah, McDonald & Barchenger, 2019).

The Formative Assessment Design Cycle (FADC) uses learning progressions as foundational tools to organize routines for teacher learning about student thinking, as well as deepening teachers' own content understandings (Heredia, Furtak, Morrison & Gröschner, 2021). It consists of five steps, each of which uses the learning progression to support teachers as they plan for, enact, and reflect upon formative assessment tasks; these steps are summarized in Table 22.1. First, teachers *explore* student thinking with the support of a learning progression, and then they *design* formative assessment tasks to embed in units of instruction. Next, they *practice* using those tasks, anticipating how to elicit and respond to student ideas related to the learning progression before *enacting* the tasks with students. Finally, teachers gather to *reflect* upon enactment, using the learning progression to interpret artifacts of student thinking and to plan the next steps for instruction.

We have enacted these routines iteratively both within and across academic years and focused on particular disciplinary domains or units of instruction. One complete cycle might

Table 22.1 Formative assessment design cycle

Step	*Description*	*Role of Learning Progression*	*Example*
Explore	Explore student thinking, including discussions of teachers' prior experience with student ideas in a given disciplinary area/practice/crosscutting concept	Unpack and discuss learning progression levels; use learning progression to interpret student work samples; add/modify language in levels based on teachers' experiences	Using sample copies of student work, teachers practice assigning responses to levels of progression and discuss disagreements
Design	Adapt existing assessment tasks or develop new tasks that create space for students to share their thinking/engagement in practice	Use learning progression to identify key 'look fors' that can inform the design of different components in the task	Teachers discuss possible prompts for tasks that would draw out ideas from learning progression, adapting existing assessments from curriculum and other item sources
Practice	Discuss participation structures for enacting tasks with students, including different levels of groupings; plan how the task will organize routines around sharing & responding to student thinking	Use learning progression to inform kinds of questions to ask, and possible responses/feedback to student ideas to help them push their thinking	Teachers complete a guide that helps them anticipate how to enact the task with students, focusing on participation structures and ways to respond to student ideas
Enact	Use tasks in the classroom with students; collect samples of student work (written responses, posters, pictures of whiteboards or posters, etc.)	Use identified look-fors as an informal guide to navigating classroom conversations/student ideas	Teachers use tasks with students, asking questions to draw out student responses in the learning progression
Reflect	Interpret student ideas and plan next steps for instruction based on enactment of the task; where applicable, note revisions for the task for subsequent academic years	Use learning progression to identify patterns in student thinking and plan next steps for instruction	Teachers create multiple piles of student work, sorting responses into patterns that align with the learning progression

happen across an academic year as teachers plan for, enact, and reflect upon a set of assessments for one instructional unit. Another possibility would be to focus the cycle on a science practice or crosscutting concept that is explicitly taught multiple times within the year; in that case, multiple cycles might be completed within one academic year. In addition, teachers may meet multiple times as they complete routines within the cycle.

The routines within each step of the cycle are supported by additional tools that help teachers to plan for, enact, and reflect on the assessments, such as checklists for quality assessment task features, planning tools for classroom enactment, and guides for interpreting student work (Henson, 2019). University- or district-based facilitators prepare for the meetings with detailed facilitation guides and agendas to support conversation within each phase of the cycle.

Three-Dimensional Learning Progressions for Formative Assessment Professional Learning: The Aspire Project

To understand how learning progressions could serve as orienting tools to support teacher professional learning around formative assessment across high school content areas, we conducted a multiple-year project in which we designed a learning progression for energy as the basis for high school teachers to plan for, enact, and reflect upon formative assessments. This project was embedded in the context of a long-term research-practice partnership (Penuel et al., 2011) with a large culturally, ethnically, linguistically, and socioeconomically diverse school district located near a large city in the Western US. In this district, physics is taught in 9th grade, chemistry in 10th, and biology in 11th, allowing for opportunities to build students' understanding across disciplines (Furtak et al., 2021).

Early phases of the Aspire project involved piloting different learning progressions within content areas and examining how they could support teachers' formative assessment design and enactment within different science courses (see Furtak & Tayne, 2018). However, this approach revealed that different learning progressions had different affordances for formative assessment design and the interpretation of student ideas; for example, some spanned multiple grade bands (e.g., Mohan, Chen & Anderson, 2009), and others were more tailored to recognizing individual student ideas.

Three-dimensional Learning Progressions

Based on this pilot work, we shifted to take the approach of following a crosscutting concept across high school science courses. This enabled us to create a common learning progression that would serve as the foundation for multiple grade levels, creating continuity for teachers and students, while also offering opportunities to examine differences across disciplines. We identified the crosscutting concept of energy as a focus that would be taught in multiple units of instruction in high school physics, chemistry, and biology. At the same time, as our partner district was moving toward NGSS adoption, we wanted to also create space for teachers and students to engage in science practices as they applied the crosscutting concept of energy to novel scenarios.

Our approach was consistent with the *Framework* and began by building on published learning progressions that had been empirically documented, including progressions for energy in Physics (Neumann et al., 2013), the science practice of modeling (Pierson et al., 2017; Schwarz et al., 2009), and studies of the flow of energy as a crosscutting concept (e.g., Park & Liu, 2016). The final learning progression traces how students model energy across systems (Buell et al., 2019; see Figure 22.1), both building on and extending beyond energy as articulated in the *Framework*.

Our intention was for this learning progression to be further articulated to bring the crosscutting concept of energy and the practice of modeling into specific disciplinary core ideas, as shown on the right side of Figure 22.1. That is, each of the five levels of the learning progression at left combines ideas about modeling crosscutting concept of energy; the science practice of modeling and crosscutting concept of energy flows was then migrated into specific disciplinary core ideas – such as cellular respiration, as shown in Table 22.2 – to create sets of performance-expectation-specific "look-fors" that would support learning and assessment in specific curricular units (Briggs & Furtak, 2019; Buell et al., 2019; Furtak et al., 2021).

Level	A Learning Progression for Modeling Energy Flows
5	• Students are able to ***generalize their model*** to unknown or multiple phenomena, and can ***explain limitations of applying the model*** to a new phenomenon.
	• Students ***develop a model*** that illustrates a mechanism that can explain or predict the phenomenon, AND ***use*** the model to make predictions about how changing one part of the model would influence energy flows elsewhere in the system. • Students can explain how the total energy of the system constrains the magnitude of change possible. • Students can ***describe limitations of the model*** in explaining or predicting the phenomenon
3	• Students ***use*** or ***develop*** a model that relates changes in the phenomenon directly to changes in energy through transfers/transformations by identifying specific indicators. • Students begin to show evidence that their model is accounting for conservation and dissipation. • Model includes energy flows into, within, and out of the system.
2	• Students ***use*** or ***develop*** a model to illustrate a relationship or pattern between the increase in one form of energy and the decrease in another form, or transferred from one location or object to another. • Students identify the most relevant components and relationships in the model *and* distinguish between the system and surroundings • Model focuses on energy flows within the system only.
1	• Students ***use*** or ***develop*** a model that shows, through drawings or labels, the components involved in a phenomenon, some (but not necessarily all relevant) energy forms, transfers, or transformations.

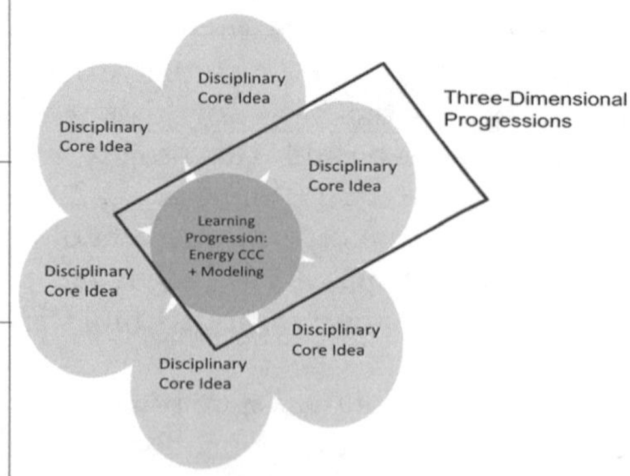

Figure 22.1 Learning progression for modeling energy.

Study Design

We partnered with six content-specific PLCs at three different high schools in our partner school district, working with two PLCs in each content area. Then, each content area worked with two PE-specific learning progressions that combined the crosscutting concept of energy with modeling in a specific disciplinary core idea, as shown in Figure 22.2.

While our collaboration with teachers spanned multiple years of the partnership, we focused our data collection on teachers' use of the learning progression for energy in one academic school year (2018–2019). We met with teachers bi-monthly on-site at each of their schools during professional learning time embedded in the school day, usually for about an hour. This meant that at each school, we met with teachers about 12–15 times. Each meeting was facilitated by a member of the university-based research team according to detailed facilitation guides that described step-by-step activities to support teachers in using the learning progression in their formative assessment design. Appendix A provides a timeline that summarizes the meetings at one of the schools, illustrating the activities teachers completed at each meeting.

To understand how the learning progression supported teacher learning about formative assessment, we collected multiple sources of data during this time, including teacher interviews, audio recordings of PLC meetings, fieldnotes, teacher-designed formative assessment tasks, student work artifacts, and video recordings of teachers' classroom enactment of those tasks.

3D Learning Progressions to Support Teacher Professional Learning

In the following sections, we draw upon prior analyses of these multiple sources of data to illustrate the different ways that the learning progression was used to guide teachers' use of formative assessment practices. Specifically, we highlight four ways that we used learning progressions to support teachers' design, enactment, and reflection upon formative assessment

Table 22.2 A learning progression for modeling energy flows (Buell et al., 2019)

Level	*A learning progression for modeling energy flows*	*What to look for*
5	• Students are able to ***generalize their model*** to unknown or multiple phenomena and can ***explain limitations of applying the model*** to a new phenomenon.	
4	• Students ***develop a model*** that illustrates a mechanism that can explain or predict the phenomenon AND ***use*** the model to make predictions about how changing one part of the model would influence energy flows elsewhere in the system. • Students can explain how the total energy of the system constrains the magnitude of change possible. • Students can ***describe the limitations of the model*** in explaining or predicting the phenomenon	• Explanation of the mechanism of cellular respiration through the breaking of glucose and oxygen and then reforming of bonds of water and carbon dioxide • Energy constraints/boundaries - limited by the amount of oxygen and glucose (lots of glucose but little oxygen= less energy flow through cellular respiration) • phenomenon (energy transfer through cellular respiration), • Limitations: … • make predictions about how changing one part of the model would influence energy flows elsewhere in the system (e.g., oxygen deprivation). • Students can explain how the total energy available to the system constrains the magnitude of change possible (e.g., oxygen or glucose availability constrains energy available) • Lactic acid buildup??
3	• Students ***use*** or ***develop*** a model that relates changes in the phenomenon directly to changes in energy through transfers/transformations by identifying specific indicators. • Students begin to show evidence that their model accounts for conservation and dissipation. • The model includes energy flows into, within, and out of the system.	• Indicators of transfer/transformation - Relationship between the number of inputs and the rate of cellular respiration; the number of outputs; energy to move/do something (mechanical/kinetic?) Muscles are not running out of ATP; muscles can continue to contract. • Indicators of conservation and dissipation are shown throughout different parts of the model (with arrow or explanation): e.g., heat, mitochondria, cells to other cells, if the system is a body, its body to the atmosphere. • Includes specific inputs and outputs of cellular respiration (oxygen, carbon dioxide, glucose, water, energy) • Food molecules and oxygen transfer energy to cells to sustain life's processes including the maintenance of body temperature despite ongoing energy transfer to the surrounding environment. • Students ***use*** or ***develop*** a model that relates changes in the phenomenon directly to changes in matter cycling and energy flow through transfers/transformations by identifying specific indicators (the person can run, less oxygen and glucose molecules are present, more CO_2 and H_2O molecules present).

(*Continued*)

Table 22.2 (Continued)

Level	*A learning progression for modeling energy flows*	*What to look for*
		• (e.g., chemical energy of inputs to mechanical/kinetic energy of outputs … eat something and breathe = run).
2	• Students ***use*** or ***develop*** a model to illustrate a relationship or pattern between the increase in one form of energy and the decrease in another form or transferred from one location or object to another. • Students identify the most relevant components and relationships in the model *and* distinguish between the system and surroundings • The model focuses on energy flows within the system only.	• Evidence of transfer/transformation (one location to another). Energy from bar to body. • Quantifying in some way: i.e., pie chart, bar chart, written explanation • Students indicate the system and surroundings • Arrows drawn connect the most relevant components of the system • System and surroundings may include the inputs, outputs, and …
1	• Students ***use*** or ***develop*** a model that shows, through drawings or labels, the components involved in a phenomenon, some (but not necessarily all relevant) energy forms, transfers, or transformations.	• General labels or drawings • Components (food molecules, oxygen) may be present, but the focus is not necessarily on the mechanisms/processes by which energy flows in cellular respiration • Some arrows

Indicator *- how changes in energy are manifested. These are the observable differences in a phenomenon, such as when an object speeds up or slows down or temperature increases or decreases, that let students know there is a transfer or transformation of energy.*
Mechanism *- the entities and activities that produce the changes in energy flow. For example, changes in kinetic energy can best be explained through particle motion. This will usually require changes in scale in a model, such as particle motion, cellular activity, or changes in fields.*
Phenomenon *- an event or state that we want to explain, in this case, through changes in energy flows.*
System/Surroundings *- the system includes the part of the universe under investigation and the surroundings include everything outside of the investigation.*

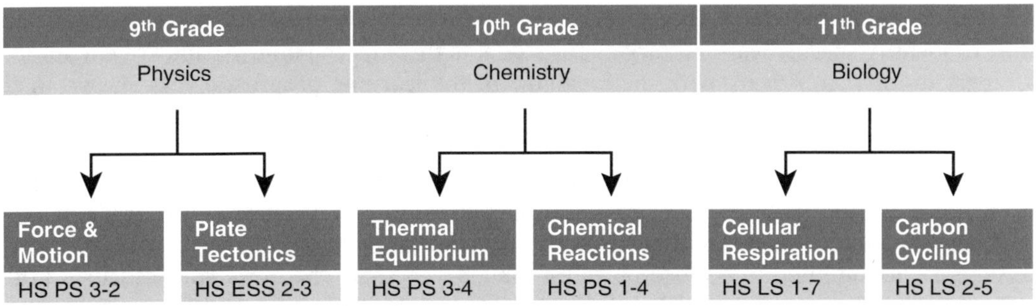

Figure 22.2 Organization of focal units on energy across high school courses.

in the professional learning experiences: supporting teacher learning about the learning progression; setting goals for formative assessment; designing tasks; and interpreting student thinking to identify subsequent supports for learning.

Learning about the Learning Progression

We began the academic year by meeting with teachers to support their own sensemaking of the newly designed learning progression. While several of the teachers had previously worked with different formats of learning progressions, these were generally one-dimensional and focused on disciplinary core ideas. Thus, initial meetings involved teachers taking the time to read and understand the core learning progression for modeling energy, and then working with university-based facilitators to interpret what the learning progression would look like in the context of a specific disciplinary core idea. In this way, the teachers both informed the development of the learning progression and developed "look fors" that they could identify in student work that would indicate when student thinking aligned with one of the levels.

Then, teachers used the learning progression to explore specific examples of student work and to practice interpreting that work according to the learning progression levels. To encourage teachers to interact directly with the learning progression, our facilitation asked teachers to "highlight evidence you can use to help locate these student ideas on the learning progression, knowing that each student work sample may have characteristics of multiple levels."

When asked how the learning progression aligned with their instructional unit, teachers reported different levels of influence of the learning progression on their planning. While some schools' curriculum materials aligned well with the curriculum, teachers noted there were still multiple structural influences on what they taught, including what they had done the previous year. When they described how the learning progression influenced their planning of units, results were similarly mixed. Several teachers noted that it provided a "baseline of where I think students need to be," and others who found their curriculum was aligned with the progression said it had been very influential for them.

Setting Goals for Formative Assessment with the Learning Progression

Next, we developed a planning tool to support teachers in using the learning progression to help them reflect upon the sequencing of their current curriculum. Our intention here was to use what we called a "staircase model" to have teachers look at how their curriculum sequentially built students' ability to model energy as shown in the learning progression, and then to identify specific learning targets for students through the unit. Again, we encouraged teachers to interact directly with the language of the learning progression, as shown in the facilitation prompts below:

1 Within this unit, where are the places students are learning about energy forms (e.g., kinetic energy, gravitational potential energy), transfers/transformation/conservation, or dissipation?
2 Where do students have an opportunity to practice modeling?
3 Where can you build in opportunities for students to share the ideas you identified in the Look-Fors? For example, in Level 1, students are learning to label energy forms and represent transfers and transformations. In what activity do you plan for this to take place? When are students going to learn how to identify systems and surroundings?
4 Let's do this for each level of the learning progression and make note of them within the staircase (Facilitation Guide 3).

Each PLC then used the planning guide to identify one or more places within the unit to embed formative assessment tasks to create space for students to share their developing understandings during that focal unit.

Designing Tasks and Drawing Out Student Thinking with the Learning Progression

An important outcome of the 2018–2019 academic year was multiple teacher-designed and curriculum-embedded formative assessment tasks intended to create space for students to model energy transfers and transformations in multiple phenomena within the same academic year. The affordances of a learning progression for teacher learning are most clearly seen at the high school which we will refer to as Mayfield. At Mayfield High School, biology, physics, and chemistry teachers all used the modeling energy learning progression to design formative assessment tasks. In a full-day meeting, teachers gathered and used the learning progression as an anchor to discuss how energy was taught in each of their disciplines and gained a greater understanding of how to support energy instruction across the disciplines. Then, during the school year, teachers worked in discipline-specific PLCs with the learning progression to map out the trajectory of their teaching units and identify places for formative assessments. The teachers used the learning progression to identify where they should introduce ideas of energy forms, transfer and transformation, and conservation and dissipation to students and to help inform the design of formative assessments.

Analysis of the tasks (Buell, 2020) shows how teachers were able to design formative assessments on modeling energy. Teachers in all three disciplinary areas co-designed at least one task asking students to model the transfer or transformation of energy to explain a phenomenon; samples of these tasks are included in Figure 22.3. For example, physics teachers asked students to model how a skateboarder goes up and down a half-pipe; chemistry teachers asked students to model how a hot liquid in a cup cools down; and biology teachers asked students to model how brine shrimp and algae survive in an enclosed ecosphere. Because teachers designed the tasks themselves, they varied greatly in structure and format, however, the learning progression did enable similarities across the different tasks.

Consistent with the focus of the learning progression, all the tasks focused on students modeling the flow of energy in a system (e.g., a skateboarder and ramp, a coffee cup, or the

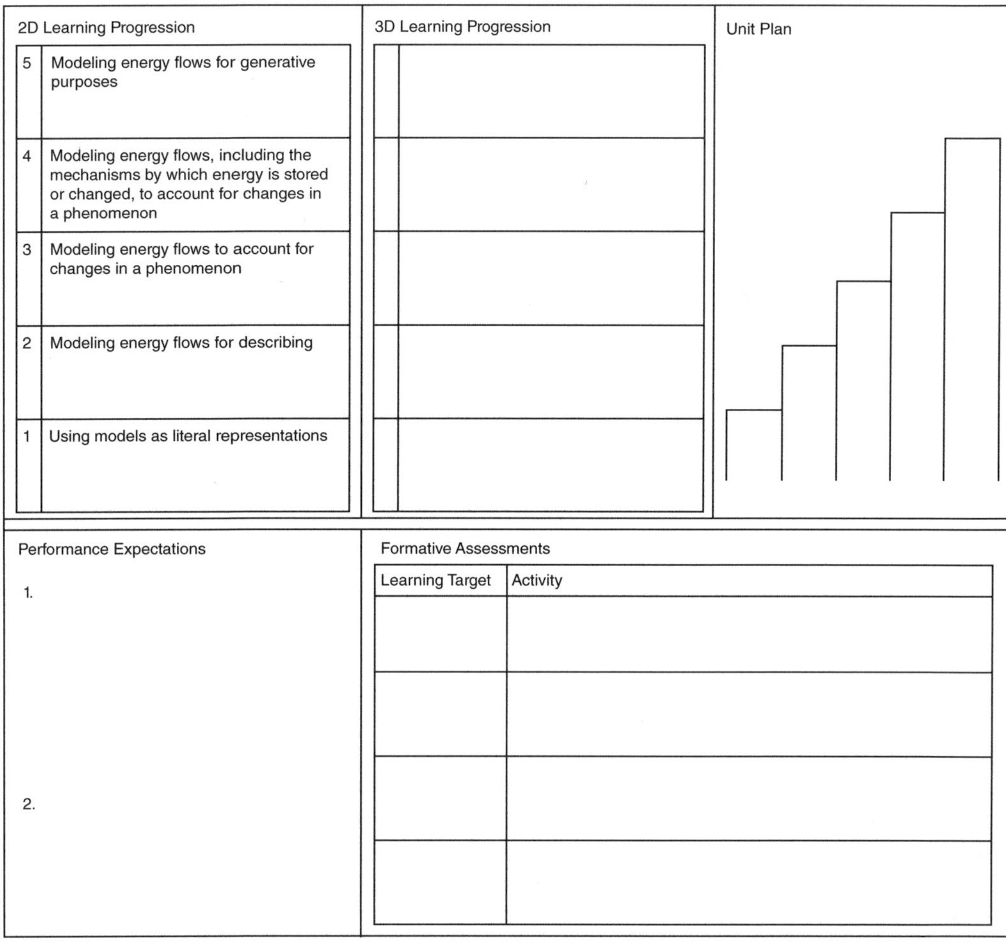

Figure 22.3 Planning guide for formative assessment.

ecosphere) but went about it with very different levels of scaffolding. In physics, students were provided with the components of the model and were asked to focus on only forms and transformation of energy. This task was placed toward the end of the unit and provided teachers with an assessment of past learning. In contrast, in chemistry, students were asked to do the representational work but provided multiple checklists of what to include in the model, including what components to include, using specific kinds of arrows to denote different energy transfers, and using a dotted line to identify the system and surroundings. The chemistry task was used to help students connect ideas about energy to their designs of cups to help them make predictions about how different cup designs would perform. The biology task was the most open-ended, only providing students with a list of components and while students were asked to model energy flows, students were not given any additional instruction on what kinds of flows or forms to use, or any representational guidance. This task was given at the beginning of instruction to use as a pre-assessment and the teachers intended to build on these initial models during the unit.

While an analysis of students' understanding of the disciplinary core ideas showed students scored highest on the more heavily scaffolded tasks (Buell, 2020), we recognize that the intentions of the tasks were different. The biology task was intended to provide information to teachers at the start of a unit, where students created a model of how possibly the ecosphere might work while the skateboarding task was intended for students to demonstrate knowledge learned over the unit.

While there were differences between the tasks, the learning progression enabled teachers across the disciplines to develop a shared practice of modeling energy across the grade levels. While the specific components and representational strategies varied, teachers across grade levels were asking students to answer questions about phenomena by modeling energy flows and using shared language in their tasks and teaching.

Interpreting Student Thinking with the Learning Progression

We found evidence of how teachers used the learning progression to interpret students' level of understanding based on the learning progressions, an activity that was also a major element of the FADC. In an in-depth case analysis of one physics PLC over the course of multiple years, including the focal year with the energy learning progression, Henson (2019) found that the routines of the formative assessment design cycle provided space for teachers to make sense of student thinking and to design better formative assessment tasks over time. Henson examined the ways that different tools associated with the learning progression – such as planning guides and different score report formats to support teacher interpretation of student responses to post-tests – both enabled and constrained teachers' collaborative activity. Of the tools examined in the Aspire study, Henson (2019) found that the learning progression, as a resource-based representation of student thinking, in addition to specific examples of student responses from each school, supported teachers' conversations around student thinking. In contrast, when teachers were working with externally developed learning progressions, they found the descriptions and examples less proximal to their practice, and ultimately less useful.

When reviewing student work with a facilitator, teachers were asked to identify what level of the learning progression each student response aligned with. Overall, engaging in this process helped teachers think more deeply about the kinds of responses they wanted the task to elicit and what those responses indicated about student understanding.

In one example, a student demonstrated points where the kinetic energy of a skateboarder would be at a minimum and at a maximum without explaining the relationship between the forms of energy in the system. (left panel of Figure 22.4). Through some back and forth, teachers looked at the student work for evidence of student thinking and then the learning

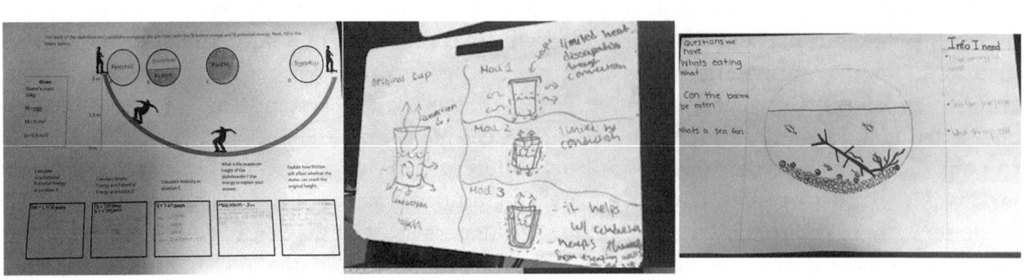

Figure 22.4 Sample assessment tasks for physics (Skateboarder), chemistry (Kinetic Cup), and biology (Ecosphere).

progression to identify the look-fors at each level. Then, they identified level 2, *using a model to illustrate the increase in one form of energy and a decrease in another*, because students were using their models to describe the points on the half pipe where kinetic and potential energy were at their minimum and maximum values.

Next, the facilitator asked the teachers to consider how the student could have been supported in developing toward a level 3 understanding where they would begin to account for energy conservation and dissipation. The teachers indicated that they would have shown the student a skateboarder on a halfpipe that didn't make it all the way back up to the top and they would have wanted to ask them why that happened. This would have prompted students to consider energy transfer between the skateboard and the halfpipe that was no longer available as kinetic energy resulting in the skateboarder not making it all the way to the top of the ramp.

In this example, the learning progression supported teachers in thinking deeply about what ideas their students were sharing and where those ideas were in the progression. More importantly, considering the next level in the learning progression helped teachers think about the kinds of instructional moves they could make to support students in developing toward a more sophisticated understanding of energy flow.

Discussion

The evidence presented in the preceding sections illustrates the ways that a central learning progression for modeling energy flows in systems provided foundations for teachers' design, use, and reflection upon formative assessments in school-based professional learning. By building on arguments that classroom assessment can be viewed as an activity system, the examples above establish how interconnected elements of classroom activity – including students' and teachers' prior knowledge and experiences, as well as the tools and tasks that organize roles and routines – can be oriented around a common representation of a learning progression. When viewed considering ongoing *Framework*-era reforms, we have found that learning progressions can be useful for the design and enactment of formative assessment tasks. From the perspective of supporting teacher learning, we have presented evidence of how teachers can learn to design tasks that create space for students to model energy across different disciplines, as well as how they can learn to use learning progressions to support their interpretations of students' responses to those assessments.

We identify several key take-aways from the study not only for teachers, but for providers of professional learning experiences. For teachers, we have illustrated how conversations around a shared representation for learning can create opportunities for better alignment across grade levels, as is the case in creating space to model energy flows in systems across high school physics, chemistry, and biology classes. We have also shown how planning for and reflecting on these assessments with the support of a learning progression can help teachers more closely interpret student thinking and focus on the resources that students bring to school.

For providers of professional learning experiences, we have shown how the dedicated time provided to teachers within the school day created space for conversations about energy. In the US, where shifts to three-dimensional standards are still taking place, these experiences can create important opportunities for making sense of the structure of the standards, designing assessments aligned to them, and learning to interpret student work. In addition, we have illustrated the benefits of using crosscutting concepts as organizing themes that span the discipline.

Our findings contribute to an emerging literature on the value learning progressions can have as direct supports for teachers to inform their instruction, both individually and through ongoing professional learning (e.g., Alonzo, this volume; Alonzo et al., 2022; Alonzo & Elby, 2019; Gunckel, Covitt & Salinas, 2018).

References

Alonzo, A. C., & Elby, A. (2019). Beyond empirical adequacy: Learning progressions as models and their value for teachers. *Cognition and Instruction*, *37*(1), 1–37.

Alonzo, A. C., Wooten, M. M., & Christensen, J. (2022). Learning progressions as a simplified model: Examining teachers' reported uses to inform classroom assessment practices. *Science Education*, *106*(4), 852–889.

Bennett, R. E. (2011). Formative assessment: A critical review. *Assessment in Education: Principles, Policy & Practice*, *18*(1), 5–25.

Black, P., & Wiliam, D. (1998). Assessment and classroom learning. *Assessment in Education*, *5*(1), 7–74.

Borko, H., Jacobs, J., Eiteljorg, E., & Pittman, M. (2008). Video as a tool for fostering productive discussions in mathematics professional development. *Teaching and Teacher Education*, *24*(2), 417–436.

Buell, J. Y. (2020). *Designing for relational science practices*. [Doctoral dissertation]. University of Colorado Boulder.

Buell, J. Y., Briggs, D. C., Burkhardt, A., Chattergoon, R., Fine, C., Furtak, E. M., Henson, K., Mahr, B., & Tayne, K. (2019). *A learning progression for modeling energy flows in systems*. Boulder, CO: Center for Assessment, Design, Research and Evaluation (CADRE).

Briggs, D. C., & Furtak, E. M. (2019). Learning progressions and embedded assessment. In S. M. Brookhart & J. H. McMillan (Eds.), *Classroom Assessment and Educational Measurement*. Routledge.

Corcoran, T., Mosher, F. A., & Rogat, A. (2009). *Learning progressions in science: An evidence-based approach to reform*. Consortium for Policy Research in Education.

Council of Chief State School Officers. (2018). *Revising the definition of formative assessment*. https://ccsso.org/sites/default/files/2018-06/Revising%20the%20Definition%20of%20Formative%20Assessment.pdf

Cowie, B., Jones, A., & Otrel-Cass, K. (2011). Re-engaging students in science: Issues of assessment, funds of knowledge and sites for learning. *International Journal of Science and Mathematics Education*, *9*(2), 347–366.

Dini, V., Sevian, H., Caushi, K., & Orduña Picón, R. (2020). Characterizing the formative assessment enactment of experienced science teachers. *Science Education*, *104*(2), 290–325.

Duschl, R., Maeng, S., & Sezen, A. (2011). Learning progressions and teaching sequences: A review and analysis. *Studies in Science Education*, *47*(2), 123–182.

Engeström, Y. (2001). Expansive learning at work: Toward an activity theoretical reconceptualization. *Journal of Education and Work*, *14*(1), 133–156.

Fine, C., & Furtak, E. M. (2020). A framework for science classroom assessment task design for emergent bilingual learners. *Science Education*, *104*(3), 393–420.

Ford, M. J., & Forman, E. A. (2006). Redefining disciplinary learning in classroom contexts. *Review of Research in Education*, *30*(1), 1–32.

Furtak, E. M. (2012). Linking a learning progression for natural selection to teachers' enactment of formative assessment. *Journal of Research in Science Teaching*. 49(9), 1181–1210.

Furtak, E. M., Heredia, S. C., & Morrison, D. M. (2019). Formative assessment in science education: Mapping a shifting terrain. In H. Andrade, R. Bennett, & G. Cizek (Eds.), *Handbook of Formative Assessment in the Disciplines*. Taylor & Francis.

Furtak, E. M., & Heredia, S. C. (2014). Exploring the influence of learning progressions in two teacher communities. *Journal of Research in Science Teaching*, *51*(8).

Furtak, E. M., Morrison, D., & Kroog, H. (2014). Investigating the link between learning progressions and classroom assessment. *Science Education*, *98*(4).

Furtak, E. M., & Penuel, W. R. (2019). Coming to terms: Addressing the persistence of "hands-on" and other reform terminology in the era of science as practice. *Science Education*, *103*(1), 167–186.

Furtak, E. M., Penuel, W. R., Badrinarayan, A., Patrick-Stuart, R. P., & Duwe, S. (2021). Crosscutting concepts: A sensemaking perspective. In J. Nordine, & O. Lee (Eds.), *Crosscutting Concepts: Strengthening Science Teaching*. Arlington, VA: NSTA Press.

Furtak, E. M., & Tayne, K. (2018). *Mobilizing learning progressions for teacher use: Examining the utility of outside learning progressions in task co-design*. In J. Kay, & R. Luckin (Eds.). *Rethinking Learning in the Digital Age: Making the Learning Sciences Count, 13th International Conference of the Learning Sciences (ICLS) 2018*, Volume 2, pp. 520–527. London, UK: International Society of the Learning Sciences.

Gunckel, K. L., Covitt, B. A., & Salinas, I. (2018). Learning progressions as tools for supporting teacher content knowledge and pedagogical content knowledge about water in environmental systems. *Journal of Research in Science Teaching*, *55*(9), 1339–1362.

Henson, K. (2019). *Designing tools to support teacher activity focused on student thinking to inform instruction*. [Doctoral dissertation]. University of Colorado Boulder.

Heredia, S. C., Furtak, E. M., Morrison, D. L., & Gröschner, A. (2021). The disciplinary nature of science teachers' talk in the process of formative assessment design. In E. de Vries, Y. Hod, & J. Ahn (Eds.). *Proceedings of the 15th International Conference of the Learning Sciences - ICLS 2021*, pp. 729–732. International Society of the Learning Sciences.

Horn, I. S., & Little, J. W. (2010). Attending to problems of practice: Routines and resources for professional learning in teachers' workplace interactions. *American Educational Research Journal*, *47*(1), 181–217.

Kang, H., & Furtak, E. M. (2021). Learning theory, classroom assessment, and equity. *Educational Measurement: Issues and Practice*, *40*(3), 73–82.

Kang, H., Thompson, J., & Windschitl, M. (2014). Creating opportunities for students to show what they know: The role of scaffolding in assessment tasks. *Science Education*, *98*(4), 674–704.

Kang, H., Windschitl, M., Stroupe, D., & Thompson, J. (2016). Designing, launching, and implementing high quality learning opportunities for students that advance scientific thinking. *Journal of Research in Science Teaching*, *53*(9), 1316–1340.

Latour, B., & Woolgar, S. (1979). *Laboratory life: The construction of scientific facts*. Princeton University Press.

Mohan, L., Chen, J., & Anderson, C. W. (2009). Developing a multi-year learning progression for carbon cycling in socio-ecological systems. *Journal of Research in Science Teaching*, *46*(6), 675–698.

Moss, P. A. (2008). Sociocultural implications for assessment I: Classroom assessment. In P. A. Moss, D. Pullin, J. P. Gee, G. Haertel, & L. J. Young (Eds.), *Assessment, equity, and opportunity to learn* (pp. 222–258). Cambridge University Press.

National Academies of Sciences, Engineering, and Medicine. (2018). *How people learn II: Learners, contexts, and cultures*. The National Academies Press.

National Academies of Sciences, Engineering, and Medicine. (2019). *Science and engineering for grades 6–12: Investigation & design at the center*. The National Academies Press.

National Research Council. (2012). *A framework for K-12 science education*. National Academies Press.

National Research Council. (2014). *Developing assessments for the next generation science standards*. National Academies Press.

Neumann, K., Viering, T., Boone, W. J., & Fischer, H. E. (2013). Towards a learning progression of energy. *Journal of Research in Science Teaching*, *50*(2), 162–188.

NGSS Lead States. (2013). *Next Generation Science Standards: For States, by States*. Achieve, Inc.

OECD. (2009). *PISA 2009 Assessment Framework—Key Competencies in Mathematics, Reading and Science*.

Park, M., & Liu, X. (2016). Assessing understanding of the energy concept in different science disciplines. *Science Education*, *100*(3), 483–516.

Penuel, W. R., Fishman, B. J., Haugan Cheng, B., & Sabelli, N. (2011). Organizing research and development at the intersection of learning, implementation, and design. *Educational Researcher*, *40*(7), 331–337.

Penuel, W. R., & Shepard, L. A. (2016). Assessment and teaching. In D. H. Gitomer, & C. Bell (Eds.), *Handbook of research on teaching* (Fifth, pp. 787–850). American Educational Research Association.

Pickering, A. (1995). *The Mangle of Practice: Time, agency, and science*. *Chicago*: University of Chicago Press.

Pierson, A. E., Clark, D. B., & Sherard, M. K. (2017). Learning progressions in context: Tensions and insights from a semester-long middle school modeling curriculum. *Science Education*, *101*(6), 1061–1088.

Ropohl, M., Nielsen, J. A., Olley, C., Rönnebeck, S., & Stables, K. (2018). *The Concept of Competence and Its Relevance for Science, Technology and Mathematics Education*. 3–25.

Schwab, J. J. (1962). The concept of the structure of a discipline. *The Educational Record*, *43*(3), 197–205.

Schwarz, C. V., Reiser, B. J., Davis, E. A., Kenyon, L., Achér, A., Fortus, D., Schwarz, Y., Hug, B., & Krajcik, J. (2009). Developing a learning progression for scientific modeling: Making scientific modeling accessible and meaningful for learners. *Journal of Research in Science Teaching*, *46*(6), 632–654.

Shepard, L. A., Penuel, W. R., & Pellegrino, J. W. (2018). Using learning and motivation theories to coherently link formative assessment, grading practices, and large-scale assessment. *Educational Measurement: Issues and Practice*, *37*(1), 21–34.

Thompson, J. J., Hagenah, S., McDonald, S., & Barchenger, C. (2019). Toward a practice-based theory for how professional learning communities engage in the improvement of tools and practices for scientific modeling. *Science Education*, *103*(6), 1423–1455.

Wertsch, J. (1998). *Mind as action*. Oxford University Press.

Appendix A

A timeline that summarizes the meetings at one of the schools, illustrating the activities teachers completed at each meeting.

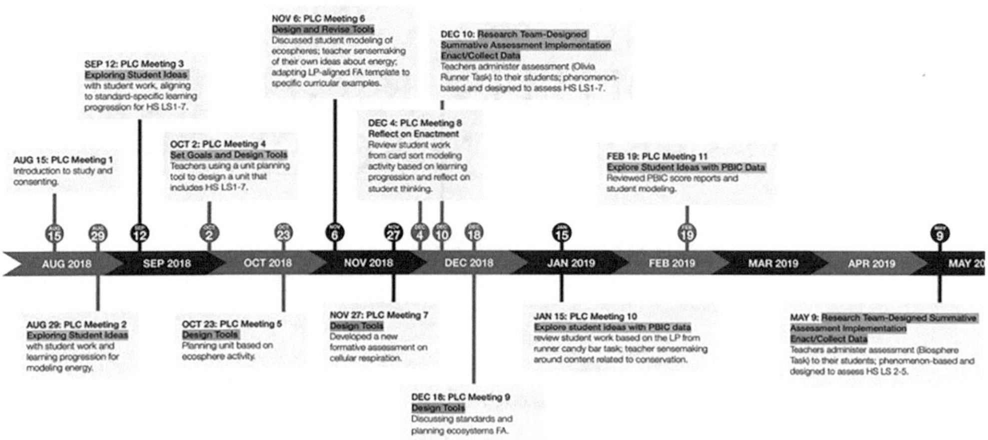

23

LEARNING PROGRESSIONS AND YOUTHS' RIGHTFUL PRESENCE IN SCIENCE

Angela Calabrese Barton and Edna Tan

Introduction

In this chapter, we explore the possibilities and limitations of the learning progressions research and practice agenda for youths' equitable presents and futures in science education. To build our argument, the chapter takes three turns. First, we explore how learning progressions research has opened new possibilities for working towards equity in science education. Second, we discuss the challenges this work will need to address in the shift from equitable to justice-oriented learning in science. Drawing from critical justice and socio-cultural perspectives of learning, we use a lens of "rightful presence" to raise questions about who and what learning progressions are for, and what this means for all youths' opportunities to learn and develop in science in meaningful and just ways. We conclude the chapter by offering fruitful directions for research and development in learning progressions research, which potentially center rightful presence and work towards just possibilities for all learners. In particular, we explore how learning progressions, in both form and function, should account for the powerful and complex ways that youth navigate both vertical (e.g., disciplinary defined movement from novice to expert) and horizontal dimensions of learning through social activity (e.g., movement across communities, with the recontextualization/hybridization of what counts as resources, ideas, and processes related to science). We offer concrete examples from the literature, and our own work, in support of these directions.

An Eye Towards Equity and the Possibilities of Learning Progressions

Learning progressions research has significantly shaped the field of science education, gaining prominence since the publication of *Taking Science to School* (NRC, 2007). Considered a theory of disciplinary learning (Kang & Furtak, 2021), and a framework and toolset for aligning and designing curriculum, instruction, and assessment (Liu and Jackson, 2019), learning progressions offer ways to make sense of learning *pathways* – including the "waypoints along the path" (Duschl, 2019, p. 3) in which students' knowledge and practices are likely to grow as students navigate towards science literacy across topics and over time (NRC, 2007).

The stance offered by the learning progressions community marks an important shift in views of learning as connected to young people's lives and development. Learning progressions focus

 DOI: 10.4324/9781003170785-26

on what children know, experience, think about, and do with a certain topic and/or practice, and how this shifts and develops as they gain new experiences and opportunities to sense-make with those experiences over time. As Shepard (2018) suggests, when built from the bottom up, learning progressions illuminate student thinking, while also providing insights into how such thinking might drive instructional practice. This approach shifts from a focus on content mastery and teacher/discipline-centered instruction to a focus on students and what they know and care about, when, and how (Alonzo, 2018). The development of learning progressions has been tied to work on learner-centered curricula and environments (Liu and Jackson, 2019).

Consider, for example, Jin et al.'s (2017) study on environmental literacy, where researchers designed a teaching unit, focused on plants, based on a learning progression they previously developed on matter and energy in socio-ecological systems. Using video-based analysis, the researchers illustrated how both teachers' teaching practices and students' learning practices mediated the impact of a learning progressions-based curriculum on learning outcomes. First, they showed that when teachers attend to students' use of epistemological frames for learning, in this case, the "mechanism frame for learning about scientific knowledge" and the "inquiry frame" for learning about scientific inquiry, students made meaningful learning gains. Second, there are specific "scaffolding" discourse moves that teachers can make to help students apply these epistemological frames. In these ways, this study points not only to the pedagogical architecture that makes learner-centered approaches to learning progressions possible, but also to the importance of expanding learning progressions to include learning at the metacognitive and epistemological levels to guide learning about inquiry.

From a sociocultural point of view, the work on learning progressions has shifted attention from learning as happening "inside the head of individual learners" to learning as "manifested through embodied experiences in activity situated in contexts" (Kang & Furtak, 2021, p. 2). This practice-based conception of learning "is conceived of not in narrow terms of inquiry skills, but rather in wider sociocultural terms that consider the particular cultural context in which the practice is enacted" (Tytler, 2018, p. 599). Furthermore, the idea of a learning progression has expanded to include not just what people know (about content or practices) but also how people engage in practice as active members of learning communities (Cordova & Balcerzak, 2016).

Thus, the move to learning progressions has been viewed as a potential lever for more equitable policy and practice in science education. In centering learners and practice-based frames of learning that happen within context, learning progressions shift attention away from reductive views of learning, teaching, and achievement towards developmental trajectories of understanding and reasoning (Nasir, et al., 2013).

Much-needed focus on a deep understanding of important concepts in a domain, as grounded in learning progressions, can counter prevailing notions of learning as accumulating bits of knowledge and testing each bit. With the increasing emphasis on high-stakes testing, education has been moving away from being guided by deep understanding as a goal, to prioritizing proficiency as measured by large-scale standardized assessments. This is a particularly problematic dynamic in under-resourced schools, where students are more likely to score "below basic" and are thus more vulnerable to being subjected to sanctions under federal education policy, exacerbating inequities.

Furthermore, learning progressions, by aiding the alignment of curriculum, instruction, and assessment potentially support more meaningful opportunities to learn in ways that show what one is learning. From a sociocultural point of view, tight cycles of learning, instruction, and assessment, with attention to enactment in practice, could improve access to high-quality teaching for all students; and again, a great improvement over current educational practice (Kang & Furtak, 2021).

The Limits of Equity and Learning Progressions

While the promises of learning progressions are exciting, many critical questions remain, especially from sociocultural and critical justice points of view. In this section, we raise questions around *whose science, whose learning,* and *whose progressions* are for consideration, and discuss how these questions inform the challenges and possibilities of learning progression research in working towards equity and justice.

Current approaches to learning progressions offer important guidance with respect to individually-oriented knowledge construction. However, they do not account for how learning and taking action with what one knows in science are social and cultural processes that shape how people learn and use what they learn in everyday living. This stance shifts attention from a focus on the cognitive/epistemic to also account for the sociocultural, ethical, and political ways people know and take action with and in science as integral to learning. Here we refer to the sociocultural aspects of learning as how people make connections to and negotiate learning in relation to their personal and cultural lives (Philip & Azevedo, 2017). By ethical we refer to how people learn and use what they learn to affect people's lives, social relations, and possibilities (Vakil, 2018). By political we refer to the contentious processes of negotiating whose ideas, experiences, and practices matter, why and how.

Asking questions such as "what does it mean to broaden what counts as valuable or necessary knowing/knowledge", especially when particular people/perspectives have been systemically and historically silenced, can make visible emergent tensions that require surfacing and disruption in working towards students' rightful presence in science (Calabrese Barton & Tan, 2020b). From this viewpoint, learning progressions only partially account for how people learn because these progressions focus only on the vertical dimensions of learning. By the vertical dimension, we refer to learning that reflects the movement from novice to expert in disciplinary-sanctioned ways. Learning and action-taking also call into question who people are, and how people, as culturally situated human beings, move and adapt ideas in social context as they navigate *horizontally* in the world, across communities and contexts. Here we refer to horizontal learning as this cross-community movement that requires people to mix and re-mix different knowledge and ways of knowing to develop useful knowledge for being and acting in the world.

Consider, for example, the following quote from Jazmyn, a 15-year-old Black girl, who participated in our community-based STEM programs for years:

> The racist stereotype is that Black people are not listening to science. But that is not true. Maybe it's the other way around, like science is not listening to us. I just wish that people could see what I could do, like what I am doing at home. Making home-made hand sanitizer, making masks, caring for my elders. I don't want to act White. I don't want people to tell me I'm not white enough. I want you to know how I feel as a young, Black girl in America and in STEM. I want to feel like I can be me in STEM and have that celebrated.

Jazmyn is calling us to recognize her for who she is: A bright, caring Black girl who has creatively leveraged STEM knowledge and practice towards the well-being of her community during a multi-pandemic. At the same time, she is also critiquing a system that has made her invisible, as a Black girl, despite her embodied presence. Jazmyn wants to be legitimately recognized and welcomed in science as a whole person with valuable knowledge, practices, and experiences that matter.

Many youths do not feel like their *whole person* is invited into science. Their feelings, experiences, histories, hopes, and fears are often not integral to what happens in science classrooms. When people are denied their whole selves in learning, their lives and histories can be partially or completely erased from science (Herrenkohl & Bevan, 2017). This erasure can result from science learning experiences that center or reinforce dominant cultural norms in science. These norms limit possibilities for seeing youth for who they are and want to be in science. They can position youth as outsiders, diminish opportunities to learn and become, and limit collective growth and transformation. Youth of Color, youth from low-income communities, youth who are undocumented, and youth who speak languages that institutional power holders do not, can face dehumanization in science on a daily basis due to its colonizing and racist histories (McGee & Stovall, 2015). Zuniga and colleagues (2005) found that in a rural high school, Latino/students who demonstrated success in standardized tests were nonetheless tracked into lower-level science courses while White students were disproportionately tracked into higher-level science courses.

Equally troubling, students who are English language learners continue to lag behind non-English language learners in dominant/normed records of science achievement (Lee & Buxton, 2013), compounded by the practice of schools prioritizing literacy and numeracy over science for their ELL students (Lee & Luykx, 2005). Such studies show *how* science learning opportunities may oppress through how power-mediated norms of being/interacting take shape in local practice.

Indeed, Jazmyn, like many minoritized youth, does not often have *their* powerful cultural experiences and expertise legitimized as integral to meaningful learning and engagement in science. This kind of marginalization or erasure can be thought of as *denying youth a rightful presence in science* (Calabrese Barton & Tan, 2020b). Rightful presence refers to youth having legitimate membership in their science learning community because of who they are, and not because of who they should be. Rightful presence is a critical mode for *making present* those who have been made *missing* by intersectional racism and other forms of oppression manifest in science and schooling. Rightful Presence orients the field towards new forms of teaching and learning that restructure learning opportunities such that the lives of all youth are made visible and integral to science, especially those made historically invisible in STEM through systemic racism and heteropatriarchy.

We thus question how youth may be more fully welcomed in science, and through learning progressions, as *rightfully present*. We see rightfulness as claimed through presence, in the sense that power relations shift so that youths' whole lives become integral to learning. In this way, the outcomes of learning focus not only on individual gains but also on the kinds of social transformations that allow for presence to thrive. This justice-oriented stance, while acknowledging the importance of access and opportunity to learn the science of scientists, attends to the ethical and political dimensions of learning. This is because this stance challenges the underlying power relationalities that shape what learning, engagement, and participation in science look like.

Rightful presence is primarily concerned with identifying and transforming underlying injustices that undergird existing structures and norms of "doing science", reflected in the canonical culture of science, and by extension, school science. Even as learning progressions account for student positionings as active members of learning communities, what counts as "active" – in terms of the kinds of resources, modes of engagements, and reasons to engage –

are constrained by the tight boundaries of a school science, classroom-based learning community. In short, the horizontal dimensions of learning, including students' temporal arc (past, present, future engagements with science) and social-spatial relationalities relevant to science (spaces across students' lives where science-related activities are undertaken in social activity with particular stakeholders), are elided.

To explicitly center such horizontal dimensions of learning in learning progressions, we suggest three critical directions for consideration:

- *Whose science* – How does a focus on how science itself as a cultural practice shape the form and function of learning progressions?
- *Whose learning* – How might a focus on the relationship between knowledge development, participation norms, and taking action with and through science reveal the heterogeneity of learners, what they might care about, and whose learning (and the forms of such learning) is thus prioritized in a learning progression?
- *Whose progressions* – How might a focus on engaging in science in everyday learning and living that involves navigating multiple epistemologies towards more robust hybrid forms of knowing and action-taking inform a more expansive orientation towards learning progressions?

These questions center on who has the power to know in science and what forms of knowing and knowledge matter. We take the stance, as have others, that the field of science education, broadly speaking, has treated epistemologies as established, agreed upon, and/or "settled" (Bang, et al., 2012, p. 302). However, such a stance "privileges epistemologies that reproduce hierarchies such as race, class, gender, sexuality, and nationality and limit other ways of seeing or imagining possibilities for equity" (Philip & Azevedo, 2017, p. 527).

Whose Science?

How might youths' lives be centered as rightfully present in the science that learning progressions map? Here we raise questions about how science itself is a cultural practice, shaped by people, context, and place. We also need to acknowledge that canonical science culture, upon which learning progressions are based, is not neutral. Rather, canonical science culture elevates White, middle-class patriarchal culture as the norm, thus making White, middle-class patriarchal culture seem objective and benign. To foster the rightful presence of historically minoritized students in science, learning progressions need to expand the cultural dimensions of science explicitly as a way to make visible youths' lives.

First, we suggest that attention needs to be paid to how science is a cultural construct, and any learning progression should actively engage science as such. What it means to progress, in what ways, and on what topics, what entry and anchor points may be, and what waypoints exist along the way, are embedded with the values, discourses, and ideologies of the cultures and histories of the people who generated them. Treating science as normative privileges of those whose experiences and ways of knowing align with the cultural underpinnings while erasing, marginalizing, and oppressing others.

As noted in this book, learning progressions have been critiqued for not adequately attending to the socially and culturally embedded nature of science. Who students are – including their lived lives, communities, and histories in context – all shape how they come to interact with and thus come to know science in ways that carry meaning. For some youth, this approach is welcoming because their families' and communities' cultural practices are reflected in the discourses and practices of science. However, for minoritized youth, this is often not the case.

Consider Ladachart et al.'s (2020) study where they sought to develop a *place-based learning progression* for haze pollution in the northern region of Thailand. These authors explain how haze air pollution affects Northern Thailand in late winter and summer due primarily to agricultural waste burning and forest fires. With increasing economic pressures to increase crop production alongside the growing population, the traditional way of life led to haze accumulation, impacting human health and well-being. Responding to the sociocultural ideal that students' lived experiences profoundly shape learning, this study explored how students' "pieces of knowledge about a place could or should be integrated into a learning progression for a topic related to a phenomenon in their own place" (p. 995).

By exploring what students understand about haze pollution in their community and region, and how their own place-based experiences contributed to their understandings, the study shows how students' place-based experiences provide a framework for a learning progression. Students' place-based experiences serve not only as lower anchors, such as how most students understood that haze pollution results from burning, and that the burning produces pollutants, such as dust and smoke, but also that students' insights, questions, and concerns could be used to identify powerful middle anchors, such as general knowledge of when haze pollution occurs, or questions about whether it matters what is being burned, by whom and for what purposes. Using these insights and questions as places for building connections between seasonal burning and traditional practices illustrates how such intersections between knowledge systems can serve as middle-level anchors for deeper exploration.

However, as these authors note, what students learn is not just tied to the production and impact of haze pollution. It is also tied to how one might teach "future generations how to live with nature in sustainable manners" which is "a basic premise of Buddhism that ones have to be mindfully living with nature as well as other people respectfully, which had been instilled in Thai cultures and schools" (p. 1011). Further, this philosophical point of view is not one introduced after scientific knowledge is developed, but rather as a part of exploring the ideas of haze pollution itself. Thus as students engage in meaning-making around the effects of seasonal burning and haze pollution, they also bring onto-epistemological lenses of how such meaning-making affects the environment and people and communities' well-being in a context larger than oneself.

Thus, this place-based approach, grounded in a sociocultural view of learning, centered students' experiences in their local contexts, including forms of indigenous knowledge, and traditional ways of thinking– "often deemed 'noises' (Ford, 2015) when a learning progression in science is being developed" (Hammer & Sikorski, 2015, p. 995). In doing so, the authors remind the field of the importance of learning progressions research that resides in knowledge-in-practice, and their onto-epistemological origins in everyday living, rather than more strict forms of Western canonical knowledge privileged in classrooms. They also remind us of the need to recognize the porous boundaries between knowledge systems, and its impact on how the field may think about learning progressions. As these authors remind us, learning progressions are "models to be modified within the context of local circumstances, community resources, and students' learning histories" (p. 994).

These authors also call attention to how a place has many dimensions, often giving socio-scientific issues great complexity in how students may come to understand them. The implication here is important. Studies in learning progressions often advance the idea that students "should learn science in ways that are consistent with those of scientists" (p. 1013). However, as meaning-making is made in a social context, and in particular for complex socioscientific issues that people encounter in everyday living, scientific ideas are imbued with relevancy and efficacy, shaping what learning actually looks like.

While this approach opens up powerful new insights about how science itself is place-based and how progressions might be mapped in ways that attend to the cultural dimensions of science, it does not fully engage how the cultural basis of science should shape the *full* progression as well as the outcomes. One question we have about this approach is how place-based knowledge formation should shape what it means to become an expert. In this particular place-based learning progression, the top anchor is still the scientific explanation for haze pollution. The lower and middle anchors account for students' indigenous and place-based ways of knowing, but these may still be viewed as on-ramps in support of scientific knowledge formation. From a justice standpoint, students may see themselves in the processes of learning about haze, and this is critically important, but they may also find their lives erased in the final outcomes of learning. Further research should be done to explore how upper anchors may account for place-based and other culturally grounded ways of knowing.

Whose Learning?

Here we raise questions about the cultural basis of learning itself, and how both the processes and outcomes of learning need to reflect how people make connections to and negotiate learning in relation to their personal and cultural lives, as well as how people use what they learn towards affecting people's lives, social relations and possibilities. One important aspect of this is how the focus on knowledge/practice development leaves out the importance of agency (or how action-taking with and in science is intimately related to coming to know science). Learning progressions need to make explicit the relationship between knowledge development and learning to participate and take action with and through science. One way to do so is to hold developing science knowledge and practices in tandem with the applicability of such in present student-driven everyday issues, instead of training applicability of scientific know-how only for the future. Such an approach prioritizes and intentionally scaffolds into learning progressions the development of student agency and immediate relevance.

One example of work aiming to do so is presented by Taylor and Brand (2021), who expand learning progressions to consider how "learning progressions could move beyond conceptual development to structuring participation in science and participation with local communities" (p. 15). In responding to Ladachart et al.'s (2020) place-based learning progressions (discussed above), these authors considered how practice-oriented progression facilitated students' epistemic agency and identity formation. Here the authors view epistemic agency as "students participating and taking ownership during the knowledge building process" to solve problems within their local community (p. 9).

These authors are concerned with how action-taking with and in science is intimately related to knowledge formation. Integrating place-based education within learning progressions requires students to engage in *participatory agency with the local community*. These authors suggest that place-based learning progressions – and other forms of learning progressions that account for social context – can be enhanced to address epistemic agency, which strategically structures students' participation, identity development, and self-efficacy as a part of the sequenced progression. As they state,

> Students possess the autonomy to construct new knowledge in their local place using science practices. While Ladachart et al.'s effectively noted that the long-term goal is to impact human behavior with respect to creating haze pollution in the northern region of

> Thailand, their focus was primarily on conceptual knowledge, which alone does not cultivate or ensure human responsibility or epistemic agency" (p. 4).

They further assert that the path of cognitive development for students is not "purely incremental" (p. 4) but rather that students learn in complex, non-linear ways as they engage in the world, and encounter new experiences.

Taylor and Brand identified a set of anchors that ranged from "entry" (level 1) where the focus was on enhancing students' science identity and self-efficacy through participation in engineering activities that examined the epistemic agency of others (in this case, the contributions of African Americans in STEM), to "target" (level 4) where the focus was on students enacting their own epistemic agency to authentically engineer for their community. While this progression is still hierarchical, it challenges the dominant approach to target anchors which have typically focused on abstracted knowledge. Here the focus is on using and adapting knowledge to solve real local problems. While the authors do not discuss the embodied aspects of their framing, it is important to note that students' enactments-in-context – their engagement with community concerns and their embodied experiences related to these concerns –are contingent on the progression.

The focus on participation in communities in this work is important. In our work, we have documented the why, where, when, how, and for whom youth engage in rigorous STEM to be central to their persistence in rigorous and iterative STEM engagement. Youth have termed such community-centered STEM "Science that matters" and labeled themselves as "community science experts" with the dual know-how of community issues and scientific expertise to take concrete actions in addressing community concerns via STEM (e.g., Calabrese Barton & Tan, 2018; 2019).

Whose Progressions?

What constitutes a learning progression should remain an open question, as well as who gets to decide that answer. As learning progressions research develops in ways that attend to these sociocultural dimensions of learning, further attention should also be paid to the new, more robust forms of knowing and practice that arise as students engage multiple epistemologies as they come to know and act in science (building on our two previous points). Incorporating students' cultural knowledge into a progression but not challenging the directionality of the progression itself limits the possibilities for justice-oriented disruption of norms. More than just an entry point into scientists' science, students' cultural knowledge should be integral to the expansion of what counts as scientists' science in thought, word, and deed, through what kinds of social activities, in which locations.

This view foregrounds how learning is a relational activity, in terms of time, place (Lee, 2006), and power (Gutierrez, 2012). Focused on more than vertical movement (e.g., novice to expert), this view illuminates how new forms of hybrid knowledge and practice arise as people move horizontally, from place to place, widening what counts as expertise. This is important because equitable practices need to broaden expertise and participation to include the culturally based assets youth bring to learning.

Accounting for the *horizontal* dimensions of learning requires more sophisticated forms of knowledge/practice that merge discourses and practices and promote boundary crossing and boundary blurring (rather than within boundary work only). Becoming more expert involves both coming to know and coming to be in ways that transgress the traditional boundaries of school science.

To be clear, we are not suggesting that "anything goes" in terms of developing expertise in science through merging vertical and horizontal dimensions of knowing and knowledge production. Merging requires critical engagement across knowledge systems, with deepening awareness within and across them. Thus, rather, we suggest two critical points. First, that learning progressions are not settled terrain, nor should they ever be. They are always in the making, as youth bring forward their lived experiences and ways of knowing to bear on what knowledge is produced, how, where, and when. Second, that learning progressions involve not just epistemological and meta-cognitive frames, but also ontological ones, as people matter in how knowledge formation happens. We saw this in the study on haze pollution where students' experiences provided both anchors and launching points for exploration along with deeper questions about how such knowledge matters. We also see this in Jazmyn's quote shared earlier as she points to her own learning and acting early on and in the pandemic. Caring for her family and learning to make science and health decisions for her own and her community's well-being shaped what she learned in and about the pandemic (see Calabrese Barton et al., 2021).

Merging her lived experiences in the pandemic with her developing knowledge of viral spread and risk mitigation, Jazmyn leveraged social media apps as a gateway to the CDC's database of COVID-19 information (e.g., clicking on banner ads the CDC posted to her YouTube homepage), as well as YouTube and TikTok's search and hashtag functions to look up more visually accessible and culturally situated instantiations of viral transmission mitigation protocols that she could not find in these big data sources. She paid particular attention to videos in which the individual talking looked like her, was wearing scrubs, and was posting using verified accounts "from hospitals and stuff." During one interview, Jazmyn distilled what she had learned from one such video:

> Maintain at least one meter or three feet distance between you and someone who's coughing or sneezing. Why? When someone coughs or sneezes, they spread small liquid droplets from their nose and mouth which may contain viruses. If you are too close, you can breathe in the droplets, including the COVID-19 virus, if the person coughing has the disease. Avoid touching eyes, nose, and mouth.

She tapped into forms of scientific data collected at inter/national levels but was also seeking more contextualized illustrations of how to make sense of, address, and potentially critique that big data, and how to use it to inform her daily actions. Jazmyn's data navigations took on particular importance as she began engaging in her city's protest efforts against racism in the Summer of 2020. Her research on viral transmission and comparative efficacy of various risk mitigation behaviors, along with her understandings of 7-day patterns of infection and spread in her region, informed her contextualized understandings of different forms of risk as she decided whether, and how, to physically partake in in-the-street actions with friends and community. She also worried about how mask-wearing harmed her black community through racial profiling. She commented that none of the data on COVID-19 and risk mitigation paired these two challenging realities she and her friends navigated. She explained in a follow-up interview: "I had to decide whether to protect myself and my family against injustice by protesting, or to protect myself and my family by not going." Her decision to protest, while wearing a mask and physically distancing herself from others, involved a complex analytical process of weighing different data inputs, including looking up predicted participants attending, current county and city infection rates, and photos of other cities' protests from that week to gauge how many participants looked like they were taking personal safety precautions.

While this example does not offer a learning progression per se, it does fundamentally shift the discourses of learning and knowing towards to also include the ethical and political in ways that were attentive to scientifically legitimatized forms of knowing alongside her onto-epistemological and social positioning. Jazmyn's example also clearly illustrated how her personal and community-anchored ways of pandemic-relevant sense-making are more than an on-ramp to canonical science knowledge. In fact, we argue, in an unprecedented science-relevant phenomenon that is the Covid-19 pandemic where literally the scientific field is grappling with a novel entity, Jazmyn's Covid-19-related inquiry and negotiations, in complex and non-linear ways, would simultaneously map onto different levels of a learning progression. Jazmyn leveraged and merged scientific data and information from different epistemological and social origins (e.g., the CDC, a medical professional with a hospital TikTok account, first-hand images of protests, and the county COVID-19 data dashboard) towards *coming-to-know* and *coming-to-act* in a pandemic. Her practices had epistemic elements (e.g., understanding tables/graphs about spread patterns), but also cultural, sociopolitical, and ethical elements (e.g., re-imagining understandings of big data to "see herself" through small data; critiquing emergent narratives from juxtaposing data about racial profiling with viral spread; projecting the tension between "health rights" & "civil rights" as a guiding interpretive dimension). Descriptions of "lower" *and* "upper" anchors, along with the middle-level waypoints, should account for how individuals adaptively and flexibly use ideas to solve interdisciplinary problems in locally situated ways, such as Jazmyn did in the example above, or the youth in northern Thailand did in the Ladachart et al. study. Furthermore, the terminology of upper and lower anchors suggests fixed starting and ending points that are hierarchical in nature. Considering how ideas are recontextualized across communities of practice suggests that this terminology fails to take into account how expert reasoning in practice operates. A fixed anchor hierarchy also fails to reflect the epistemic resources and variation both across domains and across contexts even within the same domain. For example, students participating in an engineering for sustainable communities unit are likely to define "sustainable communities" in varying ways within the domain of engineering (Tan et al., 2019, 2022). Such differing ways of identifying and operationalizing sustainability, held in dialogue with engineering epistemologies, work to open up various pathways and iterative engineering design opportunities for students. While the middle-level anchors or waypoints open up the complexity of pathways, our point is that the discourse of upper and lower anchors suggests a reality that is not responsive to all possible learning trajectories.

In our own work, we have co-designed with teachers and youth an approach to learning that supports students in navigating multiple epistemologies, along with the tensions that emerge as they do so (e.g., Tan et al., 2019; Calabrese Barton & Tan, 2019, 2020a). Using a related set of epistemic tools that supported students in recruiting and facilitating the integration of diverse community epistemologies students possess with canonical middle school STEM knowledge and practices, we were concerned with when and how these tools cultivated new forms of learning toward equitable and consequential outcomes, transforming both the processes and outcomes of learning.

When students are supported in soliciting and navigating multiple epistemologies as a part of STEM learning, they attend to how their own and their communities' lived experiences pushed back against structurally embedded ways of sense-making. As they leveraged what they were learning to take action on problems they cared about, they also were recognized for knowing the intricacies of complex problems that mattered in their community and which could be responded to with engineering design. Success in their learning and action-taking was not successful without deep engagement with how, or why, particular concerns mattered in their community. Thus, we suggest that learning progressions value multiple epistemologies towards promoting not only deeper engagement and learning in STEM, but also disrupting normed epistemologies in STEM by raising questions around what, and whose, knowledge counted, and why.

Returning to the engineering for sustainable communities unit, students and teachers have both reflected on the direct, positive-feedback relationship between the rigor of iterative engineering undertaken and a deeper dive into unpacking community social issues. In one project, 6th grader June engineered a portable school-desk cavity light-up system that consisted of two pieces of hardback paper in a foldable L shape, each with three, white LED lights in a parallel circuit powered by a hand-crank. This functional prototype was the result of three rounds of iterative design that included testing different components of the circuitry – load, power, switches—and the structural components of the innovation – how to build a portable light-up system that fits into a school desk cavity (see Tan et al., 2019). The process of iterative engineering design was contingent on June drilling down into school-identified concerns, from low school morale, to how it is related to harsh disciplinary penalties students incurred, to how a common penalty is caused by students being late to class during lesson transitions, to how being late is related to not being able to see under one's desk to quickly gather needed materials for the next class. June's light-up desk prototype illustrates how the horizontal dimensions of learning – students' everyday concerns and their embodied experiences as students moving through a school day across school spaces in social relationships with peers and teachers—are relevant and integral to how she engaged in the engineering unit. With the support of her science teacher, peers, and community members that included engineers who gave comments and feedback during the iterative design process, June navigated between multiple epistemologies that included engineering and community texts in dynamic ways.

This example also raises questions of how learning progressions might be conceived in fluid and non-linear ways where learning is viewed as a social endeavor, oriented towards collective social action.

Discussion

These considerations of whose science, learning, and progressions, as they pay attention to the place-based nature of progressions, epistemic agency, and the horizontal dimensions of learning, are of critical importance. When learning progressions center science as separate from the livelihood of minoritized communities, minoritized youth are positioned as deficient, delimiting their agency to leverage the power of science alongside other ways of knowing to take action on the questions and concerns they care about. When youths' knowledge, practices, and already-present forms of expertise they bring to science learning are made *invisible*, they, too, are made invisible (Tedesco & Bagelman, 2017). They are denied a "rightful presence" in their learning community.

One's own location in the world allows one to consider the questions, "*how* do I know what I know, *why* does that matter, and to *whom*?" Such questions are important if learning progressions are to be designed in ways "that sustainably disrupt historically shaped inequities and cultivate transformative agency from within communities" (Bang, et al., 2016, p. 2). We thus view the questions raised in this chapter – whose science? whose learning? whose progressions? – as addressing the nature of knowledge itself, as well as the sociocultural processes for achieving possible epistemic aims in both coming to know and coming to act. Without attention to these questions, the cultural basis of learning is minimized, leaving open the possibilities for reinscribing racial privilege and oppression within progressions (Nasir et al., 2013). Normative discourses about and enactments of science learning, for whom and why, can differentially position students with or without epistemic authority and/or agency. When the articulation of expertise and the construction of pathways towards that expertise neglect the role of culture, this reinscribes normative (white, middle-class) understandings of what and how students learn, making anything that deviates from this norm "outside" the learning progression.

We recommend that learning progressions research seek to make present students' whole lives by making visible and integral youths' lived experiences and ways of knowing. What this might look like includes recognizing students for what they know and have experienced, and how their expertise is used to help others, as a part of the progression itself, rather than anchors towards canonical learning. This also might look like designing research in ways that actively disrupt traditional power dynamics, including activities and discussions that actively position students as experts, recognizing them for what they know, and how their expertise is used to help others. If we return to the haze pollution example, the very learning progression itself – how it was conceptualized and framed – was grounded in the historicized lives of youth in Northern Thailand. It was not framed around the abstracted environmental or chemical knowledge that may be a part of the progression. The erasure of people and livelihoods can have a rippling effect as learning progressions are used to inform and align the entire system of learning – curriculum, instruction, and assessment. Furthermore, as the field matures, and as learning progressions become codified, they also become settled. "Settled expectations" in school science, that can be made manifest in the learning progressions which guide systems of instruction (curriculum, instruction, assessment) act as "boundaries that control the borders of acceptable meanings and meaning-making practices," and position minoritized students in deficit-oriented ways (Bang, et al., 2012, p. 303). As Kang and Furtak (2021) note, learning progressions, and the learning goals they indicate, are "expressed in ways that make them seem settled despite reflecting Whiteness and privileging the values and stories of members of the dominant communities above others (Bang, Warren, Rosebery, & Medin, 2012)" (p. 2). This forecloses possibilities for new forms of engagement with the progressions. This makes it more difficult to challenge who belongs in science as well as what constitutes a progression itself.

Despite attention to practice, learning progressions still, by and large, remain grounded in epistemological development rather than considering more expansive dimensions to learning, including onto-epistemological dimensions such as agency and identity. Note, we do not suggest that there should be progressions for agency and identity, per se, but more so that it is problematic to consider epistemologies separate from these. Further, epistemologies -what counts as knowledge that matters and how do we know -is inevitably intertwined with ontologies -who we are, can be, want to be- and axiologies -what we care about and value with our priorities.

Lastly, the methodologies and tools used to research learning progressions should be expanded to include approaches that account for the variation in experiences that individuals bring to engaging science (Nasir et al., 2013). Research designs should include attention to the heterogeneity of experiences and their role in science learning. Research designs "should take into account how culturally embedded ways of knowing and doing are not lower anchors but rather integrated across the developmental process of coming to know, do and be in science" (Nasir et al., 2013, p. 5). Methodologies and approaches, grounded in cultural and critical frameworks, should be employed to make sense of how learning progressions can be informed in place-based ways, and which make visible and account for cultural variation. The learning progressions community should also support "interdisciplinary collaborations to allow for within project debate regarding interpretive outcomes, developing research designs that seek out heterogeneity of experiences and its role in learning" (Nasir et al., 2013, p. 5) opening up new ways of thinking about and designing for learning progressions.

As the field of learning progressions continues to mature and if minoritized youths' rightful presence in science is to be of import, a way to think about progressions from lower to upper anchors is to base such hierarchy not solely on the epistemic knowledge that students might display, but in the increasing depth and pedagogical sophistication in which youth and community epistemologies are integrated, in positive feedback loop fashion, horizontally, and with community epistemologies. Such a twinned, community-science learning progression is contingent on disrupting canonical science culture for an expansive one.

References

Alonzo, A. C. (2018). An argument for formative assessment with science learning progressions. *Applied Measurement in Education*, *31*(2), 104–112.

Bang, M., Warren, B., Rosebery, A., & Medin, D. (2012). Desettling expectations in science education. *Human Development*, *55*(5–6), 302–318. 10.1159/000345322

Bang, M., Faber, L., Gurneau, J., Marin, A., & Soto, C. (2016). Community-based design research: Learning across generations and strategic transformations of institutional relations toward axiological innovations. *Mind, Culture, and Activity*, *23*(1), 28–41. 10.1080/10749039.2015.1087572

Calabrese Barton, A., & Tan, E. (2019). Designing for Rightful Presence in STEM: The Role of Making Present Practices. *Journal of the Learning Sciences*, *28*(4-5), 616–658. 10.1080/10508406.2019.15 91411

Calabrese Barton, A., Tan, E., & Birmingham, D. J. (2020a). Rethinking high-leverage practices in justice-oriented ways. *Journal of Teacher Education*, *71*(4), 477–494.

Calabrese Barton, A., & Tan, E. (2020b). Beyond equity as inclusion: A framework of "rightful presence" for guiding justice-oriented studies in teaching and learning. *Educational Researcher*, *49*(6), 433–440. 10.3102/0013189X20927363

Calabrese Barton, A., Greenberg, D., Turner, C., Riter, D., Herrenkohl, L., Tasker, T., & Davis, E. (2021). No learning lost here: Critical data practices in the COVID-19 multipandemic. *AERA Open*. 10.1177/23328584211041631

Córdova, R. A., & Balcerzak, P. (2016). Co-constructing cultural landscapes for disciplinary learning in and out of school: The next generation science standards and learning progressions in action. *Cultural Studies of Science Education*, *11*(4), 1223–1242.

Duschl, R. A. (2019). Learning progressions: Framing and designing coherent sequences for STEM education. *Disciplinary and Interdisciplinary Science Education Research*, *1*(1), 1–10.

Gutiérrez, K. D. (2012). Re-mediating current activity for the future. *Mind, Culture, and Activity*, *19*(1), 17–21.

Hammer, D., & Sikorski, T. R. (2015). Implications of complexity for research on learning progressions. *Science Education*, *99*(3), 424–431. 10.1002/sce.21165

Herrenkohl, L. R., & Bevan, B. (2017). What science and for whom?: An introduction to our focus on equity and out-of-school learning. *Science Education*, *101*(4), 517–519. 10.1002/sce.21284

Jin, H., Johnson, M. E., Shin, H. J., & Anderson, C. W. (2017). Promoting student progressions in science classrooms: A video study. *Journal of Research in Science Teaching*, *54*(7), 852–883.

Kang, H., & Furtak, E. M. (2021). Learning theory, classroom assessment, and equity. *Educational Measurement: Issues and Practice*.

Ladachart, L., Poothawee, M., & Ladachart, L. (2020). Toward a place-based learning progression for haze pollution in the northern region of Thailand. *Cultural Studies of Science Education*, *15*(4), 991–1017.

Lee, C. (2006). 'Every good-bye ain't gone': Analyzing the cultural underpinnings of classroom talk. *International Journal of Qualitative Studies in Education*, *19*(3), 305–327.

Lee, O., & Buxton, C. (2013) Integrating Science and English proficiency For English language learners. *Theory Into Practice*, *52*(1), 36–42. 10.1080/07351690.2013.743772

Lee, O., & Luykx, A. (2005). Dilemmas in scaling up innovations in science instruction with nonmainstream elementary students. *American Educational Research Journal*, *42*, 411–438.

Liu, L., & Jackson, T. (2019). A recent review of learning progressions in science: Gaps and shifts. *The Educational Review, USA*, *3*(9), 113–126.

McGee, E. O., & Stovall, D. (2015). Reimagining critical race theory in education: Mental health, healing, and the pathway to liberatory praxis. *Educational Theory*, *65*, 491–511. 10.1111/edth.12129

Nasir, N., Warren, B., Bang, M., & Calabrese Barton, A. (2013). A friendly critique of learning progressions research from a cultural perspective. *White Paper developed for Learning Progressions Conference.*

National Research Council (NRC). (2011). *A framework for K-12 science education: Practices, crosscutting concepts, and core ideas.* Washington, DC: National Academies Press.

NRC. (2007). Taking science to school: Learning and teaching science in grades K-8. In Shouse A. W.Schweingruber, H. A. Duschl, R. A. (Eds.), National Academies Press

Philip, T., & Azevedo, F. S. (2017). Everyday science learning and equity: Mapping the contested terrain. *Science Education*, *101*(4), 526–532. 10.1002/sce.21286

Shepard, L. A. (2018). Learning progressions as tools for assessment and learning. *Applied Measurement in Education*, *31*(2), 165–174.

Tan, E., Calabrese Barton, A., & Benavides, A. (2019). Engineering for sustainable communities: Epistemic tools in support of equitable and consequential middle school engineering. *Science Education*, *103*(4), 1011–1046.

Tan, E., Calabrese Barton, A., & Restrepo Nazar, C. (2022). Negotiating new social-spatial relationalities in middle school engineering: Seeing consequential learning as a justice-oriented project. *Cognition & Instruction.*

Taylor, L., & Brand, B. (2021). Enhancing place-based learning progression through epistemic agency: A response to toward a hypothetical place-based learning progression for haze pollution in the northern region of Thailand. *Cultural Studies of Science Education*, 1–18.

Tedesco, D., & Bagelman, J. (2017). The 'missing' politics of whiteness and rightful presence in settler-colonial city. *Millennium*, *45*(3), 380–402. 10.1177/0305829817712075

Tytler, R. (2018). Learning progressions from a sociocultural perspective: response to "co-constructing cultural landscapes for disciplinary learning in and out of school: the next generation science standards and learning progressions in action". *Cultural Studies of Science Education*, *13*(2), 599–605.

Vakil, S. (2018). Ethics, identity, and political vision: Toward a justice-centered approach to equity in computer science education. *Harvard Educational Review*, *88*(1), 26–52.

Zuniga, K., Olson, J. K., & Winter, M. (2005). Science education for rural Latino/a students: Course placement and success in science. *Journal of Research in Science Teaching: The Official Journal of the National Association for Research in Science Teaching*, *42*(4), 376–402. 10.1002/tea.20064

24

IMPLICATIONS OF TEACHERS' LEARNING AND USE OF LEARNING PROGRESSIONS

Commentary for Section III

Todd Campbell

While there exists a myriad of reasons for optimism about the usefulness of learning progressions (LPs) in science education research, few seem as promising as the potential role LPs might play as a resource for teaching. In this role, LPs can be both supportive of teacher learning and act as "thinking tools" that teachers might use as part of the formative assessment in interpreting, and responding to, student thinking. This section of the handbook captures not only how LPs have been leveraged productively as part of teachers' formative assessment practices (Chapter by Alonzo, Chapter by Furtak et al., and Chapter by Gunkel et al.), it also captures the factors that affect teachers' use of LPs (Chapter by Alonzo and Chapter by Luft et al.), how these LPs can be used to support teacher learning as part of professional development (PD) (Chapter by Furtak et al.), and how LPs might influence the design of curriculum (Chapter by Gunckel et al.). As well, this section recognizes the affordances of LPs for orienting the field of science education and teachers away from content mastery, toward learner-centered curricula and environments that elevate views of learning as connected to students' lives, while also pointing to the need for future LP research to consider and center students' rightful presence by taking account of the cultural dimensions of science so that science learning, and science, is connected to students' lives and communities (Chapter by Barton & Tan). Also important in this section of the handbook is the caution that the contributing scholars offer when considering the role that LPs might play in science teaching and learning. As a science teacher education researcher invited to comment on these contributions by the field-leading LP scholars, my own orientation to LPs has been, at times, hampered by questions I've had about the extent to which LPs are capable of helping teachers recognize and legitimize the plural onto-epistemological resources learners bring to science classrooms, which is something Barton & Tan also take up in-depth in this section. Other researchers (e.g., Hammer and Sikorsko, 2015; Alonzo & Elby, 2019), cited in the handbook chapters of this section, have previously levied similarly important and related, if not identical, concerns. Given what the section scholars have shared here to support our future visioning of the role of LPs in science teacher education, I've organized this commentary around the interrogation of the collective of these chapters related to the following questions:

1 What are the benefits and challenges of LPs supportive of science teacher learning and LP use?

DOI: 10.4324/9781003170785-27

2 In what ways can science teacher learning with LPs, and their use, be (un)supportive of students' plural onto-epistemological reasoning?

In the end, my aim in writing this commentary was to work beside the contributing scholars to elevate their important contributions, while at the same time looking for patterns that might be important to synthesize across their work. Such a synthesis is important to inform future directions for LP research in science teacher education. To set the stage for this commentary, I first briefly share literature to frame my interrogation of the chapters before answering each question in sequence. I conclude with implications, a conclusion, and a note of appreciation to the authors for their contributions and leadership.

Literature Framing: Learning Progressions and Onto-Epistemological Reasoning

The National Research Council (2007) conceptualizes LPs as "descriptions of the successively more sophisticated ways of thinking about a topic that can follow one another as children learn about and investigate a topic over a broad span of time" (p. 219). The aims and promise assigned to LPs include the representation of a pathway along which students might potentially move toward a more sophisticated knowledge of science over longer, defined periods of time (Alonzo & Gotwals, 2012), and the revision and alignment of curriculum, instruction, and assessments to support learning of the LP target disciplinary core ideas or practices (Neumann et al., 2013). In science teacher education, especially when supporting science teacher learning, researchers have articulated how LPs have "the potential to guide the coordination of teaching, instructional resources, and assessment with cognitive and metacognitive practices so that learning builds coherently" (Sevian & Talanquer, 2014, p. 11). As well, researchers like Duschl (2019) have also described the promising role LPs, as instructional supports for teachers, offer for informing interpretations of student thinking and subsequent instructional decision-making.

As is the case with most lines of research, scholarship focused on LPs is not without tensions, trade-offs, or limitations. Researchers like Hammer and Sikorsko (2015) and Alonzo and Elby (2019), among others (e.g., Pierson et al., 2017), have illuminated the potential tensions that emerge in the use of LPs to support teachers' learning and instruction. As an example, Hammer and Sikorski levied two critiques about the use of LPs in science education research. The first critique relates to how, early on in the conceptualization of LP research, even as the science education community recognized that:

> ... all students will follow not one general sequence, but multiple (often interacting) sequences around important disciplinary specific core ideas ... [however] despite the initial expectation of multiplicity, most published LPs describe a single, levels-based sequence from a lower to an upper anchor (Hammer and Sikorsko, 2015, p. 424).

Their second critique concerns how, "with a few notable exceptions, the analyses supporting these LPs rely on cross-sectional studies that aggregate data across large numbers of students" (p. 425). Significant to both of these critiques is the case that Hammer and Sikorski make for the importance of the idiosyncratic particularities in individual students' sensemaking as critical to their coherence seeking, a dynamic aspect of learning experiences. In making the case for the importance of idiosyncratic particularities of students' sensemaking, they expressed concerns, about how "the space of possible connections they [students] may consider in their [coherence] seeking is vast" (p. 428), and how researchers, and subsequently teachers, focusing on a single, levels-based sequence for learning might lead to instruction that prioritizes keeping students on track to meet the single sequence of the LP at the expense of their coherence seeking. More generally, there is a concern regarding analyses supportive of LP development, and refinement, that rely on the aggregation of data across large numbers of students. The specific concern highlights how "the education community needs approaches to research that can capture and consider idiosyncrasy, not as statistical noise to eliminate but as part of the phenomena of interest" (Hammer and Sikorsko, 2015, p. 429).

Similarly, while also beyond the concerns voiced by Hammer and Sikorski, Alonzo and Elby (2019), building on the framing of LPs as akin to scientific models used by other researchers, point to the limited criteria of fit with empirical data almost exclusively investigated in LP research, as well as the empirical inadequacy of LPs in relation to this limited measure. Specifically related to the empirical inadequacy of LPs, they point to the misalignment with predicted LP levels, and the failure of LPs to predict student response patterns in an extensive number of studies. Importantly, in their analysis, they also propose how the context-dependency of student thinking is greater than what can be attributed to a level of student thinking in an LP for an individual student at a specific time. Given these challenges of empirical inadequacy, Alonzo and Elby note the hopeful role researchers have for LPs, even in the context of their recognized empirical inadequacy, as researchers "often note that an LP is just a model, meaning that it may have utility even if it does not capture student thinking perfectly" (2019, p. 2). However, Alonzo and Elby also detail how there exists little to no empirical research exploring broader criteria for LPs as models, namely their utility and generativity in supporting teachers. As an example, teachers "use LPs for formative assessment purposes and to generate knowledge supportive of instructional decision-making more broadly" (Alonzo & Elby, 2019, p. 2). Their research did ultimately investigate and provide evidence of the utility and generativity of LPs in supporting teachers' formative assessment and instructional decision-making. Yet, what might be as important in their work as it relates to this section and commentary is how their critique moves past concerns with focusing solely on the empirical adequacy of LPs, to begin to explore and demonstrate how teachers can leverage their own intellectual resources to use LPs as productive models to "both to spur instructional decision-making and to generate new understandings of student thinking that go beyond the model itself" (Alonzo & Elby, 2019, p. 2)

The final critique that has shaped this commentary, my own previous concerns, relates to the extent to which LPs and their use are capable of helping teachers recognize and legitimize the plural onto-epistemological resources learners bring to science classrooms. This concern has emerged from exploring the writings and commitments of scholars like Bang and Marin (2015), Higgins (2021), Kayumova and Dou (2022), and others. Specifically, these scholars have helped me to recognize the powerful ways in which ontological (i.e., ways of being) and epistemological (i.e., ways of knowing), which are deeply entangled with axiological values, in the sciences and its offshoot, science teacher education, have traditionally been limited to Western, Eurocentric, or Cartesian perspectives in ways that foreclose the prospect of recognizing, legitimizing, and benefiting from a plurality of perspectives. In fact, failure to recognize the powerful roles these perspectives play in how we read and relate to the world not only has the potential to make science teaching and learning foreign, inequitable, and oppressive to a large portion of global K-12 science learners, but it also limits our ability to recognize and then consider different ways of orienting ourselves to, and thriving in, the world. As one example, Indigenous onto-epistemologies prioritize reciprocity, relationality, and responsibility (e.g., Bang & Marin, 2015; RunningHawk Johnson et al., 2023) and, recognized among other plural perspectives, hold promise for saving us from our continued failure to address inequities in K-12 schools (Kayumova & Dou, 2022), as well as the planetary destruction brought on by human domination, entitlement, and extraction from the natural world (Bang & Marin, 2015). This last concern may seem distant from the interrogation of LP research, however, because LPs concern themselves with the characterization of student thinking, thinking that is vast, not only because of the context-dependency within which coherence-seeking is embedded, but also because of the cultural dependency in which student thinking is rooted, it felt imperative to me to consider which student's thinking was being characterized. It also felt important to consider the extent to which characterizations captured in LPs made space for, or limited, how teachers might learn from and use LPs in ways that recognized and committed them to eliciting, interpreting, and supporting student thinking constructed from diverse onto-epistemological resources.

In the end, using this literature framing as a backdrop, and as I thought about, and learned from, the authors' contributions to this section, my aim was to synthesize from across these chapters the benefits and challenges of LP-anchored teacher learning, as well as considering the documented (un)supportive role LPs play in elevating students' plural onto-epistemological resources and ways of reasoning. This is taken up sequentially in what follows.

What Are the Benefits and Challenges of LPs Supportive of Science Teacher Learning and LP Use Captured Across the Chapters in This Section?

Benefits of LPs in Supporting Science Teacher Learning and Use

Alonzo categorized the benefits of LPs in supporting science teacher learning into two broad themes, namely: (1) as supports for orienting, to and learning about, student thinking and learning; and, (2) as a generative conceptual framework that develops from teachers making sense of LPs shared by researchers in relation to their own classroom's experiences and practice. Importantly, Gunckel et al. provide a framing for LPs similar to that offered by Alonzo. Luft et al., citing Jin et al. (2019), build on these affordances of LPs, by revealing the ways in which LPs support teachers in developing deeper understandings of how students learn ideas. In turn, these understandings can support teachers in constructing strategies for responding to the varied ideas of students. Also, most of the chapters in this section (i.e., Chapter by Alonzo; Chapter by Furtak et al., Chapter by Gunkel et al., and Chapter by Luft et al.) point to the role LPs can play in supporting teachers' assessment practices, especially in connection to formative assessments that seek to understand student knowledge, and hence aid planning the next instructional steps to supporting student learning. Related specifically to supporting teachers' assessment practices, Furtak et al. designed and studied how LPs supported teachers as they engaged in school-based professional development (PD). Furtak et al. designed PD anchored around the teaching of energy across three disciplines of high school (i.e., physics, chemistry, biology). They found that the energy LP supported teacher learning, by encouraging adaptation to the disciplinary context, the setting of learning goals, the shaping of design tasks to elicit student thinking, and responding to student thinking in ways that moved this thinking forward.

Challenges of LPs in Supporting Science Teacher Learning and Use

In this section of the handbook, one initial practical challenge identified by Gunckel et al. is how teachers struggle to find the usefulness of LPs for informing their curricular, instructional, and assessment practices, without support from, among other things, educative

LP-based curriculum materials. Related to the challenges Gunckel et al. noted, teachers experiencing using LPs, Luft et al. recognized the limitations of teachers using LPs as they referenced research by Furtak and Heredia (2014). In this, Furtak and Hererdia documented the struggle teachers experience in aligning the LP they are provided with to the experiences in their classrooms. This problem, they suggested, arises from how the LPs are not developed for any teacher's specific students or context. Beyond this, Furtak et al. identified how, even if LPs were aligned with the school curriculum, teachers identified multiple structural influences within their school contexts, like historical curricular and assessment practices, that resulted in the LP influencing teacher planning unevenly. Beyond considering the challenges in supporting teachers in using LPs, further challenges reside in how LPs are created (See Chapter by Alonzo). Here, Alonzo described how aggregation across many students that occurs in the creation of LPs represents both an affordance and a constraint. Specifically, Alonzo described how aggregating data afforded the identification of broad patterns of student thinking to illuminate information contained in the LP model, even as it simultaneously eliminated information when a particular student failed to fit within the broadly identified patterns, something Hammer and Sikorsko (2015) criticized as a practice of eliminating the statistical noise.

LPs and Students' Plural Onto-Epistemological Reasoning

Supporting Reasoning

Among other benefits of LPs, both Alonzo and Gunkel et al. pointed to their potential for helping teachers move away from what they referred to as the "gets it/doesn't get it" perspective about students' thinking toward a more nuanced view focused on identifying and using student thinking. Further, Gunkel et al., in referencing the work of Alonzo and Elby (2019), pointed to the promise of teachers using LPs as beginning resources for developing insights into the interpretation of student thinking, especially when supported by LP-based curriculum materials. While not explicitly mentioned, this perspective focused on identifying the student thinking alluded to by Alonzo and Gunkel and colleagues holds the potential for shifting teachers' aims to what Rosebery et al. (2015) referred to as developing interpretive power. This power can be described as the capacity to attune to the diverse sensemaking repertoires of students and the accompanying pedagogies needed for "encourag[ing], making visible, and intentionally building on students' ideas, experiences, and perspectives" (p. 1571). Gunkel et al. delve deeper into how LPs support teachers to intentionally build on students' ideas as they frame these ideas as stepping stones which encourage teachers to think carefully about how these can be used in developing assessments that can give insights about productive next steps in instruction.

Luft et al., citing Hamed et al. (2020) and Adadan and Oner (2014), in describing what had been learned about LPs focused on pre- and in-service teacher learning, noted how even when there was some convergence on the different levels of LPs, the pathways of teachers toward levels within LPs were varied. While Luft et al. didn't explicitly point out how these divergent pathways might be supportive of plural-ontological reasoning, what is important, is that by legitimizing the prospect of differential pathways, not as problems with LPs but as realities connected to difference, this observation has significance as a step toward supporting teachers in recognizing students' plural onto-epistemological reasoning, as part of orienting to and using LPs. Similarly, Furtak et al. identified how LPs as thinking tools for planning for, and reflecting on, formative assessments offered promise in helping teachers pay more careful attention to the interpretation of student thinking and seeing students thinking as resources for learning. While Furtak et al., in their chapter, did not distinguish the extent to which interpreting students' thinking might support teachers in recognizing plural onto-epistemological resources, and reasoning that moves beyond Western traditions, they do establish the possibility of this, even if more intentional teachers PD support is necessary to develop the facility in recognizing more plural perspectives.

Unsupportive of Reasoning

In pointing to the lack of alignment between individual students' learning and the aggregate description of student learning captured in LPs, Alonzo in her chapter asked, among other questions: "What groups of students have learning pathways that do and do not align with an LP? ... [noting] there are equity implications if students' ways of thinking and learning are not represented in an LP" (p. 18). Barton & Tan in their chapter similarly ask questions about "whose science, whose learning, and whose progressions ... [are represented in LPs and note how] particular people/perspectives have been systemically and historically silenced" (p. 6–7). These questions are of particular importance when considering the extent to which LPs may, or may not, open up space for teachers to recognize students' plural onto-epistemological resources and reasoning. Closely connected to concerns Alonzo and Barton and Tan shared about groups of students not being represented in LPs, Luft et al. and Furtak et al. also note that teachers struggle to align LPs to the work in their classrooms, since descriptions and examples in the LPs were less proximal to their practice. Gunckel et al. described how teachers' struggles using LPs could also be attributed to the abstract nature of LPs, as well as the grain size of the LPs being articulated in ways that were too coarse to inform instruction. This helps make apparent how generalizations and broad patterns of LPs fail to capture the important contextual nuances (e.g., place, culture) that are needed in working closely with students, especially students whose onto-epistemological reasoning may differ from more historically acknowledged and recognized onto-epistemologies (i.e., Western, Eurocentric, or Cartesian perspectives). Put more succinctly, if LPs are not approached cautiously, teachers who can navigate the challenges of less proximal, abstract, and overly coarse examples (or descriptions) within LPs, may still be limiting the scope of their interpretations if they only look for the student ideas and levels contained in LPs, without recognizing different possibilities beyond what is captured in LPs. Finally and importantly, especially in the context of how current approaches to LP research fall short in supporting students' plural onto-epistemological reasoning, Barton and Tan point to how LPs fail to account for

> the sociocultural, ethical and political ways people know and take action with and in science as integral to learning ... [where] [l]earning and action-taking call into question who people are, and how people, as culturally situated human beings, move and adapt ideas in social context as they navigate horizontally in the world, across communities and contexts (p. 7).

Instead, Barton and Tan point to how LP research, to date, has exclusively focused on individually oriented knowledge construction that is cognitive/epistemic in nature, such that it limits the legitimization of cultural experience and expertise inherent in socioculturally meaningful forms of learning and action-taking.

Implications, Conclusion, and Appreciation

This section of the handbook has contributed significantly in helping the science education community to think critically and expansively about the role of LPs in supporting science teacher learning and instruction. The contributors, leading scholars in the field of LP research, have revealed how LPs can help orient teachers to student thinking, how student thinking can be considered as stepping stones for more sophisticated and complex thinking over time, and how curriculum design and assessments can be informed by LPs in ways that bring coherence to science teaching and learning. The contributors to this section have also surfaced the challenges teachers might experience when working with LPs, while also offering supports that can be considered when addressing such challenges.

Important to me, and many others in the science education community, this section of the handbook has also revealed how space might be made for, and priority given to, plural onto-epistemological resources and reasoning as LPs have shown promise for shifting teachers' perspectives from extant "gets it/doesn't get it" perspectives to more nuanced interpretive inquiries about student thinking, as a potential precursor to teachers developing interpretive power (Rosebery et al., 2015). This section of the handbook also revealed how learners move along divergent pathways as they develop more sophisticated thinking, a characteristic of student (and teacher) thinking that has been recognized, not as a problem for LPs, but as a consequence of differences that deserve more attention in future LP research.

Finally, this section of the handbook, both explicitly and implicitly, asked critical questions about LP research in science education more broadly. Questions like, whose pathways and thinking are, or are not, represented in LPs? And, importantly, how can lack of representation within LPs contribute to the perpetuation of historical marginalization for minoritized learners? While not a comprehensive solution to the challenges of LP research, researchers like Hammer and Sikorsko (2015) and Alonzo and Elby (2019) point to the generativity of how the notion of LPs, as models or stepping stones, might best be understood as "thoughtful, plausible, and consensus-building intermediaries that push learners, and in our case researchers, toward new ways of thinking" (Hammer and Sikorsko, 2015, p. 429). Given this, and aligned with what Alonzo acknowledged, what might be the most important future avenues of research in thinking about LPs as a resource for teacher learning and instruction is attention to flexibility in the design and/or interpretations of LPs, along with consideration of how teachers use LPs in more expansive ways. These research opportunities come, even as we recognize, and appreciate, the affordances that the scholars contributing chapters to this section, have already made in moving forward the field of science education with LPs.

In conclusion, I, as an invited science teacher educator, am very much appreciative of scholars who contributed chapters focused on teachers' learning and use of LPs. Among other substantive contributions, LPs as tools that have demonstrated potential for orienting teachers to student thinking, which for me represents a needed transition supportive of responsive teaching. I also appreciate the critiques of LP research offered by authors like Hammer and Sikorsko (2015), Alonzo and Elby (2019), and Alonzo, as well as the solutions they offer to those questions concerning how we might think of LPs as generative models in place of focusing exclusively on the empirical adequacy of LPs. For me, this is important given the concerns I have voiced here related to supporting plural onto-epistemological reasoning. As an example, when I think of the differences in onto-epistemological resources and reasoning that researchers like Hertz et al. (2020) make apparent as they compare the differences between substance ontologies associated with Western perspectives and process ontologies associated with Indigenous relational perspectives, I can't help but pause at anything beyond LPs as generative starting places for orienting to the vastness of student thinking. More specifically, Hertz et al. point out how Western science predominately relies on dualistic substance ontologies whereby the focus of inquiries and interpretations is on isolating substances, a practice, that while important, has implications for seeing, being, and living in (and apart from) the world. When this is juxtaposed with Indigenous relational perspectives that orient to process, significantly different ways of thinking, and therefore, being, and living in (and as part of) the world arise. Capturing such ontological differences in LPs in just these two examples of differing perspectives seems beyond what is possible through aggregation across large numbers of students, a common practice used in the creation of LPs that amplifies dominant pathways and perspectives. Given how, as Hammer and Sikorsko (2015) noted, most published LP research describes only a single dominant pathway, David Hammer, in a recent communication conjectured,

> I wonder if an implication would be specifically for LP researchers to look for disaggregated statistical pockets of the phenomena, evidence of (productive) learning progressions that small numbers of students take ... it could become a review criterion, the need to see a diversity of progressions for the work to be published — even/especially if some of the progressions only show up in a minority of cases." (D. Hammer, personal communication, October 19, 2023)

This is not to say that we should downplay the benefits that can come from LPs, it is only an acknowledgement of how we, as researchers in science teacher education, should think critically about any tradeoffs we are positioning teachers to make and ways in which we can, through approaches like ensuring teachers recognize LPs as beginning generative and flexible models (Chapter by Alonzo in this volume; Alonzo & Elby, 2019; Hammer and Sikorsko, 2015) and publishing LP research that highlights diversities of pathways (D. Hammer, personal communication, October 19, 2023), ensure that we all remain oriented to this valued work with our eyes, and hearts, wide-open.

References

Adadan, E., & Oner, D. (2014). Exploring the progression in preservice chemistry teachers' pedagogical content knowledge representations: The case of "behavior of gases." *Research in Science Education, 44* (6), 829–858.

Alonzo, A. C., & Elby, A. (2019). Beyond empirical adequacy: Learning progressions as models and their value for teachers. *Cognition and Instruction, 37* (1), 1–37. 10.1080/07370008.2018.1539735

Alonzo, A. C., & Gotwals, A. W. (Eds.). (2012). *Learning progressions in science: Current challenges and future directions.* New York: Springer Science & Business Media.

Bang, M., & Marin, A. (2015). Nature–culture constructs in science learning: Human/non-human agency and intentionality. *Journal of Research in Science Teaching, 52*(4), 530–544.

Duschl, R. (2019). Learning progressions: Framing and designing coherent sequences for STEM education. *Disciplinary and Interdisciplinary Science Education Research, 1* (4), 1–10.

Furtak, E. M., & Heredia, S. C. (2014). Exploring the influence of learning progressions in two teacher communities. *Journal of Research in Science Teaching, 51*(8), 982–1020.

Hamed, S., Ezquerra, Á., Porlán, R., & Rivero, A. (2020). Exploring pre-service primary teachers' progression towards inquiry-based science learning. *Educational Research, 62* (3), 357–374.

Hammer, D., & Sikorsko, T. (2015). Implications of complexity for research on learning progressions. *Science Education, 99* (3), 424–431.

Hertz, T., Mancilla Garcia, M., & Schlüter, M. (2020). From nouns to verbs: How process ontologies enhance our understanding of social-ecological systems understood as complex adaptive systems. *People and Nature, 2* (2), 328–338.

Higgins M. (2021). *Unsettling Responsibility in Science Education Indigenous Science, Deconstruction, and the Multicultural Science Education Debate*. Palgrave Macmillan.

Jin, H., Mikeska, J. N., Hokayem, H., & Mavronikolas, E. (2019). Toward coherence in curriculum, instruction, and assessment: A review of learning progression literature. *Science Education*, *103* (5), 1206–1234.

Kayumova, S., & Dou, R. (2022). Equity and justice in science education: Toward a pluriverse of multiple identities and onto-epistemologies. *Science Education*, *106* (5), 1097–1117.

National Research Council (2007). *Taking Science to School: Learning and Teaching Science in Grades K-8*. National Academies Press.

Neumann, K., Viering, T., Boone, W. J., & Fischer, H. E. (2013). Towards a learning progression of energy. *Journal of Research in Science Teaching*, *50* (2), 162–188.

Pierson, A. E., Clark, D. B., & Sherard, M. K. (2017). Learning progressions in context: Tensions and insights from a semester-long middle school modeling curriculum. *Science Education*, *101* (6), 1061–1088.

Rosebery, A. S., Warren, B., & Tucker-Raymond, E. (2015). Developing interpretive power in science teaching. *Journal of Research in Science Teaching*, *53* (10), 1571–1600.

RunningHawk Johnson, S., Cheng, M. M. W., Karpudewan, M., Campbell, T., Melville, W., Verma, G., & Park, B. (2023). Onto-epistemological realities and assumptions beyond Western science. *Journal of Science Teacher Education*, *34* (6), 583–592.

Sevian, H., & Talanquer, V. (2014). Rethinking chemistry: A learning progression on chemical thinking. *Chemical Education Research and Practice*, *15*, 10–23.

SECTION IV

Applications of Innovative Technology and Artificial Intelligence to Support Learning Progressions

25

DYNAMIC BAYESIAN MODELS FOR LEARNING PROGRESSIONS

Russell Almond, Duanli Yan, and Hui Jin

Introduction

"Learning progressions are descriptions of successively more sophisticated ways of thinking about how students develop key disciplinary concepts and practices within a grade level and across multiple grades" (Fortus & Krajcik, 2012, p. 784). In science education, many learning progressions (LPs) describe how students develop scientific understanding through a sequence of stepping stones (e.g., Kaldaras et al., 2021; Neumann et al., 2013; Osborne et al., 2016). Such LPs can be used to monitor student learning – assessing students' competency and using the assessment results to adjust instruction to better support student learning. Several instructional decisions are involved in this process. How do we assess students' knowledge, skills, and abilities? How do we use the LP to score students' assessment responses? How do we determine students' overall proficiency (i.e., their position on the LP) based on their item scores? Given the students' current LP level, what instructional activities are appropriate? Given the students' LP level, how should instructional activities be planned to ensure that students will achieve the upper anchor of the LP at the end of the course?

This chapter discusses how dynamic Bayesian networks (DBNs) can be used to aid teachers in making evidence-based decisions when faced with these questions. More specifically, how the evidence gathered from assessments should be analyzed to predict student proficiency against the LP and how these predictions are used to inform instruction. As the evidence-centered assessment design (ECD) approach provides a framework to organize the information needed for building Bayesian networks, we use ECD to structure this chapter, focusing on the application of Bayesian networks within ECD. To illustrate this process, we use a hypothetical LP for exploratory analysis of scientific data (EASD) as an example. Netica (Norsys, Inc. 2012) can be used to create DBNs.

Evidence-Centered Design

Evidence-centered assessment design (ECD) is an approach to constructing educational assessments in terms of evidentiary arguments (Mislevy et al., 2003). An evidentiary argument infers students' knowledge, skills, and abilities from observations of what students make, say, or do in a particular context (Mislevy et al., 2003; Riconscente et al., 2015). We modified the

DOI: 10.4324/9781003170785-29

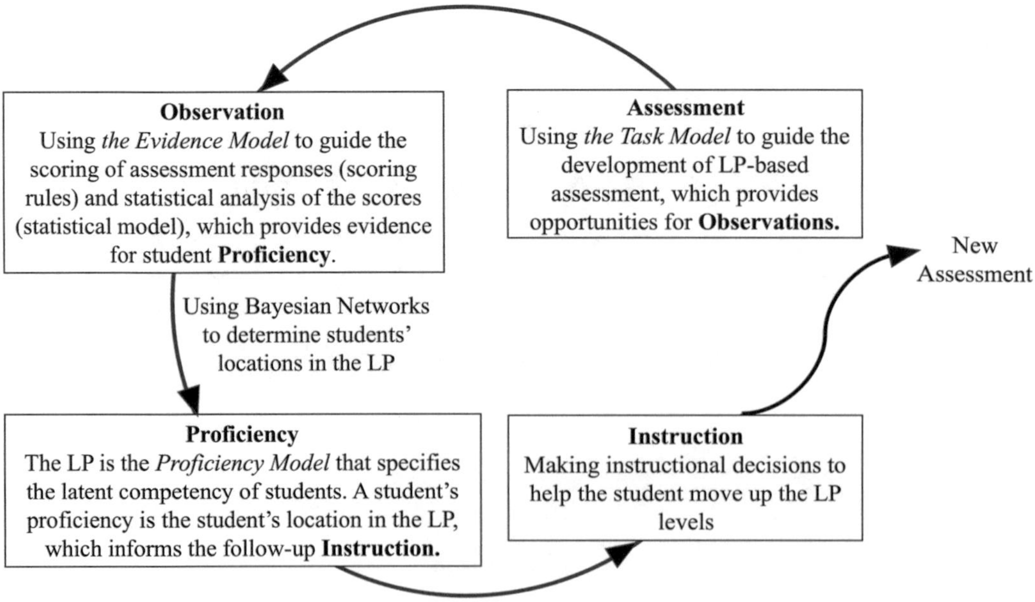

Figure 25.1 One ECD cycle for LP-based assessment and Instruction.

ECD framework (Mislevy et al., 2003) to incorporate the LP approach. The modified framework contains the following components (Figure 25.1).

- ***Assessment*** is used to gather observations: A Task Model is used to guide the development of the LP-based assessment tasks. These tasks assess students' knowledge, skills, practices, and abilities defined in the LP's upper anchor. They provide opportunities for observations as well as practice of new skills.
- ***Observation*** is used to determine proficiency: The observation includes students' assessment responses. An Evidence Model is used to guide the analyses of assessment responses. More specifically, the scoring rules guide the scoring of students' assessment responses, and the statistical model guides the statistical analysis of those scores. These analysis results provide evidence of students' proficiency. Bayesian networks are an approach to predicting students' proficiency—their locations in the LP.
- ***Proficiency*** is used to inform instruction: The LP is the Proficiency Model that specifies students' knowledge, practices, skills, and abilities. A student's proficiency can be presented as the student's location in the LP. It informs the instruction by showing where the student is and how far away the student is from the upper anchor of the LP.
- ***Instruction*** is used to move students up the LP: Based on a student's proficiency, appropriate instructional activities should be identified to help the student move up the LP levels.

Figure 25.1 shows one cycle of using assessment results to inform instruction. In science classrooms, students develop the upper anchor understanding through a sequence of cycles. Therefore, an *instructional plan* (Figure 25.2) must be laid out. The first step of the instructional plan is to figure out where the students are. Static Bayesian network can be used to locate students on an LP. The second step is to understand how instruction is likely to

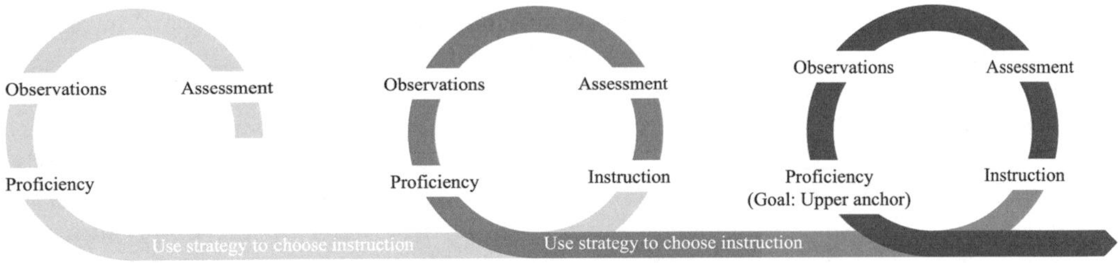

Figure 25.2 Instructional Plan: Using Assessment and Instruction to Move Student Up the LP.

advance the students. We use the DBNs to model not only where students are but also how they progress over time, which is shown as the sequence of cycles in Figure 25.2. Given an instructional plan, DBNs allow us to forecast where students will be in the LP at the end of the instructional period. A *strategy* refers to a set of rules for choosing the next instructional activity (adjusting the plan) based on the estimation of where students are on the LP.

Static Bayesian Networks

A static Bayesian network describes the relationship between a student's proficiency (location in the LP) and the observed outcomes (i.e., the LP level scores for different assessment items) at a single point in time. The original ECD model focused on a single measurement occasion (i.e., the single cycle presented in Figure 25.1). Supplied with knowledge about what observations about a student's work are likely at various levels of the LP, the Bayes net applies Bayes' rule to estimate the student's position in the LP. In addition to the likelihood of the observations, using Bayes' rule requires an initial prior probability distribution of the student at various levels. This is usually based on where the members of the target population typically are at this point in time. Given these two sources of information—the prior distribution over student locations and the likelihood of certain observations given the student's location in the LP—the Bayes net calculates a posterior (after observing the evidence) probability distribution over the student's possible locations in the LP.

The relationship between observations and LP levels produces an operational definition of the levels. Generally, at higher LP levels, the student can solve more difficult problems. They also give more sophisticated explanations for a real-world phenomenon. The most useful observations are the ones that are likely to be seen if the student is at a higher level in the progression and unlikely to be seen at a lower level. If the teacher/researcher is uncertain about which of the two levels the student is at, then the teacher/researcher should ask the student to perform tasks that are likely to show these kinds of observations. This sounds complex but is probably already done in regular practice. If the instruction is intended to move the student from one level to the next, then the assessment tasks should exercise the new knowledge, practices, and abilities introduced at the new level.

More formally, let $L(t)$ be an assignment of LP levels for a particular student at Time t. For an LP that contains multiple progress variables, $L(t)$ includes assignments to the overall LP and all its progress variables. The location of the student is unknown, so the state of knowledge about the student's location is expressed with a probability distribution $P(L(t))$ over possible positions in the LP for students at Time t; ideally, this would come from experience with past student populations similar to the target population. For example, based on the teacher's experience of teaching a student, the teacher predicts the proportion of students at each LP level at Time *t*. This *prior* distribution describes the teachers' uncertainty about the student's location *before* considering the evidence from the assessment. Let $Y(t)$ be the observed outcomes, i.e., the scores from the assessment given to that student at Time *t*. The statistical model, *P(Y(t)|L(t))*, provides the probability of students exhibiting given responses given their position in the LP; this plays the role of the likelihood in the Bayes theorem. For example, our Mathematical Thinking in Science LP (Jin et al., 2023) has four levels: Level 1 is holistic phenomenon – students describe macroscopic observations and personal feelings related to the phenomenon without identifying any qualitative attributes; Level 2 is attributes – students analyze phenomena in terms of qualitative attributes rather than measurable variables; Level 3 is measurability – students analyze phenomena in terms of measurable variables, but do not identify the correct quantitative relationships among the variables; Level 4 is relational complexity – students identify complete and correct quantitative relationships among measurable variables and use scientific ideas to conceptualize them. More detailed descriptions of each level are provided in Chapter 24 and other publications of the project (e.g., Jin et al., 2023). In this case, students with a proficiency of Level 3 have a 5% likelihood of producing responses at Level 1, a 15% likelihood of producing responses at Level 2, a 60% likelihood of producing responses at Level 3, and a 20% likelihood of producing responses at Level 4. After observing the outcomes (i.e., assessment scores showing how frequently a level score is present or absent in the student's item responses), the *posterior* (after observing the student responses) probability that a student is in a given position in the LP is *P(L(t) | Y(t))*, which can be calculated from $P(L(t))$ and *P(Y(t) | L(t))* using Bayes theorem.

Suppose that at Time *t*, experience in previous years has shown that about 50% of students are at Level 2, 40% are at Level 1, and 10% at Level 3 in a certain LP; this is the proficiency model, $P(L(t))$. Suppose that a student is asked to explain a phenomenon in terms of contrasting attributes, such as warm vs. cold and slow vs. fast, and the teacher rates the explanation as characteristic of Level 1, 2 or 3. As students sometimes produce explanations that are at a higher or lower level than their current LP level, this is expressed with the evidence model, *P(Y(t) | L(t))*. Multiplying them together produces the joint distribution as shown in Table 25.1. The marginal total below the table on the left gives the predictive probability that a randomly chosen student at this time in their instruction will exhibit the target behavior.

Now, suppose that for a particular student, the teacher rates the explanation at Level 2. Applying Bayes' rule yields the probability the student is at each level given that response, *P(L(t) | Y(t))*. This can be calculated by dividing the "L2" column in the $P(L(t), Y(t))$ table by the predictive probability on the bottom. This yields approximately 20% for Level 1,

Table 25.1 Proficiency and Statistical Models at Time t

L(t)	*P (L(t))*	*P (Y (t) | L(t))*			*P (Y (t), L(t))*		
		Response			*Response*		
		L3	*L2*	*L1*	*L3*	*L2*	*L1*
3	0.1	0.7	0.2	0.1	0.07	0.02	0.01
2	0.5	0.2	0.6	0.2	0.10	0.30	0.10
1	0.4	0.1	0.2	0.7	0.04	0.08	0.28
Total	1.0				0.21	0.40	0.49

75% for Level 2, and 5% for Level 3. Thus, observing behaviors associated with higher levels of the learning progression shifts our estimate of the student's location towards the higher levels.

Dynamic Bayesian Networks

We use a hypothetical LP for Exploratory Analysis of Scientific Data (EASD) as an example to demonstrate how DBNs can be used to monitor student progress along an LP containing multiple progress variables. In this section, we first present the hypothetical EASD LP. Next, we describe the Dynamic Bayesian Networks. Finally, we discuss how the results from DBN can be used to inform instruction.

The EASD LP

EASD refers to exploring large datasets and/or multiple datasets to identify patterns, trends, erroneous values, and outliers; exploring the scientific meanings of these analysis results; and using the results to formulate meaningful questions that are worth investigating. Based on a comprehensive literature review, we developed a hypothetical learning progression for EASD. More details of the development of the EASD learning progression and its progress variables and achievement levels are discussed in a working paper (Jin & Li, 2023). In this chapter, we focus on the use of the Bayesian networks in LP-based score reporting for teachers. Figure 25.3 illustrates the relationships among the three progress variables of EASD, which are three cognitive processes (CPs) required to conduct EASD. The CPs are:

- **CP1:** Understanding individual variables. Students use scientific content knowledge to explain the meaning of individual variables. They analyze the distribution of each variable to identify patterns, trends, and anomalies and interpret these analysis results using scientific content knowledge.
- **CP2:** Analyzing relationships between two variables. Students analyze the relationships between two variables to identify patterns, trends, and anomalies and interpret these analysis results using scientific content knowledge.
- **CP3:** Analyzing relationships among three or more variables. Students analyze the relationships among three or more variables to identify patterns, trends, and anomalies and interpret these analysis results using scientific content knowledge.

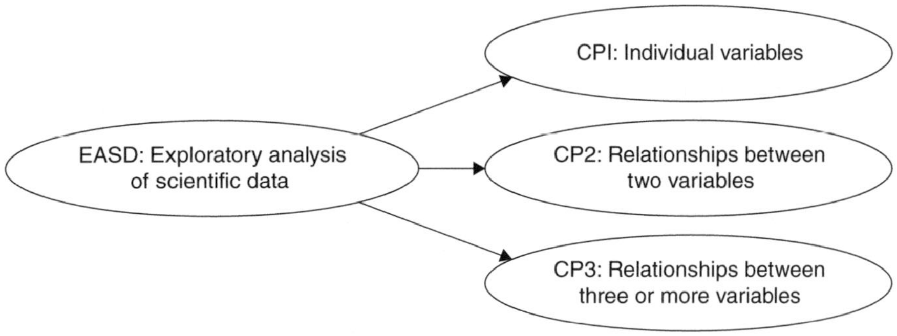

Figure 25.3 The Proficiency Model---the EASD LP.

For each of CP1, CP2, and CP3, students' proficiency is assigned to one of three levels of achievement. Level 1 is Naïve Reasoning. At this level, students do not understand the scientific meaning of data points. Neither can they identify and interpret distributions and relationships in data. Level 2 is Transitional Reasoning. At this level, students can identify some patterns, trends, and anomalies in graphs about the distribution and relationships of variables, but they often cannot use scientific ideas to interpret them. Level 3 is Analytical Reasoning. At Level 3, students can identify meaningful patterns, trends, and outliers regarding the distribution and relationships of the variables. They can also use scientific ideas to interpret those findings.

Dynamic Bayesian Network

A Bayesian network is a mechanism for representing a joint probability distribution over multiple variables. The EASD variables are represented by nodes in a graph (Figure 25.3), and the edges (arrows) mark dependencies between the nodes. In this sense, it is similar to a structural equation model or path diagram. However, a Bayesian network has an additional restriction: separation in the graph implies that the variables are conditionally independent (Pearl, 1988). This condition allows for efficient algorithms for scoring students and other computations (Almond et al., 2015). In this example, the chosen graphical structure indicates that the joint prior distribution over all the nodes will be expressed by first providing a distribution over the highest-level node (i.e., EASD) and then a distribution for each of the variables (i.e., CP1, CP2, and CP3) conditioned on the student's value for the highest node.

The nodes in Figure 25.3 represent variables in the model, and the edges (connections between the nodes) represent functional dependencies among the variables. The direction of the arrows indicates how the model is to be factored. In particular, Figure 25.3 shows a typical *part-of* pattern, where the student's level on the three component skills, CP1, CP2, and CP3, is related to the student's level on the composite skill EASD. Because of this dependency of the component skills on the composite skills, they will be correlated in the population of learners (the higher the student is on the composite skill, the higher the student is likely to be on each of the components). However, in Figure 25.3 the connection only goes through EASD, which means that given where the student is on the composite skill, the student's location on the component skills is independent. If this is not true, more edges can be added to the graph to model the relationships among the skills in the learner population. For example, if CP2 was a prerequisite for CP3, then an edge from CP2 to CP3 could indicate that this relationship needs to be modeled.

Estimates of the student location in an LP with multiple skills take into account two types of evidence: 1) direct evidence from observations from tasks that used the measured skills and 2) indirect evidence from the correlation between the skills observed in the population. In the example above, suppose a student does well on a task involving the relationship between two variables. This provides direct evidence that the student is at a higher level of the CP2 progression. That, in turn, implies that the student is higher on the EASD progression, and hence it provides indirect evidence for both CP1 (single variable representations) and CP3 (three or more variables). This use of indirect evidence is built into the Bayesian network algorithm and can improve estimates for scores representing specific dimensions of the LP (Wainer & Thissen, 2001).

Generally, teachers are not only interested in where a student is at a single time, $L(t)$, but at several different points in time, $t = 0, \ldots, T$, where $t = 0$ represents the beginning of instruction and $t = T$ represents the end. Each time point represents one trip through the learning cycle

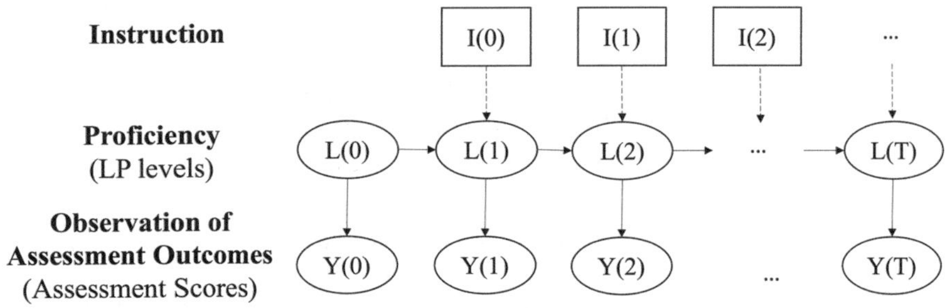

Figure 25.4 The LP across Time.

shown in Figure 25.2. The student is given instruction and then assessed to see how the student's position in the LP has changed. In each cycle, the teacher chooses instruction (based on the students' estimated positions in the LP) and assessments for the student. The student's growth depends on these choices.

A DBN (DBNs; Dean and Kanazawa, 1989) extends a Bayesian network along the time dimension. Figure 25.4 provides a schematic view of a DBN. Time flows horizontally. At each time point, the student has a given set of values for all proficiency variables, *L(t)*, which is unknown to the instructor, and a set of values on assessment tasks, *Y(t)*, which are observed (although missing values can be handled by the model). In this sense, the part of the figure at Time 0 is the same as the single measurement occasion Bayes net. What is different is the arrows between the time slices. These represent a growth model. They describe the probability distribution of where a student will be in the LP at the next time point, *L(t+1)*, given where the student was at Time *t*, *L(t)*, and the instruction given at Time *t*, *I(t)*.

The first column in Figure 25.4 is a familiar problem that ECD was designed to address: estimating the student's ability at a single time point. The ECD proficiency model provides the population distribution for the initial position in the LP $L(0)$. This includes information about the expected levels of students before any instruction begins and the correlation among the levels in a multivariate setting. Almond et al. (2015) showed how this can be done using a Bayesian network.

The ECD statistical model, $P(Y(t) \mid L(t))$, are the links between $L(t)$ and $Y(t)$. Note that all paths between $Y(t)$ and all other observations (*Ys*) pass through *L(t)*. The implication is that if the student's position in the LP at Time *t* is known, then *Y(t)* is *conditionally independent* of past and future locations in the LP, and hence past and future assessment observations. (In practice, as *L(t)* is unknown and related to both *L(t-1)* and *L(t+1)* and hence past and future performances are *marginally* dependent; it is conditioning on *L(t)*, i.e., hypothesizing that the student's location is known at Time *t* which renders them independent.) This conditional independence assumption is one of the assumptions of a *Markov* model, and it allows the estimation of student location to be factored across time points.

In a DBN, there is a new kind of edge, the horizontal arrows going from *L(t-1)* to *L(t)*. The node *L(t)* also has an arrow coming from *I(t)* representing the instruction the student engages with between Time *t-1* and Time *t*. This represents the probability that the student will advance (or regress) along the LP. This depends both on where the student was at the current time point, and what the teacher does between time points.

In general, the teacher wants to design a collection of tasks for each Time *t* that will provide optimal information about the location of the student in the LP. In particular, if the teacher is giving instruction about Level L_k the assessment tasks should be related to Levels L_{k-1} and L_{k+1}.

While the same general form of the statistical model holds for all time points, the mix of tasks in the benchmark assessments is likely to vary. In particular, early in the instruction, the assessment tasks are usually focused on the initial levels of the LP, and towards the end of the instruction, their focus shifts to the higher levels. To do good measurement, all the statistical models need to be calibrated to a common scale; this presents a vertical scaling challenge (Almond et al., 2014).

Modeling Growth over Time

While conventional ECD works for a single time slice, the dynamic ECD model also requires building a model for the transitions $L(t) \rightarrow L(t+1)$, $P(L(t+1) \mid L(t))$. This is the *growth model.* The distribution at Time 0 is given by the proficiency model, $C(0)$. At Time 1, it is given by:

$$P(L(1)) = \sum_{\ell(0) \in L(0)} P(L(1) | \ell(0)) P(\ell(0)),$$

where the sum is taken over all of the possible student locations at Time 0.

At Time t, the probability distribution over locations is given by the recursive formula:

$$P(L(t)) = \sum_{\ell(t-1) \in L(t-1)} \cdots \sum_{\ell(0) \in L(0)} P(L(t) | \ell(t-1)) \prod_{r=1}^{t-1} P(\ell(r) | \ell(r-1)) P(\ell(0)$$

Let $p_{ijt} = P(L(t+1) = j \mid L(t) = i)$, so each row of the matrix P_t is a probability distribution. This transition matrix can vary across time, and in general with the instructional activities which occur between the measurement times.

For example, consider a student at Level k of an LP. If the student reads a handout aimed at students much lower in the LP, the amount of learning will be minimal. On the other hand, if the student attempts to read a handout designed for students located much higher in the LP, then they will likely not be able to fully understand the handout, and again the amount of learning will be minimal. In general, the amount of learning is related to the activity or assignment the teacher provides at Time t, $I(t)$. So $P_t = P_{I(t)}$.

The DBN model can also forecast a student's performance at a future time, $E[Y(t+1)]$. From an information theory perspective, an assessment task offers the most information when the uncertainty about $Y(t+1)$ is maximized (e.g., there is a roughly 50–50% chance of getting the item right). These maximally uncertain tasks will generally be tasks that differentiate between what the teacher believes is the student's current level in the LP and the next level. As such, they are in Vygotsky's zone of proximal development (Vygotsky, 1978), or, using Falmagne's learning space terminology, the current level of the LP is at the student's inner fringe and the next level is the outer fringe of the student's learning space Falmagne et al. (2007). This correspondence between items with high evidentiary value and high instructional value (given the student's location in the LP) can be exploited for assessment design. Shute, Hansen, and Almond (2007) found that students randomly assigned to a group where practice items were ordered to maximize information had more learning than students with items in a fixed sequence.

In summary, to build a DBN, the experts must develop models for initial proficiency, proficiency growth as a function of instruction, and evidence models for the tasks used to assess the students. One of the advantages of Bayesian networks is that before data becomes available, the model will give predictions using expert opinion. As data becomes available, the parameters can be adjusted to reflect the observed experience (Almond et al., 2015).

Sequencing Instruction

Typically, a course is divided into a series of *periods* (these may correspond to classes, weeks, days, or in other ways that make sense). Each period will have some instructional activities (lectures, readings, assignments, experiments, etc.) designed for the level of the LP at which the students should be; this is the $I(t)$ in Figure 25.4. There is also usually some assessment of student ability done within the period, either embedded in assignments that also serve an instructional purpose or an explicit end-of-unit test; this is the $Y(t)$ in Figure 25.4. A number of these periods is $t \in \{1,...,T\}$. Time point T is the end of the course of instruction. Time point 0 is the point at which instruction starts, therefore, $Y(0)$ is the result from a pre-test. The true location in the LP at the end of each period is the unobserved value $L(t)$.

The assignment of instruction to each period is a realization of $I(t)$ for every time point. The DBN model uses this assignment of instruction for two important operations. Assume that the current time point is t, and that the series of instructions: $I(1),...I(t)$ is known as the series of assessment results $Y(0),...,Y(t)$. Then the DBNs can:

- Estimate the student's current ability, $L(t)$. This is called *filtering* in the signal processing world because it reduces the noise of the signal, or increases the reliability of the estimate.
- Given an assignment of instruction for the future, $I(t+1),...I(t')$, it can *forecast* the student's ability at a future time, $L(t')$. A particularly important application of this is that given an assignment of instruction, the DBN can forecast the student's location in the LP at the end-of-course, $L(T)$. As the instructional goal is often stated as a function of the end-of-course location, this can calculate the probability that the plan will succeed.

The assignment of instruction often has contingencies. For example, at Time point t, the teacher could continue to the next level of the LP if the $P(L(t) > c)$ is sufficiently high and review the current level if it is not.

The DBN can also forecast the end-of-course performance based on one of these assignments of instruction. In particular, it can forecast the expected result of two different assignments. In the world of artificial intelligence, this is known as planning under uncertainty. A DBN can be extended to produce a partially observed Markov decision process (POMDP; Boutilier, Dean & Hanks, 1999), which supports a number of planning algorithms. Two new pieces of information need to be added to the DBN to make it a POMDP. The first is a *utility function* that assigns relative values to the various states the student is at the end-of-course: How much worse is it to be two steps behind the goal state than one step behind? How much better is it to move beyond the goal state than to just reach the goal? The second is a *cost function*, which describes how much various activities cost. For example, leaving a student in whole-class instruction is usually less costly than assigning the student to one-on-one instruction with a specialist, but the latter may produce better results. The algorithm can then evaluate the contingency and try to search for the optimal *policy*, which is a set of rules for how to select what instruction to provide between two measurement occasions.

DBNs and Related Models in Other Research

Several researchers have developed DBNs for assessments that unfold across time (Reye, 1998; Levy, 2014; Li et al., 2016; Mislevy & Yan, 2017, Choi & Mislevy, 2022). The last chapter of Almond et al. (2015) also generally describes the DBN approach, although it does not develop an example. Reichenberg (2018) provides a review of various projects using the DBN framework.

There are very few examples of full POMDP models, with Tokac and Almond (2016) being an exception. This simulated example attempts to cast the Response-To-Intervention (Greenwood et al., 2011) process, where on the basis of periodic placement tests, students are assigned to either Tier 1 (mainstream classroom education) or Tier 2 (small group instruction). Note that the teacher might be resource-constrained as to how many students can be accommodated in Tier 2. In this case, the policy is based on not just a single POMDP, but an ensemble of such models (Boutilier & Liu, 2016).

Figure 25.4 bears a strong resemblance to a hidden Markov model (HMM). The HMM differs from the DBN in that the latent state is usually univariate, and often the observations from successive states are as well. Often the same measurement is used at each time point, for example, the same depression test could be used at each time point, and the transition model describes how a patient moves between normal and depressed moods (Eid, 2002). If the competency model consists of a single continuous variable, then the HMM can be recast as a multivariate IRT model, with the dimensions corresponding to the time points (Marvelde et al., 2006).

Consider a simple DBN, where the skill consists of a single binary variable (the student has mastered the skill or not), and at each time point, the student makes another attempt at a dichotomously scored task, which is a variant of a basic type. This *knowledge tracing* model can be characterized with just three probabilities: the probability that a non-master can *guess* the correct solution, the probability that a non-master *transitions* to a master at each attempt, and the probability that a master will *slip* and make an incorrect solution (Corbett & Anderson, 1995). Because of its simplicity, this model is easy to work with and has enjoyed great success. *Performance factor analysis* (Pavlik et al., 2009) is a generalization to more complex cases. Chang et al. (2006) sketch the relationship between knowledge-tracing models and DBNs.

If the competency model is a single (ordered) latent class, then the general framework is called *latent transition analysis* (Collins & Wugalter, 1992). This is often coupled with the use of cognitive diagnostic models for the evidence models (Sudheesh & Golden, 2021).

Note that general learning curve models (Singer & Willett, 2003) can be put into this same framework, although often the modeling focuses on estimating the future values for the observations and not the latent skill. Ye (2005) provides an example.

Using the DBNs to Inform Instruction

We use the EASD LP to demonstrate how DBNs can be used to inform instruction. Given a DBN, and a set of observations from a student, which includes both the history of instructional activities and the results of the periodic assessments, the Bayesian network estimates the probability that the student is for each of the progress variables (CP1, CP2, and CP3). Almond et al. (2009) recommended viewing these as hanging bar plots.

In Figure 25.5, a horizontal line is set to show the percentages of students who achieve the learning goal—Level 3 (Analytical Reasoning) of the EASD LP. Therefore, the line is between Level 2 (Transitional Reasoning) and Level 3 (Analytical Reasoning). The hanging bar plots make it easy to tell which students and skills have most likely reached the goal (i.e., a certain LP level). If the goal state was Level 2 (Transitional Reasoning), a simple adjustment can be made to the plot to show the percentages of students who achieved or not achieved this goal. Aggregating across students makes it easy to compare one student (yellow bars) to a class (rose bars), as is shown in Figure 25.6. The student's overall proficiency in EASD is below the class average, specifically due to the student's low proficiency in LP2.

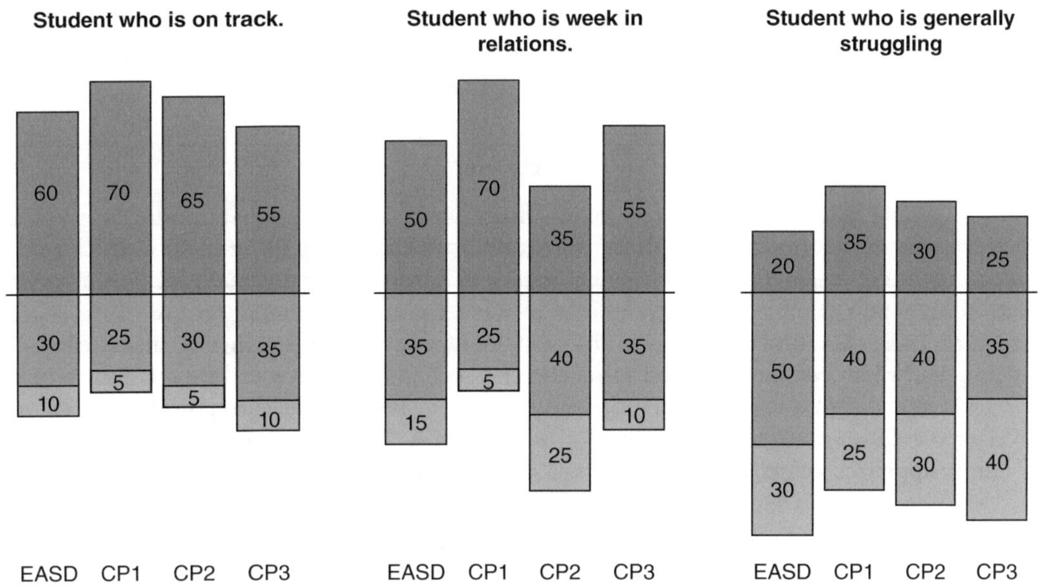

Figure 25.5 Hanging bar plots showing three different students.

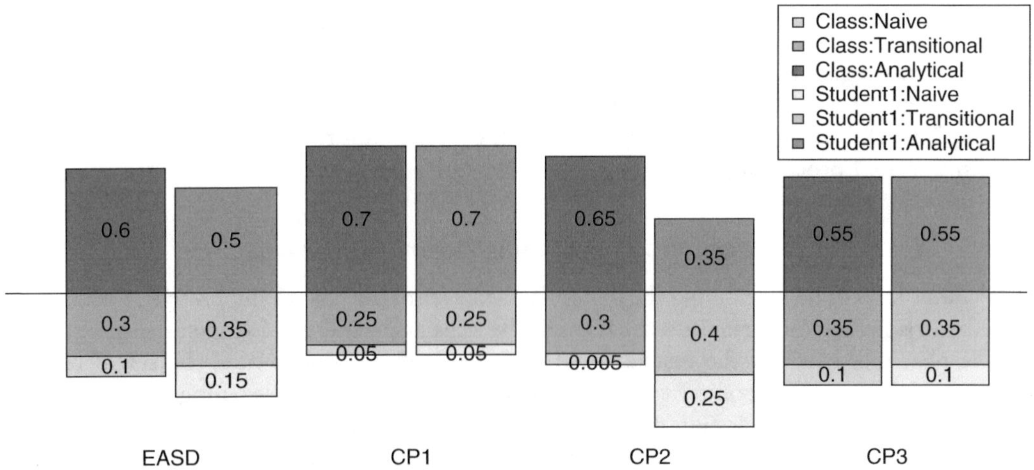

Figure 25.6 Comparing One Student to the Entire Class.

In addition to the distribution, a student can be classified into one of the states (although this loses information about the uncertainty of the classification). Often, the student is classified in the modal category (the one with the highest probability). The three students in Figure 25.5 are classified as Level 3 Analytical, Level 3 Analytical, and Level 2 Transitional for the overall EASD, respectively, based on the level that has the highest probability. As the variables are ordinal, the median and other order statistics are well-defined. Using the median, the first student clearly has a median in the Analytical category, the second one on the border between Analytical and Transitional, and the last one is clearly in the transitional state. Mitzel et al. (2001) suggest using the 67% quantile. This value is Transitional for all three students:

for each student, the Pr(*EASD* ≥ *Analytical*) < .67 and Pr(*EASD* ≥ *Transitional*) > .67. If a student is "likely" to perform at and above an achievement level, the student is categorized as a student at that level. Here, "likely" is defined as 67% probability.

Discussion

In LP research, using assessment results to inform instruction holds paramount importance. This chapter uses concrete examples to illustrate how DBNs can be used to aid teachers in decision-making when using assessment results to inform their instruction. The marginal distributions and graphical displays above do not rely on the dynamic nature of the model. Any diagnostic classification model with ordinal latent states can produce similar probability statements, which can be displayed using the displays above. However, the advantage of the dynamic model is that it can bring more information to bear on the estimation of the current level in the LP of a student (or collection of students). There are five advantages that this dynamic approach offers over separate measures at each time point.

Advantage 1: Improving Estimation of Students' Current LP Levels

A static evaluation system uses only the assessment responses at the current time point. Thus, the reliability of the estimates is limited by the amount of time devoted to testing at the current time point. The dynamic model can use assessment scores and the history of instruction to strengthen inferences about the students' current LP levels. The eventual estimates are a weighted average between the current and past scores with the past scores being down-weighted by a factor that is related to the ratio between the standard error of measurement of the current scores and the variation around the growth model (Almond, 2007b, 2007a). This borrowing strength from the past to improve estimates in the present is known in signal processing as *filtering*.

Advantage 2: Forecasting Future Performance

While filtering is done by running the dynamic model from the past to the present, *forecasting* is done by letting the model run into the future. The expected patterns of student learning depend on the plan that specifies the instructional activities and the policy (i.e., the set of rules for choosing activities). This could be a fixed plan such as a sequence of learning activities aligned with the LP, or a contingent plan which will change in response to the estimates of students' LP levels. Whatever the plan, the model can make predictions of where the student will be in the future. However, it is important to note that, because of the uncertainty in the growth model, the forecasts could get more and more uncertain as they get further and further from the present. One important application is forecasting how likely a student is to meet end-of-year goals using the current instruction. If this probability becomes too low for a few students in a class, then the teacher might consider bringing additional resources to bear for those students to get them back to the rest of the class. The use of DBNs in LP thus provides early warnings for what could be a problem with serious consequences at the end of the course or school year.

Advantage 3: Predicting Students' Performance on New and Unanswered Items

In the Bayesian network, there is no real distinction between a latent variable that cannot be observed and an observable outcome from a task that has not yet been observed. In particular, given the observed history of a particular student, the Bayesian model can go from an estimated proficiency profile to a predicted performance on a particular task or activity. Similarly, it can forecast the performance of that activity at a future time point.

There are two important applications of this. First, predicting the performance of assessment activities can be used to select assessment activities that maximize the weight of evidence from those activities (Madigan & Almond, 1995). In particular, the best activities are ones for which the model expresses a high degree of uncertainty about the result. There is also some support (Shute et al., 2008) for the idea that such items are also good for learning; they are likely to lie within Vygotsky's zone of proximal development or inner and outer fringes of Falmagne's learning space (Falmagne, 1989; Vygotsky, 1978). In the LP context, DBNs can be used to predict the scores a student may receive in an LP-based assessment. This information will allow teachers to select tasks for the students to practice.

The second application involves the construction of a market basket of standard tasks (DeVito & Koenig, 2000). This is a standardized collection of tasks calibrated to the scale used in the measurement model. The Bayesian model can forecast or predict the performance on each task, as well as on the entire set of tasks or any interesting subset. This provides an important practical illustration of what the levels of the LP mean in terms of performance and could be useful in both counseling students and their parents about the meaning of scores.

Advantage 4: Describing Student Learning Over Time

Running the model forward produces better, *filtered* estimates of ability at the present; Running the model backward produces better, *smoothed* estimates of ability in the past. This is called smoothing because it helps straighten out small variations from time-point to time-point caused by the standard error of measurement of the instrument at each time point by assuming that the student's ability is growing smoothly across time. Thus, just as filtering borrows strength from the past to improve future estimates, smoothing borrows from both the past and future to stabilize estimates in the past. These smoothed, filtered, and forecast estimates can be linked into an electronic gradebook, hiding the complexity of the model (unless the end-user wants to understand the details).

Advantage 5: Planning for the Future

Just as the dynamic model can predict whether a given student will meet an educational goal under a given policy, the model can evaluate that probability under a number of different policies. *Planning* is the process of selecting the optimal policy. Planning is the principal use of POMDPs. Two things are necessary to move from a DBN to a POMDP: some measure of what the various outcomes are worth (utilities) and some measure of the cost of the various activities.

In educational models, utilities are usually based on what level of the LP the student reaches at the end of the course of instruction. In particular, the utility function needs to express how much more valuable it is for a student to reach the Analytical level rather than the Transitional, and at least Transitional rather than being still at Naïve. Furthermore, with three skills in the model, is *CP1* more valuable than *CP3*, or are they all equally weighted?

The costs for the activities must be on the same scale as the utilities, as the goal is to maximize benefit minus cost. In the example described here, all three modeled activities are whole-class activities with similar costs. But other choices, like small group instruction or bringing in a specialist for one-on-one work with a student may consume more resources. The policy solution of a POMDP model balances the costs and expected returns.

In summary, the DBNs, unlike conventional psychometric models, have a growth component, which allows teachers to monitor student progress along the LP over time. More specifically, teachers can compare student progress with the end-of-year goals and decide what instructional activities to choose. They can develop and select assessment tasks for students to work on based on the DBNs' prediction of the students' assessment scores. They can also develop a realistic plan for instruction throughout the academic year or semester.

References

Almond, R. G. (2007a). An illustration of the use of markov decision processes to represent student growth (learning) (ETS Research Report RR-07-40). Educational Testing Service. http://www.ets.org/research/researcher/RR-07-40.html

Almond, R. G. (2007b). Cognitive modeling to represent growth (learning) using markov decision processes. *Technology, Instruction, Cognition and Learning (TICL)*, 5, 313–324. http://www.oldcitypublishing.com/TICL/TICL.html

Almond, R. G., Goldin, I., Guo, Y., & Wang, N. (2014). Vertical and stationary scales for progress maps. In J. Stamper, Z. Pardoz, M. Mavrikis, & B. M. McLaren (Eds.), Proceedings of the 7th International Conference on Educational Data Mining (pp. 169–176). http://educationaldatamining.org/EDM2014/uploads/procs2014/long\%20papers/169_EDM-2014-Full.pdf

Almond, R. G., Mislevy, R. J., Steinberg, L. S., Williamson, D. M., & Yan, D. (2015). *Bayesian Networks in Educational Assessment*. Springer New York. http://www.springer.com/statistics/social+sciences+%26+law/book/978-1-4939-2124-9

Almond, R. G., Shute, V. J., Underwood, J. S., & Zapata-Rivera, J.-D. (2009). Bayesian networks: A teacher's view. *International Journal of Approximate Reasoning*, 50, 450–460. 10.1016/j.ijar.2008.04.011

Boutilier, C., Dean, T., & Hanks, S. (1999). Decision-theoretic planning: Structural assumptions and computational leverage. *Journal of Artificial Intelligence Research*, 11, 1–94. citeseer.ist.psu.edu/viewdoc/download?doi=10.1.1.65.9397&rep=rep1&type=pdf

Boutilier, C., & Liu, T. (2016). Budget allocation using weakly coupled, constrained markov decision processes. In A. Ihler & D. Janzing (Eds.), Uncertainty in Artificial Intelligence (UAI): Proceedings of the Thirty-Second Conference (pp. 52–61). Association for Uncertainty in Artificial Intelligence (AUAI). http://auai.org/uai2016/proceedings/papers/246.pdf

Chang, K., Beck, J., Mostow, J., & Corbett, A. (2006). A Bayes net toolkit for student modeling in intelligent tutoring systems. In M. Ikeda, K. D. Ashley, & T.-W. Chan (Eds.), *Intelligent Tutoring Systems* (pp. 104–113). Springer. 10.1007/11774303_11

Choi, Y., & Mislevy, R. J. (2022). Evidence centered design framework and DBN for modeling learning progression in online assessment system. *Frontiers in Psychology*, 13, 742956. 10.3389/fpsyg.2022.742956

Collins, L. M., & Wugalter, S. E. (1992). Latent class models for level- sequential dynamic latent variables. *Multivariate Behavioral Research*, 27(1), 131–157. 10.1207/s15327906mbr2701_8

Corbett, A. T., & Anderson, J. R. (1995). Knowledge tracing: Modeling the acquisition of procedural knowledge. *User Modeling and User-Adapted Interaction*, 4, 253–278.

Dean, T., & Kanazawa, K. (1989). A model for reasoning about persistence and causation. *Computer Intelligence*, 5, 142–150.

DeVito, P. J., & Koenig, J. A. (2000). *Investigating District-Level and Market-Basket Reporting*. National Research Council. http://www.nap.edu/catalog/9891.html

Eid, M. (2002). A closer look at the measurement of change: Integrating latent state-trait models into the general framework of latent mixed Markov modeling. Methods of Psychological Research Online. http://www.mpr-online.de

Falmagne, J.-C. (1989). A latent trait model via a stochastic learning theory for a knowledge space. *Psychometrika*, 54, 283–303.

Falmagne, J.-C., Cosyn, E., Doble, C., Thiéry, N., & Uzun, H. (2007). Assessing Mathematical Knowledge in a Learning Space: Validity and/or Reliability.

Fortus, D., & Krajcik, J. (2012). Curriculum coherence and learning progressions. In B. Fraser, K. Tobin, & C. McRobbie (Eds.), *Second International Handbook of Science Education* (Vol. 24). Springer, Dordrecht. 10.1007/978-1-4020-9041-7_52

Greenwood, C. R., Bradfield, T., Kaminski, R., Linas, M., Carta, J. J., & Nylander, D. (2011). The response to intervention (RTI) approach in early childhood. *Focus on Exceptional Children*, 43(9). 10.17161/foec.v43i9.6912

Jin, X., & Li, X. (2023, May). *Preparing Students for a Data-intensive World: A Framework for Exploratory Analysis of Scientific Data.* (Working Paper)

Kaldaras, L., Akaeze, H., & Krajcik, J. (2021). Developing and validating Next Generation Science Standards -aligned learning progression to track three-dimensional learning of electrical interactions in high school physical science. *Journal of Research in Science Teaching*, 58(4), 589–618.

Levy, R. (2014). DBN modeling of game based diagnostic assessments (({CSE} Report No. 837). University of California, National Center for Research on Evaluation, Standards, Student Testing (CRESST). Retrieved from http://www.cse.ucla.edu/products/reports/R837.pdf

Li, F., Cohen, A., Bottge, B., & Templin, J. (2016). A latent transition analysis model for assessing change in cognitive skills. *Educational and Psychological Measurement*, 76(2), 181–204. 10.1177/0013164415588946

Madigan, D., & Almond, R. G. (1995). Test selection strategies for belief networks. In D. Fisher & H. J. Lenz (Eds.), *Learning from Data: AI and Statistics V* (pp. 89–98). Springer-Verlag.

Marvelde, J. M., Glas, C. A. W., Van Landeghem, G., & Van Damme, J. (2006). Application of multidimensional item response theory models to longitudinal data. *Educational and Psychological Measurement*, 66(1), 5–34.

Mislevy, R. J., Steinberg, L. S., & Almond, R. G. (2003). On the structure of educational assessment (with discussion). *Measurement: Interdisciplinary Research and Perspective*, 1(1), 3–62.

Mislevy, R. J., & Yan, D. (2017). Evidence-centered assessment design and probability-based inference to support the generalized intelligent framework for tutoring (GIFT). In R. Sottilare, A. C. Graesser, X. Hu, & G. Goodwin (Eds.), *Design Recomendations for Intelligent Tutoring Systems: Vols. Volume 5, Assessment Methods* (pp. 101–132). U.S. Army Research Laboratory (ARL). https://www.researchgate.net/publication/319330321_Design_Recommendations_for_Intelligent_Tutoring_Systems_Volume_5_-_Assessment_Methods

Mitzel, H. C., Lewis, D. M., Patz, R. J., & Green, D. R. (2001). The bookmark procedure: Psychologicalperspectives. In G. J. Cizek (Ed.), *Setting Performance Standards: Concepts, Methods, and Perspectives* (pp. 249–283). Lawrence Erlbaum Associates Publishers.

Neumann, K., & Nagy, G. (2013). Students' progression in understanding energy. Paper presented at the Annual Conference of the National Association for Research in Science Teaching (NARST), Rio Grande, Puerto Rico.

Norsys, Inc. (2012). Netica. http://www.norsys.com

Osborne, J. F., Henderson, J. B., MacPherson, A., Szu, E., Wild, A., & Yao, S. Y. (2016). The development and validation of a learning progression for argumentation in science. *Journal of Research in Science Teaching*, 53(6), 821–846. 10.1002/tea.21316

Pavlik, P. I., Cen, H., & Koedinger, K. R. (2009). Performance factors analysis – A new alternative to knowledge tracing. *Artificial Intelligence in Education*, 531–538. 10.3233/978-1-60750-028-5-531

Pearl, J. (1988). *Probabilistic Reasoning in Intelligent Systems: Networks of Plausible Inference.* Morgan Kaufmann.

Reichenberg, R. (2018). DBNs in educational measurement: Reviewing and advancing the state of the field. *Applied Measurement in Education*, 31(4), 335–350. 10.1080/08957347.2018.1495217

Reye, J. (1998). Two-Phase updating of student models based on dynamic belief networks. In G. Goos, J. Hartmanis, J. van Leeuwen, B. P. Goettl, H. M. Halff, C. L. Redfield, & V. J. Shute (Eds.), *Intelligent Tutoring Systems* (Vol. 1452, pp. 274–283). Springer Berlin Heidelberg. 10.1007/3-540-68716-5_33

Riconscente, M., Mislevy, R. J., & Corrigan, S. (2015). *Evidence-Centered Design. Handbook of Test Development.* Routledge.

Shute, V. J., Hansen, E. G., & Almond, R. G. (2007). An assessment for learning system called ACED: Designing for learning effectiveness and accessibility. *ETS Research Report Series*, 2007(2).

Shute, V. J., Hansen, E. G., & Almond, R. G. (2008). You can't fatten a hog by weighing it - or can you? Evaluating an assessment for learning system called ACED. *International Journal of Artificial Intelligence in Education*, 18(4), 289–316. http://www.ijaied.org/iaied/ijaied/abstract/Vol_18/Shute08.html

Singer, J. D., & Willett, J. B. (2003). *Applied Longitudinal Data Analysis: Modeling Change and Event Occurrence* (1st ed.). Oxford University Press, USA.

Sudheesh, A., & Golden, R. M. (2021). Exploring temporal functional dependencies between latent skills in cognitive diagnostic models. In M. Wiberg, D. Molenaar, J. González, U. Böckenholt, & J.-S. Kim (Eds.), *Quantitative Psychology* (pp. 253–262). Springer International Publishing. 10.1007/978-3-030-74772-5_23

Tokac, U., & Almond, R. G. (2016). The efficacy of the POMDP-RTI approach for early reading intervention. In K. B. Laskey & R. N. Carvalho (Eds.), Proceedings of the Thirteenth UAI Bayesian Modeling Applications Workshop (BMAW 2016) (pp. 36–43). http://ceur-ws.org/Vol-1663/bmaw2016_paper_6.pdf

Vygotsky, L. (1978). *Mind in society: The development of Higher Mental Processes*. Harvard University Press.

Wainer, H., & Thissen, D. (2001). True score theory: The traditional method. In D. Thissen, & H. Wainer (Eds.), *Test Scoring* (pp. 23–72). Lawrence Erlbaum Associates Publishers.

Ye, F. (2005). *Diagnostic Assessment of Urban Middle School Student Learning of Pre-Algebra Patterns [PhD Thesis]*. The Ohio State University.

26

THE APPLICATION OF AUTOMATED SCORING TECHNOLOGY IN LEARNING PROGRESSION-BASED ASSESSMENT

Brian Riordan, Hui Jin, Cynthia Lima, Kenneth Steimel, and Duanli Yan

Introduction

Learning progressions (LPs) are "descriptions of the successively more sophisticated ways of thinking about a topic that can follow one another as children learn about and investigate a topic over a broad span of time" (National Research Council [NRC], 2007). An LP often describes student development as a sequence of levels of achievement, including a lower anchor depicting prior knowledge and skills of students when they enter the progression, an upper anchor describing what students are expected to know and be able to do when they complete the learning, and several levels representing the intermediate steps between these two anchors. An LP may contain more than one progress variable, and each progress variable represents one specific aspect or component of the construct.

Researchers develop and validate LPs based on student responses in interviews and assessments. Constructed Response (CR) items can be an effective way of assessing students' understanding. By having students explain their reasoning, constructed responses can display individual student ideas, placing them in LPs more clearly. However, scoring CRs can be time-consuming for both researchers and teachers. Recent studies have used automated scoring technology to free teachers, instructors, and researchers from the burden of scoring and to provide scoring reports immediately after students complete the assessment (Jescovitch et al., 2020; Wang et al., 2021). Several recent reviews have addressed LPs and automated scoring in science. Zhai et al. (2020) provide a recent broad overview and taxonomy of the application of machine learning methods in science assessments. They note that most studies use supervised regression or classification to assign scores to or classify text-based responses. There has been a rapidly increasing number of studies applying machine learning methods in science over the past 15 years as tools for machine learning have matured.

In this chapter, we investigate the issue of how LP-based CR assessment items and rubrics can be constructed in ways that not only improve the validity of the scoring results but also

DOI: 10.4324/9781003170785-30

diagnose students' intuitive reasoning patterns. Using the data from an assessment for a validated Mathematical Thinking in Science (MTS) LP (Jin et al., 2023), we report on a case study of automated scoring for CR items that assess high school students' mathematical thinking in science in an online learning environment. In particular, we demonstrate how the weaknesses of holistic scoring models can be ameliorated with more detailed model validation techniques and additional models for "idea detection."

Conceptual Framework

Our conceptual framework contains two parts: (1) the key components of our approach to automated scoring – holistic scoring and idea detection; (2) the MTS LP used for automated scoring. We begin with a brief overview of holistic and analytic rubrics for automated scoring in science assessment, followed by a description of the MTS LP.

Holistic Scoring

In standard automated scoring settings, human raters score students' responses based on a set of scoring rubrics. The text of the responses and the human-assigned scores serve as the training data for the automated scoring system. In this process, two types of scoring rubrics can be used to score students' responses: holistic rubrics and analytic rubrics. These two approaches to scoring were initially developed by researchers in educational measurement (Turley & Gallagher, 2008; Mertler, 2001; Nitko, 2001; Sadler, 2009) to help teachers score students' responses and behaviors in performance assessments. A holistic rubric represents a global approach to scoring and requires simultaneous consideration of all criteria to score students' responses and behaviors. In contrast, an analytic rubric scores separate components of students' responses and behaviors. Analytic rubrics are often used to provide explicit formative feedback for students because they can be used to identify both strengths and weaknesses. In some situations, analytic scores may optionally be combined into a summary score, similar to holistic scores (e.g., by selecting the highest score, averaging all scores, or adding all scores).

Holistic rubrics have been the dominant paradigm in automated short answer scoring in education (Zesch et al., 2023), forming the basis of most research studies as well as formal competitions on short answer scoring (Shermis, 2015; Institute of Education Sciences, 2022, 2023). Analytic rubrics have received comparatively less attention for short answer scoring (e.g., Mizumoto et al., 2019; Rahimi et al, 2017) but are commonly used for general essay scoring (Wilson & Czik, 2016; inter alia).

Figure 26.1 illustrates the distinction between holistic rubrics and analytic rubrics, as well as the relationships of the rubrics with the LPs, using examples from the MTS project. One assessment item in MTS is about reindeer population growth on a remote island. Snippets of holistic and analytic rubrics for Level 3 of the LP for this item are presented in Figure 26.1. While the holistic rubric summarizes what students can or cannot do, the analytic rubric specifies the ideas students use to explain the phenomenon.

Some researchers have argued that holistic rubrics, when used by experts, are more effective than analytic rubrics because student performance is not just a sum of ideas or components (Slater & Boulet, 2001; Yune et al., 2018). However, these researchers also point out that analytic rubrics are more effective for alternative and less experienced raters (Slater & Boulet, 2001), and factors such as fatigue may affect experts' scoring (Yune et al., 2018).

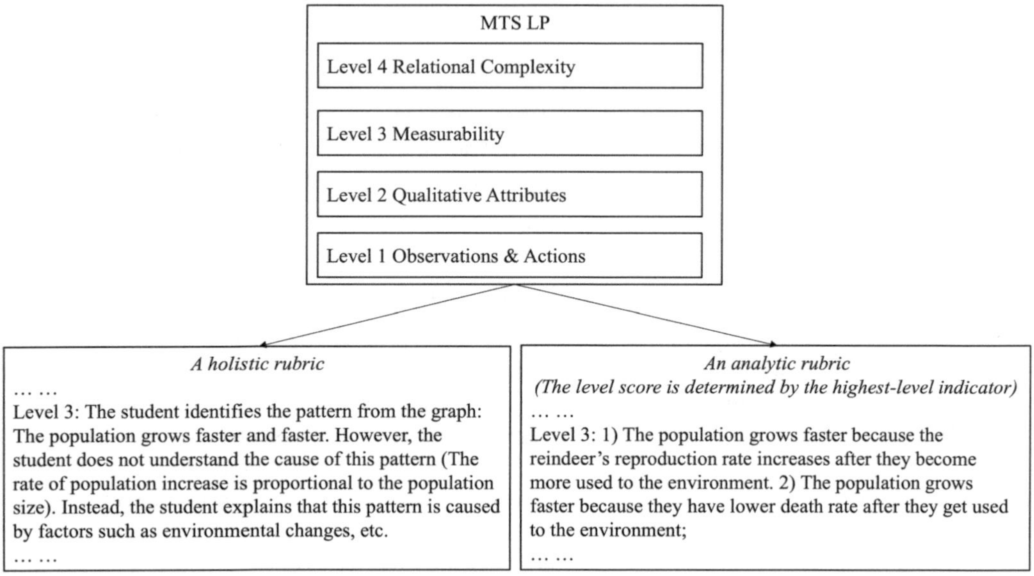

Comparison of two rubrics for an item about the exponential growth of reindeer in an isolated island.

Figure 26.1 An Example of Holistic and Analytic Rubrics for the MTS LP.

In automated scoring for science assessments, similar to general trends in automated short answer scoring, holistic scoring has been widely applied (Gerard & Linn, 2022; Kaldaras et al., 2020; *inter alia*) since automated scoring in science assessments primarily deals with short answers. However, as with general essay scoring, holistic rubrics have limitations. In particular, since holistic scoring natural language processing (NLP) systems are trained on a single score per response, they can pick up on spurious correlations of response characteristics and scores (Ding et al., 2020; Filighera et al, 2020; inter alia). This can result in issues with the validity of the scoring system (Myers & Wilson, 2022; Kaldaras & Haudek, 2022; inter alia).

Recently, analytic rubrics for short answer scoring in science assessments have garnered increasing interest (Davies et al., 2021; Jescovitch et al., 2020; Yune et al., 2018; Jonsson & Svingby, 2007). In this paradigm, each response is labeled for what ideas it contains. Holistic scoring rubrics can be "deconstructed" into tables of construct-specific ideas and (typically binary) labels. Human annotators apply these binary labels to student responses, generating multiple labels per response. The pattern of these labels for a response indicates the student's performance. Some studies comparing holistic and analytic scoring rubrics have found advantages for analytic rubrics, either in terms of higher human-machine agreement or increased validity (Jescovitch et al., 2020; Kaldaras & Haudek, 2022). However, analytic rubrics have some drawbacks. First, analytic rubrics can be difficult to develop. For example, Liu et al. (2014, 2016) found that a significant challenge in automated scoring for LPs is the difficulty of exhausting all important conceptual components in student responses and managing a large number of components. Second, while an analytical rubric depicts a more detailed relationship among the reasoning patterns and their components and can be more "mechanical" for raters, some responses, especially vague ones, may not be addressed adequately in the rubric. Third, it is uncertain whether a holistic rubric can always be decomposed into analytical ideas and labels objectively and mechanically.

Both holistic rubrics and analytic rubrics have the objective of *scoring* student responses – generating one or more numeric scores intended to represent the achievement of the student's content understanding on an ordinal scale. Such scores, even when reflecting relatively fine-grained analytical rubrics, inevitably condense and summarize students' science understanding, as the scores apply to entire student responses. As a result, the evidence for the assigned scores may not be revealed and opportunities for highly targeted feedback may be missed.

Idea Detection

To address these limitations in the standard scoring paradigms, some recent automated scoring systems for evaluating individuals' reasoning in educational contexts have begun to explore the feasibility of directly detecting ideas and relationships in responses (Schulz et al., 2018, 2019; Riordan et al., 2020). Idea detection in educational assessment attempts to both find and label "spans" of words from responses that correspond to scoring rubric components.[1] While similar to the analytic scoring approach, the idea detection approach differs in that it simultaneously focuses on detecting an idea category (and concomitant score) along with where the idea appears in the student response.[2] Both holistic scoring and analytic scoring models typically cannot guarantee that specific coherent ideas underpin their scores (without introducing additional supervision to the models).

Idea detection has been tackled for a number of educational applications. For example, the issue of finding and describing ideas has been central to research in biology education. Sripathi et al. (2019) present a thoughtful investigation of the types of ideas and their relationships in biology education. Using an analytic rubric for coding ideas at the response level, they show that students typically express multiple ideas. Ideas can be both normative and "developing". A different kind of approach to finding ideas is by "semantic matching" (Graesser et al., 2018; Carmon et al., 2019). In intelligent tutoring systems (ITSs), responses are short enough and often limited to one basic idea, making possible a matching approach to detecting a student's idea. A variety of techniques for measuring text similarity to accomplish this semantic matching have been explored for ITSs (Graesser et al., 2018; Carmon et al., 2019; Khayi et al., 2021). Hughes et al. (2018) developed an approach for detecting ideas in student essays about causal explanations. Rather than coding at the response level, this work tackles the problems of finding the extent of the ideas and labeling them correctly. This goal is shared by the work in this chapter. Schulz et al. (2019) tackle the same problem in an adult education context and demonstrate that idea detection is possible with recent neural network methods with reasonable accuracy.

In this chapter, we provide a case study of how automated scoring for LP-based CR items can be approached with holistic scoring combined with additional fine-grained model validation and NLP models for idea detection. We address validity concerns about holistic scoring models by analyzing what the models focus on in student responses and the extent to which predictions match experts' annotations. We explore the reasons for individual model score predictions leveraging *feature attribution* methods. Feature attribution scores provide a visual representation of the importance of each word in a response, computed from the trained model's parameters when making predictions. Through expert analysis of responses via feature attribution scores, we focus on the extent to which models attribute importance to words and phrases in student responses that align with item rubrics. We also demonstrate how idea detection models can provide a complementary view of student performance to holistic scoring in LP-based items, finding specific ideas in student responses along with the region of the response where they appear, in turn enabling targeted diagnosis of student performance issues and opportunities for feedback.

The MTS LP

Mathematical thinking in science refers to abstracting measurable variables from "messy" phenomena, identifying mathematical relationships among the variables, and using scientific ideas to conceptualize mathematical relationships. In the MTS project, human raters used holistic scoring rubrics to score students' responses. Validity evidence for the LP is presented in a separate report (Jin et al., 2023). In this section, we provide a brief description of the LP levels.

- Level 1. Holistic Phenomena
  At Level 1, students see the phenomenon as a whole; they do not analyze it, meaning they do not break down the phenomenon into components, aspects, and attributes. Instead, they talk about their personal feelings, tell personal stories, make human analogies, or merely describe macroscopic observations.
- Level 2. Attributes
  At Level 2, students begin to analyze the phenomenon. They identify observable and perceptible attributes of the phenomenon and use those qualitative attributes to construct explanations.
- Level 3. Measurability
  At Level 3, students begin to develop the concept of measurability: Variables can be measured and reasoned about in terms of numerical values. It is important to note that, although variables have numerical values when measured, we do not need to measure them or know what their measurements are in order to reason about them (Thompson, 2011). At this level, students abstract measurable variables from qualitative attributes of a phenomenon and use those variables to construct quantitative explanations. However, students' explanations either contain errors or are incomplete. For example, students may not identify all relevant variables, identify the quantitative relationships among the variables, or understand the scientific meanings of those quantitative relationships.
- Level 4 Relational Complexity

At Level 4, based on their understanding of measurability, the students identify and conceptualize the specific mathematical relationships between/among measurable variables. Such mathematical relationships include quadratic relationships, exponential relationships, extensive vs. intensive variables, change vs. rate of change, and so forth. Students not only identify these mathematical relationships but also understand the scientific meaning of the relationships.

Methods for Scoring Responses

In the MTS project, we conducted a field study to collect large-scale data for the development of automated scoring models. In total, 5,353 students from 22 high schools in 14 states participated in the field test. Approximately 59% of these students were white, 9% were Asian or Asian American, 8% were African American, and 8% were Hispanic or Latino; about 8% of these students were either currently English Language Learners or former English Language Learners; about 28% of these students received free or reduced lunch. Each student completed a test form that contained a set of MTS items and a set of "pure" mathematics items in 60 minutes. We focus in detail on two formative assessment items from the MTS assessments. These two items, two Boiling Situations (TBS) and Projectile, were completed by 1137 students and 798 students, respectively. More details about the field study and other studies conducted in the project can be found in other reports from the MTS project (e.g., Jin et al., 2023). In this section, we use the first item, TBS, as an example to illustrate the item design, rubrics, and idea detection details.

Item design

A CR item may generate many different responses characterized by different kinds of conceptual, realization, and nonconformity variance (Zesch et al., 2023). To categorize a wide variety of responses into score levels, the scoring rubrics can be complex. Still, many of these responses may be difficult to score even with complex rubrics. In the first round of the assessment, we found that, although the items required students to explain their reasoning, many students provided vague responses or did not explain the reasoning they used to generate their answers. To solve this problem, we revised many items into two-tier items. In a two-tier item, the first tier requires students to select an option, and the second tier is a specific question about the selected option. This approach encourages students to think deliberately and provide more detailed information about their thinking, generating more specific responses. Rubrics for those responses are also expected to be less complicated since the more targeted questions reduce the conceptual and realization variance in the responses. Details and examples of item design can be found in the chapter's online Appendix.

Figure 26.2 shows the revised TBS item that we used for the final round of assessment and automated scoring. As shown in the figure, for students who think that the temperatures in the two situations are the same, we ask them: if that is the case, then why does the degree of boiling differ? For students who think the temperatures in the two situations are different, we ask them to provide evidence to justify that claim. With these targeted questions, we were able to encourage students to think more deeply and provide specific information that better revealed their underlying reasoning.

Holistic Scoring rubrics

For the TBS item, we used the LP to develop two sets of holistic rubrics, for scoring student responses to the Option A and B questions, respectively. We used an iterative process to

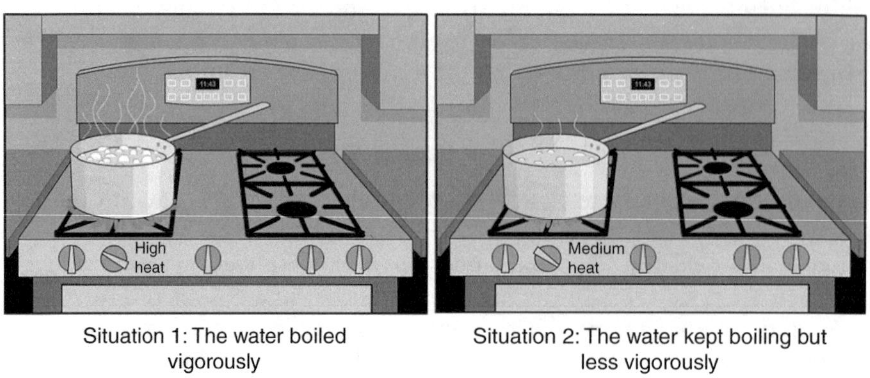

Figure 26.2 The Final Version of Two Boiling Stations (TBS) Item.

evaluate, revise, and refine the scoring rubrics (see details in Jin et al., 2023). For both rubrics, a zero score is used to code off-topic and aberrant responses (random letters, "I don't know," irrelevant responses, etc.). Responses scored at lower levels mostly selected Option B, while responses that scored at Level 4 were required to have selected Option A. Here, we use responses to Option B to illustrate Levels 1, 2, and 3 and responses to Option A to illustrate Levels 3 and 4.

For the Option A question, Level 1 responses suggest that the student does not identify any relevant attributes. Instead, the student explains their choice in terms of observations, perceptions, and actions. Some examples of Level 1 responses are: *One is boiled on high heat and one isn't; They boil at different times; He had the stove on too high.* These responses claim that the temperatures in the two situations are different because of heat settings, time, or action.

Level 2 responses suggest that the student identifies attributes relevant to the question (e.g., more or less bubbling, more or less steam, larger or smaller gas flames) and analyzes the relationships among these qualitative attributes. However, the student does not reason "quantitatively," meaning that they do not reason about measurable variables that have numerical values. For example, one response is: *The heat of the stove is changing, so the temperature of the water that it is heating would also have to be different. There is also the fact that the water boiled on high heat boiled more violently than the one on medium heat.* This response treats heat, temperature, and the intensity of boiling as interconnected qualitative attributes. It describes a chain of events where the increase or decrease of one attribute leads to the increase or decrease of the other attributes. What is missing is a sense of measurability – variables are measured by numerical values, even when not measured (Thompson, 2011).

Level 3 responses (that choose Option A) explain the boiling phenomenon in terms of measurable variables (variables with numerical values) such as mentioning temperature values and comparing the amount of heat/energy, but these responses are not accurate. For example, one response is: *Water starts to boils [boil] at around 100 degrees celceus [Celsius] and the closer it is to the 100 the more it will boil so for the water to boil more at one point means that the temperature was highest then.* This response suggests that the student thought there is a range of boiling points corresponding to boiling at varying degrees of intensity.

Among Level 3 responses that choose Option B, responses treat heat/energy, temperature, and other variables as measurable variables (variables with numerical values), but the explanation of the relationships among the variables is incomplete or inaccurate. A common error is that the relationships among the variables are not connected to phase change. For example, one response is: *Both of the water's temperature is at 100 degrees celsius [Celsius] because all of waters boiling point is that temperature. The water in situation 2 was boiling less vigorously becuase [because] less heat was being applied.* This response is incomplete because although it explains the relationship between the intensity of boiling and heat input, it does not connect this relationship with temperature remaining the same in the two situations. A comparison of the Level 3 example for Option A and Option B suggests that both students reasoned in terms of measurable variables, but they did not correctly explain the relationships among those variables (heat, temperature, and boiling intensity). In other words, although the two students selected different options, they used the same reasoning to analyze the phenomenon.

Level 4 responses explain that the temperature stays the same because the input heat is used for boiling rather than temperature increase. Students may provide explanations either at the macroscopic scale or the molecular scale. Responses at the molecular scale explain that the input heat provides the energy for the molecules of liquid water to break the intermolecular forces and become molecules of gaseous water. Responses at the macroscopic scale explain that the input heat makes the boiling more intense because more liquid water goes through a phase change and becomes gaseous water. An example of a Level 4 response is: *When a substance goes under a phase change, that is melting, boiling, freezing, condensation, sublimation, or deposition: it does not change in temperature. This is because the energy previously used to heat the substance up is now being used to break the bonds between molecules. Therefore, when Paulo puts the heat on higher, the energy being added quicker is only breaking the bonds quicker, then if he aded [added] energy slower. Because of this quirk in how phase changes affect the matter undergoing it the pots have the same temperature.*

Idea Detection

For idea detection, we examined the holistic rubrics and student responses to identify conceptual components embedded in students' responses. These conceptual components are concrete ideas that students use to explain their choices. We found that many conceptual components from responses that chose different options could be categorized into the same groups. As shown in Table 26.1, although Student 1 (S1) and Student 2 (S2) selected different options, they both thought that time was a factor affecting the water temperature. Although Student 3 (S3) and Student 4 (S4) selected different options, they both explained their answers using the idea of particle movement. The rubric used to score responses to "Question for choice A (Same Temperature)" and the rubric used to score responses for "Question for choice B (Different Temperatures)" share similar conceptual components on Levels 1, 2, and 3. An overview of the overlaps between these two rubrics can be found in the chapter's online Appendix.

Table 26.1 Examples of Student Responses

Conceptual Components	Question for choice A (Same Temperature): If the water temperatures in these two situations are the same, why does the water in situation 2 boil less vigorously than the water in situation 1?	Question for choice B (Different Temperatures): What evidence can be used to support the claim that the water temperature in situation 1 and the water temperature in situation 2 are different? Please explain why this evidence can be used to support the claim.
Time	1 A Time: S1: The temperature has been turned down, but hasn't had enough time to cool down as seen by the fact it is 11:43 on both clocks.	1B Time: S2: They boil at different times. Over time the water on medium will cool down more than the high heat.
Particle Movement	3 A Heat for particle movement: S3: Because more heat is being added to the water making the water molecules move faster.	3B Particle movement or kinetic energy: S4: When more heat is added and the temperature is higher, the water molecules moved faster, causing the water to boil vigorously. This is why the two temperatures could not be the same because when the heat was turned down, the water didn't boil as vigorously as it had been doing before

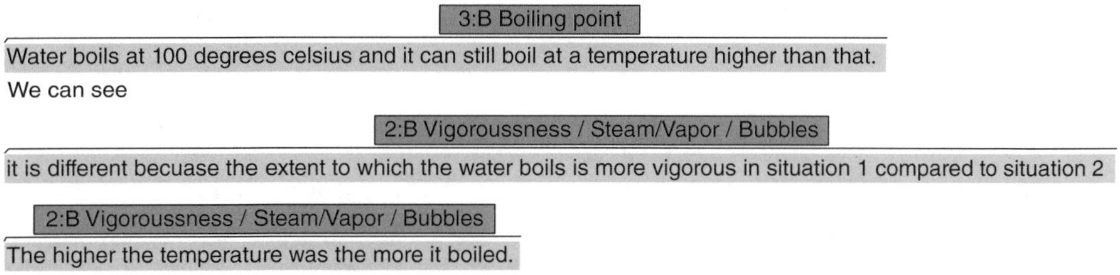

Figure 26.3 An Example of annotation.

Human Coding

Human coding of the responses to train NLP models consisted of two steps. First, two researchers used the scoring rubrics to score and annotate all item responses independently. Ideas in responses were annotated with the INCEpTION platform (Klie et al., 2018). An example annotated response is shown in Figure 26.3. It presents the thought process that the researcher used to score the response. In the response, three segments are highlighted and annotated. Each segment contains one idea that was annotated as one of the conceptual components in the rubrics. This response contains three conceptual components. One Level 2 component appeared twice, and a Level 3 component appeared once. Following our scoring rules, the highest level, Level 3, is used as the final score for this response.

Machine Learning Model Development

Holistic scoring can be formulated as a machine learning *text regression* problem: given a student response text, assign an ordinal score. While a text classification approach can also be employed (e.g., Jescovitch et al., 2020), maintaining the ordinal information in the response scores in the model can produce models with greater human-machine agreement. In this work, we employ *instance-based* machine learning models (Horbach & Zesch, 2019), which learn from {text, score} examples. This approach contrasts *with similarity-based* models that compare individual responses with a small set of *reference* responses. While similarity-based methods can be competitive for holistic scoring tasks (e.g., Bexte, Horbach, & Zesch, 2022), instance-based approaches typically yield higher accuracy.

Detecting ideas can be formulated as a word-by-word classification or *sequence labeling* task (cf. Schulz et al., 2018; Riordan et al., 2020). That is, for each word in a student response, a model assigns one or more idea categories. A contiguous span of these predictions forms a predicted idea's *span*. Since idea spans from different categories can overlap, individual words may belong to more than one idea category. Hence, the task of assigning idea categories to words is a *multi-label* classification task: for each word, the model makes multiple binary classification predictions about whether the word belongs to an idea category's span.

For both holistic scoring and idea detection, we use *pre-trained transformer-based language models* (Devlin et al., 2019). In these models, model training involves starting from a large pre-trained model and *fine-tuning* the model weights on the task-specific (i.e., holistic scoring or idea detection) training examples. Similar model architectures have achieved state-of-the-art performance on both holistic scoring (e.g., on the ASAP Short Answer Grading benchmark dataset; Steimel & Riordan, 2020) and idea detection in student writing (Alhindi & Ghosh, 2021). See the Appendix for details about the model architectures and training regimes.

Analysis and Results

In this section, we describe the analysis and results for holistic scoring and idea detection for the two items, Two Boiling Situations (TBS) and Projectile. Then we explain how we used model auditing to examine the validity of the results.

Metrics

To assess human-human inter-rater reliability (IRR) for scoring, we report Pearson's correlation, quadratic weighted kappa (QWK), and Cohen's kappa. To evaluate the agreement of human scores and machine scores, we report Pearson's correlation, quadratic weighted kappa (QWK), and root mean squared error (RMSE). QWK is a measure of agreement that ranges between 0 and 1 and is motivated by accounting for chance agreement (Fleiss & Cohen, 1973; Haberman, 2019).

To evaluate the agreement of human-labeled idea spans and machine-predicted idea spans, we use macro- and micro-averaged F1 scores (cf. the M_A metric in Schulz et al. (2018)). The F1 score is a combined measure of *precision* and *recall* of human-machine classification agreement across all word tokens in the data. The macro-average weights all idea categories equally, while the micro-average weights the idea categories by frequency.

Results

Human-human inter-rater reliability is shown in Table 26.2 and human-machine agreement is shown in Table 26.3. IRR and human-machine agreement are both very high. In particular, the human-machine QWK significantly exceeds the standard threshold of 0.7 for deploying operational scoring models (Williamson et al., 2012).

Next, human-machine agreement for idea span prediction for each item is given in Table 26.4. The idea detection models achieve F1 micro-average scores of around .6, which is consistent with other difficult discourse-level idea detection tasks (Schulz et al., 2019).

Table 26.2 Human-human Inter-rater Reliability for Holistic scoring. N = number of double-scored responses

Item	N	*Pearson*	*QWK*	*Kappa*
TBS	1137	0. 965	0. 965	0.931
Projectile	311	0. 942	0. 942	0.925

Table 26.3 Human-machine Agreement for Holistic Scoring

Item	*Pearson*	*QWK*	*RMSE*
TBS	0.844	0.840	0.471
Projectile	0.803	0.800	0.537

Table 26.4 Human-machine Agreement for Idea Span Prediction

Item	*Accuracy*	*F1 macro-avg*	*Precision macro-avg*	*Recall macro-avg*	*F1 micro-avg*	*Precision micro-avg*	*Recall micro-avg*
TBS	0.48	0.24	0.38	0.21	0.55	0.66	0.47
Projectile	0.56	0.46	0.56	0.43	0.62	0.71	0.55

Table 26.5 Descriptive Statistics for a Sample of Idea Types for Each Item

Item/Idea	*Frequency*	*Mean length*	*Type/token ratio*
TBS			
1:B Heat setting	123	19.5	0.15
2:B Vigorousness/steam/vapor/bubbles	290	22.2	0.10
3:B Boiling point	61	31.9	0.19
4:A Heat for evaporation	71	32.8	0.18
Projectile			
1 Phenomena/Actions NOT connected to a selected line graph	102	14.2	0.24
2 More/Less compression causes higher height	257	15.3	0.09
3 Quantitative relationship between H and X ONLY	142	14.3	0.16
4 Quadratic relationship AND/OR proportional relationship	24	29.1	0.29

Table 26.6 Human-machine Agreement for Idea Span Prediction for a Sample of Idea Types

Item/Idea	*F-score*	*Precision*	*Recall*
TBS			
1:B Heat setting	0.636	0.748	0.552
2:B Vigorousness/steam/vapor/bubbles	0.720	0.748	0.694
3:B Boiling point	0.601	0.673	0.543
4:A Heat for evaporation	0.594	0.636	0.557
Projectile			
1 Phenomena/Actions NOT connected to a selected line graph	0.475	0.558	0.413
2 More/Less compression causes higher height	0.775	0.811	0.742
3 Quantitative relationship between H and X ONLY	0.656	0.661	0.651
4 Quadratic relationship AND/OR proportional relationship	0.589	0.744	0.500

We further examined the human-machine agreement for idea span prediction by individual idea type. In Table 26.5, we provide descriptive statistics of the data by idea type for a sample of idea types from each item. For each item, we selected one idea type per LP score level, weighted by the frequency of the idea type. Table 26.6 provides the corresponding human-machine agreement performance statistics for the idea types.

Feature Attribution

The methods we demonstrate here, based on pre-trained transformer language models, achieve state-of-the-art human-machine agreement in conventional scoring contexts, and this performance generally transfers to tasks with LPs. However, these models are relatively opaque to both the model developer and end users. To examine the validity of the model prior to deployment in an application to end users, it is typically useful to employ "model auditing" practices for black-box NLP models.

Model auditing is an emerging set of practices that span a variety of techniques. While analyzing model performance with tools such as confusion matrices and psychometric methods such as Differential Item Functioning (DIF) have long been used, the success of large neural network-based models in recent years has spurred intense interest in new methods to probe model behavior. Raji et al. (2021), in their discussion of the role of benchmarks in NLP, provide a taxonomy of alternative evaluation methods to aggregate statistics – for example, test suites to examine model coverage of different phenomena of interest, various kinds of error analysis, ablation analysis – removing one model component to see its effect on overall model behavior, and analysis of other model properties such as stability to perturbations of the training data.

Tooling for building test suites has evolved rapidly in recent years. Ribeiro et al. (2020) presented *Checklist,* which provided ways of developing templates that allow for efficiently generating many test examples to probe model behavior. Other toolkits provide methods to test model robustness, in particular, to "adversarial" data (e.g., Robustness Gym (Goel et al., 2021) and the Adversarial Robustness Toolbox (Nicolae et al., 2018)).

Another kind of model auditing involves employing emerging methods to "explain" or "interpret" model predictions (Lyu et al., 2024; Madsen et al., 2022). In attempting to explain the reasons for model predictions in human-understandable ways, these methods can also be used for model debugging and building confidence in the reliability and validity of a model. Numerous explainability methods have been developed; see Lyu et al. (2024) for a critical overview of methods for NLP.

Examining saliency in model predictions is one kind of model audit for explainability that has been employed widely in both computer vision and natural language processing. Saliency analysis is a backpropagation-based method (Lyu et al., 2024) that requires access to model representations. Saliency-based methods have been shown to be relatively more faithful to model representations than other methods in NLP tasks (Atanasova et al., 2020). Here, we present one way of using model saliency analysis to examine what a model has learned, focusing on when successful predictions are "right for the right reasons."

Saliency estimates the importance of a word token for a model's score prediction. This "saliency" estimation procedure produces a (normalized) scalar value for each token. We employ gradient-based saliency estimation methods for neural network models. These methods access the internals of the scoring model while it makes predictions; specifically, the gradient of the model's loss from the error in its score prediction is backpropagated to the model's input level and aggregated per word token in the model output.

We visualize the saliency of each word in each response with "saliency maps." We sampled 10% of the data from the Two Boiling Situations item and generated saliency maps for each response. We use the SmoothGrad method (Smilkov et al., 2017) via AllenNLP (Wallace et al., 2019). Each sampled response was labeled by an item developer with one or more categories that represented hypotheses about what tokens the model used to make a prediction, as evidenced by the saliency scores. For this analysis, we focus on the saliency of rubric-related keywords in student responses (Table 26.7).

The distribution of saliency labels in the sample is shown in Table 26.8. Across the sample of responses, item developers judged that the model's saliency profile for responses captured the most important keywords about half of the time (46.5%). In other words, more than half

Table 26.7 Labels for model saliency behavior

Captured the most important keywords	Keywords that indicate correct relationships, i.e., accurate understanding, are marked as salient.
Non-keyword is salient	Words that are not indicative of accurate understanding are salient.

Table 26.8 Distribution of Labels for Saliency Behavior

Item	*Captured the most important keywords*	*Non-keyword is salient*
Two Boiling Situations	46.5%	20.2%

290 score=2 prediction=2
[CLS] the fact that situation 1 is put on high heat and the situation 2 pot is set to medium heat . also , in the capt ##ion it says that situation 1 boiled vigorously , the water has t ##p be hot for that to happen , the water in situation 2 barely boiled . [SEP]

Figure 26.4 A response, where the model's prediction is correct, but some keywords are not salient.

1087 score=3 prediction=2
[CLS] if the boiling was less vigorous , this means that the particles were not as active in situation 2 than in situation 1 . due to this , the situation 2 pot had a lower temperature than situation 1 . [SEP]

Figure 26.5 A response, where a non-keyword was salient ("pot").

of the time, the model did not attribute a high saliency to one or more rubric-relevant keywords.

An example is provided in Figure 26.4: the keywords "vigorously" and "boiled" are not salient. In addition, it can be observed that key *phrases* ("situation 1", "barely boiled") are not leveraged by the model. Conversely, item developers noted that a non-keyword was salient about one-fifth of the time (20.2%). Figure 26.5 shows an example: the model attributes high saliency to "pot", which, because of associations of that word with score 2, may have resulted in the model's under-prediction of the score.

Despite these issues, the item developers judged that the model performed reasonably well in attributing saliency to rubric-aligned keywords. Combined with the model's good human-machine agreement for scoring (cf. Table 26.3), the item developers were comfortable deploying the model for use in formative assessments in science classrooms.

Discussion

This chapter surveyed the different approaches to developing automated scoring in science education and for LPs in particular, focusing on the main approaches for developing scoring rubrics. The chapter then provided a case study of how one combination of methods, leveraging recently developed NLP technology, can be used to build an automated scoring system for LPs in science, achieving strong results.

Existing research generates mixed results about the effectiveness of holistic and analytic rubrics. For humans, holistic rubrics, if used by experts, can generate higher-quality scores. However, analytic rubrics may be more practical for novice raters or raters who do not have an in-depth understanding of student learning (Slater & Boulet, 2001: Yune et al., 2018). In one study, researchers found no significant differences between these two approaches in terms of inter-rater reliability (Tomas et al., 2019).

For machines, Jescovitch et al. (2020) provide a comprehensive review of the use of these two rubrics in automated scoring. In their own study of an assessment on undergraduate physiology, Jescovitch et al. (2020) decomposed holistic rubrics into analytic rubrics, in which binary labels captured single, key concepts in student responses. They found that machine learning models using analytic rubrics achieved slightly better performance for items with complex scenarios and performed similarly to models using holistic rubrics for other items. However, they also found that although both holistic and analytic scoring approaches obtained high human-human agreement, the human-human agreement for individual labels of the analytic rubric varied. Since the labels that received low human-human agreement also occurred with very low frequency, they did not affect the overall reliability of the level scores.

In brief, empirical evidence suggests that analytic rubrics can be effective for automated scoring of complex science items. However, unlike undergraduates' responses, K-12 students' responses often are short (one or two sentences) and use informal language. If many responses only have one idea/label presented in an assessment, the low human-human reliability of individual labels (Jescovitch et al., 2020) could lead to low reliability of the final level scores. In addition, to generate analytic rubrics, humans need to identify the concepts to be labeled, which can be challenging and labor-intensive. Human rubric developers may have difficulty compiling an exhaustive list of conceptual components (Liu et al., 2016). Motivated by these issues, our research contributes to overcoming this challenge by using two approaches to automated scoring: the holistic scoring approach and idea detection.

Our work demonstrates how holistic scoring can be augmented and vetted in two ways: (1) analysis of automated holistic scores with feature-attribution (specifically, saliency-based) methods; (2) training additional NLP models for detecting specific student ideas. First, the analysis of holistic scoring model predictions with saliency-based methods provided insights into the model behavior at a more detailed level than is possible with standard analysis of aggregate human-machine agreement and error analysis. In particular, the saliency analysis revealed the model may rely on forms of keyword spotting as a surrogate for a nuanced understanding of student ideas in LPs. This kind of model auditing has become increasingly important with the increasing power of machine learning-based scoring models and the increasing maturity and availability of model auditing tools.

Second, we applied separate idea detection models to the two LP-based items. We demonstrated that moderate human-machine agreement on both idea categories and idea spans (i.e., where ideas occur in responses) can be achieved with state-of-the-art NLP models. These results provide an additional, complementary perspective on student performance vis-a-vis the targeted LPs, opening up possibilities for more targeted LP-based feedback.

Saliency-based methods for auditing automated scoring models have both benefits and limitations. The main benefits of the approach include the following: saliency scores are readable by many stakeholders in model development, and some saliency methods can be relatively faithful to neural network model internal representations. At the same time, current saliency-based methods have several limitations. First, different saliency methods operating on the same model representations can produce widely varying predictions of which word tokens are salient. The degree of faithfulness depends on the saliency method and has not been well-established empirically. Second, saliency scores are continuous, while human-annotated data is typically binary (a word token is or is not important), making comparison challenging. Third, the degree to which saliency scores are actionable has been called into question (Wallace et al. 2020). That is, what the model developer should do given a set of model saliency behavior is not always clear. Fourth, tooling to use saliency analysis in model development workflows is still limited. While the basic methods are available in several libraries, it may take some technical effort to apply and visualize the outputs and to conduct experiments to compare with human annotations (Boggust et al., 2022).

Tools for model auditing and explaining model predictions are maturing rapidly. We can expect that these methods will become more usable for domain experts in education. Moreover, it has been observed that one kind of model auditing or explainability is typically not sufficient to examine model behavior and make decisions on whether to deploy models (Arya et al., 2022; Raji et al., 2021); evaluating models from different perspectives, and taking measures to address model deficiencies, should enable more reliable and valid models in education.

With the idea detection models, we showed how specific categories of student ideas that underpinned the holistic scoring rubrics and LP levels could be detected and foregrounded. In contrast with the analytic scoring approach, which focuses on binary idea categories applied to an entire student response, the idea detection approach simultaneously focuses on where the idea appears and what idea category (and score) it belongs to. This approach provides several benefits. First, while developing automated scoring models, model developers can examine what types of ideas are and are not detected well by examining actual spans of text and hence may be better able to take steps to modify the training data and NLP model to improve scoring performance. Second, identifying specific idea spans enables transparent feedback to students and teachers. Predicted idea spans allow students and teachers to focus quickly on parts of the response to improve.

While the levels of accuracy that can be achieved with the idea detection approach are promising, we saw that models for idea detection based on machine learning alone have some difficulty capturing the diversity of expression of student ideas, especially less frequent ideas. The way the NLP task is formulated – as a joint discourse-level segmentation and classification task (Schulz et al., 2019) – can be challenging for expert human annotators to apply and learn with current machine learning techniques. Different ways of formulating the NLP task may prove easier to apply. In addition, more powerful methods for capturing this "long tail" of ideas are an active area of NLP research, with many innovations in labeling and modeling data emerging. We expect that a combination of task framing and more "interaction-centric" model development workflows will continue to yield improvements in automated scoring in STEM education.

Acknowledgement

The research described in this chapter was supported by the Institute of Educational Sciences (Grant Number R305A160219).

Notes

1 Idea detection is similar to "snippet detection" for question-answering systems or entity and relationship detection in information extraction systems (cf. Jurafsky & Martin, 2023: Chapters 14, 21).

2 Figure 7 later in the chapter shows an example from a student response on a formative science assessment with idea spans labeled.

References

Alhindi, T., & Ghosh, D. (2021). "Sharks are not the threat humans are": Argument Component Segmentation in School Student Essays. *Proceedings of the 16th Workshop on Innovative Use of NLP for Building Educational Applications (BEA 2021).*

Arya, V., Bellamy, R. K. E., Chen, P.-Y., Dhurandhar, A., Hind, M., Hoffman, S. C., Houde, S., Liao, Q. V., Luss, R., Mojsilovic, A., Mourad, S., Pedemonte, P., Raghavendra, R., Richards, J. T., Sattigeri, P., Shanmugam, K., Singh, M., Varshney, K. R., Wei, D., & Zhang, Y. (2022). AI Explainability 360: Impact and Design. *AAAI.*

Atanasova, P., Simonsen, J. G., Lioma, C., & Augenstein, I. (2020). A Diagnostic Study of Explainability Techniques for Text Classification. *Conference on Empirical Methods in Natural Language Processing.*

Bexte, M., Horbach, A., & Zesch, T. (2022). Similarity-Based Content Scoring—How to Make S-BERT Keep Up With BERT. *Proceedings of the 17th Workshop on Innovative Use of NLP for Building Educational Applications (BEA 2022).*

Boggust, A., Hoover, B., Satyanarayan, A., & Strobelt, H. (2022). Shared Interest: Measuring Human-AI Alignment to Identify Recurring Patterns in Model Behavior. *CHI Conference on Human Factors in Computing Systems.*

Carmon, C. M., Hampton, A. J., Morgan, B., Cai, Z., Wang, L., & Graesser, A. C. (2019). Semantic Matching Evaluation of User Responses to Electronics Questions in AutoTutor. *Proceedings of the Sixth (2019) ACM Conference on Learning @ Scale.*

Davies, P. M., Goa, Y., Muresan, S., & Passonneau, R. (2021). Analytical Techniques for Developing Argumentative Writing in STEM: A Pilot Study. *IEEE Transactions on Education*, *65*, 373–383.

Devlin, J., Chang, M.-W., Lee, K., & Toutanova, K. (2019). BERT: Pre-training of Deep Bidirectional Transformers for Language Understanding. *Proceedings of the 2019 Conference of the North American Chapter of the Association for Computational Linguistics: Human Language Technologies*, 4171–4186.

Ding, Y., Riordan, B., Horbach, A., Cahill, A., & Zesch, T. (2020). Don't take "nswvtnvakgxpm" for an answer – The surprising vulnerability of automatic content scoring systems to adversarial input. *Proceedings of the 86th International Conference on Computational Linguistics (COLING).*

Filighera, A., Steuer, T., & Rensing, C. (2020). Fooling Automatic Short Answer Grading Systems. *Artificial Intelligence in Education*, *12163*, 177–190.

Fleiss, J. L., & Cohen, J. (1973). The equivalence of weighted kappa and the intraclass correlation coefficient as measures of reliability. *Educational and Psychological Measurement*, *33*(3), 613–619.

Gerard, L., Linn, M. C., & Berkeley, U. C. (2022). Computer-based guidance to support students' revision of their science explanations. *Computers & Education*, *176*, 104351. 10.1016/j.compedu.2021.104351

Goel, K., Rajani, N., Vig, J., Tan, S., Wu, J., Zheng, S., Xiong, C., Bansal, M., & R'e, C. (2021). Robustness Gym: Unifying the NLP Evaluation Landscape. *Proceedings of the 2021 Conference of the North American Chapter of the Association for Computational Linguistics: Human Language Technologies.*

Graesser, A. C., Hu, X., Rus, V., & Cai, Z. (2018). AutoTutor and other conversation-based learning and assessment environments. In Handbook of Automated Scoring: Theory into Practice.

Haberman, S. J. (2019). Measures of Agreement Versus Measures of Prediction Accuracy. *ETS Research Report Series.*

Horbach, A., & Zesch, T. (2019). The influence of variance in learner answers on automatic content scoring. *Frontiers in Education*, *4*, 28.

Institute of Education Sciences. (2022). *ED.gov National Assessment of Educational Progress (NAEP) Automated Scoring Challenge.* https://www.challenge.gov/?challenge=naep-automated-scoring-challenge

Institute of Education Sciences. (2023). *NAEP Math Automated Scoring Challenge.* https://www.challenge.gov/?challenge=naep-math

Jescovitch, L. N., Scott, E. E., Cerchiara, J. A., Merrill, J., Urban-Lurain, M., Doherty, J. H., & Haudek, K. C. (2020). Comparison of machine learning performance using analytic and holistic coding approaches across constructed response assessments aligned to a science learning progression. *Journal of Science Education and Technology*, 1–18.

Jin, H., Shin, H-J., & Cisterna, D. (2023). Systematic validation in science learning progression research. *International Journal of Science and Mathematics Education.*

Jonsson, A., & Svingby, G. (2007). The use of scoring rubrics: Reliability, validity, and educational consequences. *Educational Research Review*, 2, 130–144.

Jurafsky, D., & Martin, J. H. (2023). *Speech and Language Processing: An Introduction to Natural Language Processing, Computational Linguistics, and Speech Recognition* (3rd ed.). https://web.stanford.edu/~jurafsky/slp3/

Kaldaras, L., Akaeze, H. O., & Krajcik, J. S. (2020). Developing and validating Next Generation Science Standards -aligned learning progression to track three-dimensional learning of electrical interactions in high school physical science. *Journal of Research in Science Teaching.*

Kaldaras, L., & Haudek, K. C. (2022). Validation of automated scoring for learning progression-aligned Next Generation Science Standards performance assessments. *Frontiers in Education.*

Khayi, N. A., Rus, V., & Tamang, L. J. (2021). Towards Improving Open Student Answer Assessment using Pretrained Transformers. *The International FLAIRS Conference Proceedings.*

Klie, J.-C., Bugert, M., Boullosa, B., Castilho, R. E. de, & Gurevych, I. (2018). The INCEpTION Platform: Machine-Assisted and Knowledge-Oriented Interactive Annotation. *Proceedings of the 27th International Conference on Computational Linguistics: System Demonstrations*, 5–9.

Liu, O. L., Brew, C., Blackmore, J., Gerard, L., Madhok, J., & Linn, M. C. (2014). Automated Scoring of Constructed-Response Science Items: Prospects and Obstacles. *Educational Measurement: Issues and Practice*, *33*(2), 19–28.

Liu, O. L., Rios, J. A., Heilman, M., Gerard, L., & Linn, M. C. (2016). Validation of Automated Scoring of Science Assessments. *Journal of Research in Science Teaching*, *53*(2), 215–233.

Lyu, Q., Apidianaki, M., & Callison-Burch, C. (2024). Towards faithful model explanation in NLP: A survey. *Computational Linguistics.*

Madsen, A., Reddy, S., & Chandar, A. P. S. (2022). Post-hoc Interpretability for Neural NLP: A Survey. *ACM Computing Surveys.*

Mertler, C. A. (2001). Designing scoring rubrics for your classroom. *Practical Assessment, Research & Evaluation*, *7*(25).

Mizumoto, T., Ouchi, H., Isobe, Y., Reisert, P., Nagata, R., Sekine, S., & Inui, K. (2019). Analytic Score Prediction and Justification Identification in Automated Short Answer Scoring. *14th Workshop on Innovative Use of NLP for Building Educational Applications.*

Myers, M. C., & Wilson, J. (2022). Evaluating the Construct Validity of an Automated Writing Evaluation System with a Randomization Algorithm. *International Journal of Artificial Intelligence in Education.*

National Research Council. (2007). *Taking Science to School.* Washington, DC: The National Academies Press.

Nicolae, M.-I., Sinn, M., Tran, M.-N., Buesser, B., Rawat, A., Wistuba, M., Zantedeschi, V., Baracaldo, N., Chen, B., Ludwig, H., Molloy, I., & Edwards, B. (2018). Adversarial Robustness Toolbox v1.0.0. *arXiv preprint arXiv:1807.01069.*

Nitko, A. J. (2001). *Educational Assessment of Students* (3rd ed.). Prentice-Hall, Inc.

Rahimi, Z., Litman, D. J., Correnti, R., Wang, E. L., & Matsumura, L. C. (2017). Assessing Students' Use of Evidence and Organization in Response-to-Text Writing: Using Natural Language Processing for Rubric-Based Automated Scoring. *International Journal of Artificial Intelligence in Education*, *27*, 694–728.

Raji, I. D., Bender, E. M., Paullada, A., Denton, E. L., & Hanna, A. (2021). AI and the Everything in the Whole Wide World Benchmark. *Proceedings of the 35th Conference on Neural Information Processing Systems (NeurIPS 2021) Track on Datasets and Benchmarks.*

Riordan, B., Cahill, A., Chen, J. K., Wiley, K., Bradford, A., Gerard, L., & Linn, M. C. (2020). Identifying NGSS-Aligned Ideas in Student Science Explanations. *Workshop on Artificial Intelligence for Education (AI4EDU@AAAI).*

Ribeiro, M. T., Wu, T., Guestrin, C., & Singh, S. (2020). Beyond Accuracy: Behavioral Testing of NLP models with CheckList. *Proceedings of the 58th Annual Meeting of the Association for Computational Linguistics (ACL).*

Sadler, D. R. (2009). Transforming holistic assessment and grading into a vehicle for complex learning. In G. Joughin (Ed.), *Assessment, Learning and Judgement in Higher Education* (pp. 1–19). Springer Netherlands.

Shermis, M. D. (2015). Contrasting state-of-the-art in the machine scoring of short-form constructed responses. *Educational Assessment*, *20*(1), 46–65.

Schulz, C., Meyer, C.M., Sailer, M., Kiesewetter, J., Bauer, E., Fischer, F., Fischer, M. R., & Gurevych, I. (2018). Challenges in the Automatic Analysis of Students' Diagnostic Reasoning. *Proceedings of the 33rd AAAI Conference on Artificial Intelligence (AAAI).*

Schulz, C., Meyer, C. M., Kiesewetter, J., Sailer, M., Bauer, E., Fischer, M. R., Fischer, F., & Gurevych, I. (2019). Analysis of Automatic Annotation Suggestions for Hard Discourse-Level Tasks in Expert Domains. *Proceedings of the 57th Annual Meeting of the Association for Computational Linguistics (ACL)*, 2761–2772.

Slater, S. C., & Boulet, J. R. (2001). Predicting holistic ratings of written performance assessments from analytic scoring. *Advances in Health Sciences Education*, *6*(2), 103–119.

Smilkov, D., Thorat, N., Kim, B., Viégas, F. B., & Wattenberg, M. (2017). SmoothGrad: Removing noise by adding noise. *ICML Workshop on Visualization for Deep Learning.*

Sripathi, K. N., Moscarella, R. A., Yoho, R., You, H. S., Urban-Lurain, M., Merrill, J. E., & Haudek, K. C. (2019). Mixed student ideas about mechanisms of human weight loss. *CBE—Life Sciences Education, 18.*

Steimel, K., & Riordan, B. (2020). Towards Instance-Based Content Scoring with Pre-Trained Transformer Models. *Workshop on Artificial Intelligence for Education (AI4EDU@AAAI).*

Thompson, P. W. (2011). Quantitative reasoning and mathematical modeling. In L. L. Hatfield, S. Chamberlain, & S. Belbase (Eds.), *New Perspectives and Directions for Collaborative Research in Mathematics Education* (pp. 33–57). University of Wyoming.

Tomas, C., Whitt, E., Lavelle-Hill, R., & Severn, K. (2019). Modeling holistic marks with analytic rubrics. *Frontiers in Education, 4.* 10.3389/feduc.2019.00089

Tunstall, L., von Werra, L., & Wolf, T. (2022). *Natural Language Processing with Transformers.* Sebastopol, CA: O'Reilly Media, Inc.

Turley, E., & Gallagher, C. W. (2008). On the "Uses" of Rubrics: Reframing the Great Rubric Debate. *The English Journal*, *97*(4), 87–92.

Wallace, E., Gardner, M., & Singh, S. (2020). Interpreting Predictions of NLP Models. *Proceedings of the 2020 Conference on Empirical Methods in Natural Language Processing: Tutorial Abstracts*, 20–23.

Wallace, E., Tuyls, J., Wang, J., Subramanian, S., Gardner, M., & Singh, S. (2019). AllenNLP Interpret: A Framework for Explaining Predictions of NLP Models. *Proceedings of the 2019 Conference on Empirical Methods in Natural Language Processing (EMNLP).*

Wang, C., Liu, X., Wang, L., Sun, Y., & Zhang, H. (2021). Automated scoring of Chinese grades 7–9 students' competence in interpreting and arguing from evidence. *Journal of Science Education and Technology*, *30*(2), 269–282.

Williamson, D. M., Xi, X., & Breyer, F. J. (2012). A framework for evaluation and use of automated scoring. *Educational Measurement: Issues and Practice*, *31*(1), 2–13.

Wilson, J., & Czik, A. (2016). Automated essay evaluation software in English Language Arts classrooms: Effects on teacher feedback, student motivation, and writing quality. *Computers & Education*, *100*, 94–109.

Yune, S. J., Lee, S. Y., Im, S. J., Kam, B. S., & Baek, S. Y. (2018). Holistic rubric vs. Analytic rubric for measuring clinical performance levels in medical students. *BMC Medical Education*, *18*(124).

Zesch, T., Horbach, A., & Zehner, F. (2023). To Score or Not to Score: Factors Influencing Performance and Feasibility of Automatic Content Scoring of Text Responses. *Educational Measurement: Issues and Practice.*

Zhai, X., Yin, Y. M., Pellegrino, J. W., Haudek, K. C., & Shi, L. (2020). Applying machine learning in science assessment: A systematic review. *Studies in Science Education*, *56*, 111–151.

Appendix

Item Design

As shown in Figure A, to score a wide variety of responses of a Constructed Response (CR) item is challenging. Usually, complex scoring rubrics are needed. Still, many of these responses may be difficult to score even with complex rubrics. As shown in Figure B, in a two-tier item, the first tier requires students to select an option, and the second tier is a specific question about the selected option. This approach encourages students to think deliberately and provide more detailed information about their thinking, generating more specific responses. Rubrics for those responses are also expected to be less complicated since the more targeted questions reduce the conceptual and realization variance in the responses. Figure C and Figure D provide an example of the contrast between the traditional item and the two-tier item. The item, named Two Boiling Positions (TBP) is discussed in the chapter. Figure E provides an overview of the scoring rubrics for responses to questions posed to students selecting option A (green) and option B (red). It shows that the two rubrics share similar conceptual components on the lower and middle levels of the LP.

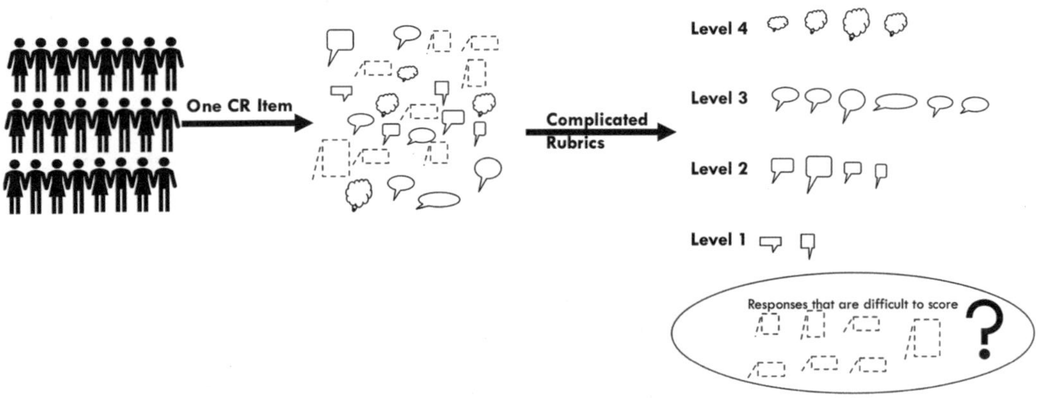

Figure A Complex Scoring Rubrics of a Traditional CR Item.

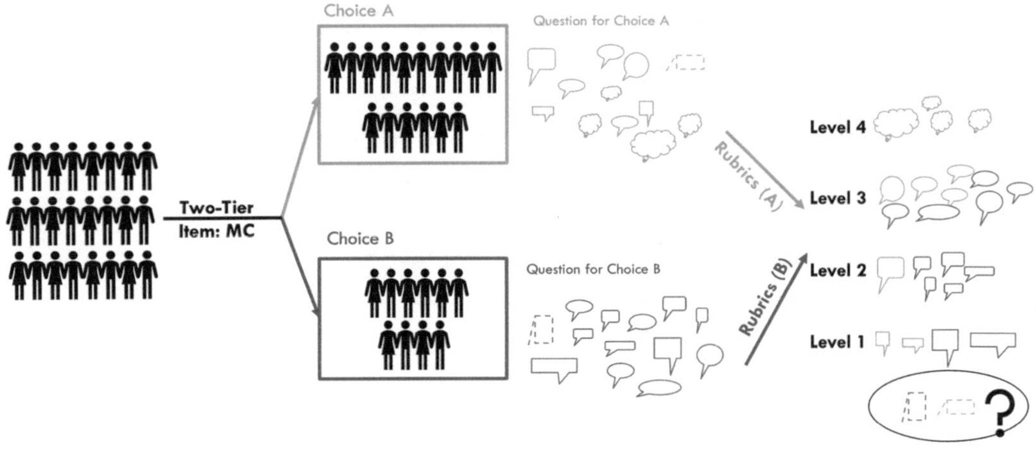

Figure B Collecting and Analyzing Responses for a Two-tier Item.

Paulo put a pot with water on the stove at high heat. After a few minutes, the water started to boil vigorously. Paulo turned down the heat setting to medium, and the water kept boiling but less vigorously.

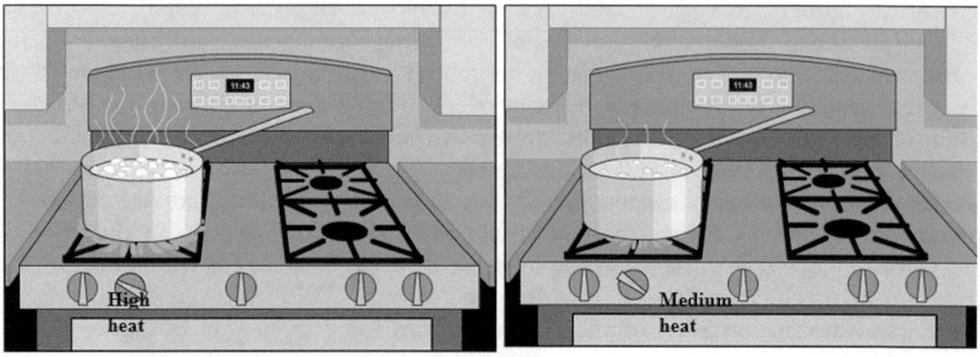

Situation 1: The water boiled vigorously

Situation 2: The water kept boiling but less vigorously

Do you think the temperature of the water is the same in these two situations? Please explain your reasoning.

Figure C An Example of CR Item – Boiling Water in Two Situations.

Paulo put a pot with water on the stove at high heat. After a few minutes, the water started to boil vigorously. Paulo turned down the heat setting to medium, and the water kept boiling but less vigorously.

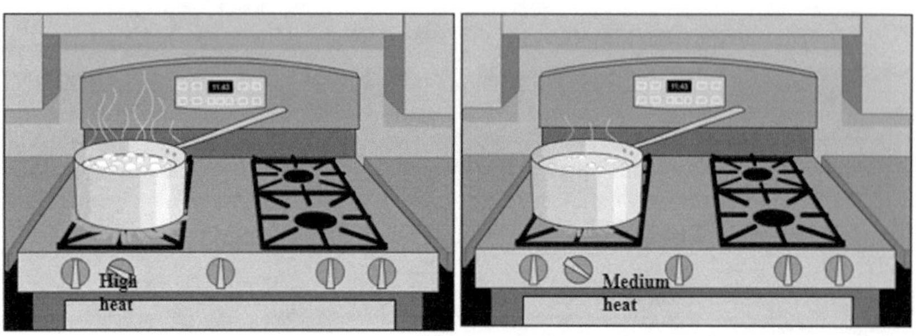

Situation 1: The water boiled vigorously.

Situation 2: The water kept boiling but less vigorously.

1. Do you think the temperature of the water is the same in these two situations?
   A. The water temperature in situation 1 and the water temperature in situation 2 are the same.
   B. The water temperature in situation 1 and the water temperature in situation 2 are different.

2. [Different sets of questions are shown when the student chooses A or B.]

   [Chooses A]
   → If the water temperatures in these two situations are the same, why does the water in situation 2 boil less vigorously than the water in situation 1?

   [Chooses B]
   → What evidence can be used to support the claim that the water temperature in situation 1 and the water temperature in situation 2 are different? Please explain why this evidence can be used to support the claim.

Figure D The Two-tier Item Revised from the Initial CR Item.

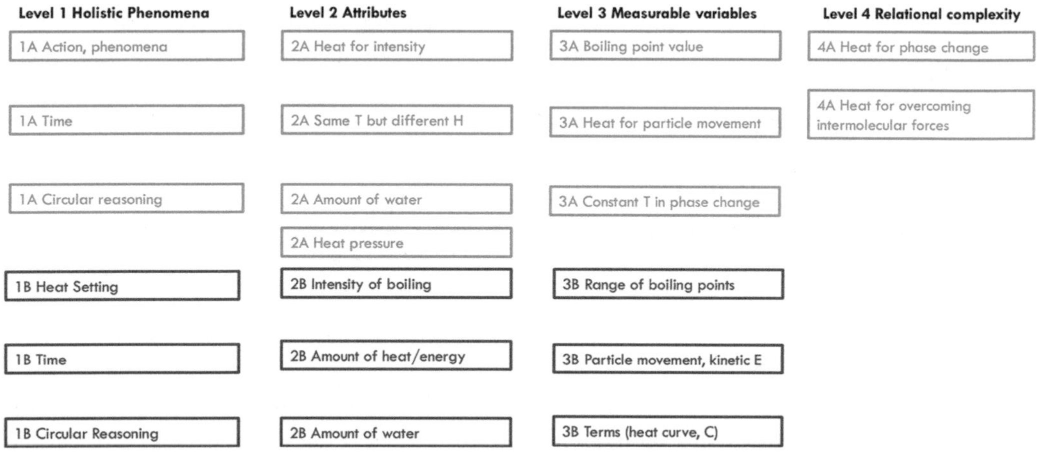

Figure E Conceptual components for different LP levels in the "Two Boiling Situations" item. Green indicates components for Option A, and red indicates components for Option B.

Model Architectures and Training Regimes

General

Pre-trained transformer models leverage word representations that are learned from language models trained on very large corpora. A language model learns to predict words in the input corpus that are masked out during training. When coupled with the transformer neural network architecture, this self-training of predicting words in the input yields word representations that are useful across many NLP tasks. Transformer networks use self-attention between word-like representations to learn word representations that consider the context of words' usage.

Holistic Scoring Model Details

Student responses were tokenized with the Spacy English tokenizer. The basic model architecture was as follows. First, a pre-trained transformer language model embedded tokens from the response. Scoring models used the "BERT-base-uncased" pre-trained instance (Tunstall et al., 2022). Second, a single vector representing the response was generated. Following standard practice, we used the output vector for the "CLS" special token. Third, dropout was applied (0.1). Last, a single linear layer with sigmoid activation generated an unnormalized scalar logit.

Holistic Scoring Model Training

Student response scores were scaled to [0, 1] to train the network. For evaluation, the scaled scores were converted back to their original range. The training objective was the minimization of a mean squared error loss function.

Idea Detection Model Details

Student responses were tokenized with the Spacy English tokenizer. For idea detection, a multi-label sequence labeling model was employed. First, response tokens were embedded with the same base language model as in holistic scoring. Next, to further contextualize tokens, the response representation was passed through a simple *sequence-to-sequence* recurrent neural network (RNN; 1-layer Gated Recurrent Unit). Then, for each token, a linear layer with sigmoid activation produced an unnormalized scalar logit per idea category.

Idea Detection Model Training

The network was trained on per-token idea category labels (see Section "Machine learning model development"). The training objective was minimizing a sequence-based cross-entropy loss.

General Model Training

Models were trained with 10-fold cross-validation with train/validation/test (80/10/10) splits. Predictions were pooled (concatenated) across folds and used for evaluation. For hyperparameter tuning, we trained on each train split. We evaluated performance on the validation split, keeping the best predictions across epochs and the epoch on which that performance was observed. Specifically, predictions were pooled from all folds on the validation sets, performance was evaluated, and the best-performing configuration of hyperparameters was selected. For final model training, models were trained on combined train and validation splits with 10-fold cross-validation to the median best epoch across folds from the hyperparameter tuning phase. The final performance was evaluated on the pooled predictions from the test splits. This training and evaluation procedure aims to increase the stability of performance estimates during both the tuning and final testing phases and to use more data for training and evaluating the final models to provide better estimates of model performance. Both models were optimized with the Adam optimizer, a learning rate tuned from {2e-5, 3e-5, 5e-5}, and a batch size 16. Hyperparameters were tuned for 20 epochs. An exponential moving average was used to smooth the model's weights across training epochs (decay rate = 0.999).

27

INTEGRATING ARTIFICIAL INTELLIGENCE INTO LEARNING PROGRESSION-BASED LEARNING SYSTEMS TO SUPPORT STUDENT KNOWLEDGE-IN-USE

Opportunities and Challenges

Peng He, Namsoo Shin, Leonora Kaldaras, and Joseph Krajcik

Introduction

Science education worldwide seeks to have all students develop knowledge-in-use to apply scientific knowledge with practices to make sense of compelling phenomena and solve real-world problems (NRC, 2012). Various nations have adapted this vision of knowledge-in-use in their national science standards (e.g., Chinese Ministry of Education, 2017; NGSS Lead States, 2013) and the Programme for International Student Assessment [PISA] (e.g., Organization for Economic Cooperation and Development, [OECD], 2019). For instance, the *Framework of K-12 Science Education* (hereafter *the Framework*, NRC, 2012) and the *Next Generation Science Standards* (NGSS, the Lead NGSS States, 2013) recommend that students apply three-dimensional knowledge, including disciplinary core ideas (DCIs), science and engineering practices (SEPs), and crosscutting concepts (CCCs), in explaining compelling phenomena and solving problems to demonstrate knowledge-in-use. However, knowledge-in-use development requires long-term science learning and coherent support (NRC, 2012). Learning progressions (LPs) guide the coherence among curriculum, assessment, instruction, and professional learning in a learning system to support science learning over time (Krajcik & Shin, 2023). In science education, an LP describes students' knowledge-in-use development of core ideas and practices over time so that knowledge becomes more sophisticated, allowing learners to apply their knowledge in new situations. To do so, researchers put efforts into developing and using LPs in learning systems to support student knowledge-in-use (e.g., Kaldaras et al., 2021; He et al., 2023b).

Some studies have found ways to design LPs for student knowledge-in-use; however, challenges remain in using LP-based learning systems to support student science learning over time effectively. For instance, classroom assessments or artifacts are critical in an LP-based learning system. Teachers usually collect student performance on classroom assessments to diagnose and track students' progress across time (NRC, 2014). That assessment information helps teachers make evidence-based instructional decisions about

DOI: 10.4324/9781003170785-31

whether and how to support their students' learning further (He et al., 2024; Krajcik, 2021). More importantly, performance-based or open-ended assessments require students' constructed responses, such as drawn models or written explanations (NRC, 2014; He et al., 2023a; 2023b). Assessing knowledge-in-use holds promise to support learning and provide information about learning. However, such assessment tasks are time-consuming for teachers to grade (Krajcik, 2021), which impedes the delivery of timely feedback to all students. Providing immediate feedback for each student is challenging. Teachers who do not have sufficient teaching experience may not provide actionable feedback for all students in class (Holstein et al., 2022). Even though students have received timely feedback, they may still need teachers' instructional support to achieve knowledge-in-use learning goals. Those challenges impede LP-based learning systems' efficiency and effectiveness in supporting student knowledge-in-use development.

With the rapid development of digital technologies, artificial intelligence (AI) can serve as a partner to provide potential solutions to support LP-based learning systems further. In science education, AI technologies, such as Machine Learning (ML), have been widely used to develop algorithmic models to automatically score students constructed responses on assessment tasks (e.g., Li et al., 2023). Researchers have developed immediate feedback systems based on students' automatic scores (e.g., Lee et al., 2021). Research teams have recently designed dashboards to provide assessment information for teachers and students (Matuk et al., 2016; Moharreri et al., 2014). These studies have made significant progress in applying AI to develop algorithms for providing teachers and students with timely information. However, few have explored how AI can support student learning across time, especially in an LP-based learning system.

While AI has been used for years (e.g., Dickler et al., 2021; Haudek et al., 2012), leveraging AI into an LP learning system to support student knowledge-in-use shows promise. This chapter addresses this essential research agenda by 1) positioning theoretical perspectives and a conceptual model for integrating AI into an LP learning system; 2) using our two LP projects (middle and high school examples) to discuss how we envision integrating AI into an LP with automatic scoring information, personalized feedback, and instructional support in learning systems; and 3) positing the future research work with recommendations for developing and using AI-empowered LP-based learning systems.

Theoretical Perspectives

This section illustrates the theoretical perspectives of integrating AI into LP research to support student science learning over time. We first introduce our vision of learning and the learning goals in contemporary science education (i.e., knowledge-in-use). We elaborate on the term learning progressions as a model to depict student development of science learning across time. Then, we discuss what AI is and how AI works to support science learning.

Knowledge-in-Use for Articulating Science Learning Goals

Individuals learn science by interacting and interpreting the world, including essential interactions with others. Students acquire new knowledge through learning experiences to transfer their understanding to other sophisticated situations (Li et al., 2024; NRC, 2000). However, understanding is not just memorizing but continuously building and refining knowledge as one uses ideas and experiences to make sense of phenomena and solve problems (He et al., 2023; Krajcik & Shin, 2023).

Science is more than just a compilation of facts. Instead, scientific knowledge contains multiple dimensions, such as disciplinary core ideas, crosscutting concepts, and science and engineering practices (NRC, 2012) that work together to support making sense of phenomena or solving problems. Moreover, learning science is a complex, challenging, and multifaceted effort. Developing sophisticated knowledge in one area of science requires an in-depth understanding of other areas, such as mathematics and engineering design. In the *Framework* (NRC, 2012), the integration of disciplinary core ideas, crosscutting concepts, and science and engineering practices is called three-dimensional learning. Just like a rope becomes stronger when individual strands are braided together, when the three dimensions of knowledge are integrated learners can use their knowledge to solve problems, make decisions, think innovatively, make sense of phenomena, and learn more when needed. Their proficiency through three-dimensional learning is called knowledge-in-use (He et al., 2023; Li et al., 2024). The NGSS expresses student knowledge-in-use performance as Performance Expectation (PE) statements (the Lead NGSS States, 2013). For instance, MS-PS1-1 is a middle school physical science PE, stating, "Develop models to describe the atomic composition of simple molecules and extended structures." The DCI is "PS1.A: Structure and Properties of Matter," the SEP is "Developing and Using Models," and the CCC is "Scale, Proportion, and Quantity." Students demonstrate their proficiency in knowledge-in-use by drawing models that include the atomic composition of simple molecules and extended structures to explain the relevant and compelling phenomena. For instance, how can pencil lead and diamond be composed of the same substance (carbon) but have different properties? Learning environments should facilitate students' learning of core ideas in and across science disciplines and develop capabilities to apply those ideas with science and engineering practices in real situations for understanding natural phenomena or solving complex problems. For instance, to explain the phenomenon "why pencil lead is soft, but diamond is hard," students could draw two atomic models that show the atomic structure of carbon in the pencil lead and the diamond and use the models to explain the phenomenon.

Three-Dimensional Learning Progressions for Depicting Knowledge-in-Use Development

At all ages, students need sustained effort and focus to learn science. Learning disciplinary core ideas, crosscutting concepts, and science and engineering practices requires introducing them to students from early grades through college. Because of their importance to various science domains, three-dimensional learning needs to be taught across grade levels and subjects (e.g., biology, physics, chemistry, and earth science) with increased depth and sophistication (NRC, 2012). Developing a useable understanding of disciplinary core ideas, crosscutting concepts, and science and engineering practices takes time for learners to transfer their capabilities to make sense of more sophisticated phenomena and solve more challenging problems. As student progress in their learning, how they explain phenomena changes and advances. For instance, why can density be used to identify substances, but mass cannot? What is the substance made of at atomic and molecular levels? How do we know a chemical reaction occurs at macroscopic and microscopic levels? Students' progressive understanding of the properties and structure of matter could begin by exploring the similarities and differences among substances, which children can see and compare; then moving to more microscopic-based identification of substances and advancing to make mechanistic explanations to explain phenomena, such as rearrangement of atoms in molecules to explain chemical reaction (e.g., Morell et al., 2017).

A three-dimensional learning progression (3DLP) describes how students' knowledge-in-use develops over time and becomes more sophisticated so that learners can use their knowledge in new and challenging situations (e.g., Krajcik & Shin, 2023; Kaldaras et al., 2021). A 3DLP is not a listing of all concepts in a domain that students need to memorize (e.g., the definition of ideas); instead, a 3DLP describes how a set of disciplinary core ideas, crosscutting concepts, and science and engineering practices integrate so learners can apply them when needed (knowledge-in-use). A vital aspect of a 3DLP is showing how to support learners in developing disciplinary core ideas, crosscutting concepts, and science and engineering practices over time (Krajcik & Shin, 2023), which provides a promising avenue to develop and monitor knowledge-in-use development (He et al., 2023; Li et al., 2024). However, mastering a level along LP depends on whether learners have the opportunities to learn. Achieving new levels requires learners to recall their prior knowledge and learning experience to select, connect, and apply core ideas (DCIs and CCCs) with science and engineering practices to explain phenomena and solve complex problems. However, learners' prior knowledge and learning experience may vary, which results in an LP with multiple learning pathways that learners might follow as they move from less to more sophisticated understanding (Krajcik & Shin, 2023). While multiple pathways are possible, LPs provide a design framework to guide the development and use of curriculum materials, instruction, and assessment to support students' knowledge-in-use development over time.

Artificial Intelligence for Enhancing Science Teaching and Learning

Artificial intelligence (AI) refers to the simulation of human intelligence in machines, especially computer systems, that are programmed to perform tasks that typically require human intelligence, such as visual perception, decision-making, and problem-solving (e.g., Chen et al., 2020; Kent, 2022). AI is grounded in several theories, such as neural-like growing networks, temporary and long-term memory, and the functional organization of the "brain" of artificial intelligent systems (Nilsson, 1982; Russell, 2010; Winston, 1992). AI systems generally work by ingesting large amounts of labeled training data, analyzing the data to identify patterns, and learning from patterns and experiences to make predictions. The primary goal of AI is to create intelligent machines that can exhibit behavior similar to human intelligence. As AI is noteworthy for its potential to change how we live, work, and play, it has been widely used in many areas, such as industry, medicine, and education. AI has also become a critical tool in education to support student learning and teaching practices in science classrooms. With the rapidly increasing advances and adoption of technologies, AI applications have begun to flourish in science education research, especially using machine learning for automatic scoring (e.g., Li et al., 2023), timely feedback (e.g., Lee et al., 2021), and data mining for establishing teacher and student platforms (e.g., Dickler et al., 2021). While the initial research stage has expanded the knowledge of using AI to provide timely support for teachers and students, more studies are urgently needed to advance our understanding of how AI can integrate into a learning system to support teaching and learning over time.

Leveraging AI to Support Science Learning: An Overview of Literature

AI can potentially advance LP use in teaching and learning with its features of analyzing large data immediately for automatically scoring and generating class and student performance reports, providing timely teacher and student feedback, and delivering instructional strategies for class and student needs. Below, we summarize the existing literature on using AI to support science learning.

Automatic Scoring

Machine learning (ML), an AI approach, has successfully automatically scored short, textual, constructed responses (CR) across various STEM contexts with reliability close to human scoring. Further, ML-based scoring is highly consistent with human scoring when measuring progression toward higher-order understanding, as supported by the work on automated scoring of LP-aligned assessments (e.g., Jescovitch et al., 2021). The results from existing studies show that using ML-based scoring along with an LP brings a promising way to efficiently support students in building deeper science understanding over time. Therefore, applying ML approaches for automatically scoring LP-aligned assessments represents a new challenge and a promising way of tackling the issues of accurate scoring and providing timely assessment feedback to teachers and students (Krajcik, 2021).

Personalized Feedback

The existing research using automatic feedback has focused on supporting students in developing written arguments in science (Zhu et al., 2017) and interacting with simulations coupled with constructed response assessments (Lee et al., 2021). The results of these studies indicate that students' ability to construct scientific arguments in the context of Earth Science improved upon receiving automatic feedback (Zhu et al., 2017). Another study examined the effect of automatic feedback received after engagement with computer simulations related to water systems and writing open-ended arguments. The study concluded that student scores on engagement with the simulations and scores of argumentations improved upon receiving automated feedback (Lee et al., 2021). However, the feedback was generic to the practice of argumentation and might not necessarily be aligned with any core ideas. These studies show the promise of automatic feedback to help students better understand science at higher levels along the LP.

Instructional Support

Automatic AI reports of individuals and whole class performance on assessments provide teachers with immediate and meaningful information to adjust teaching on time. To our knowledge, only a few dashboards have been developed for science teachers. Some dashboards, such as the WISE dashboard (Matuk et al., 2016) and the CK Biology curriculum dashboard (Acosta and Slotta, 2018), provide students with information about the understanding of science. Dickler and colleagues (2021) developed a science intelligent tutoring system, Inq-ITS, to provide teachers with data on students' inquiry practices, such as forming questions/hypothesizing, collecting data, analyzing and interpreting data, and constructing explanations. The existing dashboards are either content-focused or practice-specific. Few of them presented information about students' performance on knowledge-in-use assessments or delivered suggestions for timely instructional strategies for teachers to make instructional decisions. Moreover, the field also has limited knowledge on how to incorporate timely teacher instructional support with AI to support student knowledge-in-use development over time along with an LP.

Summary

Although the above AI application studies (e.g., Dickler et al., 2021; Lee et al., 2021; Jescovitch et al., 2021) did not build an AI-based LP for supporting student knowledge-in-use over time, they used a developmental perspective of learning to design scoring rubrics, student and teacher feedback, and instructional support for teachers to help students move to advanced levels of understanding. An LP has the potential to help track the development of student understanding over time and adjust instruction to the needs of individual learners (Kaldaras et al., 2021). However, it is only possible if students and teachers access timely and meaningful feedback and instructional support. Generating such feedback involves designing assessments capable of accurately probing student progress at all levels of the LP, assigning the LP level to individual students on these assessments in a timely fashion, providing guidance to help teachers adjust their teaching, and producing teacher and student-facing feedback based on the LP level placement. Carrying out these steps in an iterative fashion using human scoring is time-consuming, extremely expensive, and therefore not feasible in real-life classroom settings. Thus, leveraging AI into an LP will provide timely and adequate information. With this in mind, students can receive immediate and personalized feedback based on their needs, and teachers can make instructional decisions using appropriate instructional strategies to support their students to move up to upper levels aligned with an LP.

The Conceptual Model for Integrating AI into Learning Progression-Based Learning Systems

Addressing the challenge of integrating AI into a learning system, we synthesize our current project work and articulate a conceptual model to describe our vision of integrating AI into an LP to enhance a long-term and coherent learning system to efficiently support student knowledge-in-use over time (See Figure 27.1). Our conceptual model extends the existing knowledge about the components of a learning system (e.g., curriculum, assessment, instruction, and professional learning) with the guidance of learning progression and the affordance of AI technologies.

Figure 27.1 shows our conceptual model. Developing student knowledge-in-use over time needs a long-term and coherent learning system consisting of curriculum, instruction, assessment, and professional learning (He et al., 2023). We use an LP to guide the intra-unit and inter-unit coherence in a learning system and its alignment with national/local science standards (e.g., NGSS) (Fortus & Krajcik, 2012; Jin et al., 2019). Intra-unit coherence requires the designed learning activities (focal SEPs, CCCs, and DCIs) and classroom-embedded assessments to be sequentially arranged and aligned with the learning performance goals in a specific unit. Inter-unit coherence builds strong connections across the units. Student learning from the prior unit (including SEPs, CCCs, and DCIs) can form a substantial foundation for their science learning in the next unit. In an LP-based learning system, unit and lesson sequences with learning activities and assessment tasks are carefully ordered and aligned with the LP levels. Teachers provide constant instructional practices to support all students from lower to upper levels across time. Of course, teachers also need comprehensive support from professional learning sessions to implement an LP-based learning system in classrooms.

AI is a critical tool for enhancing the efficiency of an LP-based learning system by gathering timely information and providing personalized student and teacher feedback. In an AI support system, collecting and interpreting information on all students' performance in a

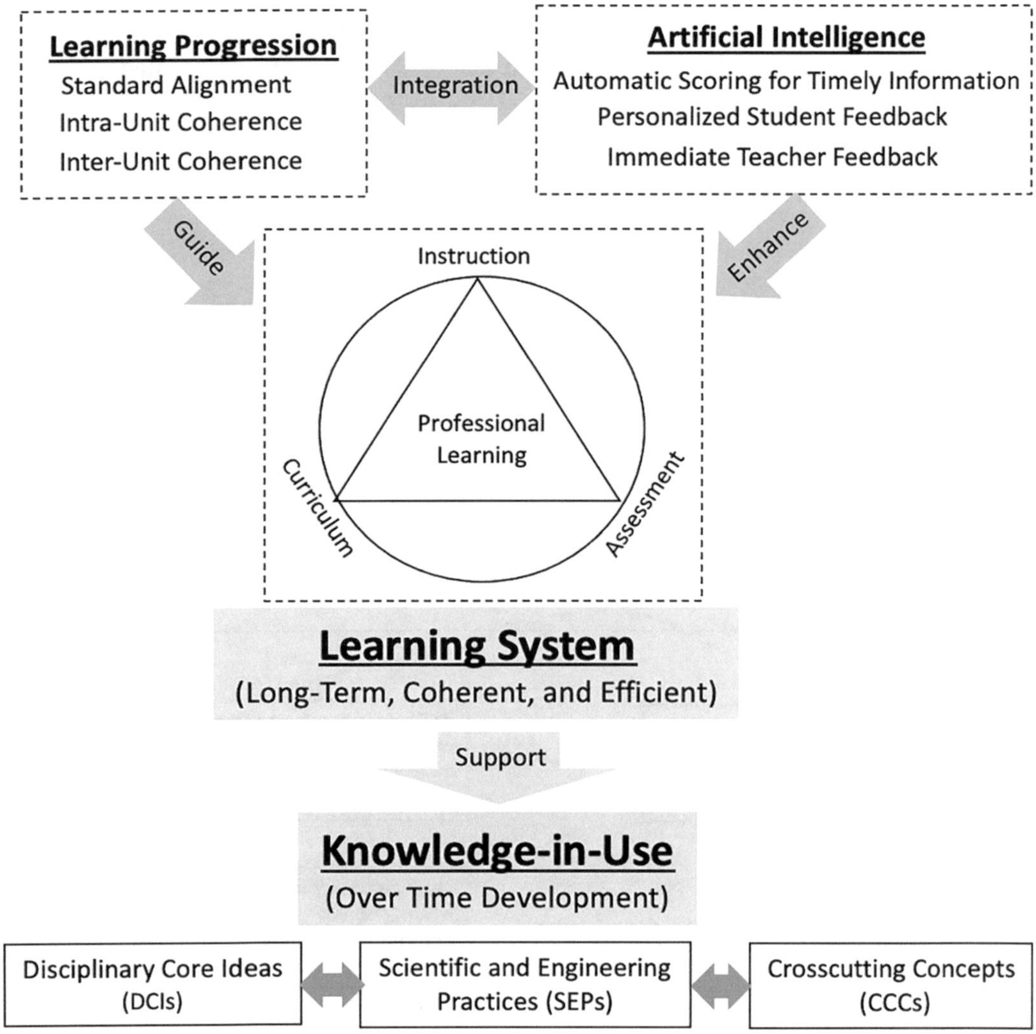

Figure 27.1 Conceptual Model for Integrating LPs and AI.

timely and adequate manner are the prerequisites for supporting students' knowledge-in-use development over time. In this conceptual model, we propose AI approaches, such as machine learning, can develop algorithms to automatically score students' constructed responses (e.g., Kaldaras et al., 2022; Zhai et al., 2022). In doing so, information on student performances on each level can be available on time for both students and teachers. Research-based personalized feedback can be pre-designed and offered for students to revise and improve their responses to further support students in achieving each level. Moreover, individual and class-level information can support teachers in making evidence-based instructional decisions (He et al., 2024). In addition, researchers and developers can pre-design research-based instructional strategies aligned with each level to facilitate inexperienced teachers' instructional actions to improve their students' understanding and support them in moving to the following levels.

Integrating AI into Learning Progression: Our View

This section employs two LPs (middle and high school examples) and the related AI work to unpack our conceptual model (See Figure 27.1) on integrating AI into the LPs to support students' knowledge-in-use development over time. This section includes three topics of using AI to 1) examine LP-based student performance, 2) provide LP-based automatic feedback, and 3) support LP-based teacher instructional decisions.

An Overview of Two LPs

Middle School LP

Figure 27.2 presents a hypothetical LP to support middle school students' knowledge-in-use to apply the disciplinary core ideas (e.g., properties of substances and chemical reactions) and crosscutting concepts (e.g., cause and effect) to construct scientific explanations about the

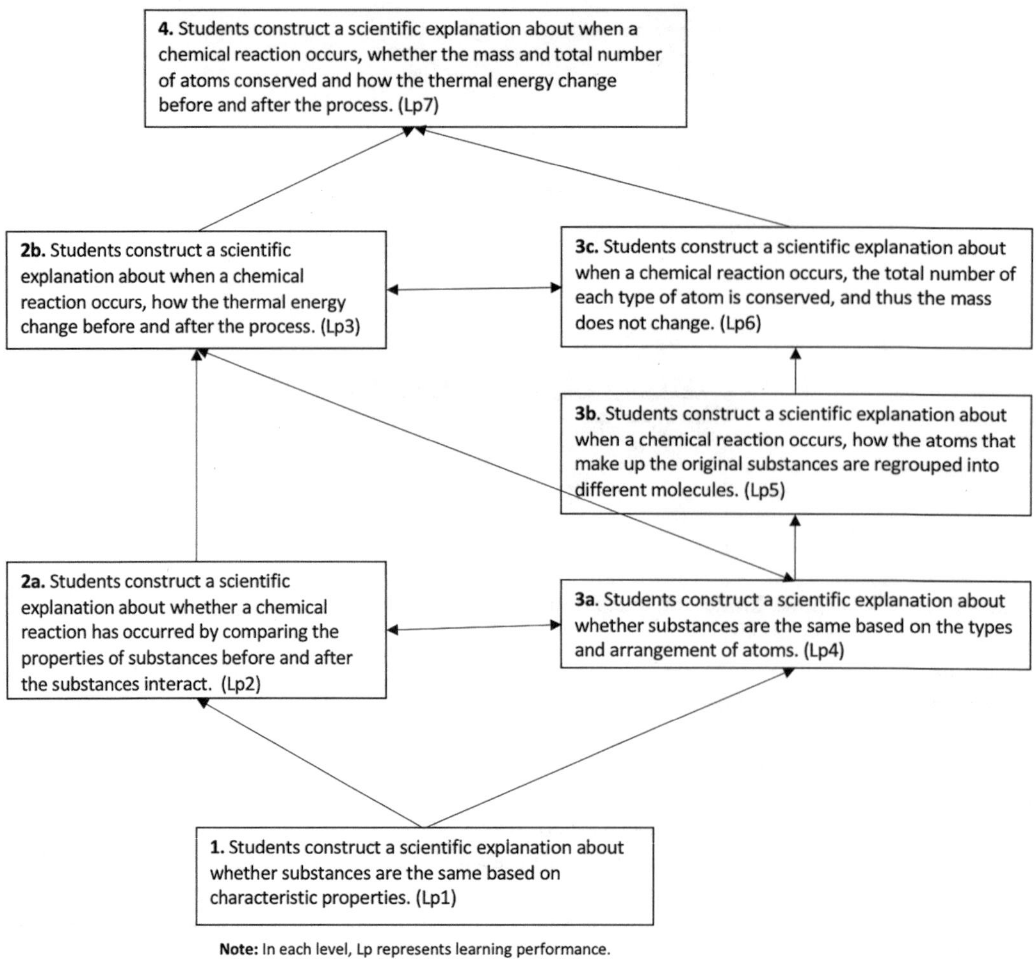

Figure 27.2 A Middle School Hypothetical Knowledge-in-Use LP for Properties of Substances and Chemical Reactions Illustrating Different Pathways.

relevant phenomena. We followed the design principles (see Figure 27.2 of Chapter 14 in this handbook, He et al. 2024) to develop a hypothetical LP aligned with NGSS PEs (MS-PS1 Matter and Its Interactions). We articulated four levels in the hypothetical LP with seven performance-based three-dimensional learning goals (i.e., DCIs, SEPs, and CCCs). Each learning goal is a small grain size PE, which serves as a knowledge-in-use statement for all students required to achieve (Harris et al., 2019). The focal SEP is constructing scientific explanations. Although the focal CCC is cause and effect, we include other CCCs, such as patterns, when cause and effect are not required at some levels. For instance, at Level 2a, the learning goal states that "students construct a scientific explanation about whether a chemical reaction has occurred by comparing the properties of substances before and after the substances interact (Lp2)." At this level, the DCI is "characteristic properties that can be used to identify whether a chemical reaction occurs," and the SEP is "constructing a scientific explanation." CCC is "Patterns (difference of characteristic properties)." The CCC of cause and effect is not required for students to explain the relevant phenomena. Moreover, student knowledge-in-use development has multiple learning pathways from lower to upper levels. Figure 27.2a shows different pathways that students might take. One pathway can be 1-2a-2b-3a-3b-3c-4; another possible way is 1-3a-2a-2b-3b-3c-4. Students' multiple learning pathways rely on their prior learning experience and teachers' curriculum materials, instruction, and assessments. Our LP intends to coherently guide middle school teachers' adaptions to their local curriculum, instruction materials, and assessment tasks to support students' knowledge-in-use development.

Performance-based assessment tasks from the *Next Generation of Science Assessment* (NGSA) project are adopted and modified to align with the levels in the LP. Each level has at least one task. We take the task of "Nami's Careful Experiment" (hereafter Nami task) as an example (see the task[1] via the ***Online Resource 26.1***) to assess students' knowledge-in-use at Level 2a. Rubrics are developed to score students' written responses (see the rubric example in Table 27.1 below). Students achieve Level 2a by demonstrating that the scores of their written responses completely meet the knowledge-in-use statement (Lp2, see Figure 27.2). The tasks are embedded into different locations across a unit or lesson sequence to collect information about how students' performance aligns with the LP over time.

High School LP

Figure 27.3 shows a previously validated NGSS-aligned LPs focusing on students' ability to integrate relevant DCIs, SEPs, and CCCs when explaining electrical interaction phenomena. Both LPs focus on the SEPs of Developing and Using Models and Constructing Explanations, the CCC of Cause and Effect, and DCIs of the Relationship between Energy and Forces and Types of Interactions. The two LPs describe students' increasing competency to develop causal models and explanations of phenomena involving electrical interactions, focusing specifically on Coulomb's law and Energy (Kaldaras et al., 2021) to explain electrical interactions at the macroscopic and atomic-molecular level (see detailed description in the ***Online Resource 26.2***). The two LPs are closely related and follow similar logic. The LPs start from Level 0, reflecting no substantive evidence of using either Coulomb's law or Energy to explain electrostatic phenomena. Level 1 reflects the ability to use Coulomb's law and Energy to develop partial causal models and explain electrostatic phenomena. Level 2 reflects the ability to use either Coulomb's law or Energy to explain electrostatic phenomena at the macroscopic level and partially at the atomic-molecular level. Finally, Level 3 reflects

Table 27.1 The Analytic Rubric and Example of Student Responses for the Nami Task

Student response:

1 Yes, because when we mixed the two liquids together, it made a new substance, so yes, on that. When we boiled the new substance, it separated the two liquids from each other.
2 One there is density, second there is boiling point's, third there is mass differences.
3 Because it helps understand what a new substance was and became a separation.

Dimensions	Criteria (C)	Score 1/0
DCI	C1. Student clearly states that a chemical reaction occurs (when Acid and Alcohol are mixed and heated).	1
SEP/CCC	C2. Student supports the claim by referring to the different boiling points of substances (before and after acid and alcohol were mixed).	1
SEP/CCC	C3. Student supports the claim by referring to the different densities of substances (before and after acid and alcohol were mixed).	1
DCI	C4. Student indicates that boiling point is the characteristic (or unique/important) property to identify substances.	0
DCI	C5. Student indicates that density is the characteristic property (or unique/important) to identify substances.	0
SEP/CCC	C6. Student indicates layers 1 and 2 are new substances by comparing the differences of characteristic properties (before and after acid and alcohol were mixed).	1
DCI	C7. Student indicates new substances formed, which means that a chemical reaction occurs.	0
Misunderstanding	C08. Option 1: Student does not mention the mass change. Option 2: Student states that the mass changed, but it does not show that a chemical reaction occurred. Option 3: Student states the mass changes and indicates that it shows a chemical reaction occurs (this does not indicate a chemical reaction).	Option 3

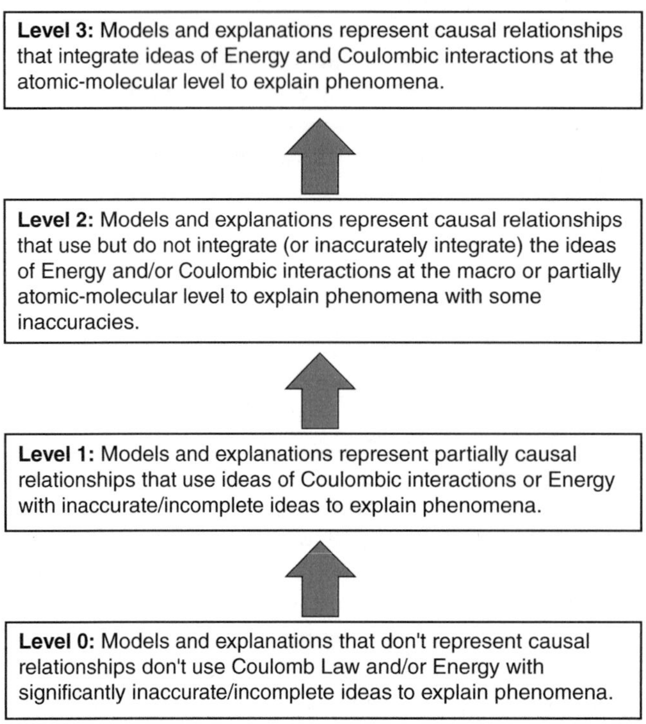

Figure 27.3 A High School Knowledge-in-Use LP for Electrical Interactions Combing Coulomb's Law and Energy Ideas (Adopted from Kaldaras et al., 2022).

the ability to use Coulomb's law and Energy ideas to explain electrostatic phenomena at the atomic-molecular level.

The performance-based assessments were designed to assess student performance at the levels of the LP shown in Figure 27.3. A detailed assessment development process has been previously described (Kaldaras et al., 2022). We use the Carts task (See it in the ***Online Resource 26.2***) as an example that is designed to assess all three levels of the LPs. The task assesses the DCI of the Relationship between Energy and Forces, SEP of Constructing Explanations, and CCC of Cause and Effect. Using a modified evidence-centered design (mECD, Kaldaras et al., 2021; Mislevy & Haertel, 2006), we provided a claim of three-dimensional learning performance and detailed evidence statements that elicit students' performances to demonstrate their achievement.

Use AI in Examining LP-based Student Performance

Examining student knowledge-in-use performance over time depends on the LP-based performance-based assessment tasks and associate rubrics (analytic or holistic perspectives). Given the time needed for teacher scoring, the field has widely employed supervised machine learning approaches to develop algorithms so that the computer can automatically score students' constructed responses. Supervised machine learning requires computers to learn from human scores for developing algorithms and validate them using the accuracy of ML scores and the inter-rater reliability between human and ML scores. Both holistic and analytic scoring approaches (Jescovitch et al., 2021) have been shown to yield well-functioning ML models for performance-based assessments. However, it is essential to develop rubrics that accurately capture student thinking at the appropriate LP levels and yield high inter-rater reliability between human and machine scores. Below, we elaborate on how AI examines LP-based students' performance using analytic and holistic rubrics.

Automatic Diagnostic Information

We contextualize the automatic diagnostic information procedure in the middle school LP. Analytic rubrics represent a series of binary dichotomous categories that identify the presence or absence of relevant ideas in student responses. Such rubrics can diagnose students' problems with knowledge-in-use at each level. Many ML studies have employed analytic rubrics to score student responses to provide teachers and students with diagnostic information (Moharreri et al., 2014). When designing such rubrics, researchers need to ensure that the rubric captures students' ability to integrate disciplinary core ideas, crosscutting concepts, and science and engineering practices at various LP levels (Kaldaras et al., 2022).

We developed diagnostic analytic rubrics using each of the three dimensions (DCIs, SEPs, and CCCs) as a criterion to score student performance with binary scores (1 correct and 0 incorrect) on assessment tasks and collect examples of actual student responses (He et al., 2023). Each criterion includes either a DCI aspect or a SEP/CCC aspect. We grouped the SEP/CCC aspect for this specific task because the SEP of analyzing and interpreting data and the CCC of Patterns (i.e., similarity and difference of data) cannot be separated. The combination of three-dimensional knowledge might vary depending on the specific context. However, the vision of knowledge-in-use requires a key principle that a SEP cannot be separated from a CCC or a DCI because we expect students to apply knowledge (CCC or DCI) with a scientific practice. A set of criteria functions together to assess student knowledge-in-use performance.

Table 27.1 shows the analytic rubric for the Nami task, including eight criteria with four DCI aspects, three SEP/CCC aspects, and one misunderstanding. Teachers can also use rubrics to diagnose students' strengths and weaknesses in achieving the knowledge-in-use goal at a specific level. For instance, the student response in Table 27.1 shows that the student understands one DCI and three SEP/CCCs but has challenges understanding the other essential DCIs (i.e., characteristic properties can be used to identify substances). Moreover, the student misunderstood that changes in mass also indicate a chemical reaction. A deep-learning application, Bidirectional Encoder Representations from the Transformers (BERT) approach (Devlin et al., 2018), is used to develop algorithms. Usually, thousands of human-labeled data are required for training computers to develop algorithms for automatically scoring students' constructed responses (Zhai et al., 2022). Using the BERT approach, diagnostic information about individual students' performance can be automatically presented to teachers and students.

For teachers, AI provides them with information about individual students' performances and whole-class performances. For instance, except for the last criteria (C08 misunderstanding), the score pattern of the student response above (see Table 27.2) is 1110010, which indicates the student had achieved the SEP/CCC aspects (C2, C3, and C6 correct) but need supports in the understanding of DCIs (C1 correct but C4, C5, and C7 incorrect). To achieve Level 2a (Figure 27.2), the student needs to improve their understanding of DCIs. All score patterns are categorized into four groups (see Table 27.2): 3D, SEP/CCC, DCI, and Missing 3D. For instance, using the example in Table 27.1, students' responses meeting the same score pattern (e.g., 1110010) are categorized into the SEP/CCC group. We categorize students' responses into the DCI group if the score patterns meet the criteria for all DCI aspects (C1, C4, C5, C7) but not all SEP/CCC aspects (C2, C3, C6), such as 1001101. AI approaches, such

Table 27.2 Diagnostic Information about Collective Student Performance

Group	*Achieved*	*Not Yet Achieved*
3D	Students have achieved understanding (DCI, SEP, and CCC): *DCI: characteristic properties can be used to identify whether a chemical reaction occurs,* *SEP: constructing explanations, and* *CCC: looking for patterns in data.*	Students have achieved three-dimensional learning (DCI, SEP, and CCC) in the task and are ready to move up to the upper levels.
SEP/CCC	Students have achieved understanding (SEP and CCC): *SEP: constructing explanations, and* *CCC: looking for patterns in data.*	Students need support in the DCI: *DCI: characteristic properties can be used to identify whether a chemical reaction occurs.*
DCI	Students have achieved understanding (DCI): *DCI: characteristic properties can be used to identify whether a chemical reaction occurs.*	Students need support in the SEP/CCC: *SEP: constructing explanations, and* *CCC: looking for patterns in data.*
3D Missing	Students have not achieved three-dimensional learning (DCI, SEP, and CCC).	Students need support in the DCI, SEP, and CCC: *DCI: characteristic properties can be used to identify whether a chemical reaction occurs,* *SEP: constructing explanations, and* *CCC: looking for patterns in data.*

as supervised and unsupervised ML (Murphy, 2012), can train the computers to categorize individual students into four groups so that teachers can receive diagnostic information about collective students' performance at the class level.

Automatic Evaluation Information

We situate the automatic evaluation information in the high school LP. Holistic rubrics aim to assess the overall quality of student responses concerning LP levels, informing teachers and students about what level the students achieved. Initially, we designed a holistic rubric aiming to diagnose student performance in response to the three levels of the LP. A holistic rubric is a multi-level rubric historically used to map student responses to specific LP levels (Haudek et al., 2012; Kaldaras et al., 2021; Li et al., 2023). However, holistic rubrics may be less effective in contexts of automatic scoring on LP-aligned assessments because they may not allow for a straightforward diagnosis of mis-scores. This can diminish the validity of the resulting automatic scores for LP of interest (Kaldaras et al., 2022). Therefore, we decomposed the original holistic rubric into analytic rubric categories (see the ***Online Resource 26.3***). Each category aligns with the corresponding mECD argument above (see the ***Online Resource 26.2***). These categories aim to reflect the smallest possible aspect of three-dimensional knowledge that can be described without compromising the three-dimensional nature of the task focused on measuring knowledge-in-use. Further, these categories ensure that all essential aspects of the phenomenon are captured in student responses. Finally, all rubric categories centered on ideas students use in their responses, not simply the presence or absence of certain words. This assures that the scores captured ideas rather than specific phrases or words. Similarly, for each category, the designers aimed to capture different ways students can express the ideas captured by a given category to verify that the rubrics capture a diverse range of student thinking. We used an ensemble-supervised machine learning approach (Haudek et al., 2012) to develop algorithms to automatically assign scores for each analytic category of the item (Kaldaras et al., 2022). We used 1252 student responses to train the ML algorithm. Human raters previously scored these responses using the analytic rubric (see the ***Online Resource 26.3***). See more details of the training process in Kaldaras and colleagues' (2022) paper.

To evaluate student LP levels, we needed a way to map specific combinations of the categories in the above analytic rubric to determine individual levels in the LP. That is, no single analytic category represents an LP level performance above the lowest level. Table 27.3 shows the alignment between specific combinations of the analytic rubric categories and the LP levels by relying on the pieces of evidence within the mECD argument. The resulting

Table 27.3 Alignment between Combinations of Analytic Rubric Categories and the LP Levels (Adopted from Kaldaras et al., 2022)

LP level	*Analytic Rubric Categories Combination*
3	Categories (1,2,3,6) or (1,2,6) or (1,3,4,5) or (1,2,3,4,5) or (1,3,6) or (1,2,4,5)
2	Categories (1,2,3,4) or (1,2,3,4,5) or (1,2,3,5,7) or (1,2,3,5) or (1,3,4) or (1,3,5)
1	Categories (1,2) or (1,2,3) or (1,2,4) or (1,2,5) or (1,4,7) or (1,3,5,7) or (1,2,3,7) or (1,2,7) or (1,5,7) or (1,4) or (1,3,4,7)
0	Categories (1,3) or (1,7) or (3,7) or (1,3,7)Category (1) or (2) or (7) or Zero on all categories

Note: Any other combination of analytic categories not defined above was assigned to level 0.

combinations of analytic categories ensured that the student's ability to use the three dimensions of NGSS together when answering the question, which reflects knowledge-in-use, is captured in the resulting final score. Combining analytic rubric categories is the basis for using AI to automatically evaluate student LP levels. To examine whether the combinations for the analytic categories led to the same LP level assignment as the original holistic rubrics, we compared scores assigned by human coders using these two coding approaches. From the entire data set, we randomly selected 200 responses by LP levels to compare codes assigned *via* the original holistic rubric with codes assigned *via* the analytic categories and combination. Holistic scoring was conducted by an experienced coder who participated in the holistic rubric development. These responses were coded independently of the analytic scores. The agreement on the final score for each response between the two approaches was almost perfect (Cohen's kappa = 0.986). We also calculated Spearman's rho (correlation) between final analytic and holistic scores. We found $r_s = 0.995$, $p < 0.01$, indicating that combining the analytic categories could lead to the same LP level using the original holistic rubrics.

Use AI in Providing LP-based Automatic Feedback

Without any learning opportunities, it is challenging for students to achieve understanding and move to the next levels in an LP (Krajcik & Shin, 2023). To support individual students' achievement at each level, they need immediate and personalized feedback based on the identical non-achieved aspects in their constructed responses to performance-based assessments. Using AI to provide automatic information on assessment performance along an LP over time brings tremendous potential for offering automatic and coherent feedback and guidance to support individual students' knowledge-in-use development.

Automatic Feedback to Achieve the LP Levels

Here, we situate this type of automatic feedback in the context of the middle school LP. At each LP level, individual students need immediate and personalized feedback to improve their understanding to achieve the goal at this level. To design personalized feedback for each student, we employed the criteria in our analytic rubrics (see Table 27.4). We transformed each criterion into two-fold feedback statements: achieved feedback (score 1) and non-achieved feedback (score 0). We thus articulated the achieved and non-achieved feedback for all criteria. In our feedback statements, we avoided discouraging wordings, such as "unfortunately," "did not correctly answer," or "however." The score patterns of students' responses are categorized using the components of achieved and non-achieved feedback so that we can design personalized and holistic feedback for an individual student. For instance, the score pattern

Table 27.4 An Example of Personalized Student Feedback

The Score Pattern of the Student Response: 1110010 **Personalized Student Feedback** [Your response]
• clearly stated that a chemical reaction occurs (achieved C1), supported by referring to the boiling points (achieved C2) and the density (achieved C3) of substances are different AND indicates layers 1 and 2 are new substances (achieved C6) based on the data patterns in the table. *[To improve your work, your response should include]* • the rationale for why boiling point (non-achieved C4) and density (non-achieved C5) provide evidence to support your claim that a chemical reaction occurred needs justification. Why can density and boiling point be used to identify if substances are the same or different?

of the student's response above is 1110010. The student will receive personalized feedback on the achievement and the further improvement needed in the response (see Table 27.4). With AI support, each student can receive personalized feedback to improve their work to achieve the current level and move up to the next levels of understanding.

Researchers also employed innovative AI approaches (e.g., deep learning, Goodfellow et al., 2016) to detect students' science ideas (Li, Gerard, et al., 2023; Riordan et al., 2020) so that it can automatically send probing prompts in response to performance-based tasks. As the example in Table 27.1 shows, we detect students' ideas such as "new substance," "densities are different," "boiling points are different," and "masses are different." Teachers can ask probing questions related to SEP/CCC aspects or DCI to support student's reasoning. For SEP/CCC aspects, the probing questions can be: 1) Can you use the data table to describe the similarities and differences of Density, Boiling Points, and Mass you found? 2) Can the data on Density, Boiling Point, and Mass support your claim? If yes, explain why. For DCI aspects, potential probing questions include: 1) what does it mean that the properties of the four substances (Acid, Alcohol, Layer 1, and Layer 2) differ? 2) Why can changes in properties be used to identify if a chemical reaction occurs? Based on the automatically detected students' ideas, such probing prompts can potentially be provided to engage students step-by-step in working out the challenges they experienced in responding to the tasks at this level. Once a student submits the first response to a task, the AI system can detect the ideas in the response and automatically offer a probing prompt to deepen the student's reasoning and ideas. If the student still does not achieve the task, the AI system offers a follow-up prompt until the student achieves the goal at this level.

Automatic Feedback to Move Up to Next Levels

We elaborate our ideas in the context of the high school LP. Table 27.5 shows sample student responses for each LP level and the corresponding combinations of analytic scoring categories that led to the response being assigned to that level. As one advantage of using analytic scores to provide a more fine-grained view of student performance on a task, we considered possible feedback for this task and aligned it to LP levels. Table 27.5 provides possible feedback statements that aim to guide students to think about what ideas are missing from their answers to help them advance in the LP. Notice that each feedback statement builds on specific ideas in student responses to help elicit the ideas needed to support them in moving to the next LP level. We are currently developing and testing the automatic feedback system capable of achieving this goal. We further tested multiple feedback types to study the feedback most helpful to learners with diverse backgrounds and support them in moving to the next LP level.

Use AI in Supporting LP-based Teacher Instructional Decisions

The above section describes how students can improve their understanding using individual and personalized feedback. However, such feedback relies heavily on individual students' self-regulated learning skills (e.g., Butler & Winne, 1995). Students may still need timely instructional support from their teachers. Even though teachers receive essential AI-based information about their students' performances individually and collectively, we argue that challenges still exist for teachers to take instructional actions to support their students' achievements on a specific LP level, especially for teachers who do not have sufficient NGSS teaching experience. Instructional strategies aligned with each LP level can benefit teachers'

Table 27.5 Sample Combinations of Analytic Categories in Student Responses, the Associated LP Levels, and Suggested Feedback

NGSS-aligned LP levels	*Sample Student Responses and Associated Analytic Scoring*	*Possible Automatic Feedback*
Level 3: Student models and explanations represent causal relationships that integrate ideas of energy and Coulombic interactions at the atomic-molecular level to explain phenomena.	*Response: The cars will move in opposite directions [C1] because they are of the same charge [C2]. There is much energy when they are close together because they want to repel. When the wedges are moved, and the cars go away from each other [C6], they will move until [C3] there is no more repulsive force between them. The farther they move, the less energy they have and the less force they have between each other[C6].* *Score:* Category 1 + Category 2 + Category 3+ Category 6	*Feedback:* Combining ideas of electric force and energy to explain when the carts will stop is a useful strategy.
Level 2: Student models and explanations represent causal relationships that use but do not integrate (or inaccurately integrate) the ideas of energy and/or Coulombic interactions at the macro or atomic-molecular level to explain phenomena with some inaccuracies	*Response: Once the wedges are removed, I predict the cars will move away from each other [C1] because the metal sheets attached to both cars are negatively charged, and objects with the same charge repel away[C2].* *I predict that the cars will stop moving once [C3] they are out of each other's electric field because once the cars are no longer in the other car's electric field, there will be no more repelling forces from the other car's electric field, so both cars will then stop since there is no force causing them to move [C4].* *Score:* Category 1 + Category 2 + Category 3 + Category 4	*Feedback:* Consider adding ideas related to energy to your explanation of when the carts will stop. *Purpose*: elicit categories 5 or 6, which are required to move to level 3 of the LP
Level 1: Student models and explanations represent partially causal relationships that use ideas of Coulombic interactions or energy with inaccurate/incomplete ideas to explain phenomena	*Response: They will move away [C1] because the metal sheets are neg, and the same charges push each other away [C2]. They will stop once they are far enough apart [C3] because they will not sense each other [C7].* *Score:* Category 1 + Category 2 + Category 3 + Category 7	*Feedback:* Consider explaining why the carts will stop using ideas of energy and force. *Purpose*: elicit categories 4, 5, or 6 required to move to levels 2 and 3 of the LP)
Level 0: Student models and explanations do not represent causal relationships and use ideas of Coulombic interactions and/or energy with significantly inaccurate/incomplete ideas	*Response: the objects would move away from each other [C1].* *Score:* Category 1 only	*Feedback:* Consider adding the explanation for why the two carts will move in opposite directions and when they will stop. *Purpose: elicit categories 2 and 3, which are required to move to level 1 of the LP.*

Note: [C1] in the texts (2nd column) stands for Category 1 where the preceding phrase or sentence was coded, and so do the others.

instructional decision-making. In our prior work (He et al., 2024), we developed four types of instructional strategies (e.g., first-hand, second-hand, simulation, and multimodal experiences) based on the theoretical foundations of content-specific (Magnusson et al., 1999), inclusive (Thomas et al., 2018), and constructive and interactive (Chi, 2009) strategies. We argue that our instructional strategies can engage students, support equity, and promote knowledge-in-use for a wide range of students from diverse backgrounds, particularly from minoritized and marginalized racial and ethnic groups.

We present an example of first-hand instructional strategies (see the ***Online Resource 26.4***) to support students' achievement at Level 1b (See Figure 27.2). Here, we illustrate how our strategies could work. For instance, a teacher implements the Nami task (see the ***Online Resource 26.1***) to examine whether students have achieved the learning goal at Level 1b. AI-based information and individual feedback support students to improve their work. After that, the teacher may find some students struggling to understand the DCIs (characteristic properties that can be used to identify substances) or SEP/CCCs (constructing explanations and data patterns). At this point, the teacher may need support with well-designed instructional strategies. Based on the previous lessons and the understanding of student learning experience, the teacher may have students conduct a lab experiment on whether a chemical reaction occurs when mixing a drying agent ($CaCl_2$) with baking soda (Na_2CO_3). Then, the teacher asks students to collect and analyze data to find patterns in the data. The teacher prompts students to work individually and collaboratively to learn how to construct a scientific explanation with a clear claim responding to the question, sufficient and appropriate evidence, and an organized reasoning process using scientific principles that justify why the data can serve as evidence to support the claim (e.g., McNeill et al., 2006). Such instructional strategies are available to support students with diverse backgrounds since we carefully selected phenomena relevant to their daily lives. The learning materials, modalities, and use of language are accessible, which makes our instructional strategies foster fairness and equity in learning science. As an LP is a long journey of student learning development, teachers need continual and coherent support with such well-designed instructional strategies.

Recommendations for Future Research and Conclusion

This chapter expands on previous work on using AI in teaching and learning science (e.g., Lee et al., 2021; Kaldaras et al., 2022; Matuk et al., 2016; Zhai et al., 2022) to support student science learning over time aligned with three-dimensional LPs. Although we have made some progress, we call for more research to expand the field's understanding of integrating AI into LPs and enhance learning systems to better support student knowledge-in-use over time. To move forward, we provide several recommendations for future research.

Recommendation 1. Evidence-centered design approaches should guide the coherence between AI support materials (e.g., automatic information, individual feedback) and LPs. Creating a systematic approach using evidence-centered design (Mislevy & Haertel, 2006) is encouraged to ensure coherence between the AI support materials and LPs. In our study, both the AI support and the LP materials are under the construct of knowledge-in-use. To ensure the coherence of integrating AI into an LP, we recommend collaborations between researchers with diverse expertise (e.g., science education, learning sciences, AI, equity, and social justice) and research-practice partnerships (Coburn & Penuel, 2016). Researchers conduct design-based research with a design-test-revise cycle (Barab & Squire, 2004; Krajcik & Shin, 2023) to investigate the coherence between AI materials and LPs. In doing so, researchers with diverse expertise contribute their intellectual knowledge to ensure the AI-empowered LPs are NGSS-aligned, AI-empowered, and equitable to support a range of students with diverse backgrounds. Researchers must collaborate with teachers and students to develop and test AI systems. For instance, we work with teachers from minority and marginalized racial and ethnic groups to use rubrics to score students' responses on assessment tasks, potentially reducing the human data bias for developing AI algorithms. We recommend that empirical studies be conducted to see how their feedback and users' experience improve the feasibility of AI-LP materials in classrooms.

Recommendation 2. LPs integrated with AI can enhance a learning system to support student knowledge-in-use better. An LP does not provide students with the opportunity to learn. Instead, a coherent learning system with LP guidance supports all students' learning opportunities. Given that, using an AI-empowered LP to guide the adaption of a well-designed learning system has substantial potential to support students better learning over time. Such AI-empowered learning systems (e.g., Acosta & Slotta, 2018; Dickler et al., 2021; Lee et al., 2021) may dramatically change the existing paradigm of classroom learning environments. However, a learning system with multiple components (e.g., curriculum, assessment, instruction, and professional learning) is complex. Bringing an AI-empowered LP to enhance the efficiency of the existing learning system brings tremendous challenges for researchers and partitioners. Implementing and testing an AI-empowered learning system in classrooms also brings challenges. For instance, policymakers and school districts may not be ready to use AI in classrooms; not all districts nationwide have the infrastructure to equip AI tools; and teachers may need sustainable professional learning to support implementing AI-empowered LPs in classrooms.

Recommendation 3. Rigorous research should be conducted to generate empirical evidence for employing and improving AI-empowered LP-based learning systems. We call for researchers to conduct qualitative and quantitative research to explore the effect of an AI-empowered LP in a learning system and the mechanism of how such learning systems support diverse students' learning with multiple learning pathways. In addition, researchers need to conduct substantial studies, including 1) teachers' and students' views of using AI in classrooms, 2) teacher knowledge, understanding, and practices of using AI-empowered LPs in classrooms, and 3) professional learning for supporting teachers' use of an AI-empowered LP-based in classrooms. Collecting and analyzing multiple data sources, including interviews, observations, log file data, questionnaires, assessments, and artifacts, is preferred to provide empirical evidence to address the above questions.

Recommendation 4. Ethics and transparency issues should be addressed when applying AI-empowered LP-based learning systems. Researchers and practitioners should consider various aspects of ethics and transparency to ensure that AI-empowered LP-based learning systems are accountable, fair, and trustworthy (Holmes et al., 2022). To do so, we need to minimize the potential biases in generating algorithm processes, such as automatic scoring and feedback. As teachers and students access an AI-empowered LP-based learning system using a web portal, all information they take will be stored there. Their data should be protected and stored securely. In addition to the ethics issues, the AI-empowered LP-based learning systems should be transparent and understandable to teachers, students, and other stakeholders. For instance, we should explain to teachers how the AI algorithms work and let teachers and students know the accuracies of automatic scoring and feedback for their decision-making.

In conclusion, this chapter discusses our research on integrating AI into an LP-based learning system to support student knowledge-in-use better. We believe that building an equitable and efficient AI-empowered LP-based learning system, including automatic and timely information, personalized feedback, and instructional support, would be a promising research agenda and potential solution for better-supporting teacher teaching practices and student learning across time.

Acknowledgments

This study is supported by the National Science Foundation (Grant Number DRL-2201068; DRL- 2200757; DRL-2100964). Any opinions, findings, conclusions, or recommendations expressed in this material are those of the authors and do not necessarily reflect the views of the National Science Foundation.

Note

1 Also see the "Nami's Careful Experiment" assessment via the NGSA project website: https://ngss-assessment.portal.concord.org/resources/198/nami-s-careful-experiment-id-015-03-c02

References

Acosta, A., & Slotta, J. D. (2018). CKBiology: An active learning curriculum design for secondary biology. *Frontiers in Education*, *3*, 52. 10.3389/feduc.2018.00052

Barab, S., & Squire, K. (2004). Design-based research: Putting a stake in the ground. *The Journal of the Learning Sciences*, *13*(1), 1–14. 10.1207/s15327809jls1301_1

Butler, D. L., & Winne, P. H. (1995). Feedback and self-regulated learning: A theoretical synthesis. *Review of Educational Research*, *65*(3), 245–281. 10.3102/00346543065003245

Chen, L., Chen, P., & Lin, Z. (2020). Artificial intelligence in education: A review. *IEEE Access*, 8, 75264–75278. 10.1109/ACCESS.2020.2988510

Chi, M. T. (2009). Active-constructive-interactive: A conceptual framework for differentiating learning activities. *Topics in Cognitive Science*, *1*(1), 73–105. 10.1111/j.1756-8765.2008.01005.x

Coburn, C. E., & Penuel, W. R. (2016). Research–practice partnerships in education: Outcomes, dynamics, and open questions. *Educational Researcher*, *45*(1), 48–54. 10.3102/0013189X16631750

Devlin, J., Chang, M. W., Lee, K., & Toutanova, K. (2018). Bert: Pre-training of deep bidirectional transformers for language understanding. *arXiv preprint arXiv:1810.04805*.

Dickler, R., Gobert, J., & Sao Pedro, M. (2021). Using Innovative Methods to Explore the Potential of an Alerting Dashboard for Science Inquiry. *Journal of Learning Analytics*, *8*(2), 105–122. 10.18608/jla.2021.7153

Fortus, D., & Krajcik, J. (2012). Curriculum coherence and learning progressions. In Fraser, B., Tobin, K., & McRobbie, C. (Eds.) *Second international handbook of science education* (pp. 783–798). Springer International Handbooks of Education, vol 24. Springer, Dordrecht. 10.1007/978-1-4020-9041-7_52

Goodfellow, I., Bengio, Y., & Courville, A. (2016). *Deep learning*. MIT Press.

Harris, C. J., Krajcik, J. S., Pellegrino, J. W., & DeBarger, A. H. (2019). Designing Knowledge-In-Use Assessments to Promote Deeper Learning. *Educational Measurement: Issues and Practice*, *38*(2), 53–67. 10.1111/emip.12253

Haudek, K. C., Prevost, L. B., Moscarella, R. A., Merrill, J., & Urban-Lurain, M. (2012). What are they thinking? Automated analysis of student writing about acid–base chemistry in introductory biology. *CBE—Life Sciences Education*, *11*(3), 283–293. 10.1187/cbe.11-08-0084

He, P., Chen, I.-C., Touitou, I., Bartz, K., Schneider, B., & Krajcik, J. (2023a). Predicting student science achievement using post-unit assessment performances in a coherent high school chemistry project-based learning system. *Journal of Research in Science Teaching*, *60*(4), 724–760. 10.1002/tea.21815

He, P., Shin, N., Zhai, X., & Krajcik, J. (2024). A design framework for integrating artificial intelligence to support teachers' timely use of knowledge-in-use assessments. In Zhai, X., & Krajcik, J. (Eds.) *Uses of artificial intelligence in STEM education*. Oxford University Press.

He, P., Zhai, X., Shin, N., & Krajcik, J. (2023b). Applying Rasch measurement to assess knowledge-in-use in science education. In: Liu, X., & Boone, W. J. (Eds.) *Advances in applications of Rasch measurement in science education*. Contemporary Trends and Issues in Science Education, vol. 57. Springer, Cham. 10.1007/978-3-031-28776-3_13

Holmes, W., Porayska-Pomsta, K., Holstein, K., Sutherland, E., Baker, T., Shum, S. B., ... & Koedinger, K. R. (2022). Ethics of AI in education: Towards a community-wide framework. *International Journal of Artificial Intelligence in Education*, *32*, 504–526. 10.1007/s40593-021-00239-1

Holstein, A., Weber, K. E., Prilop, C. N., & Kleinknecht, M. (2022). Analyzing pre-and in-service teachers' feedback practice with microteaching videos. *Teaching and Teacher Education*, *117*, 103817. 10.1016/j.tate.2022.103817

Jescovitch, L. N., Scott, E. E., Cerchiara, J. A., Merrill, J., Urban-Lurain, M., Doherty, J. H., & Haudek, K. C. (2021). Comparison of machine learning performance using analytic and holistic coding approaches across constructed response assessments aligned to a science learning progression. *Journal of Science Education and Technology*, *30*(2), 150–167. 10.1007/s10956-020-09858-0

Jin, H., Mikeska, J. N., Hokayem, H., & Mavronikolas, E. (2019). Toward coherence in curriculum, instruction, and assessment: A review of learning progression literature. *Science Education*, *103*(5), 1206–1234. 10.1080/09500693.2019.1606471

Kaldaras, L., Akaeze, H., & Krajcik, J. (2021). Developing and validating Next Generation Science Standards-aligned learning progression to track three-dimensional learning of electrical interactions in high school physical science. *Journal of Research in Science Teaching*, *58*(4), 589–618. 10.1002/tea.21672

Kaldaras, L., Yoshida, N. R., & Haudek, K. C. (2022). Rubric development for AI-enabled scoring of three-dimensional constructed-response assessment aligned to NGSS learning progression. *Frontiers in Education*, *7*, 983055. 10.3389/feduc.2022.983055

Kent, D. (2022). *Artificial intelligence in education: Fundamentals for educators*. Kotesol DCC.

Krajcik, J. S. (2021). Commentary—Applying machine learning in science assessment: Opportunity and challenges. *Journal of Science Education and Technology*, *30*(2), 313–318. 10.1007/s10956-021-09902-7

Krajcik, J., & Shin, N. (2023). Student conceptions, conceptual change, and learning progressions. In Abell, S. K., & Lederman, N. G. (Eds.) *Handbook of research on science education* (pp. 121–157). Routledge.

Lee, H. S., Gweon, G. H., Lord, T., Paessel, N., Pallant, A., & Pryputniewicz, S. (2021). Machine learning-enabled automated feedback: supporting students' revision of scientific arguments based on data drawn from simulation. *Journal of Science Education and Technology*, *30*(2), 168–192. 10.1007/s10956-020-09889-7

Li, T., Chen, I., Miller, E., Miller, C., Schneider, B., & Krajcik, J. (2024). The relationships between elementary students' knowledge-in-use performance and their science achievement. *Journal of Research in Science Teaching*. 10.1002/tea.21900.

Li, T., Liu, F., & Krajcik, J. (2023). Automatically Assess Elementary Students' Hand-Drawn Scientific Models Using Deep Learning of Artificial Intelligence. *Proceedings of the Annual meeting of the International Society of the Learning Sciences (ISLS)*.

Li, W., Gerard, L., Lim-Breitbart, J., Bradford, A., Riordan, B., Steimel, K., & Linn, M. C. (2023). Explaining thermodynamics: Impact of an adaptive dialog based on a natural language processing idea detection model. *Proceedings of the 17th International Conference of the Learning Sciences (ICLS 2023)*.

Magnusson, S., Krajcik, J., & Borko, H. (1999). Nature, sources, and development of pedagogical content knowledge for science teaching. In Gess-Newsome, J., & Lederman, N. G. (Eds.) *Examining pedagogical content knowledge*. Science & Technology Education Library, 6. (pp. 95–132). Springer, Dordrecht. 10.1007/0-306-47217-1_4

Matuk, C., Gerard, L., Lim-Breitbart, J., & Linn, M. (2016). Gathering requirements for teacher tools: Strategies for empowering teachers through co-design. *Journal of Science Teacher Education*, *27*(1), 79–110. 10.1007/s10972-016-9459-2

McNeill, K. L., Lizotte, D. J., Krajcik, J., & Marx, R. W. (2006). Supporting students' construction of scientific explanations by fading scaffolds in instructional materials. *Journal of the Learning Sciences*, *15*, 153–191. 10.1207/s15327809jls1502_1

Ministry of Education, P. R. China. (2017). *Chemistry curriculum standards for senior high school* [普通高中化学课程标准]. People's Education Press.

Mislevy, R., & Haertel, G. (2006). Implications of evidence-centered design for educational testing. *Educational Measurement: Issues and Practice*, *25*(4), 6–20. 10.1111/j.1745-3992.2006.00075.x

Moharreri, K., Ha, M., & Nehm, R. H. (2014). EvoGrader: an online formative assessment tool for automatically evaluating written evolutionary explanations. *Evolution: Education and Outreach*, *7*(1), 15. 10.1186/s12052-014-0015-2

Morell, L., Collier, T., Black, P., & Wilson, M. (2017). A construct-modeling approach to develop a learning progression of how students understand the structure of matter. *Journal of Research in Science Teaching*, *54*(8), 1024–1048. 10.1002/tea.21397

Murphy, K. P. (2012). *Machine learning: A probabilistic perspective*. MIT Press.

National Research Council. (2000). *How people learn: Brain, mind, experience, and school: Expanded edition* (Vol. 1). National Academies Press. 10.17226/9853

National Research Council. (2012). *A framework for K–12 science education: Practices, crosscutting concepts, and core ideas*. National Academies Press. 10.17226/13165

National Research Council. (2014). *Developing assessments for the next generation science standards*. Washington, DC: The National Academies Press. 10.17226/18409

NGSS Lead States (2013). *Next generation science standards: For states, by states*. National Academies Press. 10.17226/18290

Nilsson, N. J. (1982). *Principles of artificial intelligence*. Springer Science & Business Media.

Organization for Economic Cooperation and Development (2019). *PISA 2018 assessment and analytical framework*. OECD Publishing.

Riordan, B., Cahill, A., Chen, J. K., Wiley, K., Bradford, A., Gerard, L., & Linn, M. C. (2020). Identifying NGSS-aligned ideas in student science explanations. *Paper presented at the Workshop on Artificial Intelligence for Education (AI4EDU@AAAI)*.

Russell, S. J. (2010). *Artificial intelligence a modern approach*. Pearson Education, Inc.

Thomas, C., Van Garderen, D., Sadler, K., Decker, M., & Hanuscin, D. (2018). Applying a universal design for learning framework. In Koomen, M., Kahn, S., Atchison, C., & Wild, T. (Eds.) *Towards inclusion of all learners through science teacher education* (pp. 91–103). Brill Sense.

Winston, P. H. (1992). *Artificial intelligence* (3rd Edition). Addison-Wesley Longman Publishing Co., Inc.

Zhai, X., He, P., & Krajcik, J. (2022). Applying machine learning to automatically assess scientific models. *Journal of Research in Science Teaching*, *59*(10), 1765–1794. 10.1002/tea.21773

Zhu, M., Lee, H. S., Wang, T., Liu, O. L., Belur, V., & Pallant, A. (2017). Investigating the impact of automated feedback on students' scientific argumentation. *International Journal of Science Education*, *39*(12), 1648–1668. 10.1080/09500693.2017.1347303

IV27_He_Chapter_Online Resources

Online Resource 27.1 "Nami's Careful Experiment" Task (Adopted from NGSA, 2023)

Learning performance: Students construct a scientific explanation about whether a chemical reaction has occurred by comparing the properties of substances before and after the substances interact.

Task: Nami wondered if mixing an acid with alcohol would cause a chemical reaction. She did the following experiment:

Step 1	Step 2	Step 3	Step 4	Step 5
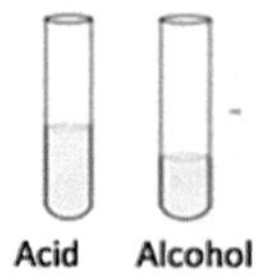	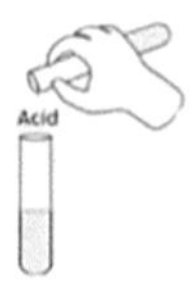		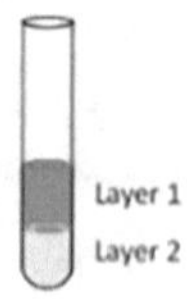	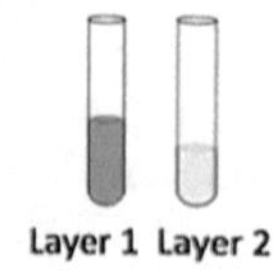
Measured and tested properties of acid and alcohol at room temperature. Recorded data in Table 1.	Mixed the acid and alcohol in a test tube.	Heated the test tube with the mixture.	After heating, observed Layer 1 and Layer 2 form.	Measured and tested properties of substance in Layer 1 and Layer 2 at room temperature. Recorded data in Table 1.

She measured the boiling point and mass, calculated each substance's density, and then recorded the data in the table below.

Data table for liquids **before** and **after** the experiment

Substance	*Density*	*Boiling Point*	*Mass*
Acid	1.2 g/cm^3	100 °C	6.9 g
Alcohol	0.80 g/cm^3	98 °C	9.0 g
Layer 1	0.91 g/cm^3	82 °C	13.2 g
Layer 2	1.0 g/cm^3	100 °C	2.7 g

Based on Nami's results in the data table, write a scientific explanation about whether or not a chemical reaction occurred. Your scientific explanation should include the three parts below.

1 Make a claim about whether a chemical reaction occurred by relating the substance **before** and **after** heating.
2 Describe similarities and differences between the properties of substances from the data table as evidence to support your claim.
3 Provide a reason or reasons for why the evidence you used supports your claim.

Online Resource 27.2 *The Carts Task (Adopted from* Kaldaras et al., 2022)

NGSS PE	HS-PS3-5. Develop and use a model of two objects interacting through electric or magnetic fields to illustrate the forces between objects and the changes in energy of the objects due to the interaction.
Claim (3D Learning Performance)	Students apply a model of electrostatic interactions that includes the relationship between electric force and potential energy to explain and predict phenomena involving changes in the distance between two charged objects.
Evidence Statements	1 Students will use electrical interactions between two objects to predict their movement in a system where objects can move freely. a When two objects have the same charge, they repel each other so that they will move apart. b As the similarly charged objects get further apart, the interactions between their respective electric fields and charges decrease, and the repulsive force between them decreases (*Coulomb's law*). 2 In their explanations and predictions, students will relate the relative position of two objects within a system to the potential energy. a When the distance between two interacting objects with the same charge decreases, the potential energy increases. 3 In a system where objects are allowed to move freely, students will predict that the objects will move to a more stable state. They will justify their prediction with the idea that the final stable state should have a lower potential energy than the initial state. a Students will predict that the potential energy will increase when the system moves to a less stable state. b Students will relate an increase in stability to a decrease in potential energy.*Note: limited to electric or gravitational potential energy* 4 Students will track the energy as the system moves toward a more stable state using explanations or models (for example, bar graphs) to indicate the relative changes. *Note: the energies and energy changes are only relative (qualitative, not quantitative).* 5 Account for changes in the system's potential and kinetic (and thermal) energies through transfer and conversion.
Task	The picture shows two wooden carts with metal sheets attached. Both metal sheets are negatively charged. The wedges prevent the carts from moving. 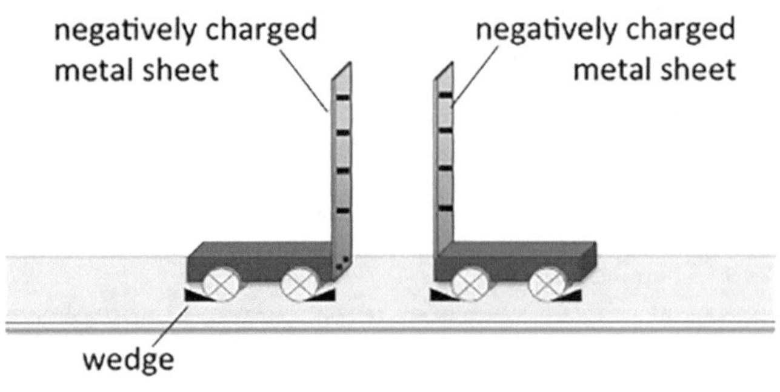 When the wedges are removed, the carts will move. Predict which direction they will move and when they will stop. Use ideas about forces and energy as appropriate.

Online Resource 27.3 *Alignment between Analytic Rubric Categories and the mECD Argument for the Carts Task (Adopted from* Kaldaras et al., 2022)

Rubric Category	*Description*	*mECD components*
1	Prediction about the movement of the carts 1 They will repel OR they will move away– 1 point 2 Anything else- 0 points	1a
2	Use the fundamental property of electric charges to **construct a causal account** that supports the prediction for which direction the carts will move: 1 Carts will repel because similar charges repel- 1 point OR 2 Cars will repel because two "-" cannot attract- 1 point 3 Anything else- 0 points	1a
3	Prediction about when the carts will stop 1 The carts will stop when … - 1 point 2 The carts will keep moving away (or repelling) until … - 1 point 3 The carts will eventually slow down … - 1 point 4 Anything else- 0 points	1b
4	Use the Coulombic **relationship to construct a causal explanation** that supports the prediction for when the carts will stop: 1 Distance between two plates related to charges OR electric field OR electric force- 1 point 2 Carts will stop moving when: a The charges are far enough away so they no longer interact b The distance between electric fields is large enough so they no longer interact. c The distance is far enough away that electric forces are too weak to move the carts 3 Use Coulombic relationship without connecting to the phenomenon (without Category 3 or without saying when the carts will stop) 4 Anything else- 0 points	1b
5	Construct a causal relationship using Energy only to explain when the carts will stop by either talking about energy conversion between potential and kinetic (1 point) OR systems moving to the lowest energy state (1 point) and relating to the phenomenon: 1 When the wedges are removed, the potential energy will be converted to kinetic energy (or PE goes down while KE goes up), kinetic energy is transferred to the surroundings and the carts will stop when all kinetic energy has been transferred to the surroundings OR there is not additional potential energy transferred to the kinetic energy of the cart -> conversion between energy forms 2 Systems move to the lowest energy state or potential energy is at its lowest state. Therefore, the carts will stop when the potential energy of the system is at the minimum -> systems moving to the lowest energy state 3 Use Energy ideas without connecting to the phenomenon (no Category 3) 4 Anything else- 0 points	Part 1) relates to 4a of the mECD; Part 2) relates to 3a,b of the mECD

(*Continued*)

Rubric Category	*Description*	*mECD components*
6	Construct a causal relationship between Energy and Coulombic interactions between charged plates; use this relationship to explain in which direction the carts will move and when they will stop: All of the following ideas should be present in the answer: 1 Potential energy is high when two similar charges are close because of the high repulsive force between two similar charges. Therefore, the carts will move away from each other -> **integration of energy and Coulomb law and use the relationship to explain which directions the carts will move.** 2 As the carts move away from each other after the wedges are removed, the system's repulsive force* and energy will decrease as distance increases. The carts will stop when the distance is far enough that the repulsive forces are very weak, and the system's energy is minimal-> **integration of energy and Coulomb law and use the relationship to explain when the carts will stop.***Students can also say that as the distance between the two carts increases, charges no longer interact, or electric fields no longer interact. 3 Integrate Coulomb's Law and Energy without connecting to the phenomenon (without category 3). 4 Anything else- 0 points.	2a
7	Either Coulomb's law or Energy is mentioned, but usable knowledge (meaning causal explanation and relation to the phenomenon in question) is not evident from the answer, or there are too many inaccuracies.	

Online Resource 27.4 An Example of Instructional Strategies (First-Hand)

Scenario: Baking soda is a popular baking ingredient for making bread, cakes, and cookies. Baking soda absorbs moisture if the box is left open for a long time. Can a drying agent such as calcium chloride remove the moisture from baking soda? The teacher shares his/her experience that the baking soda sample becomes worse when the two powders are mixed. Some of the mixed powders have turned into small solid chunks.

Learning Activity 1: Did a chemical reaction occur when mixing a drying agent ($CaCl_2$) with baking soda (Na_2CO_3)? To test that, students work in small groups to conduct the lab experiment of mixing baking soda and calcium chloride solutions.

Step 1: Data collection. [Purpose: Students conduct the lab experiment, observe the phenomena, and collect data.]

1.1 dissolve each substance in water to make different solutions. Students make observations and record data on the colors and solubilities of *dissolved baking soda and calcium chloride*.
1.2 conduct a lab experiment by mixing baking soda and calcium chloride solutions. They observe what happens.
1.3 filter the precipitation from the mixture and observe it.

Then, the teacher prepares a data table with information on substances before and after mixing. Students fill out the other data in the table below.

Data table about properties of substances

		Substances	*Solubility*	*Density (g/cm^3)*	*Color*
Before mixing	Calcium chloride solution	$CaCl_2$	[Yes]	2.15	[White]
	Baking soda solution	Na_2CO_3	[Yes]	2.54	[White]
After mixing	Product 1	Precipitation	[No]	2.71	[White]
	Product 2	Leftover and water	[Yes]	2.16	[White]

Note: the content inside "()" is for teacher information; the content inside "[]" should be filled out by students.
Prompt 1-1: What are the colors and solubilities of substances before and after mixing?
Prompt 1-2: What happened after mixing the baking soda and calcium chloride solutions? Make an initial claim about whether or not a chemical reaction occurred.

Learning Activity 2: How do you know if a chemical reaction occurs when mixing a drying agent ($CaCl_2$) with baking soda (Na_2CO_3)?

Step 2: Identify patterns of data. [Purpose: Students identify data patterns and use them as evidence to support or revise their initial claim of whether or not a chemical reaction occurred.]

Prompt 2-1: What similarities and differences exist among the substances before and after mixing baking soda and calcium chloride?

Prompt 2-2: Did a chemical reaction occur? Use the data patterns as evidence to support or revise your initial claim. Please share your ideas with your classmates.

Step 3: Reasoning that connects the evidence to the claim. [Purpose: Students apply core ideas and evidence to determine whether or not a chemical reaction occurred.]

Prompt 3-1: Based on the data patterns above, what properties help determine whether or not a chemical reaction occurred? And why? Please share your reasoning with your classmates.

28

USING AN AI-BASED DASHBOARD TO HELP TEACHERS SUPPORT STUDENTS' LEARNING PROGRESSIONS FOR SCIENCE PRACTICES

Janice D. Gobert, Rachel Dickler, and Amy Adair

Science reform documents such as the *Framework for K-12 Science Education* (NRC, 2012) and the Next Generation Science Standards (2013; NGSS, henceforth) have been adopted or adapted by most states in the US. Authentic scientific (and engineering) practices, one of the three strands of the NGSS, are purported, in part, to help students develop a deep understanding of scientific phenomena. However, teachers have found inquiry challenging to implement in their science classrooms for many reasons (Pruitt, 2014), including: the time-intensive, subjective nature of grading inquiry, the lack of immediate, actionable data for identifying students' difficulties, and the lack of authentic, empirically tested resources.

A recent technological development, namely, teacher dashboards, can address many of teachers' assessment and instructional needs for science. Dashboards are monitoring tools that can provide data on students' progress, enabling teachers to quickly ascertain who needs immediate assistance, and, in turn, increase the number of interactions teachers have with students (van Leeuwen & Rummel, 2018). Although there are a number of dashboards that provide data on students' content understandings (WISE dashboard, Matuk et al., 2016; SAIL Smart Space's tablet tool, Tissenbaum & Slotta, 2019), these cannot support teachers' instruction of scientific practices (SEPs, henceforth; NGSS, 2013) because they were *not* designed or instrumented to assess students' competencies on SEPs. We propose that our dashboard, instrumented and paired with our learning platform, can be used by teachers to actively monitor and promote students' learning progressions for SEPs as they conduct inquiry in Inq-ITS. We first introduce Inq-ITS and Inq-Blotter, explaining how they can be used to help teachers monitor student learning. Next, we describe an empirical study using Inq-ITS and Inq-Blotter to promote students' learning of SEPs.

Inq-ITS Combined with Inq-Blotter to Help Teachers Monitor Student Learning

Inq-ITS (inqits.com) is a platform with virtual, lab-based formative assessment activities for middle and high school for life, earth, and physical science. With Inq-ITS, students form scientific questions, collect data, interpret data, do relevant mathematics, and write

DOI: 10.4324/9781003170785-32

explanations in claim-evidence-reasoning format (McNeill & Krajcik, 2011). Inq-ITS uses patented AI algorithms (Gobert et al., in press) to assess students on their competencies on key SEPs (NGSS, 2013). In addition, these algorithms trigger Rex (our virtual agent) to provide scaffolds to students in realtime when students struggle with SEPs, leading to robust learning of SEPs that can be applied both to new science topics and over long periods of time (tested up to 170 days).

In our more recent research and development work, we sought to design a tool that would: provide teachers with *real-time data* about their students' science competencies on SEPs, *identify which students were struggling* with various SEPs and in what specific ways (i.e., the sub-components of the practices), and *give teachers real-time, actionable alerts* so they could monitor students as they conduct virtual science experiments in Inq-ITS (Inquiry Intelligent Tutoring System; Gobert et al., in press). With Inq-Blotter, teachers receive real-time alerts for our 150+ Inq-ITS formative assessment labs in the form of textual descriptions and visualizations about who in the class is struggling with SEPs. Each alert provides a *diagnosis* of how students are struggling with the specific sub-components of the SEPs (see vignette for an example of a sub-component) based on Inq-ITS' automatic assessment capabilities. In addition to supporting individual students, the dashboard provides data on how many students in the class are struggling with each practice and their respective sub-components so that the teacher can provide differentiated instruction to small groups or do whole-class instruction if a large percentage of the class is struggling with the same practice (when a class-wide alert is triggered). Other key features of Inq-Blotter include: a) *scores on each of the SEPs* for the student for whom an alert was triggered, b) *the amount of Rex feedback* the student has received on the lab on which they are currently working, and c) *configurable settings* so teachers can customize alerts to their specific pedagogical needs at different points in the curriculum or at different grade levels, etc. (see Gobert et al., 2023).

Inq-ITS and Inq-Blotter are currently used in 50 states and many countries to support teachers' formative assessment and instruction of the practices. Further, many schools have seen marked improvement in students' competencies at SEPs and improved state summative test scores when pairing Inq-ITS formative assessments and alerts with rich science curricula (https://www.openscied.org/; Gobert & Lane, 2023). Figure 28.1 below outlines the full system: Inq-ITS (our student platform), its algorithms, Rex (our virtual tutor), and Inq-Blotter (our teacher dashboard) with its alerts and TIPS (Teacher Inquiry Practice Supports, not discussed in this chapter; Gobert et al., 2023). Figure 28.2 provides a detailed view of Inq-Blotter with an example of an alert. The vignette illustrates how Inq-ITS and Inq-Blotter are used in classrooms.

Vignette

8th-grade teacher, Mrs. Johnson, is having her class do a formative assessment with Inq-ITS' States of Matter virtual lab in which students conduct investigations to determine the effects of the independent variables (e.g., level of heat, amount of substance) on the dependent variables (e.g., time to melt, temperature when melted) by forming a question, carrying out an investigation to test their question, analyzing their data, and communicating their findings in claim-evidence-reasoning format (McNeill & Krajcik, 2011). A student, Paul is carrying out an investigation (NGSS Practice 3) with the Inq-ITS simulation, and though he is trying to test the question he posed (a sub-component of NGSS Practice 2), he is changing too many variables at once, which will not generate the data he needs for data analysis (NGSS

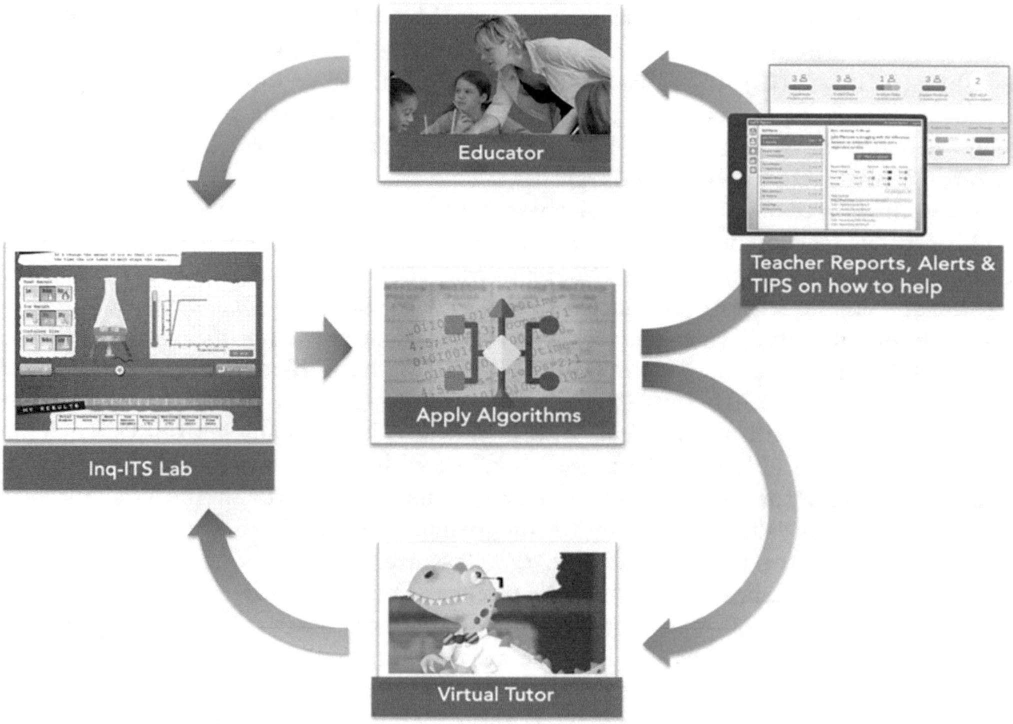

Figure 28.1 Components of the Inq-ITS and Inq-Blotter Systems.

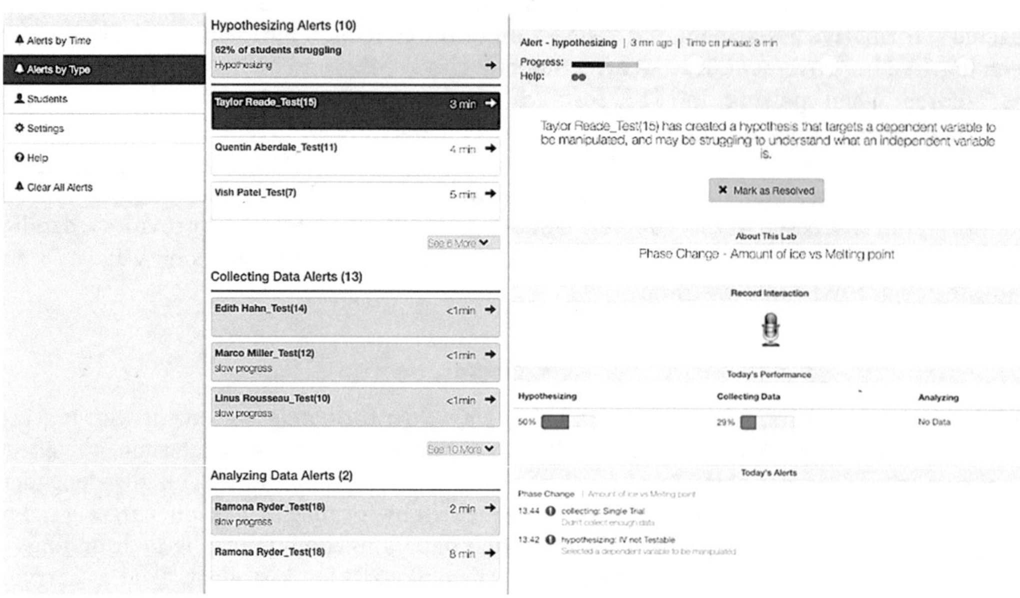

Figure 28.2 Detailed View of Inq-Blotter with an Example of an Alert.

Practice 4) or his C-E-R statement (NGSS Practices 6 & 7). The system's assessment algorithms (Gobert et al., in press) detect Paul's problem and automatically send an alert to Mrs. Johnson's Inq-Blotter saying, "Paul is having difficulty collecting controlled trials." Mrs. Johnson is then able to walk over to converse with Paul (or chat with him via video screen (Dickler, 2021; Dickler et al., 2021b)) to help him understand that multiple controlled trials are needed in which he targets the variable he said he was going to test during the question formation phase. After helping him, Mrs. Johnson clicks on the "Mark as resolved" button on Inq-Blotter and can move on to the next student, Taylor, for whom an alert has been automatically triggered regarding a sub-component for question formation (NGSS Practice 1), i.e., he is targeting a dependent rather than an independent variable (Figure 28.2).

An Empirical Study Using Inq-ITS and Inq-Blotter to Monitor Student Learning of SEPs

We argue that to support teachers' instruction, they need rigorous, fine-grained, formative data on their students' competencies on SEPs. By providing alerts on Inq-Blotter that include fine-grained data identifying *which students are struggling and on what specific sub-components for each SEP*, teachers can maximize the benefits of the automated assessment data generated in realtime by Inq-ITS, respond to students' difficulties in realtime, and better support students on their developing competencies for each SEP (NGSS, 2013). A number of studies conducted both in-class (Dickler et al., 2021a) and in remote settings (Dickler, 2021; Dickler et al., 2021b) have demonstrated that students' competencies for SEPs were improved compared to students whose teachers did not use Inq-Blotter (Sao Pedro et al., 2019); further, such effects have been shown for a variety of SEPs (Dickler, 2021), including question formation, carrying out investigations, as well as graphing data, a sub-component associated with using mathematics (NGSS Practice 5). In this section, we report on an empirical study that used Inq-ITS and Inq-Blotter to provide teachers with rigorous, fine-grained formative assessment data and alerts regarding students' difficulties with SEPs.

Methods

Participants and Procedure

Two 8th-grade teachers at different middle schools in the United States used Inq-Blotter while their students completed Inq-ITS virtual lab activities during their regular science class. A variety of Inq-ITS formative assessment labs (Gobert et al., 2023) – including collisions, waves, lunar phases, and phase change – were used as per teachers' respective needs. The teachers helped 35 students who went on to complete at least two Inq-ITS labs after being helped by the teacher based on Inq-Blotter alerts. More specifically, after the students completed a lab, the automated AI assessments in Inq-ITS sent alerts to teachers' Inq-Blotters identifying the specific sub-component of each SEP students struggled with, and other prior difficulties the student had while completing that activity. When teachers responded to alerts, they pressed the microphone button on their screen that captured the voice recordings using the internal microphones on the teachers' devices (i.e., iPads), and recordings were automatically triangulated with the log file data from both Inq-ITS and Inq-Blotter by our system's infrastructure. After the students received the teachers' feedback, they continued working and then moved to the next lab.

Data Collection

All students' actions while using Inq-ITS labs were logged. Student competencies on the SEPs were assessed for the first three stages of the Inq-ITS lab via machine-learned and knowledge-engineered algorithms (Gobert et al., in press), namely, question formation (e.g., identifying variables to explore based on a given goal), investigating/collecting data (e.g., running controlled trials targeted towards the target variable), and analyzing and interpreting data (e.g., making a claim and selecting trials that support/refute the claim). Inq-ITS has since been expanded to automatically assess students' written C-E-R statements as well as mathematical practices). Third, voice recordings of teacher support were extracted, transcribed, and split into clauses according to the speaker (i.e., teacher or student).

Data Analyses

We analyzed the three datasets. First, using the data automatically triangulated from Inq-ITS and Inq-Blotter, we identified: the SEPs (and respective sub-components) on which students were helped by the teacher, students' competencies on SEPs prior to being helped (i.e., first lab), and students' competencies on the practice on which they were helped in their next two Inq-ITS activities (i.e., their 2nd and 3rd labs, respectively). A repeated measures ANOVA was used to identify whether student performance on the sub-component on which they were helped significantly improved over labs. Paired samples t-tests were used for follow-up analyses.

Second, for teachers' voice recordings, the teacher clauses (N = 190) were coded for whether *general support* or *inquiry support* was provided. As shown in Table 28.1, seven types of supports were identified. Three types of supports belong to general support. They included whether teachers provided content help either in the form of general comments, evaluative comments, or questions. Four types of supports belong to inquiry support. They included *orienting* (i.e., directing attention to a particular practice), *procedural* (i.e., information on the steps involved in a practice), *conceptual* (i.e., definition/explanation of

Table 28.1 Seven Types of Teacher Support

Categories	*Types of Teacher Support*	*Example*
General Support that provides content help	General comments	"So that's saying, ya know, if you have a shorter string the wave speed will be faster, if you have a longer string it'll be faster."
	Evaluative Comments	"According to this, you're struggling with your hypothesis."
	Questions	"So you're gonna change the speed of the wave and that's gonna affect the loudness?"
Inquiry Support that provides help with SEPs	Orienting	"Let's look at analyzing your data … Let's look at your data"
	Procedural	"You always run a control and you always run it more than one time. Why do you think you need to run it more than one time?"
	Conceptual	"The independent variable is what's manipulated. So what are you changing here?"
	Instrumental	"It's saying to you that it didn't change … So you need to click no change."

a practice), and *instrumental* (i.e., the exact actions to take to complete a practice). Two researchers, trained on the coding scheme, scored 100 clauses independently. The raters agreed on 97% of codes, disagreements were discussed, and agreed-upon codes were used for analyses. One rater coded the remaining clauses (N = 90) independently.

Finally, we analyzed the data by examining the patterns, e.g. co-occurrences, etc., of the types of teacher support (see above) using ENA (Shaffer et al., 2016) with respect to the patterns of support for which students correspondingly improved (or not) on the practice on which they were helped in their first lab compared to that practice on their 2nd, and then again for their 3rd lab; that is two separate ENAs were conducted to examine what types of support promoted learning that was maintained in lab 3). As discussed in the results section, such evidence illuminates teachers' pedagogical strategies that promote students' competencies on SEPs along a progression, as expected by NGSS (2013), as well as informs future research, strategies, etc., on LPs.

Results

Descriptive analyses indicated that the majority of students improved on SEPs after being helped by their teachers using Inq-Blotter when comparing their 1st lab opportunity (prior to being helped by the teacher in response to an Inq-Blotter alert) to their 2nd lab opportunity on which 80% of students improved on the practice(s) on which they received help, and comparing their 1st lab opportunity to their 3rd lab opportunity, on which 77% of students improved on the practice(s) on which they were helped. Results of the repeated measures ANOVA revealed a significant main effect for the number of labs/opportunities, $F(2, 68) = 22.91$, $p < .001$, $\eta^2 = .40$. Follow-up paired-sample t-tests yielded significant gains: students improved from their 1st lab opportunity ($M = .52$, $SD = .20$) to their 2nd lab opportunity ($M = .83$, $SD = .24$), $t(34) = -5.56$, $p < .001$, $d = -.94$.

Students also significantly improved from their 1st lab opportunity ($M = .52$, $SD = .20$) to their 3rd lab opportunity ($M = .85$, $SD = .26$; $t(34) = -5.84$, $p < .001$, $d = -.99$). There were no significant differences between students' performance on their 2nd lab opportunity ($M = .83$, $SD = .24$) and 3rd lab opportunity ($M = .85$, $SD = .26$; $t(34) = -.489$, $p = .63$, $d = -.08$), indicating that students *maintained* their improved performance after being helped (see Figure 28.3).

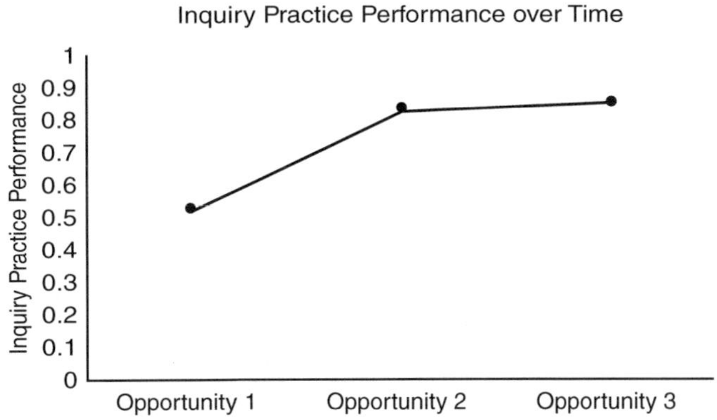

Figure 28.3 Students' performance in the practice in which they were helped across labs.

Since most students improved (in their 2nd lab) and then maintained their performance (in their 3rd lab) on the SEP(s) on which they were helped by the teacher in lab 1, we examined the patterns of teachers' support to identify which instructional strategies teachers used to support students' learning of the SEPs along a progression as expected by the NGSS (2013).

First, ENA compared the types of support related to whether students improved or not on their *2nd opportunity* after being helped. Each node of the network was representative of one of the seven codes for types of support (described earlier). The strength/width of the connection between nodes is representative of the frequency with which two codes occur consecutively or sequentially. This permitted us to identify the most common types of

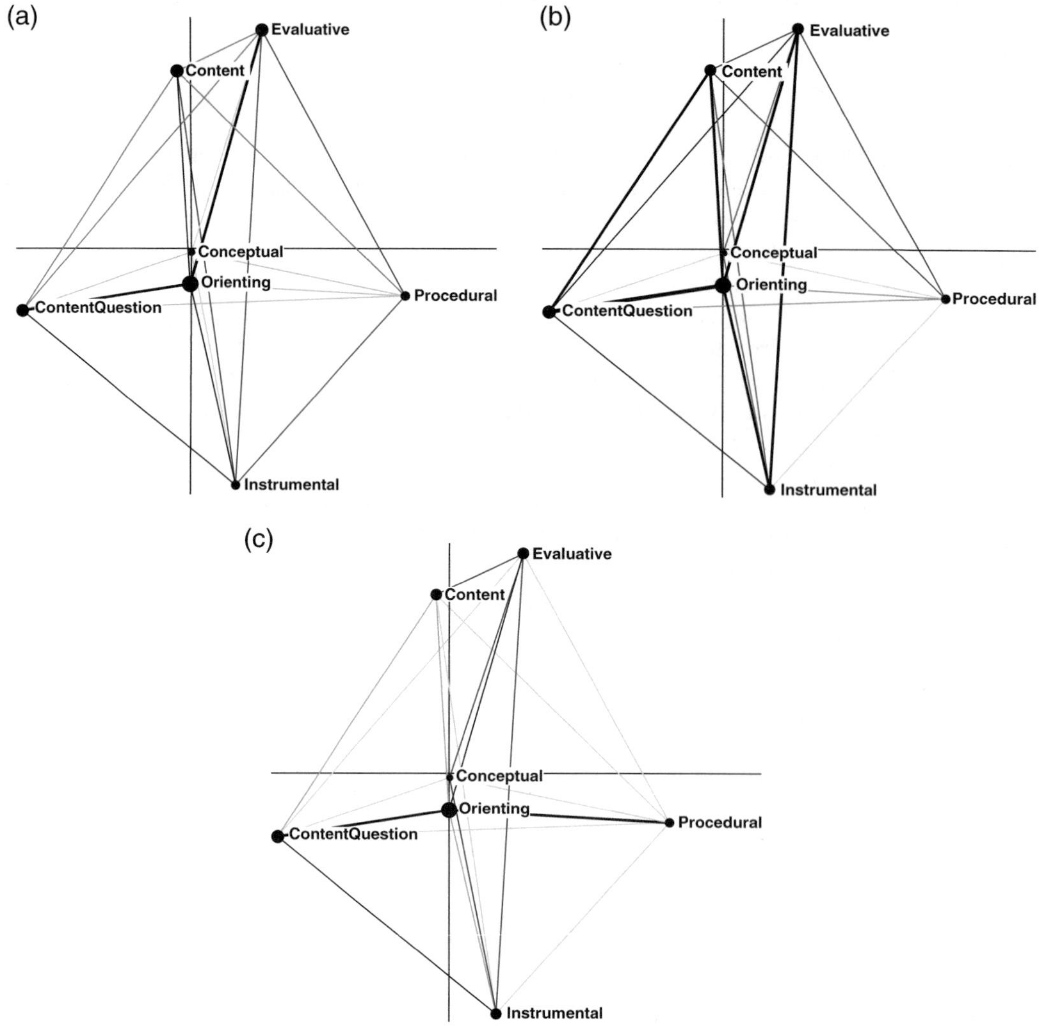

Figure 28.4 a. Epistemic Networks for Student Improvement on Second Lab Opportunity. b. Epistemic Networks for No Student Improvement on Second Lab Opportunity. c. Subtracted Epistemic Networks for Student Improvement and No Student Improvement on Second Lab Opportunity.

support that were verbalized together and to subtract networks to compare differences in the strengths of connections between networks. Additionally, mean centroid values are computed in relation to the weights of connections within a network and allow for comparing networks. The ENA revealed a significant difference in the patterns of teacher support when students improved (M = .11, SD = .49) versus those who did not improve (M = –.46, SD = .56; $t(34)$ = 2.45, p = .04, d = 1.12). In particular, the subtracted network revealed that students who improved on their 2nd lab are those who received mostly inquiry practice supports (including lower-level inquiry support such as orienting and procedural scaffolds) in combination with evaluative comments (i.e., 61% of support associated with improvement). However, students who did *not* improve on their next lab opportunity received primarily content-related support (both comments and questions) as well as low-level orienting hints (i.e., 68% of support associated with no improvement; see Figure 28.4c).

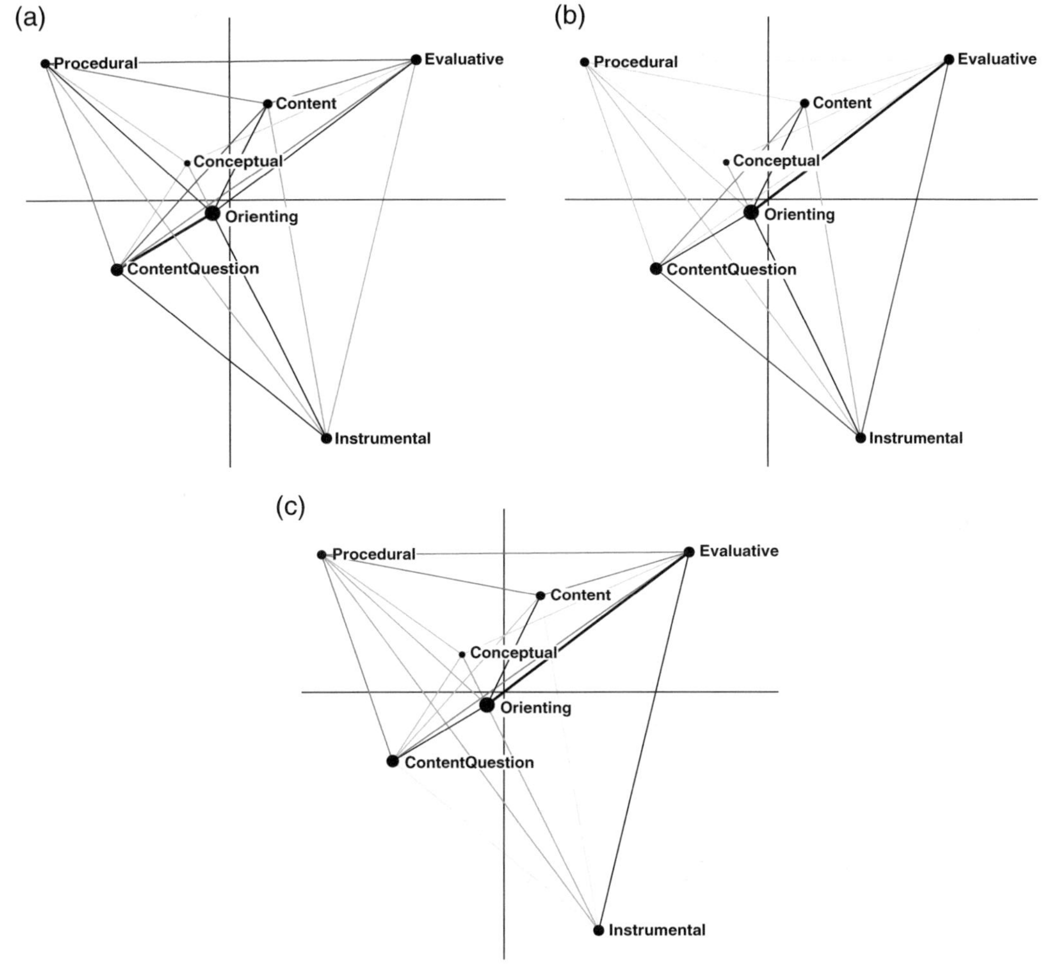

Figure 28.5 a. Epistemic Networks for Student Improvement on Third Opportunity. b. Epistemic Networks for No Student Improvement on Third Opportunity. c. Subtracted Epistemic Networks for Student Improvement and No Student Improvement on Third Opportunity.

We also conducted ENA to compare the types of support related to whether students improved or not from their *1st* to their *3rd* opportunity. Again, we found that there was a significant difference in the patterns of teacher support elicited by Inq-Blotter's alerts for the students who improved ($M = -.20$, $SD = .54$) versus those who did not improve ($M = .66$, $SD = .84$; $t(34) = 2.73$, $p = .02$, $d = 1.40$). Specifically, the strength of connections in the subtracted network revealed that teacher support involving a combination of inquiry practice support (both high and low-level scaffolds) and content-related support was more strongly associated with student improvement on the 3rd lab (i.e., 88% of support associated with improvement), whereas evaluative comments in combination with low-level orienting and instrumental support did *not* appear to yield robust student learning of SEP(s) that could be used on the 3rd lab (i.e., 68% of support associated with no improvement; see Figure 28.5c). These findings make sense intuitively, i.e., lower-level supports may lead to initial improvement, but they are unlikely to result in robust learning that is maintained across labs. For example, telling a student, "No that's not right, remember you are supposed to run a controlled experiment, so only change the amount of salt you are adding in the water" may help the student to complete the current or a very similar activity. However, a combination of inquiry supports such as "Remember you are supposed to run a controlled experiment and controlled trials means only changing one variable at once, so which variable should you change?" seems to be more beneficial to students' learning of SEPs such that they can apply it in their next lab. See Figures 28.5a, b, and c.

Conclusions

As per policymakers' vision statements (NGSS (2013), students need to do authentic science practices (SEPs), and teachers need to both provide instruction and be able to assess students' developing competencies on these SEPs. Without rigorous, real-time data, teachers would need to gauge students' developing competencies on SEPs as students work, which is nearly impossible (Pruitt, 2014). To support teachers so that they can better support students' learning, we unpacked the SEPs into sub-components, used Inq-ITS' assessment algorithms to diagnose students' difficulties on SEPs, and developed Inq-Blotter to alert teachers about students' difficulties on SEPs (Gobert et al., 2023; in press). In this chapter, we address how our system helps teachers monitor students' progression in learning SEPs over time.

Alerting teachers on the sub-components of the SEPs on which students are struggling, as we do with Inq-Blotter, provides specificity to the NGSS practices (Shea & Duncan, 2013) at a grain-size that can support teachers' instruction (Harris et al., 2016), which leads to robust learning of SEPs such that students can apply what they have learned across investigations, as shown in the study described earlier (and in our other studies, cited earlier).

Although our research does not specify LPs, our research provides implications for other researchers who seek to develop automated systems that help teachers assess and scaffold students in moving up the LP levels. Further, we recognize the potential of providing teachers with further instructional support aligned to LPs for each SEP. Our recently added TIPS (Teacher Inquiry Practice Supports; Gobert et al., 2023; Adair et al., 2022) are a step in this direction, i.e., they provide targeted instructional guidance for teachers so that they do not default to giving solely content support because this will not help students in future investigations on different science topics (Dickler et al., 2021a). In the future, we could examine how our alerts and TIPS might be iteratively refined to provide specific guidance for each SEP along a learning progression. Further, fine-grained log files collected as students work in Inq-ITS could provide empirical data to inform and validate which sub-components should be foregrounded in an LP for each SEP. This is important (Liu & Jackson, 2019) because relatively little work has been done to date on LPs for SEPs with the exception of LPs for communication practices (NGSS practices 6 and 7; Berland and McNeill, 2010; Osborne et al., 2016).

In closing, researchers working in the areas of assessment and learning progressions seek to close the formative assessment loop, permitting teachers to monitor students in realtime (Liu & Jackson, 2019), orchestrate whole-class learning, and have assessment conversations (Duschl & Gitomer, 1997) with small groups (for differentiated instruction) or individual students, including those in the messy middle (Gotwals & Songer, 2010). In all our work, we seek to develop technologies to support teachers and students, thus, Inq-ITS and Inq-Blotter provide fine-grained formative data, scaffolds, alerts, and TIPS, which have been shown to support teachers in helping their students develop competencies at SEPs that are robust; furthermore, since our technologies are undergirded by AI (Gobert et al., 2023; in press), they support teachers and students *in realtime, and at scale*, helping to realize the vision of NGSS (2013).

Author Note

Acknowledgment. This research was funded by the National Science Foundation (NSF-IIS-1629045, NSF-IIS- 1902647) and the US Department of Education (R305A210432, EDIES16C0014, EDIES15C0018). Any opinions, findings, etc. expressed are those of the authors and do not necessarily reflect those of the funding agencies.

References

Adair, A., Owens, J., & Gobert, J. (2022). Using epistemic network analysis to explore discourse patterns across design iterations of a teacher dashboard. In C. Chinn, E. Tan, C. Chan, & Y. Kali (Eds.), *Proceedings of the 16th International Conference of the Learning Sciences - ICLS 2022* (pp. 297–304). International Society of the Learning Sciences.

Berland, L. K., & McNeill, K. L. (2010). A learning progression for scientific argumentation: Understanding student work and designing supportive instructional contexts. *Science Education, 94*(5), 765–793.

Corcoran, T., Mosher, F. A., & Rogat, A. (2009). *Learning progressions in science: An evidence-based approach to reform (CPRE Research Report #RR-63).* Consortium for Policy Research in Education.

Dickler, R. (2021). *An intelligent tutoring system and teacher dashboard to support students on mathematics in science inquiry* (Publication No. 28643659). [Doctoral dissertation, Rutgers The State University of New Jersey, School of Graduate Studies].

Dickler, R., Adair, A., Gobert, J., Hussain-Abidi, H., Olsen, J., O'Brien, M., & Sao Pedro, M. (2021b). Examining the use of a teacher alerting dashboard during remote learning. In I. Roll, D. McNamara, S. Sosnovsky, R. Luckin, & V. Dimitrova (Eds.), *International Conference on Artificial Intelligence in Education* (pp. 134–138). Springer, Cham.

Dickler, R., Gobert, J., & Sao Pedro, M. (2021a). Using innovative methods to explore the potential of an alerting dashboard for science inquiry. *Journal of Learning Analytics, 8*(2), 105–122.

Duschl, R. A., & Gitomer, D. H. (1997). Strategies and challenges to changing the focus of assessment and instruction in science classrooms. *Educational Assessment, 4*(1), 37–73. 10.1207/s15326977ea0401_2

Gobert, J., & Lane, J. (2023, October). *Using Inq-ITS AI to assess and support students on OpenSciEd [Conference presentation].* New Jersey Science Convention, Princeton, NJ.

Gobert, J. D., Li, H., Dickler, R., & Lott, C. (in press). Can AI-based scaffolding promote students' robust learning of authentic science practices? To appear In X. Zhai, & J. Krajcik (Eds.), *Uses of artificial intelligence in STEM education.* Oxford University Press.

Gobert, J. D., Sao Pedro, M. A., & Betts, C. G. (2023). An AI-based teacher dashboard to support students' inquiry: Design principles, features, and technological specifications. In N. Lederman, D. Zeidler, & J. Lederman (Eds.), *Handbook of research on science education* (pp. 1011–1044). Routledge.

Gotwals, A. W., & Songer, N. B. (2010). Reasoning up and down a food chain: Using an assessment framework to investigate students' middle knowledge. *Science Education, 94*(2), 259–281.

Harris, C. J., Krajcik, J. S., Pellegrino, J. W., & McElhaney, K. W. (2016). *Constructing assessment tasks that blend disciplinary core ideas, crosscutting concepts, and science practices for classroom formative applications*. SRI International.

Liu, L., & Jackson, T. (2019). A recent review of learning progressions in science: Gaps and shifts. *The Educational Review, USA, 3*(9), 113–126.

Matuk, C., Gerard, L., Lim-Breitbart, J., & Linn, M. (2016). Gathering requirements for teacher tools: Strategies for empowering teachers through co-design. *Journal of Science Teacher Education, 27*(1), 79–110.

McNeill, K. L., & Krajcik, J. S. (2011). *Supporting grade 5-8 students in constructing explanations in science: The claim, evidence, and reasoning framework for talk and writing*. Pearson.

National Research Council. (2012). *A framework for K-12 science education: Practices, crosscutting concepts, and core ideas*. National Academies Press.

NGSS Lead States. (2013). *Next Generation Science Standards: For states, by states*. National Academies Press.

Osborne, J. F., Henderson, J. B., MacPherson, A., Szu, E., Wild, A., & Yao, S. Y. (2016). The development and validation of a learning progression for argumentation in science. *Journal of Research in Science Teaching, 53*(6), 821–846.

Pruitt, S. L. (2014). The next generation science standards: The features and challenges. *Journal of Science Teacher Education, 25*(2), 145–156.

Sao Pedro, M. A., Gobert, J., & Dickler, R. (2019, April). Can an alerting teacher dashboard improve how teachers help their students learn science inquiry practices? [Conference presentation]. American Educational Research Association Annual Meeting: Learning and Instruction, Toronto, ON, Canada.

Shaffer, D. W., Collier, W., & Ruis, A. R. (2016). A tutorial on epistemic network analysis: Analyzing the structure of connections in cognitive, social, and interaction data. *Journal of Learning Analytics, 3*(3), 9–45. 10.18608/jla.2016.33.3

Shea, N. A., & Duncan, R. G. (2013). From theory to data: The process of refining learning progressions. *Journal of the Learning Sciences, 22*(1), 7–32.

Tissenbaum, M., & Slotta, J. (2019). Supporting classroom orchestration with real-time feedback: A role for teacher dashboards and real-time agents. *International Journal of Computer-Supported Collaborative Learning, 14*(3), 325–351. 10.1007/s11412-019-09306-1

van Leeuwen, A., & Rummel, N. (2018). Considering teachers' informational needs and the role division between teacher and orchestration tool in the context of collaborative learning. In *Proceedings of the International Conference of the Learning Sciences* (pp. 1230–1231). International Society of the Learning Sciences.

29

THE ROLE OF AI AND TECHNOLOGIES IN SCIENCE LEARNING PROGRESSION

Commentary for Section IV

Xiaoming Zhai

Introduction

The complexity of thinking in the context of engaging students in scientific practices is a multifaceted phenomenon that warrants nuanced exploration. As students navigate the intricacies of scientific practices, they are required to employ a range of cognitive, metacognitive, and epistemic skills. These include, but are not limited to, hypothesis formulation, data interpretation, and the application of scientific principles to real-world problems (National Research Council, 2012). Moreover, the interplay between Disciplinary Core Ideas (DCIs), Science and Engineering Practices (SEPs), and Crosscutting Concepts (CCCs) adds layers of complexity to their cognitive engagement. Students must not only grasp the foundational knowledge within one discipline but also understand how to apply this knowledge through scientific practices, while recognizing the broader themes that cut across different scientific domains. This complex orchestration of skills and knowledge is further influenced by students' prior experiences, cultural backgrounds, and preconceptions about science, which can either facilitate or hinder their engagement in scientific practices. Therefore, any attempt to assess or support student thinking in this context must take into account this rich tapestry of cognitive, social, and cultural factors.

One of these nuanced endeavors to understand these complexities proposed in the *Framework for K-12 Science Education* (National Research Council, 2012) is to employ structured frameworks such as learning progressions (LPs). LPs delineate the increasingly complex thinking regarding how students evolve in grasping core disciplinary ideas and practices, both within a specific grade and across several grades (Krajcik, 2012). LPs offer a sequenced, developmental trajectory of students' conceptual understanding, skills, and competencies, thereby serving as an invaluable tool for both instructional planning and assessment (Alonzo & Gotwals, 2012). In the context of science education, the utility of LPs is particularly pronounced, as they allow for an integrated approach to teaching and learning that aligns DCIs, CCCs, and SEPs in a coherent and developmentally appropriate manner (Jin et al., 2019a). This alignment not only enhances the effectiveness of instruction but also ensures that assessments are more accurately targeted (Furtak, 2012), providing actionable insights into the intricacies of students' scientific thinking. Moreover, LPs enable differentiated instruction by allowing educators to pinpoint the specific stage each student has reached,

DOI: 10.4324/9781003170785-33

thereby facilitating more personalized and effective teaching strategies (Zhai et al., 2018). Beyond the classroom, LPs have broader implications for curriculum design and educational policy, offering a research-based structure that can guide the development of curricular materials and instructional methods to ensure they are attuned to the developmental stages of student thinking in specific scientific domains (Jin et al., 2019b). Therefore, LPs serve as an indispensable cornerstone in the quest to improve science education, particularly when leveraged alongside technological advancements like artificial intelligence (AI), which hold the promise of further enriching our understanding of student learning trajectories (Kubsch et al., 2022).

Assessing students' LPs in science education presents a formidable challenge due to the multifaceted and dynamic nature of scientific understanding and skills. Traditional assessment methods often fall short of capturing the complexity and nuance of students' evolving conceptual frameworks, scientific practices, and crosscutting themes. Moreover, the sheer volume of data that needs to be processed and interpreted to make meaningful educational decisions often exceeds human analytical capabilities (Gao et al., 2020). In this context, advanced technologies like Bayesian models, cognitive diagnostic models, and AI offer transformative potential. Bayesian models can provide probabilistic frameworks that accommodate the uncertainty and variability inherent in educational data, thereby offering more nuanced insights into student learning. AI, particularly machine learning algorithms, can handle large-scale, high-dimensional data and can be trained to recognize intricate patterns in student responses, thereby automating the labor-intensive process of assessment and providing real-time, personalized feedback (Zhai et al., 2020b). These technologies not only enhance the efficiency and accuracy of assessments but also enable a more granular analysis of LPs, thereby advancing our understanding of how students acquire complex scientific knowledge and skills. The integration of these advanced technologies into educational assessment thus represents a critical step toward a more effective, data-driven science education.

In the sections to follow, I first reviewed the needs and contributions of LP research, laying out the foundation of LP research for the audience, and then discussed the roles of technology and AI in assessing and implementing LPs. From a technical and practical perspective, I presented the strengths and limitations of each chapter included in this section and discussed their implications for science learning and assessment practices. Synthesizing the information across the four chapters, I reported the emergent themes in this evolving field of science education. At last, I discussed practical implications and future research directions in the field.

Theoretical Foundations

Learning Progressions in Science Education

The concept of LPs in science education has gained considerable traction in the past more than ten years, serving as a pivotal framework for understanding the trajectory of students' conceptual understanding and competence development in scientific domains. LPs offer a nuanced, developmental view of learning, contrasting sharply with more static approaches to assessment and instruction (Neumann et al., 2013). They provide educators, researchers, and policymakers with a roadmap that delineates the incremental steps through which students typically move as they advance toward more sophisticated levels of understanding and competence. This roadmap is invaluable for curriculum development, as it allows for the alignment of instructional materials and activities with specific stages of conceptual development (Jin et al., 2019b), thereby ensuring that students are neither overwhelmed with complexity too early nor stymied by a lack of challenge.

Moreover, LPs have the potential to democratize science education by serving as a tool to meet the needs of students with diverse backgrounds. By making explicit the milestones and benchmarks that signify different levels of understanding, they offer a transparent mechanism for identifying where additional resources and interventions are needed, particularly for students from marginalized or underrepresented groups (Furtak, 2023). This level of diagnostic insight is crucial for tailoring instruction to meet diverse learning needs, thereby narrowing achievement gaps.

Furthermore, the utility of LPs extends to the realm of educational assessment. Traditional assessments often provide a snapshot of student performance at a single point in time, which may not accurately reflect their developmental trajectory. LPs, by contrast, offer a dynamic framework for assessment, allowing for the measurement of growth and change over time (Yin et al., 2014). This is particularly important in science education, where understanding often builds cumulatively, with new knowledge and skills scaffolding upon prior ones.

Despite the importance of LPs in science education, assessing students' progress and implementing LPs in real classroom settings has never been easy. Developing a comprehensive hypothesized LP has to be based on learning theories, while understanding how students develop scientific thinking around three-dimensional science learning is even more challenging (Hadenfeldt et al., 2016). A hypothesized LP needs extensive data and long-term efforts to validate, which can be challenging due to student retention and other practical concerns. The diversity and complexity of students' thinking and learning paths further challenge the assessment and validation of LPs (Jin & Anderson, 2012).

Roles of Technology and AI in Modern Assessment and Learning Progressions

The advent of technology and AI has ushered in a transformative era for modern educational paradigms, fundamentally altering how teaching, learning, and assessment are conceptualized and executed. Particularly in assessment practices, technologies may help target complex constructs, assist in evidentiary inferences about student progressions, and ease human efforts. These technologies extend far beyond the mere digitization of educational content or the automation of routine tasks, offering a reconceptualization of educational processes. Such novel transformation enables a shift from simple multiple-choice questions to performance-based constructed response questions in assessments and facilitates personalized, adaptive learning experiences that can be finely tuned to individual needs, preferences, and developmental trajectories (Gerard et al., 2015).

AI, in particular, has the potential to revolutionize educational assessment and analytics. Traditional assessment models often suffer from a lack of real-time feedback, limiting their utility for immediate instructional adjustment. AI-driven assessment systems can analyze student responses in real-time, providing instant, targeted feedback that is aligned with LPs. For example, Wilson et al. (2023) employed machine learning to assess students' argumentation practices that aligned with a learning progression. The team further employed cognitive diagnostic modeling to uncover students' cognitive patterns in argumentation practices (Zhai et al., 2022). These efforts not only provide educators with insights into each student's learning pathway but also enhance the actionable uses of LPs in classrooms, thereby enabling more effective, differentiated instruction. Furthermore, AI algorithms can sift through vast amounts of data to identify patterns and trends that may not be readily apparent, offering the potential for predictive analytics that can flag at-risk students before they fall critically behind.

Moreover, technology and AI can elicit performance in scientific practices to assess complex thinking. Through intelligent tutoring systems, virtual labs, and other AI-driven instructional platforms, students across diverse geographical and socio-economic contexts can have access to high-quality assessment tasks and personalized learning experiences. This is particularly salient for science education, where hands-on experimentation and data analysis are crucial for deep understanding. Virtual labs powered by AI can simulate complex scientific phenomena, allowing students to engage in authentic scientific inquiry irrespective of the resources available in their physical learning environment, thus providing educators with insights into students' learning progression on various competencies.

To assess and validate hypothesized LPs, researchers have applied many technologies and methods, such as the Rasch modeling (Fulmer et al., 2014), Latent Cluster Analysis and Rule Space Models (Gao et al., 2021), Rasch-Cognitive Diagnostic Modeling (Gao et al., 2023), etc. These methods have facilitated the validation of students' progression for various big ideas, yet more indispensable tools and technologies are needed to infer students' progression. Dynamic Bayesian Networks (DBNs) allow teachers to trace student learning progression over time, presenting a more flexible and comprehensive approach than traditional models like the Rasch model. A significant advantage of DBNs lies in their ability to model the dependencies and interactions between various science concepts over time. This contrasts with other models, which often assume a degree of independence among items and traits and typically do not account for the sequential nature of learning. Predictive in nature, DBNs extend beyond merely describing student abilities at a given time to forecast future performance based on longitudinal data. This capability allows educators to anticipate learning outcomes and intervene proactively, which is a substantial leap from the static assessment snapshots provided by models like Rasch. Therefore, DBNs can potentially contribute to the validation of LPs, providing opportunities to delve into students' learning trajectories.

The integration of technology and AI in LP studies is not without challenges. Ethical considerations around data privacy, potential biases in AI algorithms, and the digital divide are pressing concerns that require thoughtful attention. Moreover, the effective implementation of these technology-involved LPs requires substantial professional development for educators, as well as a rethinking of curriculum design and assessment practices to align with the affordances and limitations of these LPs. Thus, although technology and AI hold the promise of redefining the LPs, offering more personalized, adaptive, and equitable learning experiences, and have the potential to serve as powerful catalysts for the development and implementation of LPs, their integration demands careful planning, ethical considerations, and a commitment to ongoing research and professional development. The four chapters presented in this section evidenced how technologies play significant roles in delving into students' thinking and LPs around several big ideas in science.

In the following sections, I applied this framework as an analytical tool to review the four chapters presented in this section. For each chapter, I summarized the major takeaways for the audience and then presented the strengths and limitations of each chapter. I conclude each section by discussing the implications for science learning and assessment practices.

Dynamic Bayesian Models for Learning Progressions

Chapter 25 explores how DBNs are used to monitor student progression along science LPs. The chapter outlines how DBNs can be employed to estimate individual students' locations in an LP. It also discusses how DBNs can help teachers trace students' progression across multiple time points to offer a more nuanced understanding of student learning trajectories.

This is particularly evident in the chapter's discussion of how DBNs can be used for both "filtering" to improve current estimates and "forecasting" to predict future performance. The chapter also discusses the extension of DBNs to Partially Observed Markov Decision Processes (POMDPs), which allows for more sophisticated planning under uncertainty. This is crucial for making informed decisions about instructional interventions and resource allocation.

Strengths and Limitations

The chapter's strengths lie in its thorough and detailed exposition of the theoretical foundations and practical applications of DBNs in a specific educational setting--LPs. It goes beyond merely introducing the concept, offering concrete examples and graphical representations to elucidate how DBNs can be operationalized to monitor how students move up the achievement levels of an LP. Thereby, the chapter provides a robust theoretical framework. This is particularly beneficial for researchers and practitioners who are looking for a rigorous understanding of how DBNs can be applied in predicting students' LPs.

Another notable strength is the chapter's focus on monitoring learning over time. Traditional assessment models often provide a snapshot of student performance at a single point in time, which can be limiting when trying to understand learning as a developmental process. The chapter addresses this limitation head-on by discussing how DBNs can extend across multiple time points. This allows for a more nuanced understanding of how students progress along an LP, thereby providing educators with the tools to make more informed instructional decisions. As far as the author knows, there is very little research among LP publications (e.g., Gao et al., 2021) that attempts to tackle this challenge. The chapter's discussion on "filtering" and "forecasting" is a testament to this strength, offering concrete methodologies for improving current assessments and predicting student future performance based on model inferences.

While this chapter focuses its discussions on Bayesian theories, it provides an approach that can be easily applied to many assessment systems. It offers concrete examples and graphical representations and even discusses extensions like POMDPs to illustrate how DBNs can be operationalized. This makes the chapter not just a theoretical discourse but a practical guide for implementation. The discussion on how DBNs can be used for planning instructional activities and resource allocation is particularly noteworthy. It moves the conversation from assessment to actionable insights, which is a critical aspect often missing in similar scholarly works. This contribution testifies to the said potential of LPs in aligning the assessment with the curriculum and instructions (Fortus & Krajcik, 2012).

Furthermore, the chapter is comprehensive in its coverage of the advantages that DBNs offer over traditional models. It enumerates five specific advantages, ranging from improving the estimation of students' current LP levels to planning for future instructional activities. Each advantage is discussed in detail, providing a full picture of what DBNs bring to the table. This exhaustive list serves as a compelling argument for the adoption of DBNs in educational assessment and planning.

However, one important limitation of this chapter is that it falls short of providing empirical evidence to substantiate the claims made. The absence of real-world case studies or empirical validations makes the arguments less compelling. Furthermore, the chapter's technical complexity could be a barrier for educators without a background in statistics. This limits the immediate applicability of the concepts presented, as the chapter assumes a level of expertise that may not be universally present among its readers.

Implications for Science Learning and Assessment

The implications of the chapter for science learning and assessment are manifold and significant. The dynamic nature of DBNs allows for a more nuanced and temporally sensitive understanding of student learning, which is particularly crucial in science education, where concepts are often interrelated and build upon one another. The chapter's discussion of "forecasting" is particularly relevant here. By employing DBNs, educators can make predictions about students' future performance, thereby allowing for timely interventions. This is especially important in science education, where early foundational knowledge is critical for grasping more complex, later-stage concepts.

The chapter opens avenues for the integration of advanced statistical techniques into educational assessment practices. This could lead to the development of more diagnostical and predictive assessment systems, thereby revolutionizing how educational assessments are designed and implemented. However, this also raises ethical questions about transparency and fairness in model inferences, a point that the chapter does not delve into but is crucial for real-world applications. Moreover, this chapter provides a compelling, albeit technically complex, argument for the integration of DBNs in LPs. While the chapter could benefit from empirical validations and a discussion of practical and ethical considerations, it nonetheless offers a rich theoretical framework that has the potential to significantly impact the field of science learning and assessment.

In summary, the chapter's strengths lie in its theoretical rigor, focus on temporal dynamics, practical applicability, and comprehensive coverage of the advantages of using DBNs in the context of LPs. These strengths make it a valuable contribution to the literature, offering both depth and breadth in its exploration of a complex but critical subject.

The Application of Automated Scoring Technology in Learning Progression-based Assessment

Chapter 26 reports a study that used a Natural Language Processing (NLP) approach to automatically score student responses in LP-based assessments. The authors employed a multi-faceted approach, utilizing holistic rubrics for scoring and introducing a novel "idea detection" methodology. Through the use of pre-trained transformer models, the chapter demonstrates high human-machine agreement, evidenced by Pearson correlation and Quadratic Weighted Kappa (QWK) scores that exceed standard thresholds for operational deployment. Additionally, the chapter emphasizes the importance of model auditing, incorporating saliency-based methods to ensure the validity and reliability of the deployed models.

Strengths and Limitations

One of the most salient strengths of the chapter is the empirical rigor that underpins the entire study. The authors go to great lengths to validate their methodologies and findings, employing a range of statistical measures such as Pearson correlation, QWK, and F1 scores. These metrics are not just impressive; they exceed the standard thresholds for operational deployment in educational settings, as cited from the meta-analysis by Zhai et al. (2021). This level of empirical validation is crucial for the credibility of automated scoring systems, especially when these systems are to be deployed in educational contexts where stakes are demanding for both students and educators.

Another significant strength lies in the chapter's innovative approach to "idea detection." This is not merely a theoretical proposition but is substantiated with empirical evidence. The chapter, for example, shows F1 micro-average scores of around 0.6, which is consistent with other difficult discourse-level idea detection tasks. The importance of this cannot be overstated. Traditional scoring methods often focus solely on the correctness of a response, but "idea detection" goes a step further by identifying and categorizing the specific ideas within a student's response. This nuanced approach opens up new avenues for formative assessment, with the potential to dig into students' fundamental ideas (Zhai & Li, 2021), allowing educators to provide more targeted, idea-specific feedback. It also has the potential to revolutionize our understanding of how students progress along LPs, offering a more granular view of student conceptual development.

The chapter's focus on model auditing is another commendable strength. In an era where machine learning models are becoming increasingly complex and opaque, the authors rightly emphasize the need for transparency and interpretability. They discuss a range of model auditing practices, from traditional methods like confusion matrices to more recent innovations like saliency-based methods. Their work evidences that the field is rapidly evolving to address the challenges of model interpretability. This focus on model auditing is not just a theoretical exercise; it has real-world implications for the ethical and responsible deployment of machine learning models in educational settings.

However, the chapter also acknowledges limitations, particularly the challenge of capturing the diversity of student expressions for less frequent ideas. While the authors discuss model auditing as a way to address the many challenges, the chapter could delve deeper into alternative machine learning techniques that might be more effective in capturing this diversity. For example, recent research uses text data augmentation to enrich the opportunity to more accurately assess the less frequent ideas (Fang et al., 2023). Another limitation is the complexity of the models used, which may not be easily interpretable by educators or students. The chapter does discuss the use of model auditing and saliency maps to make the models more understandable, but the practicality of these methods in a classroom setting remains an open question.

Implications for Science Learning and Assessment

The high levels of human-machine agreement, as evidenced by the Pearson and QWK scores, have significant implications for the scalability of automatically scored science assessments. Automated scoring systems that meet or exceed the standard accuracy thresholds can be confidently deployed, reducing the manual workload for educators and allowing for quicker, more targeted feedback. This contribution adds to the ongoing discussions about using AI in formative assessment practices.

The "idea detection" methodology also has far-reaching implications. By identifying specific idea spans in student responses, this approach allows for a more nuanced understanding of student learning. This is particularly important in the context of LPs, where understanding the progression of student ideas is challenging. Moreover, the chapter's focus on model auditing is a timely and important contribution. As the authors point out, tooling for model auditing has evolved rapidly. This suggests that as these tools become more mature, they will become more accessible to educators, thereby making the deployment of machine learning models in educational settings more responsible and ethical.

In summary, the chapter's strengths are deeply rooted in its empirical rigor, innovative methodologies, and focus on ethical considerations. Each of these strengths is not just discussed in the abstract but is substantiated with detailed empirical evidence, making the chapter a significant contribution to the field of automated scoring in science education. The chapter's findings have significant implications for the future of science education, particularly in the context of LPs, and contribute meaningfully to the discourse on the ethical and responsible use of machine learning in education.

Integrating Artificial Intelligence into Learning Progression-Based Learning Systems

Chapter 27 offers an exhaustive exploration of the integration of AI into LPs for the advancement of science education. The authors employ evidence-centered design as a framework to construct diagnostic analytic rubrics, which are aligned with the three foundational dimensions of the Next Generation Science Standards (NGSS Lead States, 2013): DCIs, SEPs, and CCCs. This alignment ensures a holistic approach to science education that is both rigorous and comprehensive. The chapter also delves into the practical application of AI algorithms, specifically highlighting the use of the BERT approach for the automated scoring of student responses and the provision of diagnostic feedback. The chapter also delves into the potential of AI in providing personalized, immediate feedback to students and in aiding teachers in making informed instructional decisions. Furthermore, the chapter does not shy away from discussing the ethical and practical challenges of implementing AI in educational settings, including issues related to data security, transparency, and potential biases in algorithmic processes. The chapter concludes with a series of thoughtful recommendations for future research, emphasizing the necessity for a coherent alignment between AI support materials and LPs, the ethical considerations that come with the use of AI in educational settings, and the need for empirical studies to substantiate the proposed approaches.

Strengths and Limitations

The chapter provides a comprehensive scope, which serves as a one-stop resource for understanding the integration of AI into LPs for science education. This comprehensive approach is particularly beneficial for educators and researchers who are new to either AI or LPs, as it provides a thorough grounding in both, as well as in their intersection. The chapter introduces the key concepts of AI and LPs, and delves deep into their complexities, offering a nuanced understanding that is often missing in academic literature that tackles either subject in isolation.

Secondly, the methodological rigor of the chapter is evidenced by using the evidence-centered design as a framework, lending credibility and academic rigor to the chapter's propositions. This is particularly important given the state of research at the intersection of AI and science education. By grounding their approach in a well-established research methodology, the authors provide a blueprint for future research in this area, enhancing the chapter's utility as a seminal resource.

Another strength lies in the chapter's practical applicability. The authors go beyond theoretical discussions to include concrete examples and algorithmic approaches that can be directly applied in educational settings. This focus on real-world applicability is invaluable for practitioners who are interested in implementing AI in their educational assessments but may not know where to start. The chapter serves as a practical guide, offering not just theoretical knowledge of AI and LPs, but also actionable insights into the multiple aspects and considerations of applying LPs.

The innovation factor is also a significant strength. The chapter refers to using AI, specifically the BERT approach, for automated scoring and diagnostic feedback, which aligns with cutting-edge technologies in automatic scoring (Liu et al., 2023). This application of AI exemplifies the potential of AI in revolutionizing educational assessments, making the chapter timely and forward-looking. The authors are keeping pace with technological advancements; they are also staying ahead of the curve, pushing the boundaries of what is currently possible in educational technology. Moreover, the chapter highlights equity as both commendable and crucial. In an educational landscape that is increasingly diverse, the authors' commitment to making science education accessible to students from all backgrounds is not just socially responsible but also academically rigorous. They pay lip service to the idea of equity and embed it into their methodology and recommendations, making it a core component of their research approach.

However, the chapter is not without its limitations. Its comprehensive nature, while a strength, could also be viewed as a weakness, as the breadth of topics covered might dilute the focus and make it challenging for readers to fully grasp the core contributions. Moreover, the chapter could be strengthened by the inclusion of more empirical data to substantiate its claims. While the authors discuss the use of AI algorithms like BERT, the chapter lacks empirical evidence to demonstrate the effectiveness, accuracy, and fairness of these algorithms in educational settings. Additionally, the chapter could benefit from a more nuanced discussion on the ethical implications of using AI in education, particularly concerning data privacy and algorithmic bias.

Implications for Science Learning and Assessment

The implications of this chapter for science learning and assessment are manifold and significant. The integration of AI into LPs opens up new vistas for a more personalized, timely, and data-driven approach to education. The use of automated scoring and diagnostic feedback could potentially alleviate some of the burdensome aspects of assessment for teachers, thereby allowing them to channel their energies more productively into instructional strategies tailored to individual student needs. This is not merely a theoretical proposition; the authors provide concrete examples to illustrate how AI can be used to categorize student performance, thereby enabling more targeted instructional interventions.

Moreover, the highlight of equity and inclusion in this chapter is both timely and critical. By explicitly considering the needs of students from diverse and marginalized backgrounds, the authors underscore the potential for AI-integrated LPs to democratize science education. This focus on equity is not just a nod to current educational trends but is deeply embedded in the chapter's methodology and recommendations. The authors argue for the development of instructional strategies that are not only timely but also crucial, thereby making a significant contribution to ongoing discussions about social justice in educational settings.

In summary, the chapter's strengths lie in its comprehensive scope, methodological rigor, practical applicability, innovative focus, and commitment to equity. These strengths make it a valuable contribution to the field of science education. This chapter serves as both a primer and a comprehensive guide to the potential for integrating AI into science education through LPs. While the chapter could benefit from further empirical validation and a more nuanced discussion of ethical considerations, it nonetheless lays a robust and innovative foundation for future research in this interdisciplinary and rapidly evolving field.

Using an AI-based Dashboard to Support Teachers

Chapter 28 presents a comprehensive exploration of the role of AI-based dashboards, specifically Inq-ITS and Inq-Blotter, in facilitating formative assessment and instruction in science education. The authors argue that these dashboards can significantly aid teachers in the real-time monitoring and assessment of students' competencies in scientific practices. The chapter is grounded in the context of the *Framework for K-12 Science Education* (NRC, 2012) and the Next Generation Science Standards (NGSS Lead States, 2013), emphasizing the need for effective tools to support inquiry-based learning. Through a detailed vignette and subsequent analyses, the authors demonstrate that the AI-based dashboard not only helps teachers identify struggling students but also provides actionable insights for immediate intervention.

Strengths and Limitations

The empirical rigor of Chapter 28 stands out as a particularly noteworthy strength. The authors employ a mixed-methods approach that combines quantitative statistical analyses with qualitative epistemic network analyses (ENA). This multi-pronged methodological framework allows for a nuanced and comprehensive understanding of students' learning both prior to and after using Inq-ITS and Inq-Blotter. By providing both numerical data and contextual insights into teachers' pedagogical strategies, the study achieves a level of empirical depth that is often lacking in similar research endeavors.

Another significant strength lies in the real-world applicability of the study's findings. The tools under investigation, namely the Inq-ITS and Inq-Blotter dashboards, are not merely theoretical constructs but are operational systems already deployed in educational settings. This lends an immediate practical relevance to the study, making its findings actionable for educators, administrators, and policymakers who are interested in leveraging technology to enhance science education. The study thus bridges the often-observed gap between academic research and practical application, contributing to its broader impact.

Furthermore, the study is deeply grounded in the context of current educational frameworks, specifically the *Framework for K-12 Science Education* (NRC, 2012) and the Next Generation Science Standards (NGSS Lead States, 2013). This contextual grounding not only enhances the relevance of the study but also makes it timely and directly applicable to the challenges and needs of contemporary educational practices and standards. By aligning the study with these frameworks, the authors ensure that their findings speak directly to ongoing debates and initiatives of three-dimensional learning in the field of science education.

Additionally, the chapter offers actionable insights that extend beyond the mere identification of challenges in teaching scientific inquiry. The real-time data and alerts generated by the dashboard are designed for immediate instructional intervention, thereby serving both diagnostic and prescriptive functions. This dual utility makes the tool exceptionally valuable for educators who aim to dynamically adapt their teaching strategies in response to real-time student performance data.

The chapter presents valuable insights, yet there is room to maximize its impact. By establishing a stronger linkage with the LPs literature and actively participating in the scholarly discourse on LPs' construction and implementation, the theoretical base of the proposed approach could be further enhanced. Moreover, situating the study's major findings within the framework of LPs would clarify their relevance and contributions to the evolution of LPs, thus increasing the chapter's informational depth and academic contribution. While the development and implementation of Inq-ITS and Inq-Blotter represent notable advancements in the domain, there is an opportunity to strengthen the connection to the existing corpus of research on AI applications in science education.

Implications for Science Learning and Assessment

The chapter has significant implications for the future of science education, particularly in the realm of formative assessment and real-time instructional support. The Inq-ITS and Inq-Blotter dashboard represent a paradigm shift in how educators can approach teaching scientific inquiry, offering a more dynamic and responsive pedagogical strategy. The real-time data and actionable insights provided by the dashboard can potentially revolutionize the way teachers interact with students, allowing for immediate intervention and personalized instruction. This could lead to more effective teaching strategies and, ultimately, a deeper understanding of scientific practices for students.

Moreover, the chapter's findings could catalyze further research into the integration of AI and machine learning technologies in intelligent tutor systems. Given the promising results, there is a compelling case for broader implementation and further study, including larger and more diverse samples, to explore the full range of capabilities and limitations of such dashboards.

In summary, Chapter 28 offers a compelling look at the potential of AI-based dashboards to significantly enhance formative assessment and instruction in science education. While there are limitations that warrant the consistency with the theme of the volume, the chapter nonetheless provides valuable insights that could shape the future of science education and educational technology. In summary, the strengths of Chapter 28 lie in its methodological rigor, practical applicability, contextual relevance, and its provision of actionable insights for science educators. These strengths collectively contribute to the study's significance and its potential to affect meaningful improvements in the field of science education.

Emergent Themes from the Cross-Chapter Synthesis

In a comprehensive examination of the interplay between technology, pedagogy, and assessment, Chapters 25 through 28 collectively offer a nuanced understanding of the evolving landscape of LPs of science education (Krajcik & Shin, 2023). A recurring theme across these chapters is the transformative potential of technology, particularly AI, in enhancing the efficacy and reach of science teaching and learning, which aligns with the ongoing trends in science education (Zhai et al., 2020a). Each chapter, while rooted in its unique context and objectives, underscores the importance of evidence-based approaches, be it through evidence-centered design in Chapters 25 and 27 or through empirical studies advocated in Chapters 26 and 28. The chapters collectively emphasize the need for a multi-dimensional approach to LPs in science education that integrates DCIs, SEPs, and CCCs, aligning closely with the Next Generation Science Standards (NGSS Lead States, 2013).

Despite these commonalities, the chapters exhibit divergent views and methodologies that enrich the discourse. For instance, while Chapter 25 delves into the details of validating LPs using AI-based automatic measures, Chapter 27 takes a broader view, focusing on the role of technology in fostering collaborative learning environments. The methodological frameworks employed also vary, ranging from design-based research to more traditional qualitative and quantitative approaches. These divergences are not contradictions but rather complementary perspectives that offer a multi-faceted understanding of the complex ecosystem of science education.

Technological advancements serve as a unifying thread that weaves these disparate elements into a coherent narrative. Whether it is the use of machine learning algorithms for automated assessment or the application of virtual laboratories, technology emerges as an enabler that amplifies the impact of pedagogical strategies and assessment frameworks. It serves as a catalyst for innovation, pushing the boundaries of what is possible in science education, from diagnosing students' learning pathways to real-time analytics and feedback for instructions. However, it is crucial to note that the chapters collectively caution against the uncritical adoption of technology. Ethical considerations, particularly concerning data security and algorithmic bias, are consistently highlighted, emphasizing the need for responsible implementation. Moreover, the chapters evidenced the need for interdisciplinary collaboration involving experts in science education, AI, statistics, and ethics to navigate the complexities of this evolving landscape.

In summary, Chapters 25 through 28 offer a rich tapestry of insights, methodologies, and perspectives that, while diverse, converge on the transformative potential of technology in science education. They collectively argue for a balanced, evidence-based approach that harnesses technological advancements while remaining cognizant of the ethical and practical challenges that such integration entails.

Practical Implications and Future Research Directions

As a scholar whose research delves into both AI and LPs, I view the findings from Chapters 25 through 28 as significant—they collectively offer a wealth of practical implications for various stakeholders in the educational ecosystem, including educators, policymakers, and researchers. For educators, the integration of technology, particularly AI, into science learning and assessment provides a robust framework for personalized, diagnostic, and formative assessment practices. The use of LPs, as detailed in Chapter 27 for instance, allows for a nuanced understanding of student performance, thereby enabling targeted interventions. Educators can leverage these technological tools to not only assess but also to adapt their instructional strategies in real time, thereby enhancing the learning outcomes. For policymakers, the chapters make a compelling case for the need to invest in technological infrastructure and professional development for educators in an era of AI. The potential of technology to democratize access to quality science education, especially for marginalized communities, is a recurring theme that policymakers would do well to heed (Zhai & Nehm, 2023).

The research presented in these chapters also highlights several gaps that need to be addressed. One such gap is the ethical considerations surrounding the use of AI in educational settings, including issues of data privacy and algorithmic bias. Another gap is the need for interdisciplinary research that brings together experts from the fields of science education, technology, and ethics to develop holistic solutions. These gaps present fertile ground for future research that is well aligned with prior findings (Latif et al., 2023). Studies could explore, for example, the long-term impacts of technology-integrated LPs on student outcomes or investigate the ethical implications of data-driven personalized learning. There is also a need for empirical studies that evaluate the efficacy of these technological LPs-supported interventions in diverse educational settings.

The overarching themes that emerge from these chapters include the transformative potential of technology, the importance of evidence-based approaches, and the need for ethical considerations in the validation and implementation of LPs in science education. These themes collectively contribute to a nuanced understanding of the complex landscape of science learning, as well as effective pedagogies to address the learning needs of students with diverse backgrounds. They offer a balanced perspective that recognizes the potential of technological advancements while also cautioning against their uncritical adoption in LP studies.

In conclusion, the future of technology in facilitating LPs appears to be promising yet fraught with challenges that require careful navigation. The integration of technology into LP studies is not merely a trend but a paradigm shift that has the potential to redefine how science is taught and learned. However, this transformation must be guided by empirical evidence, ethical considerations, and a commitment to equity and inclusion. As we move forward, the insights from these chapters serve as both a roadmap and a cautionary tale, highlighting the immense possibilities and the significant responsibilities that come with the integration of technology into science education.

Acknowledgment

The author is a Visiting Humboldt Professor at IPN, Germany, at the time of crafting this paper. He is grateful for the sponsor—IPN, and the host professor Knut Neumann, as well as the funder, Alexander Humboldt Foundation. The opinions and ideas do not reflect the sponsor and the funder.

References

Alonzo, A. C., & Gotwals, A. W. (2012). *Learning progressions in science: Current challenges and future directions*. Sense Publishers.

Fang, L., Lee, G.-G., & Zhai, X. (2023). Using Gpt-4 to Augment Unbalanced Data for Automatic Scoring. *arXiv preprint: 2310.18365v1*.

Fortus, D., & Krajcik, J. (2012). Curriculum coherence and learning progressions. *Second international handbook of science education*, 783–798.

Fulmer, G. W., Liang, L. L., & Liu, X. (2014). Applying a force and motion learning progression over an extended time span using the force concept inventory. *International Journal of Science Education*, *36*(17), 2918–2936.

Furtak, E. M. (2012). Linking a learning progression for natural selection to teachers' enactment of formative assessment. *Journal of Research in Science Teaching*, *49*(9), 1181–1210.

Furtak, E. M. (2023). *Formative Assessment for 3D Science Learning: Supporting Ambitious and Equitable Instruction*. Teachers College Press.

Gao, Y., Zhai, X., Andersson, B., Zeng, P., & Xin, T. (2020). Developing a learning progression of buoyancy to model conceptual change: A latent class and rule space model analysis. *Research in Science Education*, *50*(4), 1369–1388.

Gao, Y., Zhai, X., Bae, A., & Ma, W. (2023). Rasch-CDM: A combination of rasch and cognitive diagnosis models to assess a learning progression. In *Advances in Applications of Rasch Measurement in Science Education* (pp. 97–122). Springer.

Gao, Y., Zhai, X., Cui, Y., Xin, T., & Bulut, O. (2021). Re-validating a learning progression of buoyancy for middle school students: A longitudinal study. *Research in Science Education*, 1–29. 10.1007/s11165-021-10021-x

Gerard, L., Matuk, C., McElhaney, K., & Linn, M. C. (2015). Automated, adaptive guidance for K-12 education. *Educational Research Review*, *15*, 41–58.

Hadenfeldt, J. C., Neumann, K., Bernholt, S., Liu, X., & Parchmann, I. (2016). Students' progression in understanding the matter concept. *Journal of Research in Science Teaching*, *53*(5), 683–708. https://onlinelibrary.wiley.com/doi/abs/10.1002/tea.21312

Jin, H., & Anderson, C. W. (2012). A learning progression for energy in socio-ecological systems. *Journal of Research in Science Teaching*, *49*(9), 1149–1180.

Jin, H., Delgado, C., Bauer, M. I., Wylie, E. C., Cisterna, D., & Llort, K. F. (2019a). A hypothetical learning progression for quantifying phenomena in science. *Science & Education*, *28*, 1181–1208.

Jin, H., Mikeska, J. N., Hokayem, H., & Mavronikolas, E. (2019b). Toward coherence in curriculum, instruction, and assessment: A review of learning progression literature. *Science Education*, *103*(5), 1206–1234.

Krajcik, J., & Shin, N. (2023). Student conceptions, conceptual change, and learning progressions. *Handbook of Research on Science Education: Volume III*.

Krajcik, J. S. (2012). The importance, cautions and future of learning progression research: Some comments on Richard Shavelson's and Amy Kurpius's "Reflections on Learning Progressions". In *Learning progressions in science* (pp. 27–36). Brill Sense.

Kubsch, M., Czinczel, B., Lossjew, J., Wyrwich, T., Bednorz, D., Bernholt, S., Fiedler, D., Strauß, S., Cress, U., & Drachsler, H. (2022). Toward learning progression analytics—Developing learning environments for the automated analysis of learning using evidence centered design. *Frontiers in Education*,

Latif, E., Mai, G., Nyaaba, M., Wu, X., Liu, N., Lu, G., Li, S., Liu, T., & Zhai, X. (2023, April 01, 2023). Artificial General Intelligence (AGI) for Education. arXiv:2304.12479. https://arxiv.org/abs/2304.12479

Liu, Z., He, X., Liu, L., Liu, T., & Zhai, X. (2023). Context matters: A strategy to pre-train language model for science education. In N. Wang & et al. (Eds.), *AI in Education 2023* (Vol. CCIS 1831, pp. 1–9). Springer. 10.1007/978-3-031-36336-8_103

National Research Council. (2012). *A framework for K-12 science education: Practices, crosscutting concepts, and core ideas*. National Academies Press.

Neumann, K., Viering, T., Boone, W. J., & Fischer, H. E. (2013). Towards a learning progression of energy. *Journal of Research in Science Teaching*, *50*(2), 162–188. https://onlinelibrary.wiley.com/doi/abs/10.1002/tea.21061

NGSS Lead States. (2013). *Next generation science standards: For states, by states*. National Academies Press.

Wilson, C., Haudek, K., Osborne, J., Stuhlsatz, M., Cheuk, T., Donovan, B., Bracey, Z., Mercado, M., & Zhai, X. (2023). Using automated analysis to assess middle school students' competence with scientific argumentation. *Journal of Research in Science Teaching*, 1–32. 10.1002/tea.21864

Yin, Y., Tomita, M. K., & Shavelson, R. J. (2014). Using formal embedded formative assessments aligned with a short-term learning progression to promote conceptual change and achievement in science. *International Journal of Science Education*, *36*(4), 531–552.

Zhai, X., Haudek, K., & Ma, W. (2022). Assessing argumentation using machine learning and cognitive diagnostic modeling. *Research in Science Education*, *53*, 405–424. 10.1007/s11165-022-10062-w

Zhai, X., Haudek, K. C., Shi, L., Nehm, R., & Urban-Lurain, M. (2020a). From substitution to redefinition: A framework of machine learning-based science assessment. *Journal of Research in Science Teaching*, *57*(9), 1430–1459.

Zhai, X., & Li, M. (2021). Validating a partial-credit scoring approach for multiple-choice science items: an application of fundamental ideas in science. *International Journal of Science Education*, *43*(10), 1640–1666.

Zhai, X., Li, M., & Guo, Y. (2018). Teachers' use of learning progression-based formative assessment to inform teachers' instructional adjustment: a case study of two physics teachers' instruction. *International Journal of Science Education*, *40*(15), 1832–1856.

Zhai, X., & Nehm, R. (2023). AI and formative assessment: The train has left the station. *Journal of Research in Science Teaching*, *60*(6), 1390–1398.

Zhai, X., Shi, L., & Nehm, R. (2021). A meta-analysis of machine learning-based science assessments: Factors impacting machine-human score agreements. *Journal of Science Education and Technology*, *30*(3), 361–379.

Zhai, X., Yin, Y., Pellegrino, J. W., Haudek, K. C., & Shi, L. (2020b). Applying machine learning in science assessment: A systematic review. *Studies in Science Education*, *56*(1), 111–151.

30

REFLECTIONS ON THE EDUCATION VALUE OF LEARNING PROGRESSIONS

Joseph Krajcik, Hui Jin, and Duanli Yan

Attention to improving science education has grown nationally and globally as science and policy communities find themselves challenged by complex real-world problems without straightforward solutions. The following ten years will continue to bring about unforeseen changes and challenges in how we live and interact with each other and the environment. To find solutions to pressing problems of our era, all students need to develop knowledge-in-use, which refers to the ability to use disciplinary knowledge and practices to find solutions to novel and challenging problems, make sense of compelling and complex phenomena, make informed decisions about their lives, and learn more when needed. Otherwise, we will not be able to resolve and implement solutions to some of the most pressing problems of our times, including climate change, clean energy, clean water, pandemics, and sustainable and healthy food.

Unfortunately, as an education community, we have failed most students in the K–12 education system, particularly those from underrepresented and underserved communities, to develop knowledge-in-use. Curriculum materials, instruction, and assessment lack the coherence and depth to provide the necessary learning opportunities for students to develop and apply their knowledge (OECD, 2019). Students of color and traditionally underserved students from poor communities and countries suffer the most. According to the reports of the National Center for Science and Engineering Statistics (NCSES, 2023), the proportion of Black or African American individuals, American Indian or Alaska Native individuals, and Hispanic or Latine individuals in the US do not participate in engineering education and occupations at the same proportion as their white counterparts. While policy and other attempts have tried to bring about change, the percentage of STEM bachelor's degrees awarded to Black students has historically been lower than their representation in the general population. Throughout the globe, including in the US, a persistent achievement gap among whites and other ethnic groups exists, highlighting an equity gap and injustice in our world. These disparities in STEM education must be addressed because they can have long-lasting impacts on economic and social equity as well as the environment and well-being of individuals. Efforts to close the achievement gap in STEM education have been a focal point for educators, policymakers, and stakeholders, given the importance of STEM fields throughout the globe.

As the chapters in this book have illustrated, Learning Progressions (LPs) and LP research offer hope in bringing about change to provide more students with opportunities to learn

DOI: 10.4324/9781003170785-34

challenging scientific ideas and practices. More specifically, implementation of LPs across K-16 science education holds the potential to create coherent and equitable curriculum, instruction, and assessment systems. Such systems support a wide range of students to achieve challenging learning goals necessary in a knowledge-production society while providing different pathways contingent upon individual student backgrounds. The LP approaches therefore hold promise to close the achievement gap while keeping academic achievement at the level necessary to live and thrive in a world where the ability to use existing knowledge and generate new knowledge is required. In this concluding chapter, we begin with a discussion of the nature of LP—depicting student development over time in a purposefully designed learning environment. We subsequently explore how multi-dimensional LPs can be developed to fit the needs of a variety of students, encompassing underrepresented students, and examine how LP and LP-based systems can support teachers in classroom teaching. Lastly, we wrap up this chapter by outlining potential avenues for future research.

LP to Depict Development Over Time in a Learning Environment

A vital aspect of an LP is the development over time within a specific learning environment: An LP shows how to support students in developing knowledge of big ideas and practices that students will need throughout their lives. Many adults, unfortunately, have not developed the depth of expertise to use big ideas and scientific practices to understand everyday phenomena. For example, most adults do not understand the transfer of energy that occurs in a car accident and how car designs minimize the transfer of energy away from the passengers. Adults with high school and higher education struggled with interpreting various data representations of COVID-19 data to make everyday decisions (Tabak & Dubovi, 2022). This lack of understanding of disciplinary ideas and scientific practices is unsurprising. Although topics like energy, natural selection, interaction among molecules, growth, and development of organisms, infectious disease, and plate tectonics are typically part of the K-12 science education curriculum and taught across different grade levels, they often are dealt with superficially and not over time. Many students do not engage in learning big ideas of science and scientific practices systematically across grade levels in contexts that require them to apply the ideas more deeply and in more sophisticated situations. Although no single trajectory or pathway exists that all students will follow, all students need to develop certain scientific ideas before they have the background knowledge to learn other, more sophisticated ideas, as some ideas depend on others. Imagine a student attempting to develop proficiency in a high school standard without the proficiencies of the prior learning developed in earlier years. Without prior knowledge and experience, attempting to develop high school-level proficiency in complex content would be exceptionally challenging. This often results in students memorizing ideas and then quickly forgetting those ideas. The lack of building on students' prior knowledge and experiences provides just one reason why the achievement gap exists.

As highlighted throughout this Handbook, although students develop an understanding of science in different ways and at different rates, opportunities to learn entail that all students have the chance to create knowledge-in-use, so that they can apply ideas to explain compelling phenomena, find solutions to challenging problems, and learn more when needed. Such pathways can potentially close the persistent achievement gap in the US and globally. The learning opportunities should occur throughout an individual's lifespan because learning challenging big ideas (e.g., evolution, energy, sustainability, or how matter reacts) takes time to develop and is not developmentary inevitable but dependent on experiences and education (Krajcik, 2023; National Research Council, 2007). Because students' learning depends on appropriate instruction, researchers need to develop learning environments that provide rich opportunities for a wide range of students to learn big ideas and scientific and engineering practices.

LPs provide a way to rethink, restructure, and push forward the content taught in schools and that students experience (Krajcik, 2023; Lehrer & Schauble, 2015). As conceived in this Handbook, LPs can serve as models to shift teachers' and students' views of what it means to understand a big idea and engage in scientific practice to make sense of phenomena and solve problems. This is achieved through two fundamental ways. Firstly, substantial evidence suggests that humans hold specific models in mind, which they use to construct language and interpret their environment (Pinker, 2007). It follows that students, as human beings, tend to rely on a few (not many) intuitive reasoning patterns to analyze and understand the material world. These intuitive reasoning patterns are often used in LPs as the lower anchors. The other LP levels and the instruction associated with the LPs must address these lower anchors. Secondly, an LP depicts a pathway a student follows as a result of purposefully designed instruction. In essence, LPs and curriculum, instruction, and assessment systems go hand in hand. As discussed in the chapters by Duncan et al., and Fortus & Nordine, even for the same scientific idea or practice, different LP and curriculum, instruction, and assessment system pairs can be developed; their effectiveness can be tested and compared in real classrooms.

Therefore, developing LPs using traditional curricula or business-as-usual approaches takes a step in the wrong direction. Instead, researchers need to design LPs through iterative development and testing, using what we know works to promote student learning. This requires that educational researchers, teachers, administrators, and policymakers shift their view of what it means to learn big ideas along with scientific practices. As several chapters in this volume discuss, this shift will occur if LP development occurs in design-based research settings. Research teams composed of educational researchers, content experts, and practitioners need to work together to develop meaningful contexts, instructional strategies that teachers can adapt to students with different backgrounds, and learning tasks and assessments that will support them in developing usable understanding across time. Researchers working collaboratively with teachers need to design LPs through iterative development and testing, using what we know works to promote student learning. Validation plays a crucial role in this design-based research process. Moreover, validation activities must be carried out in a systematic manner to ensure the effectiveness of the LP and associated curriculum, instruction, and assessment. In this handbook, the chapter by Kant discusses how validity theories can be used to evaluate and validate claims about LPs and the use of LP-based curriculum, instruction, and assessment; the chapters by Kaldaras and colleagues and Wilson colleagues use empirical research to illustrate LP-based validation activities.

Building and testing LPs that depict how students' knowledge of disciplinary core ideas (DCIs), crosscutting concepts (CCCs), and scientific and engineering practices (SEPs) grows over a long period of time (K–16) – several weeks, a semester, a year, or several years – will require a concerted effort by researchers, teachers, and policymakers. Because LPs focus on complex DCIs essential for understanding the world and require the integration of SEPs and CCCs, it will take time to develop them, and the road to designing and testing LPs across various domains will cause challenges as the road is steep and rocky. Yet, as pointed out in several chapters, we need to develop LPs that integrate the 3-dimensions of scientific knowledge. LPs are developmentally appropriate, as they build over time based on students' prior knowledge and experiences, cultural background, and gender (National Research

Council, 2007; Smith et al., 2006). Unfortunately, most individuals across the globe do not understand this developmental approach or how to implement it. As such, it will take time to educate educators to understand a developmental approach to learning and how to implement a developmental approach in classrooms. As a science education community, we still have much to learn. This road, too, will provide many challenges. The importance of building on prior knowledge to develop new and deeper understandings has been recognized for many years, yet in educational settings, particularly in science education, we continue to ignore what students know and move forward to the detriment of student learning, or we don't take the time to slowly build complex ideas in meaningful ways. Today, the situation presents more complications because not only do we need to build on prior knowledge, but we also now know that to reach the greatest number of learners, educational environments need to build on a learner's cultural experiences, race, and gender.

Multi-dimensional LPs to Support a Variety of Learners

A significant premise underlying LP theory asserts that although "individual differences" exist and students develop understanding in different ways (e.g., visual, verbal, and psychomotor) and at different rates, every student needs equal opportunities and resources to develop deep, usable knowledge of scientific ideas and practices; moreover, this must happen through meaningful experiences in a range of learning contexts (National Academies of Sciences Engineering and Medicine, 2019; National Research Council, 2000). We now know that the one textbook or the same experience for all students will fail to develop knowledge-in-use for the majority of students. Yet, we have in our means the ability to develop LPs that take multiple pathways and experiences and to adapt curriculum resources and instruction to support a variety of learners. As such, implementation of LPs throughout K-16 education holds the potential to create equitable curriculum materials to support a wide range of students achieving the same learning goals while providing different pathways contingent upon individual student backgrounds, culture, race, and gender, helping to close the persistent achievement gap that exists while keeping academic achievement at the level necessary to live and thrive in a society where students need to use knowledge and learn more when needed.

As discussed throughout this volume, LPs provide developmental descriptions of student learning of scientific big ideas (i.e., disciplinary core ideas and crosscutting concepts) and practices, delineating the pathways that students travel as they progress toward higher proficiency levels over time when given appropriate opportunities to learn. LPs should describe how students develop a more sophisticated understanding of three-dimensional knowledge – the integration of DCIs, SEPs, and CCCs over an extended period. LPs need to clearly define what it means to move towards a more advanced understanding of using DCIs integrated with SEPs and CCCs to make sense of complex phenomena and challenging problems. Yet, as this volume indicates, the research needed to develop multi (2 or more)-dimensional learning progression has just begun. While researchers have known about the importance of integrating DCIs and SEPs for over two decades (National Research Council, 2007), and while researchers developed some LPs, more researchers need to put a more concerted effort into building LPs with multiple dimensions.

Learning Progressions delineate evidence-based learning pathways about how big ideas in science develop over time and become more sophisticated so that learners can use their knowledge. The multiple pathways in LPs, which consider students' prior knowledge, experiences, cultural background, and race, can guide teachers in promoting useable knowledge. While this volume has shown that some areas of various science disciplines have developed LP, including energy, the structure, and interaction of matter, genetics, water quality, ecosystems, and earth systems, additional research needs to occur to expand these LPs across the grade level and to expand into new areas – like photosynthesis, matters interaction with light, and natural selection and adaptation. However, many previously designed LPs were one-dimensional, focusing only on how the content idea developed. However, the research is clear that to make sense of phenomena and solve challenging problems, researchers and teachers need a multi-dimensional approach to knowledge and LP development. While a few examples of multi-dimensional (i.e., 3-dimensional) LPs exist, they are few.

Supporting Teachers in Using the LP Approaches

To successfully implement LPs, teachers will need equitable resources (curriculum, instructional strategies, assessment tasks) that they can use to support the development of understanding scientific ideas and practices throughout a grade level and revisiting those ideas in subsequent grades so that knowledge becomes more sophisticated and usable. Designers and researchers need to develop these research-based resources. Teachers will also need long-term LP professional learning to establish communities of learners where they can grapple with important ideas related to students' learning over time. Professional learning will need to provide teachers with an understanding of how learning develops over time, how to use LPs during instruction, and how to use the resources to move students further along the learning progression (see Luft, this volume). Teachers will need to learn how to provide appropriate feedback to learners to move them further along the learning progression (see Peng and colleagues, this volume). A critical aspect moving forward is that the LP community must engage in research to address these challenges so that a diverse range of students develop usable knowledge.

Once researchers develop and test 3-dimensional learning progressions and teachers learn how to use these LPs in their teaching, teachers will face challenges in implementing LPs. Teachers face challenges in analyzing student responses to open-ended tasks such as developing scientific explanations and constructing models. These tasks take time and are challenging to analyze, but students need opportunities to participate in them to help them build knowledge in use. More importantly, student learning requires timely and specific feedback on such tasks to help them learn and progress along the LP. Analyzing and then using that feedback to promote learning present challenges. Therefore, professional development that helps teachers understand the development perspective of LPs and how LPs can be used for classroom assessments should be provided (see chapters by Along and by Furtak and colleagues).

In addition to teachers' conceptual understanding of how LPs work to promote student learning, their workload should also be considered. Imagine a good teaching situation in which a teacher has four classes with only 20 students in each class (an unlikely scenario in most US classrooms). Suppose this teacher assigns students to develop a model to explain how a chemical reaction produces new substances with different properties, even if students worked in pairs to build the models. In that case, the teacher will need to analyze and provide appropriate and developmental feedback on 40 models that evening. Even the most dedicated teacher would be overwhelmed to accomplish this task. Yet, we know appropriate and timely feedback is essential to promote learning. Fortunately, artificial intelligence (AI) technologies, in particular, Natural Language Processing (NLP), used as a tool, have shown promise in the automatic scoring of students' constructed responses to provide timely feedback for creating equitable opportunities for students to learn (see the chapter by Riordan and colleagues, and the chapter by Peng and colleagues this volume). The recent advancement of large language models (i.e., GPT and BERT) holds even more promise, but research is needed to provide evidence of this use.

More importantly, even if teachers could receive the results of how students did on open-ended tasks that placed students on a learning progression, teachers would need to know how to move students further along the LP. Providing appropriate feedback and adapting learning experiences to students' responses presents challenges, particularly if students need support on different aspects (i.e., the DCI, SEP, or CCCs) and learn using different modalities. AI can help in this area as well. Leveraging cutting-edge AI technology (i.e., large language models and data analytics), LPs will provide teachers and students with automatic, prompt performance reports and appropriate suggestions on the type of feedback to provide individual learners and the class. This will enable teachers to make essential adjustments to instruction while also allowing students to seek help in meeting their learning needs and progress toward targeted proficiency levels.

The integration of AI tools has the transformative potential to revolutionize the learning experience for students. AI can provide tailored, personalized feedback that addresses each student's unique needs and learning preferences by harnessing a wealth of information about students, including their backgrounds, interests, and academic performance (See the chapter by Russell and colleagues). This personalized approach extends to adaptive learning, where cutting-edge AI techniques continuously analyze students' real-time performance and engagement. The result is the dynamic adaptation of teaching materials and feedback to align with each student's learning preferences, supporting them to move along the LP. These AI-driven systems create digital profiles of students' educational journeys, regularly updated with new data, serving as comprehensive repositories of a student's academic development across time.

These digital profiles offer valuable performance reports, which provide teachers, students, and parents with crucial insights into students' learning and engagement. These insights guide educators in adjusting their instructional strategies to cater to individual student's needs, ensuring that every student has opportunities to thrive and deepen their understanding of the disciplinary core ideas and scientific practices. However, integrating AI resources into education also necessitates investment in professional learning for teachers, students, and parents. This education is essential to empower stakeholders with the knowledge and skills required to leverage AI tools effectively.

Moreover, it's crucial to remain vigilant against introducing bias into AI systems. Ensuring that these resources offer equitable learning opportunities for all students is critical. This vigilance involves continuous monitoring and mitigation of bias, emphasizing fairness and impartiality in the educational journey. AI's role in education promises personalized, adaptive learning experiences, but it also demands a commitment to professional development and bias mitigation to maximize its benefits for student learning.

Concluding Comment and Future Research Needs

Grasping complex ideas and practices in such a way that learners can apply them in novel situations is rooted in the depth and sequence of their learning experiences. For truly effective learning, students must encounter scenarios demanding the layering of ideas in coherent sequences. Presenting stimulating and intellectually engaging contexts where learners experience the intrigue of the phenomena propels them to build and connect new knowledge. However, this intricate process of knowledge construction is not inevitable or instantaneous. It takes time that spans over extended periods, grounded in continuous exposure and instruction.

The framework of Learning Progressions (LPs) has emerged as a cornerstone in science education. Researchers need to anchor the development of LPs in research and an understanding of student developmental trajectories. Learning, as a phenomenon, is multifaceted. It unfolds in non-linear patterns, charting numerous pathways influenced heavily by a student's unique background, culture, and prior knowledge. Consequently, to design LPs to promote learning, ongoing research efforts must delve deep into the myriad ways students engage with and process scientific ideas and practices.

The digital revolution over the past decade, especially recently, has triggered a transformative shift in how we interact, seek information, and manage daily life. This transformation compels us to reconsider and reconstruct traditional frameworks. Formulating and implementing LPs is critical to bringing about changes in our world. LPs serve as roadmaps, guiding students through a structured journey of knowledge development, ensuring that what they learn is not just stored information and inert facts but knowledge that is applicable, adaptable, and relevant in the rapidly changing world in which students live. LPs can help to empower learners with knowledge structures that are both deep-rooted and versatile, equipping them for the challenges and opportunities of the modern world.

Learning progression theory and research provide an inspiring vision to narrow the achievement gap that exists in our society. To make this possible, there are several pivotal research efforts that the science education and learning science communities should actively undertake. Policymakers need to support this effort by providing the appropriate resources to complete this research. Some of the research that needs to occur includes the following.

1 Three-dimensional Learning Progressions for Disciplinary Core Ideas Integrated with CCCs and SEPs: Researchers need to develop comprehensive learning progressions that tackle the disciplinary core ideas – which include, but are not limited to, energy, ecology, the nature of matter and its interaction, genetics, and force – integrated with crosscutting concepts and scientific practices. Creating three-dimensional LPs ensures depth, breadth, and contextual relevance, allowing for a more holistic understanding of the subject matter and supporting the development of useable knowledge.
2 Coherent Curriculum, Instruction, and Assessment Systems aligned with LPs: Researchers need to create curriculum materials, instructional approaches, resources, and assessment tasks that align with the designed LPs. Coherent resources ensure that teaching and assessment strategies support the trajectories outlined in the LPs, leading to more consistent and effective learning opportunities.
3 Long-term Impact of LP-based Education: Perhaps one of the most critical and immediate research efforts that needs to occur include conducting longitudinal studies to determine the long-term outcomes of students taught through LP-based curricula versus traditional methods. Assessing the long-term impacts of student learning with LPs can provide insights into the sustainability and effectiveness of LPs.
4 Professional Learning for Effective LP Utilization: Researchers, along with teachers, need to develop, test, and offer ongoing professional learning programs for teachers across all grade levels. These programs should elucidate the nuances of LPs and provide guidance on using LP resources effectively. Equipping science teachers, including teachers who teach science in K-12, with knowledge and the appropriate tools and how to use them can enhance their ability to guide students progressively through the LP and advance student learning.
5 AI-driven Analysis and Feedback on Student Tasks: Using automated scoring technology to evaluate students' constructed responses and provide real-time feedback to educators and students can significantly promote student learning. Automated scoring models can

process large amounts of data quickly, identifying patterns and gaps in student understanding that may be difficult for educators to spot manually. Researchers need to learn how to harness this power.

6 AI and Data Analytics for Personalized Learning Experiences: Researchers need to leverage AI and data analytics to tailor and adapt learning experiences and resources based on individual student needs, backgrounds, learning styles, culture, race, and gender. Personalized learning experiences cater to students' unique strengths, weaknesses, and interests, potentially accelerating their learning journey.
7 Professional learning for teachers on how to utilize AI Feedback: Professional learning also needs to help teachers integrate immediate feedback from AI systems into their teaching and provide appropriate strategies for moving students along the LP. Immediate feedback serves as a potent tool for promoting student learning. When teachers are adept at interpreting and acting on this feedback, they can make real-time adjustments to their teaching methods to cater to student needs.
8 Integrating of Digital Learning Environments: Research needs to investigate the effectiveness of merging traditional classroom learning with digital tools and platforms. Such efforts have the potential to enhance learning. As digital platforms become more integrated into educational settings, understanding how to leverage their strengths can improve the efficacy of LPs and student learning.
9 Environmental and Contextual Learning. Researchers need to investigate the potential benefits of project-based, outdoor, and experiential learning environments as curricula that support the use of LPs. Real-world, tangible experiences that environments like project-based learning can provide can deepen understanding and provide context to learning challenging science ideas and practices (Krajcik et al., 2022; Schneider et al., 2022).
10 Interdisciplinary LPs: Researchers need to explore the development and implementation of LPs that bridge multiple disciplines, such as science, language arts, history, and art. Interdisciplinary learning has the potential to provide a richer and more interconnected understanding of the world.

Improving science education rests heavily on the effective construction and implementation of learning progressions. By intertwining technological advancements like AI with LPs and by investing in sustained professional learning, educators and researchers can pave the way for a more informed, dynamic, and responsive educational landscape that promises to reduce the achievement gap that has continued to persist in our world even though the various stakeholders have discussed and policy made to help remedy this situation.

All students, regardless of their backgrounds, race, and culture, need the opportunity to develop usable knowledge of scientific ideas and practices if we hope to find solutions to the challenges facing our world. LPs can serve as vehicles for allowing this to happen. With the integration of AI resources, we have the opportunity to reduce the persistent learning gap that exists in our world. LP research provides a promising venue to support a wide range of learners in developing and monitoring usable knowledge. Learning environments should facilitate students' learning of big ideas of science and scientific practices in and across disciplines and develop capabilities to apply those ideas in real situations. While we can't consider LPs the miracle cure of education, implemented appropriately, LPs provide one promising avenue to reduce the persistent achievement gap among learners that can create a more equity-focused, inclusive, and just educational system throughout the globe. The road will present challenges as the road is steep and rocky, but we have the tools; we need the will and resources to work together to build such a system.

References

Krajcik, J., Schneider, B., Miller, E. A., Chen, I. C., Bradford, L., Baker, Q., Bartz, K., Miller, C., Li, T., Codere, S., & Peek-Brown, D. (2022). Assessing the effect of project-based learning on science learning in elementary schools. *American Educational Research Journal*, *60*(1), 70–102. 10.3102/00028312221129247

Krajcik, J. S., & Namsoo (2023). Student concepts, conceptual change and learning progressions. In N. G. Lederman, Zeidler, D.L. & Lederman, J.S. (Ed.), *Handbook of Research on Science Education, Vol III* (Vol. 3, pp. 121–157). Taylor and Francis group.

Lehrer, R., & Schauble, L. (2015). Learning progressions: The whole world is NOT a stage. *Science Education*, *99*(3), 432–437.

National Academies of Sciences Engineering and Medicine. (2019). *Science and Engineering for Grades 6-12: Investigation and Design at the Center*. The National Academies Press. 10.17226/25216

National Center for Science and Engineering Statistics [NCSES]. (2023). *Diversity and STEM: Women, minorities, and persons with disabilities 2023*. National Science Foundation; Special Report NSF 23-315. https://ncses.nsf.gov/pubs/nsf23315/

National Research Council. (2000). *How people learn: Brain, mind, experience, and school: Expanded edition*. National Academies Press.

National Research Council. (2007). *Taking science to school: Learning and teaching science in grades K-8*. National Academies Press.

OECD, O. f. E. C. a. D. (2019). *PISA 2018 assessment and analytical framework*.

Pinker, S. (2007). *The stuff of thought: Language as a window into human nature*. Penguin.

Schneider, B., Krajcik, J., Lavonen, J., Salmela-Aro, K., Klager, C., Bradford, L., Chen, I.-C., Baker, Q., Touitou, I., Peek-Brown, D., Dezendorf, R. M., Maestrales, S., & Bartz, K. (2022). Improving science achievement—is it possible? Evaluating the efficacy of a high school Chemistry and Physics project-based learning intervention. *Educational Researcher*, *51*(2), 109–121. 10.3102/0013189x211067742

Smith, C. L., Wiser, M., Anderson, C. W., & Krajcik, J. (2006). Implications of research on children's learning for standards and assessment: A proposed learning progression for matter and the atomic-molecular theory. *Measurement: Interdisciplinary Research & Perspective*, *4*, 1–98. 10.1080/15366367.2006.9678570

Tabak, I. , & Dubovi, I. (2022). *What drives the public's use of data? The mediating role of trust in science and data literacy in functional scientific reasoning concerning COVID-19*. Science Education. https://doi.org/10.1002/sce.21789

INDEX

Note: Locators in *italics* represent figures and **bold** indicate tables in the text. Page numbers followed by "n" refer to notes.

Made in the USA
Monee, IL
13 September 2026

62978988R00304